Computational Actuarial Science with R

Chapman & Hall/CRC
The R Series

Series Editors

Aims and Scope

This book series reflects the recent rapid growth in the development and application of R, the programming language and software environment for statistical computing and graphics. R is now widely used in academic research, education, and industry. It is constantly growing, with new versions of the core software released regularly and more than 5,000 packages available. It is difficult for the documentation to keep pace with the expansion of the software, and this vital book series provides a forum for the publication of books covering many aspects of the development and application of R.

The scope of the series is wide, covering three main threads:

- Applications of R to specific disciplines such as biology, epidemiology, genetics, engineering, finance, and the social sciences.
- Using R for the study of topics of statistical methodology, such as linear and mixed modeling, time series, Bayesian methods, and missing data.
- The development of R, including programming, building packages, and graphics.

The books will appeal to programmers and developers of R software, as well as applied statisticians and data analysts in many fields. The books will feature detailed worked examples and R code fully integrated into the text, ensuring their usefulness to researchers, practitioners and students.

Published Titles

Event History Analysis with R, *Göran Broström*

Computational Actuarial Science with R, *Arthur Charpentier*

Statistical Computing in C++ and R, *Randall L. Eubank and Ana Kupresanin*

Reproducible Research with R and RStudio, *Christopher Gandrud*

Displaying Time Series, Spatial, and Space-Time Data with R, *Oscar Perpiñán Lamigueiro*

Programming Graphical User Interfaces with R, *Michael F. Lawrence and John Verzani*

Analyzing Baseball Data with R, *Max Marchi and Jim Albert*

Growth Curve Analysis and Visualization Using R, *Daniel Mirman*

R Graphics, Second Edition, *Paul Murrell*

Customer and Business Analytics: Applied Data Mining for Business Decision Making Using R, *Daniel S. Putler and Robert E. Krider*

Implementing Reproducible Research, *Victoria Stodden, Friedrich Leisch, and Roger D. Peng*

Dynamic Documents with R and knitr, *Yihui Xie*

Computational Actuarial Science with R

Edited by
Arthur Charpentier
University of Québec at Montreal
Canada

CRC Press is an imprint of the
Taylor & Francis Group, an **informa** business
A CHAPMAN & HALL BOOK

CRC Press
Taylor & Francis Group
6000 Broken Sound Parkway NW, Suite 300
Boca Raton, FL 33487-2742

First issued in paperback 2016

Version Date: 20140623

ISBN 13: 978-1-138-03378-8 (pbk)
ISBN 13: 978-1-4665-9259-9 (hbk)

Library of Congress Cataloging-in-Publication Data

Computational actuarial science with R / [edited by] Arthur Charpentier.
pages cm. -- (Chapman & Hall/CRC the R series)
Includes bibliographical references and index.
ISBN 978-1-4665-9259-9 (hardback)
1. Actuarial science. I. Charpentier, Arthur, editor of compilation.

HG8781.C637 2014
368'.0102855133--dc23 2013049493

Visit the Taylor & Francis Web site at
http://www.taylorandfrancis.com

and the CRC Press Web site at
http://www.crcpress.com

Contents

III Finance 407

IV Non-Life Insurance 473

Preface

This book aims to provide a broad introduction to computational aspects of actuarial science in the R environment. We assume that the reader is either learning or is familiar with actuarial science. It can be seen as a companion to standard textbooks on actuarial science. This book is intended for various audiences: students, researchers, and actuaries.

As explained in Kendrick et al. (2006) (discussing the importance of computational economics):

> *Our thesis is that computational economics offers a way to improve this situation and to bring new life into the teaching of economics in colleges and universities [...] computational economics provides an opportunity for some students to move away from too much use of the lecture-exam paradigm and more use of a laboratory-paper paradigm in teaching undergraduate economics. This opens the door for more creative activity on the part of the students by giving them models developed by previous generations and challenging them to modify those models.*

Based on the assumption that the same holds for computational actuarial science, we decided to publish this book.

As claimed by computational scientists, *computational actuarial science* might simply refer to modern actuarial science methods. Computational methods probably started in the 1950s with Dwyer (1951) and von Neumann (1951). The first one emphasized the importance of linear computations, and the second one the importance of massive computations, using random number generations (and Monte Carlo methods), while (at that time) access to digital computers was not widespread. Then statistical computing and computational methods in actuarial science intensified in the 1960s and the 1970s, with the work of Wilkinson (1963) on rounding errors and numerical approximations, and Hemmerle (1967) on computational statistics. Then, the S language started, initiating R.

R

R includes a wide range of standard statistical functions and contributed packages that extend the range of routine functions, including graphical methods. And once we get used to it, R is an easy and intuitive programming language. As explained in Becker (1994): "*from the beginning,* S *was designed to provide a complete environment for data analysis. We believed that the raw data would be read into* S, *operated on by various* S *functions, and results would be produced. Users would spend much more time computing on* S *objects than they did in getting them into or out of* S." In the statistical literature, new methods are often accompanied by implementation in R, so that R became a sort of *lingua franca* in the statistical community (including connected areas, from social science to finance, and actuarial science).

The best way to learn computational actuarial science is to do computational actuarial science. And one of the best ways to do computational actuarial science is probably to start with existing models, and to play with them and experiment with them. This is what we tried to do in this book. The focus in the book is on implementation rather than theory, and we hope to help the reader understand the concepts without being burdened by the theory. Nevertheless, references will be mentioned for those willing to go back to the theory.

A long time ago, before becoming the CEO of AXA, Claude Bébéar published an article on *clarity and truth on the accounts of a life-insurance company*, starting with a short Persian tale:

> *Once upon a time, there was a very clever, prominent and respected life insurer. He was regarded by all as a great technician, perhaps even a savant. There was no one but him to estimate the premium to charge the insured, and to value commitments of the insurer. But sometimes he had nagging doubts: was his science really valued? What if mere commoners got their hands on this science, insinuating that there was nothing admirable about it and in so doing, destroying his reputation he had so patiently built? He had to act swiftly to protect himself. And so our esteemed insurer, respected and ever so clever, invented an esoteric language he called* "actuarial science". *Thanks to him, generations of contented actuaries lived and still live in the shelter of potential critics, adored by all. Each year, bolstered by the solid respect of accountants, they develop impressive Annual Reports which those who are insured, shareholders and controllers of all kinds contemplate without understanding and without daring to ask questions for fear of sounding stupid. The system was perfect.*

If actuarial science can be seen as an esoteric language to the vulgar, then computational aspects are usually seen as magic tricks. This is what Clarke (1973) mentioned in his so-called *third law of prediction*, when claiming that "*any sufficiently advanced technology is indistinguishable from magic.*" The ambitious goal of this book is to demystify computational aspects of actuarial science, and to prove that even complex computations can usually be done without too much pain. I hope that after reading this book, everyone will be able to become a magician.

— Introduction to the R language —

Chapter 1 will provide an introduction to the R language, for Windows, Mac OS X, and Linux users. We will see how to manipulate standard objects, how to write a function, how to import a dataset, how to deal with dates, how to plot graphs, etc. The difficult part of books on computational techniques is to find a balance between

- a code efficient and fast, but where the algorithm is not explicit; and
- a simple code, where iterations or sums can be visualized easily, but which might be slow.

The goal of the book is to provide an introduction to computational aspects of actuarial science, so we will focus on simple codes. Nevertheless, more advanced methods will be mentioned, such as parallel computing, and C/C++ embedded codes. Several technical aspects of R will be mentioned in this introduction, but can be skipped the first time the reader goes through the book. A dedicated section will introduce graphics with R. As Yogi Berra (baseball player and/or philosopher) said:

> *You can see a lot by just looking.*

Visualization is extremely important, and simple R code can help to plot (almost) anything. After reading those pages, anyone will be able to understand the computational aspects of the algorithms dedicated to actuarial computations described in the following chapters.

— Statistical Models with R —

From Chapters 2 to 6, we will get back to the **methodology** and **statistical modeling** issues, with R. Because these chapters focus on methodological aspects, several packages will be used in each chapter (while later on, for each specific application, one dedicated package will be used). In Chapter 2, Christophe Dutang will discuss standard inference, with a description of univariate loss distribution, parametric inference, and goodness of fit. A section will be dedicated to the collective model and aggregate loss distribution, and a section will also mention multivariate distribution and copulas. In Chapter 3, Arthur Charpentier and Benedict Escoto will introduce the Bayesian philosophy, insisting on Bayesian computation. Two sections will focus on important applications: regression modeling from a Bayesian perspective and credibility models. Then, in Chapter 4, Arthur Charpentier and Stéphane Tufféry will give an overview of credit scoring and statistical learning. The goal of this chapter is to describe techniques used to model a binary variable, starting with a logistic regression (from the standard version to ridge and lasso regression techniques), classification trees, random forests, and concluding with boosting of trees. Then, in Chapter 5, Renato Assunção, Marcelo Azevedo Costa, Marcos Oliveira Prates, and Luís Gustavo Silva e Silva will introduce spatial analysis, with an application to car accidents. And finally, in Chapter 6, Eric Gilleland and Mathieu Ribatet will recall extreme value theory, with an application on reinsurance pricing.

— Life Insurance and Mortality with R —

Chapters 7 to 10 discuss the computational aspects of **life insurance**. In Chapter 7, Giorgio Spedicato will discuss life contingencies calculations using R. Then, in Chapter 8, Heather Booth, Rob J. Hyndman, and Leonie Tickle will introduce prospective life tables, and extend computations of Chapter 7 by incorporating a dynamic component on population datasets. In Chapter 9, Julien Tomas and Frédéric Planchet will focus on prospective life tables from an insurer's perspective, and willing to use portfolio experience. And finally, in Chapter 10, Frédéric Planchet and Pierre-E. Thérond will recall techniques used in survival analysis when dealing with censored data (with partial information).

— Actuarial Finance with R —

In Chapters 11 to 13, we will focus on **finance** from an actuarial perspective. In Chapter 11, Yohan Chalabi and Diethelm Würtz will discuss stock price models and nonlinear time series. In Chapter 12, Sergio S. Guirreri will describe standard techniques used to model yield curves and interest rates models, again from an actuarial perspective. And finally, in Chapter 13, Yohan Chalabi and Diethelm Würtz will present techniques used on portfolio optimization problems.

— Nonlife Insurance with R —

Finally, in Chapters 14 to 16, we will see how to use R to deal with computational aspects of **nonlife insurance**. In Chapter 14, Jean-Philippe Boucher and Arthur Charpentier will discuss motor insurance pricing, using generalized linear models, to model claims frequency, and average costs of motor claims. In Chapter 15, Katrien Antonio, Peng Shi, and Frank van Berkum will discuss extension on longitudinal models, introducing dynamics. In this chapter, they mention no-claim bonus systems, and make connections with credibility models described in Chapter 3. And finally, in Chapter 16, Markus Gesmann will show how to use R for IBNR computations and loss reserving.

— Before Starting —

To read the book, keep in mind that all chapters are independent (at least from a computational point of view). We suggest starting with an empty workspace to ensure that no lurking objects can affect code execution. Emptying the workspace is easy using

```
> rm(list = ls())
```

Then datasets used in this book data can be obtained in an R package called `CASdatasets` that accompanies this book. It is available from the CRAN servers at http://CRAN.R-project.org/.

On a computer connected to the Internet, its installation is as simple as typing

```
> install.packages("CASdatasets")
```

at the prompt. Note that additional references can be downloaded from

https://github.com/CASwithR

— Acknowledgements —

As the editor of the book, I am honored and proud that all the contributors agreed to spend time on this book. All of them know more about the topic they write about than I do, and most of them have written a package (sometimes several) used in computational actuarial science. One of the constraints mentioned when I asked them to write their chapter is related to the reproducibility concept mentioned in the first paragraph of this preface: every reader should be able to reproduce what contributors have done. But probably more important, I do not expect the readers to believe—blindly—that what they read is valid. I specifically asked the contributors to show the core of the algorithm, so that readers can reproduce what they have done with their own packages, as I truly believe in the power of the do-it-yourself strategy, especially to understand algorithms.

All programs and data in this book are provided in good faith. The authors do not guarantee their accuracy and are not responsible for the consequences of their use.

Finally, I should also mention that we tried in this book to mention important references on actuarial models and computational aspects. Readers not (yet) familiar with R will find there are dozens of recent books published on the R language. But they should keep in mind that the strength of R is probably the community of R users. Most large cities in the world now have their own R groups, and the community is extremely active online. For instance, stackoverflow (a popular question-and-answer site "*for professional and enthusiast programmers*") has a tag for R related questions, http://stackoverflow.com/questions/tagged/r. See also http://r.789695.n4.nabble.com/ for another large forum. I should also mention blogs, as some of the contributors have one (and additional information related to what they have can be found on their own blog). A nice blog aggregator is r-bloggers where almost 500 bloggers contribute, see http://www.r-bloggers.com/.

To conclude, I want to thank colleagues and friends for their valuable comments on preliminary drafts of this book (Michel Denuit, Ewen Gallic, José Garrido, Hélène Guérin, Stuart Klugman, Dan Murphy, Emiliano Valdez); students who asked for more computational aspects in actuarial courses; actuaries who asked for more details on how to implement theoretical methods they have seen in conferences; and all the staff from Chapman & Hall/CRC, starting with John Kimmel (without forgetting Marcus Fontaine), for their constant support of this project.

Arthur Charpentier, Montréal

Contributors

Katrien Antonio
Universiteit van Amsterdam
Amsterdam, Netherlands

Renato Assunção
Universidade Federal de Minas Gerais
Belo Horizonte, Brazil

Frank van Berkum
Universiteit van Amsterdam
Amsterdam, Netherlands

Heather Booth
Australian National University
Canberra, Australia

Jean-Philippe Boucher
Université du Québec à Montréal
Montréal, Québec, Canada

Yohan Chalabi
ETH Zürich
Zürich, Switzerland

Arthur Charpentier
Université du Québec à Montréal
Montréal, Québec, Canada

Marcelo Azevedo Costa
Universidade Federal de Minas Gerais
Belo Horizonte, Brazil

Christophe Dutang
Université de Strasbourg & Université du Maine
Le Mans, France

Benedict Escoto
Aon Benfield
Chicago, Illinois, USA

Markus Gesmann
ChainLadder Project
London, United Kingdom

Eric Gilleland
National Center for Atmospheric Research
Boulder, Colorado, USA

Sergio S. Guirreri
Accenture S.p.A.
Milan, Italy

Rob J. Hyndman
Monash University
Melbourne, Australia

Rob Kaas
Universiteit van Amsterdam
Amsterdam, Netherlands

Frédéric Planchet
Université de Lyon 1
Lyon, France

Marcos Oliveira Prates
Universidade Federal de Minas Gerais
Belo Horizonte, Brazil

Mathieu Ribatet
Université de Montpellier 2
Montpellier, France

Peng Shi
University of Wisconsin - Madison
Madison, Wisconsin, USA

Luís Gustavo Silva e Silva
Universidade Federal de Minas Gerais
Belo Horizonte, Brazil

Giorgio Spedicato
UnipolSai Assicurazioni
Bologna, Italy

Pierre-E. Thérond
Université de Lyon 1
Lyon, France

Leonie Tickle
Macquarie University
Sydney, Australia

Julien Tomas
Université de Lyon 1
Lyon, France

Stéphane Tufféry
Ecole Nationale de la Statistique et de l'Analyse de l'Information (ENSAI) &
Université de Rennes 1
Rennes, France

Diethelm Würtz
ETH Zürich
Zürich, Switzerland

List of Figures

List of Tables

1

Introduction

Arthur Charpentier
Université du Québec à Montréal
Montréal, Québec, Canada

Rob Kaas
Amsterdam School of Economics, Universiteit van Amsterdam
Amsterdam, Netherlands

CONTENTS

1.1 R for Actuarial Science?

As claimed on the CRAN website, http://cran.r-project.org/, R is an "*open source software package, licensed under the GNU General Public License*" (the so-called GPL). This simply means that R can be installed for free on most desktop and server machines. This platform independence and the open-source philosophy make R an ideal environment for reproducible research.

Why should students or researchers in actuarial science, or actuaries, use R for computations? Of primary interest, as suggested by Daryl Pregibon, research scientist at Google—quoted in Vance (2009)—is that R "*allows statisticians to do very intricate and complicated analyses without knowing the blood and guts of computing systems.*"

In this chapter, we will briefly introduce R, compare it with other standard programming languages, explain how to link R with them (if necessary), give an overview of the language, and show how to produce graphs. But as stated in Knuth (1973), "*premature optimization is the root of all evil (or at least most of it) in programming,*" so we will first describe intuitive (and hopefully easy to understand) algorithms before introducing more efficient ones.

1.1.1 From Actuarial Science to Computational Actuarial Science

In order to illustrate the importance of computational aspects in actuarial science (as introduced in the preface), consider a standard actuarial problem: computing a quantile of a compound sum, such as the 99.5% quantile (popular for actuaries dealing with economic capital, and the Value-at-Risk concept). There exists an extensive literature on computing the distribution in collective risk models. From a probabilistic perspective, we have to calculate, for all $s \in \mathbb{R}$,

$$F(s) = \mathbb{P}(S \leq s), \text{ where } S = \sum_{i=1}^{N} X_i.$$

By independence, a straightforward convolution formula (see e.g. Chapter 3 in Kaas et al. (2008)) can be used:

$$F(s) = \sum_{n=0}^{\infty} F_X^{*n}(x) \cdot \mathbb{P}(N = n).$$

From a statistician's perspective, the distributions of N and X_i are unknown but can be estimated using samples. To illustrate the use of R in actuarial science, consider the following (simulated) sample, with 200 claims amounts

```
> set.seed(1)
> X <- rexp(200,rate=1/100)
> print(X[1:5])
[1]  75.51818 118.16428  14.57067  13.97953  43.60686
```

From now on, forget how we generate those values, and just keep in mind that we have a sample, and let us use statistical techniques to estimate the distribution of the X_i's. A standard distribution for loss amounts is the Gamma(α, β) distribution. As we will see in Chapter 2, other techniques can be used to estimate parameters of the Gamma distribution, but here we can solve the normal equations, see for example Section 3.9.5 in Kaas et al. (2008). They can be written as

$$\log \widehat{\alpha} - \frac{\Gamma'(\widehat{\alpha})}{\Gamma(\widehat{\alpha})} - \log \overline{X} + \overline{\log X} = 0 \text{ and } \widehat{\beta} = \frac{\widehat{\alpha}}{\overline{X}}.$$

To solve the first equation, Greenwood & Durand (1960) give tables for $\widehat{\alpha}$ as a function of $M = \log \overline{X} - \overline{\log X}$, as well as a rational approximation. But using the digamma function $\Gamma'(\alpha)/\Gamma(\alpha)$, nowadays solving it is a trivial matter, by means of the function `uniroot()`. This goes as follows:

```
> f <- function(x) log(x)-digamma(x)-log(mean(X))+mean(log(X))
> alpha <- uniroot(f,c(1e-8,1e8))$root
> beta <- alpha/mean(X)
```

We now have a distribution for the X_i's. For the counting process, assume that N has a Poisson distribution, with mean 100. A standard problem is to compute the quantile with probability 99.5% (as requested in Solvency II) of this compound sum. There are hundreds of academic articles on that issue, not to mention chapters in actuarial textbooks. But the computational aspect of this problem can actually be very simple. A first idea can be to compute numerically the sum (with a given number of terms) in the convolution formula above, using that a sum of independent Gamma random variables still has a Gamma distribution,

```
> F <- function(x,lambda=100,nmax=1000) {n <- 0:nmax
+ sum(pgamma(x,n*alpha,beta)*dpois(n,lambda))}
```

Once we have a function to compute the cumulative distribution function of S, we just need to invert it. Finding x such that $F(x) = .995$ is the same as finding the root of function $x \mapsto F(x) - .995$.

```
> uniroot(function(x) F(x)-.995,c(1e-8,1e8))$root
[1] 13654.43
```

A second idea can be to use fast Fourier transform techniques, since it is (at least from a theoretical perspective) much more convenient to use the generating function when computing the distribution of a compound sum, if we use a discretized version of X_i's. Here the code to compute the probability function of S (after discretizing the Gamma distribution, on $\mathbb{N} = \{0, 1, 2, \ldots\}$, with an upper bound, here 2^{20}) is simply

```
> n <- 2^20; lambda <- 100
> p <- diff(pgamma(0:n-.5,alpha,beta))
> f <- Re(fft(exp(lambda*(fft(p)-1)),inverse=TRUE))/n
```

This is possible because R has a function, called `fft()`, to compute either the Fourier transform, or its inverse. To compute the quantile of level α, we just have to find x_α such that $F(x_\alpha - 1) < \alpha \leq F(x_\alpha)$ (since we use a discretization on $\mathbb{N}$). The R code to compute that value is

```
> sum(cumsum(f)<.995)
[1] 13654
```

All those methods will be discussed in Chapter 2. The point here was to prove that using an appropriate computational language, many actuarial problems can be solved easily, with simple code. The most difficult part is now to understand the grammar of the R language.

1.1.2 The S Language and the R Environment

R is a scripting language for data manipulation, statistical analysis and graphical visualization. It was inspired by the S environment (the letter S standing for *statistics*), developed by John Chambers, Douglas Bates, Rick Becker, Bill Cleveland, Trevor Hastie, Daryl Pregibon and Allan Wilks from the AT&T research team in the 1970s. In 1998 the Association for Computing Machinery (ACM) gave John Chambers the Software System Award, for "*the S system, which has forever altered the way people analyze, visualize, and manipulate data.*"

R was written by Ross Ihaka and Robert Gentleman, at the Department of Statistics of the University of Auckland. John Chambers contributed in the early days of Rand later became a member of the core team. The current R is the result of a collaborative project, driven by this core team, with contributions from users all over the world.

R is an *interpreted language*: when expressions are entered into the R console, a program within the R system (called the interpreter) executes the code, unlike C/C++ but like JavaScript (see Holmes (2006) for a comparison). For instance, if one types 2+3 at the command prompt and presses Enter, the computer replies with 5. Note that it is possible to recall and edit statements already typed, as R saves the statements of the current session in a buffer. One can use the ↑ and the ↓ keys either to recall previous statements, or to move down from a previous statement. The active line is the last line in the session window (called console), and one can use the ← and the → keys to move within a statement in this active line. To execute this line, press Enter.

In this book, illustrations will be based on copies of the console. Of course, we strongly recommend the use of an editor. The user will type commands into the editor, select them (partially or totally), and then run selected lines. This can be done in Windows RGui, via File and New script menu (or Open script for an existing one). The natural extension for R script is a .R file (the workspace will be stored in a .RData file, but we will get back to those objects later on). It is also possible to use any R editor, such as RStudio, Tinn-R or JGR (which stands for Java Gui for R). In all the chapters of this book, R codes will be copies of prompts on the screen: the code will follow the > symbol (or the + symbol when the command is not over). If we go back to the very first example of this book, in the script file there was

```
set.seed(1)
X <- rexp(200,rate=1/100)
print(X[1:5])
```

If we select those three lines and press Run, the output in the console is

```
> set.seed(1)
> X <- rexp(200,rate=1/100)
> print(X[1:5])
[1]  75.51818 118.16428  14.57067  13.97953  43.60686
```

To go further, R is an Object-Oriented Programming language. The idea is to create various *objects* (that will be described in the next sections) that contain useful information, and that could be called by other functions (e.g. graphs). When running a regression analysis with SAS® or SPSS®, output appears on the screen. By contrast, when calling the `lm` regression function in R, the function returns an object containing various information, such as the estimated coefficients $\widehat{\boldsymbol{\beta}}$, the implied residuals $\widehat{\boldsymbol{\varepsilon}}$, estimated variance matrix of the coefficients $\text{var}(\widehat{\boldsymbol{\beta}})$, etc. This means that in R, the analysis is usually done in a series of consecutive steps, where intermediate results are stored in objects. Those objects are then further manipulated to obtain the information required, using dedicated functions. As we will see later (in Section 1.2.1), "*everything in S is an object*," as claimed in the introduction of Chapter 2 of Venables & Ripley (2002). Not only are vectors and matrices objects that can be passed to and returned by functions, but functions themselves are also objects, even function calls.

In R, the preferred assignment operator (by the user community) is the arrow made from two characters `<-`, although the symbol `=` can be used instead. The main advantage of the `<-` is that it allows one to assign objects within a function,. For instance

```
> summary(reg <- lm(Y~X1+X2))
```

will not only print the output of a linear regression, but will also create the object `reg`. Again, Section 1.2 will be dedicated to R objects.

An mentioned previously, in this book, R code will follow the `>` symbol, or the `+` symbol when the line is not over. In R, comments will follow the `#` tag (everything after the `#` tag will not be interpreted by R).

```
> T <- 10    # time horizon
> r <- .05   # discount rate = 5%
> (1+r)^(-
+ T)         # 1$ in T years
[1] 0.6139133
```

R computations will follow the `[1]` symbol. Observe that it is possible to use spaces. It might help to read the code,

```
> (1 + r)^( -T )
[1] 0.6139133
```

and it will not affect computations (unless we split operators with a space; for instance, `<-` is not the same as `< -` : just compare `x<-1` and `x< -1` if you are not convinced). We also encourage the use of parentheses (or braces, but not brackets, used to index matrices) to get a better understanding.

```
> {1+r}^(-T)
[1] 0.6139133
```

If you are not convinced, try to understand what the output will be

```
>  -2 ^ .5
```

It is also possible to use a `;` to type two commands on the same line. They will be executed as if they were on two different lines.

```
> {1+r}^(-T); {1+r/2}^(-T)
[1] 0.6139133
[1] 0.7811984
```

As mentioned previously, instead of typing code in the console, one can open a script window, type a long code, and then run it, partially or totally. It is also possible to load functions stored in a text file using the `source()` function:

```
> source("Rfunctions.txt")
```

1.1.3 Vectors and Matrices in Actuarial Computations

The most common objects in R are vectors (vectors of integers, reals, even complex, or TRUE-FALSE boolean tests). Vectors can be used in arithmetic expressions, in which case the operations are performed element by element: a*b will return the vector $[a_i \cdot b_i]$. In actuarial science, most quantities are either vectors or matrices. For instance, the probability that a person aged x will be alive at age $x+k$ is ${}_kp_x$, which is a function of x and k (both integers) and can be stored in a matrix p. Then any actuarial quantity can be computed using matrix arithmetics. For instance, the curtate expectation of life, given by

$$e_x = \sum_{k=1}^{\infty} {}_kp_x,$$

can be computed using

```
> life.exp <- function(x){ sum(p[1:nrow(p),x]) }
```

Several actuarial activities (from ratemaking to claims reserving) are simply using past data to create models that should describe future behavior. A set of techniques, called *predictive modeling*, became widespread among actuaries. The goal is to infer from the data some factors that better explain the risk, in order to compute the premium for different policyholders, or to calculate reserves for different types of claims. As we will see, using dedicated functions on those data, it will be possible to compute any actuarial quantity. But before discussing datasets, let us spend some time on coding functions in R.

1.1.4 R Packages

A *package* is a related set of functions, including help files and data files, that have been bundled together and is shared among the R community. Those packages are similar to *libraries* in C/C++ and *classes* in Java. To get the list of packages loaded by default, the getOption command can be used:

```
> getOption("defaultPackages")
[1] "datasets" "utils"    "grDevices" "graphics" "stats"     "methods"
> (.packages(all.available=TRUE))
[1] "AER"     "evd"     "sandwich" "lmtest"   "nortest"
```

and many more packages, previously installed on the machine. All these packages are available; you just have to load them. But before, we have to *install* a package from the Internet (for instance quantreg to run quantile regressions) we use

```
> install.packages("quantreg", dependencies=TRUE)
```

and then we select a mirror site to download. The option dependencies=TRUE is used because the quantreg package might be using functions coming from other packages (here, the MatrixModels package has to be installed too). See Zhang & Gentleman (2004) for interactive exploration of R packages. Note that if a package has not been *loaded*, it is not possible to call associated functions:

```
> fit <- rq(Y ~ X1 + X2, data = base, tau = .9)
Error: could not find function "rq"
```

To *load* a package in R, one should use either the library() command, or the require() one,

```
> library(quantreg)
```

As mentioned in Fox (2009), the number of packages on CRAN has grown roughly exponentially, by almost 50% each year (this was confirmed on updated counts). With thousands of packages, inevitably, functions with similar names can be loaded. For instance, the default `stats4` package (containing statistical functions related to the S4 class that will be discussed in the next paragraph) contains a `coef` function,

```
> coef
function (object, ...)
UseMethod("coef")
<bytecode: 0x16d1d80>
<environment: namespace:stats>
```

Note the use of ellipsis (`...`) in the function. This is an interesting feature of R-functions, that allows functions to have a variable number of arguments. But a function with the same name exists in the `VGAM` package (for additive models, see Yee (2008))

```
> library(VGAM)
The following object(s) are masked from 'package:stats4':

    coef
```

Now function `coef` will be called from this package,

```
> coef
standardGeneric for "coef" defined from package "VGAM"

function (object, ...)
standardGeneric("coef")
<environment: 0x2981fcb0>
Methods may be defined for arguments: object
Use  showMethods("coef")  for currently available ones.
```

If the `VGAM` package is loaded, and we need to run function `coef()` from library `stats4`, it is either possible to unload it, using

```
> detach(package:VGAM, unload=TRUE)
```

or to call the function using `stats4::coef`.

Finally, observe that packages are regularly updated. In order to run the latest version, it might be a good thing to run, frequently, the following line

```
> update.packages()
```

The update should be run before loading packages, and one should keep in mind that updates might be possible only on the latest version of R (see Ripley (2005) for a discussion). It is also possible to get an overview of existing packages on different topics on the *task views* page, see http://cran.r-project.org/web/views/, such as *empirical finance* or *computational econometrics* (see Zeileis (2005) for additional information).

Another difficult task, for new users, is to find which package will be appropriate to deal with a specific problem. Consider the problem of longitudinal mixed models (developed in Chapter 15). Consider the following hierarchical model

$$Y_i = \beta_1 X_i + \underbrace{\gamma_0 + \gamma_1 Z_{j[i]} + u_{j[i]}}_{\text{random component } \alpha_{j[i]}} + \varepsilon_i,$$

where individual i belongs to a group (company, region, car type, etc.) j. Several packages can be used to deal with this model, or more generally nonlinear mixed-effect models, such as nlme or lme4. One can also use plm for panel regression models. At least two functions can be used to estimate the model above, lme() and lmer(). More details about function lmer() can be found in Bates (2005), and from linear and nonlinear mixed effects models described in Pinheiro & Bates (2000). For instance, to run a regression, use either

```
> reg <- lme(fixed = Y ~ X, random = ~ X | Z, method='ML' )
```

or

```
> reg <- lmer( Y ~ X+ Z + (1 | Z), method='ML' )
```

Note that the syntax can be different, as well as the output. Gelman & Hill (2007) suggest the use of lmer() for instance, as lme() only accepts nested random effects, while lmer() handles crossed random effects. On the other hand, lme() can handle heteroscedasticity (while lmer() does not). Further, lme() returns p-values, while lmer() does not. Choosing a package is not a simple task.

1.1.5 S3 versus S4 Classes

In order to get a good understanding of R functions and packages, it is necessary to understand the distinction between S3 and S4 classes. The basic difference is that S3 objects were created with an *old* version of R (or S, the so-called third version), while S4 objects were created with a more recent version of R (or S, the so-called fourth version). But it is still possible to create S3 objects with the latest version of R. For example, in regression models, lm, glm and gam are S3 objects, while lmer (for mixed effects models) and VGAM (for vector generalized linear and additive models) are S4. See Bates (2003) for a discussion about those two classes.

To illustrate the distinction between the two, consider the case of health insurance, where we have characteristics of some individuals. It is possible to define a person object that will contain all important information. And then we can define functions on such an object.

S3 is a primitive concept of classes, in R. To define a class that will contain characteristics of a person, use

```
> person3 <- function(name,age,weight,height){
+  characteristics<-list(name=name,age=age,weight=weight,height=height)
+  class(characteristics)<-"person3"
+  return(characteristics)}
```

To create a person, we just have to give proper arguments:

```
> JohnDoe <- person3(name="John",age=28, weight=76, height=182)
> JohnDoe
$name
[1] "John"

$age
[1] 28

$weight
[1] 76
```

```
$height
[1] 182

attr(,"class")
[1] "person3"
```

Observe the use of the `$` symbol, to get attributes of that list,

```
> JohnDoe$age
[1] 28
```

Then it is possible to define a function on a `person3` object. If we want a function that returns the BMI (Body Mass Index), use

```
> BMI3 <- function(object,...) {return(object$weight*1e4/object$height^2)}
```

Then, we can call that function on `JohnDoe`

```
> BMI3(JohnDoe)
[1] 22.94409
```

As mentioned in the previous paragraph, `lm` objects are in the S3 class.

```
> reg3 <- lm(dist~speed,data=cars)
> reg3$coefficients
(Intercept)       speed
 -17.579095    3.932409
> coef(reg3)
(Intercept)       speed
 -17.579095    3.932409
> plot(reg3)
```

Note that with S3 objects, a function is usually defined with a certain list of arguments (see the `lm` function), and then to define a generic function, there is a `UseMethod` call with the name of the generic function. For example, the generic `summary`

```
> summary
function (object, ...)
UseMethod("summary")
<environment: namespace:base>
```

The latest version, S4, allows object-oriented programming with S (see Chambers & Lang (2001)). An object is a set (or a list) of functions, and it should be associated with functions dealing with that object. Object programming might appear more complex, as the code should be thought through in advance. As mentioned in Lumley (2004), "*as a price for this additional clarity, the* S4 *system takes a little more planning.*"

To create an object, use

```
> setClass("person4", representation(name="character",
+ age="numeric", weight="numeric", height="numeric"))
[1] "person4"
```

To create a person, use the `new` function, for example,

```
> JohnDoe <- new("person4",name="John",age=28, weight=76, height=182)
> JohnDoe
An object of class "person"
Slot "name":
[1] "John"

Slot "age":
[1] 28

Slot "weight":
[1] 76

Slot "height":
[1] 182
```

Attributes of those objects are obtained using the @ symbol (and no longer the `$`)

```
> JohnDoe@age
[1]  28
```

Then, it is possible to define functions on those objects, for example, to compute the BMI. The first argument will be the name of the function (here `BMI4`), the second one will be the object name (here a `person4`), and then the code of the function. But first, it is necessary to define the method `BMI4` using the `setGeneric` function

```
> setGeneric("BMI4",function(object,separator) return(standardGeneric("BMI")))
[1] "BMI4"
> setMethod("BMI4", "person4",
function(object){return(object@weight*1e4/object@height^2)})
[1] "BMI4"
```

Then, we can use this function on all individuals we have,

```
> BMI4(JohnDoe)
[1] 22.94409
```

As mentioned previously, `VGAM` objects are in the S4 class.

```
> library(VGAM)
> reg4 <- vglm(dist~speed,data=cars,family=gaussianff)
> reg4@coefficients
(Intercept)       speed
 -17.579095    3.932409
> coefficients(reg4)
(Intercept)       speed
 -17.579095    3.932409
```

For those two examples, we can see that S3 and S4 classes are rather similar, and actually, both classes coexist (so far peacefully) in R. See Genolini (2008) for more details, especially on the concept of inheritance that can be used only with S4 classes, Leisch (2009) on R packages or Lumley (2004) which compares ROC curves (defined in Chapter 4) with S3 and S4 classes, respectively.

It is necessary to understand that those two classes both exist, as some libraries use only S3 classes (described in this introductory chapter), while others do use S4, such as several probability distributions used in Chapter 2 (see Ruckdeschel et al. (2006) for a discussion on S4 classes for distributions) or `lifetable` objects in Chapter 8.

1.1.6 R Codes and Efficiency

In this book, we will discuss computational aspects of actuarial sciences. So there will be a tradeoff between a very efficient code, where the structure of the algorithm might not be explicit, and a simple code, to illustrate how to compute quantities, but which might be quite slow. Efficient techniques will be mentioned in this chapter, and to illustrate efficiency, instead of using flop counting (we might refer to Knuth (1973), Press et al. (2007) or Cormen et al. (1989) for details), we will use R functions that measure how long a particular operation takes to execute, such as `system.time`, which returns CPU times, and others. For instance, if we consider the product of two matrices $1,000 \times 1,000$,

```
> n <- 1000
> A <- matrix(seq(1,n^2),n,n)
> B <- matrix(seq(1,n^2),n,n)
> system.time(A%*%B)
       user      system      elapsed
      1.040       0.020        1.226
```

we can see that on a standard machine, it took around 1 second to compute the product (time is here in seconds). And around the same time to see that this matrix could not be inverted

```
> system.time(solve(A%*%B))
Error in solve.default(A %*% B) :
  Lapack routine dgesv: system is exactly singular
Timing stopped at: 1.466 0.069 1.821
```

For deterministic computations, if we want to compare computation time, use function `benchmark` from `library(rbenchmark)`

```
> benchmark(A*B,A%*%B,replications=1)[,c(1,3,4)]
     test elapsed relative
1   A * B   0.006    1.000
2 A %*% B   0.989  164.833
```

where we can see compare element by element and matrices product computations. In the context of random number generation, it might be interesting to use the `library(microbenchmark)`. Here, based on ten computations of the same quantity,

```
> microbenchmark(A*B,A%*%B,times=10)
Unit: milliseconds
    expr        min         lq      median          uq          max neval
   A * B   3.369552    3.49644    3.763942    5.085174     8.988348    10
 A %*% B 970.899445 979.78303 994.234688 999.586548 1024.985193    10
```

Here, on deterministic computations for the matrices product, it took 1 second, each time. Those times might be very different when using Monte Carlo simulations, with much more variability (especially when rejection techniques are used).

1.2 Importing and Creating Various Objects, and Datasets in R

After this general introduction, let us spend some time to discover R's grammar.

1.2.1 Simple Objects in R and Workspace

As claimed in the introduction of Chapter 2 of Venables & Ripley (2002),

"Everything in S is an object."
"Every object in S has a class."

For instance, to assign a value to an object (and to create that object if it does not exist yet), we use the assignment operator <-. Command

```
> x <- exp(1)
> x
[1] 2.718282
```

will create a `numeric` variable denoted `x`, with value e.

```
> class(x)
[1] "numeric"
```

Almost all names can be used, except a small list of taken words, such as `TRUE` or `Inf`. The latter is defined such that

```
> 1/0
[1] Inf
```

Observe that the largest number (before reaching infinity) is

```
> .Machine$double.xmax
[1] 1.797693e+308
```

So here,

```
> 2e+307<Inf
[1] TRUE
> 2e+308<Inf
[1] FALSE
```

And finally observe that

```
> 0/0
[1] NaN
```

where `NaN` stands for Not A Number (which makes sense because 0/0 is not properly defined). If `pi` has a default value when starting R (π, ratio of a circle's circumference to its diameter, numerically $3.141593\cdots$ etc.), it is possible to create an object named `pi`. But it might be a bad idea. Keep in mind also that, when loading R, `T` and `TRUE` are equivalent (and so are `F` and `FALSE`), which explains why, in some codes, we can find `mean(x,na.rm=T)`, for instance. Nevertheless, we strongly recommend to avoid using `T` instead of `TRUE`, and to keep `T` as a possible variable (that can be used for time). The list of all objects stored in R memory can be obtained using the `ls()` function. It is possible to define objects `x2`, `x_2`, or `x.2`, but not `2x`. Object names cannot start with a numeric value.

Using

```
> y <- x+1
```

will create another object, `y` with value $e+1$. From a technical point of view, R uses copying semantics , which makes R a pass by value language: `y` stores a numeric value and is not a function of `x`. Thus, if the value of `x` changes, this will not affect the value of `y`,

```
> x <- pi
> y
[1] 3.718282
```

A `numeric` class object has values in $\mathbb{R}$, while the class `integer` refers to values in $\mathbb{Z}$. The `logical` class is for values `TRUE` and `FALSE`.

The objects we created can be stored in a file called .RData (in the directory where we started R). Our workspace is one of the several locations where R can find objects.

```
> find("x")
[1] ".GlobalEnv"
```

The workspace is just an environment in R (a mapping between names, and values). Note that predefined objects are stored elsewhere

```
> find("pi")
[1] "package:base"
```

Objects can be stored in several locations,

```
> search()
[1]  ".GlobalEnv" "tools:RGUI" "package:stats"
[4]  "package:graphics" "package:grDevices" "package:utils"
[7]  "package:datasets" "package:methods" "Autoloads"
[10] "package:base"
```

1.2.2 More Complex Objects in R: From Vectors to Lists

1.2.2.1 Vectors in R

The most natural way to define and store more than one value in R is to create a vector, which is probably the simplest R object. This can be done using the `c()` function (to *concatenate* or *combine*)

```
> x <- c(-1,0,2)
> x
[1] -1  0  2
> y <- c(0,2^x)
> y
[1]  0.0 0.5 1.0 4.0
```

Here, `[1]` states that the answer is starting at the first element of a vector: when displaying a vector, R lists the elements, from the left to the right, using (possibly) multiple rows (depending on the width of the display). Observe that if an object is followed by the assignment operator `<-`, then a value (or a dataframe, or a function, etc.) will be assigned to the object. If we simply type the name of the object, and then Enter, the value of the object will appear in the console (or the code of the function, if the object is a function). Each new row includes the index of the value starting that row, that is,

```
> u <- 1:50
> u
 [1]  1  2  3  4  5  6  7  8  9 10 11 12 13 14 15 16 17
[18] 18 19 20 21 22 23 24 25 26 27 28 29 30 31 32 33 34
[35] 35 36 37 38 39 40 41 42 43 44 45 46 47 48 49 50
```

Note that there is a NULL symbol in R.

```
> c(NULL,x)
[1] -1  0  2
```

Such an object can be useful to create an object used in a loop.

```
> x <- NULL
> for(i in 1:10){ x <- c(x,max(sin(u[1:i]))) }
> x
 [1] 0.8414710 0.9092974 0.9092974 0.9092974 0.9092974
 [6] 0.9092974 0.9092974 0.9893582 0.9893582 0.9893582
```

We have seen function c(), used to concatenate series of elements (having the same type), but one can also use seq() to generate a sequence of elements evenly spaced:

```
> seq(from=0, to=1, by=.1)
[1] 0.0 0.1 0.2 0.3 0.4 0.5 0.6 0.7 0.8 0.9 1.0
> seq(5,2,-1)
[1] 5 4 3 2
> seq(5,2,length=9)
[1] 5.000 4.625 4.250 3.875 3.500 3.125 2.750 2.375 2.000
```

or rep(), which replicates elements

```
> rep(c(1,2,6),3)
[1] 1 2 6 1 2 6 1 2 6
> rep(c(1,2,6),each=3)
[1] 1 1 1 2 2 2 6 6 6
```

It is important to keep in mind that R is case sensitive, so x is not the same as X. Observe also that there are no pointers in R. If we use the sort function, it will print the sorted vector. But the order will not change.

```
> x <- c(-1,0,2)
> sort(x,decreasing=TRUE)
[1]  2  0 -1
> x
[1] -1  0  2
```

If we want to sort vector x, then we should reassign the vector (which is possible in R),

```
> x <- sort(x,decreasing=TRUE)
> x
[1]  2  0 -1
```

Observe that it is possible to assign names to the elements of the vector,

```
> names(x) <- c("A","B","C")
> x
 A  B  C
-1  0  2
```

and then components of the vectors can be called using brackets [], with either

```
> x[c(3,2)]
C B
2 0
```

or

```
> x[c("C","B")]
C B
2 0
```

which is a shorter version for `x[names(x)%in%c("C","B")]`.

With `runif()`, we can generate random variables, uniformly distributed over the unit interval $[0,1]$ (most standard distributions can be generated with standard functions in R, as discussed later on in this chapter and in Chapter 2),

```
> set.seed(1)
> U <- runif(20)
```

By setting the seed of the generating algorithm, so-called random numbers will always be the same, and we will have examples that can be reproduced.

By default, seven digits are displayed

```
> U[1:4]
[1] 0.2655087 0.3721239 0.5728534 0.9082078
```

It is possible to display more digits, or less, by setting the number of digits to print, from 1 to 22:

```
> options(digits = 3)
> U[1:4]
[1] 0.266 0.372 0.573 0.908
> options(digits = 22)
> U[1:4]
[1] 0.2655086631420999765396 0.3721238996367901563644
[3] 0.5728533633518964052200 0.9082077899947762489319
```

Only the display is affected, not the way numbers are stored, in R.

R has a recycling rule when working with vectors. The recycling rule is implicit when we use expression `x+2`, where `2` does not have the size of `x`. More generally, shorter vectors are recycled as often as needed, until they match the length of the longest one. Hence,

```
> x <- c(100,200,300,400,500,600,700,800)
> y <- c(1,2,3,4)
> x+y
[1] 101 202 303 404 501 602 703 804
```

This works also when vector lengths are not multiples,

```
> y <- c(1,2,3)
> x+y
[1] 101 202 303 401 502 603 701 802
```

Note that `NA` values are used to represent missing values, as NA stands for *not available*

```
> age <- seq(0,90,by=10)
> length(age) <- 12
> age
[1] 0 10 20 30 40 50 60 70 80 90 NA NA
```

But R will not always assign values when lengths are not appropriate (and some error message might also appear).

The strength of a vector-based language is that it is very simple to access parts of vectors, specifying subscripts. To return the values of `U` strictly larger than 0.8, we use

```
> U[U>.8]
[1] 0.9082078 0.8983897 0.9446753 0.9919061
```

or if we want value(s) between 0.4 and 0.5,

```
> U[(U>.4)&(U<.5)]
[1] 0.4976992
```

Here, a boolean test is made, and the value of `U` is returned only if the test is `TRUE`:

```
> (U>.4)&(U<.5)
 [1] FALSE FALSE FALSE FALSE FALSE FALSE FALSE FALSE FALSE
[10] FALSE FALSE FALSE FALSE FALSE FALSE  TRUE FALSE FALSE
[19] FALSE FALSE
```

If the test is `FALSE` for all components, then a `numeric(0)` is returned:

```
> U[(U>.4)&(U<.45)]
numeric(0)
```

From a mathematical point of view, the set $\{U > .4\} \cap \{U < .45\}$, here, is empty. Thus, the vector here, two times has a zero length

```
> length(U[(U>.4)&(U<.45)])
[1] 0
```

It is also possible to return a vector containing the subscripts of the vector for which the logical test was true:

```
> which((U>.4)&(U<.6))
[1]  3 16
```

In order to get the complementary, one can use operators which stands for *and*, while `|` stands for *or*:

```
> which((U<=.4)|(U>=.6))
 [1]  1  2  4  5  6  7  8  9 10 11 12 13 14 15 17 18 19 20
```

or one can use the negation operator `!`

```
> which(!((U>.4)&(U<.6)))
 [1]  1  2  4  5  6  7  8  9 10 11 12 13 14 15 17 18 19 20
```

For integer values, it is possible to use `==` to compare values:

```
> y
[1] 1 2 3 4
> y==2
[1] FALSE  TRUE FALSE FALSE
```

But this symbol can yield tricky situations when comparing non-integers

```
> (3/10-1/10)
[1] 0.2
> (3/10-1/10)==(7/10-5/10)
[1] FALSE
```

Those two fractions are not equal, for R, as

```
> (3/10-1/10)-(7/10-5/10)
[1] 2.775558e-17
```

In that case, it might be more judicious to use `all.equal()`,

```
> all.equal((3/10-1/10),(7/10-5/10))
[1] TRUE
```

Similarly, $(\sqrt{2})^2$ is slightly different from 2,

```
> sqrt(2)^2 == 2
[1] FALSE
```

To go further on floating-point numbers (that can be represented exactly by a computer), see Goldberg (1991). In R, the smallest positive floating-point number, ϵ, such that $(1+\epsilon) \neq 1$ is

```
> print(eps<-.Machine$double.eps)
[1] 2.220446e-16
> 1+eps==1
[1] FALSE
```

It does not mean that R cannot deal with smaller numbers, only that `==` cannot be used. The smallest number is actually

```
 .Machine$double.xmin
[1] 2.225074e-308
```

As mentioned earlier, the important idea with vectors is that they are (ordered) collections of elements of the same type, which can be `numeric` (in $\mathbb{R}$), `complex` (in $\mathbb{C}$), `integer` (in $\mathbb{N}$), `character` for characters or strings, `logical`, that is `FALSE` or `TRUE` (or in $\{0,1\}$). If we try to concatenate elements that are not of the same type, R will coerce elements to a common type, for example,

```
> x <- c(1:5,"yes")
> x
[1] "1" "2" "3" "4" "5" "yes"
> y <- c(TRUE,TRUE,TRUE,FALSE)
> y
[1] TRUE TRUE TRUE FALSE
> y+2
[1] 3 3 3 2
```

1.2.2.2 Matrices and Arrays

A matrix is just a vector of data, along with an additional vector, accessible by the `dim()` function, that contains the dimensions (i.e. number of rows `nrow` and columns `ncol`).

```
> M <- matrix(U,nrow=5,ncol=4)
> M
          [,1]       [,2]      [,3]      [,4]
[1,] 0.2655087 0.89838968 0.2059746 0.4976992
[2,] 0.3721239 0.94467527 0.1765568 0.7176185
[3,] 0.5728534 0.66079779 0.6870228 0.9919061
[4,] 0.9082078 0.62911404 0.3841037 0.3800352
[5,] 0.2016819 0.06178627 0.7698414 0.7774452
```

The dimension of matrix `M` is obtained using

```
> dim(M)
[1] 5 4
```

If we want to reshape the matrix, to get a 4×5 matrix, instead of a 5×4, it is possible to change the *attribute* of the object. Here, we can specify ex-post the dimension of the matrix, using

```
> attributes(M)$dim=c(4,5)
```

All the elements are sorted the same way (per column), but the matrix has been reshaped,

```
> M
          [,1]      [,2]       [,3]      [,4]      [,5]
[1,] 0.2655087 0.2016819 0.62911404 0.6870228 0.7176185
[2,] 0.3721239 0.8983897 0.06178627 0.3841037 0.9919061
[3,] 0.5728534 0.9446753 0.20597457 0.7698414 0.3800352
[4,] 0.9082078 0.6607978 0.17655675 0.4976992 0.7774452
```

The dimension of matrix `M` is

```
> dim(M)
[1] 4 5
```

Here again, it is possible to use logical subscripts. If we want the lines of `M` for which the element in the last column is larger than 0.8, we use

```
> M[M[,5]>0.8,]
[1] 0.37212390 0.89838968 0.06178627 0.38410372 0.99190609
```

If we want the columns for which the element in the last row is larger than 0.8, we use

```
> M[,M[4,]>0.8]
[1] 0.2655087 0.3721239 0.5728534 0.9082078
```

A lot of functions can be used to manipulate matrices. For instance, `sweep()` can be used to apply a function either to rows (`MARGIN=1`) or columns (`MARGIN=2`). For instance, if we want to add i to row i of matrix `M`, the code will be

```
> sweep(M,MARGIN=1,STATS=1:nrow(M),FUN="+")
         [,1]     [,2]     [,3]     [,4]     [,5]
[1,] 1.265509 1.201682 1.629114 1.687023 1.717619
[2,] 2.372124 2.898390 2.061786 2.384104 2.991906
[3,] 3.572853 3.944675 3.205975 3.769841 3.380035
[4,] 4.908208 4.660798 4.176557 4.497699 4.777445
```

Here, we will (mainly) work with matrices containing numeric values. But it is possible to have matrices of (almost) any format, for instance logical values,

```
> M>.6
      [,1]  [,2]  [,3]  [,4]  [,5]
[1,] FALSE FALSE  TRUE  TRUE  TRUE
[2,] FALSE  TRUE FALSE FALSE  TRUE
[3,] FALSE  TRUE FALSE  TRUE FALSE
[4,]  TRUE  TRUE FALSE FALSE  TRUE
```

Observe that the recycling rule of R affects also matrices (a matrix—or an array—being just a rectangular collection of elements of the same type)

```
> M <- matrix(seq(1,8),nrow=4,ncol=3,byrow=FALSE)
Warning :
In matrix(seq(1,8), nrow = 4, ncol = 3) :
data length [8] is not a sub-multiple or multiple of the number of rows [3]
> M
     [,1] [,2] [,3]
[1,]    1    5    1
[2,]    2    6    2
[3,]    3    7    3
[4,]    4    8    4
> M+c(10,20,30,40,50)
     [,1] [,2] [,3]
[1,]   11   55   41
[2,]   22   16   52
[3,]   33   27   13
[4,]   44   38   24
Warning :
In M + c(10,20,30,40,50) :
longer object length is not a multiple of shorter object length
```

Note that the standard way of storing a vector as a matrix in R is not by row but by column, so in this case we could have omitted the `byrow=FALSE` argument.

Observe that if there was a `c()` function to concatenate vectors, there are two functions to concatenate matrices, `rbind()` to concatenate matrices by adding rows, or `cbind()` to concatenate by adding columns, for suitable dimensions

```
> A <- matrix(0,3,6)
> B <- matrix(1,2,6)
> C <- rbind(B,A,B)
> C
     [,1] [,2] [,3] [,4] [,5] [,6]
[1,]    1    1    1    1    1    1
[2,]    1    1    1    1    1    1
[3,]    0    0    0    0    0    0
[4,]    0    0    0    0    0    0
[5,]    0    0    0    0    0    0
[6,]    1    1    1    1    1    1
[7,]    1    1    1    1    1    1
```

or

```
> A <- matrix(0,6,4)
> B <- matrix(1,6,3)
> C <- cbind(B,A,B)
> C
     [,1] [,2] [,3] [,4] [,5] [,6] [,7] [,8] [,9] [,10]
[1,]    1    1    1    0    0    0    0    1    1     1
[2,]    1    1    1    0    0    0    0    1    1     1
[3,]    1    1    1    0    0    0    0    1    1     1
[4,]    1    1    1    0    0    0    0    1    1     1
[5,]    1    1    1    0    0    0    0    1    1     1
[6,]    1    1    1    0    0    0    0    1    1     1
```

Finally, recall that even if two quantities are (formally) equal, computation times can be quite different. For instance, if A, B and C are $k\times m$, $m\times n$ and $n\times p$ matrices, respectively, then

$$(A\times B)\times C = A\times(B\times C)$$

Computing the simplest matrices first will be more efficient

```
> n <- 1000
> A<-matrix(seq(1,n^2),n,n)
> B<-matrix(seq(1,n^2),n,n)
> C<-1:n
> benchmark((A%*%B)%*%C,A%*%(B%*%C),replications=1)[,c(1,3,4)]
                test elapsed relative
1 (A %*% B) %*% C   0.945      135
2 A %*% (B %*% C)   0.007        1
```

A matrix-vector multiplication goes much faster than a matrix-matrix multiplication.

Note that for matrix crossproducts ($A^\top \times B$), using the function `crossproduct()` might lead to faster computations. More general than matrices, arrays are multidimensional extensions of vectors (and like vectors and matrices, all the objects of an array must be of the same type). The matrix can be seen as a two-dimensional array.

```
> A <- array(1:36,c(3,6,2))
> A
, , 1

     [,1] [,2] [,3] [,4] [,5] [,6]
[1,]    1    4    7   10   13   16
[2,]    2    5    8   11   14   17
[3,]    3    6    9   12   15   18

, , 2

     [,1] [,2] [,3] [,4] [,5] [,6]
[1,]   19   22   25   28   31   34
[2,]   20   23   26   29   32   35
[3,]   21   24   27   30   33   36
```

Several operators are used in R for accessing objects in a data structure,

```
> x[i]
```

to return objects from object `x` (a vector, a matrix, or a dataframe). `i` may be an integer, an integer vector, a logical vector (with `TRUE` or `FALSE`), or characters (of object names).

```
> x[[i]]
```

returns a single element of `x` that matches `i` (`i` is either an integer or a character).

1.2.2.3 Lists

Finally, note that it is possible to store a variety of objects into a single one using a `list` object.

```
> stored <- list(submatrix=M[1:2,3:5],sequenceu=U,x)
> stored
$submatrix
          [,1]      [,2]      [,3]
[1,] 0.8298559 0.3947363 0.1446575
[2,] 0.9771057 0.3233137 0.9415277

$sequenceu
 [1] 0.2948071 0.2692372 0.4756646 0.4196496 0.5012345 0.6599705
 [7] 0.4496317 0.6041229 0.8298559 0.9771057 0.2093930 0.2677681
[13] 0.3947363 0.3233137 0.5937027 0.6698777 0.1446575 0.9415277
[19] 0.4466402 0.8573008

[[3]]
 A  B  C
-1  0  2
```

Names of the list elements can be obtained using the `names()` function

```
> names(stored)
[1] "submatrix" "sequenceu"  ""
```

The various list elements can be called using the `$` character, when objects have names: `stored$sequenceu` is the vector stored in the list. It is also possible to use `stored[[2]]` if we want to use the second element of that list.

Keep in mind that lists are important in R, as most functions use them to store a lot of information, without necessarily displaying them in the console.

```
> f <- function(x) { return(x*(1-x)) }
> optim.f <-  optimize(f, interval=c(0, 1), maximum=TRUE)
```

Here, `optim.f` is a list with the following information:

```
> names(optim.f)
[1] "maximum"   "objective"
> optim.f$maximum
[1] 0.5
```

To get further information on the `optimize()` function (and more generally on any function from a documented package), the command is `help(optimize)`, or for a faster alternative `?optimize`.

For lists, or dataframes,

```
> x$n
```

returns the object with name n from object x. Finally,

```
> x@n
```

is used when x is an S4 object (that will be mentioned in several chapters later on). It returns the element stored in the slot named n.

1.2.3 Reading csv or txt Files

Sometimes we do not want to *create* objects that we might be using, but we wish to *import* them. In life insurance, we might need to import life tables, or yield curves, while datasets with claims information as well as details on insurance contracts will be necessary for motor insurance ratemaking. In R terminology, we need a *dataframe*, which is a list that contains multiple named vectors, with the same length. It is like a spreadsheet or a database table. If the dataframe is too large to be printed, it is still possible to use function head() to view the first few data rows and tail() to view the last few. The read.table() function is used to read data into R, and to create a dataframe object. This function expects all variables in the input source to be separated by a character defined by the sep argument (using quote signs, such as sep=";"). The default is spaces and/or tabs (to specify that the variable is changing) or a carriage return (to specify that the individual is changing). Missing values are either an empty section, or defined using a specific notations, for example, -9999 or ?. In that case, one has to specify argument na.strings.

The default location is the working directory, obtained using

```
> getwd()
```

It is possible to change the working directory or to specify the location of the text files. But the specification is different for Windows users and Mac-Linux users (see Ripley (2001)). On a Windows platform, use

```
> setwd("c:\\Documents and Settings\\user\\Rdata\\")
```

to relocate the working directory, and then

```
> db <- read.table("file.txt")
```

to create the db object, or directly

```
> db <- read.table("c:\\Documents and Settings\\user\\Rdata\\file.txt")
```

to read a file at a specific directory. The backslash symbols used to specify location in Windows have to be preceded by R's backslash used to introduce special symbols in character strings (see Section 1.2.5 for more details on strings and text). For a Mac-Linux platform, the syntax is

```
> setwd("/Users/Rdata/")
```

and then

```
> db <- read.table("file.txt")
```

or

```
> db <- read.table("/Users/Rdata/file.txt")
```

Note that on a Windows platform one can also use a single forward slash notation /, but it is not the common way to specify locations.

It might be convenient to specify the location and the name of the file as a string object, as several functions can be used to debug some codes. Consider the list of all tropical cyclones in the NHC best track record over the period 1899–2006, as in Jagger & Elsner (2008). We want to investigate if there is an upward trend in the number of cyclones in the Atlantic Ocean, Gulf of Mexico, and Caribbean Sea (including those that have made landfall in the United States). Consider the following csv file, available in the Github folder.

```
> file <- "extreme2datasince1899.csv"
> StormMax <- read.table(file, header=TRUE, sep=",")
Error in scan(file, what, nmax, sep, dec, quote, skip, nlines, ,  :
line 5 did not have 11 elements
```

Observe that `read.csv()` could be used and should have the right defaults. Some parts of the sixth line of the csv file have been dropped. It is possible to use the `count.fields()` function to discover whether there are other errors as well (and if there are, to identify where they are located), our benchmark being here the number of variates for 90% of the dataset,

```
>  nbvariables <- count.fields(file,sep=",")
>  which(nbvariables !=quantile(nbvariables,.9))
[1] 6
```

The `header=TRUE` argument in the `read.table()` function is used to identify names of variables in the input file, if any. It is also possible to skip some early lines of the text file using `skip` and to specify the number of lines to be read using `nrow` (which can be used if the file is too large to be read completely). Finally, as discussed previously, one can also specify strings that should be read as missing values, `na.string=c("NA"," ")`.

```
> file <- "extremedatasince1899.csv"
> StormMax <- read.table(file,header=TRUE,sep=",")
> tail(StormMax,3)
       Yr Region     Wmax       sst sun        soi split naofl naogulf
2098 2009  Basin 90.00000 0.3189293 4.3 -0.6333333     1  1.52   -3.05
2099 2009     US 50.44100 0.3189293 4.3 -0.6333333     1  1.52   -3.05
2100 2009     US 65.28814 0.3189293 4.3 -0.6333333     1  1.52   -3.05
> str(StormMax)
'data.frame':        2100 obs. of  11 variables:
 $ Yr     : int  1899 1899 1899 1899 1899 1899 1899 1899 1899 1899 ...
 $ Region : Factor w/ 5 levels "Basin","East",..: 1 1 1 1 3 1 4 5 5 5 ...
 $ Wmax   : num  105.6 40 35.4 51.1 87.3 ...
 $ sst    : num  0.0466 0.0466 0.0466 0.0466 0.0466 ...
 $ sun    : num  8.4 8.4 8.4 8.4 8.4 8.4 8.4 8.4 8.4 8.4 ...
 $ soi    : num  -0.21 -0.21 -0.21 -0.21 -0.21 -0.21 -0.21 -0.21 -0.21 ...
 $ split  : int  0 0 0 0 0 0 0 0 0 0 ...
 $ naofl  : num  -1.03 -1.03 -1.03 -1.03 -1.03 -1.03 -1.03 -1.03 -1.03 ...
 $ naogulf: num  -0.25 -0.25 -0.25 -0.25 -0.25 -0.25 -0.25 -0.25 -0.25 ...
```

The object here is called a dataframe. This dataframe has a (unique) name (here `StormMax`), each column within this table has a unique name (the second one for instance is `Region`), and each column has a unique type associated with it (a column is a vector). It is possible to use the third column either using its name, with a `$` symbol (here `StormMax$Wmax`) or using a matrix notation (here `StormMax[,3]`). But instead of accessing this vector with `StormMax$Wmax`, it is possible to attach the dataframe, using `attach(StormMax)`, and then simply indicate the column name, as now, vector `Wmax` does exist. An alternative can be to use function `with()` when running functions in variables on that dataset.

Several functions can be used to manipulate dataframes. For instance, it is possible to sort a database according to some variable. Consider

```
> set.seed(123)
> df <- data.frame(x1=rnorm(5),x2=sample(1:2,size=5,replace=TRUE),x3=rnorm(5))
> df
           x1 x2         x3
1 -0.56047565  2  1.2805549
2 -0.23017749  1 -1.7272706
3  1.55870831  2  1.6901844
4  0.07050839  2  0.5038124
5  0.12928774  1  2.5283366
```

If we want to sort according to `x2` (increasing), and `x1` (decreasing), use

```
> df[ order(df$x2, -df$x1), ]
           x1 x2         x3
5  0.12928774  1  2.5283366
2 -0.23017749  1 -1.7272706
3  1.55870831  2  1.6901844
4  0.07050839  2  0.5038124
1 -0.56047565  2  1.2805549
```

Let us get back to our previous example: observe that the `read.table` function automatically converts character variables to factors, in the dataframe. This can be avoided using the `stringsAsFactors` argument.

For extremely large datasets, one strategy can be to select only some columns to be imported, either manually or by using a function from `library(colbycol)`, a package intended for reading big datasets into R.

```
> mycols <- rep("NULL",11)
> mycols[c(1,2,3)] <- NA
> StormMax <- read.table(file,header=TRUE,sep=",",colClasses=mycols)
> tail(StormMax,3)
       Yr Region     Wmax
2098 2009  Basin 90.00000
2099 2009     US 50.44100
2100 2009     US 65.28814
```

Here, `colClasses` simply has non-null elements to specify columns of interest. It is actually faster to specify the class of the elements to import the dataset,

```
> mycols <- rep("NULL",11)
> mycols[c(1,2,3)] <- c("integer","factor","numeric")
```

For large datasets, it might be faster to import a zipped dataset, for instance using

```
> read.table(unz("file.zip",filename="file.txt"))
```

or if the dataset is online,

```
> import.url.zip <- function(file,name="file.txt"){
+ temp = tempfile()
+ download.file(file,temp);
+ read.table(unz(temp,name ),sep=";",header=TRUE,encoding="latin1")
+ }
```

(the `unz` function works only on files located on our computer, so we have to download the file first, and then unzip it). If we consider the contract database used in Chapter 14, then it is two times faster to open a zipped file,

```
> system.time(read.table("CONTRACTS.txt",sep=";",header=TRUE))
      user      system     elapsed
     5.200       0.122       5.319
> system.time(read.table(unz("CONTRACTS.txt.zip",
+ filename="CONTRACTS.txt"),sep=";",header=TRUE))
      user      system     elapsed
     2.679       0.053       2.722
```

Because R uses the computer RAM, it can handle only small sets of data. But some packages might allow one to work with much larger volumes, like `ff` or `bigmemory`. It is also possible to use R within Python using the `rpy2` package, as Python reads data much more efficiently than R. And both have well established means of communicating with Hadoop, mainly leveraging Hadoop. In R, it is also possible to use the `mapReduce` package.

In databases, there might be missing values. The value `NA` represents a missing value. To test whether there is a missing value or not, we use the `is.na()` function. This function will return either `TRUE` or `FALSE`. It is then possible to work with components of a vector which are not missing,

```
> Xfull <- X[is.na(X)==FALSE]
```

or equivalently

```
> Xfull <- X[!is.na(X)]
```

Note that most of the statistical functions have an option to specify how to deal with missing values. With the `mean` function (to compute the mean), if the argument `na.rm` is `TRUE`, then missing values are removed and the mean is computed on the sub-vector. Similarly, with the `lm()` function (to estimate a linear model), it is possible to specify the `na.action` argument.

To speed up dataframe operations when working with (very) large datasets, it is possible to use the library `data.table` (this format will be used, and discussed, in Chapter 16). Subsetting the dataset is here two times faster.

```
> library(data.table)
> DF <- data.frame(matrix(rnorm(100000), 10000, 10)); DF$index <- 1:nrow(DF)
> DT <- data.table(DF)
> benchmark(DF[DF$X1 >2, ], DT[DT$X1 >2, ])[,c(1,3,4)]
             test elapsed relative
1 DF[DF$X1 > 2, ]   0.254    3.098
2 DT[DT$X1 > 2, ]   0.082    1.000
```

Note that the function `write.table()` can be used to export an R matrix, or dataframe, as a text file. For more complex objects, it is possible to use `cat()`

```
> cat(object,file="namefile.txt", append=FALSE)
```

where append means that we can either add the object to the existing file (if append=TRUE) or overwrite the file (if append=FALSE). It is convenient to use the cat() function to write sentences in the R console:

```
> cat("File DF contains",nrow(DF),"rows \n")
File DF contains 10000 rows
```

One can also use the sink() function, usually seen as the complement of the source() function. We can create a text file, and store any kind of object inside:

```
> sink('DT.txt')
> DT
> sink()
```

There exists a more elementary function, named scan(), to import data not conforming to the matrix layout required by read.table(). The scan can be used to read html pages, for example,

```
> scan("http://cran.r-project.org/",what="character",encoding="latin1")
Read 69 items
```

When working with dataframes, it is also possible to use SQL queries, using function sqldf() (from the eponyme library) for instance. For instance, to merge two dataframes df1 and df2, based on a common Id variable, use

```
> library(sqldf)
> df3 <- sqldf("SELECT Id, X1, X2 FROM df1 JOIN df2 USING(Id)")
```

for a standard inner join. There is also a join() function in the plyr package,

```
> library(plyr)
> df3 <- join(df1, df2, type="inner")
```

Another application of SQL queries will be mentioned in the next section, to extract data from Excel® files.

1.2.4 Importing Excel® Files and SAS® Tables

In Section 1.2.3 we saw how to load a txt or a csv file. Note that R provides a package called foreign to read (and write) files in formats that are commonly used by other software tools (see Table 1.1 and Murdoch (2002) for more details).

Sometimes, datasets are stored in spreadsheets, Excel spreadsheets, for instance. In finance and insurance, spreadsheets are very common as data storage and for communication purposes. But one should keep in mind that there may be problems when working with spreadsheets. Because some spreadsheets might contain subsheets with interconnected formulas, macros and so on, it might be difficult to read spreadsheets with multiple sheets, to extract proper information, and not to be confused by other contents within the file. Further, some spreadsheets are encoded in proprietary formats that encrypt the data, or make it hard to read (think of dates, or amounts with $ symbol or commas). The most convenient way to import data from a spreadsheet is to extract the data from the file using one or more text files.

But if one still wants to read Excel spreadsheets directly, it is possible. On a Windows platform, one can use the ODBConnectExcel() function of the library(RODBC). The first step is to connect the file, using

TABLE 1.1
Reading datasets in other formats, using library `foreign`.

Base function	Format
`read.dbf`	Read a DBF file
`read.dta`	Read a Stata binary file
`read.epiinfo`	Read a Epi Info data file
`read.mtp`	Read a Minitab worksheet
`read.octave`	Read a Octave text data files
`read.spss`	Read an spss data file
`read.ssd`	Read a dataframe from a SAS permanent dataset, via `read.xport`
`read.systat`	Read a Systat dataframe
`read.xport`	Read a SAS XPORT file

```
> sheet <- "c:\\Documents and Settings\\user\\excelsheet.xls"
> connection <- odbcConnectExcel(sheet)
> spreadsheet <- sqlTables(connection)
```

The `sqlTables()` function is helpful in case sheets have different names. Now `spreadsheet$TABLE_NAME` will return sheet names. Then, we can make an SQL request:

```
> query <- paste("SELECT * FROM",spreadsheet$TABLE_NAME[1],sep=" ")
> result <- sqlQuery(connection,query)
```

This function can also be used to import Access tables. An alternative, available on all platforms, is to use the `read.xls()` function of `library(gdata)` , the syntax being

```
> result <- read.xls("excelsheet.xls", sheet="Sheet 1")
```

It is possible to use more advanced SQL functions with `library(RMySQL)` . The generic function here is

```
> drv <- dbDriver("MySQL")
```

To read SAS databases, namely files with extension sas7bdat, the most convenient way (especially if we do not have SAS on our computer, and therefore can cannot export the file in a more appropriate format) is to use function `read.sas7bdat` from libary `sas7bdat`. For more details, SAS users should read Kleinman & Horton (2010), while Spector (2008) gives a lot of more general information on database management with R. And Wei (2012) describes the PROC_R macro, which enables native R programming in SAS.

1.2.5 Characters, Factors and Dates with R

1.2.5.1 Strings and Characters

Several functions can be used when dealing with strings. For instance, define object

```
> city <- "Boston, MA"
> city
[1] "Boston, MA"
```

One can count the number of characters in that object

```
> nchar(city)
[1] 10
```

It is possible to extract parts of a string, at specific locations,

```
> substr(city,9,10)
[1] "MA"
```

or to add characters

```
> city <- paste(city,"SSACHUSETTS",sep="")
> city
[1] "Boston, MASSACHUSETTS"
```

even to split strings into a list of elements

```
> (strsplit(city, ", "))
[[1]]
[1] "Boston"          "MASSACHUSETTS"
```

The output of this function is a list. It is possible to obtain a vector using function `unlist()`. Of course, all those operations can be done on vectors (as R is a vector language)

```
> cities <- c("New York, NY", "Los Angeles, CA", "Boston, MA")
> substr(cities, nchar(cities)-1, nchar(cities))
[1] "NY" "CA" "MA"
```

or

```
> unlist(strsplit(cities, ", "))[seq(2,6,by=2)]
[1] "NY" "CA" "MA"
```

Strings of characters can be inputs in actuarial modelling (with location, disease, names, etc.) but also output, as it might be preferable to write sentences instead of simply reporting a number. Function `cat()` can. be used to output objects (including strings):

```
> cat("Number of available packages = ",length(available.packages()[,1]))
Number of available packages =  4239
```

If we want to see how many packages start with an 'e' (or an 'E' if we use function `tolower()`) , we can use

```
> packageletter <- "e"
> cat("Number of packages \n starting with a \"",packageletter,"\" is ",
+ sum(tolower(substr(available.packages()[,1],1,1))==packageletter),sep="")
Number of packages
 starting with a "e" is 154
```

Note that if we actually want to print a quote symbol, it is necessary to put a backslash symbol in front, \" (as for location of files in Windows format, where \\ was used). See Feinerer (2008) for more information about character and string manipulation, as well as an introduction to textmining.

1.2.5.2 Factors and Categorical Variables

In statistical modeling, characters (or character sequences) are usually used as factors. It is always possible to convert names using

```
> x <- c("A", "A", "B", "B", "C")
> x
[1] "A" "A" "B" "B" "C"
```

It is also possible to use the `letters` object for lower-case letters of the Roman alphabet, or `LETTERS` for upper-case letters,

```
> x <- c(rep(LETTERS[1:2],each=2),LETTERS[3])
> x
[1] "A" "A" "B" "B" "C"
```

One can transform those letters for factors, or levels of some qualitative categorical variable,

```
> x <- factor(x)
> x
[1] A A B B C
Levels: A B C
```

Factors are labelled observations with a predefined set of labels:

```
> unclass(x)
[1] 1 1 2 2 3
attr(,"levels")
[1] "A" "B" "C"
```

As we can see from this example, a factor is stored in R as a set of codes, taking values in $\{1, 2, \ldots, n\}$, where n is the predefined number of categories that can be interpreted as levels. Observe that those levels are sorted using an alphabetic ordering,

```
> factor(rev(x))
[1] C B B A A
Levels: A B C
```

It is possible to change those labels easily (the order will then be the one specified with the `labels` parameter):

```
> x <- factor(x, labels=c("Young", "Adult", "Senior"))
> x
[1] Young  Young  Adult  Adult  Senior
Levels: Young Adult Senior
```

If the variable `x` is used in a regression context, then level `Young` will be the reference (as the first level). In order to specify another reference level, one should use

```
> relevel(x,"Senior")
[1] Young  Young  Adult  Adult  Senior
Levels: Senior Young Adult
```

From that vector with different categories, it is possible to create dummy-coded variables (sometimes called contrasts) that represent the levels of the factor,

```
> model.matrix(~0+x)
  xYoung xAdult xSenior
1      1      0       0
2      1      0       0
3      0      1       0
4      0      1       0
5      0      0       1
```

This symbolic notation `~0+x` will be discussed in Section 1.2.6. Finally, we can also mention that levels can be ordered,

```
> x <- factor(x, labels=c("Young", "Adult", "Senior"),ordered=TRUE)
> x
[1] Young  Young  Adult  Adult  Senior
Levels: Young < Adult < Senior
```

(that might be interesting in the context of multinomial ordered regression).

```
> cut(U,breaks=2)
 [1] (0.0609,0.527] (0.0609,0.527] (0.527,0.993]  (0.527,0.993]  (0.0609,0.527]
 [6] (0.527,0.993]  (0.527,0.993]  (0.527,0.993]  (0.527,0.993]  (0.0609,0.527]
[11] (0.0609,0.527] (0.0609,0.527] (0.527,0.993]  (0.0609,0.527] (0.527,0.993]
[16] (0.0609,0.527] (0.527,0.993]  (0.527,0.993]  (0.0609,0.527] (0.527,0.993]
Levels: (0.0609,0.527] (0.527,0.993]
```

When `breaks` is specified as a single number (here 2), the range of the data is divided into two pieces of equal length. Observe that the outer limits are moved away by 0.1% of the range. One can rename those two pieces,

```
> cut(U,breaks=2,labels=c("small","large"))
 [1] small small large large small large large large large small small small large
[14] small large small large large small large
Levels: small large
```

The cutoff point here depends on the range of the initial data. In order to have a fixed split, consider

```
> cut(U,breaks=c(0,.3,.8,1),labels=c("small","medium","large"))
 [1] small  medium medium large  small  large  large  medium medium small  small
[12] small  medium medium medium medium medium large  medium medium
Levels: small medium large
```

To get the frequency for each factor, we use `table()`:

```
> table(cut(U,breaks=c(0,.3,.8,1),labels=c("small","medium","large")))

 small medium  large
     5     11      4
```

To generate vectors of factors, it is possible to use function `gl()`:

```
> gl(2, 4, labels = c("In", "Out"))
[1] In  In  In  In  Out Out Out Out
Levels: In Out
```

1.2.5.3 Dates in R

Among simple functions to create dates are the `strptime()` and `as.Date()` functions, used to convert character chains into dates. The `strptime()` function creates a `POSIXct` or `POSIXlt` object (based on—signed—number of seconds since the beginning of 1970, in the UTC timezone for `POSIXct` objects, while a `POSIXlt` object is a list of day, month, year, hour, minute, second, etc.). As it is a date/time class, one can specify the hour and the time zone (using the `tz` option).

```
> some.dates <- strptime(c("16/Oct/2012:07:51:12","19/Nov/2012:23:17:12"),
+ format="%d/%b/%Y:%H:%M:%S")
> some.dates
[1] "2012-10-16 07:51:12" "2012-11-19 23:17:12"
```

To find how many days have elapsed, do

```
> diff(some.dates)
Time difference of 34.68472 days
```

but it is also possible to use the dedicated function `difftime`

```
> difftime(some.dates[2],some.dates[1],units = "hours")
Time difference of 832.4333 hours
```

Function `as.Date()` converts character chains into objects of class `Date` (representing calendar dates)

```
> some.dates <- as.Date(c("16/10/12","19/11/12"),format="%d/%m/%y")
> some.dates
[1] "2012-10-16" "2012-11-19"
```

It is possible to use the `seq()` function to create date sequences:

```
> sequence.date <- seq(from=some.dates[1],to=some.dates[2],by=7)
> sequence.date
[1] "2012-10-16" "2012-10-23" "2012-10-30" "2012-11-06" "2012-11-13"
```

Consider the following function, that generates a date from the month, the day and the year:

```
> mdy = function(m,d,y){
+ d.char = as.character(d); d.char[d<10]=paste("0",d.char[d<10],sep="")
+ m.char = as.character(m); m.char[m<10]=paste("0",m.char[m<10],sep="")
+ y.char = as.character(y)
+ return(as.Date(paste(m.char,d.char,y.char,sep="/"),"%m/%d/%Y"))
+ }
> mdy(c(12,6),5,c(1975,1976))
[1] "1975-12-05" "1976-06-05"
```

One can also convert those dates using the `format()` function,

```
> format(sequence.date,"%b")
[1] "oct" "oct" "oct" "nov" "nov"
```

or use a more specific functions, like `weekdays()`, to know the weekday,

```
> weekdays(some.dates)
[1] "Tuesday" "Monday"
```

But in that case, we did not create the objects. In order to extract the month, and define a Months object, we use

```
> Months <- months(sequence.date)
> Months
[1] "october"  "october"  "october"  "november" "november"
```

To create a vector that contains the year, we can use

```
> Year <- substr(as.POSIXct(sequence.date), 1, 4)
> Year
[1] "2012" "2012" "2012" "2012" "2012"
```

Note that the use of as.POSIXct function to extract the year is slow, and strftime(,"%Y") is actually much faster

```
> randomDates <- as.Date(runif(100000,1,100000))
> system.time(year1 <- substr(as.POSIXct(randomDates), 1, 4))
   user  system elapsed
  8.112 0.039 8.112
> system.time(year2 <- strftime(randomDates,"%Y"))
   user  system elapsed
   0.128 0.003 0.130
```

See Ripley & Hornik (2001) or Grothendieck & Petzoldt (2004) for more information about date and time classes.

One should keep in mind that outputs are related to the language R is using, which can be changed. If we want outputs in German, use

```
> Sys.setlocale("LC_TIME", "de_DE")
[1] "de_DE"
```

and weekdays are now

```
> weekdays(some.dates)
[1] "Dienstag" "Montag"
```

while in French,

```
> Sys.setlocale("LC_TIME", "fr_FR")
[1] "fr_FR"
```

the output of function weekdays is

```
> weekdays(some.dates)
[1] "Mardi" "Lundi"
```

and with a Spanish version

```
> Sys.setlocale("LC_TIME", "es_ES")
[1] "es_ES"
```

the output of months is

```
> months(some.dates)
[1] "octubre"   "noviembre"
```

1.2.6 Symbolic Expressions in R

In some cases, it is necessary to write symbolic expressions in R, using a version of the commonly used notations of Wilkinson & Rogers (1977), for example when running a regression where a *formula* has to be specified. In a call of the `lm()` function, formula `y ~ x1 + x2 + x3` means that we consider a model

$$Y_i = \beta_0 + \beta_1 X_{1,i} + \beta_2 X_{2,i} + \beta_3 X_{3,i} + \varepsilon_i.$$

The function `lm()` returns an object of class `lm` and the generic call is

```
> fit <- lm(formula = y ~ x1 + x2 + x3, data=df)
```

where `df` is a dataframe which contains variables named `x1`, `x2`, `x3` and `y`. In a formula, `+` stands for inclusion (not for summation), and `-` for exclusion. To run a regression on X_1 and the variable $X_2 + X_3$, we use

```
> fit <- lm(formula = y ~ x1 + I(x2+x3), data=df)
```

For categorical variables, possible interactions between X_1 and X_2 can be obtained using `x1:x2`. To get a better understanding of symbolic notations, consider the following dataset

```
>  set.seed(123)
>  df <- data.frame(Y=rnorm(50), X1=as.factor(sample(LETTERS[1:4],size=50,
+  replace=TRUE)), X2=as.factor(sample(1:3,size=50,replace=TRUE)))
> tail(df)
        Y X1 X2
45  1.030  B  3
46  0.684  C  3
47  1.667  B  3
48 -0.557  B  2
49  0.950  C  2
50 -0.498  A  3
```

The default model, with a regression on `X1+X2` will generate the following matrix:

```
> reg <- lm(Y~X1+X2,data=df)
> model.matrix(reg)[45:50,]
   (Intercept) X1B X1C X1D X22 X23
45           1   0   0   1   0   1
46           1   0   0   0   1   0
47           1   0   0   0   1   0
48           1   0   0   0   1   0
49           1   0   0   0   0   0
50           1   0   1   0   1   0
```

with an `(Intercept)` vector, and then, indicator variables, except for the first level, namely `A` for `X1` and `1` for `X2`. It is a model with $1 + (4 - 1) + (3 - 1) = 6$ explanatory variables. Now, if we add `x1:x2` in the regression, the model matrix will be

```
> reg <- lm(Y~X1+X2+X1:X2,data=df)
> model.matrix(reg)[45:50,]
   (Intercept) X1B X1C X1D X22 X23 X1B:X22 X1C:X22 X1D:X22 X1B:X23 X1C:X23 X1D:X23
45           1   1   0   0   0   1       0       0       0       1       0       0
46           1   0   1   0   0   1       0       0       0       0       1       0
47           1   1   0   0   0   1       0       0       0       1       0       0
```

```
48              1   1   0   0   1   0       1       0       0       0       0       0
49              1   0   1   0   1   0       0       1       0       0       0       0
50              1   0   0   0   0   1       0       0       0       0       0       0
```

Thus, cross products of all remaining variables are considered, here, namely {B, C, D} × {2, 3}. The code above is (strictly) equivalent to

```
> reg <- lm(Y~X1*X2,data=df)
> model.matrix(reg)[45:50,]
   (Intercept) X1B X1C X1D X22 X23 X1B:X22 X1C:X22 X1D:X22 X1B:X23 X1C:X23 X1D:X23
45           1   1   0   0   0   1       0       0       0       1       0       0
46           1   0   1   0   0   1       0       0       0       0       1       0
47           1   1   0   0   0   1       0       0       0       1       0       0
48           1   1   0   0   1   0       1       0       0       0       0       0
49           1   0   1   0   1   0       0       1       0       0       0       0
50           1   0   0   0   0   1       0       0       0       0       0       0
```

It is a model with $1+(4-1)+(3-1)+(4-1)\times(3-1)=12$ explanatory variables. Note that cross interactions, of {A, B, C, D} × {1, 2, 3}, can be obtained using

```
> reg <- lm(Y~X1:X2,data=df)
```

which contains $4\times 3+1=13$ columns,

```
> ncol(model.matrix(reg))
[1] 13
```

For more subtle interpretations of regressions, it is possible to use %in% with some nested models. For instance,

```
> reg <- lm(Y~X1+X2%in%X1,data=df)
> model.matrix(reg)[45:50,]
   (Intercept) X1B X1C X1D X1A:X22 X1B:X22 X1C:X22 X1D:X22 X1A:X23 X1B:X23 X1C:X23 X1D:X23
45           1   1   0   0       0       0       0       0       0       1       0       0
46           1   0   1   0       0       0       0       0       0       0       1       0
47           1   1   0   0       0       0       0       0       0       1       0       0
48           1   1   0   0       0       1       0       0       0       0       0       0
49           1   0   1   0       0       0       1       0       0       0       0       0
50           1   0   0   0       0       0       0       0       1       0       0       0
```

where variable X1 is here (without the first level because the constant is included here), as well as cross interactions of {A, B, C, D} × {2, 3}. It is a model with $1+3+4\times(3-1)=12$ explanatory variables. See Pinheiro & Bates (2000) or Kleiber & Zeileis (2008), among many others, for more details on symbolic expressions in regressions.

To conclude this section, observe that a *formula* is a string, so it is possible to use dedicated functions to run arbitrary regressions

```
> stringformula <- paste("Y ~",paste(names(df)[2:3],collapse=" + "))
> stringformula
[1] "Y ~ X1 + X2"
> fit <- lm(formula= stringformula, data=df)
```

1.3 Basics of the R Language

In this section, we introduce briefly how to use R, more precisely how to use functions from R libraries and how to create our own functions. We will also discuss how to visualize outputs. For more details, we refer to Matloff (2011), Teetor (2011), Kabacoff (2011) or Craley (2012), among many others.

1.3.1 Core Functions

Because R is open-source, it is possible to see what R is actually computing, even for core functions. Just type the name of a function will return the code of the function. For instance, the `factorial()` function,

```
> factorial
function (x)
gamma(x + 1)
<bytecode: 0x1708aa7c>
<environment: namespace:base>
```

where we see that $n! = \Gamma(n+1)$, where $\Gamma(\cdot)$ is the standard gamma function[1]. Now, if we want to see what this `gamma()` function is

```
> gamma
function (x)  .Primitive("gamma")
```

which looks up for a primitive (internally implemented) function.

Standard statistical functions are already defined in R, such as `sum()`, `mean()`, `var()` or `sd()`. Note that `var()` is the (standard) unbiased estimator of the variance, defined as

$$\frac{1}{n-1}\sum_{i=1}^{n}(x_i - \overline{x})^2 \text{ where } \overline{x} = \frac{1}{n}\sum_{i=1}^{n} x_i$$

See for instance

```
> x <- 0:1
> sum((x-mean(x))^2)
[1] 0.5
> var(x)
[1] 0.5
```

All those functions can be used on vectors,

```
> x <- c(1,4,6,6,10,5)
> mean(x)
[1] 5.333333
```

but it also works if `x` is a matrix, which will be considered a vector

```
> m <- matrix(x,3,2)
> m
     [,1] [,2]
[1,]    1    6
[2,]    4   10
[3,]    6    5
> mean(m)
[1] 5.333333
```

[1] Note that this function will return numbers slightly different from the ones obtained using `prod(1:x)` for large values of `x`. In that case, one might use `library(gmp)` and either `factorialZ(x)` or `prod(as.bigz(1:x))` (which are now identical).

To compute means per row (or per column), we can use the `apply()` function, argument `1`, meaning that the function will be applied on each *row* (the first index):

```
> apply(m,1,mean)
[1] 3.5 7.0 5.5
```

Nevertheless, to compute sums per row, or column, `rowSums()` and `colSums()` can be much faster. With the `apply()` command, it is possible to use more complex functions that return not a numeric value, but a vector. For instance, `cumsum()` can be used to return cumulated sums per columns, or rows,

```
> apply(m,2,cumsum)
     [,1] [,2]
[1,]    1    6
[2,]    5   16
[3,]   11   21
```

We might also be interested to compute means given another variate (a factor)

```
> sex <- c("H","F","F","H","H","H")
> base <- data.frame(x,sex)
> base
   x sex
1  1   H
2  4   F
3  6   F
4  6   H
5 10   H
6  5   H
```

Then we use the `tapply()` function:

```
> tapply(x,sex,mean)
  F   H
5.0 5.5
```

Note that if the function of interest is the sum, per factor, then the `rowsum()` function can also be used

```
> rowsum(x,sex)
  [,1]
F   10
H   22
```

Consider now a second categorical variable

```
> base$hair <- c("Black","Brown","Black","Black","Brown","Blonde")
```

One can compute a two-way contingency table using

```
> table(base$sex,base$hair)

    Black Blonde Brown
  F     1      0     1
  H     2      1     1
```

that may also include sums by row, and column,

```
> addmargins(table(base$sex,base$hair))

      Black Blonde Brown Sum
  F       1      0     1   2
  H       2      1     1   4
  Sum     3      1     2   6
```

We will discuss later on how to speed up R codes (for example, using C/C++, as discussed in Section 1.4.3).

1.3.2 From Control Flow to "Personal" Functions

We have seen, so far, many R functions. To get some help on functions, it is possible to use either `help` or `?`. For instance,

```
> ?quantile
```

will display a help page for the `quantile()` function, with details on the input values, the output, the algorithm, and some examples.

But most of the time, there is no function to compute what we need, so we have to write our own functions. But let us start with a short paragraph about some interesting commands, related to control flow, and then try to code our own functions.

1.3.2.1 Control Flow: Looping, Repeating and Conditioning

Computers are great to repeat the same task a lot of times. Sometimes, it is necessary to repeat some statements a given number of times, or until a condition becomes true (or false). This can be done using either `for()` or `while()`.

A `for` loop executes a statement (between two braces, `{` and `}`), repetitively, until a variable is no longer in a given sequence. In `for(i in 2^(1:4)){...}`, some statements will be repeated four times, and `i` will take values `2`, `4`, `8` and `16`. Consider a folder where several csv files are stored; you wish to import all of them and store them in a list. You can use

```
> listdf <- list()
```

to create an empty list object, and then a vector with all the names of the files,

```
> listcsv <- dir(pattern = "*.csv")
```

You can also use `Sys.glob("*.csv")`. Finally, you can use a loop to import all the files,

```
> for(filename in listcsv){ listdf[filename] <- read.csv(filename)}
```

A `while` loop executes a statement (again between two braces), repetitively, until a condition is no longer true. In

```
> T <- NULL
> while(sum(T)<=10){ T <- c(T,rexp(1)) }
```

we generate a Poisson process (more precisely, inter-arrival time, see Chapter 2), until the total time exceeds 10 (for the first time). `while` loops can be dangerous, when badly specified: R loop will perhaps never end. Use `?control` for more information. But one should keep in mind that loops can be time consuming, and inefficient, and a better alternative is to use an `apply()` type of function (see next section). For instance, a faster way to import all the csv files in a list is to use

```
> listdf <- lapply(dir(pattern = "*.csv"), read.csv)
```

A common use of those functions can be found in the gradient descent, to derive maximum likelihood estimators (see Chapter 2). The looping procedure is here

1. Start from some initial value $\boldsymbol{\theta}_0$
2. At step $k \geq 1$, set $\boldsymbol{\theta}_k = \boldsymbol{\theta}_{k-1} - H[\log \mathcal{L}(\boldsymbol{\theta}_{k-1})]^{-1} \nabla \log \mathcal{L}(\boldsymbol{\theta}_{k-1})$

and loop this second item. Here, $\nabla \log \mathcal{L}(\boldsymbol{\theta})$ is the gradient of the log-likelihood, and $H[\log \mathcal{L}(\boldsymbol{\theta})]$ the Hessian matrix. We can either decide to use a finite loop, using `for`, with 100 iterations for instance, or a loop that ends only when the change is too small using `while`, where we repeat the iterative algorithm while $\|\boldsymbol{\theta}_k - \boldsymbol{\theta}_{k-1}\| > \epsilon$ for some small ϵ. An application (using a `for` loop) is given in Section 4.2.1 to derive maximum likelihood estimators for the logistic regression. Actually, computing the Hessian matrix can be long, and for numerical reasons, working with an approximation is usually sufficient. This is the idea underlying the so-called Broyden–Fletcher–Goldfarb–Shanno (BFGS) algorithm; see Broyden (1970), Fletcher (1970), Goldfarb (1970) and Shanno (1970), using an R optimisation routine.

Conditional statements are also possible, using `if()` or `ifelse()`.

```
> set.seed(1)
> u <- runif(1)
> if(u>.5) {("greater than 50%")} else {("smaller than 50%")}
[1] "smaller than 50%"
> ifelse(u>.5,("greater than 50%"),("smaller than 50%"))
[1] "smaller than 50%"
> u
[1] 0.2655087
```

The main difference is that `ifelse()` is vectorizable, but not `if()`.

```
> u <- runif(3)
> if(u>.5) {print("greater than 50%")} else {("smaller than 50%")}
[1] "smaller than 50%"
Warning message:
In if (u > 0.5) { :
  the condition has length > 1 and only the first element will be used
> ifelse(u>.5,("greater than 50%"),("smaller than 50%"))
[1] "smaller than 50%" "smaller than 50%" "greater than 50%"
> u
[1] 0.2655087 0.3721239 0.5728534
```

1.3.2.2 Writing Personal Functions

In R, writing a function is rather simple. For instance, consider function $\mathbb{R}^n \times [0,1]^n \times \mathbb{R}_+^n \to \mathbb{R}$ defined as

$$(\boldsymbol{x}, \boldsymbol{p}, \boldsymbol{d}) \mapsto \sum_{i=1}^{n} \frac{p_i \cdot x_i}{(1+d_i)^i}$$

The code to define this function can be

```
> f <- function(x,p,d){
+ s <- sum(p*x/(1+d)^(1:length(x)))
+ return(s)
+ }
```

As mentioned previously, if we type `f` in the console, the code of the function will appear:

```
> f
 function(x,p,d){
 s <- sum(p*x/(1+d)^(1:length(x)))
 return(s)
 }
```

If we ask for `f()`, then R will return an error message

```
> f()
Error in p * x : 'p' is missing
```

because parameters of the function were not specified. To call that function, the syntax is the same as for R core functions,

```
> f(x=c(100,200,100),p=c(.4,.5,.3),d=.05)
[1] 154.7133
```

or equivalently

```
> f(c(100,200,100),c(.4,.5,.3),.05)
[1] 154.7133
```

Functions with named arguments also have the option of specifying default values for those arguments, for example `f <- function(x,p,d=.05)` (and in that case, it may be left out in a call).

```
> f(c(100,200,100),c(.4,.5,.3))
[1] 154.7133
```

It is not necessary to name the argument when calling a function: if names of arguments are not given, R assumes they appear in the order of the function definition. A standard example is probably the `log()` function. This function has two arguments: `x` (a positive scalar, or a vector) and `base` (the base with respect to which the logarithm is computed). Thus, the following five calls are equivalent:

```
> log(base = 2, x = 16)
> log(x = 16, base = 2)
> log(16, 2)
> log(x = 16, 2)
> log(16, base = 2)
[1] 4
```

It is the same for most R functions; for instance, the `qnorm()` function computes quantiles of the $\mathcal{N}(0,1)$ distribution

```
> qnorm(.95)
[1] 1.644854
```

To get quantiles of a $\mathcal{N}(\mu,\sigma^2)$ distribution, we use

```
> qnorm(.95,mean=1,sd=2)
[1] 4.289707
```

Quantities related to standard statistical distributions will be studied later on in this chapter, and more intensively in Chapter 2.

It is also possible to define functions within functions. For instance, if we want to compute

$$f : x \mapsto \frac{H(x)}{\int_x^\infty H(t)dt} \text{ where } H(t) = 1 - \Phi_{\mu,\sigma}(t) = \int_t^\infty \varphi_{\mu,\sigma}(t)dt,$$

where $\varphi_{\mu,\sigma}()$ is the density function of the $\mathcal{N}(\mu, \sigma^2)$ distribution,

```
> f <- function(x,m=0,s=1){
+  H<-function(t) 1-pnorm(t,m,s)
+  integral<-integrate(H,lower=x,upper=Inf)$value
+  res<-H(x)/integral
+  return(res)
+ }
```

The (first) argument of function `f` is not a vector. Using one yields a warning=

```
> f(x <- 0:1)
[1] 1.2533141 0.3976897
Warning :
In if (is.finite(lower)) { :
the condition has length > 1 and only the first element will be used
```

If we want to compute a vector $[f(x_i)]$ for some x_i's, we should *vectorize* the function, using

```
> Vectorize(f)(x)
[1] 1.253314 1.904271
```

Similarly, we can also use a loop

```
> y <- rep(NA,2)
> x <- 0:1
> for(i in 1:2) y[i] <- f(x[i])
> y
[1] 1.253314 1.904271
```

or use the `sapply` function

```
> y <- sapply(x,"f")
> y
[1] 1.253314 1.904271
```

It is also possible to use recursion to define functions, in R. Consider the popular towers of Hanoi example (see Section 1.1 in Graham et al. (1989)). The function used in this example is $h(n) = 2h(n-1) + 1$ if $n \geq 2$ and $h(1) = 1$. Then it is possible to define recursively a`hanoi()` function naturally

```
> hanoi <- function(n) if(n<=1) return(1) else return(2*hanoi(n-1)+1)
> hanoi(4)
[1] 15
```

The R function can use local variables that might have the same name as non-local ones. Consider the following code to illustrate this point:

```
> beta <- 0
> slope <- function(X,Y){
+ beta <- coefficients(lm(Y~ X))[2]
+  return(as.numeric(beta))
+ }
> attach(cars)
> slope(speed,dist)
[1] 7.864818
> beta
[1] 0
```

It is possible to use the `cat()` function to print comments or values:

```
>  slope <- function(X,Y){
+  beta <- coefficients(lm(Y~ X))[2]
+  cat("The slope is",beta,"\n")
+  return(as.numeric(beta))
+  }
> slope(speed,dist)
The slope is 7.864818
[1] 7.864818
```

Note that the `<<-` operator can be used to assign in the global environment; this might be important in some functions.

In the syntax of those functions, we did not specify either the class or the size of input and output variables. The bivariate Gaussian density function with zero means and unit variances is

$$\varphi(x,y) = \frac{1}{2\pi\sqrt{1-\rho^2}} \exp\left(-\frac{1}{2(1-\rho^2)}\left[x^2+y^2-2\rho xy\right]\right), \forall x,y \in \mathbb{R}^2.$$

```
> binorm <- function(x1,x2,r=0){
+ exp(-(x1^2+x2^2-2*r*x1*x2)/(2*(1-r^2)))/(2*pi*sqrt(1-r^2))
+ }
```

Note that such a function exists in the `mnormt` package (see Hothorn et al. (2001)). If the input values are real vectors of length n, so will be the output,

```
> u <- seq(-2,2)
> binorm(u,u)
[1] 0.002915024 0.058549832 0.159154943 0.058549832 0.002915024
```

It is also possible to return a matrix (or more generally an array) with generic values $\varphi(u_i, v_j)$, from two vectors $\boldsymbol{u}$ and $\boldsymbol{v}$.

```
> outer(u,u,binorm)
            [,1]       [,2]       [,3]       [,4]        [,5]
[1,] 0.002915024 0.01306423 0.02153928 0.01306423 0.002915024
[2,] 0.013064233 0.05854983 0.09653235 0.05854983 0.013064233
[3,] 0.021539279 0.09653235 0.15915494 0.09653235 0.021539279
[4,] 0.013064233 0.05854983 0.09653235 0.05854983 0.013064233
[5,] 0.002915024 0.01306423 0.02153928 0.01306423 0.002915024
```

Observe that the previous vector is the diagonal of this matrix. This function will be used to plot surfaces in dimension 2. An alternative is to use `expand.grid()`,

```
> (uv<-expand.grid(u,u))
> head(uv)
  Var1 Var2
1   -2   -2
2   -1   -2
3    0   -2
4    1   -2
5    2   -2
6   -2   -1
> matrix(binorm(uv$Var1,uv$Var2),5,5)
            [,1]       [,2]       [,3]       [,4]        [,5]
[1,] 0.002915024 0.01306423 0.02153928 0.01306423 0.002915024
[2,] 0.013064233 0.05854983 0.09653235 0.05854983 0.013064233
[3,] 0.021539279 0.09653235 0.15915494 0.09653235 0.021539279
[4,] 0.013064233 0.05854983 0.09653235 0.05854983 0.013064233
[5,] 0.002915024 0.01306423 0.02153928 0.01306423 0.002915024
```

Here, function `binorm` can be defined on vectors, but sometimes it might be more complicated. For instance in function `f`, parameter `x` appears as a lower bound in an integral. If we wish to plot $x \mapsto f(x)$, we need to compute `f` for a sequence of `x` values. We can use the `Vectorize()` function to do so:

```
> Vectorize(f)(u)
[1] 0.4865593 0.7766387 1.2533141 1.9042712 2.6794169
```

It is also possible to define new binary operators, surrounded by percent signs. For instance, R recognizes `%*%` as the standard product for matrices. But define, for instance, the following function, to derive easily confidence intervals,

```
> "%pm%" <- function(x,s) x + c(qnorm(.05),qnorm(.95))*s
```

which will return the 5% and 95% bounds of a (standard) confidence interval,

```
> 100 %pm% 10
[1]  83.55146 116.44854
```

As an illustration in this section on functions, assume that we want to compute the inverse of a strictly increasing function, for instance the quantile function associated to a continuous random variable. Consider the cumulative distribution function of a mixture model (discussed in Chapter 2), denoted $F = p_1F_1 + p_2F_2$.

```
> p <- .4; m1 <- 0; m2 <- 1; s1 <- 1; s2 <- 2
> F <- function(x) p*pnorm(x,m1,s1)+(1-p)*pnorm(x,m2,s2)
```

R has its own function, `uniroot()`, that can be used to compute the inverse of a function, as discussed in the introduction of this chapter.

```
> uniroot(function(x) F(x)-.95,interval=c(0,10))$root
[1] 3.766705
```

But we wish, here, to write our own function. A natural idea is to use a bisection method to compute F^{-1} at some point $u \in (0, 1)$.

```
> Finv1 <- function(H,u,xinf=-100,xsup=100){
+ cond <- FALSE
```

```
+ while(!cond){
+   xmid <- (xinf+xsup)/2
+ if(H(xmid)<u) xinf <- xmid else xsup <- xmid
+ cond <- abs(H(xmid)-u)<1e-6
+ }
+ return(xmid)}
```

but we need to have a lower and an upper bound for the quantile. Here, we consider extremely large values to have a code as general as possible.

```
> Finv1(F,.95)
[1] 3.766727
```

An alternative can be to use a vectorized version of the function, and to use the $\inf\{x \in [x_{\text{inf}}, x_{\text{sup}}], F(x) > u\}$ expression of the quantile function. Thus, the code is

```
> Finv2 <- function(H,u,n=10000,xinf=-100,xsup=100){
+ vx <- seq(xinf,xsup,length=n+1)
+ vh <- Vectorize(H)(vx)
+ return(min(vx[vh>=u]))}
```

But in order to have an accurate estimate, the code might be extremely long to run,

```
> Finv2(F,.95)
[1] 3.766708
```

A third idea is to combine those two algorithms, as the latter gave us an upper bound for the quantile (and because $\sup\{x \in [x_{\text{inf}}, x_{\text{sup}}], F(x) < u\}$ will give us a lower bound).

```
> Finv <- function(H,u,n=1000,xinf=-100,xsup=100){
+  vx <- seq(xinf,xsup,length=n+1)
+  vh <- Vectorize(H)(vx)
+  xsup <- min(vx[vh>=u])
+  xinf <- max(vx[vh<=u])
+ cond <- FALSE
+ while(!cond){
+  xmid <- (xinf+xsup)/2
+  if(H(xmid)<u) xinf <- xmid else xsup <- xmid
+  cond <- abs(H(xmid)-u)<1e-9
+}
+ return(xmid)}
```

This algorithm is faster than the previous two (even if we ask for a higher precision):

```
> Finv(F,.95)
[1] 3.766708
```

1.3.3 Playing with Functions (in a Life Insurance Context)

Consider a vector corresponding to the number of people alive at age x,

```
> alive <- TV8890$Lx
```

To construct the vector `death` corresponding to curtate life times, that is, containing the number of people who died at a particular age $x = 0, 1, 2, \ldots$, do

```
> death <- -diff(alive)
```

A standard mortality law is the one suggested by Makeham, with survival probability function

$$S(x) = \exp\left(-ax - \frac{b}{\log c}[c^x - 1]\right), \forall x \geq 0,$$

for some parameters $a \geq 0, b \geq 0$ and $c > 1$. The R function to compute this function can be defined as

```
> sMakeham <- function(x,a,b,c){ ifelse(x<0,1,exp(-a*x-b/log(c)*(c^x-1))) }
```

The use of the function `ifelse()` ensures that $S(x) = 1$ if $x < 0$. The probability function associated to this survival function can be computed as

```
> dMakeham <- function(x,a,b,c){
+ ifelse(x>floor(x),0,sMakeham(x,a,b,c)-sMakeham(x+1,a,b,c))
+ }
```

Using this function, it is possible to use standard maximum likelihood techniques (see Chapter 2) to estimate those parameters, based on the sample where deaths at birth are removed (Makeham's distribution cannot capture this feature), as well as above age 105,

```
> death <- death[-c(1,107:111)]
> ages <- 1:(length(death))
> loglikMakeham <- function(abc){
+ - sum(log(dMakeham(ages,abc[1],abc[2],abc[3]))*death[ages])
+ }
```

The `optim()` function can be used to obtain maximum likelihood estimators for parameters in Makeham's survival function (assuming that we can find adequate starting values for the algorithm)

```
> mlEstim <- optim(c(1e-5,1e-4,1.1),loglikMakeham)
> abcml <- mlEstim$par
```

Based on observed ages of deaths, it is possible to compute the average age-at-death

```
> sum((ages+.5)*death)/sum(death)
[1] 81.1998
```

which can be compared to the one obtained using Makeham's survival function

```
> integrate(sMakeham,0,Inf,abcml[1],abcml[2],abcml[3])
81.16292 with absolute error < 0.0034
```

1.3.4 Dealing with Errors

It is rather common to obtain errors when running intensive computations. For instance, if we use Monte Carlo techniques to generate a sample and we fit automatically a parametric distribution, optimization might fail. To avoid that a function fails, it is possible to use function `try()` to allow the user's code to handle error recovery. Consider the estimation of a mixture distribution. One way to deal with boundary conditions is to transform the parameters, so optimization can be done over non-bounded spaces. Let `X` denote a sample of observations,

```
> X <- rnorm(100)
```

here, a $\mathcal{N}(0,1)$ sample. The density is here

```
> mixnorm <- function(x,p,m1,s1,m2,s2){
+         p <- exp(p)/(1+exp(p));
+         s1 <- exp(s1); s2 <- exp(s2)
+         (p*dnorm(x,m1,s1)+(1-p)*dnorm(x,m2,s2))*(m1<m2)}
```

where a logistic transformation of the probability is considered, as well as the logarithm of the volatility parameters. We also order the two distributions to ensure identifiability. The log-likelihood (with a minus sign as most optimization functions seek the *minimum*) is

```
> llmix <- function(p,m1,s1,m2,s2){ -sum(log(Vectorize(mixnorm)
(X,p,m1,s1,m2,s2)))}
```

The standard code to obtain the minimum is based on the `mle()` function. Unfortunately, if initial values are not chosen correctly, we might obtain an error

```
> startparam <- list(p=.5,m1=-1,m2=1,s1=1,s2=10000)
> estmix <- mle(minuslog=llmix, start=startparam)
Error in solve.default(oout$hessian) :
Lapack routine dgesv: system is exactly singular
```

If we want to print the parameters only if the algorithm did converge, it is possible to use the `try()` function:.

```
> if(!inherits(estmix <- mle (minuslog = llmix, start = startparam),
                 "try-error")){
> if(!inherits(estmix,"try-error")){
+         param<-stats4::coef(estmix)
+         cat("Probability = ",exp(param[1])/(1+exp(param[1])),
+             "Mean = ",param[2]," Std Dev = ",exp(param[3]),"\n")
+         cat("Probability = ",1/(1+exp(param[1])),
+             "Mean = ",param[4]," Std Dev = ",exp(param[5]),"\n")
+ }
```

This code will return estimates only if there were no errors in the optimization procedure.

1.3.5 Efficient Functions

In order to illustrate how to code R functions, consider the following problem: We want to create a function to generate compound random variables, $S = X_1 + \cdots + X_N$, with convention $S = 0$ when $N = 0$, where N is a counting variable, and the X_i's are i.i.d. random variables. Assume that N can be generated using some generic function `rN`, and the X_i's using `rX`. For instance, consider a compound Poisson variable, with exponential sizes.

```
> rN.Poisson <- function(n) rpois(n,5)
> rX.Exponential <- function(n) rexp(n,2)
```

The first natural idea is to use loops (see Ligges & Fox (2008) for a discussion on making loops faster),

```
> rcpd1 <- function(n,rN=rN.Poisson,rX=rX.Exponential){
+ V <- rep(0,n)
+ for(i in 1:n){
+  N <- rN(1)
+  if(N>0){V[i] <- sum(rX(N))}
+ }
+ return(V)}
```

TABLE 1.2
Splitting and combining data.

Base Function	plyr Function	Input	Output
aggregate	ddply	dataframe	dataframe
apply	aaply (or alply)	array	array (or list)
by	dlply	dataframe	list
lapply	llply	list	list
mapply	maply (or mlply)	array	array (or list)
sapply	laply	list	array

Actually, as discussed previously, the sum of an empty vector is null,

```
> sum(NULL)
[1] 0
```

so the `if()` condition is not necessary here

```
> rcpd1 <- function(n,rN=rN.Poisson,rX=rX.Exponential){
+ V <- rep(0,n)
+ for (i in 1:n) V[i] <- sum(rX(rN(1)))
+ return(V)}
```

Functions based on the `apply` functions (including `tapply`, `sapply` or `lapply`) can be used to get sums per line, per column, or cross tables. Note that `lapply` and `sapply` take a list or vector as first argument and a function to be applied to each element as second argument. The difference between the two functions is that `lapply` will return its result in a list, while `sapply` will simplify its output to a vector (or matrix); see Table 1.2 for a summary (with standard functions, as well as the one from the `plyr` package, see, for example, Wickham (2011) for a discussion).

Those functions can be interesting to avoid loops (described in Section 1.3.2.1). Consider two simple situations to illustrate those functions. In Chapter 6, on extreme value theory, we will look for the distribution of the maximum from an i.i.d. sample. To generate such values, we can use loops. Consider the maximum of 10 $\mathcal{E}(1)$ variables, for instance,

```
> set.seed(1)
> M <- rep(NA,5)
> for(i in 1:5) M[i] <- max(rexp(10))
> M
[1] 2.894969 4.423934 3.958933 1.435285 2.007832
```

The same output can be obtained using `replicate()`

```
> replicate(5, max(rexp(10)))
[1] 2.894969 4.423934 3.958933 1.435285 2.007832
```

or `apply`,

```
> apply(matrix(rexp(10*5),10,5),2,max)
[1] 2.894969 4.423934 3.958933 1.435285 2.007832
```

(where a 10×5 matrix is generated, and the maximum per column is returned).

Another popular example is the Jackknife procedure, introduced by Quenouille (1949) and Tukey (1958); see also Shao & Tu (1995) for more details. Given a sample $\boldsymbol{X} = \{x_1, \ldots, x_n\}$, and some statistic s, the idea is to use all subsamples $\boldsymbol{X}_{-i} = \{x_1, \ldots, x_{i-1}, x_{i+1}, \ldots, x_n\}$ to derive an estimate for the bias of the estimator, for instance.

```
> set.seed(1)
> n <- 10
> X <- rnorm(n)
> s <- function(x) sd(x)
```

A natural algorithm is

```
> for (i in 1:n) { u[i] <- s(X[-i])}
> shat <- s(X)
> (jck.bias <- (n - 1) * (mean(u) - shat))
[1] -0.02007206
```

for an estimator of the bias, and

```
> (jck.se <- sqrt(((n - 1)/n) * sum((u - mean(u))^2)))
[1] 0.1768931
```

for the standard error. But it is possible to avoid the `for` loop using `sapply`,

```
> u <- sapply(1:n,function(i) s(X[-i]))
```

The output is exactly the same

```
> (jck.bias <- (n - 1) * (mean(u) - shat))
[1] -0.02007206
> (jck.se <- sqrt(((n - 1)/n) * sum((u - mean(u))^2)))
[1] 0.1768931
```

Now, we can use those `apply` functions to generate compound random variables,

```
> rcpd2 <- function(n,rN=rN.Poisson,rX=rX.Exponential){
+ N <- rN(n)
+ X <- rX(sum(N))
+ I <- factor(rep(1:n,N),levels=1:n)
+ return(as.numeric(xtabs(X ~ I)))}

> rcpd3 <- function(n,rN=rN.Poisson,rX=rX.Exponential){
+ N <- rN(n)
+ X <- rX(sum(N))
+ I <- factor(rep(1:n,N),levels=1:n)
+ V <- tapply(X,I,sum)
+ V[is.na(V)] <- 0
+ return(as.numeric(V))}

> rcpd4 <- function(n,rN=rN.Poisson,rX=rX.Exponential){
+  return(sapply(rN(n), function(x) sum(rX(x))))}

> rcpd5 <- function(n,rN=rN.Poisson,rX=rX.Exponential){
+ return(sapply(Vectorize(rX)(rN(n)),sum))}

> rcpd6 <- function(n,rN=rN.Poisson,rX=rX.Exponential){
+  return(unlist(lapply(lapply(t(rN(n)),rX),sum)))}
```

As seen earlier, unlist prints the result as a numeric vector, with named components.

In order to get more details about time computation, we can run 1,000 computations,

```
> n <- 100
> library(microbenchmark)
> options(digits=1)
> microbenchmark(rcpd1(n),rcpd2(n),rcpd3(n),rcpd4(n),rcpd5(n),rcpd6(n))
Unit: microseconds
       expr  min   lq median   uq   max
1 rcpd1(n)  674  712    742  797  1896
2 rcpd2(n) 1324 1405   1487 1557 15586
3 rcpd3(n)  724  762    792  841  1102
4 rcpd4(n)  365  386    399  421  1557
5 rcpd5(n)  525  556    590  622  1692
6 rcpd6(n)  320  339    358  377  1513
```

As mentioned earlier, those functions can be used also on dataframes. But if a matrix representation is possible (instead of a dataframe), it might speed things up.

We have seen previously that function Vectorize() could be nice to optimize code, and to avoid loops. But trying to avoid loops at all cost is probably not optimal. In the paragraphs above, while writing a function to generate a compound Poisson random variable, we have seen that the code based on loops was quite fast actually. But more important, sometimes loops cannot be avoided. And that is probably not a big deal. Ligges & Fox (2008) mentioned the following example. Consider the following list matriceslist which contains 100,000 matrices $n \times n$, denoted M_k

```
> matriceslist <- vector(mode = "list", length = 100000)
> for (i in seq_along(matriceslist)) matriceslist[[i]] <- matrix(rnorm(n^2),n,n)
```

The goal is to compute the $n \times n$ matrix such that $M = \sum M_k$. From Section 1.2, it would seem natural to store matrices in an object larger than a matrix (an array), and then use the apply function:

```
> M <- apply(array(unlist(matriceslist),dim=c(n,n,10000)),1:2,sum)
```

Even if the code runs, it will take a while when n is large. More precisely, we will create a very large array. Why not use a simple loop?

```
> M <- NULL; for(i in 1:100000) M=M+matriceslist[[i]]
```

We simply create an object which is only an $n \times n$ matrix.

Recall finally that simple (old) tricks can also be used to speed up computations. In Section 1.1.1, we gave a straightforward code to compute an estimate for the α parameter of a gamma distribution

```
> f <- function(x) log(x)-digamma(x)-log(mean(X))+mean(log(X))
> alpha <- uniroot(f,c(1e-8,1e8))$root
```

This was based on the example of Section 3.9.5 in Kaas et al. (2008) but the algorithm given there was actually

```
> constant <- -log(mean(X))+mean(log(X))
> fastf <- function(x) log(x)-digamma(x)+constant
> alpha <- uniroot(fastf,c(1e-8,1e8))$root
```

In the original implementation, each time the function `f()` is called within `uniroot()`, the same quantity `-log(mean(X))+mean(log(X))` is computed. This is clearly inefficient. Using the auxiliary scalar really makes the code run a lot faster:

```
> benchmark(uniroot(f,c(1e-8,1e8))$root,uniroot(fastf,c(1e-8,1e8))$root,
+ replications=100)[,c(1,3,4)]
                                 test elapsed relative
1     uniroot(f, c(1e-08, 1e+08))$root   0.097   10.778
2 uniroot(fastf, c(1e-08, 1e+08))$root   0.009    1.000
```

1.3.6 Numerical Integration

In this section, we will briefly see how to compute integrals (which is a standard actuarial problem). Consider the case where we would like to compute $\mathbb{P}[X > 2]$, where X has a given distribution, say a Cauchy one. Thus, we want to compute

$$\theta = \mathbb{P}[X > 2] = \int_2^\infty \frac{dx}{\pi\,(1 + x^2)}.$$

R has mathematical functions to compute standard quantities, such as integrals. Here, define the density of X,

```
> f <- function(x) 1/(pi*(1+x^2))
```

To compute integrals numerically, on finite support $[a, b]$, several techniques can be used, such as Gauss quadrature (see Chapter 25.4 in Abramowitz & Stegun (1970)). The idea is to write

$$\int_a^b f(x)dx \approx \sum_{i=1}^n \omega(x_i) f(x_i),$$

where the x_i's are associated with zeros of orthogonal polynomials (the so-called integration points), and ω is a weighting function. Those functions can be obtained using `gauss.quad()` For instance, the ChebyshevGauss formula yields

```
> library(statmod)
> GaussChebyshev <- function(f, a, b, n) {
+ x <- gauss.quad(n, kind = "chebyshev1")$nodes
+ y <- f((x + 1) * (b - a)/2 + a) * (1 - x^2)^(0.5) *pi * (b - a)/(2 * n)
+ return(sum(y))}
> GaussChebyshev(f,2,1000,10000)
[1] 0.1472654
```

while Gauss-Legendre is here

```
> GaussLegendre <- function(f, a, b, n) {
+ qd <- gauss.quad(n, kind = "legendre")
+ x <- qd$nodes
+ w <- qd$weights
+ y <- w * f((x + 1) * (b - a)/2 + a) * (b - a)/2
+ return(sum(y))}
> GaussLegendre(f,2,1000,10000)
[1] 0.1472653
```

The natural R function to compute integrals is `integrate`:

```
> integrate(f,2,1000)
0.1472653 with absolute error < 4e-07
```

Observe that the lower and upper bound can be `Inf`,

```
> integrate(f,lower=2,upper=Inf)
0.147584 with absolute error < 1.3e-10
```

It is also possible to use Monte Carlo simulations to approximate that integral (see Jones et al. (2009) or Robert & Casella (2010)), using the law of large numbers and the fact that the Cauchy distribution is the ratio of independent $\mathcal{N}(0,1)$ variables,

$$\frac{1}{n}\sum_{i=1}^{n}\frac{X_i}{Y_i}\to\theta,$$

where the convergence should be understood either in probability (weak version) or almost surely (strong version), where X_i and Y_i's are independent $\mathcal{N}(0,1)$ variables. The code to compute the left part above is

```
> n <- 1e7
> set.seed(123)
> mean(rnorm(n)/rnorm(n)>2)
[1] 0.1475814
```

Here, we do not have the absolute error, but from the central limit theorem, it is possible to control the error, as the variance here is $\theta(1-\theta)/n \sim 0.1275n^{-1}$. To increase accuracy, a first idea is to run more simulations. But here, `1e7` is probably large enough. A more interesting idea would be to use variance reduction techniques. For instance, observe that because the Cauchy distribution is symmetric,

$$\frac{1}{2n}\sum_{i=1}^{n}\left|\frac{X_i}{Y_i}\right|\to\theta,$$

```
> set.seed(123)
> mean(abs(rnorm(n)/rnorm(n))>2)/2
[1] 0.1475936
```

Here the variance is $\theta\,(1-2\theta)\,/2n = 0.0525n^{-1}$. It is also possible to use importance sampling techniques. Here, observe that

$$\theta=\frac{1}{2}-\mathbb{P}[0\leq X\leq 2]=\frac{1}{2}-\int_0^2\frac{dx}{\pi\,(1+x^2)}=\frac{1}{2}-\mathbb{E}[g\,(V)]\text{ where }g\,(x)=\frac{2}{\pi\,(1+x^2)}$$

and $V\sim\mathcal{U}([0,2])$. Thus,

$$\frac{1}{2}-\frac{1}{n}\sum_{i=1}^{n}g(2\cdot U_i)\to\theta,\text{ where }U_i\sim\mathcal{U}([0,1]),$$

which can be computed using

```
> g <- function(x) 2/(pi*(1+x^2))
> set.seed(123)
> .5-mean(g(runif(n)*2)
[1] 0.1475538
```

One can prove that the variance of this quantity is $0.0092n^{-1}$. Another transformation can be to observe that

$$\theta = \int_0^{1/2} \frac{y^{-2}}{\pi\left(1+y^{-2}\right)} dy = \mathbb{E}\left[h\left(V\right)\right] \text{ where } h\left(y\right) = \frac{1}{2\pi\left(1+y^2\right)}$$

and $V \sim \mathcal{U}([0,1/2])$. Thus,

$$\frac{1}{n}\sum_{i=1}^{n} h\left(\frac{U_i}{2}\right) \to \theta, \text{ where } U_i \sim \mathcal{U}([0,1]),$$

where the left term can be computed using

```
> h <- function(x) 1/(2*pi*(1+x^2))
> set.seed(123)
> mean(h(runif(n)/2))
[1] 0.1475856
```

Here, the variance is $0.00095n^{-1}$ (which is one thousand times smaller than the first one).

Observe that it is possible to use quasi-Monte Carlo techniques, as introduced in Niederreiter (1992),

```
> library(randtoolbox)
```

The idea is that the law of large numbers can still be valid for non-i.i.d. sequences, and that sequence with less discrepancy can be used. Instead of generating variables using `rnorm(n)`, it is possible to generate different sequences, using the Torus algorithm, the Sobol and Halton sequences, respectively, with functions `torus(n,normal=TRUE)`, `sobol(n,normal=TRUE)` (from Sobol (1967)) and `halton(n,normal=TRUE)` (from Halton (1960)). In the case of a uniform distribution on the unit interval, those four generators can be visualized in Figure 1.1. Observe that those sequences have lower discrepancy than the standard random number generator.

```
> plot(runif(250),rep(.2,250),ylim=c(0,1),axes=FALSE,xlab="",ylab="")
> points(torus(250),rep(.4,250))
```

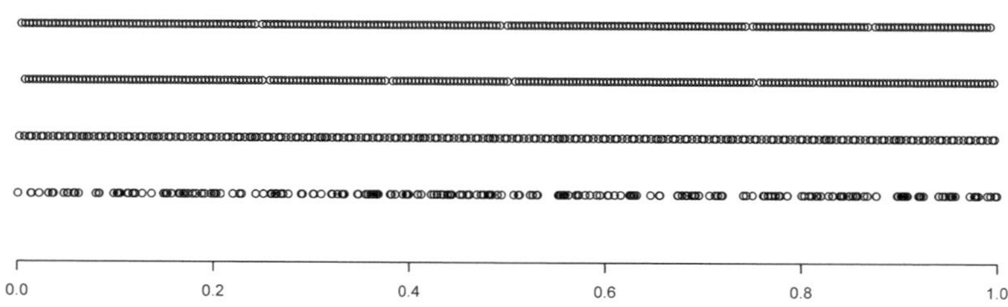

FIGURE 1.1
Three quasi-Monte Carlo generators, `halson`, `sobol` and `torus` (from top to bottom), and one random generator, `runif`.

```
> points(sobol(250),rep(.6,250))
> points(halton(250),rep(.8,250))
> axis(1)
```

(More details about plotting functions will be given in the next section.)

If we compare a random number generator with any of those sequences, we can see that, indeed, they have less discrepancy:

```
> set.seed(123)
> runif(10)
 [1] 0.2875775 0.7883051 0.4089769 0.8830174 0.9404673
 [6] 0.0455565 0.5281055 0.8924190 0.5514350 0.4566147
> halton(10)
 [1] 0.5000 0.2500 0.7500 0.1250 0.6250 0.3750 0.8750 0.0625
 [9] 0.5625 0.3125
```

It is possible to use codes described above with those sequences, to obtain a more accurate approximation of the integral. For instance,

```
> mean(h(sobol(n)/2))
[1] 0.1475836
```

The drawback of quasi-Monte Carlo techniques is that it is usually more complicated to quantify the error (see Niederreiter (1992) or Lemieux (2009) for a discussion).

1.3.7 Graphics with R: A Short Introduction

If S stands for *Statistics*, as mentioned earlier, observe that early publications such as Becker & Chambers (1984) or Ihaka & Gentleman (1996) (describing S as "*An Interactive Environment for Data Analysis and Graphics*" and R as "*A Language for Data Analysis and Graphics*", respectively) emphasize the importance of the graphic interface.

The starting point to produce a graph is the `plot()` function, which will cause a window to pop up and then plot the points. As always, functions that produce graphical output rely on a series of arguments, for example, to specify the range for the x-axis with the optional parameter `xlim`, or to have a log-scale on the y-axis using `log="y"`; `lty` to specify the type of line used, `pch` for the plotting character and `col` for the color. Then, it is possible to add objects, such as straight lines with `abline()`, curves with `lines()`, points with `points()` and colored areas with `polygon()`. The `legend()` function adds a legend. But before playing with those functions, let us see other graphs, ready-made for specific types of data.

1.3.7.1 Basic Ready-Made Graphs

Based on the `StormMax` dataframe mentioned above, let us count the number of major storms per decade.

```
> table(trunc(StormMax$Yr/10)*10)[-1]

1900 1910 1920 1930 1940 1950 1960 1970 1980 1990 2000
 183  153  135  204  191  199  179  186  178  191  281
```

Some simple functions can be used to plot this series, such as `barplot()`:

```
> barplot(table((trunc(StormMax$Yr/10)*10))[-1])
```

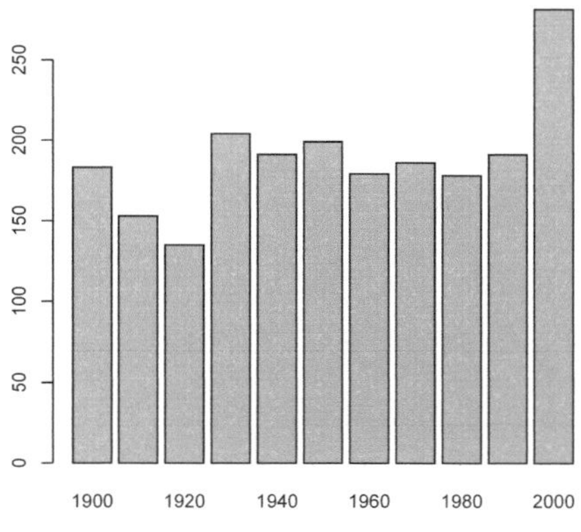

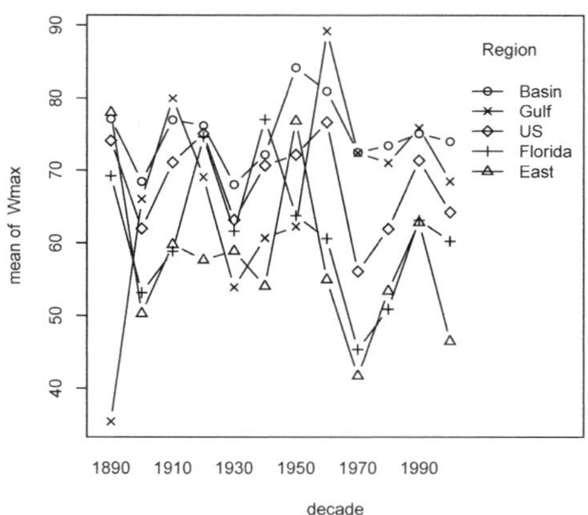

FIGURE 1.2
Some ready-made graphs, `barplot` and `interaction.plot`.

or—a bit more complex—`interaction.plot()`

```
> attach(StormMax)
> decade=trunc(Yr/10)*10
> interaction.plot(decade,Region,Wmax,type="b",pch=1:5)
```

Those ready-made graphs are easy to use, with simple options, such as the color of bars, or lines, or the shape of symbols (the `pch` parameter)

○ △ + × ◇ ▽ ⊠ ✳ ◇ ⊕ ✡ ⊞ ⊠ ⊠ ■ ● ▲ ◆ ● •
1 2 3 4 5 6 7 8 9 10 11 12 13 14 15 16 17 18 19 20

But the strength of R is that it is possible to draw almost anything.

1.3.7.2 A Simple Graph with Lines and Curves

Consider an i.i.d. sample, with observations from a normal distribution,

```
> X <- rnorm(37)
```

In Figure 1.3 are plotted a histogram with estimated densities on the left, and the empirical cumulative distribution function on the right.

Using the `par` parameter, it is possible to divide the graphic window in subparts and plot different graphs in it.

```
> par(mfrow = c(1, 2))
```

or equivalently,

```
> op <- par(mfrow=c(1,2))
```

gives one graph on the left and one graph on the right (the graphical window is divided in two, symmetrically): `c(1, 2)` means one row and two columns. By saving the current parameters, one can restore the previous parameters using `par(op)`. The `hist` function is

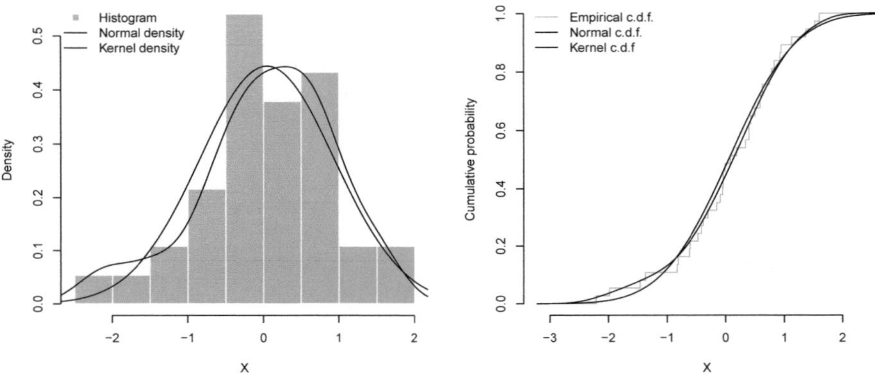

FIGURE 1.3
Histogram and empirical cumulative distribution function, from a $\mathcal{N}(0,1)$ sample.

used to compute and plot the histogram of the sample. With option `probability=TRUE`, the axis is not a count of observations with a partition, but the density, so that the histogram has a total area of 1.

```
> hist(X,xlab="X",ylab="Density",main="",col="grey",
+ border="white",probability=TRUE)
```

It is possible to add lines and curves on the graph, for instance the density of a normal distribution fitted using moment estimators,

```
> u <- seq(min(X)-1,max(X)+1,by=.01)
> lines(u,dnorm(u,mean(X),sd(X)),lty=2)
```

or a kernel-based estimator of the density, with a Gaussian kernel (this option can be changed through `kernel`), and the standard rule-of-thumb for choosing the bandwidth from Silverman (1986),

```
> d <- density(X)
> lines(d$x,d$y)
```

Actually, because the output of the `density()` function contains two vectors named `x` and `y`, it is possible to use instead

```
> lines(density(X))
```

Finally, we can add a legend in the upper left corner (using the `"topleft"` option, but it might be located anywhere):

```
> legend("topleft",c("Histogram","Normal density","Kernel density"),
+ col=c("grey","black","black"),lwd=c(NA,1,1),lty=c(NA,2,1),
+ pch=c(15,NA,NA),bty="n")
```

Note that to locate an area of the graph which might be appropriate for the legend (the text can be placed interactively via a mouseclick), it is possible to use

```
> legend(locator(1), ...)
```

Next we plot the empirical cumulative distribution function, computed using

```
> F.empirical <- function(y) mean(X<=y)
```

Then we plot it using the standard `plot` function. Note that to plot a step-function, we use a `type="s"` style

```
> plot(u,Vectorize(F.empirical)(u),type="s",lwd=2,col="grey",
+ xlab="X",ylab="Cumulative probability",main="",axes=FALSE)
```

So far, with the `axes=FALSE` option, there are no axes on the graph. We can add the x-axis below and the y-axis on the left using

```
> axis(1); axis(2)
```

Then, we can add two lines, the distribution function obtained by cumulating the kernel density estimator, and the distribution function of a Gaussian random variable where μ is the empirical mean and σ the empirical standard deviation,

```
> lines(d$x,cumsum(d$y)*diff(d$x)[1])
> lines(u,pnorm(u,mean(X),sd(X)),lty=2)
```

Again, in the upper left corner, we can add a legend,

```
> legend("topleft",c("Empirical c.d.f.","Normal c.d.f.","Kernel c.d.f"),
+ col=c("grey","black","black"),lwd=c(2,1,1),lty=c(1,2,1),bty="n")
```

If we do not know where to plot the legend, it is possible to use the `locator` function,

```
> legend(locator(1),c("Empirical c.d.f.","Normal c.d.f.","Kernel c.d.f"),
+ col=c("grey","black","black"),lwd=c(2,1,1),lty=c(1,2,1),bty="n")
```

In that case, R reads the position of the graphics cursor when the mouse button is pressed.

Note that graphs are displayed in the graphics window of R. It is possible to export them in a standard image format (bmp, jpeg, png and also pdf). One should add

```
> png(filename = "rplot.png", width = 480, height = 240,
+ units = "px", bg = "white")
```

before the `plot` command (several options can be included), and then add

```
> dev.off()
```

at the end (indicating that we are done creating the plot). An alternative, if we want to save a graph that has been plotted, without running again the code, is to use save the output of the graphic window in the .eps format by right-clicking the mouse (or any other format, see Murrell & Ripley (2006) for more details).

1.3.7.3 Graphs That Can Be Obtained from Standard Functions

It is possible to use graphical functions in R that extract information from complex objects such as the ones obtained from a regression model:

```
> data(StormMax)
> StormMaxBasin <- subset(StormMax,(Region=="Basin")&(Yr>1977))
> attach(StormMaxBasin)
```

Note here the use of function `subset()`. Indeed, to keep only some variables in a dataset; one can use the following generic code

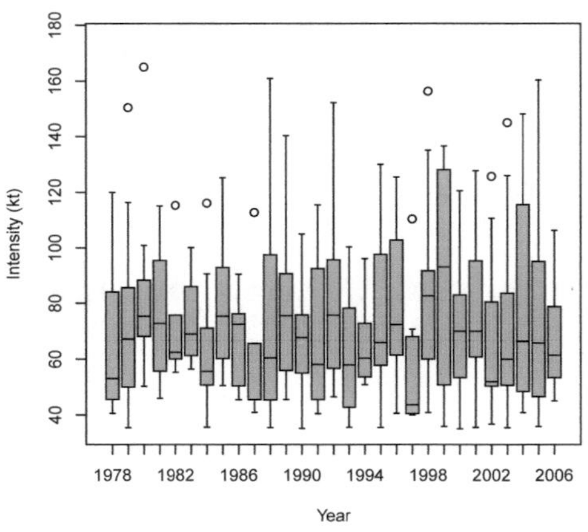

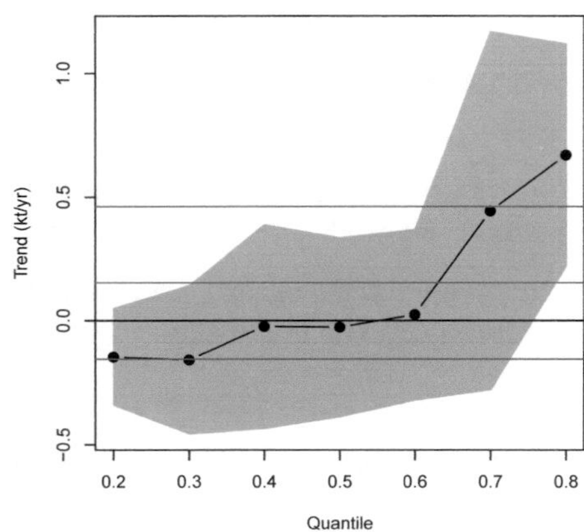

FIGURE 1.4
Boxplots of cyclone intensity (kt), per year, and slope of quantile regressions (intensity versus years), for different probability levels.

```
> df <- subset(df, select = c(X1,X2))
```

or, to drop some variables,

```
> df <- subset(df, select = -c(X1,X2))
```

The first graph, on the left of Figure 1.4, contains boxplots, per year. We can use `boxplot(Wmax~as.factor(Yr))` to display several boxplots on the same graph (vertically here):

```
> boxplot(Wmax~as.factor(Yr),ylim=c(35,175),xlab="Year",
+ ylab="Intensity (kt)",col="grey")
> library(quantreg)
> model <- rq(Wmax~Yr,tau=seq(0.2,0.8,0.1))
> model
Call:
rq(formula = Wmax ~ Yr, tau = seq(0.2, 0.8, 0.1))

Coefficients:
               tau= 0.2    tau= 0.3     tau= 0.4     tau= 0.5
(Intercept) 189.8900242 391.2367884 140.80069596 210.10132020
Yr           -0.0715789  -0.1696345  -0.04040035  -0.07219432
                 tau= 0.6   tau= 0.7     tau= 0.8
(Intercept)  7.500000e+01 -221.78568 -1169.6654437
Yr          -1.075529e-16    0.15365     0.6361049
```

Here, seven regressions have been computed and stored in the `model` object. Using the `summary()` function on this object, many interesting quantities can be computed (for example, confidence intervals on parameters) and then plotted. The following graph is on the right of Figure 1.4.

```
> plot(summary(model,alpha=0.05,se="iid"),
+ parm=2,pch=19,cex=1.2,mar=c(5,5,4,2)+0.1,
+ ylab="Trend (kt/yr)",xlab="Quantile")
```

1.3.7.4 Adding Shaded Area to a Graph

Consider the case where we want to sketch the possible range for a quantile function. Assume that X and Y are two $\mathcal{N}(0,1)$ random variables, and let us plot the upper and the lower bound for the quantile function of $X+Y$. If `Finv` (that is, F^{-1}) denotes the quantile function of X, and `Ginv` (that is, G^{-1}) the quantile function of Y, then those bounds can be computed using (see Williamson (1989) for theoretical details, as well as algorithms):

```
> n <- 1000
> Qinf <- Qsup <- rep(NA,n-1)
> for(i in 1:(n-1)){
+ J <- 0:i; Qinf[i] <- max(Finv(J/n)+Ginv((i-J)/n))
+ J <- (i-1):(n-1); Qsup[i] <- min(Finv((J+1)/n)+Ginv((i-1-J+n)/n))
+ }
```

Then, lines are drawn using the `lines` function, for example,

```
> x <- seq(1/n,1-1/n,by=1/n)
> lines(x,Qsup,lwd=2)
```

and to add colored area, we use the `polygon` function. The color is chosen within the gray palette,

```
> gray.col <- gray.colors(n=100, start = 0, end = 1)
```

from `library(RColorBrewer)`. To colorize between the upper bound `Qsup` and the sum of quantile functions, `qnorm(.,sd=2)`, the code is

```
> polygon(c(x,rev(x)),c(Qsup,rev(qnorm(x,sd=2))),col=gray.col[45],border=NA)
```

It is also possible to look at constraint bounds, when we assume that X and Y are positively dependent (with quadrant positive dependence, i.e. for all x,y, $\mathbb{P}(X>x,Y>y)\geq\mathbb{P}(X>x)\cdot\mathbb{P}(Y>y)$, or equivalently $\mathbb{P}(X>x|Y>y)\geq\mathbb{P}(X>x)$) In that case, those bounds can be computed using

```
> Qinfind <- Qsupind <- rep(NA,n-1)
> for(i in 1:(n-1)){
+ J <- 1:(i); Qinfind[i] <- max(Finv(J/n)+Ginv((i-J)/n/(1-J/n)))
+ J <- (i):(n-1); Qsupind[i] <- min(Finv(J/n)+Ginv(i/J))
+ }
```

Then, we use

```
> polygon(c(x,rev(x)),c(y.lower,rev(y.upper)),col=gray.col[45],density=20,border=NA)
```

where `y.upper` and `y.lower` are the upper and the lower curves for the area that should be colorized.

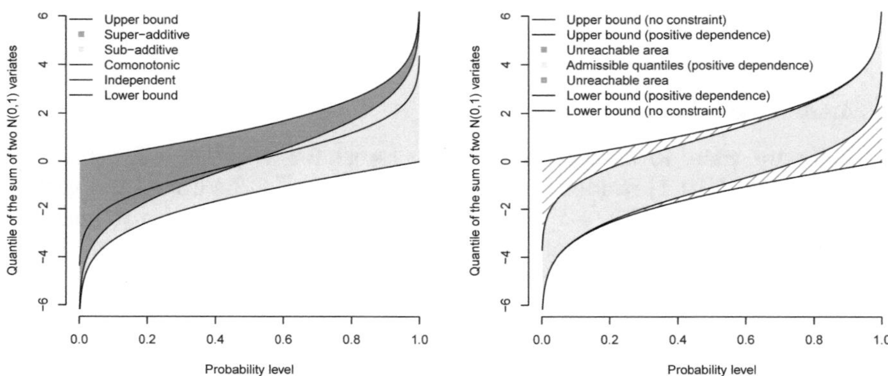

FIGURE 1.5
Admissible values for the quantile function of the sum of two $\mathcal{N}(0,1)$ random variables, with, on the right, the restriction when X and Y are positively dependent.

1.3.7.5 3D Graphs

Finally, note that 3D graphs can also easily be constructed. Use here

```
> x <- y <- seq(-2.5,2.5,by=.25)
> z <- outer(x,y,function(u,v) binorm(u,v,r=.4))
```

To visualize level curves, use

```
> image(x,y,z,col=rev(gray.col))
> contour(x,y,z,add=TRUE)
```

Here, contour lines are added to the colorized graph. For surfaces of densities, use

```
> persp(x,y,z)
```

Because we keep visualizing graphs on a screen, a 3D graph is just a 2D visualization. It is possible to define an object, storing the transformation matrix used for that representation

```
> pmat <- persp(x,y,z,theta=210, col=gray.col[45],shade=TRUE)
> pmat
              [,1]        [,2]       [,3]       [,4]
[1,] -3.464102e-01  0.05176381 -0.1931852  0.1931852
[2,] -2.000000e-01 -0.08965755  0.3346065 -0.3346065
[3,]  3.526252e-16 11.12516046  2.9809778 -2.9809778
[4,] -3.061800e-17 -0.96598365 -2.9908853  3.9908853
```

We can still use 2D graphical functions to add lines or points on the figure. For instance, to draw a line, that is, $(x_t, y_y, z_t)_{t\in[a,b]}$, use function `trans3d()`,

```
> u <- x; v <- rep(1,length(y)); w <- binorm(u,v,r=.4)
> lines(trans3d(u,v,w, pmat),lwd=4,col="black")
```

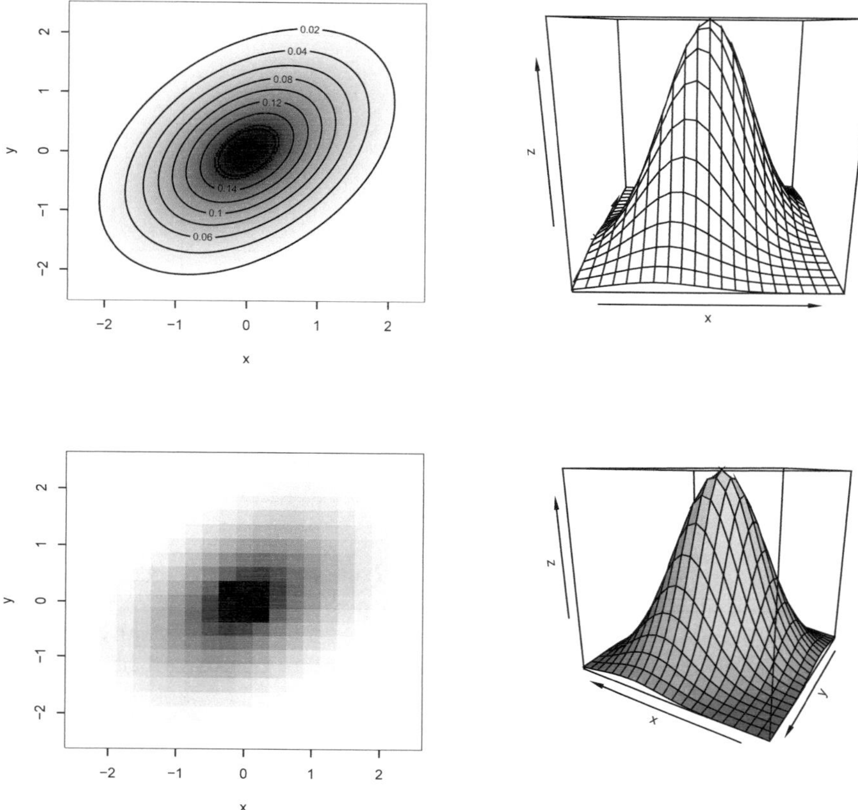

FIGURE 1.6
Representation of the density of a bivariate Gaussian random vector.

1.3.7.6 More Complex Graphs

In many applications, we might like to enhance plots, in order to visualize various interesting features in one single graph. In previous sections we have seen how to plot several curves (and lines) on the same graph. With the following code we can plot two different regression lines, for two different factors (here `Z` takes values `M` or `F`), the output being the graph on the left of Figure 1.7,

```
> attach(linearmodelfactor)
> plot(X,Y,pch=(Z=="M")*2+1)
> abline(lm(Y~X,data=B,subset=which(B$Z=="M")))
> abline(lm(Y~X,data=B,subset=which(B$Z=="F")),lty=2)
> legend("topright",c("M","F"),pch=c(3,1),lty=c(1,2),bty="n")
```

Another way of visualizing those regression lines is by trellis-type graphics (as introduced in Cleveland (1993)) using the `lattice` library (see also Sarkar (2002)).

```
> xyplot(Y ~ X| Z, panel = function(x, y) {panel.xyplot(x, y);panel.lmline(x, y)})
```

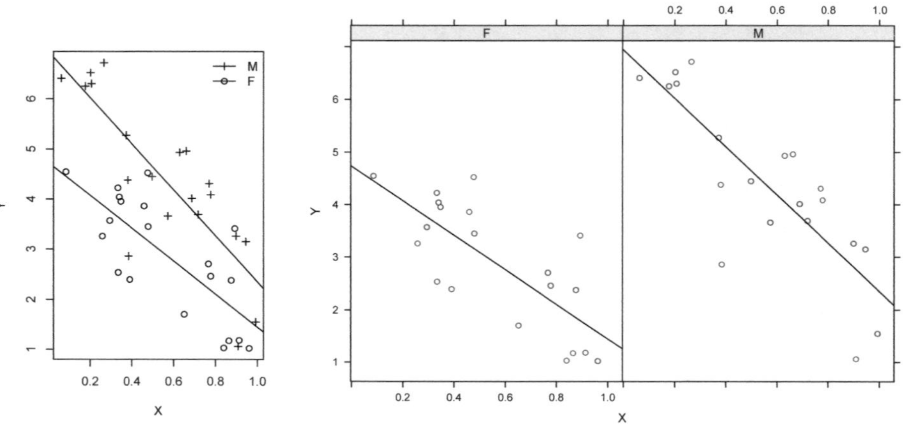

FIGURE 1.7
Regression lines of Y against X, for different values of a factor Z, with trellis functions on the right.

In the first part, we ask for graphs for different values of Z, of Y against X. In the second part, we specify a generic function that will be added to the scatterplot (here, a linear regression line). Those two graphs are on the right of Figure 1.7. Observe that—automatically—the same y-axis is used, to allow for comparisons. More details on trellis graphs can be found in Sarkar (2008).

With standard functions, it is also possible to plot two or more figures on a single graph. Graphics parameters mfrow and mfcol subdivide the plotting region into arrays of figure regions, as discussed previously. But graphics will have the same size. It is possible to have windows of different shapes on a single graph using either mai or mar in the par description (margin sizes are either in inches with mai, or in text line units with mar).

Consider the case where we wish to plot surfaces and contour plots (with level curves) to visualize dependencies, as well as histograms to describe marginal distributions. Consider for instance the popular loss-ALAE dataset (see e.g. Klugman & Parsa (1999)), with indemnity payment (loss), and the allocated loss adjustment expense (ALAE), for 1,500 individual claims, in U.S. dollars:

```
> library(evd); data(lossalae); library(MASS)
> X <- lossalae[,1]; Y <- lossalae[,2]
> xhist <- hist(log(X), plot=FALSE)
> yhist <- hist(log(Y), plot=FALSE)
> top <- max(c(xhist$counts, yhist$counts))
```

In order to visualize the distribution of the points, consider the bivariate kernel estimator of the joint density of (X, Y):

```
> kernel <- kde2d(log(X),log(Y),n=201)
```

Those histograms (in the xhist and yhist object) and surfaces (in the kernel object) can be plotted on the same graph using the following code:

```
> par(mar=c(3,3,1,1))
> layout(matrix(c(2,0,1,3),2,2,byrow=TRUE),
> c(3,1), c(1,3), TRUE)
```

The window will be split into four parts, $\frac{2 \mid 0}{1 \mid 3}$, in that specific order, and if the length of the last column and the first row is 1, the length of the first column and the last row will be 3.

```
> plot(X,Y, xlab="", ylab="",log="xy",col="grey25")
> contour(exp(kernel$x),exp(kernel$y),kernel$z,add=TRUE)
> par(mar=c(0,3,1,1))
> barplot(xhist$counts, axes=FALSE, ylim=c(0, top),space=0,col="grey")
> par(mar=c(3,0,1,1))
> barplot(yhist$counts, axes=FALSE, xlim=c(0, top),space=0,
+ horiz=TRUE,col="grey")
```

This graph is on the left of Figure 1.8. Color `grey25` has been used to get a wider palette of colors. But other functions can be used, for instanced `gray.colors()` (from the `grDevices` package) can be used to generate a palette:

```
> gray.colors(8)
[1] "#4D4D4D" "#737373" "#8E8E8E" "#A4A4A4"
[5] "#B7B7B7" "#C8C8C8" "#D7D7D7" "#E6E6E6"
```

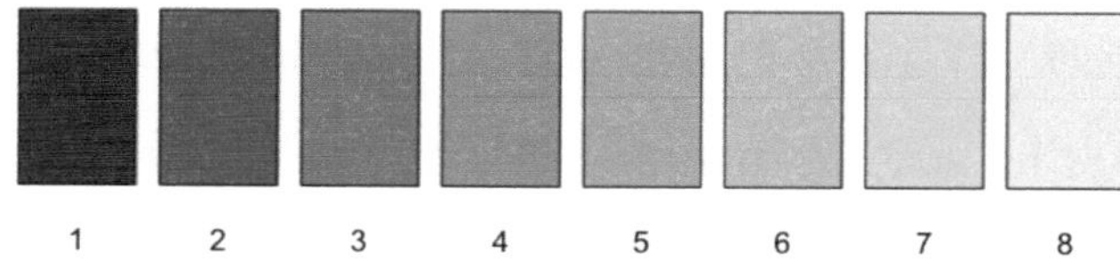

For other palettes, one can use library `RColorBrewer`, :

```
> brewer.pal(n=8, "Greys")
[1] "#FFFFFF" "#F0F0F0" "#D9D9D9" "#BDBDBD"
[5] "#969696" "#737373" "#525252" "#252525"
```

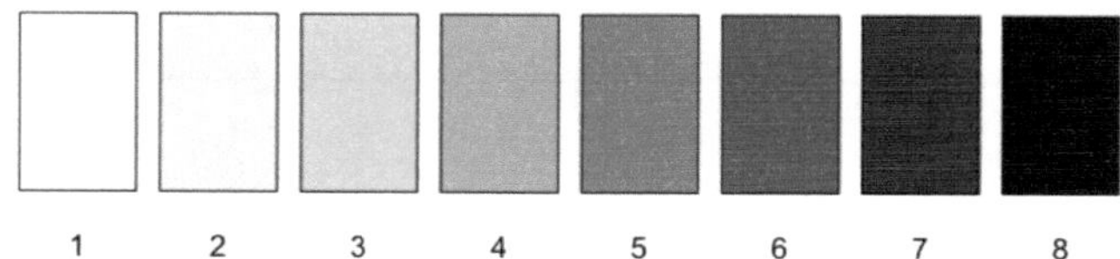

Use `display.brewer.all()` to visualize a lot of possible palettes.

Finally, to get *elegant* graphs, an alternative can be to use the `ggplot2` library, where the first two letters stand for *grammar of graphics*; from Wilkinson (1999) and Wickham (2009). This library is based on the `reshape` and `plyr` packages. The idea is to use different layers to generate a complex graph. For instance, consider two financial indices:

```
> library(tseries)
> SP500 <- get.hist.quote("^GSPC")
> Nasdaq <- get.hist.quote("^NDX")
> BS <- data.frame(Date=time(SP500),SP500=SP500$Close)
> BSN <- data.frame(Date=time(Nasdaq),Nasdaq=Nasdaq$Close)
> B <- merge(BS,BSN)
> indices <- melt(B, id.vars = "Date")
```

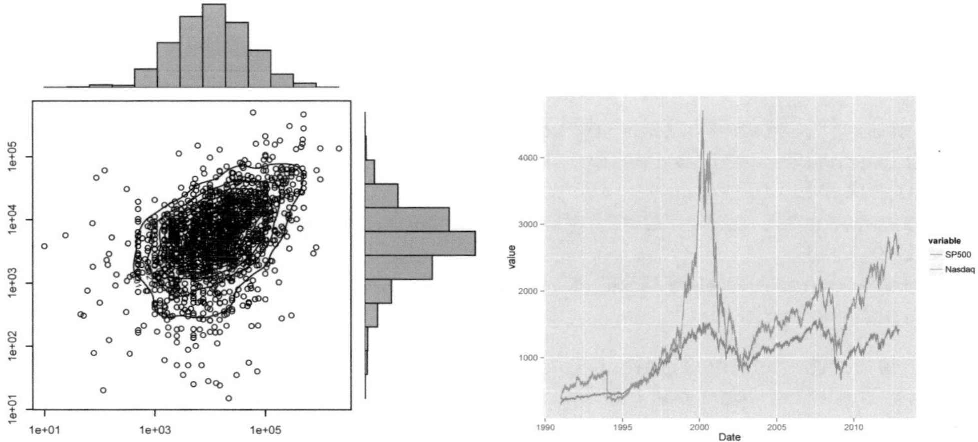

FIGURE 1.8
Scatterplot of the loss and allocated expenses dataset, including histograms, on the left, with a graph obtained using the `ggplot2` library.

The graph is on the right of Figure 1.8. More details on `ggplot2` graphs can be found in Wickham (2009).

Note here the use of the `merge()` function. This function merges two dataframes by common columns (or row) names. This function can perform one-to-one, many-to-one and many-to-many merges. If we want to plot the two series, the code is close to the one used in the `lattice` package:

```
> zp1 <- ggplot(indices)
> zp1 <- zp1 + geom_line(aes(x = Date, y = value, colour = variable))
> plot(zp1)
```

1.4 More Advanced R

As mentioned earlier, the aim of this chapter is to introduce basics about the R language. Of course, it is possible to optimize the code. But it is clearly a matter of personal taste, and priorities. It might be important to have codes that will make a procedure run faster, but it might also be important to have *readable* codes, or to have codes that run on a desktop or a laptop, not necessarily a server.

So far, we have seen a lot of things that R can do. But R has limitations too. R is a bit slow in some computations (as discussed earlier), and sometimes R cannot define objects that are too large. In this section, we will briefly show how to code advanced functions in R.

1.4.1 Memory Issues

The memory R can use depends on several factors:

- The RAM physically in the computer
- The RAM used by the computer (by other programs)

- The processor (32 or 64 bits) as the allocated RAM is bounded at 2^{32} bytes (i.e. 4×1024 KB, or 4 GB) for 32 bits, while it might be 2^{64} bytes for 64 bits (but in that case, the limit comes from the architecture of the processor)

Recall that to get numerical characteristics of the computer, one should use the `.Machine` object. For instance,

```
> .Machine$sizeof.pointer
[1] 4
```

returns 8 on 64-bit builds, and 4 for 32-bit. On a 32-bit computer, some operations are not possible; for instance,

```
> D<-matrix(0,12000,12000)
Error: cannot allocate vector of size 1.1 GB
```

but

```
> D <- matrix(0,10000,10000)
```

does work

```
> object.size(D)
800000112 bytes
```

If the dataset is a very large one, we might not have enough memory to run standard regression functions. Some functions (slower than the standard regression functions) will work when there is not enough memory to use the standard functions. Namely,

```
> library(biglm)
> fit <- biglm(Y ~ X1 + X2, data=base)
```

1.4.2 Parallel R

It is possible to do calculations faster using several processors, all of them cooperating on a given task. The idea of parallel processing is to break up the task, to split it among multiple processors, and then to put the components back together. This is very useful and easy if the task can be split up (especially without any communication between them, like using a value computed on one processor on another one, as discussed in Yu (2002) or more recently Schmidberger et al. (2009) and Hoffmann (2011)). In the `plyr` package, it is possible to use option `.parallel=TRUE` in the `aaply()` function, for instance. It is also possible to use libraries dedicated to parallelization; see Schmidberger et al. (2009) for a description of the state of the art. The following code was run on a 4 core computer:

```
> library(parallel)
> (allcores <- detectCores(all.tests=TRUE))
[1] 4
```

With the `snow` package, tasks are sent to the processors (that can also be clients on a complex server), then results are sent back and assembled into the final output. One can also use package `doMC`:

```
> library(doMC)
> registerDoMC()
```

where `register` determines the number of cores to use. Consider here the following function, where a loop is used and at each iteration, outputs are stored in a list:

```
> F<-function(n,m=100){
+   L<-vector("list", n)
+   for(i in 1:n)
+     {
+      X<-rnorm(m,0,sd=log(1+i))
+      L[[i]]<-max(eigen(X%*%t(X))$values)
+     }
+     return(L)
+ }
```

It is possible to make the algorithm faster using a parallelized `foreach %dopar%` loop, that always returns a list:

```
> parF<-function(n,m=100){
+      foreach(i = 1:n) %dopar%
+      {
+      X<-rnorm(m,0,sd=log(1+i))
+      L[[i]]<-max(eigen(X%*%t(X))$values)
+     }
+ }
```

Here, the parallelized function is slightly faster on a desktop:

```
> microbenchmark(F(), parF(), times = 10)
Unit: milliseconds
    expr      min       lq   median       uq      max
1    F() 285.3781 288.3071 290.9585 306.8264 340.7759
2 parF() 218.4247 224.5321 226.0498 237.5514 255.7262
```

Remark 1.1 *Random number generation might be a good example of tasks that can be split among multiple processors, but one should keep in mind that random number generators are iterative, and thus theoretically not splittable.*

The idea is that if we work on a typical desktop, then implementing parallel processing in some R code might speed up your program. But not every task runs better in parallel. Further, distributing processes among the cores may cause computation overhead: we might lose time (and memory) firstly by distributing, secondly by gathering the patches shared out among the processing units. So, sometimes parallel computing can be rather inefficient. Consider the case where we generate datasets with a possibly large number of rows, and we have to compute, say, quantiles per column. To compare the use of (standard) `lapply()` with `mclappy()` from library `multicore`, `parLapply()` from library `snow` or `sfLapply()` of library `snowfall`, we generate a dataframe with 100 columns and n =250,000 rows.

```
> n <- 250000
> base <- data.frame(matrix(rnorm(n*100),n,100))
```

The goal can be to compute quantiles per column. Consider here the following function:

```
> microbenchmark(
+ mclapp=data.frame(mclapply(base, quantile, probs = 1:3/4 ,
+ mc.cores = allcores)),
```

```
+ mlapp=data.frame(lapply(base, quantile, probs = 1:3/4 )),
+ sflapp=data.frame(sfLapply(base, quantile, probs = 1:3/4 )),
+ times=100)
Unit: milliseconds
    expr     min lq median uq  max
1 mclapp  710  744    785  811  951
2  mlapp 1444 1465   1524 1562 1788
3 sflapp 1443 1470   1530 1567 1787
```

In this example, `mclapp()` is (only) twice as fast as simply `lapply()`. Even with 250,000 rows, parallel codes on a small number of cores are not very efficient.

Remark 1.2 *In order to go even faster, it is also possible to run computations not on the CPU (*central processing unit*), but on the GPU (*graphical processing unit*), using the* gputools *on a Linux system. But in the sequel, we will simply dispatch each task to a different CPU core.*

1.4.3 Interfacing R and C/C++

As mentioned earlier, R code is interpreted when it is run, unlike some other languages that are first compiled. But R actually does have compiling abilities that might speed up functions by a factor of 3 or 4. A first strategy is to use `.Internal`, which may speed up some functions but is only recommended for R-wizards:

```
> u <- runif(100)
> benchmark(mean(u),sum(u)/length(u),.Internal(mean(u)),replications=1e5)[,c(1,3,4)]
                  test elapsed relative
3 .Internal(mean(u))   0.320    1.000
1            mean(u)   0.870    2.719
2   sum(u)/length(u)   0.339    1.059
```

Another strategy is to use the `compiler` library. Consider the case where we want to compute the sum per row, of a given matrix:

```
> library(compiler)
> f1 <- function(x) apply(x,1,sum)
> g1 <- cmpfun(f1)
```

Note that it might be more efficient to use a loop, instead of using a R function:

```
> f2 <- function(x) {
+ v <- rep(NA,nrow(x))
+ for(i in 1:length(v)) v[i] <- sum(x[i,])
+ return(v)}
> g2 <- cmpfun(f2)
```

We can now compare those four functions:

```
> n <- 100
> M <- matrix(runif(n^2),n,n)
> benchmark(f1(M),f2(M),g1(M),g2(M),
+ replications=1000)[,c(1,3,4)]
   test elapsed relative
1 f1(M)   0.519    1.736
2 f2(M)   0.462    1.545
3 g1(M)   0.485    1.622
4 g2(M)   0.299    1.000
```

Observe that the compiled function of the second code, obtained using `cmpfun`, is almost twice as fast as the `apply()` function here. One could also have used `rowSums()` or the even faster internal function `.rowSums`.

Finally, if R is not fast enough, to improve performance, it is possible to write key functions in C/C++, by ourselves. Use of C/C++is more appealing than Fortran, for instance, because of the Rcpp API (application programming interface), as discussed in Lang (2001). Note that to use the `Rcpp` package on a Windows platform, it is necessary to install (properly) the Rtools kit (but this is straightforward on Mac or Linux).

```
> library(Rcpp)
> evalCpp("7+3")
[1] 10
```

It is possible to create ad-hoc R functions, such as a function `square()` that will compute the square of any integer:

```
> cppFunction("int square(int x) {return x*x;}")
> square(9)
[1] 81
```

or `power()` that will compute; more generally; the power n of any (positive) number x:

```
> cppFunction("double power(double x, int n){double xn = pow(x,n);return xn;}")
> power(4,2)
[1] 16
```

Following Eddelbuettel & François (2011), consider the construction of the Fibonacci sequence, define recursively by

$$u_n = u_{n-1} + u_{n-2}, \text{ with } \begin{cases} u_0 = 0 \\ u_1 = 1 \end{cases}$$

The R code to define that function is

```
> FibonacciR <- function(n){
+ if(n<2) return(n)
+ else return(FibonacciR(n-1)+FibonacciR(n-2))
+ }
> FibonacciR(10)
[1] 55
```

(see the previous example of the towers of Hanoi for recursive programming) but it can be very slow to run (even for a small n),

```
> system.time(FibonacciR(35))
      user      system      elapsed
    35.431       0.128       35.538
```

The C/C++code to compute that sequence is

```
int g(int n) {if(n<2) return(n) ; return(g(n-1)+g(n-2));}
```

It is possible to use that function in the following R code:

```
> library(Rcpp)
> cppFunction("
```

```
+ int FibonacciC(int n){
+ if (n<2) return(n)    ;
+ else return(FibonacciC(n-1)+FibonacciC(n-2));
+ }")
```

The output is hopefully, the same as the R function:

```
> FibonacciC(10)
[1] 55
```

but here, the code is much faster to run (here, 1,000 times faster):

```
> library(rbenchmark)
> benchmark(FibonacciR(30),FibonacciC(30),replications=1)[,c(1,3,4)]
            test elapsed relative
2 FibonacciC(30)   0.004        1
1 FibonacciR(30)   4.136     1034
```

It is possible to use a standard R function to compute vectorized versions of that function. Using `Vectorize()` or `sapply()` yields comparable computation times. But again, using C/C++ will make our code run 1,000 times faster!

```
> N=1:35
> benchmark(sapply(N,FibonacciR),Vectorize(FibonacciR)(N),
+ sapply(N,FibonacciC),Vectorize(FibonacciC)(N),replications=1)[,c(1,3,4)]
                      test elapsed relative
3    sapply(N, FibonacciC)   0.094    1.000
1    sapply(N, FibonacciR)  91.888  977.532
4 Vectorize(FibonacciC)(N)   0.099    1.053
2 Vectorize(FibonacciR)(N)  91.566  974.106
```

For further information about the C/C++ language, see Kernighan (1988) or Stroustrup (2013).

It is also possible to use more complex formats than integers (`int`) or reals (`double`). For instance, from a matrix, it is possible to return a vector,

```
> cppFunction("
+   NumericVector SumRow(NumericMatrix M) {
+     int nrow = M.nrow(), ncol = M.ncol();
+     NumericVector out(nrow);
+     for (int i = 0; i < nrow; i++) {
+       double total = 0;
+       for (int j = 0; j < ncol; j++) {
+         total += M(i, j);
+       }
+       out[i] = total;
+     }
+     return out;
+   }
+ ")
> SumRow(matrix(1:6,3,2))
[1] 5 7 9
```

which returns the sum per row.

For those not familiar with C/C++, using package `RPy`, it is also possible to call R using Python.

1.4.4 Integrating R in Excel®

Though actuaries might be interested in advanced methods to run smoothing techniques on large datasets (that will be memory—and time—consuming), in practice they also like to use Excel and Visual Basic for simple reporting. An interesting feature of R, mentioned in Baier & Neuwirth (2003), is that "R *can completely control any of the Office applications—at least as far as Office allows this*".

To import data from Excel spreadsheets, run computations in R and then export R outputs in Excel spreadsheets; it is possible to use the XLConnect package,.

```
> library(XLConnect)
> writeWorksheetToFile(file='test.xlsx',data=M,sheet='test',
+ startRow = 10, startCol = 3)
```

for some matrix M.

An alternative is to use the following nice tool, described in Courant & Hilbert (2009): the Excel add-in RExcel.xla that allows one to use R from within Excel®, using the `library(rcom)`. R can be used in a *scratchpad mode* that consists of writing R code directly in an Excel worksheet and transferring scalar, vector, and matrix variables between R and Excel. But R can also be called directly in functions in worksheet cells. An alternative is to use a *macro mode* where the user can write macros using Visual Basic and R codes. For instance, within a cell of a worksheet, one can compute the value of an R function `rfunction` (with one parameter) by typing =RApply('rfunction',A1) in a cell, assuming the parameter is the value given in cell A1.

1.4.5 Going Further

The most difficult task for R users is probably to follow the active community of R users. Academic publishers now have their own R collections, with 'The R Series' of Chapman & Hall/CRC, and the 'UseR!' of Springer Verlag. There are several books on the R language, such as Cohen & Cohen (2008), Dalgaard (2009), Krause (2009), Zuur et al. (2009), Teetor (2011), Kabacoff (2011), Maindonald & Braun (2007), Craley (2012), or Gandrud (2013), to mention only some of them. On top of those general books, one can easily find books dedicated to specific topics, such as Wickham (2009), Murell (2012), Lawrence & Verzani (2012), Højsgaard et al. (2012) on graphics (mainly, not to say only). Hundreds of books related to S and R are mentioned at http://www.r-project.org/doc/bib/R-books.html, including books that are related to S or R. Free ebooks can be found at http://cran.r-project.org/manuals.html or http://cran.r-project.org/other-docs.html.

It is also possible to follow updates through the R *Journal*, available online at http://journal.r-project.org/. This journal contains short introductions to R packages, and hints for newcomers as well as more advanced programmers. The journal contains updates about R conferences, and R user groups, that can be found now in almost any (large) city. R bloggers also provide examples, codes, and hints that can be useful to anyone willing to learn more about R.

1.5 Ending an R Session

This introduction was a bit long. And it might be time to close R. One way to exit (properly) R is to use the `q()` function:

```
> q()
```

R will then ask whether to save the workspace image, or not.

```
Save workspace image? [y/n/c]:
```

Answering `n` (for 'no') will exit R without saving anything, whereas answering `y` (for 'yes') will save all defined objects in a file .RData, and the command history will be stored in a file called .Rhistory, both in the working directory. Note that those two files are text files and can be opened using any text editor.

1.6 Exercises

1.1. What is the difference between a `data.frame` and a `data.table` object?

1.2. What is the difference between `read.table()` and `read.ftable()` functions?

1.3. Display the value of log(4) with fifteen digits.

1.4. What is function `intersect` for? What would `intersect(seq(4,28,by=7), seq(3,31,by=2))` return?

1.5. What would `c(TRUE,TRUE,FALSE,FALSE) & c(TRUE,FALSE,FALSE,TRUE)` return?

1.6. Import file extremedatasince1899.csv with years from 1900 to 2000 only.

1.7. Sort the previous variable according to variable `Wmax`.

1.8. From database extremedatasince1899.csv, compute the average value of variable `Wmax` when `Region` is equal to `Basin`.

1.9. Using the `xtable` package, generate an html page that contains a table, obtained from the sub-dataset of extremedatasince1899.csv where `Region` is equal to `Basin`, and `Yr` is between 1950 and 1959.

1.10. Still from the same file, create a sub-dataset that contains only variables that are numeric.

1.11. What is function `attach` used for?

1.12. What will the following lines return?

```
> n<<--1
> if (n==0) "yes" else "no"; n
> if (n < - 0) "yes" else "no"; n
> if (n<-0) "yes" else "no"; n
> if (n=0) "yes" else "no"; n
> if (n<-2) "yes" else "no"; n
```

1.13. Explain the results of the statements below:

```
> 9*3 ^-2; (9*3)^-2
> -2^-.5; (-2)^-.5
> 1:4^2; 1:4*4
> 2^2^3; 2-2-3
> n <- 5; for (i in 1 : n+1) print(i)
> k <- 1:3; k[3]^2; k^2; k^2[3]
```

1.14. Are the results of the following statements as you expected?

```
> c(1, 7, NA, 3) == NA; is.na(c(1, 7, NA, 3))
> Inf*Inf; 0*Inf; Inf*-Inf; Inf>Inf/2; sqrt(Inf)
> 2^1023.999<Inf; 2^1024<Inf
> 1/0; 0/0
> log(0); log(-1)
> sqrt(-1); sqrt(-1+0i)
> integrate(dnorm, 0, 20)
> integrate(dnorm, 0, +Inf)
> integrate(dnorm, 0, 20000)
```

1.15. What will `1:10*1:5` return?

1.16. Given a vector `x`, write a function which returns only elements of `x` larger than `mean(x)`.

1.17. Write a function `seqrep(n)` which returns vector $(1, 2, 2, 3, 3, 3, \ldots, n)$ where integer k is repeated k times. How long is this vector when $n = 50$?

1.18. Write a function that counts the number of `NA`'s in a vector.

1.19. Create a function `second.diag(M)` which returns the second diagonal of squared matrix `M`.

1.20. What will `M[,2]` return if `M <- matrix(1:5,3,3)`?

1.21. Get the help page on command `%%`. What does that mean for `x` if `x %% 3 == 0` is `TRUE`?

1.22. Given matrix `m <- matrix(1:20,5,4)`, what will `which(m %% 3 == 0, arr.ind=TRUE)` produce?

1.23. Write a function that computes the power of any square matrix, `power(M,n)`.

1.24. Which function should you use to compute M^{-1}?

1.25. Create the identity matrix of size 5×5.

1.26. Write a function to compute $\sin^2(x) + x/10$. Find its minimum in $(0, 2\pi)$.

1.27. Using commands `!` and `%in%`, return the subvector `x <- sample(1:15)` where values `c(3,7,12)` are removed.

1.28. Given a matrix M, write a function which returns the following Kronecker product:

$$\begin{pmatrix} 1 & 3 & 4 \\ 2 & 0 & 5 \end{pmatrix} \otimes M = \begin{pmatrix} M & 3M & 4M \\ 2M & 0 & 5M \end{pmatrix}$$

1.29. Compute $\sum_{i=10}^{20}(i^2 + 4/i)$.

1.30. Solve numerically the following system

$$\begin{cases} 3x + 2y - z = 1 \\ 2x - 2y + 4z = -2 \\ -x + \frac{1}{2}y - z = 0 \end{cases}$$

1.31. Given a vector $\boldsymbol{x}$ in $\mathbb{R}^n$ and a function $f : \mathbb{R} \to \mathbb{R}$, create a function `sum.function(x,f)` which computes $\sum_{i=1}^{n} i \cdot f(x_i)$.

1.32. Compute $\sum_{i=1}^{10}\sum_{j=i}^{10} i^2/(5 + i * j)$.

1.33. Create a function `mat(n)` which returns the $n \times n$ matrix, such that $M_{i,i} = 2$, $M_{i+1,i} = M_{i,i+1} = 1$ and 0 elsewhere.

1.34. Are `sqrt(7)` and `7^.5` equal?

1.35. Given `a <- c(-0.2,0.2,0.49,0.5,0.51,.99,1.2)`, what is the difference between `trunc(a)`, `floor(a)`, `ceiling(a)` and `round(a)`?

1.36. Given a vector `x`, use function `ifelse()` to generate a vector with the same length as `x` with the logarithm of elements of `x` that are positive, and `NA` when elements are negative.

1.37. Create a function `anagram(word1,word2)` which returns `TRUE` if `word1` and `word2` are anagrams.

1.38. Find roots of polynomial $x^2 + x = 1$.

1.39. What is function `pretty` used for?

1.40. Given a vector `x`, write a function `which.closest(x,x0)` which returns the element in `x` that is the closest to `x0`.

1.41. Given two vectors `x` and `y`, write a function `subcount(y,x,k)` which returns the number of elements of `y` smaller than `x[k]`.

1.42. Create vector of length 100 `'Ins1','Ins2', ..., 'Ins100'`.

1.43. Create vector `c("London (2012)","Beijing (2008)","Athens (2004)", "Sydney (2000)")` from `c("London","Beijing","Athens","Sydney")`.

1.44. What will the two functions return, `paste("a", c("b c","d"), sep="")` and `paste("a", c("b c","d"), collapse="")`?

1.45. What will `grep("ab",c("abc","b","a","ba","cab"))` return?

1.46. What is function `apropos()` used for?

1.47. Given a vector `x`, what function(s) can be used to return the location of the largest element?

1.48. Define `Z <- ts(rnorm(240), start=c(1960,3), frequency=12)`. Compute the sum of elements of time series `Z` related to `January`.

1.49. Create a matrix with four columns, that contains all combination of four terms in `c(1,2,7,6,12,37,59)`, each row being a combination.

1.50. Get the help page on command `%o%`.

1.51. What will the following lines return?

```
> as.integer(c(TRUE,FALSE))
> (-2:2)==TRUE
> (-2:2)==FALSE
> as.logical(-2:2)
```

1.52. Generate a vector `x <- rpois(7,4)`. What does `unique(sort(x))[1:3]` return? What about `sort(unique(x))[1:3]`?

1.53. In function `density()`, what is parameter `n` used for?

1.54. Given a matrix `M`, write a function `range.row` which returns a vector whose ith entry is the difference between the largest and the smallest value on the ith row of `M`.

1.55. Find the dataset `accident` from package `hmmm`. How many rows does the dataset have?

1.56. On dataset `accident`, what is the average of variable `Freq` if the dataset is restricted to `Type` equal to `uncertain`?

1.57. What is function `save.image()` for?

1.58. Write a `qqplotQ(x,q)` function which plots the standard QQ plot function, for a sample x, and a quantile function `q()`.

1.59. Given a vector `x` and a cumulative distribution function `F`, write a function which plots the associated PP-plot.

1.60. Compute the integral of $f(x) = x \cdot \log(x)$ on $[0,1]$.

1.61. Plot $x \mapsto \int_0^x f(t)dt$, where $f(x) = x \cdot \log(x)$, on $[0,1]$.

1.62. What is the minimum of function f on $[0,1]$?

1.63. When plotting `y` against `x` using `plot()`, what will option `asp=1` do?

1.64. When plotting `y` against `x` using `plot()`, what will options `xaxs="i"` and `yaxs="i"` do?

1.65. When plotting `y` against `x` using `plot()`, what will option `las=3` do?

1.66. When plotting `y` against `x` using `plot()`, what will option `xlim=rev(range(x))` do?

1.67. Before a `plot()` call, what happens if we add `par(bg = "thistle")`? How do we restore the initial parameters?

1.68. To save a graph in `pdf` format, one can use function `dev.print()`. How should it be used?

1.69. What is the quantile of the $\mathcal{N}(0,1)$ distribution, for probability 95%?

1.70. Using Monte Carlo simulations, approximate $\mathbb{E}[\cos(X)]$, where $X \sim \mathcal{N}(0,1)$.

1.71. Using Monte Carlo simulations, approximate the density of $\max\{B_s, s \in [0,1]\}$, where (B_s) is a standard Brownian motion.

1.72. Plot the quantile function of the $\mathcal{N}(0,1)$ distribution, on $[1\%, 99\%]$. With a red line.

1.73. In function `persp()`, what is the difference between `col="red"` and `border="red"`?

1.74. Using several calls of function `contour` option `levels`, reproduce the graph in the upper left corner of Figure 1.6 where levels curves are in red when density is lower than 0.1, and in blue when density is over 0.1.

1.75. What will `Df <-D(expression(cos(x)/sin(x)), "x")` return? If we execute command `x <- pi/4`, what will `eval(Df)` return?

1.76. Find a package that can help you compute the kurtosis from a vector.

1.77. What is the following function testing, for some integer `n`, `function(n) sum(n%%(1:n)==0)==2`?

1.78. What is `Rprof()` used for?

Part I

Methodology

2

Standard Statistical Inference

Christophe Dutang
Université de Strasbourg and Université du Maine, Le Mans, France
Strasbourg, France

CONTENTS

2.1 Probability Distributions in Actuarial Science

Let X be our quantity of interest. Actuarial models rely on particular assumptions on the probability distribution of X. When X represents the claim amount or the life length of an individual, one expects X to have a distribution on $\mathbb{R}_+$, whereas when X represents the claim number, we deal with distribution on $\mathbb{N}$. But, characterizing the support of the random variable X is a necessary but not a sufficient step to characterize our quantity of interest.

In the discrete case, probability distributions are generally characterized by the mass probability function p_X or the "elementary" probabilities: $p_X(x) = \mathbb{P}(X = x)$ for $x \in \mathbb{N}$. In the continuous case, we define the probability distribution by its density $f_X(x)$, being the infinitesimal version of p_X such that $f_X(x)dx = \mathbb{P}(X \in [x, x + dx[)$. A third case is when the random variable has both continuous and discrete parts, for which there is no proper density. In such a case, we define the distribution with the cumulative distribution function $F_X(x) = \mathbb{P}(X \leq x)$. We recall that for a discrete distribution on $\mathbb{N}$, $F_X(x) = \sum_{n=0}^{\lfloor x \rfloor} p_X(n)$, while for a continuous distribution, $F_X(x) = \int_{-\infty}^{x} f_X(y)dy$.

The purpose of this section is to present the most common distributions used in actuarial sciences, being continuous, discrete or mixed-type. As always in this book, a special emphasis is put on how this topic is implemented and can be extended in R.

2.1.1 Continuous Distributions

There are a lot of ways to classify and to distinguish distributions. We present here the Pearson system and the exponential family, the latter being used, for instance, in generalized linear models (GLM, see Chapter 14). Pearson (1895) considers the family of continuous distributions such that the density function f_X verifies the following ordinary differential equation:

$$\frac{1}{f_X(x)} \frac{df_X(x)}{dx} = -\frac{a + x}{c_0 + c_1 x + c_2 x^2},$$

where a, c_0, c_1, c_2 are constants. Let $p(x) = c_0 + c_1 x + c_2 x^2$. The solution is defined up to a constant K which is derived by the constraint $\int_{\mathbb{R}} f_X(x)dx = 1$. Type 0 is obtained when $c_1 = c_2 = 0$: we get $f_X(x) = Ke^{-(2a+x)x/(2c_0)}$, which is the the normal distribution. Type 1 is the case where the polynomial function $c_0 + c_1 x + c_2 x^2$ has two distinct real roots a_1 and a_2 such that $a_1 < 0 < a_2$: we get $f_X(x) = K(x - a_1)^{m_1}(a_2 - x)^{m_2}$. We recognize the beta distribution. Type 2 corresponds to the case where $m_1 = m_2 = m$.

Type 3 is obtained when $c_2 = 0$ leading a first-order polynomial function $c_0 + c_1 x$. In this case, we get the gamma distribution with $f_X(x) = K(c_0 + c_1 x)^m e^{x+c_1}$. Type 4 corresponds to the case where the polynomial function $p(x) = c_0 + c_1 x + c_2 x^2$ has no real roots, in which case $p(x) = C_0 + c_2(x + C_1)^2$. We get $f_X(x) = K(C_0 + c_2(x + C_1)^2)e^{k \tan^{-1}((x+c_1)/\sqrt{c_0/c_2})}$, which is closely linked to the generalized inverse Gaussian distribution of Barndoff-Nielsen.

We get type 5 when p is a perfect square, that is, $p(x) = (x + C_1)^2$. The associated density is $f_X(x) = K(x + C_1)^{-1/c_2} e^{k/(x+C_1)}$. Two special cases are obtained when $k = 0$, $c_2 > 0$ for type 8 and $c_2 < 0$ for type 9.

Type 6 is obtained when p has two real roots a_1, a_2 of the same sign for which we get $f_X(x) = K(x - a_1)^{m_1}(x - a_2)^{m_2}$, a generalized Beta distribution. Finally, type 7 is obtained when $a = c_1 = 0$, leading to $f_X(x) = K(c_0 + c_2 x^2)^{-1/(2c_2)}$.

Those distributions are implemented in the package `PearsonDS`. In Figure 2.1, we plot the densities for the first seven types in order to compare the different possible shapes.

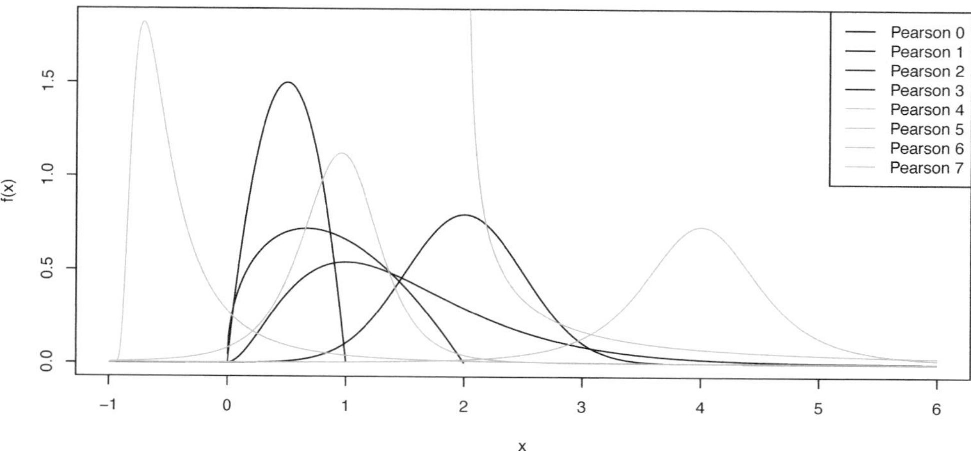

FIGURE 2.1
Pearson's distribution system.

```
> library(PearsonDS)
> x <- seq(-1, 6, by=1e-3)
> y0 <- dpearson0(x, 2, 1/2)
> y1 <- dpearsonI(x, 1.5, 2, 0, 2)
> y2 <- dpearsonII(x, 2, 0, 1)
> y3 <- dpearsonIII(x, 3, 0, 1/2)
> y4 <- dpearsonIV(x, 2.5, 1/3, 1, 2/3)
> y5 <- dpearsonV(x, 2.5, -1, 1)
> y6 <- dpearsonVI(x, 1/2, 2/3, 2, 1)
> y7 <- dpearsonVII(x, 3, 4, 1/2)
> plot(x, y0, type="l", ylim=range(y0, y1, y2, y3, y4, y5, y7),
+ ylab="f(x)", main="The Pearson distribution system")
> lines(x[y1 != 0], y1[y1 != 0], lty=2)
> lines(x[y2 != 0], y2[y2 != 0], lty=3)
> lines(x[y3 != 0], y3[y3 != 0], lty=4)
> lines(x, y4, col="grey")
> lines(x, y5, col="grey", lty=2)
> lines(x[y6 != 0], y6[y6 != 0], col="grey", lty=3)
> lines(x[y7 != 0], y7[y7 != 0], col="grey", lty=4)
> legend("topright", leg=paste("Pearson", 0:7), lty=1:4,
+ col=c(rep("black", 4), rep("grey", 4)))
```

Another important class of distribution is the exponential family, that Andersen (1970) traces back to the work of Pitman, Darmois and Koopman in the mid-1930s. This family contains distributions where the density function can be written as

$$f_X(x) = \exp\left(\sum_{j=1}^{d} a_j(x)\alpha_j(\boldsymbol{\theta}) + b(x) + \beta(\boldsymbol{\theta})\right),$$

where $\boldsymbol{\theta} \in \mathbb{R}^d$ is the d-dimensional parameter vector, and a_j, α_j, b and β are known functions (see Bickel & Doksum (2001) for more details). The exponential family includes

TABLE 2.1
Continuous distributions in R.

Probability Distribution	Root	Probability Distribution	Root
Beta	`beta`	Logistic	`logis`
Cauchy	`cauchy`	Lognormal	`lnorm`
Chi-2	`chisq`	Normal	`norm`
Exponential	`exp`	Student t	`t`
Fisher F	`f`	Uniform	`unif`
Gamma	`gamma`	Weibull	`weibull`

many familiar distributions. We recover the exponential distribution $f_X(x) = \lambda e^{-\lambda x}$ with $d = 1$, $a(x) = x$, $\alpha(x) = \lambda$, $b(x) = 0$ and $\beta(\lambda) = \log(\lambda)$, or the normal distribution, $f_X(x) = e^{-(x-\mu)^2/(2\sigma^2)}/\sqrt{2\pi\sigma^2}$ with $d = 2$, $a_1(x) = x^2$, $\alpha_1(m, \sigma^2) = -1/(2\sigma^2)$, $a_2(x) = x$, $\alpha_2(m, \sigma^2) = m/\sigma^2$, $b(x) =$ and $\beta(m, \sigma^2) = -m/(2\sigma^2) - \log\sqrt{2\pi\sigma^2}$. In the exponential family, the gamma and the inverse Gaussian distributions are also examples of particular interest in actuarial science.

In R, each probability distribution is implemented by a set of four functions and a particular root name `foo`: `dfoo` computes the density function $f_X(x)$ or the mass probability function $p_X(x)$, `pfoo` the cumulative distribution function $F_X(x)$, `qfoo` the quantile function $F_X^{-1}(x)$ and `rfoo` a random number generator. For instance, the gamma distribution with density $f_X(x) = \lambda^\alpha x^{\alpha-1}e^{-\lambda x}/\Gamma(\alpha)$ is implemented in `dgamma`, `pgamma`, `qgamma` and `rgamma`; see example below.

```
> dgamma(1:2, shape=2, rate=3/2)
[1] 0.5020429 0.2240418
> pgamma(1:2, shape=2, rate=3/2)
[1] 0.4421746 0.8008517
> qgamma(1/2, shape=2, rate=3/2)
[1] 1.118898
> set.seed(1)
> rgamma(5, shape=2, rate=3/2)
[1] 0.553910 2.380504 2.308780 1.367208 2.590273
```

In Table 2.1, the continuous distributions implemented in Rare listed. This set of distributions is rather limited, and in practice, other distributions such as Pareto are particularly relevant in actuarial science. Most of distributions are generally implemented in a dedicated package. The full list of non R-base distributions are listed on the corresponding task view http://cran.r-project.org/web/views/Distributions.html. Among the numerous packages, two packages focus on distributions relevant to actuarial science : `actuar` and `ActuDistns`. Note that `actuar` provides the raw moment $\mathbb{E}(X^k)$, the limited expected values $\mathbb{E}(\min(X, l)^k)$ and the moment generating functions $\mathbb{E}(e^{tX})$ for many distributions in three dedicated functions `mfoo`, `levfoo` and `mgffoo`.

When on a particular problem all classical distributions have been exhausted, it is sometimes appropriate to create new probability distributions. Typical transformations of a random variable X are listed:

(i) Translation $X + c$ (e.g. the shifted lognormal distribution),

(ii) Scaling λX,

(iii) Power X^α (e.g. the generalized beta type 1 distribution),

(iv) Inverse $1/X$ (e.g. the inverse gamma distribution),

(v) The logarithm $\log(X)$ (e.g. the loglogistic distribution),

(vi) Exponential $\exp(X)$ and

(vii) The odds ratio $X/(1-X)$ (e.g. the beta type 2 distribution).

With the small code below, we can visualize all those transformations (except the last one) on gamma-distributed variables, using

$$f_Y(y) = \left|\frac{d}{dy}(g^{-1}(y))\right| \cdot f_X(g^{-1}(y)), \text{ where } Y = g(Y),$$

g being a monotonic transformation.

```
> f  <- function(x) dgamma(x,2)
> f1 <- function(x) f(x-1)
> f2 <- function(x) f(x/2)/2
> f3 <- function(x) 2*x*f(x^2)
> f4 <- function(x) f(1/x)/x^2
> f5 <- function(x) f(exp(x))*exp(x)
> f6 <- function(x) f(log(x))/x
> x=seq(0,10,by=.025)
> plot(x,f(x), ylim=c(0, 1.3), xlim=c(0, 10), main="Theoretial densities",
+ lwd=2,  type="l", xlab="x", ylab="")
> lines(x,f1(x), lty=2, lwd=2)
> lines(x,f2(x), lty=3, lwd=2)
> lines(x,f3(x), lty=4, lwd=2)
> lines(x,f4(x), lty=1, col="grey", lwd=2)
> lines(x,f5(x), lty=2, col="grey", lwd=2)
> lines(x,f6(x), lty=3, col="grey", lwd=2)
> legend("topright", lty=1:4, col=c(rep("black", 4), rep("grey", 3)),
+ leg=c("X","X+1","2X", "sqrt(X)", "1/X", "log(X)", "exp(X)"))
```

We can also run simulations and visualize kernel-based densities:

```
> set.seed(123)
> x <- rgamma(100, 2)
> x1 <- x+1
> x2 <- 2*x
> x3 <- sqrt(x)
> x4 <- 1/x
> x5 <- log(x)
> x6 <- exp(x)
> plot(density(x), ylim=c(0, 1), xlim=c(0, 10), main="Empirical densities",
+ lwd=2, xlab="x", ylab="f_X(x)")
> lines(density(x1), lty=2, lwd=2)
> lines(density(x2), lty=3, lwd=2)
> lines(density(x3), lty=4, lwd=2)
> lines(density(x4), lty=1, col="grey", lwd=2)
> lines(density(x5), lty=2, col="grey", lwd=2)
> lines(density(x6), lty=3, col="grey", lwd=2)
```

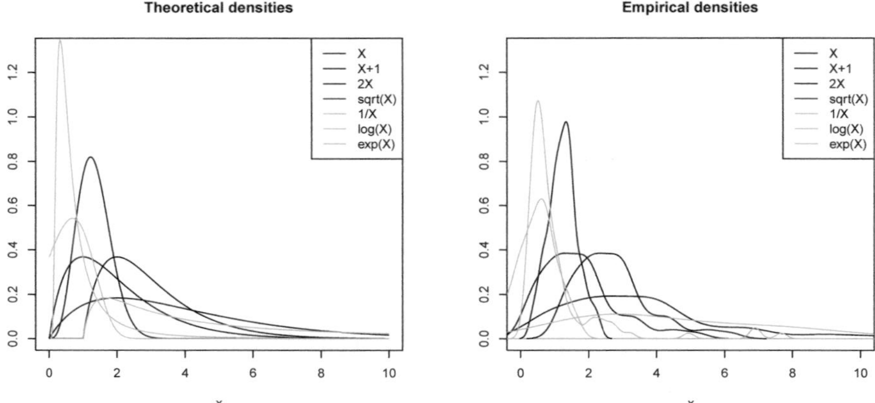

FIGURE 2.2
Transformation of random variables (from a gamma distribution).

In Figure 2.2, we plot the empirical densities (as estimated by the `density()` function, using a kernel approach). Note that the exponential transformation has a heavy-tailed distribution and only the right-tail is shown on the graphic. With these transformations in mind, we can now list the set of distributions generally used in actuarial science.

The most important distribution with finite-support is the uniform distribution with a density $f_X(x) = \mathbf{1}_{[0,1]}(x)$. The uniform distribution is always used for non-uniform random generation as the random variable $F_X(U)^{-1}$ with U a uniform variable has distribution F_X.

Another popular distribution is the beta distribution, defined as

$$f_X(x) = \frac{x^{a-1}(1-x)^{b-1}}{\beta(a,b)}\mathbf{1}_{[0,1]}(x) \text{ and } F_X(x) = \frac{\beta(a,b,x)}{\beta(a,b)}.$$

where $\beta(.,.)$ is the beta function and $\beta(.,.,.)$ is the incomplete lower beta function; see Olver et al. (2010). When $a = b = 1$, we get back to the uniform distribution, that is, f_X is constant. When $a, b < 1$, the density f_X is U-shaped, whereas for $a, b > 1$, the density is unimodal. A monotone density is obtained when a and b have opposite signs. Both of these distributions are implemented in R; see `?dunif` and `?dbeta`. By appropriate scaling and shifting, that is, $c + (d-c)X$, a distribution on any interval $[c, d]$ can be obtained. Finally, another important distribution $\mathcal{T}r(a, b, c)$ is the triangular distribution given by

$$f_X(x) = \frac{2(x-a)}{(b-a)(c-a)}\mathbf{1}_{[a,c]}(x) + \frac{2(x-a)}{(b-a)(c-a)}\mathbf{1}_{]c,b]}(x),$$

which as its name suggests has a triangular-shaped density. When $b = (a+c)/2$, the triangular is the sum of two uniform variates on interval $[a, b]$. The triangular distribution is available in `triangle`.

As presented in Klugman et al. (2009), the two main families of (unbounded) positive continuous distributions are the gamma-transformed family and the beta-transformed family. Let X follow a gamma distribution $\mathcal{G}(\alpha, 1)$. The gamma-transformed family is the distribution of $Y = X^{1/\tau}/\lambda$ for $\tau > 0$, which has the following density and distribution functions

$$f_Y(y) = \frac{\lambda^{\tau\alpha}}{\Gamma(\alpha)}\tau y^{\alpha\tau-1}e^{-(\lambda y)^\tau} \text{ and } F_Y(y) = \Gamma(\alpha, (\lambda y)^\tau)/\Gamma(\alpha),$$

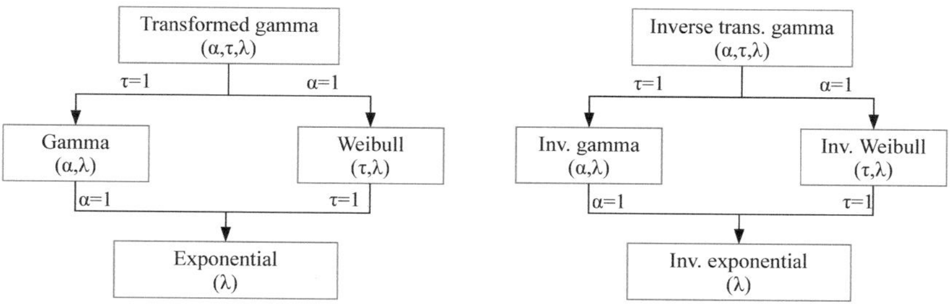

FIGURE 2.3
Transformed gamma family.

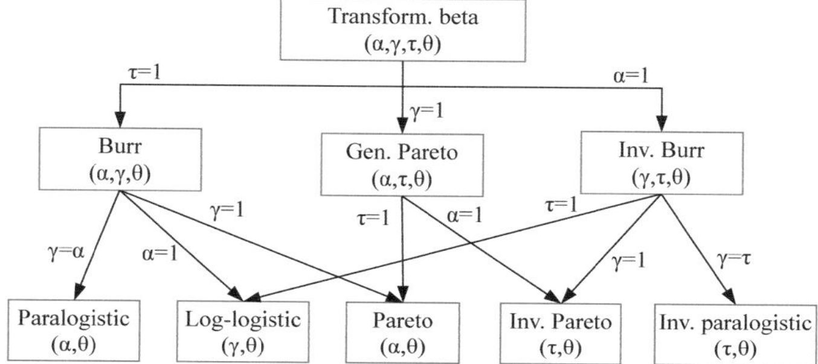

FIGURE 2.4
Transformed beta family.

where $\Gamma(.,.)$ denotes the incomplete lower gamma function, see, for example, Olver et al. (2010). When $\tau < 0$, we get the inverse gamma-transformed family. Let $\tau^\star = -\tau$. The density and the distribution function of $Y = 1/(\lambda X^{1/\tau^\star})$ are given by

$$f_Y(y) = \frac{\tau^\star e^{-(\lambda y)^{-\tau^\star}}}{\lambda^{\tau^\star \alpha} y^{\alpha\tau^\star+1}\Gamma(\alpha)} \text{ and } F_Y(y) = 1 - \Gamma(\alpha, (\lambda y)^{-\tau^\star})/\Gamma(\alpha).$$

On Figure 2.3, we list the different special cases of the transformed gamma distribution and their relationships.

The beta-transformed family is based on the beta distribution of the second kind (or type II), that is, the distribution of $X/(1-X)$ when X follows a beta distribution of type I, see the previous subsection. The beta distribution of type II has a density $f_X(x) = \frac{x^{a-1}}{\beta(a,b)(1+x)^{a+b}}$. Renaming $a = \alpha$ and $b = \tau$, the transformed beta is the distribution of $Y = \theta X^{1/\gamma}$ and has the following density and distribution function:

$$f_Y(y) = \frac{1}{\beta(\alpha,\tau)} \frac{\gamma(y/\theta)^{\gamma\tau}}{y\,(1+(y/\theta)^\gamma)^{\alpha+\tau}} \text{ and } F_Y(y) = \frac{\beta(\alpha,\tau,\frac{x}{1+x})}{\beta(\alpha,\tau)},$$

where $x = (y/\theta)^\gamma$ and $\beta(.,.,.)$ denotes the incomplete lower beta function. These two families are available in `actuar`.

2.1.2 Discrete Distributions

The Sundt $(a, b, 0)$ family of distributions is the set of distributions verifying

$$\frac{\mathbb{P}(X = k+1)}{\mathbb{P}(X = k)} = a + \frac{b}{k},$$

for $k \in \mathbb{N}$ and $a, b \geq 0$ positive parameters. This recurrence equation can be seen a simplified discrete equation of the Pearson system (see Johnson et al. (2005)). We get back to the binomial distribution $\mathcal{B}(n, p)$ with $a = -p/(1-p)$ and $b = p(n+1)/(1-p)$, the Poisson distribution $\mathcal{P}(\lambda)$ with $a = 0$ and $b = \lambda$, and the negative binomial distribution $NB(m, p)$ with $a = 1 - p$ and $b = (1-p)(m-1)$. A generalization of the $(a, b, 0)$ family is obtained by truncating the values smaller than n. Thus, the (a, b, n) family verifies

$$p_X(k) = p_X(k-1)\left(a + \frac{b}{k}\right)\mathbf{1}_{(k \geq n)}.$$

Furthermore, the exponential family also models discrete distributions by considering the mass probability function p_X that verifies

$$p_X(k) = \exp\left(\sum_{j=1}^{d} a_j(k)\alpha_j(\theta) + b(k) + \beta(\theta)\right).$$

It includes many familiar distributions: the Bernoulli distribution with $d = 1$, $a(x) = x$, $\alpha(p) = \log(p/(1-p))$, $b(x) = 0$ and $\beta(p) = \log(1-p)$, and the Poisson distribution with $d = 1$, $a(x) = x$, $\alpha(\lambda) = \lambda$, $b(x) = -\log(x!)$, and $\beta(\lambda) = -\lambda$. See Chapter 14 for a discussion of the negative binomial distribution and the exponential family.

As for continuous distributions, discrete distributions are implemented in four functions: `dfoo` computes the mass probability function p_X, `pfoo` the cumulative distribution function F_X, `qfoo` the quantile function F_X^{-1} and `rfoo` the random number generator. For instance, the Poisson distribution is implemented in `dpois`, etc. Here is a standard call:

```
> dpois(0:2, lambda=3)
[1] 0.04978707 0.14936121 0.22404181
> ppois(1:2, lambda=3)
[1] 0.1991483 0.4231901
> qpois(1/2, lambda=3)
[1] 3
> rpois(5, lambda=3)
[1] 2 2 3 5 2
```

Typical transformations of an integer-valued random variable X are listed: (i) translation $X + m$ for a non-null interger m (e.g. the shifted Poisson distribution), (ii) scaling mX, (iii) zero-inflation $(1-B)X$ where B follows a Bernoulli distribution $\mathcal{B}(q)$ and (iv) zero-modification $(1-B)(X+1)$ where B follows a Bernoulli distribution. The resulting mass probability function for the transformed variable Y is

(i) $\mathbb{P}(Y = k) = \mathbb{P}(X = k - m)$ for $k \geq m$,

(ii) $\mathbb{P}(Y = k) = \mathbb{P}(X = k/m)$ for $k = 0, m, 2m, 3m, \ldots$,

(iii) $\mathbb{P}(Y = 0) = q + (1-q)\mathbb{P}(X = 0)$ and $\mathbb{P}(Y = k) = (1-q)\mathbb{P}(Y = k)$ for $k \geq 1$,

(iv) $\mathbb{P}(Y = 0) = q$ and $\mathbb{P}(Y = k) = (1-q)\mathbb{P}(Y = k-1)$ for $k \geq 1$.

The zero-modification and the zero-inflation are useful to add a parameter to standard discrete distributions, for example, the Poisson distribution. A particular of the zero-modification is the zero-truncation when the variable B equals almost surely 0. Those transformations will be considered in Chapter 14, in the context of modeling claims frequency in motor insurance.

Such transformations are implemented in special packages, see the task view, but can be easily implemented.

```
> dpoisZM <- function(x, prob, lambda)
+ prob*(x == 0) + (1-prob)*(x > 0)*dpois(x-1, lambda)
> ppoisZM <- function(q, prob, lambda)
+ prob*(q >= 0) + (1-prob)*(q > 0)*ppois(q-1, lambda)
> qpoisZM <- function(p, prob, lambda)
+ ifelse(p <= prob, 0, 1+qpois((p-prob)/(1-prob), lambda))
> rpoisZM <- function(n, prob, lambda)
+ (1-rbinom(n, 1, prob))*(rpois(n, lambda)+1)
> x <- rpoisZM(100, 1/2, 3)
> plot(ecdf(x), main="Zero-modified Poisson(prob=1/2, lam=3)")
> lines(z <- sort(c(0:12, 0:12-1e-6)),
+ ppoisZM(z, 1/2, 3), col="grey", lty=4, lwd=2)
> legend("bottomright", lty=c(1,4), lwd=1:2,
+ col=c("black","grey"), leg=c("empir.","theo."))
```

In Figure 2.5, we plot the empirical cumulative distribution function of a zero-modified Poisson distribution.

The main discrete distributions are the binomial $\mathcal{B}(n,p)$, the Poisson $\mathcal{P}(\lambda)$ and the negative binomial $NB(m,p)$ distributions, for which we recall the mass probability function $p_X(k)$

$$\mathbb{P}(X=k)=\binom{n}{k}p^k(1-p)^{n-k},$$

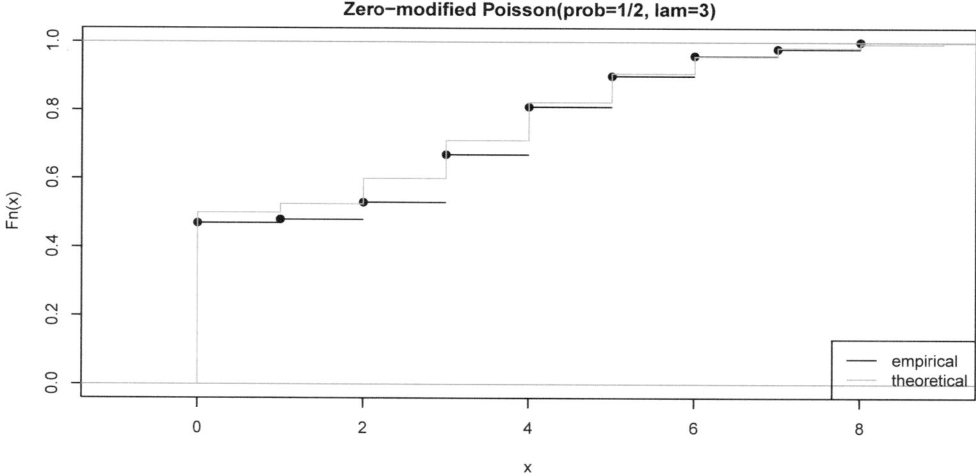

FIGURE 2.5
Cumulative distribution function of a zero-modified Poisson sample.

for $p \in [0,1]$ (the Bernoulli distribution is obtained with $n = 1$).

$$\mathbb{P}(X = k) = \frac{\lambda^k}{k!}e^{-\lambda},$$

for $\lambda > 0$, and

$$\mathbb{P}(X = k) = \binom{m+k-1}{k} p^m (1-p)^k,$$

for $p \in [0,1]$. The discrete analog of the Pareto distribution is the Zipf distribution whose mass probability function is given by

$$\mathbb{P}(X = k) = \frac{k^\eta}{\zeta(\eta)},$$

where $\zeta(.)$ is the zeta's Rieman function; see Olver et al. (2010).

2.1.3 Mixed-Type Distributions

Mixed-type distributions are distributions of random variables that are neither continuous nor discrete, that is, $0 < \sum_{x\in D_X}(F_X(x) - F_X(x_-)) < 1$ for D_X the set of discontinuities where the lower bound corresponds to continuous distributions and the upper bound discrete distributions. Thus, the distribution function has discontinuities and continuous parts. A first example of mixed-type distribution is the zero-modified gamma distribution which has the distribution function

$$F_X(x) = p\mathbf{1}_{x\geq 0} + (1-p)\frac{\Gamma(\alpha, \lambda x)}{\Gamma(\alpha)},$$

where $\Gamma(.,.)$ denotes the incomplete gamma function. X has an improper density function $f_X(x) = (1-p)\lambda^\alpha x^{\alpha-1}e^{-\lambda x}/\Gamma(\alpha)$. In a similar way, zero-modified Pareto or zero-modified lognormal distributions can be defined.

An application of mixed-type distributions to destruction rate models is now presented. Destruction rate models focus on the distribution $X = L/d$ where L is the loss amount and d the maximum possible loss (as defined in the insurance terms). By definition, X is bounded to the interval $[0,1]$, and may have a mass at 1 when the object insured is entirely destroyed. In the application that will follow, we will consider the one-modified beta and the MBBEFD distributions. The one-modified beta is the distribution of $X = BY$ where Y follows a beta distribution $\mathcal{B}(a,b)$ and B follows a Bernoulli distribution $\mathcal{B}(q)$. The distribution function is given by

$$F_X(x) = \frac{\beta(a,b,x)}{\beta}(1-q) + q\mathbf{1}_{x\geq 1},$$

for which the improper density is $f_X(x) = (1-q)x^{a-1}(1-x)^{b-1}/\beta(a,b)$. In R, we define it as

```
> dbetaOM <- function(x, prob, a, b)
+   dbeta(x, a, b)*(1-prob)*(x != 1) + prob*(x == 1)
> pbetaOM <- function(q, prob, a, b)
+   pbeta(q, a, b)*(1-prob) + prob*(q >= 1)
```

The Maxwell-Boltzmann Bore-Einstein Fermi-Dirac (MBBEFD) distribution was introduced and popularized by Bernegger (1997) in the context of reinsurance treaties. The distribution function is given by

$$F_X(x) = a\left(\frac{a+1}{a+b^x} - 1\right)\mathbf{1}_{[0,1[}(x) + \mathbf{1}_{[1,+\infty[}(x),$$

where $(a,b)\in]-1,0[\times]1,+\infty[$ or $(a,b)\in(\mathbb{R}\backslash[-1,0])\times]0,1[$. Note that there is a probability mass at 1, since $\mathbb{P}(X=1)=(a+1)b/(a+b)=q$. The improper density function is

$$f_X(x)=\frac{-a(a+1)b^x\log(b)}{(a+b^x)^2}\mathbf{1}_{]0,1[}(x).$$

At the time this book is written, there is no package implementing the MBBEFD distribution, but this can be remedied by the following lines:

```
> dMBBEFD <- function(x, a, b)
+ -a*(a+1)*b^x*log(b)/(a + b^x)^2 + (a+1)*b/(a+b)*(x == 1)
> pMBBEFD <- function(x, a, b)
+ a*((a+1)/(a+b^x)-1)*(x<1)+1*(x>=1)
```

Those two distributions will be used in the subsequent section on destruction rate data.

Mixing distributions consists of randomly drawing a distribution among a finite set of distributions. Consider a set of distribution functions $F_1,\dots,F_p$ and a set of weights $\omega_1,\dots,\omega_p\in[0,1]$. The choice Θ of a distribution is such that $\mathbb{P}(\Theta=i)=\omega_i$ for $i=1,\dots,p$. The random generation process given by (i) draw Θ according to ω_i's and (ii) draw according to F_c knowing Θ is the mixture distribution among $(F_1,\dots,F_p)$ according to $\omega_1,\dots,\omega_p$. This is characterized by the following distribution function

$$F_X(x)=\sum_{i=1}^{p}\omega_iF_i(x)$$

for all $x\in\mathbb{R}$. If distributions F_i are differentiable, then the density function of the mixture variable X is simply $f_X(x)=\sum_{i=1}^{p}\omega_if_i(x)$.

A first simple example is the mixture of two normal distributions $\mathcal{N}(m_1,s_1^2)$, $\mathcal{N}(m_2,s_2^2)$ with the following density:

$$f_X(x)=p\frac{e^{-(x-m_1)^2/(2s_1^2)}}{\sqrt{2\pi s_1^2}}+(1-p)\frac{e^{-(x-m_2)^2/(2s_2^2)}}{\sqrt{2\pi s_2^2}},$$

with a proportion $p\in[0,1]$ and $x\in\mathbb{R}$. This distribution is implemented in the package `mixtools` and `norm1mix`. A second example of more interest in actuarial science is the mixture of a light-tailed and heavy-tailed claim distribution. Say, for example, the mixture of a gamma distribution $\mathcal{G}(\nu,\lambda)$ and a Pareto distribution $\mathcal{P}(\alpha,\theta)$. The density is given by

$$f_X(x)=p\frac{\lambda^\nu x^{\nu-1}e^{-\lambda x}}{\Gamma(\nu)}+(1-p)\alpha/\theta\left(\frac{\theta}{\theta+x}\right)^{\alpha+1},$$

with a proportion $p\in[0,1]$ and $x\in\mathbb{R}_+$. In R, we implement it as

```
> library(actuar)
> dmixgampar <- function(x, prob, nu, lambda, alpha, theta)
+ prob*dgamma(x, nu, lambda) + (1-prob)*dpareto(x, alpha, theta)
> pmixgampar <- function(q, prob, nu, lambda, alpha, theta)
+ prob*pgamma(q, nu, lambda) + (1-prob)*ppareto(q, alpha, theta)
```

where `dpareto` is implemented in the `actuar` package.

Another important family obtained using mixtures are the so-called phase-type distributions, obtained as mixtures of exponential distributions. Given $\boldsymbol{p}$ a vector of probabilities of length k, and $\boldsymbol{M}$ a $k\times k$ matrix, X is said to be phase-type distributed, with parameters $\boldsymbol{p}$ and M if

$$F_X(x)=\mathbb{P}(X\le x)=1-\boldsymbol{p}\exp[\boldsymbol{M}x]\mathbf{1}$$

where exp denotes here the matrix exponential; see Moler & Van Loan (1978) for a recent survey. The phase-type distribution can be seen as the distribution of the time to absorption (in the state 0) of a Markov jump process on the set $\{0, 1, \ldots, n\}$ with initial probability $(0, \boldsymbol{p})$ and intensity matrix

$$\Lambda = \left(\begin{array}{c|c} 0 & 0 \\ \hline m_0 & \boldsymbol{M} \end{array} \right),$$

where the vector $m_0 = -\boldsymbol{M}\mathbf{1}_{\mathbb{R}^k}$. Observe that X has density

$$f_X(x) = -\boldsymbol{p} \exp[\boldsymbol{M}x] m_0.$$

In package `actuar`, `phtype` distributions do exist, `prob` being vector $\boldsymbol{p}$ and `rates` being matrix $\boldsymbol{M}$. One particular case is the Erlang distribution: Erlang (k, λ) distribution, with density

$$f(x; k, \lambda) = \frac{\lambda^k x^{k-1} e^{-\lambda x}}{\Gamma(k)}$$

is obtained when $\boldsymbol{p} = (1, 0, \ldots, 0)$ (of length k), and $\boldsymbol{M}$ is the $k \times k$ matrix with $-\lambda$ on the diagonal, λ above, and 0 elsewhere (see O'Cinneide (1990) and Bladt (2005) for more details). To generate an Erlang distribution with $k = 3$ and $\lambda = 2$, one can use

```
> M <- matrix(0,3,3)
> diag(M) <- -2
> diag(M[1:(nrow(M)-1),2:ncol(M)]) <- 2
> M
     [,1] [,2] [,3]
[1,]   -2    2    0
[2,]    0   -2    2
[3,]    0    0   -2
>  set.seed(123)
>  rphtype(5, prob=c(1,0,0), rates=M)
[1] 0.3311529 2.3017693 0.5631011 2.7375481 2.0612129
```

Other distributions such as mixture of generalized Erlang distributions are phase-type, but it is shown that a phase-type distribution does have Laplace transform $\widehat{f}(s) = \pi(-sI_k - \boldsymbol{M})^{-1} m_0$. Therefore, the phase-type family does not include heavy-tailed distribution.

2.1.4 S3 versus S4 Types for Distribution

In the previous chapter, the distinction between S3 and S4 objects was introduced. Some packages allow one to use S4 objects to deal with distributions. For instance, using

```
> library(distr)
> library(distrEx)
```

we define an object, which is a *distribution*, and then various functions can be used to get the density, the quantile function or a random number generator, based on that distribution. Consider, for instance, the $\mathcal{N}(5, 2^2)$ distribution:

```
> X <- Norm(mean=5,sd=2)
> X
Distribution Object of Class: Norm
 mean: 5
 sd: 2
```

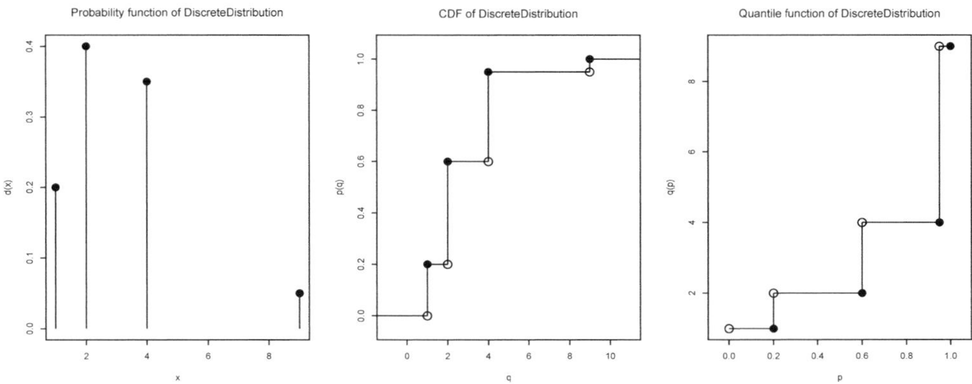

FIGURE 2.6
Using `plot` of a `distr` object for a discrete distribution.

If we want to compute quantiles associated to that distribution, we use the quantile `q()` function: `q(X)` is then the function $u \mapsto F_X^{-1}(u)$,

```
> q(X)
function (p, lower.tail = TRUE, log.p = FALSE)
{
    qnorm(p, mean = 5, sd = 2, lower.tail = lower.tail, log.p = log.p)
}
<environment: 0x10d796c98>
```

And if we want to evaluate that function, for instance to get the value of $F_X^{-1}(0.25)$, we use

```
> q(X)(0.25)
[1] 3.65102
```

(which is the same as the standard `qnorm(0.25,mean=5,sd=2)`). Various functions can also be used to derive simple quantities associated to that distribution, such as moments

```
> mean(X)
[1] 5
```

We can also create discrete distributions, such as

```
> N <- DiscreteDistribution(supp=c(1,2,4,9) , prob=c(.2,.4,.35,.05))
```

where the support `supp` and the associate probabilities `prob` are mentioned. One can then use `r()` to generate random numbers, `d()` to compute the density function, `p()` to compute the cumulative distribution function, and `q()` to compute the quantile function. We can also visualize that distribution using

```
> plot(N)
```

An interesting feature of this `S4` class is that simple arithmetics on distributions can be performed. Consider two distributions `X1` and `X2`:

```
> X1 <- Norm(mean=5,sd=2)
> X2 <- Norm(mean=2,sd=1)
```

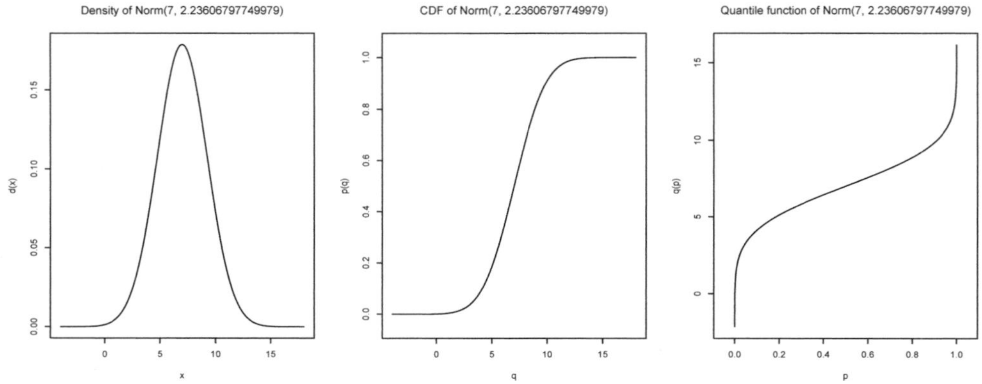

FIGURE 2.7
Using `plot` of a `distr` object for a sum of two variables.

then operator `+` can be used to define a distribution which will be the sum of two independent Gaussian random variables,

```
> S <- X1+X2
> plot(S)
```

If we look at the titles on Figure 2.7, we can see that `S` is recognized as a Gaussian distribution (the sum of two independent Gaussian distributions being also a Gaussian distribution).

Other operators can be used, such as `-`, `*` or `/`, even `^`,

```
> U <- Unif(Min=0,Max=1)
> N <- DiscreteDistribution(supp=c(1,2,4,9) , prob=c(.2,.4,.35,.05))
> Z <- U^N
> plot(Z)
```

Such a function is (absolutely continuous), and is recognized as an `AbscontDistribution` object. A more complex object is obtained if N can take value 0. Then, the distribution is no longer absolutely continuous:

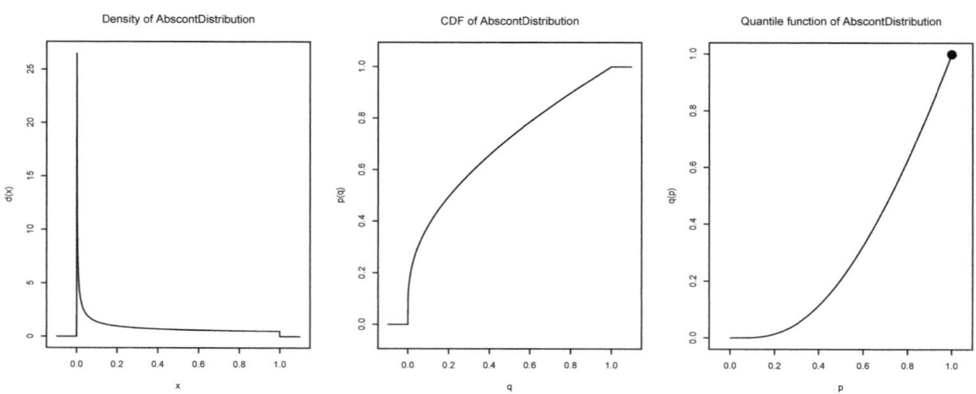

FIGURE 2.8
Using `plot` of a `distr` object for U^N, N being a discrete distribution.

```
> N <- DiscreteDistribution(supp=c(0,1,2,4) , prob=c(.2,.4,.35,.05))
> Z <- U^N
> Z
An object of class "UnivarLebDecDistribution"
 --- a Lebesgue decomposed distribution:

    Its discrete part (with weight 0.200000) is a
 Distribution Object of Class: Dirac
 location: 1
 This part is accessible with 'discretePart(<obj>)'.

    Its absolutely continuous part (with weight 0.800000) is a
 Distribution Object of Class: AbscontDistribution
 This part is accessible with 'acPart(<obj>)'.
```

Using `plot(Z)`, we can see that this distribution has two components: a (absolutely) continuous one, and a Dirac mass, in 1.

Observe finally that compound distributions can also be generated easily. The standard compound Poisson distribution is obtained using

```
> CP <- CompoundDistribution(Pois(), Gammad())
> CP
An object of class "CompoundDistribution"

 The frequency distribution is:
Distribution Object of Class: Pois
lambda: 1
 The summands distribution is/are:
Distribution Object of Class: Gammad
shape: 1
scale: 1
```

2.2 Parametric Inference

Parametric inference deals with the estimation of an unknown parameter of a chosen distribution. The experimenter assumes that $(x_1, \ldots, x_n)$ are realizations of a random sample $(X_1, \ldots, X_n)$ such that X_i are independent and identically distributed randoms variables according to a generic random variable X (this is the blanket assumption). The random variable X has a distribution function $F(.;\boldsymbol{\theta})$ for $\boldsymbol{\theta} \in \Theta \subset \mathbb{R}^d$. For example, $F(x;\theta) = (1 - e^{-\theta x})\mathbf{1}_{\mathbb{R}_+}(x)$ when considering an exponential distribution $\theta \in \mathbb{R}_+$. In the following subsequent sections, classical estimation methods are presented and provide criteria to establish an estimator $\hat{\boldsymbol{\theta}}$ of $\boldsymbol{\theta}$. Once a model is fitted, the experimenter can derive its quantities of interest (mean, variance, quantiles, survival probabilities,...), derived from the fitted distribution $F(x;\hat{\boldsymbol{\theta}})$. In most applications, X has either a continuous or a discrete distribution. Therefore, we work either the density function $f_X(.;\boldsymbol{\theta})$ or the mass probability function $p_X(.;\boldsymbol{\theta})$. For a general introduction to statistical inference, we refer to Casella & Berger (2002).

2.2.1 Maximum Likelihood Estimation

As its name suggests, maximum likelihood estimation consists of maximizing the likelihood with respect to $\boldsymbol{\theta}$, which is defined as

$$\mathcal{L}(\boldsymbol{\theta}, x_1, \ldots, x_n) = \prod_{i=1}^{n} f_X(x_i; \boldsymbol{\theta}) \text{ or } \prod_{i=1}^{n} p_X(x_i; \boldsymbol{\theta}),$$

depending on the type of the random variable X. For many reasons, it is more convenient to maximize the log-likelihood $\log \mathcal{L}$ with respect to $\boldsymbol{\theta}$. A school example is to consider the exponential distribution $\mathcal{E}(\theta)$, that is, $\log \mathcal{L}(\theta) = n\log(\theta) - \sum_{i=1}^{n} \theta x_i$. The maximizer is $1/\bar{x}_n$, leading to the estimator $\hat{\theta} = 1/\bar{X}_n$. In practice, closed-form formulas of the maximizers may not exist, thus we use a numeric optimization. The `fitdistrplus` package provides routines to compute the maximum likelihood estimator for most standard distributions.

We consider a claim dataset `itamtplcost` which contains large losses (in excess of 500,00 euros) of an Italian Motor-TPL company since 1997. For pedagogical purposes (despite that the distribution is not really appropriate), we choose to fit a gamma distribution $\mathcal{G}(\alpha, \lambda)$ defined as $f_X(x) = \lambda^\alpha x^{\alpha-1} e^{-\lambda x}/\Gamma(\alpha)$, with parameter $\boldsymbol{\theta} = (\alpha, \lambda) \in \mathbb{R}_+^2$.

```
> data(itamtplcost)
> library(fitdistrplus)
> x <- itamtplcost$UltimateCost / 10^6
> summary(x)

    Min.  1st Qu.   Median     Mean  3rd Qu.     Max.
0.002161 0.627700 0.844000 1.015000 1.224000 6.639000

> fgamMLE <- fitdist(x, "gamma", method="mle")
> fgamMLE

Fitting of the distribution ' gamma ' by maximum likelihood
Parameters:
      estimate Std. Error
shape 2.398655  0.1489696
rate  2.362486  0.1631542

> summary(fgamMLE)

Fitting of the distribution ' gamma ' by maximum likelihood
Parameters :
      estimate Std. Error
shape 2.398655  0.1489696
rate  2.362486  0.1631542
Loglikelihood:  -385.1474   AIC:  774.2947   BIC:  782.5441
Correlation matrix:
          shape      rate
shape 1.0000000 0.8992915
rate  0.8992915 1.0000000
```

Without a scaling of cost from euros to millions of euros, the call to `fitdist` raises an error, thus we divide the ultimate cost by 10^6. In this example, $\hat{\boldsymbol{\theta}}$ is estimated as $(2.398655, 2.362486)$. Note that the `fitdist` function returns an S3-object of class `fitdist`, for which `print`, `summary` and `plot` methods have been defined. In addition to the estimation of standard errors of $\hat{\boldsymbol{\theta}}$, the `summary` method gives an estimation of the asymptotic correlation matrix as well as the (optimal) log-likelihood. This is based on the asymptotic normality of the maximum likelihood estimators (under the hypotheses of the Cramer–Rao model; see Casella & Berger (2002)).

2.2.2 Moment Matching Estimation

The moment matching estimation is also commonly used to fit parametric distributions. This consists of finding the value of the parameter $\boldsymbol{\theta}$ that matches the first theoretical raw moments of the parametric distribution to the corresponding empirical raw moments as

$$\mathbb{E}(X^k|\boldsymbol{\theta}) = \frac{1}{n}\sum_{i=1}^{n} x_i^k,$$

for $k = 1, \ldots, d$, with d the number of parameters to estimate and x_i the n observations of variable X. For moments of order greater than or equal to 2, it may be relevant to match centered moments defined as

$$\mathbb{E}(X|\boldsymbol{\theta}) = \bar{x}_n \ , \ \mathbb{E}\left((X - \mathbb{E}(X))^k|\boldsymbol{\theta}\right) = m_k, \text{ for } k = 2, \ldots, d,$$

where $m_k = \frac{1}{n}\sum_{i=1}^{n}(x_i - \bar{x}_n)^k$ denotes the empirical centered moments. For instance, consider the gamma distribution $\mathcal{G}(\alpha, \lambda)$. The moment matching estimation solves

$$\begin{cases} \alpha/\lambda = \bar{x}_n \\ \alpha/\lambda^2 = m_2 \end{cases} \Leftrightarrow \begin{cases} \alpha = (\bar{x}_n)^2/m_2 \\ \lambda = \bar{x}_n/m_2 \end{cases}$$

In general, there are no closed-form formulas for this estimator and use a numerical method. Still considering the gamma distribution fit on the MTPL dataset, we use the `fitdistrplus` package.

```
> fgamMME <- fitdist(x, "gamma", method="mme")
> cbind(MLE=fgamMLE$estimate, MME=fgamMME$estimate)
```

```
           MLE      MME
shape 2.398655 2.229563
rate  2.362486 2.195851
```

2.2.3 Quantile Matching Estimation

Fitting of a parametric distribution may also be done by matching theoretical quantiles of the parametric distribution (for some specified probabilities) against the empirical quantiles (see Tse (2009) among others). The equation below is very similar to the previous equations for matching moments

$$F^{-1}(p_k; \boldsymbol{\theta}) = Q_{n,p_k},$$

for $k = 1, \ldots, d$ and Q_{n,p_k} the empirical quantiles for specified probabilities p_k. Empirical quantiles Q_{n,p_k} are computed on observations $x_1, \ldots, x_n$ using the `quantile` function of the `stats` package. When $n \cdot p_k$ is an integer, the empirical quantile is uniquely defined as $Q_{n,p_k} = x^\star_{p_k n}$, where $(x^\star_1, \ldots, x^\star_n)$ is the sorted sample. Otherwise, the empirical quantile is the convex combination of $x^\star_{\lfloor p_k n \rfloor}$ and $x^\star_{\lceil p_k n \rceil}$; see `?quantile` and Hyndman & Fan

(1996). The theoretical quantile $F^{-1}(.;\boldsymbol{\theta})$ can have a closed-form formula. For example, when considering the exponential distribution $\mathcal{E}(\lambda)$, the quantile function is

$$F^{-1}(p;\theta) = -\frac{\log(1-p)}{\lambda}.$$

Solving the d equations $F^{-1}(p_k;\theta) = Q_{n,p_k}$ is achieved by a numeric optimization in the `fitdist` function.

Continuing the MTPL example, we fit a gamma distribution against the probabilities $p_1 = 1/3$ and $p_2 = 2/3$.

```
> fgamQME <- fitdist(x, "gamma", method="qme", probs=c(1/3, 2/3))
> cbind(MLE=fgamMLE$estimate, MME=fgamMME$estimate,
+        QME=fgamQME$estimate)
```

```
           MLE      MME     QME
shape 2.398655 2.229563 4.64246
rate  2.362486 2.195851 4.95115
```

Note that compared to the method of moments and the maximum likelihood estimation, the estimate paramater $\hat{\boldsymbol{\theta}}$ differs significantly when using the quantile matching estimation, despite considering probabilities in the heart of the distribution.

2.2.4 Maximum Goodness-of-Fit Estimation

A last method of estimation called maximum goodness-of-fit estimation or (minimum distance estimation) is presented here; see D'Agostino & Stephens (1986) or Dutang et al. (2008) for more details. In this section, we focus on the Cramér–von Mises distance and refer to Delignette-Muller & Dutang (2013) for other distances (i.e. Kolmogorov–Smirnov and Anderson–Darling). The Cramer–von Mises looks at the squared difference between the candidate distribution $F(x;\boldsymbol{\theta})$ and the empirical distribution function F_n, the latter being defined as the percentage of observations below x: $F_n(x) = \sum_{i=1}^n \mathbf{1}_{x_i \leq x}$. The Cramer–von Mises distance is defined as

$$D(\theta) = \int_{-\infty}^{\infty} (F_n(x) - F(x;\boldsymbol{\theta}))^2 dx,$$

and is estimated in practice by

$$\widehat{D}(\boldsymbol{\theta}) = \frac{1}{12n} + \sum_{i=1}^{n} \left(F(x_i;\boldsymbol{\theta}) - \frac{2i-1}{2n} \right)^2.$$

The maximum goodness-of-fit estimation consists of finding the value of θ minimizing $\widehat{D}(\boldsymbol{\theta})$. The name comes from the fact that the Cramer–von Mises distance measures the goodness-of-fit of $F(.;\boldsymbol{\theta})$ against F_n. There is no closed-form formula for $\arg\min\{\widehat{D}(\boldsymbol{\theta})\}$, and a numerical optimization is used in the `fitdist` function.

Finally, we fit a gamma distribution by maximum goodness-of-fit estimation:

```
> fgamMGE <- fitdist(x, "gamma", method="mge", gof="CvM")
> cbind(MLE=fgamMLE$estimate, MME=fgamMME$estimate,
+ QME=fgamQME$estimate, MGE=fgamMGE$estimate)
```

```
           MLE      MME     QME      MGE
shape 2.398655 2.229563 4.64246 3.720546
rate  2.362486 2.195851 4.95115 3.875971
```

As for quantile matching estimation, the value of $\hat{\boldsymbol{\theta}}$ differs widely. A practitioner approach could be take the average by component irrespectively of the methods tested. This leads to the question of how to choose between fitted parameters and between fitted distributions.

2.3 Measures of Adequacy

This section focuses on measures of adequacy either graphical methods or numerical methods.

2.3.1 Histogram and Empirical Densities

A typical plot to assess the adequacy of a distribution is the histogram. We recall that for plotting a histogram, observed data are divided into k classes $]a_{j-1}, a_j]$ for $j = 1, \dots, k$ (generally k is proportional to $\log(n)$); the number of observation in each class is computed, that is, frequencies f_j; finally rectangles are drawn such that the basis is a class $]a_{j-1}, a_j]$ and the height is the absolute f_j or the relative f_j/n frequencies. Thus, the histogram is an estimator of the empirical density, as the area of a rectangle is proportional to $\mathbb{P}(X \in]a_{j-1}, a_j])$. This graph is generically provided in the `plot` function of a `fitdist` object, but does not allow multiple fitted distributions. So in the example of a gamma fit to the MTPL dataset, we use the `denscomp` function:

```
> txt <- c("MLE","MME","QME(1/3, 2/3)", "MGE-CvM")
> denscomp(list(fgamMLE, fgamMME, fgamQME, fgamMGE), legendtext=txt,
+ fitcol="black", main="Histogram and fitted gamma densities")
```

Alternatively, we can estimate directly the density function by the popular kernel density estimation. This is implemented in the `density` function as shown below

```
> hist(x, prob=TRUE, ylim=c(0, 1))
> lines(density(x), lty=5)
```

In order to better assess the fitted gamma densities, the two above graphs are plotted on separate graphics. We observe that the MLE and the MME fits best approximates the density between $x \in [0.5, 1.5]$, while the QME and the MGE fit best assess the density between $x \in [1.5, 4]$. However, it is clear that the gamma distribution cannot appropriately fit the whole distribution, mainly due to its light-tailedness.

2.3.2 Distribution Function Plot

Another typical graph is to plot the fitted distribution $F(.;\hat{\theta})$ and the empirical cumulative distribution function F_n. As already given, the computation of F_n is simpler than for the empirical density $F_n(x) = \sum_{i=1}^{n} \mathbf{1}_{x_i \leq x}$. A new claim dataset is considered to illustrate this type of plot: we use the popular Danish dataset, used in McNeil (1997). The dataset is stored in `danishuni` for the univariate version and contains fire loss amounts collected at Copenhagen Reinsurance between 1980 and 1990. We consider three distributions: a gamma distribution

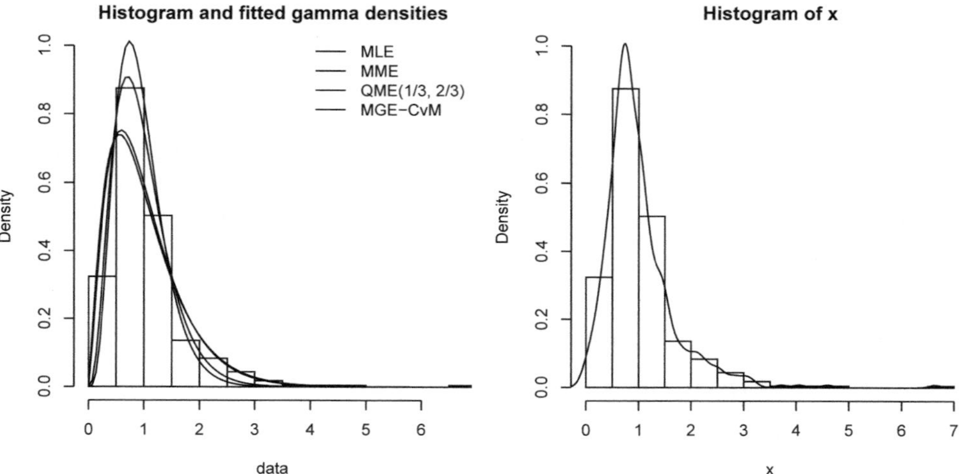

FIGURE 2.9
Comparison of fits on a MTPL dataset.

$\mathcal{G}(\alpha, \lambda)$, a Pareto distribution $\mathcal{P}(\alpha, \theta)$, a Pareto-gamma mixture $\mathcal{PG}(p, \alpha_1, \theta_1, \alpha_2, \lambda_2)$ defined in Section 2.1.3 and a Burr distribution $\mathcal{B}u(\gamma, \tau, \theta)$.

```
> data(danishuni)
> x <- danishuni$Loss
> fgam <- fitdist(x, "gamma", lower=0)
> fpar <- fitdist(x, "pareto", start=list(shape=2, scale=2), lower=0)
> fmixgampar <- fitdist(x, "mixgampar", start=
+ list(prob=1/2, nu=1, lambda=1, alpha=2, theta=2), lower=0)
> cbind(SINGLE= c(NA, fgam$estimate, fpar$estimate),
+ MIXTURE=fmixgampar$estimate)
```

```
           SINGLE    MIXTURE
               NA  0.6849568
shape   1.2976150 10.8706430
rate    0.3833335  6.5436349
shape   5.3689277  5.4182746
scale  13.8413207 30.0700544
```

When fitted alone, the parameters of the gamma distribution are estimated as $(\hat{\alpha}, \hat{\lambda}) = (1.2976, 0.3833)$ and the parameters of the Pareto distribution are estimated as $(\hat{\alpha}, \hat{\theta}) = (5.3689, 13.8413)$. When used in the mixture, we get

$$(\hat{p}, \hat{\alpha}_1, \hat{\theta}_1, \hat{\alpha}_2, \hat{\lambda}_2) = (0.6849, 10.8706, 6.5436, 5.4182, 30.0700).$$

As only the shape parameter $\hat{\alpha}_2$ is of similar amplitude, only heavy-tailed distributions (like the Pareto) are appropriate for this dataset. Finally, we fit a Burr distribution:

```
> fburr <- fitdist(x, "burr", start=list(shape1=2, shape2=2,
+   scale=2), lower=c(0.1,1/2, 0))
> fburr$estimate
```

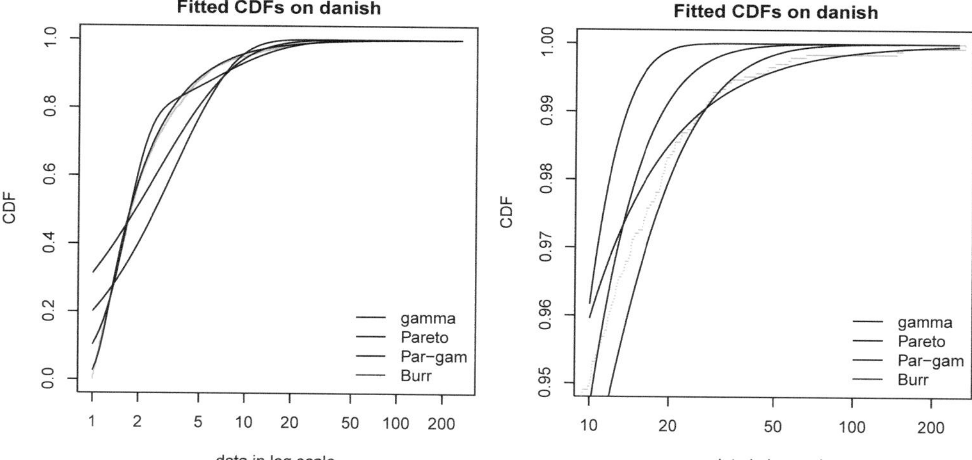

FIGURE 2.10
Comparison of fits on the Danish fire dataset.

```
  shape1   shape2    scale
 0.10000 14.44286  1.08527
```

Comparing the fitted densities is then carried out using the `cdfcomp` function:

```
> cdfcomp(list(fgam, fpar, fmixgampar, fburr), xlogscale=TRUE,
+   datapch=".", datacol="grey", fitcol="black", fitlty=2:5,
+   legendtext=c("gamma","Pareto","Par-gam","Burr"),
+   main="Fitted CDFs on danish")
```

We also plot the tail of the distribution function on Figure 2.10 When using the maximum likelihood estimation, the best fit is provided by the Burr distribution, yet the first shape parameter hits the lower bound of 0.1.

2.3.3 QQ-Plot, PP-Plot

On the two previous graphs, we consider the plot of the empirical density (respectively the empirical distribution function) and the fitted density (respectively the fitted distribution function). The PP-plot (respectively the QQ-plot) consists of plotting (directly) the empirical distribution function F_n against the fitted distribution function $F(.;\hat{\boldsymbol{\theta}})$ (respectively the empirical quantile function $Q_{n,.}$ against the fitted quantile function $F^{-1}(.;\hat{\boldsymbol{\theta}})$). Those quantities are computed at the observations, which leads to further simplifications $F_n(x_i) = \text{rank}(x_i)/n$ and $Q_{n,i/n} = x_i^\star$. This is illustrated on the Danish fire dataset `danishuni` and the four distributions considered by using the `ppcomp` and `qqcomp` functions:

```
> qmixgampar <- function(p, prob, nu, lambda, alpha, theta)
+ {
+ L2 <- function(q, p)
+ (p - pmixgampar(q, prob, nu, lambda, alpha, theta))^2
+ sapply(p, function(p) optimize(L2, c(0, 10^3), p=p)$minimum)
```

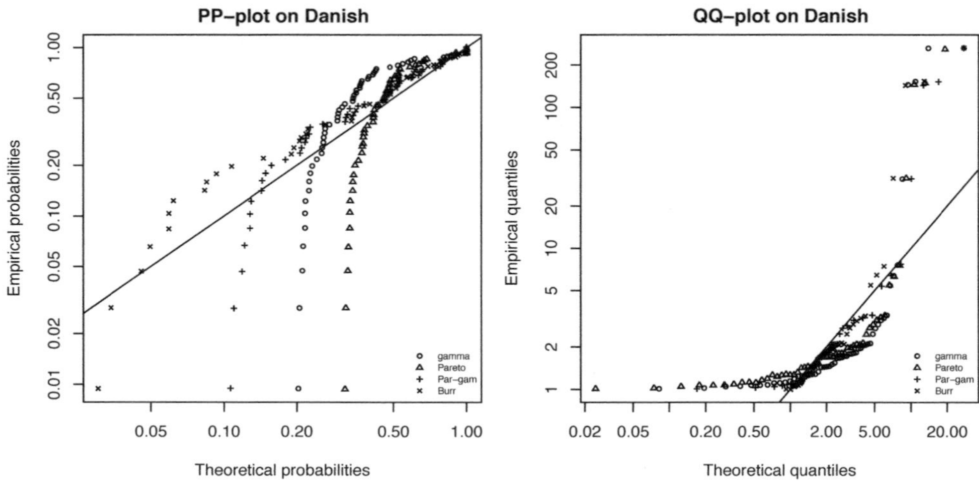

FIGURE 2.11
Comparison of fits on the Danish fire dataset.

```
+ }
> ppcomp(list(fgam, fpar, fmixgampar, fburr), xlogscale=TRUE,
+ ylogscale=TRUE, fitcol="black", main="PP-plot on danish",
+ legendtext=c("gamma","Pareto","Par-gam","Burr"), fitpch=1:4)
> qqcomp(list(fgam, fpar, fmixgampar, fburr), xlogscale=TRUE,
+ ylogscale=TRUE, fitcol="black", main="QQ-plot on danish",
+ legendtext=c("gamma","Pareto","Par-gam","Burr"), fitpch=1:4)
```

As there is no closed-form formula for the quantile function of the mixture distribution (i.e. inverse of $x \mapsto p\gamma(\alpha, \nu x)/\Gamma(\nu)+(1-p)(1-\theta^\alpha/(\theta+x)^\alpha)$), a numerical optimization is carried out using the Golden line-search (implemented in `optimize`). On Figure 2.11, quantiles and probabilities are plotted as a point, while the straight line corresponds to the identity function. The more points that are close to the line, the better fit the distribution. The pp-plot reveals that only the Burr distribution sufficiently fits the data, whereas the qq-plot shows that both the Pareto-gamma mixture and the Burr distributions best approximate the data.

The `plot` method of a `fitdist` object provides the four above graphs (for the fitted distribution) in the following order: a histogram with the fitted density, an `ecdf`-plot with the fitted distribution function, a (theoretical) quantile—(empirical) quantile plot and a (theoretical) probability—(empirical) probability plot.

2.3.4 Goodness-of-Fit Statistics and Tests

We turn our attention to goodness-of-fit statistics to complement the four previous goodness-of-fit graphs. For continuous distributions, the three statistics presented in Section 2.2.4 can be computed, that is, Cramér–von Mises, Kolmogorov–Smirnov and Anderson–Darling statistics. For discrete distributions, the more common statistic is the chi-square statistic defined as

$$\Delta^2 = \sum_{i=0}^{m} \frac{(n_i - np_i)^2}{np_i}$$

where n_i is the empirical frequency count for the level i, n is the total number of observations, the theoretical probability $p_i = P(X = i; \theta)$ (i.e. np_i the theoretical frequency count), and m is the number of cells. In practice, the number of cells is either fixed by the experimenter or chosen so that empirical frequencies are greater than five and p_i is replaced by $\hat{p}_i = \mathbb{P}(X = i; \hat{\theta})$. The chi-square statistic is linked to the Pearson's hypothesis test of goodness-of-fit, for which under the null hypothesis, Δ^2 converges in law to a chi-square distribution $\chi^2(m-d-1)$ (where d is the number of parameters). For all distributions, we consider also the information criteria (AIC and BIC) proportional to the opposite of the log-likelihood. All of this is provided in the `gofstat` function (of the `fitdistrplus` package). A numerical illustration is proposed on a TPL claim number dataset, for which a Poisson, a negative binomial and a zero-modified Poisson distribution are fitted using maximum likelihood techniques.

```
> data(tplclaimnumber)
> x <- tplclaimnumber$claim.number
> fpois <- fitdist(x, "pois")
> fnbinom <- fitdist(x, "nbinom")
> fpoisZM <- fitdist(x, "poisZM", start=list(
+   prob=sum(x == 0)/length(x), lambda=mean(x)),
+   lower=c(0,0), upper=c(1, Inf))
> gofstat(list(fpois, fnbinom, fpoisZM), chisqbreaks=c(0:4, 9),
+   discrete=TRUE, fitnames=c("Poisson","NegBinomial","ZM-Poisson"))

Chi-squared statistic:  Inf 11765679 Inf
Degree of freedom of the Chi-squared distribution:  5 4 4
Chi-squared p-value:  0 0 0
   the p-value may be wrong with some theoretical counts < 5
Chi-squared table:
     obscounts theo Poisson theo NegBinomial theo ZM-Poisson
<= 0    653047 6.520559e+05     6.530606e+05    6.530411e+05
<= 1     23592 2.545374e+04     2.353633e+04    2.351466e+04
<= 2      1299 4.968076e+02     1.326372e+03    1.413873e+03
<= 3        62 6.464481e+00     8.372804e+01    4.250619e+01
<= 4         5 6.308707e-02     5.568862e+00    8.519276e-01
<= 9         5 4.957574e-04     4.104209e-01    1.296158e-02
> 9          3 0.000000e+00     7.649401e-07    0.000000e+00

Goodness-of-fit criteria
                                Poisson NegBinomial ZM-Poisson
Aikake's Information Criterion 226880.4    225375.1   225585.7
Bayesian Information Criterion 226891.8    225398.0   225608.5
```

From the chi-square statistic and the chi-square table $(n_i, n\hat{p}_i)_i$, the negative binomial distribution is clearly the best distribution. This is also confirmed by the AIC and the BIC criteria.

2.3.5 Skewness–Kurtosis Graph

When selecting a distribution, depending on the type of applications, the experimenter may give particular attention to the tail, some quantiles or the body of the distribution for which a natural way of choosing the "best" distribution emerges. In actuarial science, a great care is given to the tail of distribution, and also on first moments. The code below provide values of quantiles (plotted before) as well as the first two raw moments.

```
> p <- c(.9, .95, .975, .99)
> rbind(
+ empirical= quantile(danishuni$Loss, prob=p),
+ gamma= quantile(fgam, prob=p)$quantiles,
+ Pareto= quantile(fpar, prob=p)$quantiles,
+ Pareto_gamma= quantile(fmixgampar, prob=p)$quantiles,
+ Burr= quantile(fburr, prob=p)$quantiles)

                p=0.9    p=0.95  p=0.975   p=0.99
empirical    5.541526  9.972647 16.26821 26.04253
gamma        7.308954  9.261227 11.18907 13.71207
Pareto       7.412375 10.341301 13.67386 18.79426
Pareto_gamma 7.093200 12.164896 17.92871 26.77259
Burr         5.344677  8.636739 13.95655 26.32118

> compmom <- function(order)
+ c(empirical= sum(danishuni$Loss^order)/length(x),
+ gamma=mgamma(order, fgam[[1]][1], fgam[[1]][2]),
+ Pareto=mpareto(order, fpar[[1]][1], fpar[[1]][2]),
+ Pareto_gamma= as.numeric(fmixgampar[[1]][1]*
+ mgamma(order, fmixgampar[[1]][2], fmixgampar[[1]][3])+
+ (1-fmixgampar[[1]][1])*
+ mpareto(order, fmixgampar[[1]][4], fmixgampar$estimate[5])),
+ Burr=mburr(order, fburr[[1]][1], fburr[[1]][2], fburr[[1]][3]))
> rbind(Mean=compmom(1), Mom2nd= compmom(2))

        empirical     gamma    Pareto Pareto_gamma    Burr
Mean   0.01081909  3.385081  3.168128      3.28202 2.98562
Mom2nd 0.26784042 20.289412 26.032657     39.78745     Inf
```

For higher moments, it is typical to look at the skewness and the kurtosis coefficients defined as

$$sk(X) = \frac{\mathbb{E}[(X - \mathbb{E}(X))^3]}{\text{var}(X)^{\frac{3}{2}}}\ ,\ kr(X) = \frac{\mathbb{E}[(X - \mathbb{E}(X))^4]}{\text{var}(X)^2},$$

for a random variable X. For heavy-tailed distributions, such coefficients may not exist, yet empirically they always exist. The `descdist` function provides the so-called Cullen and Frey graph, which plots the empirical estimates of $sk(X)$ and $kr(X)$ as well as the possible values for some classic distributions (including the gamma family for continuous distributions and the Poisson distribution for discrete distributions) This is illustrated on the `danishuni` and the `tplcaimnumber` datasets on Figure 2.12, using so-called Cullen and Frey graphs, from Cullen & Frey (1999). The fit analysis can also be completed by looking at the uncertainty of parameter estimate with a bootstrap analysis. This is possible with the `bootdist` function of `fitdistrplus`.

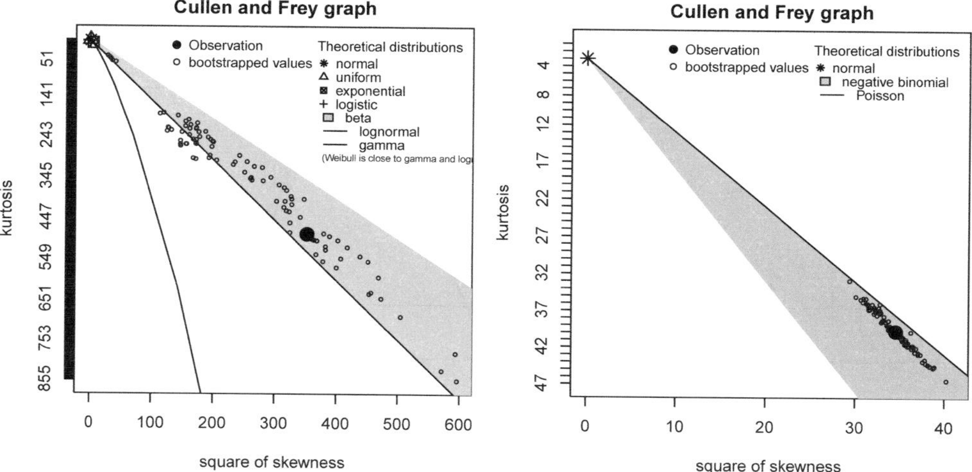

FIGURE 2.12
Cullen and Frey graph for `danish` and `tplclaimnumber`, on the left and on the right, respectively.

2.4 Linear Regression: Introducing Covariates in Statistical Inference

In the first part of this chapter, we did mention the normal distribution. If it is still a popular distribution in financial models (see Chapter 11), it is also frequently used in actuarial science because of its connection with linear regression.

2.4.1 Using Covariates in the Statistical Framework

So far, we have assumed that observations were i.i.d., for example, with distribution $\mathcal{N}(\theta, \sigma^2)$. But in most assumptions, it can yield a very restrictive model. For instance, consider dataset `Davis` and let `X` denote the height of a person (in centimeter):

```
> X <- Davis$height
```

We can fit a Gaussian distribution to the weight

```
> (param <- fitdistr(X,"normal")$estimate)
     mean          sd
170.56500   8.90987
```

If we plot the distribution (see Figure 2.13), we can see that using a mixture of two Gaussian distributions is much better than using only a single model,

$$X \sim p \cdot \mathcal{N}(\mu_1, \sigma^2) + [1-p] \cdot \mathcal{N}(\mu_2, \sigma^2)$$

(where the + sign should be understood in the context of mixtures, as described in Section 2.1.3). This model can be estimated using maximum likelihood techniques. Let us define the log-density as

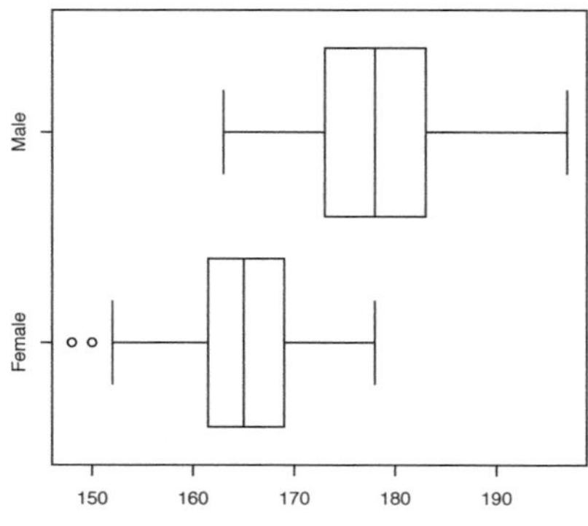

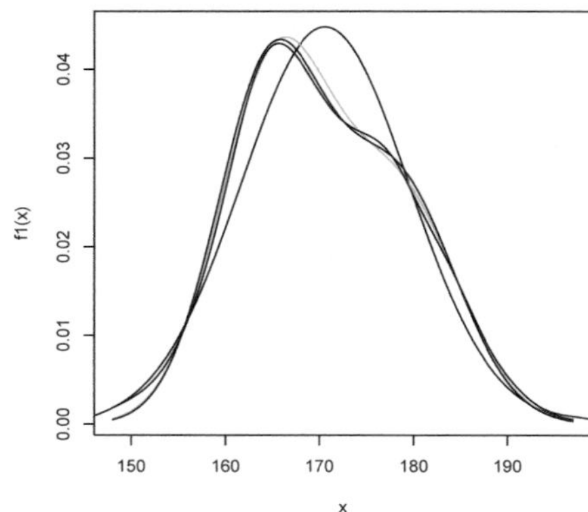

FIGURE 2.13
Distribution of the height, using a normal distribution, and mixtures of normal distributions: One with a non-observable latent factor, one where mixture is related to the sex.

```
> logdf <- function(x,parameter){
+ p  <- parameter[1]
+ m1 <- parameter[2]
+ s1 <- parameter[4]
+ m2 <- parameter[3]
+ s2 <- parameter[5]
+ return(log(p*dnorm(x,m1,s1)+(1-p)*dnorm(x,m2,s2)))
+ }
```

and in order to take into account various constraints, namely $p \in (0,1)$ and $\sigma_1, \sigma_2 \in (0,\infty)$, that can be written

$$\underbrace{\begin{pmatrix} 1 & 0 & 0 & 0 & 0 \\ -1 & 0 & 0 & 0 & 0 \\ 0 & 0 & 0 & 1 & 0 \\ 0 & 0 & 0 & 0 & 1 \end{pmatrix}}_{\boldsymbol{A}} \boldsymbol{\theta} + \underbrace{\begin{pmatrix} 0 \\ 1 \\ 0 \\ 0 \end{pmatrix}}_{-\boldsymbol{b}} \geq \boldsymbol{0}$$

Function `constrOptim` will seek the minimum of a function, so we will consider here the opposite of the log-likelihood:

```
> logL <- function(parameter) -sum(logdf(X,parameter))
> Amat <- matrix(c(1,-1,0,0,0,0,
+ 0,0,0,0,1,0,0,0,0,0,0,0,0,1), 4, 5)
> bvec <- c(0,-1,0,0)
> constrOptim(c(.5,160,180,10,10), logL, NULL, ui = Amat, ci = bvec)$par
[1]   0.5996263 165.2690084 178.4991624   5.9447675   6.3564746
```

Because we use a (finite) normal mixture here, it is also possible to use the EM algorithm, from the `mixtools` package,

```
> library(mixtools)
> mix <- normalmixEM(X)
```

```
number of iterations= 391
> (param12 <- c(mix$lambda[1],mix$mu,mix$sigma))
[1]   0.5995197 165.2676186 178.4951348   5.9448806   6.3579494
```

The two methods yield rather similar outputs.

If previously we assumed that the mixing variable Θ was a latent unobservable random variable, here it would make sense to assume that a good proxy of this variable can be the sex of the individuals. And here,

$$X \sim p_M \cdot \mathcal{N}(\mu_M, \sigma^2) + p_F \cdot \mathcal{N}(\mu_F, \sigma^2)$$

Here, p_M and p_F are known, and are, respectively, the proportion of males and females in the population.

```
> sex <- Davis$sex
> (pM <- mean(sex=="M"))
[1] 0.44
> (paramF <- fitdistr(X[sex=="F"],"normal")$estimate)
      mean          sd
164.714286   5.633808
> (paramM <- fitdistr(X[sex=="M"],"normal")$estimate)
      mean          sd
178.011364   6.404001
```

If we compare the three models, including a kernel-based estimator, we obtain the graph of Figure 2.13.

```
> f1 <- function(x) dnorm(x,param[1],param[2])
> f2 <- function(x) param12[1]*dnorm(x,param12[2],param12[4])+
+ (1-param12[1])*dnorm(x,param12[3],param12[5])
> f3 <- function(x) pM*dnorm(x,paramM[1],paramM[2])+(1-pM)*dnorm(x,paramF[1],paramF[2])
> boxplot(X~sex,horizontal=TRUE,names=c("Female","Male"))
> x <- seq(min(X),max(X),by=.1)
> plot(x,f1(x),lty=2,type="l")
> lines(x,f2(x),col="grey",lwd=2)
> lines(x,f3(x),col="black",lwd=2)
> lines(density(X))
```

Actually, this factor-based mixture is a particular case of what is known as the linear model.

2.4.2 Linear Regression Model

To use standard notions in regression modeling, let Y denote the variable of interest. And assume that some additional variables can be used, denoted $\boldsymbol{X} = (X_1, \cdots, X_k)$'s. This means that for each observation Y_i, we observe also $\boldsymbol{X}_i = (X_{1,i}, \cdots, X_{k,i})$. As discussed in Chapter 14, X_k's can be either numeric (also called continuous) or categorical (also called factor) variables.

In the `Davis` dataset, the varible of interest is the height of a person, denoted `height` (our variable Y), and two additional variables can be used, `sex` (variable X_1) and `weight` (variable X_2).

```
> Y  <- Davis$height
> X1 <- Davis$sex
> X2 <- Davis$weight
> df <- data.frame(Y,X1,X2)
```

Instead of assuming that

$$Y \sim \mathcal{N}(\theta, \sigma^2),$$

we will assume, in a regression model, that

$$Y|\boldsymbol{X} = \boldsymbol{x} \sim \mathcal{N}(\theta(\boldsymbol{x}), \sigma^2),$$

where $\theta()$ is now a function of the explanatory variables. Consider here the case where we observe two covariates, the sex and the weight (in kilograms) of the individuals. If we restrict ourselves to linear models, then

$$\theta(x_1, x_2) = \beta_0 + \beta_{1,H}\mathbf{1}(x_1 = H) + \beta_{1,F}\mathbf{1}(x_1 = F) + \beta_2 x_2.$$

From properties of the Gaussian distribution, it is also possible to write this model as

$$Y = \beta_0 + \beta_{1,H}\mathbf{1}(x_1 = H) + \beta_{1,F}\mathbf{1}(x_1 = F) + \beta_2 x_2 + \varepsilon,$$

where ε is an error term, usually called residuals, centered, and normally distributed,

$$\varepsilon \sim \mathcal{N}(0, \sigma^2).$$

The unknown parameters (that should be estimated) are now β_0, $\beta_{1,H}$, $\beta_{1,F}$, β_2 and σ^2.

2.4.3 Inference in a Linear Model

Recall, first of all, that the previous model cannot be identified: We cannot have the intercept and the two factors (`M` and `F`) at the same time. The standard procedure is to keep the intercept and to remove one of the two factors. The factor that was dropped will become the reference.

The maximum likelihood estimator is obtained by maximizing

$$\mathcal{L}((\boldsymbol{\beta}, \sigma); \boldsymbol{y}, \boldsymbol{x}) = \prod_{i=1} \varphi(y_i; \beta_0 + \beta_1 x_1 + \beta_2 x_2, \sigma^2),$$

where here $\varphi(y; \mu, \sigma^2)$ is the density of the $\mathcal{N}(\mu, \sigma^2)$ distribution.

Here, the maximum likelihood estimators of the set of parameters can be written explicitly. When writing the problem using the logarithm of the likelihood, we can observe that the optimal value of $\boldsymbol{\beta} = (\beta_0, \beta_1, \beta_2)$ should satisfy

$$\widehat{\boldsymbol{\beta}} = \text{argmin}\left\{\sum_{i=1}^{n} \underbrace{(y_i - [\beta_0 + \beta_1 x_1 + \beta_2 x_2])}_{\text{residuals } \varepsilon_i}{}^2\right\}.$$

Thus, using maximum likelihood techniques in a Gaussian linear model is the same as minimizing the sum of squares of residuals (known as Ordinary Least Squares estimation).

Fitting a linear model is done using function `lm()`. Using symbolic notions (introduced in Chapter 1), we write here

```
> lin.mod <- lm(Y~X1+X2,data=df)
```

As mentioned in Chapter 1, `lin.mod` is a `S3` object, and many functions can be used to extract information from that object. To visualize the standard output of a linear regression, use

```
> summary(lin.mod)

Call:
lm(formula = Y ~ X1 + X2, data = df)

Residuals:
    Min      1Q  Median      3Q     Max
-85.204  -4.183   0.446   5.224  19.009

Coefficients:
             Estimate Std. Error t value Pr(>|t|)
(Intercept) 175.26607    3.30681  53.002  < 2e-16 ***
X1M          17.86160    1.66941  10.699  < 2e-16 ***
X2           -0.19917    0.05503  -3.619 0.000376 ***
---
Signif. codes:  0 *** 0.001 ** 0.01 * 0.05 . 0.1   1

Residual standard error: 9.424 on 197 degrees of freedom
Multiple R-squared:  0.3903,        Adjusted R-squared:  0.3841
F-statistic: 63.05 on 2 and 197 DF,  p-value: < 2.2e-16
```

It is also possible to get predictions, using `predict`. Keep in mind that we should have the same input format as in the `lm` call: The regression was run on a dataframe, so `predict` should also be called on a dataframe (with the same variable names),

```
> new.obs <- data.frame(X1=c("M","M","F"),X2=c(100,70,65))
> predict(lin.mod,newdata=new.obs)
       1        2        3
173.2110 179.1860 162.3202
```

which will return $\boldsymbol{x}'\widehat{\boldsymbol{\beta}}$ for any observation $\boldsymbol{x}$.

Linear models with R have been intensively described, so we refer to Venables & Ripley (2002), Fox & Weisberg (2011) or Kleiber & Zeileis (2008) for more details. Extensions of this model will be given in the next chapters (the logistic regression of binary responses in Chapter 4, and GLMs in Chapter 14, among others) as well as a Bayesian interpretation of this model (in Chapter 3).

2.5 Aggregate Loss Distribution

This section deals with the aggregate loss amount distribution, which is the distribution of the compound sum

$$S = \sum_{i=1}^{N} X_i, \text{ with } S = 0 \text{ if } N = 0,$$

where N is the claim number and (X_i)'s are the claim severities (which are assumed to be strictly positive). Firstly, the computation of the distribution function F_S of S is studied. Then, an application to a TPL motor dataset is carried out. Finally, a continuous-time version of this problem is analyzed via the ruin theory framework.

2.5.1 Computation of the Aggregate Loss Distribution

A classical assumption on the aggregate amount S is to require that N is independent of claim amounts $(X_i)_i$. Another common assumption is that $(X_i)_i \overset{i.i.d.}{\sim} X$. Therefore, the distribution function simplifies to

$$F_S(s) = \sum_{n=0}^{+\infty} \mathbb{P}(N=n)\mathbb{P}(X_1 + \cdots + X_n \leq s) = \sum_{n=0}^{+\infty} \mathbb{P}(N=n) F_X^{\star n}(s),$$

where $F_X^{\star n}$ is the n-order convolution product of F_X. In a small number of distributions of X, the distribution of the sum $X_1 + \cdots + X_n$ is easy. For instance, when X follows a gamma distribution $\mathcal{G}(\alpha, \lambda)$, then the sum $X_1 + \cdots + X_n$ follows a gamma distribution $\mathcal{G}(n\alpha, \lambda)$. This can be implemented as

```
> pgamsum <- function(x, dfreq, argfreq, shape, rate, Nmax=10)
+ {
+ tol <- 1e-10; maxit <- 10
+ nbclaim <- 0:Nmax
+ dnbclaim <- do.call(dfreq, c(list(x=nbclaim), argfreq))
+ psumfornbclaim <- sapply(nbclaim, function(n)
+ pgamma(x, shape=shape*n, rate=rate))
+ psumtot <- psumfornbclaim %*% dnbclaim
+ dnbclaimtot <- dnbclaim
+ iter <- 0
+ while( abs(sum(dnbclaimtot)-1) > tol && iter < maxit)
+ {
+         nbclaim <- nbclaim+Nmax
+         dnbclaim <- do.call(dfreq, c(list(x=nbclaim), argfreq))
+         psumfornbclaim <- sapply(nbclaim, function(n)
+                 pgamma(x, shape=shape*n, rate=rate))
+         psumtot <- psumtot + psumfornbclaim %*% dnbclaim
+         dnbclaimtot <- c(dnbclaimtot, dnbclaim)
+         iter <- iter+1
+ }
+ as.numeric(psumtot)
+ }
```

In general, the distribution of the sum $X_1 + \cdots + X_n$ does not necessarily have the same distribution as X. Alternative computations are possible. The Panjer recursion provides a recursive method to compute the mass probability function of S in the case that X has a discrete distribution and N belongs to the (a, b, n) family; see Panjer (1981). The recursion formula for the mass probability function p_S is

$$p_S(s) = \frac{[p_X(1) - (a+b)p_X(0)]p_X(s) + \sum_{y=1}^{s \wedge m} (a + by/x) p_X(y) p_S(s-y)}{1 - a p_X(0)},$$

where $s \in \mathbb{N}$, X has a discrete distribution on $\{0, 1, \ldots, m\}$ with a mass probability function p_X, N belongs to $(a, b, 0)$ family and starting at $p_S(0) = G_N(p_X(0))$ with G_N the probability generating function. The recursion is stopped when the sum of elementary probabilities $P(S = 0, 1, \ldots)$ is arbitrarily close to 1. In practice, the distribution of the claim amount is not discrete but can be discretized. The upper discretization is the forward difference

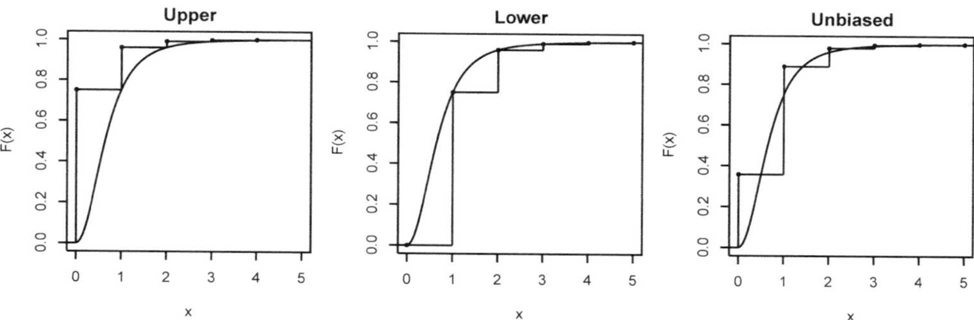

FIGURE 2.14
Comparison of three discretization methods (a paralogistic distribution).

$\tilde{f}(x) = F_X(x+h) - F_X(x)$, the lower discretization is the backward difference $\tilde{f}(x) = F_X(x) - F_X(x-h)$, and the unbiased discretization is $\tilde{f}(x) = (2\mathbb{E}(X \wedge x) - \mathbb{E}(X \wedge x - h) - \mathbb{E}(X \wedge x + h))/h$, where h is the step of discretization; see Figure 2.14 and Dutang et al. (2008) for further details.

Approximations based on the normal distribution are also available: (i) the normal approximation is given by

$$F_S(x) \approx \Phi\left(\frac{x - \mathbb{E}(S)}{\sigma(S)}\right), \text{ and}$$

(ii) the normal-power approximation is given by

$$F_S(x) \approx \Phi\left(-\frac{3}{sk(S)} + \sqrt{\frac{9}{sk(S)^2} + 1 + \frac{6}{sk(S)}\frac{x - \mathbb{E}(S)}{\sigma(S)}}\right),$$

where $\sigma(S)$ is the standard deviation of S and $sk(S)$ is the skewness coefficient of S. The skewness coefficient can be written as

$$sk(S) = \frac{sk(N)\text{var}(N)^{3/2}\mathbb{E}(X)^3 + 3\text{var}(N)\mathbb{E}(X)Var(X) + \mathbb{E}(N)sk(X)\text{var}(X)^{3/2}}{\text{var}(S)^{3/2}}.$$

An approximation based on the gamma distribution is also possible; see Gendron & Crepeau (1989). These approximations are reasonably correct at the heart of the distribution but not at the tails of the distribution. A last alternative to exact computation is the simulation procedure. It consists of simulating n claim numbers $N_1, \ldots, N_n$, N_i's claim severities $(X_{i,j})_j$ in order to get n realizations $S_1, \ldots, S_n$. All these alternative methods are available in the `aggregateDist` function of the `actuar` package. We consider two examples: the gamma case (N follows a Poisson distribution $\mathcal{P}(10)$ and X follows a gamma distribution $\mathcal{G}(3, 2)$) and the Pareto case (N follows a Poisson distribution $\mathcal{P}(10)$ and X follows a Pareto distribution $\mathcal{P}(3.1, 4.2)$). The following code computes the gamma case:

```
> parsev <- c(3, 2); parfreq <- 10
> meansev <- mgamma(1, parsev[1], parsev[2])
> varsev <- mgamma(2, parsev[1], parsev[2]) - meansev^2
> skewsev <- (mgamma(3, parsev[1], parsev[2]) -
+              3*meansev*varsev - meansev^3)/varsev^(3/2)
> meanfreq <- varfreq <- parfreq[1]; skewfreq <- 1/sqrt(parfreq[1])
> meanagg <- meanfreq * meansev
```

```
> varagg <- varfreq * (varsev + meansev^2)
> skewagg <- (skewfreq*varfreq^(3/2)*meansev^3 + 3*varfreq*meansev*
+               varsev + meanfreq*skewsev*varsev^(3/2))/varagg^(3/2)
> Fs.s <- aggregateDist("simulation", model.freq = expression(y =
+   rpois(parfreq)), model.sev = expression(y =
+    rgamma(parsev[1], parsev[2])), nb.simul = 1000)
> Fs.n <- aggregateDist("normal", moments = c(meanagg, varagg))
> Fs.np <- aggregateDist("npower", moments = c(meanagg, varagg, skewagg))
> Fs.exact <- function(x) pgamsum(x, dpois, list(lambda=parfreq),
+                                 parsev[1], parsev[2], Nmax=100)
> x <- seq(25, 40, length=101)
> plot(x, Fs.exact(x), type="l",
+      main="Agg. Claim Amount Distribution", ylab="F_S(x)")
> lines(x, Fs.s(x), lty=2)
> lines(x, Fs.n(x), lty=3)
> lines(x, Fs.np(x), lty=4)
> legend("bottomright", leg=c("exact", "simulation",
+ "normal approx.", "NP approx."), col = "black",
+ lty = 1:4, text.col = "black")
```

Similarly, we have the Pareto case. We show here only the recursive computation calls.

```
> parsev <- c(3.1, 2*2.1) ; parfreq <- 10
> xmax <- qpareto(1-1e-9, parsev[1], parsev[2])
> fx2 <- discretize(ppareto(x, parsev[1], parsev[2]), from = 0,
+ to = xmax, step = 0.5, method = "unbiased",
+ lev = levpareto(x, parsev[1], parsev[2]))
> Fs2 <- aggregateDist("recursive", model.freq = "poisson",
+ model.sev = fx2, lambda = parfreq, x.scale = 0.5, maxit=2000)
> fx.u2 <- discretize(ppareto(x, parsev[1], parsev[2]), from = 0,
+ to = xmax, step = 0.5, method = "upper")
> Fs.u2 <- aggregateDist("recursive", model.freq = "poisson",
+ model.sev = fx.u2, lambda = parfreq, x.scale = 0.5, maxit=2000)
> fx.l2 <- discretize(ppareto(x, parsev[1], parsev[2]), from = 0,
+ to = xmax, step = 0.5, method = "lower")
> Fs.l2 <- aggregateDist("recursive", model.freq = "poisson",
+ model.sev = fx.l2, lambda = parfreq, x.scale = 0.5, maxit=2000)
```

The two graphs are displayed on Figure 2.15. Despite the expectation that $E(X)$ is identical in both cases, high-level quantiles of the aggregate claim distribution are significantly different. For the gamma case, the normal-power approximation suitably fits the exact distribution function, while for the Pareto case, the normal-power approximation overestimates as the skewness $sk(S)$ is very high and not representative of the shape of the distribution. As their name suggests, the upper and the lower recursive computations surround the true distribution function. The simulation number is voluntarily chosen low (1,000), but can be set to a much larger number. If convergence is achieved for a high number of simulations, parallelization, GPU computation and quasi-Monte Carlo sampling methods can be used to fasten the process; see Chapter 1, http://cran.r-project.org/web/views/HighPerformanceComputing.html and http://cran.r-project.org/web/views/Distributions.html for more details.

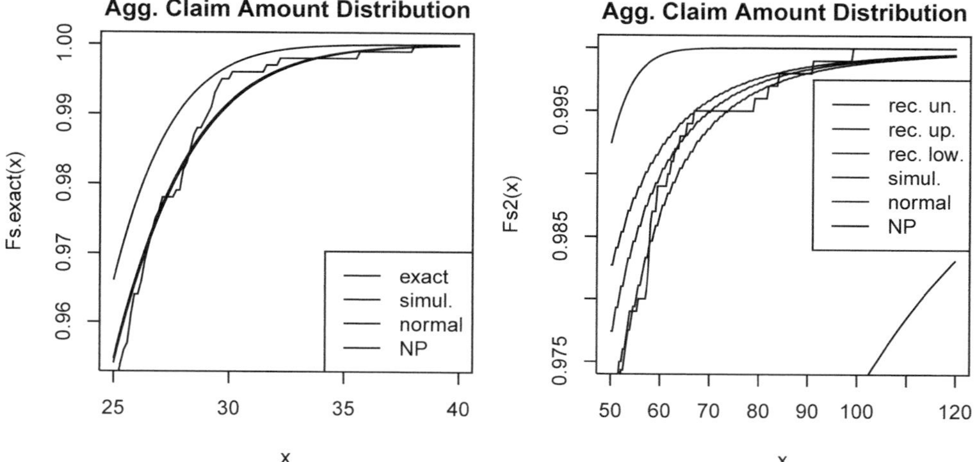

FIGURE 2.15
Aggregate claim distribution (gamma case/Pareto case).

2.5.2 Poisson Process

The Poisson process is probably the most important stochastic process in general insurance. It is used to describe the number of claims that occurred in a time interval. In its basic form, the homogeneous Poisson process is a counting process (N_t), with independent and stationary increments. At time t, N_t, the number of claims that occurred from time 0 until time t, has a Poisson distribution, and the distribution of waiting time until the next claim is an exponential distribution. Let (T_i) denote the ith arrival time, so that $\{N_t = 0\}$ can equivalently be represented by $\{T_1 > t\}$ and more generally

$$\{N_t = n\} = \{T_n \leq t \text{ and } T_{n+1} > t\} \text{ for all } n \geq 1.$$

In the case where inter-arrival times $W_i = T_i - T_{i-1}$ are i.i.d. random variables, then (N_t) is called a renewal process, and it is fully characterized by the distribution of inter-arrival times, $F(x) = \mathbb{P}(W \leq x)$. The Poisson process is obtained when F is the distribution of an exponential random variable. The code to generate a renewal process up to time T, or more precisely an arrival time sequence, is

```
> rate <- 1
> rFexp <- function() rexp(1,rate)
> rRenewal <- function(Tmax=1,rF=rFexp){
+ t <-0
+ vect.W<- NULL
+ while(t<Tmax){
+   W<-rF()
+   t<-t+W
+   if(t<T) vect.W=c(vect.W,W)}
+ return(list(T=cumsum(vect.W),W=vect.W,N=length(vW)))}
> set.seed(1)
> rRenewal(Tmax=2)
$T
[1] 0.7551818 1.9368246
```

```
$W
[1] 0.7551818 1.1816428

$N
[1] 2
```

An interesting alternative, in the case of the Poisson process with intensity λ, is to use a uniform property of the process: For all $n \geq 1$, given $\{N_t = n\}$, the joint distribution of the n arrival time $T_1, \ldots, T_n$ is the same as the joint distribution of $U_{1:n}, \ldots, U_{n:n}$, the order statistics of n i.i.d. random variables uniformly distributed on $[0, t]$. Thus, a natural algorithm to generate the Poisson process is the following:

```
> rPoissonProc <- function(Tmax=1,lambda=rate){
+ N <- rpois(n=1,lambda*Tmax)
+ vect.T <- NULL
+ if(N>0) vect.T=sort(runif(N))*lambda*Tmax
+ return(list(T=vect.T,W=diff(c(0,vect.T)),N=N))}

> set.seed(1)
> rPoissonProc(T=5)
$T
[1] 1.008410 1.860619 2.864267 4.541039

$W
[1] 1.0084097 0.8522098 1.0036473 1.6767721

$N
[1] 4
```

An homogeneous Poisson process, with intensity $\lambda \geq 0$, satisfies

$$\mathbb{P}(N_{t+h} - N_t = k) = \frac{1}{k!} e^{\lambda h} [\lambda h]^k.$$

It is possible to consider a non-homogeneous Poisson process with intensity (λ_t). Then

$$\mathbb{P}(N_{t+h} - N_t = k) = \frac{1}{k!} e^{\int_t^{t+h} \lambda_s ds} \left[\int_t^{t+h} \lambda_s ds \right]^k.$$

See Rolski et al. (1999) for more details.

To generate a Poisson process, several algorithms can be considered. In order to illustrate, consider a cyclical Poisson process with intensity $\lambda_t \propto (1 + \sin(\pi t))$,

```
> lambda <- function(t) 100*(sin(t*pi)+1)
```

so that the cumulated intensity Λ is

```
> Lambda <- function(t) integrate(f=lambda,lower=0,upper=t)$value
```

Given that the last claim occurred at time t, let F_t be the conditional distribution function of the waiting time before the next claim. Then

$$F_t(x) = 1 - \exp[\Lambda(t) - \Lambda(x + t)] = 1 - \exp\left[\int_t^{x+t} \lambda_s ds\right].$$

From a computational aspect, we just have to invert this function and use a rejection technique algorithm,

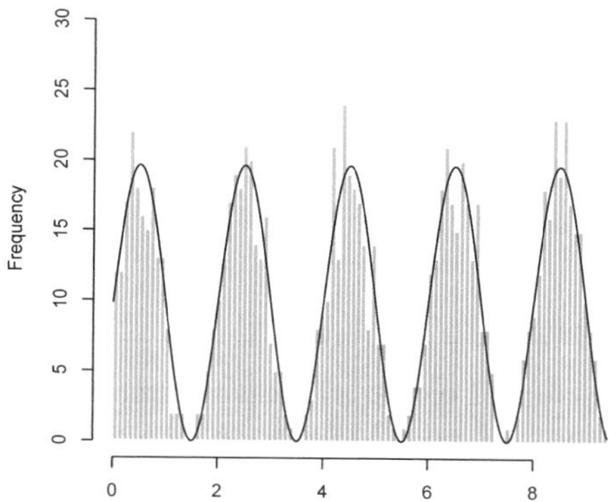

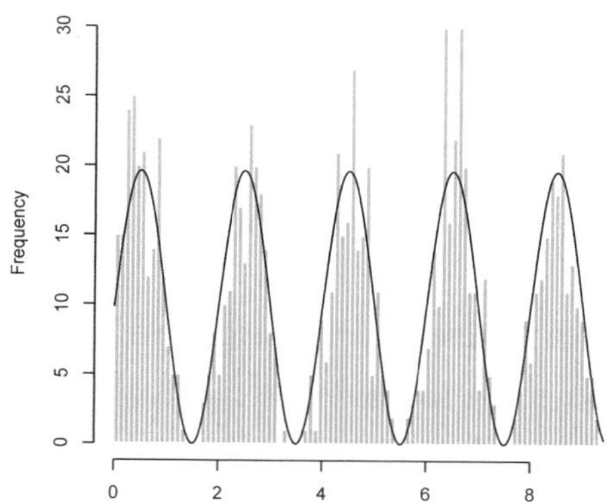

FIGURE 2.16
Histogram associated with a non-homogeneous Poisson process, with cyclical intensity λ.

```
> Tmax <- 3*pi
> set.seed(1)
> t <- 0; X <- numeric(0)
> while(X[length(X)] <= Tmax){
+ Ft <- function(x) 1-exp(-Lambda(t+x)+Lambda(t))
+ x <- uniroot(function(x) Ft(x)-runif(1),interval=c(0,Tmax))$root
+ t <- t+x
+ X <- c(X,t)}
> X <- X[-which.max(X)]
```

To visualize the cycle of occurrences, let us consider the following histogram:

```
>  hist(X,breaks=seq(0,3*pi,by=pi/32),col="grey",
+ border="white",xlab="",main="")
> lines(seq(0,3*pi,by=.02),lambda(seq(0,
+ 3*pi,by=.02))*pi/32,lwd=2)
```

See Figure 2.16. Pasupathy (2010) suggested also to use a rejection technique as an alternative. What we need is an upper bound for the intensity process. A natural upper bound is a constant one, obtained using max $\{\lambda_s, s \in \mathbb{R}\}$,

```
> lambda.up <- 200
```

The code to generate a Poisson process is

```
> set.seed(1)
> t <- 0; X <- t
> while(X[length(X)]<=Tmax){
+   u <- runif(1)
+   t <- t-log(u)/lambda.up
+  if(runif(1)<=lambda(t)/lambda.up) X <- c(X,t)}
> X <- X[-c(1,which.max(X))]
```

The two algorithms can be visualized in Figure 2.16.

Consider a Poisson process with intensity λ. If we keep each point according to some Bernoulli distribution $\mathcal{B}(p)$, then the new point process is also a Poisson process, with intensity $p\lambda$. A standard application is obtained when we consider some deductible d. If the occurrence of claims, for some reinsurer is driven by a Poisson process, with intensity λ, and if individual losses have distribution F, then the process of claims above the deductible is a Poisson process with intensity $[1 - F(d)] \cdot \lambda$. This property will be extremely important in compound Poisson processes (introduced in the next section).

2.5.3 From Poisson Processes to Lévy Processes

A natural extension to the Poisson process is the compound Poisson process, extremely useful to model a surplus process of an insurance company. Given a Poisson process (N_t) and a collection of i.i.d. random variables $X_1, X_2, \ldots$, define

$$S_t = \sum_{i=1}^{N_t} X_i.$$

To generate such a process on time interval $[0, T_{\max}]$, we need to generate a collection of variables, for claims arrival, and claim sizes given a function `randX` that generates independent variables X_i's, such as

```
> randX <- function(n) rexp(n,1)
```

The code can be the following:

```
> rCompPoissonProc <- function(Tmax=1,lambda=rate,rand){
+ N <- rpois(n=1,lambda*Tmax)
+ X <- randX(N)
+ vect.T <- NULL
+ if(N>0) vect.T=sort(runif(N))*lambda*T
+ return(list(T=vect.T,W=diff(c(0,vect.T)),X=X,N=N))}
> set.seed(1)
> rCompPoissonProc(Tmax=5,rand=randX)
$T
[1] 0.3089314 0.8827838 1.0298729 3.4351142

$W
[1] 0.3089314 0.5738524 0.1470891 2.4052414

$X
[1] 1.1816428 0.1457067 0.1397953 0.4360686

$N
[1] 4
```

Based on such a simulation, it is possible to define a function $t \mapsto S_t$:

```
> set.seed(1)
> compois <- rCompPoissonProc(Tmax=5,rand=randX)
> St <- function(t){sum(compois$X[compois$T<=t])}
```

and we can visualize this trajectory (left part of Figure 2.17) using

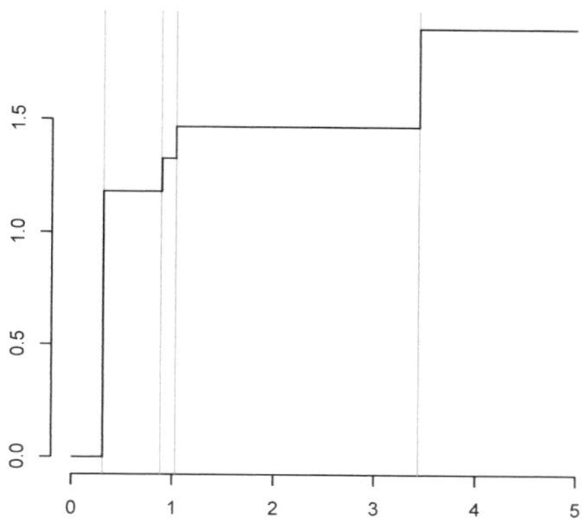

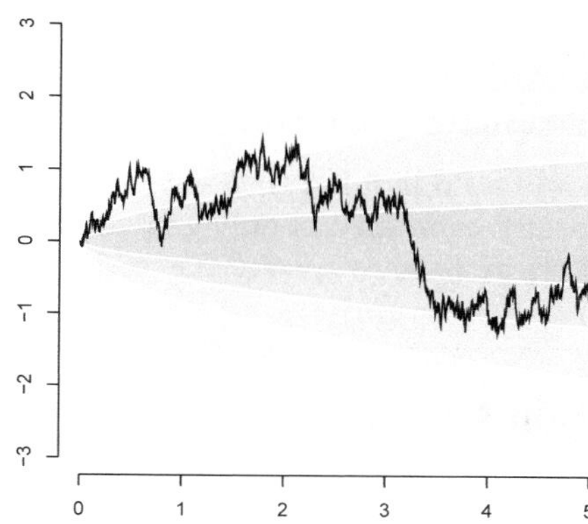

FIGURE 2.17
Sample path of a compound Poisson process on the left, and a Brownian motion on the right.

```
> time <- seq(0,5,length=501)
> plot(time,Vectorize(St)(time),type="s")
> abline(v=compois$T,lty=2,col="grey")
```

To generalize to Lévy processes, we simply have to have a random part, corresponding to a Brownian motion,

$$L_t = -\mu \cdot t + B_t + \sum_{i=1}^{N_t} X_i.$$

The Brownian (B_t) motion satisfies, for all n,

$$B_t = \sum_{i=1}^{[nt/T]} dB_i, \text{ for all } t = \frac{T}{n}, \frac{2T}{n}, \dots, \frac{(n-1)T}{n}, T,$$

where increments dB_i are i.i.d. Gaussian random variables, centered, with variance T/n. Thus, to generate a trajectory of the Brownian motion on $[0, T_{\max}]$, we have to discretize, and given n, use the function above,

```
> n <- 1000
> h <- Tmax/n
> set.seed(1)
> B <- c(0,cumsum(rnorm(n,sd=sqrt(h))))
```

The analogous function `St` would be, here,

```
> Bt <- function(t){B[trunc(n*t/Tmax)+1]}
```

and we can visualize this trajectory (right part of Figure 2.17) using

```
> time <- seq(0,5,length=501)
> plot(time,Vectorize(Bt)(time),type="s")
```

(where level curves related to quantiles of Gaussian random variables were added).

Based on these two functions, it is then possible to generate a trajectory for the Lévy process,

```
> mu <- lambda*rate
> L <- function(t) -mu*t+St(t)+Bt(t)
```

but one should keep in mind that if we can generate the first continuous time process (compound Poisson), then we can only generate an approximation of the second one (Brownian motion); first, we have to specify the grid (choosing `n`), and then, on that grid, we generate a path.

2.5.4 Ruin Models

Ruin theory deals with the study of stochastic processes linked to the wealth of an insurer; see Asmussen & Albrecher (2010) or Dickson (2010) for a recent survey. A reserve risk process $(U_t)_{t\geq 0}$ is considered. The initial model of Cramér–Lundberg assumes that the surplus $(U_t)_{t\geq 0}$ of an insurance company at time t is represented by

$$U_t = u + ct - \sum_{i=1}^{N_t} X_i,$$

where u is the initial surplus, c is the premium rate, $(X_i)_{i\geq 1}$ are i.i.d. successive claim amounts and $(N_t)_{t\geq 0}$ is the claim arrival process assumed to be a Poisson process of intensity λ (see Rolski et al. (1999) for more details on the Poisson process). Andersen (1957) generalized this model by proposing a renewal process for the claim arrival process $(N_t)_{t\geq 0}$ (the claim waiting times are denoted by $(T_i)_{i\geq 1}$). When claim severities and claim waiting times follow a phase-type distribution, closed-form formulas exist for the ruin probability,

$$\psi(u) = \mathbb{P}(\exists t > 0 : U_t < 0 | U_0 = u);$$

see Asmussen & Rolski (1991). We provide below examples of that article.

```
> psi <- ruin(claims = "e", par.claims = list(rate = 1/0.6),
+ wait   = "e", par.wait   = list(rate = 1/0.6616858))
```

Consider Phase-type claims, exponential inter-arrival times:

```
> p <- c(0.5614, 0.4386)
> r <- matrix(c(-8.64, 0.101, 1.997, -1.095), 2, 2)
> lambda <- 1/(1.1 * mphtype(1, p, r))
> psi2 <- ruin(claims = "p", par.claims = list(prob = p, rates = r),
+ wait   = "e", par.wait   = list(rate = lambda))
```

Consider Phase-type claims, a mixture of two exponentials for inter-arrival times:

```
> a <- (0.4/5 + 0.6) * lambda
> psi3 <- ruin(claims = "p", par.claims = list(prob = p, rates = r),
+ wait   = "e", par.wait   = list(rate = c(5 * a, a), weights =
+ c(0.4, 0.6)), maxit = 225)
> plot(psi, from = 0, to = 50)
> plot(psi2, add=TRUE, lty=2)
> plot(psi3, add=TRUE, lty=3)
> legend("topright", leg=c("Exp - Exp", "PH - Exp",
+ "PH - MixExp"), lty=1:3, col="black")
```

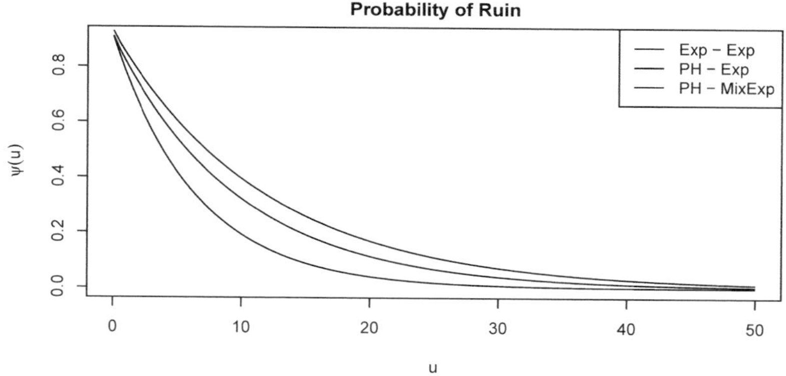

FIGURE 2.18
Ruin probability as a function of the initial surplus of the company, u.

2.6 Copulas and Multivariate Distributions

This final section deals with distributions of multivariate random vectors $\boldsymbol{X} = (X_1, \ldots, X_d)$. Due to the growing literature (see Frees & Valdez (1998), Embrechts et al. (2001), Frees & Wang (2006), among others) on copulas during the past decade (defined as multivariate distribution functions of random vector with uniform marginals), we focus on copulas in this section.

2.6.1 Definition of Copulas

Let $F_{\boldsymbol{X}}$ be the distribution function of $\boldsymbol{X}$ with marginals F_{X_j} that is,

$$F_{\boldsymbol{X}}(\boldsymbol{x}) = \mathbb{P}(X_1 \leq x_1, \ldots, X_d \leq x_d).$$

As $F_{X_j}^{-1}(U)$ has the same distribution as X_j for U a uniform variate, it is easily checked that $\mathbb{P}(X_1 \leq F_{X_1}^{-1}(u_1), \ldots, X_d \leq F_{X_d}^{-1}(u_d)) = P(F_{X_1}(X_1) \leq u_1, \ldots, F_{X_d}(X_d) \leq u_d)$. A copula function C is a multivariate distribution function such that $C(u_1, \ldots, u_d) = \mathbb{P}(F_{X_1}(X_1) \leq u_1, \ldots, F_{X_d}(X_d) \leq u_d)$ for $\boldsymbol{u} \in [0,1]^d$. The C function is bounded by the so-called Fréchet bound as

$$\left(\sum_{i=1}^{d} u_i - (d-1)\right)_+ \leq C(\boldsymbol{u}) \leq \min(u_1, \ldots, u_d)$$

generally denoted by $W(\boldsymbol{u})$ and $M(\boldsymbol{u})$; see Nelsen (2006) for a recent introduction. By the Sklar theorem (from Sklar (1959)) for any random vectors $\boldsymbol{X}$ with marginals F_{X_j}, there exists a copula function C such that

$$\mathbb{P}(\boldsymbol{X} \leq \boldsymbol{x}) = \mathbb{P}(X_1 \leq x_1, \ldots, X_d \leq x_d) = C(F_{X_1}(x_1), \ldots, F_{X_d}(x_d)),$$

for all $\boldsymbol{x} \in \mathbb{R}^d$. Note that the copula C is unique on the support of $\boldsymbol{X}$, and not otherwise. Let us note that in the independent case, the copula function is simply $C(\boldsymbol{u}) = u_1 \cdots u_d = \Pi(\boldsymbol{u})$. As described below, classical multivariate distributions such as the multivariate Gaussian distribution and the multivariate Pareto distribution can be represented using a copula function. Note further that there exists a copula function $C^\star$ such that

$$\mathbb{P}(\boldsymbol{X} > \boldsymbol{x}) = \mathbb{P}(X_1 > x_1, \ldots, X_d > x_d) = C^\star(1 - F_{X_1}(x_1), \ldots, 1 - F_{X_d}(x_d)),$$

for all $\boldsymbol{x} \in \mathbb{R}^d$. This copula $C^\star$ will be called *survival* or *dual* of C. If $\boldsymbol{U}$ has distribution function C, then $\mathbf{1} - \boldsymbol{U}$ has distribution function $C^\star$.

2.6.2 Archimedean Copulas

A wide class of copulas is given by the family of Archimedean copulas. An Archimedean copula is characterized by a generator function $\phi : [0,1] \mapsto [0,\infty]$ such that

$$C(\boldsymbol{u}) = \phi^{-1}\left(\sum_{i=1}^{d} \phi(u_i)\right),$$

where $\boldsymbol{u} \in [0,1]^d$ and ϕ is infinitely differentiable, completely monotone and invertible (weaker conditions can be required for specific dimensions d). We refer to Theorem 2.1 of Marshall & Olkin (1988) for the construction of Archimedean copulas. In this family, the three most classical copulas are the Gumbel copula $\phi(t) = (-\log(t))^{-\alpha}$, the Frank copula $\phi(t) = \log(e^\alpha - 1) - \log(e^{\alpha t} - 1)$ and the Clayton copula $\phi(t) = t^{-\alpha} - 1$ for a parameter α. We get the following copula function:

- Gumbel: $C_{Gu}(\boldsymbol{u}) = \exp(-((-\log(u_1))^\alpha + \cdots + (-\log(u_d))^\alpha)^{1/\alpha})$, for $\alpha \geq 1$.
- Frank: $C_F(\boldsymbol{u}) = -\log(1 + (e^{-\alpha u_1} - 1)\cdots(e^{-\alpha u_d} - 1)/(e^\alpha - 1))/\alpha$, for $\alpha \neq 0$.
- Clayton: $C_C(\boldsymbol{u}) = (1 - d + u_1^{-\alpha} + \cdots + u_d^{-\alpha})^{-1/\alpha}$, for $\alpha > 0$ (or $\alpha \in [-1, +\infty[\setminus\{0\}$ in dimension 2).

The survival Clayton defined as $\mathbb{P}(\boldsymbol{U} > \boldsymbol{u}) = C_C(1 - \boldsymbol{u})$ is linked to the multivariate Pareto distribution. According to Arnold (1983), the multivariate Pareto distribution is characterized by the following survival function:

$$\mathbb{P}(\boldsymbol{X} > \boldsymbol{x}) = \left(1 + \sum_{i=1}^{d} x_i^{\frac{1}{\gamma_i}}\right)^{-\alpha}.$$

The marginal distribution of X_i is also Pareto distributed, as $\mathbb{P}(X_i > x_i) = (1 + x_i^{1/\gamma_i})^{-\alpha} = 1 - F_{X_i}(x_i)$. It is easy to check that $\mathbb{P}(X_1 > F_{X_1}^{-1}(u_1), \ldots, X_d > F_{X_d}^{-1}(u_d)) = C_C(1 - \boldsymbol{u})$.

2.6.3 Elliptical Copulas

Before introducing elliptic copulas, we define elliptical distributions. A random variable X has an elliptical distribution if its characteristic function φ_X satisfies $\varphi_X(t) = e^{it\mu}\psi(t^2\sigma)$ for some parameters μ, σ and some function ψ. Generally a random vector $\boldsymbol{X}$ follows an elliptical distribution if its characteristic function verifies

$$\varphi_{\boldsymbol{X}}(\boldsymbol{t}) = e^{i\boldsymbol{t}^\top \mu}\psi(\boldsymbol{t}^\top \Sigma \boldsymbol{t}), \text{ where } \boldsymbol{t} \in \mathbb{R}^d,$$

for some vector μ, some positive definite matrix Σ, and some function ψ. For such a distribution, the density function is given by

$$f_{\boldsymbol{X}}(\boldsymbol{x}) = \frac{c_d}{\sqrt{\det(\Sigma)}}\phi((\boldsymbol{x} - \mu)^\top \Sigma^{-1}(\boldsymbol{x} - \mu)/2), \text{ where } \boldsymbol{x} \in \mathbb{R}^d,$$

some function $\phi : \mathbb{R}_+ \mapsto \mathbb{R}$ such that $\int_0^\infty x^{d/2-1}\phi(x)dx < \infty$ and some normalizing constant c_d. We get the multivariate normal distribution when $\phi(t) = e^{-t}$ with mean vector μ and covariance matrix Σ, the multivariate Student distribution with m degrees of freedom when $\phi(t) = (1+t/m)^{(d+m)/2}$. See Fang et al. (1990) or Genton (2004) for more details on elliptical distributions.

An elliptical copula is defined as

$$C(\boldsymbol{u}) = H(H_1^{-1}(u_1), \ldots, H_d^{-1}(u_d)),$$

where H is a multivariate distribution with marginals H_i belonging to the elliptical family. In particular for a symmetric positive definite matrix Σ, the Gaussian and the student copulas are defined as

- Gaussian

$$C_{Ga}(\boldsymbol{u}) = \int_{-\infty}^{z_1} \cdots \int_{-\infty}^{z_d} \tilde{c}_d e^{-\boldsymbol{x}^\top \Sigma^{-1} \boldsymbol{x}/2} dx_1 \ldots dx_d,$$

where $z_i = \Phi^{-1}(u_i)$ and Φ^{-1} is the quantile function of the standard normal distribution.

- Student

$$C_{St}(\boldsymbol{u}) = \int_{-\infty}^{z_1} \cdots \int_{-\infty}^{z_d} \tilde{c}_d \left(1 + \frac{\boldsymbol{x}^\top \Sigma^{-1} \boldsymbol{x}}{2m}\right)^{\frac{d+m}{2}} dx_1 \ldots dx_d,$$

where $z_i = F_{St}^{-1}(u_i)$ is the quantile of a Student distribution with $m > 0$ degrees of freedoms.

2.6.4 Properties and Extreme Copulas

Copulas presented in the previous subsections have a density function $c : [0,1]^d \mapsto [0,1]$ because the copula function is differentiable with respect to all variables on the unit hypercube. The dependence induced by a particular copula can be quantified through the theory of concordance measures introduced by Scarsini (1984). The two main measures of concordance are Kendall's tau and Spearman's rho. Kendall's tau for a bivariate vector (X, Y) is defined as

$$\tau(X,Y) = \mathbb{P}((X - \tilde{X})(Y - \tilde{Y}) > 0) - \mathbb{P}((X - \tilde{X})(Y - \tilde{Y}) < 0),$$

where $(\tilde{X}, \tilde{Y})$ is an independent replicate of (X, Y). Similarly, Spearman's rho for (X, Y) is defined as

$$\tau(X,Y) = 3\mathbb{P}((X - \tilde{X})(Y - \bar{Y}) > 0) - 3\mathbb{P}((X - \tilde{X})(Y - \bar{Y}) < 0),$$

where $(\tilde{X}, \tilde{Y})$ and $(\bar{X}, \bar{Y})$ are independent replicates of (X, Y). As these two measures satisfy the criteria of concordance measures, $\tau(X,Y) \in [0,1]$, $\tau(X,Y) = 1$ means that the copula of (X, Y) is the upper Fréchet bound, and $\tau(X,Y) = 0$ means that the copula of (X,Y) is the independent copula (the same holds for $\rho(X,Y)$). In the bivariate case, closed-form formulas are available for the copulas previously presented; see, for example, Nelsen (2006) and Joe (1997).

A desirable feature of copulas lies in the fact that they can model dependence between two or more variables with or without a tail dependency. This is characterized by the tail dependance coefficients. The upper tail coefficient of (X, Y) is defined as

$$\lambda_U(X,Y) = \lim_{t \to 1^-} \mathbb{P}(Y > F_Y^{-1}(t) | X > F_X^{-1}(t)),$$

while the lower tail coefficient $\lambda_L(X,Y)$ is obtained considering

$$\lambda_L(X,Y) = \lim_{t \to 0^+} \mathbb{P}(Y \leq F_Y^{-1}(t) | X \leq F_X^{-1}(t))$$

When X, Y have a continuous distribution with a dependency given by a copula $C_{X,Y}$, those coefficients can be rewritten as

$$\lambda_U(X,Y) = \lim_{t\to 1^-} \frac{1-2t-C_{X,Y}(t,t)}{1-t} \text{ and } \lambda_L(X,Y) = \lim_{t\to 0^+} \frac{C_{X,Y}(t,t)}{t}.$$

For the copulas presented here, we have $\lambda_U = 0$ except for the Gumbel copula $\lambda_U = 2-2^{1/\alpha}$ and the Student copula $\lambda_U = 2F_{St}(\sqrt{m+1}\sqrt{1-\rho}/\rho 1+\rho)$, whereas $\lambda_L = 0$ except for the Clayton copula $\lambda_L = 2^{-1/\alpha}$ and the Student copula $\lambda_L = \lambda_U$. In other words, copulas with $\lambda_U = 0$ cannot model dependence at the right-hand tail.

Another desirable property of copulas can be the max-stability. A copula function C is max-stable if

$$C(u_1,\dots,u_d) = \left(C(u_1^{1/k},\dots,u_d^{1/k})\right)^k,$$

for all $k > 0$. This property is linked to the extreme value theory because the right-hand side is the copula of component-wise maxima of a random vector sample $(\boldsymbol{X}_1,\dots,\boldsymbol{X}_k)$, where the $\boldsymbol{X}_i$'s have copula C. Copulas verifying this property are called extreme copulas: The Gumbel and the Hüsler–Reiss copulas belong to this family. The Hüsler–Reiss copula is defined as follows in the bivariate case,

$$C_{HR}(u_1,u_2) = \exp\left(\log(u_1)\Phi(d_+) + \log(u_2)\Phi(d_-)\right),$$

where $d_\pm = 1/\alpha \pm \alpha/2 \log(\log(u_1)/\log(u_2))$ and ϕ is the distribution function of the standard normal distribution.

2.6.5 Copula Fitting Methods

There are four main methods to calibrate copulas which differ on how the marginals are considered in the fitting process. Consider a sample of random vectors $(\boldsymbol{X}_1,\dots,\boldsymbol{X}_n)$ and corresponding observations $\boldsymbol{x}_1,\dots,\boldsymbol{x}_n$ where the ith marginal has a density $f_i(.;\theta_i)$ and a distribution function $F_i(.;\theta_i)$. A (full) maximum likelihood estimation is the first option, which consists of maximizing the likelihood

$$\begin{aligned}\mathcal{L}(\alpha,\theta_1,\dots,\theta_d,\boldsymbol{x}_1,\dots,\boldsymbol{x}_n) &= \prod_{i=1}^{n} c(F_1(x_{1,i};\theta_1),\dots,F_d(x_{d,i};\theta_d);\alpha)\cdot f_1(x_{1,i};\theta_1)\cdots f_{d,i} \\ &= f_j(x_{d,i};\theta_d),\end{aligned}$$

α being the parameter of the copula C and θ_i being the parameter for the ith marginal distribution. The optimization is carried out over the whole parameter space.

The second estimation method is the method of moments which consists, as in the univariate, of matching theoretical moments and empirical moments. Marginal parameters θ_i are set by equalizing the empirical moments of the sample $(X_{i,1},\dots,X_{i,n})$, while the copula parameters α are determined by matching Kendall's tau or Spearman's rho.

The third estimation, called inference for margins, is a two-step procedure. First, marginal distributions are fitted by maximum likelihood, and then a pseudo sample is defined as

$$\hat{\boldsymbol{u}}_i = (\hat{u}_{1,i},\dots,\hat{u}_{d,i}) = (F(x_{1,i},\hat{\theta}_1),\dots,F(x_{d,i},\hat{\theta}_d))$$

for $i = 1,\dots,n$. Then the copula is fitted on $\hat{\boldsymbol{u}}_1,\dots,\hat{\boldsymbol{u}}_n$ by maximizing the likelihood

$$\mathcal{L}(\alpha,\hat{\boldsymbol{u}}_1,\dots,\hat{\boldsymbol{u}}_n) = \prod_{i=1}^{n} c(\hat{u}_{1,i},\dots,\hat{u}_{d,i};\alpha).$$

The inference for margins method takes advantage of the two steps to reduce the dimension of the likelihood from $(\alpha, \theta_1, \dots, \theta_d)$ to α. Finally, the canonical maximum likelihood method is similar to the inference for margins and consists of replacing the parametric estimate by the non-parametric estimates in the pseudo data. That is to say, $\hat{\boldsymbol{u}}_i = (F_n(x_{1,i}), \dots, F_n(x_{d,i}))$, which further simplifies to $\hat{\boldsymbol{u}}_i = (\text{rank}(x_{1,i})/n, \dots, \text{rank}(x_{d,i})/n)$. In the following section, we only consider the inference for margins method.

2.6.6 Application and Copula Selection

Numerical illustrations of copulas and their estimation are carried out on the loss-ALAE dataset used in Frees & Valdez (1998) and Klugman & Parsa (1999). The dataset consists of 1,500 general liability claims (expressed in USD) where each claim is a two-component vector: an indemnity payment (loss) and an allocated loss adjustment expense (ALAE).

```
> data(lossalae)
> par(mfrow=c(1,2))
> plot(lossalae, log="xy", main="Scatterplot of loss-ALAE")
> plot(apply(lossalae, 2, rank)/NROW(lossalae),
+   main="rank transform of loss-ALAE")
```

In Figure 2.19, we plot the scatterplots of the data (x_i, y_i) and the empirical distributions evaluated at (x_i, y_i), that is, $(F_{n,X}(x_i), F_{n,Y}(y_i)) = (\text{rank}(x_i)/n, \text{rank}(y_i)/n)$.

On this dataset, we choose to fit the following bivariate copulas:

(i) Gaussian copula $C_{Ga}(.,.;\rho)$,

(ii) Student copula $C_{St}(.,.;\rho, m)$,

(iii) Gumbel copula $C_{Gu}(.,.;\alpha)$,

(iv) Frank copula $C_F(.,.;\alpha)$

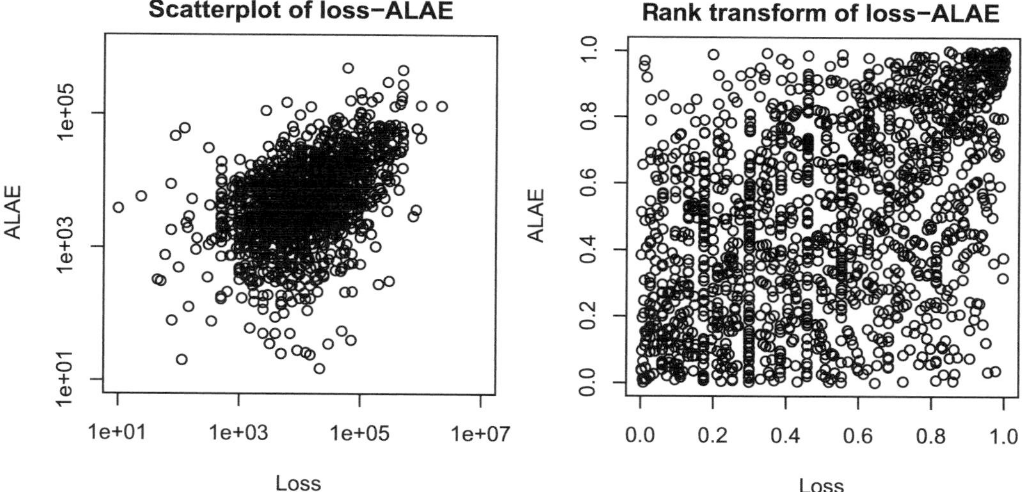

FIGURE 2.19
Loss-ALAE dataset.

(v) Hüsler–Reiss copula $C_{HR}(.,.;\alpha)$.

We use the implementation done in the `fCopulae` package (part of the Rmetrics project, see https://www.rmetrics.org/; see also Chapters 11 and 13). For convenience, we define the following functions

```
> dnormcop <- function(U, param)
+ as.numeric(dellipticalCopula(U, rho=param[1], type="norm"))
> dtcop <- function(U, param)
+ as.numeric(dellipticalCopula(U, rho=param[1], type="t",
+ param=param[2]))
> dgumcop <- function(U, param)
+ as.numeric(devCopula(U, type="gumbel", param=param[1]))
> dHRcop <- function(U, param)
+ as.numeric(devCopula(U, type="husler.reiss", param=param[1]))
> dfrankcop <- function(U, param)
+ as.numeric(darchmCopula(U, type="5", alpha=param[1]))
```

In addition to finding an appropriate copula, a choice of distribution must be done for marginals. A Pareto chart on both marginals shows that they follow heavy-tailed distributions.

```
> paretochart <- function(x, ...)
+   plot(-log((1:length(x))/(length(x)+1)), log(sort(x)), ...)
> paretochart(lossalae$Loss)
> paretochart(lossalae$ALAE)
```

Therefore, we choose a Pareto type II distribution and a lognormal distribution for candidate distributions of marginals. As there is no package fitting copulas for any kind of copula, we implement the inference for margins method in the following function:

```
> fit.cop.IFM.2 <- function(obs, copula, marg, arg.margin=list(),
+ method.margin="mle", arg.cop=list(), initpar, ...)
+ {
+ Obs1 <- obs[,1]
+ Obs2 <- obs[,2]
+ if(marg %in% c("exp","gamma","lnorm","pareto","burr")){
+         Obs1 <- Obs1[Obs1 > 0]
+         Obs2 <- Obs2[Obs2 > 0] }
+ marg1 <- do.call(fitdist, c(list(data= Obs1, distr=marg,
+       method=method.margin), arg.margin))
+ marg2 <- do.call(fitdist, c(list(data= Obs2, distr=marg,
+       method=method.margin), arg.margin))
+ comput.cdf <- function(fit, obs) {
+     para <- c(as.list(fit$estimate), as.list(fit$fix.arg))
+     distname <- fit$distname
+     pdistname <- paste("p", distname, sep = "")
+     do.call(pdistname, c(list(q = obs), as.list(para)))}
+ pseudomarg1 <- comput.cdf(marg1, Obs1)
+ pseudomarg2 <- comput.cdf(marg2, Obs2)
+ U <- cbind(pseudomarg1, pseudomarg2)
+ copLogL <- function(x) {
+ if(arg.cop$lower <= x && arg.cop$upper >= x)
```

```
+   res <- -sum(remove.naninf(log(copula(U, param=x))))
+ else res <- Inf
+   return(res)}
+ resopt <- optim(par=initpar, fn=copLogL, method="L-BFGS-B",
+                  lower=arg.cop$lower, upper=arg.cop$upper, ...)
+   list(marg1=marg1, marg2=marg2, copula=
+          list(name=arg.cop$name, alpha=resopt$par))}
> remove.naninf <- function(x)
+   x[!is.nan(x) & is.finite(x)]
```

The copulas are now fitted using the function `fit.cop.IFM.2` by defining the corresponding `arg.cop` argument. Note that the marginal distributions are fitted using `fitdist`.

```
> library(fCopulae)
> argnorm <- list(length=1, lower=0, upper=1, name="Gaussian")
> argt <- list(length=2, lower=c(0,0), upper=c(1,1000),
+ name="Student")
> arggum <- list(length=1, lower=1, upper=100, name="Gumbel")
> argHR <- list(length=1, lower=0, upper=1000, name="Husler-Reiss")
> argfrank <- list(length=1, lower=-1000, upper=1000, name="Frank")
> fgausspareto <- fit.cop.IFM.2(lossalae, copula= dnormcop,
+ marg="pareto", arg.margin=list(start=list(shape=10, scale=100),
+ lower=c(1, 1/2)), arg.cop= argnorm, initpar=1/2)
> ftpareto <- fit.cop.IFM.2(lossalae, copula= dtcop,
+ marg="pareto", arg.margin=list(start=list(shape=10, scale=100),
+ lower=c(1, 1/2)), arg.cop= argt, initpar=c(1/2, 4))
> fgumbelpareto <- fit.cop.IFM.2(lossalae, copula= dgumcop,
+ marg="pareto", arg.margin=list(start=list(shape=10, scale=100),
+ lower=c(1, 1/2)), arg.cop= arggum, initpar=10)
> fHRpareto <- fit.cop.IFM.2(lossalae, copula= dHRcop,
+ marg="pareto", arg.margin=list(start=list(shape=10, scale=100),
+ lower=c(1, 1/2)), arg.cop= argHR, initpar=10)
> ffrankpareto <- fit.cop.IFM.2(lossalae, copula= dfrankcop,
+ marg="pareto", arg.margin=list(start=list(shape=10, scale=100),
+ lower=c(1, 1/2)), arg.cop= argfrank, initpar=10)
> recap <- function(x){
+   res <- c(alpha=x$copula$alpha, x$marg1$estimate, x$marg2$estimate)
+         if(length(res) < 6)
+                 res <- c(res[1], NA, res[2:5])
+         res <- as.matrix(res)
+         colnames(res) <- x$copula$name
+         res}
> round(cbind(recap(fgausspareto), recap(ftpareto),
+ recap(fHRpareto), recap(fgumbelpareto),
+ recap(ffrankpareto)        ), 4)

        Gaussian    Student Husler-Reiss     Gumbel      Frank
alpha     0.4783     0.4816       1.1133     1.4444     3.1140
              NA     9.6475           NA         NA         NA
shape     1.2377     1.2377       1.2377     1.2377     1.2377
scale 16228.2572 16228.2572   16228.2572 16228.2572 16228.2572
```

```
shape      2.2230     2.2230      2.2230     2.2230     2.2230
scale 15133.3463 15133.3463  15133.3463 15133.3463 15133.3463
```

The level of dependency seems low as the value of the first parameter is either close to 1 (for non-elliptic copulas) or close to 0 for elliptic copulas. Unsurprisingly, the fitted parameters of the marginal distributions are identical. In order to assess the quality of the fit, we look at the tail coefficients $\lambda_U(X,Y)$ and $\lambda_L(X,Y)$, which are computable given a copula. They can be estimated on data by using a non-parametric estimate of $C(t,t)$, that is, the empirical bivariate distribution function $C_n(t,t) = \sum_{i=1}^n \mathbf{1}_{x_i \leq t}\mathbf{1}_{y_i \leq t}/n$. This is done by the following function:

```
> Lemp <- function(u, obs)
+ sapply(1:length(u), function(i)
+ 1/NROW(obs)*sum(obs[,1] <= u[i] & obs[,2] <= u[i]) )/u
> Uemp <- function(u, obs)
+ (1-2*u+sapply(1:length(u), function(i)
+ 1/NROW(obs)*sum(obs[,1] <= u[i] & obs[,2] <= u[i]) ))/(1-u)
> Lcop <- function(u, pcop, param=param)
+   pcop(cbind(u, u), param=param)/u
> Ucop <- function(u, pcop, param=param)
+ (1-2*u+pcop(cbind(u, u), param=param))/(1-u)
```

The plot of the two tail coefficients is now possible.

```
> u <- seq(0, 0.4, length=101)
> rklossalae <- apply(lossalae, 2, rank)/NROW(lossalae)
> plot(u, Lemp(u, rklossalae), type="l", main="Lower coefficient",
+ ylim=c(0,.6), xlab="u", ylab="L(u)")
> lines(u, Lcop(u, pgumcop, fgumbelpareto$copula$alpha), lty=2)
> lines(u, Lcop(u, pHRcop, fHRpareto$copula$alpha), lty=3)
> lines(u, Lcop(u, ptcop, ftpareto$copula$alpha), lty=4, col="grey25")
> lines(u, Lcop(u, pnormcop, fgausspareto$copula$alpha), lty=5, col="grey25")
> lines(u, Lcop(u, pfrankcop, ffrankpareto$copula$alpha), lty=6, col="grey25")
> legend("bottomright", lty=1:6, col=c(rep("black", 3), rep("grey25", 3)),
+ leg=c("emp.", "Gumbel","Husler-Reiss","Student","Gaussian","Frank"))
```

As shown on Figure 2.20, the tail coefficients are best approximated by the extreme copulas: Gumbel and Hüsler–Reiss. When considering Kendall's tau or Spearman's rho, these two copulas are also reasonably good. Kendall's tau is here:

```
> cbind(emp=cor(lossalae, method="kendall")[1,2],
+ Frank=taufrankcop(ffrankpareto$copula$alpha),
+ Gumbel= taugumcop(fgumbelpareto$copula$alpha),
+ HR= tauHRcop(fHRpareto$copula$alpha),
+ Gauss = taunormcop(fgausspareto$copula$alpha),
+ Student = tautcop(ftpareto$copula$alpha))

          emp     Frank    Gumbel        HR     Gauss   Student
Tau 0.3154175 0.3171107 0.3076705 0.3008525 0.3174771 0.3198691
```

while Spearman's rho is

```
> cbind(emp=cor(lossalae, method="spearman")[1,2],
+ Frank=rhofrankcop(ffrankpareto$copula$alpha),
+ Gumbel= rhogumcop(fgumbelpareto$copula$alpha),
```

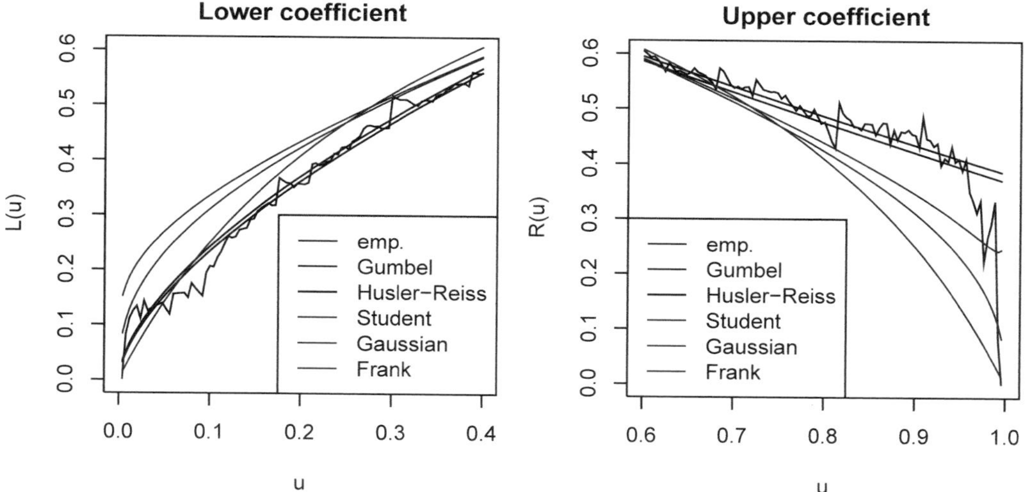

FIGURE 2.20
Tail coefficients.

```
+ HR= rhoHRcop(fHRpareto$copula$alpha),
+ Gauss = rhonormcop(fgausspareto$copula$alpha),
+ Student = rhotcop(ftpareto$copula$alpha))
```

```
         emp     Frank    Gumbel        HR      Gauss      Student
Rho 0.451872 0.4622724 0.4424486 0.4360027  0.4500000   0.4500000
```

Therefore, we continue the analysis only with the Gumbel and Hüsler–Reiss copulas.

Given a copula, quantities of interest can be estimated by a Monte–Carlo method. Focusing on the distribution of the total expense, that is, the sum of loss and ALAE, we simulate this sum with the following template function:

```
> simul.cop.2 <- function(n, rcopula, fit){
+ U <- rcopula(n, fit$copula$alpha)
+ qmarg1 <- paste("q", fit$marg1$distname, sep="")
+ qmarg2 <- paste("q", fit$marg2$distname, sep="")
+ cbind(
+ X1=do.call(qmarg1, c(list(p=U[,1]), fit$marg1$estimate)),
+ X2=do.call(qmarg2, c(list(p=U[,2]), fit$marg2$estimate))
+ )}
```

where `rcopula()` is the copula random generator defined as

```
> rgumcop <- function(n, param)
+   revCopula(n, type="gumbel", param=param[1])
> rHRcop <- function(n, param)
+   revCopula(n, type="husler.reiss", param=param[1])
> rindep <- function(n,param)
+   cbind(runif(n), runif(n))
```

Choosing a sample size of $n = 1e4$, we plot the empirical distribution functions of the sum for the Gumbel, the Husler–Reiss and the independent copulas.

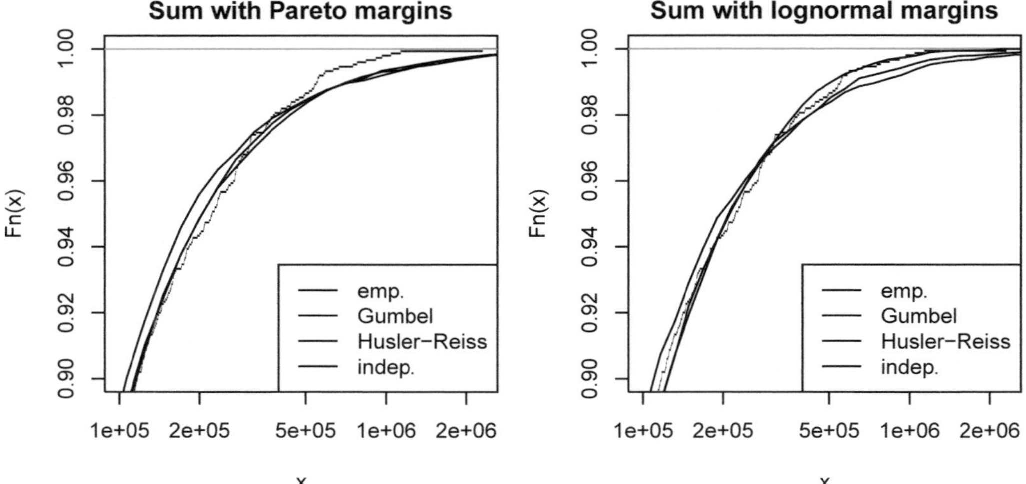

FIGURE 2.21
Tails of the distribution of the total claim (loss + ALAE).

```
> sumlossalae <- rowSums(lossalae)
> n <- 10^4
> sumgumpareto <- rowSums(simul.cop.2(n, rgumcop, fgumbelpareto))
> sumHRpareto <- rowSums(simul.cop.2(n, rHRcop, fHRpareto))
> sumindeppareto <- rowSums(simul.cop.2(n, rindep, fgumbelpareto))
> plot(ecdf(sumlossalae), log="x",  main="", xlim=range(sumlossalae))
> z <- 10^seq(1, log(max(sumlossalae)), length=201)
> lines(z, ecdf(sumgumpareto)(z), lty=2)
> lines(z, ecdf(sumHRpareto)(z), lty=3)
> lines(z, ecdf(sumindeppareto)(z), lty=4)
> legend("bottomright", lty=1:4, col="black",
+ leg=c("emp.", "Gumbel", "Husler-Reiss", "indep."))
```

On Figure 2.21, we observe that both extreme copulas are particularly adapted for the `lossalae` dataset, irrespective of the choice of the marginal distribution. Yet, we notice that the choice of the lognormal distribution better suits the tail of the empirical distribution. However, as extreme claims above 1 million dollars are less observed (by definition) than large claims (between 100 thousand and 1 million), the experimenter must take a prudential view of the right-hand tail without overfitting data. The independent copula seems to be a reasonable approximation of the tail of the distribution (yet not of the rest of the distribution).

2.7 Exercises

For these exercises, consider the following vectors,

```
> set.seed(123)
> X1 <- trunc(rlnorm(n=753,5))
```

```
> X2 <- rgamma(n=267,1.2,.25)
fitdistr(X2,"lognormal")
```

2.1. Benford's law is defined on $\{1,2,\ldots,8,9\}$ as follows:

$$\mathbb{P}(N=n)=\frac{\log\left(1+\dfrac{1}{n}\right)}{\log(10)}.$$

Write functions `dBenford`, `pBenford` and `rBenford` that return the density, the (cumulative) distribution function and generate random values according to this distribution.

2.2. Given a vector of integers, `x`, write a function that counts frequency of first digits of elements of `x`. Write a function, based on Person's statistics, that tests if Benford's distribution is relevant, or not (using functions of Exercise 1.). Try this function on `X1`.

Hint: Use function `chisq.test` to run the test.

2.3. Using `fitdistr` from library `MASS`, find maximum likelihood estimators if we fit a lognornal distribution to `X2`.

2.4. According to the Kolmogorov–Smirnov test, can we model `X2` using a $LN(1,1)$ distribution?

2.5. Assume that $N \sim \mathcal{P}(10)$. Compute $\mathbb{P}(N \in [8,12])$. Find (numerically) $\max\{\mathbb{P}(N\in[a:a+4]), a\in\mathbb{N}\}$.

2.6. Use function `barplot` to visualize the probability function associated to Sibuya's distribution, defined on $\mathbb{N}$ by

$$\mathbb{P}(N\leq n)=1-(1-n)^r, \text{ for some } r\in(0,1).$$

2.7. Without using a loop, write a function `rmix(n,p,rf1,rf2,...)` that generates `n` random variables, i.i.d., with distribution the mixture of `f1`, ..., `fn`, where `p` is a vector of probabilities of length `n`.

2.8. Compute the 95% quantile of a compound sum, when N has a negative binomial distribution, with mean 2 and variance 3, and when X_i have a gamma distribution, with mean 100 and variance 150, using the Panjer algorithm.

2.9. Compute the 95% quantile of a compound sum, when N has a negative binomial distribution, with mean 2 and variance 3, and when X_i have a gamma distribution, with mean 100 and variance 150, using Monte Carlo simulations.

2.10. Fit a gamma and a lognormal distribution to `data(danishuni)` from library `CASdatasets`. Which model gives the highest 95% quantile. Compare it with the empirical one.

2.11. Fit a gamma and a lognormal distribution to `lossalae$ALAE` from library `CASdatasets`. Which model gives the highest 95% quantile. Compare it with the empirical one.

2.12. Using a gamma distribution on `lossalae$ALAE`, what would be the expected value, given that `lossalae$ALAE` is larger than 10,000.

2.13. From results mentioned in Section 2.1.3 and functions associated to `phtype`, write functions `dErlang(x,k,lambda)`, `pErlang(x,k,lambda)` and `rErlang(n,k,lambda)`.

2.14. Fit a copula on the following sample:

```
> n <- 1500
> set.seed(123)
> theta <- rgamma(n,.5)
> U <- cbind(rank(rexp(1500,rate=theta))/(n+1),
+ rank(rexp(1500,rate=theta))/(n+1))
```

2.15. Given $a > 0$, fit a truncated distribution to the following sample:

```
> n <- 1500
> a <- 1
> set.seed(123)
> X <- rep(NA,n)
> for(i in 1:n){
> x <- a; v<- 2
+ while(v*x>a) v<- runif(1); x<-sqrt(a^2-2*log(runif(1)))
+ X[i] <- x
+ }
```

2.16. Fit a distribution to the following sample:

```
> n <- 1500
> set.seed(123)
> f <- function(v) -v[1]*log(v[2]*v[3])
> X <- apply(matrix(runif(3*n),n,3),1,f)
```

2.17. Fit a distribution to the following sample:

```
> n <- 1500
> set.seed(123)
> f <- function(v) log((v[1]*v[2])/(v[3]*v[4]))
> X <- apply(matrix(runif(4*n),n,4),1,f)
```

Hint: It is the density of the sum of two independent and identically distributed random variables.

2.18. Fit a distribution to the following sample:

```
> n <- 1500
> set.seed(123)
> a <- .4; b <- .2
> U <- runif(n)^(1/a)
> V <- runif(n)^(1/b)
> X <- (U/(U+V))[U+V <= 1]
```

2.19. Fit a distribution to the following sample:

```
> n <- 1500
> set.seed(123)
> a <- 17; b <- 12
> B <- rbeta(n,a,b)
> X <- b*B/(a*(1-B))
```

2.20. Fit a distribution to the following sample:

```
> n <- 1500
> set.seed(123)
> B <- rbeta(n,a/2,a/2)
> X <- sqrt(a)*(B-.5)/sqrt(B*(1-B))
```

2.21. Consider the following counting process, with sequence of arrival times `T`:

```
> Tmax <- 20
> set.seed(123)
> T <- 0
> while(max(T)<=Tmax) T<-c(T,sqrt(max(T)^2+2*rexp(1)))
```

Assuming that this sequence is generated from a non-homogeneous Poisson process, suggest values for an affine intensity λ_t.

3

Bayesian Philosophy

Benedict Escoto
AON Benfield
Chicago, Illinois, United States

Arthur Charpentier
Université du Québec à Montréal
Montréal, Québec, Canada

CONTENTS

3.1 Introduction

Bayesian philosophy has a long history in actuarial science, even if it was sometimes hidden. Liu et al. (1996) claim that "*Statistical methods with a Bayesian flavor* [...] *have long been used in the insurance industry.*" And according to McGrayne (2012), actuaries and Arthur L. Bailey played an important rule to prove that it might be relevant to have prior opinions about the next experiment, not to say credible.

> "[Arthur] *Bailey spent his first year in New York* [in 1918] *trying to prove to himself that* 'all of the fancy actuarial [Bayesian] procedures of the casualty business were mathematically unsound.' *After a year of intense mental struggle, however, he realized to his consternation that actuarial sledgehammering worked. He even preferred it to the elegance of frequentism. He positively liked formulae that described* 'actual data ... I realized that the hard-shelled underwriters were recognizing certain facts of life neglected by the statistical theorists.' *He wanted to give more weight to a large volume of data than to the frequentists small sample; doing so felt surprisingly 'logical and reasonable.' He concluded that only a* 'suicidal' *actuary would use Fisher's method of maximum likelihood, which assigned a zero probability to nonevents. Since many businesses file no insurance claims at all, Fisher's method would produce premiums too low to cover future losses.*"

3.1.1 A Formal Introduction

Lindley (1983) explained the Bayesian paradigm as follows. The interest here is a quantity θ, which is unknown. But we might have some personal idea about its distribution that should express our relative opinion as to the likelihood that various possible values of θ are the true value. This will be the prior distribution of θ, and it will be denoted $\pi(\theta)$. This represents the state of our knowledge, somehow, prior to conducting an experiment, or observing the data. Further, there is a probability distribution $f(x|\theta)$, which describes the relative likelihood of values x, given that θ is the true parameter value, for a random variable X. But there is nothing new here. As discussed in Chapter 2, this is the parametric distribution of a random variable X, that was sometimes denoted $f_\theta(x)$. The main difference might be that θ is a random variable here, and thus $f_\theta(x)$ is now a conditional distribution, of X, given the information that θ is the true value.

Based on those two functions, we use Bayes theorem to compute

$$\pi^\star(\theta|x) = \frac{f(x|\theta)\pi(\theta)}{\int f(x|\theta)\pi(\theta)d\theta},$$

which is the posterior distribution of θ. It can be seen as a revised opinion, once the results of the experiment are known. Once we have this posterior distribution, which contains all the knowledge we have (our subjective a priori, and the sample, which can be seen as more objective), we can compute any quantity of interest. This can be done either using analytical formulas, when the later are tractable, or using Monte Carlo simulations to generate possible values θ, given the sample that was observed.

Consider, for instance, the case where θ is a position measure that can be the mode, the median, the mean, etc. In the classical and standard setting (see Chapter 2), we usually compute confidence intervals, so that we can claim that with probability $1-\alpha$, θ belongs to

TABLE 3.1
Objective and subjective probabilities.

	Common Example	Insurance Example
Objective	The coin has a 25% probability of coming up heads twice in a row.	A policyholder has a 1% chance of a property loss.
Subjective	There's a 90% chance that life exists on other planets.	There's a 50% chance that national liability costs per exposure decrease next year.

some interval. In the Bayesian framework, we compute a $1-\alpha$ credibility interval, so that

$$\mathbb{P}(\theta \in [\ell, u]|\boldsymbol{X}) = \int_{\ell}^{u} \pi^{\star}(\theta|\boldsymbol{X})d\theta = 1-\alpha.$$

We can also use a Bayesian version of the Central Limit Theorem (see Berger (1985)) to derive a credibility interval: Under suitable conditions, the posterior distribution can be approximated with a Gaussian distribution, and an approximated credibility interval is then

$$\mathbb{E}(\theta|\boldsymbol{X}) \pm \Phi^{-1}(1-\alpha/2)\sqrt{\text{Var}(\theta|\boldsymbol{X})}.$$

3.1.2 Two Kinds of Probability

There are two kinds of probability: objective and subjective. Roughly speaking, objective probabilities are defined by reality via scientific laws and physical processes. Subjective probabilities are created by people to help them quantify their beliefs and analyze the consequences of their beliefs.

For example, take the sentences in Table 3.1. The sentences in the first row are objective. Whether or not the coin is fair (50% probability of heads) or biased depends not on what anyone believes about the coin, but on the physical properties of the coin (its shape and distribution of mass).

On the other hand, the sentences in the second row seem less objective. Life exists on other planets or not; it is not clear what would make it objectively true that the probability of this is 90%. A probability like this varies across people and reflects strength of belief.

Although the exact boundary between subjective and objective probability is controversial, the basic distinction itself is somewhat obvious. It is important to this chapter because this distinction is what distinguishes Bayesian from classical statistics. Bayesians find it useful to deal with subjective probabilities. On the other hand, classical statisticians think only objective probabilities are worthwhile, either because subjective probabilities are nonsensical or because they are too subjective.

3.1.3 Working with Subjective Probabilities in Real Life

How would subjective probabilities that are just about beliefs have any practical value? Consider the following argument:

1. There is a 50% chance that liability costs per exposure will decrease.
2. Independent of liability costs, there is a 50% chance that our CEO will decide to reduce expenses.

3. Our company will only be profitable if and only if liability costs decrease and the CEO reduces expenses.
4. Therefore, there is less than a 25% chance we will make a profit.

Although statements (1) and (2) may be purely subjective probabilities, an argument like the one above can still be powerful. Even though the first two statements may not be objectively true or false, someone who endorses (1)–(3) and yet rejects (4) seems to be in error.

Bayesian statistics really gets going when you add the principle of conditionalization:

> If your beliefs are associated with probability function $\mathbb{P}$, and you learn evidence e, then your new beliefs should now be associated with probability function $\mathbb{P}_e$, where for all x, $\mathbb{P}_e(x) = \mathbb{P}(x|e)$. Either by definition or axiom, $\mathbb{P}(x|e) = \dfrac{\mathbb{P}(x \wedge e)}{\mathbb{P}(e)}$.

We call $\mathbb{P}(x)$ the *prior probability* of x because it is the probability of x prior to learning the evidence e. Similarly, $\mathbb{P}_e(x) = \mathbb{P}(x|e)$ is called the *posterior probability* of x.

With the principle of conditionalization, Bayesian statistics provides a simple yet powerful suggestion about belief dynamics: how a person's beliefs should change when he or she learns new things. For instance, suppose we add to the sentences (1)–(4) above:

5. The CEO has decided to reduce expenses.
6. The new probability that we will make a profit is 50%.

When (5) is learned, then (6) describes the new probablitiy of a profit via the conditionalization rule.

3.1.4 Bayesianism for Actuaries

More than a hundred years ago, actuaries observed that when setting insurance premiums, the best ratemaking was obtained when the premium was somewhere between the actual experience of the insured and the overall average for all insureds. The mathematical formulation yields to the credibility theory, which has strong connections with the Bayesian philosophy.

Bayesianism seems a natural fit for actuaries. As claimed in Klugman (1992), "*within the realm of actuarial science there are a number of problems that are particularly suited for Bayesian analysis.*" Actuaries are known for their numerical literacy, so it may be natural for them to quantify the strength of their beliefs via probabilities. Actuaries are valued for their extensive industry knowledge judgment, so their considered beliefs make useful starting points in statistical analyses.

On the other hand, actuaries also prize objectivity. They like to communicate facts and unimpeachable analyses to their employers and clients, and are not content merely because their beliefs are probabilistically consistent. The rest of this chapter will introduce several applications of Bayesian reasoning. The chapter's conclusion will revisit the pros and cons of Bayesianism for actuaries.

3.2 Bayesian Conjugates

As we will see shortly, although Bayesianism has a simple statement, its implementation often requires complex computations. However, this complexity can often be sidestepped using *Bayesian conjugates*, also called *conjugate distributions*.

3.2.1 A Historical Perspective

Pierre-Simon Laplace observed, from 1745 to 1784, 770,941 births: 393,386 boys and 377,555 girls. Assuming the sex of a baby is a Bernoulli random variable, so that p denotes the probability of having a boy, Pierre-Simon Laplace wanted to quantify the credibility of the hypothesis $p \geq 0.5$. His idea was to assume that a priori, p might be uniformly distributed on the unit interval, and then, given statistics observed, to compute the a posteriori probability that $p \geq 0.5$. If N is the number of boys, out of n_0 births, then

$$f(p|N=n) = \frac{\pi(p)\cdot\mathbb{P}(N=n|p)}{\mathbb{P}(N=n)} \propto \binom{n}{n_0} p^n(1-p)^{n_0-n}.$$

Observe that

$$f(p|N=n) \propto \underbrace{\pi(p)}_{\text{prior}}\cdot\underbrace{\mathbb{P}(N=n|p)}_{\text{likelihood}},$$

as mentioned in the introduction. A more general model is to assume that a priori p has a beta distribution, $p \sim \mathcal{B}(\alpha,\beta)$. Then

$$f(p|N=n) \propto p^{\alpha+n}(1-p)^{\beta+n_0-n}$$

which is the density of a beta distribution, $\mathcal{B}(\alpha+n, \beta+n_0-n)$. As a consequence,

$$\mathbb{E}(p|N=n) = \frac{\alpha+n}{\alpha+\beta+n_0} \approx \frac{n}{n_0}$$

when n is large enough, and

$$\text{Var}(p|N=n) = \frac{(\alpha+n)(\beta+n_0-n)}{[\alpha+\beta+n_0]^2(\alpha+\beta+n_0+1)} \approx \frac{n(n_0-n)}{n_0^3} = \frac{1}{n_0}\cdot\frac{n}{n_0}\left(1-\frac{n}{n_0}\right).$$

In this example, the distribution of the variable of interest is in the exponential family (here a binomial distribution), and the a priori distribution is the conjugate distribution (here a beta distribution).

Consider another example, where after discussing with 81 agents, 51 answered positively to some proposition. We would like to test whether this proposition exceeds 2/3, or not. Let $\boldsymbol{X} = (X_1,\cdots,X_n)$ denote the answers of the n agents, X_i is 1 when they answered positively, otherwise 0. Here, $X_i \sim \mathcal{B}(\theta)$, and the Bayesian interpretation of the test is based on the computation of $\mathbb{P}(\theta > 2/3|\boldsymbol{X})$. Consider a priori beta distribution for θ, so that $\theta \sim \mathcal{B}(\alpha,\beta)$. Then

$$\theta|\boldsymbol{X}=\boldsymbol{x} \sim \mathcal{B}\left(\alpha+\sum x_i, \beta+n-\sum x_i\right)$$

To visualize the prior and the posterior distribution, consider the following code:

```
> n <- 81
> sumyes <- 51
> priorposterior <- function(a,b){
+ u <- seq(0,1,length=251)
+ prior <- dbeta(u,a,b)
+ posterior <- dbeta(u,sumyes+a,n-sumyes+b)
+ plot(u,posterior,type="l",lwd=2)
+ lines(u,prior,lty=2)
+ abline(v=(sumyes+a)/(n+a+b),col="grey")}
```

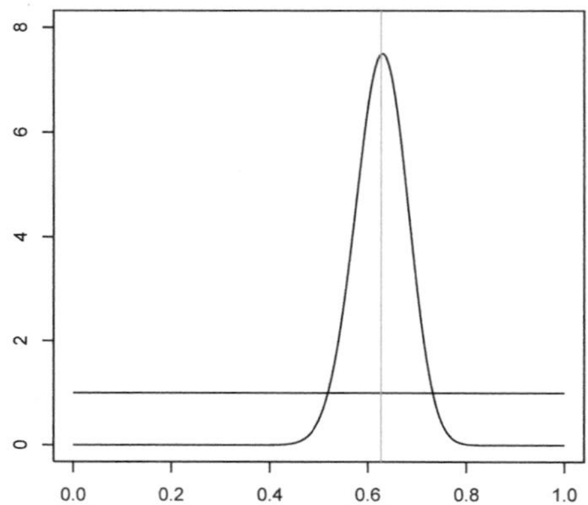

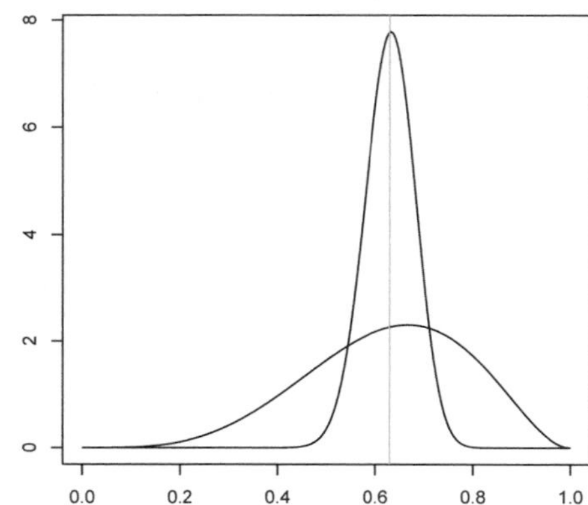

FIGURE 3.1
Prior (dotted line) and posterior (solid line) distributions for the parameter of a Bernoulli distribution.

Assuming a flat prior, $\pi(\theta) \propto 1$, corresponds to a uniform distribution, obtained when $\alpha = \beta = 1$.

```
> priorposterior(1,1)
```

An alternative can be to consider a more informative prior, for instance, $\alpha = 5$ and $\beta = 3$:

```
> priorposterior(5,3)
```

Those distributions can be visualized on Figure 3.1, the dotted line being the prior distribution, and the solid one the posterior. The vertical straight line represents the posterior mean.

If we get back to the initial question, $\mathbb{P}(\theta > 2/3|\boldsymbol{X})$ can be obtained easily as

```
> 1-pbeta(2/3,1+51,1+81-51)
[1] 0.2273979
> 1-pbeta(2/3,5+51,3+81-51)
[1] 0.2351575
```

with the two priors considered.

3.2.2 Motivation on Small Samples

Another simple example can motivate this Bayesian approach. Our company deployed a new product 5 years ago and would like a prospective estimate of the loss ratio for next year. Here is the data available:

- From our feel of the market and knowledge of similar products, we believe that this product's expected loss ratios is around $70\% \pm 10\%$.
- We believe the product's loss ratio will vary $\pm 20\%$ from year to year around its actual mean.

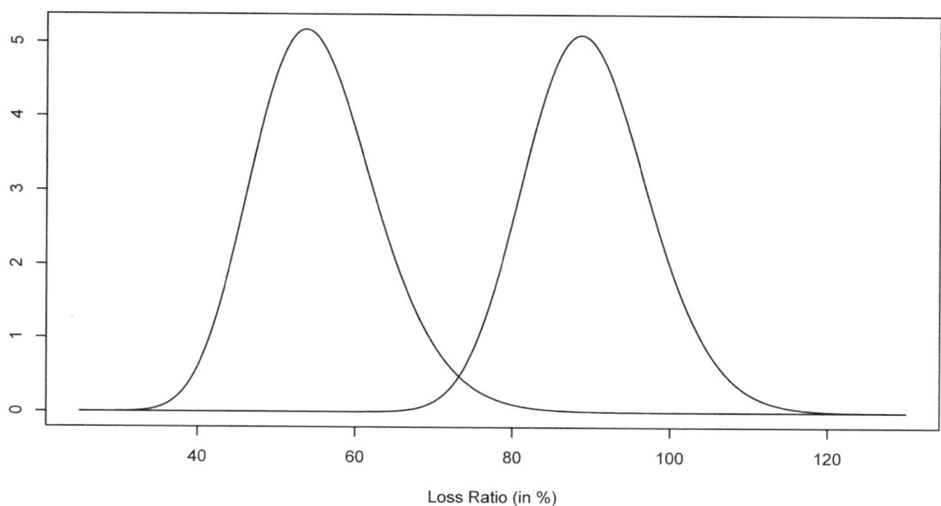

FIGURE 3.2
Prior and posterior distribution of a loss ratio.

- After the last 5 years of losses are trended and developed, and premium on-leveled, the historical loss ratios are 95.8%, 61.4%, 97.7%, 96.1%, and 75.6%.

What could be the distribution of next year's loss ratio?

We can model this problem by assuming that the actual random process yields a lognormally distributed loss ratio L:

$$\log(\mathrm{L}) \sim \mathcal{N}(\mu, \sigma).$$

For σ we represent our beliefs using the constant 0.25, so that no matter what μ is, l is usually between 0.75μ and 1.25μ. If μ is close to the 70% mean, then this implies a typical range of $0.75 \times 70\% = 52.5\%$ to $1.25 \times 70\% = 87.5\%$, which is similar to the $70\% \pm 20\%$ range of our beliefs.

For μ, we need to recognize that the true loss ratio is unknown, but we believe it to be $70\% \pm 10\%$. A simple way to model this belief is by assuming the true median loss ratio, e^μ, is itself lognormally distributed with mean 70% and standard deviation 10%. Then

$$\log(e^\mu) = \mu \sim \mathcal{N}(\mu_0, \sigma_0),$$

where $\mu_0 = -0.367$ and $\sigma_0 = 0.142$. Because μ_0 and σ_0 determine a distribution of μ, and μ itself parametrizes a distribution, μ_0 and σ_0 are *hyperparameters*. Using hyperparameters is common in Bayesian statistics because we frequently have beliefs about what the actual parameters of some process might be.

Using the terminology of the previous section, the equations above determine our prior probability function $\mathbb{P}(x)$. Now we need to conditionalize on our evidence e (the historical loss ratios) to determine our posterior beliefs $\mathbb{P}_e(x) = \mathbb{P}(x|e)$.

To calculate the posterior distribution of μ, we can us *Bayes' theorem*. It has a variety of statements, but this version is common for propositions and discrete variables, where T can be thought of as standing for "theory" and e for "evidence."

$$\mathbb{P}(T|e) = \frac{\mathbb{P}(e|T)\mathbb{P}(T)}{\mathbb{P}(e)} = \frac{\mathbb{P}(e|T)\mathbb{P}(T)}{\Sigma_{i=1}^{n}\mathbb{P}(e|T_i)\mathbb{P}(T_i)}.$$

for continuous variables, the following version is more common. θ can be a scalar or a vector, and usually represents parameters to a theory or distribution.

$$p(\theta|e) = \frac{p(e|\theta)p(\theta)}{p(e)} = \frac{p(e|\theta)p(\theta)}{\int p(e|\phi)p(\phi)d\phi}.$$

All versions of Bayes' theorem follow from basic probably and the definition of conditional probability.

In this case, Bayes' theorem implies that our posterior distribution for μ is given by

$$p(\mu|e) = \frac{p(e|\mu)p(\mu)}{\int p(e|\phi)p(\phi)d\phi},$$

where e is now the five observed historical loss ratios. We can now theoretically solve this equation because $p(e|\mu)$ and $p(e|\phi)$ are given by the lognormal distribution, and $p(\phi)$ is normally distributed.

Luckily, the equation has a simple solution in this case. Because our prior probability for μ is normal, and the likelihood of each loss ratio e (on a log-scale) given μ is also normal, the posterior distribution for μ is also normal, no matter what e is. When the prior and posterior are guaranteed to be in the same family, they are known as conjugate distributions or Bayesian conjugates. There are two other reasons this is convenient:

1. The posterior will remain in the same family as the prior when one data point is observed. Thus, this reasoning can be repeated to show that the posterior stays in the same family after any number of observations.
2. The parameters of the posterior distribution are typically simple functions of the parameters of the prior distribution and the observed data.

Our problem is an example of the *normal/normal conjugate pair.* See Table 3.2 for a table of conjugate pairs and their equations. In the normal/normal case, under the posterior probability distribution,

$$\mu_1 \sim \mathcal{N}\left(\left(\frac{1}{\sigma_0^2}+\frac{5}{\sigma^2}\right)^{-1}\left(\frac{\mu_0}{\sigma_0^2}+\frac{1}{\sigma^2}\sum_{i=1}^{5}\log(r_i)\right),\left(\frac{1}{\sigma_0^2}+\frac{5}{\sigma^2}\right)^{-1}\right),$$

where r_i represents the five observed loss ratios. The posterior mean of μ is $\mu_1 = -0.113$. and the posterior standard deviation is $\sigma_1 = 0.0879$. The situation can be represented graphically in the chart. The observed loss ratios are marked with vertical lines. Here is the set of prior parameters:

```
> sigma <- .25
> prior.mean <- .7
> prior.sd <- .1
> sigma0 <- sqrt(log((prior.sd / prior.mean)^2 + 1))
> mu0 <- log(prior.mean) - prior.sd^2/2
```

Record observations are here:

```
> r <- c(0.958, 0.614, 0.977, 0.921, 0.756)
> n <- length(r)
```

TABLE 3.2
Conjugate priors for distributions in the exponential family.

$X\|\theta$	θ	Conjugate Prior	Prior Parameter	Posterior Parameters
Bernoulli	p	Beta	(α,β)	$(\alpha+\sum x_i, \beta+n-\sum x_i)$
Poisson	λ	Gamma	(α,β)	$(\alpha+\sum x_i, \beta+n)$
Geometric	p	Beta	(α,β)	$(\alpha+n, \beta+\sum x_i)$
Normal	μ	Normal	(μ_0,σ_0^2)	$\left(\frac{\sigma^2\mu_0+\sigma_0^2\sum x_i}{\sigma^2+n\sigma_0^2}, \frac{\sigma^2\cdot\sigma_0^2}{\sigma^2+n\sigma_0^2}\right)$
Normal	σ^2	Inverse gamma	(α,β)	$\left(\alpha+\frac{n}{2}, \beta+\frac{1}{2}\sum(x_i-\mu)^2\right)$
Normal	$\boldsymbol{\mu}$	Normal	$(\boldsymbol{\mu}_0,\boldsymbol{\Sigma}_0)$	$\left(\boldsymbol{\Sigma}_0^{-1}+n\boldsymbol{\Sigma}^{-1}\right)^{-1}\left(\boldsymbol{\Sigma}_0^{-1}\boldsymbol{\mu}_0+n\boldsymbol{\Sigma}^{-1}\bar{\boldsymbol{x}}\right)$, $\left(\boldsymbol{\Sigma}_0^{-1}+n\boldsymbol{\Sigma}^{-1}\right)^{-1}$
Normal	$\boldsymbol{\Sigma}$	Wishart	$(\nu, \boldsymbol{\Psi})$	$\left(n+\nu, \boldsymbol{\Psi}+\sum(\mathbf{x_i}-\boldsymbol{\mu})(\mathbf{x_i}-\boldsymbol{\mu})^\top\right)$
Exponential	λ	Gamma	(α,β)	$\left(\alpha+n, \beta+\sum x_i\right)$
Gamma	β	Gamma	(a_0,b_0)	$\left(a_0+n\alpha, b_0+\sum x_i\right)$

Using previous discussion, we can compute posterior probabilities

```
> sigma1 <- (sigma0^-2 + n / sigma^2)^-.5
> mu1 <- (mu0 / sigma0^-2 + sum(log(r)) / sigma^2) * sigma1^2}
```

Those distribution can be visualized using

```
u <- seq (from = 0.3, to = 1.3, by = 0.01)
> prior <- dlnorm(u,mu0,sigma0)
> posterior <- dlnorm(u,mu1,sigma1)
> plot(u,posterior,type="l",lwd=2)
> lines(u,prior,lty=2)
```

We conclude this example with a few observations:

- Our distribution around the true loss ratio μ increased after observing the 5 years of data. This was to be expected because the observed loss ratios were, in general, larger than our prior mean of 70%.
- The posterior distribution of the true loss ratio is narrower (has a smaller standard deviation). This was also to be expected—after observing the data, we know more, and thus our uncertainty about the true loss ratio has decreased.
- The conceptual distinction between the *parameter* and *predictive* distributions is quite important. The parameter distributions are distributions over the parameters to other distributions (μ in this case). If we want to predict new observations, we must use the predictive distribution. The parameter distribution only quantifies parameter risk.
- The predictive distributions are significantly more uncertain than the parameter distributions. This is also quite intuitive—much of the variability in insurance losses is contributed by process risk.

- In this case, our posterior predictive distribution is actually more uncertain than our prior predictive distribution (32.9% versus 30.5%). Although there is less parameter risk, there is more process risk because of the larger expected loss ratio.

3.2.3 Black Swans and Bayesian Methodology

The use of Bayes's approach is interesting when you have no experience, at all. Consider the following simple case. Consider some Bernoulli sample, $\{0, 0, 0, 0, 0\}$. What could you say about the probability parameter ? As a frequentist, nothing. This was actually the question asked to Longley-Cook in the 1950s: is it possible to predict the probability of two planes colliding in midair? There had never been any (serious) collision of commercial planes by that time. Without any past experience, statisticians could not answer that question.

Assume that X_i's are i.i.d. $\mathcal{B}(p)$ variables. With a beta prior for p, with parameters α and β, then p given $\boldsymbol{X}$ has a beta distribution with parameters α and $\beta + 5$. Thus, with a flat prior ($\alpha = \beta = 1$),

```
> qbeta(.95,1,6)
[1] 0.3930378
```

the 95% confidence interval for p is $[0\%; 40\%]$. With Jeffrey's non-informative prior, (see Section 3.2.6.)

```
> qbeta(.95,.5,.5+5)
[1] 0.3057456
```

the 95% confidence interval for p is $[0\%; 30\%]$. It is possible to visualize the posterior transformation, with information coming, on Figure 3.3. For the uniform prior,

```
> pmat =persp(0:5,0:1,matrix(0,6,2),zlim=c(0,5), ticktype=
+ "detailed",box=FALSE,theta=-30)
> title("Uniform prior",cex.main=.9)
> y=seq(0,1,by=.01)
```

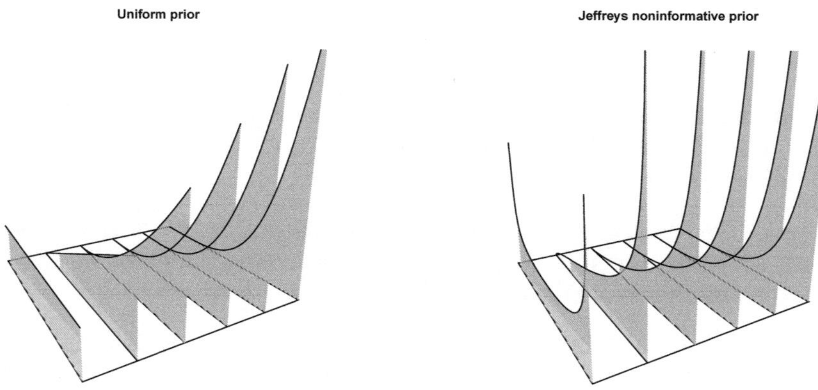

FIGURE 3.3
Distortion of the posterior distribution, with information coming.

```
> for(k in 0:5){
+ z=dbeta(y,1,1+k)
+ indx=which(y<=qbeta(.95,1,1+k))
+ xy3d=trans3d(rep(k,2*length(indx)),c(rev(y[indx]),y[indx]),
+ c(rep(0,length(y[indx])),z[indx]),pmat)
+ polygon(xy3d,col="grey",density=50,border=NA,angle=-50)
+ xy3d=trans3d(rep(k,length(y)),y,z,pmat)
+ lines(xy3d,lwd=.5)
+ xy3d=trans3d(rep(k,length(indx)),y[indx],z[indx],pmat)
+ lines(xy3d,lwd=3)}
```

To get back to the airplane collision, in 1955, Longley-Cook predicted "*anything from 0 to 4 [...] collisions over the next ten years,*" without any experience at all (so purchasing reinsurance might be a good idea). Two years later, 128 people died over the Grand Canyon, and 4 years after that, more than 133 people died over New York City.

3.2.4 Bayesian Models in Portfolio Management and Finance

Following Frost & Savarino (1986), it is possible to use a Bayesian methodology to discuss portfolio selection (see Chapter 13 for more details). Consider one assset, and a monthly sequence of returns for a given asset (a time series, see Chapter 11), $\boldsymbol{r} = (r_1, \dots, r_T)$. Let μ and σ denote the mean and the volatility of that return. Assume that returns are independent draws from a Gaussian distribution, $f(\boldsymbol{r}|\mu)$. Consider the conjugate prior, $\mu \sim \mathcal{N}(\mu_0, \tau_0)$. Then the posterior distribution is Gaussian, $\mathcal{N}(\mu_T, \tau_T)$, where

$$\mu_T = \left(\frac{\mu_0}{\tau_T^2} + \frac{T}{s^2}\frac{1}{T}\sum_{t=1}^{T} r_i\right)\left(\frac{1}{\tau_T^2} + \frac{T}{s^2}\right)^{-1},$$

where

$$\tau_T^2 = \left(\frac{1}{\tau_T^2} + \frac{T}{s^2}\right)^{-1}.$$

The posterior mean μ_T is here a weighted average of the priori mean μ_0 and the historical average Observe that the less volatile the information, the greater its precision. And as more data become available, the prior information becomes increasingly irrelevant (and the posterior mean approaches the historical average).

Of course, it is possible to extend that model in higher dimension, with multiple assets. Consider a series of monthly returns, $(\boldsymbol{r}_1, \dots, \boldsymbol{r}_T)$. Let $\boldsymbol{\mu}$ and $\boldsymbol{\Sigma}$ denote the vector of means, and the volatility matrix, of those returns. Assume that returns are independent draws from a Gaussian distribution, $f(\boldsymbol{r}|\mu)$. Consider the conjugate prior for the means $\boldsymbol{\mu} \sim \mathcal{N}(\boldsymbol{\mu}_0, \boldsymbol{\tau}_0)$. Here, $\boldsymbol{\tau}_0$ reflects our priors on the covariance of the means, not the covariance of the asset returns. Actually, we can also consider priors on the covariance of the returns, too. The prior distribution of the covariance matrix has an inverse Wishart distribution, with ν degrees of freedom, $\boldsymbol{\Sigma} \sim IW(\nu, \boldsymbol{\Sigma}_0)$. Then the posterior distribution is Gaussian, $\mathcal{N}(\boldsymbol{\mu}_T, \boldsymbol{\Sigma}_T)$ where

$$\boldsymbol{\mu}_T = \left(\boldsymbol{\tau}_0^{-1} + T\hat{\boldsymbol{\Sigma}}^{-1}\right)^{-1}\left(\boldsymbol{\tau}_0^{-1}\boldsymbol{\mu}_0 + \hat{\boldsymbol{\Sigma}}^{-1}T\hat{\boldsymbol{\mu}}\right)$$

and

$$\boldsymbol{\Sigma}_T = \frac{T+1+\nu}{T+\nu-d-2}\left(\frac{\nu}{\nu+T}\boldsymbol{\Sigma}_0 + \frac{T}{\nu+T}\hat{\boldsymbol{\Sigma}}\right),$$

where d is the number of assets considered. Posterior means and covariances are linear mixtures between the prior and the empirical (historical) estimates. If we have non-informative prior $\nu = 0$ and $\boldsymbol{\tau}_0^{-1} = \mathbf{0}$, then

$$\boldsymbol{\mu}_T = \hat{\boldsymbol{\mu}} \text{ and } \boldsymbol{\Sigma}_T = \frac{T+1}{T-d-2}\hat{\boldsymbol{\Sigma}}.$$

An extension is the so-called Black–Litterman model, introduced in Black & Litterman (1992) and discussed in Satchell & Scowcroft (2000). This model is popular among actuaries (and more generally investors) because it allows one to distinguish forecasts from conviction (or beliefs) about uncertainty associated to the forecast. This model will be discussed in Chapter 13.

3.2.5 Relation to Bühlmann Credibility

In the credibility framework, we assume that insurance portfolios are heterogeneous. There is an underlying (and non-observable) risk factor θ, for all insured, and we would like to model individual claim frequency for ratemaking issues. Let $Y_{i,t}$ denote the number of claims for year t and insurer i. In Chapters 14 and 15, the goal will be to use covariates $\boldsymbol{X}_{i,t}$ as a proxy of θ_i and to use $\mathbb{E}(Y_{i,t+1}|\boldsymbol{X}_{i,t})$ as an approximation for $\mathbb{E}(Y_{i,t+1}|\theta_i)$. Here, instead of using covariates, we would like to use past historical observations. Namely, use $\mathbb{E}(Y_{i,t+1}|\boldsymbol{Y}_{i,t})$ before as an approximation for $\mathbb{E}(Y_{i,t+1}|\theta_i)$, where $\boldsymbol{Y}_{i,t} = (Y_{i,t}, Y_{i,t-1}, Y_{i,t-2}, \ldots)$.

Consider a contract, observed during T years. Let Y_t denote the number of claims, and assume that $Y_t \in \{0,1\}$, so that $Y_t \sim \mathcal{B}(p)$. From the previous section, if we assume a priori beta distribution $\mathcal{B}(\alpha,\beta)$ for p, then

$$\mathbb{E}(p|\boldsymbol{Y}) = \frac{\alpha + T\bar{Y}_T}{\alpha+\beta+T} = \underbrace{\frac{T}{\alpha+\beta+T}}_{Z}\bar{Y}_T + \underbrace{\left(1 - \frac{T}{\alpha+\beta+T}\right)}_{1-Z}\underbrace{\frac{\alpha}{\alpha+\beta}}_{\mathbb{E}(p)},$$

where $Z \in [0,1]$. Because

$$\mathbb{E}(Y_t|p) = p \text{ and } \text{Var}(Y_t|p) = p(1-p),$$

then

$$\begin{aligned}\frac{\mathbb{E}(\text{Var}(Y_t|p))}{\text{Var}(\mathbb{E}(Y_t|p))} &= \frac{\mathbb{E}(p(1-p))}{\text{Var}(p)} \\ &= \frac{\mathbb{E}(p) - [\text{Var}(p) + \mathbb{E}(p)^2]}{\text{Var}(p)} = \alpha + \beta.\end{aligned}$$

Thus,

$$Z = \frac{T}{\alpha+\beta+T} = \frac{T}{k+T} \text{ with } k = \frac{\mathbb{E}(\text{Var}(Y_t|p))}{\text{Var}(\mathbb{E}(Y_t|p))}.$$

This will be known as the Bühlmann credibility approach: If insured i was observed T years, with past experience $\boldsymbol{Y}_T = (Y_1, \ldots, Y_T)$, then $\mathbb{E}(Y_{T+1}|\boldsymbol{Y}_T)$ can be approximated by

$$Z\overline{Y}_{i,T} + (1-Z)\mu, \text{ where } \overline{Y}_T = \frac{1}{T}\sum_{t=1}^{T} Y_{i,t},$$

and μ is the average on the whole portfolio. Z is then the share of the premium that is based on past information for a given insured, and is related to the amount of credibility we give to past information. From Bühlmann (1969), inspired by the Bayesian approach described previously, Z satisfies

$$Z = \frac{T}{T+k} \text{ where } k = \frac{\mathbb{E}(\text{Var}(Y_t|\theta))}{\text{Var}(\mathbb{E}(Y_t|\theta))}.$$

As mentioned in Bühlmann & Gisler (2005), in a Bayesian model, we wish to compute $\mathbb{E}(Y_{T+1}|\boldsymbol{Y}_T)$, but here, we restrict ourselves to projections on linear combinations of past experience. We can actually prove that this Bühlmann credibility estimator is the best linear least-squares approximation to the Bayesian credibility estimator. And when $X_{i,t}$'s are i.i.d. random variables in the exponential family, and if the prior distribution of θ is conjugate to this family, then Bühlmann credibility estimator is equal to the Bayesian credibility estimator. This was the case previously, where $X_{i,t}$'s were Bernoulli and θ has a beta prior. Another popular example is obtained when $X_{i,t}$'s are Poisson and θ has a gamma prior.

In order to illustrate Bühlmann credibility, consider claim counts, from Norberg (1979), $Y_{i,t}$, for contract $i = 1, \ldots, m$ and time $t = 1, \ldots, T$.

```
> norberg
      year00 year01 year02 year03 year04 year05 year06 year07 year08 year09
 [1,]      0      0      0      0      0      0      0      0      0      0
 [2,]      0      0      0      0      0      0      0      0      0      0
 [3,]      1      0      1      0      0      0      0      0      0      0
 [4,]      0      0      0      0      0      0      0      0      0      0
 [5,]      0      0      0      0      0      1      0      0      1      0
 [6,]      0      0      0      0      0      0      0      0      0      0
 [7,]      0      1      1      0      0      0      0      0      0      0
 [8,]      0      0      0      0      0      0      0      0      0      0
 [9,]      0      1      1      0      1      1      1      0      0      1
[10,]      0      0      1      0      0      0      0      0      0      0
[11,]      1      1      0      0      1      0      0      0      1      0
[12,]      0      0      0      0      0      1      0      1      0      1
[13,]      0      0      0      0      0      0      0      0      1      0
[14,]      0      0      0      0      0      0      0      1      0      0
[15,]      0      0      0      0      0      0      0      0      0      0
[16,]      0      0      0      0      0      0      0      0      0      0
[17,]      1      1      0      1      0      0      1      0      0      1
[18,]      1      0      0      0      0      0      0      0      0      0
[19,]      0      0      0      0      0      0      0      0      0      1
[20,]      0      0      0      0      0      0      0      0      0      0
```

For each contract, we wish to predict the expected value $X_{i,T+1}$ given past observations. Bühlman's estimate is

$$Z \cdot \bar{Y}_i + (1 - Z)\bar{Y}, \text{ with } \frac{T}{T+k},$$

where

$$\bar{Y}_i = \sum_{t=1}^{T} Y_{i,t} \text{ while } \bar{Y} = \sum_{i=1}^{m} \bar{Y}_i,$$

and where $k = s^2/a$,

$$s^2 = \frac{1}{m(T-1)} \sum_{i=1}^{m} \sum_{t=1}^{T} (X_{i,t} - \bar{X}_i)^2,$$

and

$$a = \frac{1}{m-1} \sum_{i=1}^{m} (\bar{X}_i - \bar{X})^2 - \frac{1}{n} s^2;$$

see Herzog (1996) for a discussion of those estimators. The code to compute Bühlman's estimate for year `11` will be:

```
> T<- ncol(norberg)
> (m <- mean(norberg))
[1] 0.145
> (s2 <- mean(apply(norberg,1,var)))
[1] 0.1038889
> (a <- var(apply(norberg,1,mean))-s2/T)
[1] 0.02169006
```

Bühlman's credibility factor is based on

```
> (Z <- T/(T+s2/a))
[1] 0.6761462
```

and Bühlman's estimates are

```
> Z*apply(norberg,1,mean)+(1-Z)*m
 [1] 0.0469588 0.0469588 0.1821880 0.0469588 0.1821880
 [6] 0.0469588 0.1821880 0.0469588 0.4526465 0.1145734
[11] 0.3174173 0.2498027 0.1145734 0.1145734 0.0469588
[16] 0.0469588 0.3850319 0.1145734 0.1145734 0.0469588
```

3.2.6 Noninformative Priors

In the first part of this section, we will use conjugate priors because of computational ease. As we will see in the next section, not having simple analytical expression is not a big deal. So it can be interesting to have more credible priors, even if it is not a simple problem, as discussed formally in Berger et al. (2009).

A simple example to start with can be the case where X given θ has a Bernoulli distribution. As we have seen previously, on births, the idea of Pierre Simon Laplace was to assume a uniform prior distribution, $\pi(\theta) = 1$. Assuming $\pi(\theta) \propto 1$ (for more general distribution) will be called a flat prior. If Pierre-Simon Laplace thought it would be as neutral as possible, or non-informative, this is actually not the case. A more modern idea of formalizing the idea of non-informative priors (see Yang & Berger (1998) for a discussion) is that we should get an equivalent result when considering a transformed parameter. Given some one-to-one transformation $h()$, so that the parameter is no longer θ, but $\tau = h(\theta)$, we want here to have invariant distributions. More formally, recall that the density of τ is here

$$p(\tau) = \pi(\theta) \cdot \left| \frac{\partial \theta}{\partial \tau} \right| . \tag{3.1}$$

Recall that this Jacobian has an interpretation in terms of information, as Fisher information of τ is

$$I(\tau) = -\mathbb{E}\left(\frac{\partial^2 \log f(x|\tau)}{\partial \tau^2} \right) = I(\theta) \cdot \left| \frac{\partial \theta}{\partial \tau} \right|^2 .$$

Equation (3.1) can then be written as $p(\tau)\sqrt{I(\tau)} = \pi(\theta)\sqrt{I(\theta)}$. Thus, it is natural to have a prior density which is proportional to the square root of Fisher information,

$$\pi(\theta) \propto \sqrt{I(\theta)}.$$

This is called Jeffrey's principle. For the Poisson distribution, it means that $\pi(\theta) \propto \theta^{-\frac{1}{2}}$, and if we get back to the initial problem, on births, the non-informative prior will be a beta distribution with parameters 1/2, as

$$\pi(\theta) \propto \sqrt{\theta \left(\frac{1}{\theta} - \frac{0}{1-\theta}\right)^2 + [1-\theta]\left(\frac{0}{\theta} - \frac{1}{1-\theta}\right)^2} = \frac{1}{\sqrt{\theta(1-\theta)}}.$$

3.3 Computational Considerations

In many applications, we wish to compute

$$\mathbb{E}(g(\theta)|x) = \frac{\int g(\theta)\pi^\star(\theta|x)d\theta}{\int \pi^\star(\theta|x)d\theta},$$

where $\pi^\star$ is proportional to the posterior density. One possible method is to use the Gauss-Hermite approximation, discussed in Klugman (1992) in detail. An alternative is based on Monte Carlo integration (see Chapter 1 for an introduction).

3.3.1 Curse of Dimensionality

Unfortunately, aside from Bayesian conjugates, most Bayesian problems are hard to compute. The problem is that the posterior distribution is often continuous and multidimensional and has the format

$$\frac{p(e|\boldsymbol{\theta})p(\boldsymbol{\theta})}{\int p(e|\boldsymbol{\phi})p(\boldsymbol{\phi})d\boldsymbol{\phi}}$$

for n-dimensional vector $\boldsymbol{\theta}$. Although $p(e|\boldsymbol{\theta})p(\boldsymbol{\theta})$ are often easy to compute, the normalizing constant $\int p(e|\boldsymbol{\phi})p(\boldsymbol{\phi})d\boldsymbol{\phi}$ is an n-dimensional integral. Typically, unless conjugate distributions are used, this integral is not analytically soluble and must be approximated numerically.

Here, Bayesian statistics suffer from the *curse of dimensionality*: the rapid increase in computational difficulty of integration as the number of dimensions increases.

In general, if we need to integrate a function over the d-dimensional unit hypercube $[0,1]^d$, then this thinking suggests that we will need about 100^d evaluations. Thus, the number of function [calls] will increase quickly with dimensionality. Computers are much faster than they used to be, but even moderately small problems become computationally infeasible.

There are numerical integration algorithms such as Gaussian quadrature which can reduce the number of points required somewhat, but their error still decreases at the slow rate of $O(d^{-\frac{1}{n}})$. To make matters worse, Bayesian posterior probability distributions are often "spiky" (i.e. they have large areas with low probability and relatively small areas with lots of probability) like the distributions shown in Figure 3.1.

From a computational point of view, our interest is to approximate the value of an integral, which is an area for an $\mathbb{R} \to \mathbb{R}$ function, or a volume for an $\mathbb{R}^d \to \mathbb{R}$ function. A simple idea is to use a box, around the volume, to partition the box, and to count the proportion of points in the considered volume. For instance, to compute the area of a disk with radius 1, consider a box $[-1,+1] \times [-1,+1]$, and compute the proportion of points on a grid in the box that belongs to the disk,

```
+ diskincube <- function(n){
+ gridn <- seq(-1,+1,length=n)
+ gridcube <- expand.grid(gridn,gridn)
+ inthedisk <- apply(gridcube^2,1,sum)<=1
+ mean(inthedisk)
+ }
> diskincube(200)
[1] 0.7766
```

Here, 77.66% of the points in the square belong to the disk, and because the area of the cube is 2^2, it means that the area of the disk should be close to

```
> diskincube(200)*2^2
[1] 3.1064
> diskincube(2000)*2^2
[1] 3.138388
```

(the true value is π). Note that we always underestimate the true value with this technique. This method can be extended to higher dimensions. For instance, in dimension 4,

```
> diskincube.dim4 <- function(n){
+ gridn <- seq(-1,+1,length=n)
+ gridcube <- expand.grid(gridn,gridn,gridn,gridn)
+ inthedisk <- apply(gridcube^2,1,sum)<=1
+ mean(inthedisk)
+ }
```

The volume of the unit sphere in dimension 4 is larger than

```
> diskincube.dim4(40)*2^4
[1] 4.4488
```

(the true value is $\pi^2/2 \sim 4.9348$)

More formally, we wish to estimate a volume $\mathcal{V} \subset \mathbb{R}^d$, using $m = k^d$ points, for some integer k. If $\mathcal{V} \in [\boldsymbol{a}, \boldsymbol{b}]$, consider for any i a uniform partition of $[a_i, b_i]$ in k segments. Thus, define $x_{i,j} = a_i + (j-1)(b_i - a_i)/(k-1)$, for all $j = 1, ..., k$. An approximation of volume $\mathcal{V}$ is then

$$V(\mathcal{V}) \sim \left(\prod_{i=1}^{d}(b_i - a_i)\right) \cdot \frac{1}{n} \sum_{j_1,\ldots,j_d \in\{1,\ldots,k\}} \mathbf{1}((x_{1,j_1}, ..., x_{d,j_k}) \in \mathcal{V}).$$

Let $V_n(\mathcal{V})$ denote this approximation. If $L(\mathcal{V})$ is the length of the contour of the volume, in dimension 2, then

$$|V_n(\mathcal{V}) - V(\mathcal{V})| \leq \frac{L(\mathcal{V})}{k} = \frac{L(\mathcal{V})}{n^{1/d}}.$$

The point here is that the approximation error is $O(1/n^{1/d})$, which will increase with dimension.

This is more or less what we do when computing integrals,

$$\int_{[0,1]} h(u)du \sim \sum_{j=1}^{n} \frac{1}{n} \cdot h\left(\frac{j}{n}\right)$$

and in higher dimension,

$$\int_{[0,1]^d} h(\boldsymbol{u})d\boldsymbol{u} \sim \sum_{j_1,\ldots,j_d \in\{1,\ldots,k\}} \left(\frac{1}{k}\right)^d \cdot h\left(\frac{j_1}{k}, ..., \frac{j_d}{k}\right).$$

In dimension 1, we can prove that

$$\left| \int_0^1 h(u)du - \sum_{j=1}^{n} \frac{1}{n} \cdot h\left(\frac{j}{n}\right) \right| \leq \frac{1}{2n} \|f'\|_\infty.$$

and in higher dimension,

$$\left| \int_{[0,1]^d} h(\boldsymbol{u})d\boldsymbol{u} - \sum_{j_1,\dots,j_d} \left(\frac{1}{k}\right)^d \cdot h\left(\frac{j_1}{k}, \dots, \frac{j_d}{k}\right) \right| \leq \frac{1}{n^d} \|\partial^d f\|_\infty.$$

Of course, this numerical algorithm can be improved, using trapezoids instead of rectangles, or Simpson's polynomial interpolation. But the order will remain unchanged, and the error is of order $O(n^{-1/d})$. This is—more formally—what is called the curse of dimensionality.

3.3.2 Monte Carlo Integration

Now, if we get back to the initial example, it is possible to see what Monte Carlo methods are: Instead of considering a homogeneous grid, one idea can simply be to randomly draw n points, uniformly, in the box. An approximation of the volume will be

$$V(\mathcal{V}) \sim \left(\prod_{i=1}^{d} (b_i - a_i)\right) \times \frac{1}{n} \cdot \sum_{i=1}^{n} \mathbf{1}(\boldsymbol{X}_i \in \mathcal{V}).$$

Let $\tilde{V}_n(\mathcal{V})$ denote this approximation. To discuss and quantify errors, observe the fact that a point is, or is not, inside volume $\mathcal{V}$ can be modeled using Bernoulli random variables. It is then possible to prove that $\mathbb{E}(V_n(\mathcal{V})) = V(\mathcal{V})$ and

$$\text{Var}(V_n(\mathcal{V})) = \frac{1}{n} \times \frac{V(\mathcal{V})}{\prod_{i=1}^{d} (b_i - a_i)} \times \left(1 - \frac{V(\mathcal{V})}{\prod_{i=1}^{d} (b_i - a_i)}\right).$$

Now, this idea can be used to approximate integrals, as

$$\int_{[0,1]^d} h(\boldsymbol{u})d\boldsymbol{u} = \mathbb{E}(h(\boldsymbol{U})) \text{ where } \boldsymbol{U} \text{ has independent components with } U_i \sim \mathcal{U}([0,1]),$$

and from the law of large numbers

$$\frac{1}{n} \sum_{i=1}^{n} h(\boldsymbol{U}_i) \overset{a.s.}{\to} \mathbb{E}(h(\boldsymbol{U})).$$

To quantify uncertainty, define

$$V_n = \frac{n}{n-1} \left(\frac{1}{n} \sum_{i=1}^{n} h(\boldsymbol{U}_i)^2 - \left(\frac{1}{n} \sum_{i=1}^{n} h(\boldsymbol{U}_i) \right)^2 \right),$$

and from the central limit theorem, the error (in a more probabilistic way) is then of order $O(1/\sqrt{n})$. Note that the order does not depend on the dimension d here (actually, it does, in the constant term). Thus, Monte Carlo simulations can be more interesting in high dimension than deterministic methods.

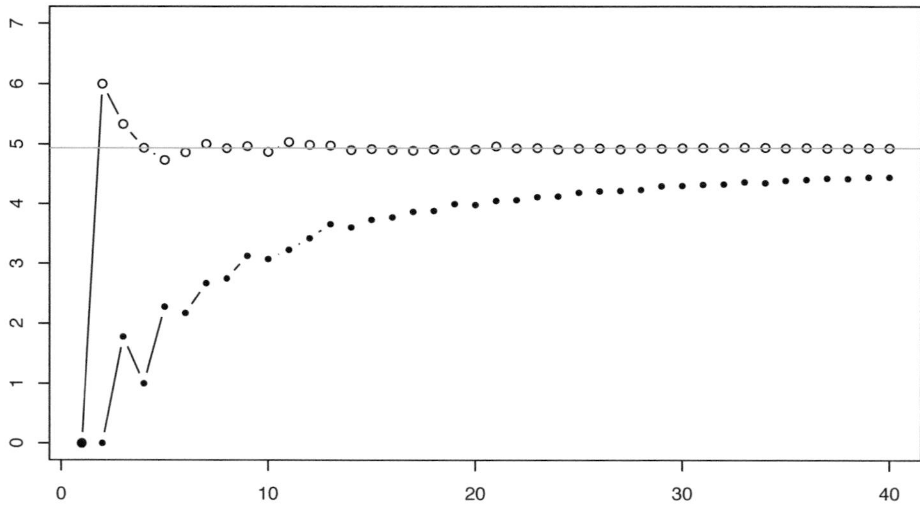

FIGURE 3.4
Curse of dimensionality, computing the volume of the unit sphere in $\mathbb{R}^4$, with a deterministic method • and using Monte Carlo simulations ○, as a function of the number of points, $n = 4^k$.

```
> sim.diskincube.dim4(40^4)*2^4
[1] 4.943763
```

(the true value is $\pi^2/2 \sim 4.9348$). Recall that in using determinstic methods, we were still far from the true value

```
> diskincube.dim4(40)*2^4
[1] 4.4488
```

To compare the deterministic algorithm and the Monte Carlo approach, we can plot the approximations obtained using the two algorithms (see Figure 3.4):

```
> v <- 1:40
> plot(v,2^4*Vectorize(sim.diskincube.dim4)(v^4),type="b",)
> lines(v,2^4*Vectorize(diskincube.dim4)(v),type="b",pch=19,cex=.6,lty=2)
> abline(h=pi^2/2,col="grey")
```

3.3.3 Markov Chain Monte Carlo

Currently the most popular way in Bayesian statistics to sidestep the curse of dimensionality is to use Markov Chain Monte Carlo (MCMC) techniques. Interestingly, the first MCMC technique, the Metropolis algorithm, was developed by the eponymous physicist in 1953. It was run on some of the very first computers.

Despite this early beginning, the Metropolis algorithm and related MCMC techniques only gained popularity in the statistical community in the 1990s. The resurgence of Bayesian statistics since then owes much to the effectiveness and practicality of these algorithms.

MCMC techniques do not even attempt to calculate the integral $\int p(e|\boldsymbol{\phi})p(\boldsymbol{\phi})d\boldsymbol{\phi}$. Rather, they provide samples of $\boldsymbol{\theta}$ from the posterior distribution of $p(\boldsymbol{\theta}|e)$. MCMC techniques are Monte Carlo, meaning that they depend on random sampling, but they also produce Markov chains, meaning that the samples they produce are not independent. Each sample depends on the previous one.

The Metropolis algorithm uses the insight that we do not need to compute $\int p(e|\boldsymbol{\phi})p(\boldsymbol{\phi})d\boldsymbol{\phi}$ to get samples; rather, we can work with the unnormalized probability distribution $p(e|\boldsymbol{\theta})p(\boldsymbol{\theta})$, which is usually easy to compute. The algorithm can be informally summarized like this:

1. Start with an arbitrary value in the posterior distribution $\boldsymbol{\theta}_0$.
2. Given an existing sample $\boldsymbol{\theta}_i$, randomly choose a "nearby" candidate sample $\boldsymbol{\theta}'$. This is done by sampling from another probability distribution $Q(\cdot|\boldsymbol{\theta}_i)$. Often, Q is chosen to be multinormal, so that $\boldsymbol{\theta}' \sim \mathcal{N}(\boldsymbol{\theta}_i, \Sigma)$, for some fixed Σ.
3. Compare $p(e|\boldsymbol{\theta}_i)p(\boldsymbol{\theta}_i)$ and $p(e|\boldsymbol{\theta}')p(\boldsymbol{\theta}')$ by computing the *acceptance ratio*

$$R = \frac{p(e|\boldsymbol{\theta}')p(\boldsymbol{\theta}')}{p(e|\boldsymbol{\theta}_i)p(\boldsymbol{\theta}_i)}.$$

4. Move to $\boldsymbol{\theta}'$ if R is big enough: with probability $\min(1, R)$ set $\boldsymbol{\theta}_{i+1} = \boldsymbol{\theta}'$ (this is called accepting $\boldsymbol{\theta}'$); with probability $1 - \min(1, R)$ set $\boldsymbol{\theta}_{i+1} = \boldsymbol{\theta}_i$ (this is called rejecting $\boldsymbol{\theta}'$).
5. Go to step (2) and continue until enough samples are drawn.

This chapter will not prove that the samples thus drawn can be considered samples from the posterior distribution $p(\cdot|e)$. However, it is hopefully clear from the algorithm that it will spend more time where $p(e|\boldsymbol{\theta}')p(\boldsymbol{\theta}')$ is large and less time where it is small. This is because the acceptance ratio will always be 1 when the candidate moves us into an area of higher probability. When the candidate sample moves us into an area of lower probability, the acceptance ratio will be less than 1 and we may just stay where we are. Because $p(e|\boldsymbol{\theta}')p(\boldsymbol{\theta}')$ is large in exactly the same places that the posterior distribution

$$p(\boldsymbol{\theta}'|e) = \frac{p(e|\boldsymbol{\theta}')p(\boldsymbol{\theta}')}{\int p(e|\boldsymbol{\phi})p(\boldsymbol{\phi})d\boldsymbol{\phi}}$$

is large, more samples will be drawn from the areas of probability space where the posterior density is large.

Once we have samples from the posterior distribution, we can estimate the answer to most questions that come up in Bayesian statistical problems. For instance, suppose we have n samples of $\boldsymbol{\theta}$ and want to know the expected value of the first parameter θ_0. Then we can use classical statistics to approximate it:

$$\mathbb{E}(\theta_0) \approx \frac{1}{n}\sum_{i=1}^{n} \theta_{0,i}.$$

It is ironic but many or most Bayesian techniques in the end depend on classical statistics to understand the behavior produced by MCMC chains.

3.3.4 MCMC Example in R

Let us examine a simple Bayesian problem that cannot be solved using Bayesian conjugates, but can be quickly analyzed with an MCMC technique. Recall the example in Section 3.2. The goal was to find the posterior mean and standard deviation (parameter risk) of the expected loss ratio. We assumed there that the process risk was lognormally distributed with fixed $\sigma = 0.25$, thus overlooking an opportunity to learn about our process risk from the data.

This time let us use the same numbers but model uncertainty about σ. We will again assume that the process risk of actual loss ratios is lognormally distributed, with parameters μ and σ. The prior distribution of μ will again be normally distributed with the same parameters μ_0 and σ_0. However, now our prior distribution of σ will be inverse gamma with parameters k_0 and θ_0 (σ cannot be negative, so the inverse gamma distribution is commonly used to model a prior variance distribution). We can set k_0 and θ_0 using the method of moments.

```
> prior.mean.mean <- .7
> prior.mean.sd <- .1
> prior.sd.mean <- .25
> prior.sd.sd <- .1
```

Using the fact that for a lognormal distribution, the coefficient of variation is $\sqrt{(}e^{s^2} - 1)$,

```
> sigma0 <- sqrt(log((prior.mean.sd / prior.mean.mean)^2 + 1))
> mu0 <- log(prior.mean.mean) - sigma0^2 / 2
```

and for an inverse Gamma distribution, $E(X) = \dfrac{\Theta}{(k-1)}$ and the coefficient of variation is $\dfrac{1}{\sqrt{k-2}}$,

```
> k0 <- 2 + (prior.sd.mean / prior.sd.sd)^2
> theta0 <- (k0 - 1) * prior.sd.mean
```

Record observations are here

```
> r <- c(0.958, 0.614, 0.977, 0.921, 0.756)
> r.log <- log(r)
> n <- length(r)
```

In order to use the `dinvgamma()` function, use

```
> library{MCMCpack}
```

The code to run MCMC simulations is here

```
> RunSim <- function(M, delta, mu, sigma) {
+    output.df <- data.frame(mu=rep(NA, M), sigma=NA)
+    set.seed(0)
+    cur.prior.log <- (dnorm(mu, mu0, sigma0, log=TRUE)
+                      + log(dinvgamma(sigma, k0, theta0)))
+    cur.like.log <- sum(dnorm(r.log, mu, sigma, log=TRUE))
+    for (i in seq_len(M)) {
+        mu.cand <- mu + rnorm(1, sd=delta)
+        sigma.cand <- max(1e-5, sigma + rnorm(1, sd=delta/2))
```

```
+         cand.prior.log <- (dnorm(mu.cand, mu0, sigma0, log=TRUE)
+                            + log(dinvgamma(sigma.cand, k0, theta0)))
+         cand.like.log <- sum(dnorm(r.log, mu.cand, sigma.cand, log=TRUE))
+         cand.ratio <- exp(cand.prior.log + cand.like.log
+                           - cur.prior.log - cur.like.log)
+         if (runif(1) < cand.ratio) {
+             mu <- mu.cand
+             sigma <- sigma.cand
+             cur.prior.log <- cand.prior.log
+             cur.like.log <- cand.like.log
+         }
+         output.df[i, ] <- c(mu, sigma) # write one sample to output
+     }
+     return(output.df)
+ }

> delta005.df <- RunSim(6000, .005, log(1.2), 0.5)
> delta20.df  <- RunSim(6000, .20, log(1.2), 0.5)
> delta80.df  <- RunSim(6000, .80, log(1.2), 0.5)
```

The comments in the code largely explain what is happening. Lines here compute the acceptance ratio

$$R = \frac{p(e|\boldsymbol{\theta}')p(\boldsymbol{\theta}')}{p(e|\boldsymbol{\theta}_i)p(\boldsymbol{\theta}_i)},$$

and line 49 updates the current values for μ and σ by moving only if the acceptance ratio is high enough. Note that the step size `delta` (δ), the number of samples to run `M`, and the starting values are purely computational decisions—they are not determined by the probabilistic structure of the problem—so above we have made them functional inputs.

The last few lines run three simulations, each with 6,000 samples. The three use the same arbitrarily chosen inputs; the difference between the three is the step size δ. Examining the three output dataframes, we find that the acceptance ratio when $\delta = .005$ is about 95%; when $\delta = 0.10$, it is 26%, and when $\delta = 0.80$, it is only 2.4%. The following graphs in Figure 3.5 of the first samples from each simulation illustrate the phenomenon. These are sometimes called *traceplots*. They can be obtained using

```
> library(coda)
> traceplot(mcmc(delta20.df[1001:6000, ]))
```

for instance, or

```
> lr <- function(m) exp(m[,1]+.5*m[,2]^2)
> lrsd <- function(m) sqrt((exp(m[,2]^2)-1)*exp(2*m[,1]+m[,2]^2))
> plot(100*lr(delta005.df[1001:6000,]),type="l",ylim=c(62,145),
+ ylab="Mean of Loss Ratio (in %)")
> abline(h=100,col="grey",lty=2)
> plot(100*lrsd(delta005.df[1001:6000,]),type="l",ylim=c(0,75),
+ ylab="Std. Dev. of Loss Ratio (in %)")
```

The goal of our simulation is to crawl around probability space so that we reach all the important parts of the space, and that the length of time we spend there is proportional to $p(e|\boldsymbol{\theta})p(\boldsymbol{\theta})$. When $\delta = 0.005$, we are moving too slowly through the space, and it takes us a long time (many processor cycles) to travel around. When $\delta = 0.8$, our jumps are too big and

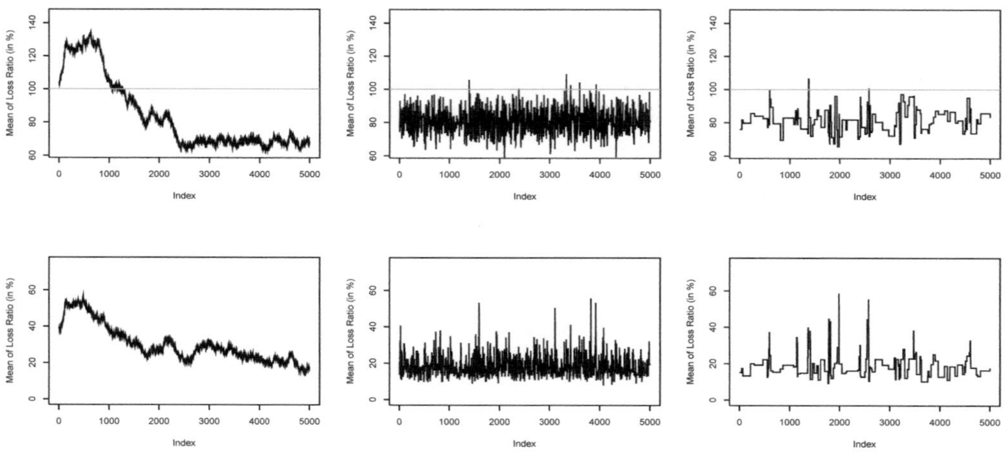

FIGURE 3.5
Traceplots with posterior mean and standard deviation of the loss ratio.

keep getting rejected, so we again do not travel efficiently. $\delta = 0.2$ achieves an acceptance ratio of 26%, which looks reasonable. Indeed, theoretical results suggest that acceptance ratios in the 20%–50% range are optimal. An MCMC simulation that is sampling efficiently from the posterior distribution is said to have good *mixing*.

Another detail to consider before using the output is the initial value of μ and σ. Ideally, the starting values would have high posterior probability, but typically this information is not available as learning the posterior probability is the whole point of the MCMC calculation. Instead, the starting values often have very low probability, as in the case in this example. If we included the starting samples, the initial probabilities would get too much weight. We can fix this by excluding the first part of the chain. These discarded results are often called *burn-in* samples. From the graphs we can see that the $\delta = 0.2$ simulation has burned in after 1,000 samples, so we will only use samples 1001–6000 in our analysis.

A final detail to worry about when using MCMC techniques is *convergence*. Because we are using a Monte Carlo technique, any answers provided will only be asymptotically correct, and the error may be very large if we do not use enough samples. Because MCMC samples will be autocorrelated, one approach is to fit a simple time series to the simulation results and estimate the error using results proven about time series. This can be done using the `coda` package in R:

```
> library(coda)
> summary(mcmc(delta20.df[1001:6000, ]))
```

Below are the `coda` results. Note that the time series error estimate is almost three times higher than the "naive" error estimate because of autocorrelation; much higher ratios are possible. This is important to keep in mind. Do not just assume that your error is low because you have 100,000 samples, because they might just be equivalent to 100 truely independent samples!

```
Iterations = 1:5000
```

```
Thinning interval = 1
Number of chains = 1
Sample size per chain = 5000

1. Empirical mean and standard deviation for each variable,
   plus standard error of the mean:

         Mean        SD  Naive SE Time-series SE
mu    -0.2450 8.387e-02 1.186e-03       0.003176
sigma  0.2220 5.754e-02 8.138e-04       0.002783

2. Quantiles for each variable:

         2.5%     25%     50%     75%    97.5%
mu    -0.4160 -0.3013 -0.2452 -0.1848 -0.08963
sigma  0.1354  0.1801  0.2116  0.2527  0.35980
```

Although mixing, burn-in are convergence important considerations when running MCMC analyses, they are still only "practical" issues. In theory, all of these simulations provided correct samples from the posterior distribution without burn-in, even if we start at a low (non-zero) probability region, and even if our mixing is slow. However, in practice, a problem with any of these issues can make MCMC results unusuable.

Aside from using the summary statistics produced by `coda`, we can use the samples of μ and σ directly. For instance, if we wanted to simulate future loss ratios in a way that reflected both process risk and parameter risk, we could do that by first sampling μ and σ from `delta20.df` and then drawing from a lognormal distribution with those parameters.

3.3.5 JAGS and Stan

The previous subsection was hopefully instructive, but it is usually a mistake to hand-code MCMC algorithms in R for practical projects. There are a few reasons for this:

1. It can be difficult to choose the step size, especially with high-dimensional problems. Notice above that I judgmentally set the σ step size to be half of the μ step size. In practice, all the dimensions may be different sizes, and the optimum step size may even be a function of the region of probability space.
2. There are better versions of the Metropolis algorithm exhibited above. The Metropolis–Hastings and Gibbs sampling algorithms are popular basic algorithms. More recently, there have been a variety of sophisticated algorithms proposed.
3. The R code can obscure the probabilistic structure of the problem—even with the simple Metropolis algorithm, there no very clear connection between the code and probabilistic assumptions. This gap would be even wider for a more complex algorithm.
4. R is slow for computations that cannot be vectorized, like the example above. It is not uncommon for a calculation coded in C/C++ to be 20 (or even 100) times faster than the equivalent coded in R; see Chapter 1.

To solve these problems, a few general-purpose MCMC engines have been developed. The two described in this chapter are JAGS (Just Another Gibbs Sampler; see Plummer

(2011)) and Stan (see Stan Development Team (2012), named after Stanislaw Ulam, one of the inventors of the Monte Carlo technique; see Metropolis & Ulam (1949)). Both of them allow you to specify your prior distribution and data in a special language designed for Bayesian statistics. They then analyze the probabilistic structure of the problem and design and run an MCMC algorithm tailored to solve it. The output is samples from the posterior distribution as in our manual example above. When it works, it is pretty magical.

This chapter will show and briefly explain sample JAGS and Stan code that solves the earlier Bayesian problem. Both JAGS and Stan come with extensive reference manuals, and example models for both are available online. However, both tools have a bit of a learning curve to install and use.

JAGS (see Plummer (2003))is an open-source, enhanced, cross-platform version of an earlier engine BUGS (Bayesian inference Using Gibbs Sampling). Here is sample JAGS code (interfaced to R using the `runjags` package) that computes the prior problem:

```
> library(runjags)
+ jags.model <- "
+ model {
+     mu ~ dnorm(mu0, 1/(sigma0^2))
+     g ~ dgamma(k0, theta0)
+     sigma <- 1 / g
+     for (i in 1:n) {
+         logr[i] ~ dnorm(mu, g^2)
+     }
+ }"
```

Then,

```
> jags.data <- list(n=n, logr=log(r), mu0=mu0, sigma0=sigma0,
+                   k0=k0, theta0=theta0)
> jags.init <- list(list(mu=log(1.2), g=1/0.5^2),
+                   list(mu=log(.8), g=1/.2^2))
> model.out <- autorun.jags(jags.model, data=jags.data, inits=jags.init,
+                           monitor=c("mu", "sigma"), n.chains=2)
> traceplot(model.out$mcmc)
> summary(model.out)
```

The actual model code is very straightforward; the only complication is that there is no built-in inverse gamma function, and the JAGS normal distribution wants the spread quantified using the precision rather than the standard deviation. We compensate for that above by calculating precision from standard deviation and the inverse gamma from the gamma distribution.

Here is the output produced by the last line, which matches our manual Metropolis algorithm to within error:

```
Iterations = 5001:15000
Thinning interval = 1
Number of chains = 2
Sample size per chain = 10000

1. Empirical mean and standard deviation for each variable,
   plus standard error of the mean:

         Mean      SD  Naive SE Time-series SE
```

```
mu     -0.2439 0.08382 0.0005927       0.0006377
sigma   0.2264 0.06182 0.0004372       0.0005879

2. Quantiles for each variable:

           2.5%     25%     50%     75%    97.5%
mu     -0.4185 -0.2974 -0.2408 -0.1877 -0.08711
sigma   0.1389  0.1830  0.2158  0.2575  0.37555
```

Stan, is a newer tool that uses the Hamiltonian Monte Carlo (HMC) sampler. HMC uses information about the derivative of the posterior probability density to improve mixing. These derivatives are supplied by algorithm differentiation, a neat computational trick for finding derivatives quickly. Unlike JAGS, which is interpreted, Stan produces C/C++ code which is then compiled by gcc (a standard open-source compiler). Stan is newer and less well tested than JAGS, and lacks a few features compared to JAGS (such as dealing with discrete distributions cleanly) but it works great on many problems and has a lot of potential.

Here is R code which uses Stan via the `rstan` package to solve our example:

```
> library(rstan)
> stan.model <- "
+ data {
+     int<lower=0> n; // number of data points
+     vector[n] r; // observed loss ratios
+     real mu0;
+     real<lower=0> sigma0;
+     real<lower=0> k0;
+     real<lower=0> theta0;
+ }
+ parameters { // Values we want posterior distribution of
+     real mu;
+     real<lower=0> sigma;
+ }
+ model {
+     mu ~ normal(mu0, sigma0);
+     sigma ~ inv_gamma(k0, theta0);
+     for (i in 1:n)
+         log(r[i]) ~ normal(mu, sigma);
+ }"

...

> stan.data <- list(n=n, r=r, mu0=mu0, sigma0=sigma0, k0=k0, theta0=theta0)
> stan.out <- stan(model_code=stan.model, data=stan.data, seed=2)
> traceplot(stan.out)
> print(stan.out, digits_summary=2)
```

The `traceplot` line suggests that Stan's MCMC chains converge quickly and have good mixing. Here is the output produced by the last line:

```
Inference for Stan model: stan.model.
4 chains, each with iter=2000; warmup=1000; thin=1;
post-warmup draws per chain=1000, total post-warmup draws=4000.
```

```
       mean se_mean   sd  2.5%   25%   50%   75% 97.5% n_eff Rhat
mu    -0.25    0.00 0.08 -0.42 -0.30 -0.25 -0.19 -0.09  1648 1.00
sigma  0.23    0.00 0.06  0.14  0.19  0.22  0.26  0.39  1358 1.00
lp__   8.64    0.03 1.06  5.77  8.23  8.96  9.39  9.67  1177 1.01
```

As you can see, the mean values for μ and σ agree with our manual Metropolis algorithm to within error. The Stan MCMC samples also show better mixing than our attempt.

3.3.6 Computational Conclusion and Specific Packages

Despite the conceptual simplicity of Bayesian statistics, computing the posterior distributions required by Bayesian statistics can still be an unfortunately complicated ordeal. This complexity has historically held Bayesianism back, but in recent decades the popularity of Bayesian statistics has been bolstered by increasing computer speed and the development of powerful MCMC algorithms and engines.

The options for computing Bayesian posteriors are usually ranked as follows in decreasing order:

1. Use a simple Bayesian conjugate if available,
2. Find an appropriate package such as `MCMCpack` or `arm` that implements the solution for you,
3. Write a custom model and let JAGS or Stan produce samples for you, and
4. As a last resort, write a custom sampler, perhaps using `Rcpp` for speed.

3.4 Bayesian Regression

To illustrate potential application of Bayesian methodology, let us describe Bayesian regression (starting with the linear model, and discussing briefly Generalized Liner Models).

3.4.1 Linear Model from a Bayesian Perspective

In the standard Gaussian linear, we assume that $Y = \boldsymbol{X}\boldsymbol{\beta} + \varepsilon$, where ε is a Gaussian i.i.d. noise, centred, with variance σ^2, where $\boldsymbol{X} = (\boldsymbol{1}, X_1, \ldots, X_k)$. The likelihood of this model, given a sample $\{(Y_i, \boldsymbol{X}_i)\}$ for $i = 1, \ldots, n$, is then

$$\mathcal{L}(\boldsymbol{\beta}, \sigma^2 | Y, \boldsymbol{X}) = [2\pi\sigma^2]^{-\frac{n}{2}} \exp\left[-\frac{1}{2\sigma^2}(Y - \boldsymbol{X}\boldsymbol{\beta})^\top (Y - \boldsymbol{X}\boldsymbol{\beta})\right].$$

As mentioned in Chapter 2, maximum likelihood estimator of $\boldsymbol{\beta}$ is $\widehat{\boldsymbol{\beta}} = (\boldsymbol{X}^\top \boldsymbol{X})^{-1}\boldsymbol{X}^\top Y$, and an unbiased estimator of the variance parameter σ^2 is

$$\widehat{\sigma}^2 = \frac{(Y - \boldsymbol{X}\boldsymbol{\beta})^\top (Y - \boldsymbol{X}\boldsymbol{\beta})}{n-k-1} = \frac{s^2}{n-k-1}.$$

Observe that the likelihood can be written as

$$\mathcal{L}(\boldsymbol{\beta}, \sigma^2 | Y, \boldsymbol{X}) = [2\pi\sigma^2]^{-\frac{n}{2}} \exp\left[-\frac{1}{2\sigma^2}(Y - \boldsymbol{X}\widehat{\boldsymbol{\beta}})^\top (Y - \boldsymbol{X}\widehat{\boldsymbol{\beta}}) - \frac{1}{2\sigma^2}(\boldsymbol{\beta} - \widehat{\boldsymbol{\beta}})^\top (\boldsymbol{X}^\top \boldsymbol{X})(\boldsymbol{\beta} - \widehat{\boldsymbol{\beta}})\right],$$

or equivalently,

$$\mathcal{L}(\boldsymbol{\beta},\sigma^2|Y,\boldsymbol{X}) = \pi(\sigma^2)\cdot\pi(\boldsymbol{\beta}|\sigma^2),$$

where the first distribution is an inverse gamma distribution, and the second is a multivariate normal with covariance matrix proportional to σ^2. For computational ease, conjugate prior can be constructed with those families. More specifically, assume that prior distributions are

$$\boldsymbol{\beta}|\sigma^2 \sim \mathcal{N}(\boldsymbol{\beta}_0, \sigma^2 M_0^{-1}),$$

for some positive definite $(k+1)\times(k+1)$ symmetric matrix M_0, while

$$\sigma^2 \sim IG(a_0, b_0).$$

In that case, the (conditional) posterior distribution of $\boldsymbol{\beta}$ given σ^2 and Y is

$$\boldsymbol{\beta}|\sigma^2, Y \sim \mathcal{N}\left([M_0+(\boldsymbol{X}^\top\boldsymbol{X})]^{-1}[M_0\boldsymbol{\beta}_0+(\boldsymbol{X}^\top\boldsymbol{X})\widehat{\beta}], \sigma^2[M_0+(\boldsymbol{X}^\top\boldsymbol{X})]^{-1}\right),$$

while the posterior distribution for σ^2 is

$$\sigma^2|Y \sim IG\left(\frac{n}{2}a_0, b_0 + \frac{[\boldsymbol{\beta}_0-\widehat{\boldsymbol{\beta}}]^\top(M_0+(\boldsymbol{X}^\top\boldsymbol{X}))^{-1}[\boldsymbol{\beta}_0-\widehat{\boldsymbol{\beta}}]}{2}\right).$$

With this framework, prior and posterior distributions are in the same family. Simple computations show that

$$\mathbb{E}(\boldsymbol{\beta}|Y,\boldsymbol{X}) = (M_0+(\boldsymbol{X}^\top\boldsymbol{X}))^{-1}\left[(\boldsymbol{X}^\top\boldsymbol{X})\widehat{\boldsymbol{\beta}}+M_0\boldsymbol{\beta}_0\right]$$

and

$$\mathbb{E}(\sigma^2|Y,\boldsymbol{X}) = \frac{2b_0+s^2+[\boldsymbol{\beta}_0-\widehat{\boldsymbol{\beta}}]^\top\left(M_0^{-1}+(\boldsymbol{X}^\top\boldsymbol{X})^{-1}\right)^{-1}[\boldsymbol{\beta}_0-\widehat{\boldsymbol{\beta}}]}{n+2(a_0-1)}$$

The interesting point is that, since we have explicitly posterior distributions, it is possible to derive credibility intervals for parameters, as well as predictions, $\widehat{Y}$ for a given $\boldsymbol{x}$. If we use the Bayesian Central Limit theorem (mentioned in the introduction of this chapter), and

$$\mathbb{E}(\widehat{Y}|Y,\boldsymbol{X}) = \boldsymbol{x}[(\boldsymbol{X}^\top\boldsymbol{X})+M_0]^{-1}\left[(\boldsymbol{X}^\top\boldsymbol{X})\widehat{\boldsymbol{\beta}}+M_0\boldsymbol{\beta}_0\right]$$

and

$$\text{Var}(\widehat{Y}|Y,\boldsymbol{X}) = \sigma^2\left(1+\boldsymbol{x}[(\boldsymbol{X}^\top\boldsymbol{X})+M_0]^{-1}\boldsymbol{x}^\top\right),$$

we get bounds for an approximated credibility interval.

To illustrate this model, consider the same example as in Section 2.4 in Chapter 2.

```
> Y <- Davis$height
> X1 <- Davis$sex
> X2 <- Davis$weight
> df <- data.frame(Y,X1,X2)
> X <- cbind(rep(1,length(Y)),(X1=="M"),X2)
> beta0 <- c(150,7,1/3)
> beta <- as.numeric(lm(Y~X1+X2,data=df)$coefficients)
> sigma2 <- summary(lm(Y~X1+X2,data=df))$sigma^2
> M0 <- diag(c(100,100,1000^2))
```

If we consider a prediction for a male with weight 70 kg,

```
> Xi <- c(1,1,70)
```

then the prediction is

```
>  (m <- Xi%*%solve(M0+t(X)%*%X) %*%(( t(X)%*%X)%*%beta +M0%*%beta0))
         [,1]
[1,] 176.7536
```

and a 95% credibility interval based on the Gaussian approximation is

```
> s2 <- sigma2*(1+ (t(Xi)%*% (solve(M0+t(X)%*%X)))%*%(Xi))
> qnorm(c(.025,.975),mean=m,sd=sqrt(s2))
[1] 166.6124 186.8947
```

Those values can be compared with the lower and upper bound obtained using function `predict()` (using a standard statistical approach)

```
> predict(lm(Y~X1+X2,data=df),newdata=
+ data.frame(X1="M",X2=70),interval="prediction")
      fit      lwr      upr
1 176.057 165.8188 186.2951
```

3.4.2 Extension to Generalized Linear Models

In Generalized Linear Models (see McCullagh & Nelder (1989) or Chapter 14), two functions must be specified:

- The (conditional) distribution of Y given $\boldsymbol{X}$, that belongs to the exponential family, and let f denote the associate density (or probability function for discrete variables)
- A link function g that relates the (conditional) mean $\mu(\boldsymbol{x}) = \mathbb{E}(Y|\boldsymbol{X} = \boldsymbol{x})$ and the covariates, as $g(\mu(\boldsymbol{x})) = \boldsymbol{x}^\top\boldsymbol{\beta}$.

This link function should be bijective, for identifiability reasons, and then, $\mu(\boldsymbol{x}) = g^{-1}(\boldsymbol{x}^\top\boldsymbol{\beta})$.

Two popular models are the logit (or probit) and the Poisson models. In the logit model, Y is a binary response, $Y|\boldsymbol{X} \sim \mathcal{B}(\pi(\boldsymbol{X}))$, where the link function is the logit transform, $g(x) = \log[x/(1-x)]$. An alternative is the probit model, where the link function is the quantile function of the $\mathcal{N}(0,1)$ distribution, $g(x) = \Phi^{-1}(x)$. For the Poisson model, Y is a counting response, with a Poisson distribution $Y|\boldsymbol{X} \sim \mathcal{P}(\lambda(\boldsymbol{X}))$, where the link function is the log transform, $g(x) = \log[x]$.

The Probit Hastings–Metropolis sampler, in the context of a Probit model can be the following: To initialize, start with maximum likelihood estimators, $\boldsymbol{\beta}_{(0)} = \widehat{\boldsymbol{\beta}}$ with (asymptotic) variance function matrix $\widehat{\boldsymbol{\Sigma}}$. At stage k,

1. Generate $\boldsymbol{\beta} \sim \mathcal{N}(\boldsymbol{\beta}_{(k-1)}, \tau^2\widehat{\boldsymbol{\Sigma}})$
2. With probability $\min\left\{1, \dfrac{\pi(\boldsymbol{\beta}|Y)}{\pi(\boldsymbol{\beta}_{(k-1)}|Y)}\right\}$, set $\boldsymbol{\beta}_{(k)} = \boldsymbol{\beta}$, otherwise $\boldsymbol{\beta}_{(k-1)}$.

A natural idea is to use a flat prior, so that the posterior distribution is

$$\pi(\boldsymbol{\beta}|Y) \propto \prod_{i=1}^{n} \Phi(\boldsymbol{X}_i^\top \boldsymbol{\beta})^{Y_i} \left[1 - \Phi(\boldsymbol{X}_i^\top \boldsymbol{\beta})\right]^{1-Y_i}.$$

Consider the attorney dataset, `autobi` (from Frees (2009)). We will consider here only fifty observations (with no missing values):

```
> index <- which((is.na(autobi$CLMAGE)==FALSE)&
+ (is.na(autobi$CLMSEX)==FALSE))
> set.seed(123)
> index <- sample(index,size=50)
```

The variable of interest is whether an insured got an attorney, or not,

```
> Y <- autobi$ATTORNEY[index]
```

Consider the following regression, where covariates are the cost of the claim, the sex of the insured and the age of the insured. The posterior density is here

```
> X <- cbind(rep(1,nrow(autobi)),
+ autobi$LOSS,autobi$CLMSEX==2,autobi$CLMAGE)
> X <- X[index,]
```

The posterior density is here

```
> posterior <- function(b)
+ prod(dbinom(Y,size=1,prob=pnorm(X%*%b)))
```

The Hastings–Metropolis alogrithm is here

```
> model <- summary(glm(Y~0+X,family=binomial(link="probit")))
> beta <- model$coeff[,1]
> vbeta <- NULL
> Sigma <- model$cov.unscaled
> tau <- 1
> nb.sim <- 10000
> library(mnormt)
> for(s in 1:nb.sim){
+ sim.beta <- as.vector(rmnorm(n=1,mean=beta,varcov=tau^2*Sigma))
+ prb <- min(1,posterior(sim.beta)/posterior(beta))
+ beta <- cbind(beta,sim.beta)[,sample(1:2,size=1,prob=c(1-prb,prb))]
+ vbeta=cbind(vbeta,beta)
+ }
```

In order to visualize generated sequences, let us remove the first 10% and plot the sequence of β_2's (parameter associated to the cost of the claim),

```
> mcmc.beta <- vbeta[,seq(nb.sim/10,nb.sim)]
> plot(mcmc.beta[2,],type="l",col="grey")
> abline(h=model$coeff[2,1],lty=2)
```

We can also visualize the distribution of the β_2's:

```
> hist(mcmc.beta[2,],proba=TRUE,col='grey',border='white')
> vb <- seq(min(mcmc.beta[2,]),max(mcmc.beta[2,]),length=101)
> lines(vb,dnorm(vb,model$coeff[2,1],model$coeff[2,2]),lwd=2)
> lines(density(mcmc.beta[2,]),lty=2)
```

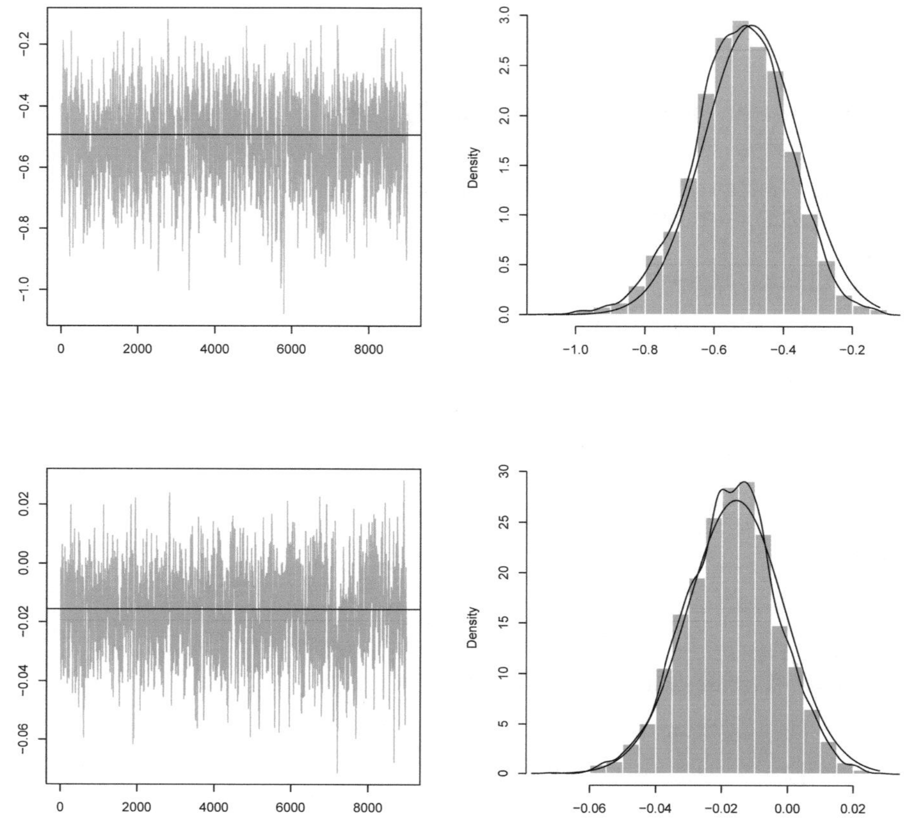

FIGURE 3.6
Hastings–Metropolis in a logistic regression, sequences of β_k's and their density.

The plain line is the standard asymptotic Gaussian distribution (in GLMs, the distribution of the parameters is only asymptotic), and the dotted line, the kernel-based estimator. The code can also be used for the age variable, with sequence of β_4's; see Figure 3.6.

3.4.3 Extension for Hierarchical Structures

In some applications, data are slightly more complex and might exhibit some hierarchical structures. Consider the case where several companies wish to share their experience through an (independent) ratemaking bureau, in order to enhance the predictive power of all companies (as discussed in Jewell (2004)). Then observations are $(Y_i, \boldsymbol{X}_i)$, related to insured i, from company j. There might be market factors, company risk factors and individual risk factors. Those models will be discussed in Chapter 15. Such complex models can be difficult to estimate using standard techniques, but a Bayesian approach can be extremely powerful.

Consider a simple linear model, where the intercept may vary by group,

$$Y_i = \alpha_{j[i]} + \beta_1 X_{1,i} + \beta_2 X_{2,i} + \varepsilon_i.$$

This is a fixed effects model, and it can be estimated using dummy variables. A more flexible alternative is to use a so-called hierarchical model, or multilevel model, also called mixed (or random) effects model. Here,

$$Y_i = \alpha_{j[i]} + \beta_1 X_{1,i} + \beta_2 X_{2,i} + \varepsilon_i, \text{ with } \alpha_j \sim \mathcal{N}(\gamma, \tau^2).$$

While previously we assumed that there were different intercepts for each group, now we assume that intercepts are drawn from an identical random variable. One can also write

$$Y_i = \alpha_{j[i]} + \beta_1 X_{1,i} + \beta_2 X_{2,i} + \varepsilon_i, \text{ with } \alpha_j = \alpha + \eta_j,$$

and it is possible to incorporate group-level covariates, such as

$$Y_i = \alpha_{j[i]} + \beta_1 X_{1,i} + \beta_2 X_{2,i} + \varepsilon_i, \text{ with } \alpha_j = \gamma_0 + \gamma_1 Z_{1,j} + \gamma_2 Z_{2,j} + \eta_j,$$

Using standard likelihood-based methods, this is a difficult model to estimate. But here, we can use a Bayesian model. Write here

$$f(\alpha, \beta, \gamma | Y) \propto f(Y | \alpha, \beta, \gamma) \cdot \pi(\alpha|\gamma)\pi(\gamma)\pi(\beta).$$

And one can use Gibbs sampling, or Hastings–Metropolis to generate sequences of parameters. See Chapter 7 in Klugman (1992) for more details, with actuarial applications, or Chapter 15 of this book.

3.5 Interpretation of Bayesianism

By now we have seen some examples of Bayesian statistics and how it differs from classical statistics. Suppose we accept the basic structure of data analysis described earlier:

1. When we first analyse a problem, before getting the data, we start with prior beliefs about the problem.
2. These prior beliefs naturally have the structure of a (subjective) probability function P.
3. When we encounter data e, probabilistic consistence compels us to move to a new belief state, reflected in the updated probability function $\mathbb{P}'(x) = \mathbb{P}(x|e)$.

If we accept (1)–(3) above, Bayesianism seems to have an airtight case going for it—doing anything else would be silly. However, there are problems with each of these assumptions above:

1. Do we always start with prior beliefs before getting the data? Two objections to this are:

   (a) Sometimes we actually get the data before we have formulated our prior beliefs about the data. For instance, suppose we need to test the theory that the sky is blue because blue light is scattered more than red or green light. The fact that the sky is blue should presumably increase our probability of this theory, but we have known that the sky is blue long before we could formulate a theory about light being scattered. This is known as the *problem of old evidence.*

   (b) It is not clear what it means to believe something. Psychologists have found that the human brain is disjointed and composed of several systems which are not always consistent. We may not have beliefs about a theory until we are forced to think about it. Even when the theory is known ahead of time, it may be the analysis of the data itself which forces us to come to a determinate belief about a theory.

2. Human beliefs are not similar to probability functions. Three arguments for their dissimilarity are:

   (a) Ordinary people, and even statisticians occasionally, are horrible at doing intuitive probability. Athough the concept of probability seems comfortable to most people, when forced to estimate probability, they tend to underestimate base rates, overestimate flashy or recent events, and confuse availability with probability. This seems to show that people are not naturally using any system that conforms to the axioms of probability.

   (b) Historically, probability took a surprisingly long time to develop. It was only treated formally by Pascal, Fermat and Huygens in the mid-seventeenth century, about the same time that Newton proposed the universal law of gravitation. By contrast, about 2,000 years earlier, Euclid's *Elements* had already proven complex theorems in a style that stands as a model of rigor and coherence even today. If probability were "natural," then it would have been formalized much earlier.

   (c) Having a full prior probability function requires superhuman intelligence, in both scope and consistency. Scope, because a full prior probability function would provide probabilities for every logically possible theory—no one would ever propose a novel theory that caused us to think, "huh, I never thought of that." Consistency, because, for instance, the laws of probability require that all logical truths be assigned probability 1 (and logical falsehoods probability 0). Mathematicians' jobs would be trivial because we would all know ahead of time whether any given mathematical statement followed from accepted axioms.

3. Finally, conditionalization is intuitively appealing, but it is not clear whether it is mandatory for a rational thinker.

   (a) Prosaically, the posterior probability distribution can be hard to compute or even approximate, despite methods like Bayesian conjugates and MCMC simulations mentioned earlier in this chapter. Even when a full prior probability function is available, conditionalization may require superhuman intelligence or more technology than is currently available.

   (b) More abstractly, an agent that does not conditionalize may not be irrational. There is a certain inconsistency over time to the agent's beliefs, but perhaps this inconsistency can be better described as the agent "changing its mind" rather than it being outright irrational.

Thus, Bayesianism cannot be considered a model of how rational people think in the same way that physics is a model of how objects move. Rather, Bayesianism is more of an idealized model of how a super-intelligent being with very specific beliefs would reason about a certain problem.

In general, Bayesianism as used in practice is less of a mandatory approach for solving all problems than one approach, albeit an extremely useful and general one, that can be used when responding to certain problems.

3.5.1 Bayesianism and Decision Theory

The practical problems that most actuaries and practioners of actuarial science face are not ultimately about the actuaries' probabilities or even their beliefs; rather, they are about the work products that practioners must deliver.

One strength of Bayesianism is its connection to classical decision theory—subjective probability functions required by Bayesianism are exactly the inputs to rational decision making according to expected utility theory. This chapter will not discuss decision theory, but classical decision theory is subject to some of the same criticisms of Bayesian theory listed above: that it is psychologically unrealistic, that it demands superhuman intelligence, and that it is not as logically inescapable as it first appears.

Bayesian analysis and classical decision theory make a great one-two punch, but they can also fall together. For instance, suppose the actuary is called upon to analyse a large book of policies for underwriting purposes, and categorize them into disjoint "interesting" classes by analysing policyholder information. The policyholder information consists of a couple dozen dimensions (e.g. age of policyholder, location of exposure, exposed value, etc.). The goal is to give the underwriters a better sense of the risks they are writing, and give them a better vocabulary to talk about changes to their book.

In this case, the full Bayesian + decision theory analysis might have to define a prior probability over which partitions are likely to exist in the data, then define a utility function on which of these to show the underwriters to maximize "interest." By contrast, a more practical approach might be to use k-means clustering or another common technique from unsupervised machine learning. It would be unclear what posterior probability distribution and utility function would be implied by that selection, if any, but the underwriters might still be satisfied with the results.

3.5.2 Context of Discovery versus Context of Justification

Because the ultimate output of the practicing actuary is a work product and not a probability distribution, it is also worth considering the distinction between the methods presented and explained in the work product and the process used to create the work product. To describe this distinction, we can borrow from the philosophy of science the labels *context of discovery* versus *context of justification.*

The context of justification would include the work product itself and any supplementary documentation or supporting material. If the actuary's boss or a regulator had a question about the work product, the answer would be drawn in the context of justification. Practicing actuaries strive to appear fair, logical, unbiased, and conservative (often even boring or conformist) in the context of justification.

The context of discovery, by contrast, involves whatever causal processes led the actuary to create the work product that he or she did. For instance, if the actuary chose a number or statistical method purely to please the CEO of the company, that would obviously belong only in the context of discovery. Less problematically, the actuary might have tried several alternate statistical methods but found that they were computationally infeasible for the problem. These failed methods are important in the causal story of why the actuary went with the method he or she picked, but would probably not be detailed in the work product.

Under a first reading of Bayesian statistical procedure, the context of discovery and context of justification coincide—the actuary simply encodes his/her prior beliefs and conditionalizes. However, the issues raised in Section 3.5 create an interplay between the two. Because humans do not actually have prior beliefs in the form of probabilities, picking a prior probability distribution is more of a choice or act than a simple result of introspection.

In fact, in practice, the choice of prior probability often depends on the results of the Bayesian analysis! At worst, this can be construed as the practioner simply trying to get the answer he or she wants. But a fairer way to view it is that humans, actuaries included, lack a probabilistically coherent set of beliefs. In order to produce a prior distribution, they need to explicitly consider all the consequences of the prior distribution, which of course includes its behavior under conditionalization.

Interestingly, when asked to justify their salaries, many practicing actuaries will not emphasize their technical ability, but rather their experience and "feel" for the numbers. For instance, when a complex technical method produces a number, an actuary may immediately reject it out of hand and know that either the method was inappropriate or faulty, or the method was applied on a bad set of assumptions. Any practioner has probably caught hundreds of mistakes through this kind of reasonability checking and considers it an invaluable component of actuarial skill. Nevertheless, this same reasonability checking could also be described as actuaries routinely changing their inputs and methods to get the answers that fit their preconceived viewpoints!

Thus, the choice of prior distribution, considered an input to the Bayesian process in the context of justification, can also be considered an output of the process in the context of discovery, in the sense that an actuary may only truly confirm the choice of prior after understanding the results of conditionalizing on the data.

Another practical issue for Bayesian statistics relating to the distinction between the two contexts is that the prior probabilities required for Bayesian analyses may look too subjective in the context of justification.

For example, suppose a consulting actuary is assigned to estimate reserves for a company. She uses an advanced Bayesian reserving method to encode her broad and deep industry experience into the prior, and computes the posterior using an appropriate numerical technique. In her work product, she details these assumptions. The client, perhaps pushing for a different reserve estimate, might then nitpick all of her prior probability assumptions ("Why did you assume a loss ratio coefficient of variance 25% in your prior instead of 23%?"). All her assumptions may have been very fair, but are necessarily fuzzy because humans do not store their beliefs in precise probabilistic form. Thus in the clients' eyes, the actuary may not be able to defend herself adequately from these criticisms. By contrast, an actuary who simply used the chain ladder method on all the historical experience might have the work accepted without comment. Carveth Read may have been right when he said that "*it is better to be vaguely right than exactly wrong.*" However, a work product that is exactly wrong may be easier to defend for an actuary.

In terms of the context of justification, Bayesian analyses may necessarily depend on more quantitative, debatable inputs. Classical statistics is often "better" at smuggling assumptions into the choice of methods, or at providing arbitrary thresholds (such as the $p = 0.05$ significance level) that are considered unimpeachable because they are "standard."

A hard-core Bayesianist might say that the appropriate response is still to conduct the correct Bayesian reserve analysis. Once the truth is known, the actuary might confine this analysis to the context of discovery, and use more common methods to get the same answer in the actual work product. This is a laudable response, but may not be possible in practice. Instead, the actuary might just use the simple, accepted technique, and count on actuarial feel to warn her if the simple method is not approximately correct.

3.5.3 Practical Classical versus Bayesian Statistics Revisited

The two diagrams in Figure 3.7 are not intended to be complete diagrams of using Bayesian and classical statistics. Rather, they are just intended to highlight some of the important dynamics when applying either type of statistics to actual actuarial problems.

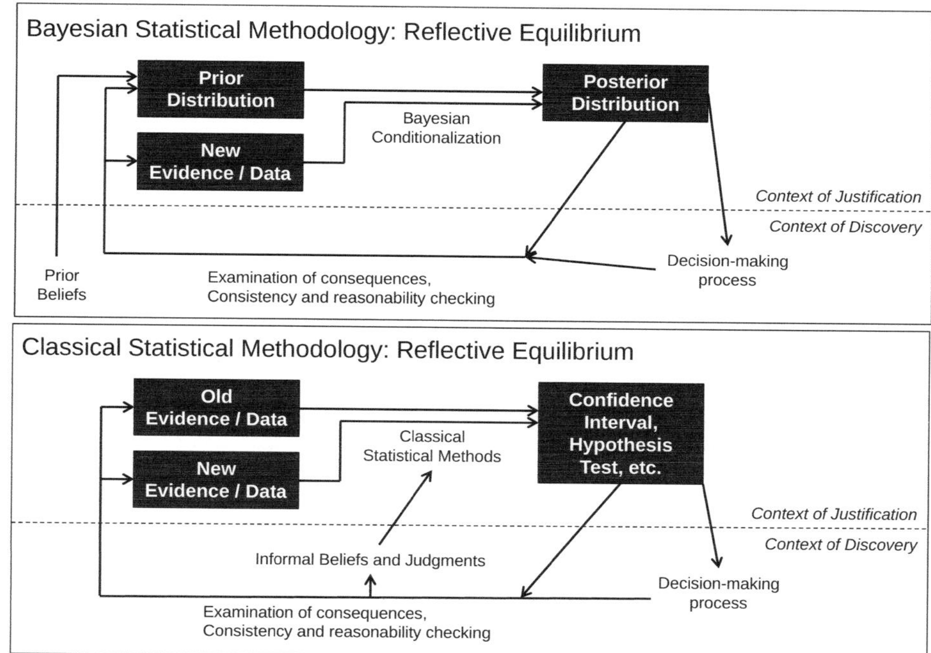

FIGURE 3.7
Diagram of statistical processes.

For both schools of statistics, problems to be solved are not isolated textbook cases, but rather originate in informal, non-statistical beliefs and inputs. And in both cases, the finished statistical analysis is not immediately entered into a blank to be graded, but must provide output couched in concepts and vocabulary that integrated into existing decision-making processes.

With that in mind, here are some quick pros and cons of Bayesianism statistics. Of course which approach is best depends on the exact problem and circumstance in question, but some generalizations seem useful. First the pros:

- It makes sense – Bayesianism is based on a consistent and beautiful idealization of rationality. When Bayesian priors are available and the computation is feasible, Bayesian analyses may be the obvious choice.
- Coherence – Unlike classical statistics which can seem like a grab bag of random ideas, Bayesianism techniques follow from only a few core ideas.
- Documentation – Bayesian analyses typically formalize more of their assumptions, instead of smuggling them in like classical statistics. This allows more information to be communicated in work products.
- Intuitive outputs – Bayesian analyses typically produce posterior probabilities for theories (or *Bayes factors* comparing two theories), and credible intervals or posterior distributions for parameters. These are much more useful and less error-prone than their counterparts, classical hypothesis testing and confidence intervals.

- Decision theory – Bayesianism shares a tight connection to decision theory. Most insurance decisions are not made formally, but decision theory may occasionally be useful.

Now for some general cons of Bayesian statistics:

- Priors can be difficult – As mentioned above, people do not actually have prior probability functions in their heads. For many problems, it can be very hard to construct an appropriate and practical prior.

- It is subjective and nitpickable – This is the flipside of self-documenting above. Sometimes it is not a good idea to expose certain assumptions and the arbitrariness of certain decisions.

- Computationally difficult – Even relatively simple Bayesian problems may require MCMC methods. Convenient computation is extremely important; the less time spent computing, the more time can be spent doing additional analyses, doing data validation and exploration, etc.

- Unwieldy outputs – A full posterior distribution may be conceptually intuitive and may contain all relevant information, but may be harder to process than the classical equivalent. For instance, compare the posteriors from Bayesian regression to the simple point estimates and covariance matricies from classical regression.

- Unfamiliar language – Apart from their intrinsic merit, many people may be more familiar with classical concepts. For instance, when talking to someone who frequently does Kolmogorov-Smirnov tests, simply communicating the K-S statistic may be the best option.

3.6 Conclusion

The foundations of Bayesianism are based on the idea of subjective probabilities. Although considered philosophically problematic by frequentists, they are both mathematically solid and lead to beautiful and productive theories of statistics, game theory and decision theory.

Conditionalization is core to Bayesian statistics. When you learn evidence e, you should condition on it by moving from your prior subjective probability distribution $\mathbb{P}(\cdot)$ to the posterior distribution $\mathbb{P}(\cdot|e)$.

Despite the simple notation, $\mathbb{P}(\cdot|e)$ can be hard to compute in practice because it involves calculating the integral $\int p(e|\phi)p(\phi)d\phi$. In the case of Bayesian conjugates, there is an exact solution, but typically this integral must be sidestepped or approximated. The most popular approach is to use MCMC methods to sample from $\mathbb{P}(\cdot|e)$ rather than computing it directly. There are a variety of R packages and MCMC engines to help with this.

Bayesian statistics is based on an elegant theory of rationality and can be applied to almost any statistical problem. However, it is not always the best choice in practice. Solving actuarial problems involves producing work products, not posterior distributions. Bayesian statistics provides actuaries with a rich and useful set of tools, but techniques from classical statistics, machine learning or other fields will fit certain situations better.

3.7 Exercises

3.1. Consider a simple mixture model, with two components, $p \cdot \mathcal{N}(\mu_1, 1) + [1-p] \cdot \mathcal{N}(\mu_2, 1)$ where the probability p is known. Assume further that $p < 0.5$ to make sure that the model is identifiable (and μ_1 cannot be confused with μ_2). We do have a sample $\{x_1, \dots, x_n\}$. Assume independent priors $\mathcal{N}(a, 1/b)$ on both μ_1 and μ_2.

Write the code of the following algorithm (so-called Gibbs sampler for mean mixtures), function of vector `x` (the observations) and of vector `c(a,b)`, the fixed hyper-parameters of the prior distribution. To initialize, chose $\mu_{1,0}$ and $\mu_{2,0}$. Then, at iteration k (with $k \geq 1$),

(a) Generate Θ_i, which will allocate i to one group, either 1 or 2, with probabilities

$$\begin{cases} \mathbb{P}(\Theta_i = 1) \propto p \cdot \exp\left(-\frac{1}{2}\left[x_i - \mu_{1,k-1}\right]^2\right) \\ \mathbb{P}(\Theta_i = 2) \propto [1-p] \cdot \exp\left(-\frac{1}{2}\left[x_i - \mu_{2,k-1}\right]^2\right). \end{cases}$$

(b) Compute $\beta_k = \sum_{i=1}^{n} \mathbf{1}(\Theta_i = 1)$, the number of observations in class 1, the total sum for observations in class 1 $\alpha_{1,k} = \sum_{i=1}^{n} \mathbf{1}(\Theta_i = 1) \cdot x_i$ and the total sum for observations in class 2 $\alpha_{2,k} = \sum_{i=1}^{n} \mathbf{1}(\Theta_i = 2) \cdot x_i$.

(c) Generate $\mu_{1,k}$ from $\mathcal{N}\left(\frac{ab + \alpha_{1,k}\beta_k}{b + \beta_k}, \frac{1}{b + \beta_k}\right)$.

(d) Generate $\mu_{2,k}$ from $\mathcal{N}\left(\frac{ab + \alpha_{2,k}(n - \beta_k)}{b + n - \beta_k}, \frac{1}{b + n - \beta_k}\right)$.

4

Statistical Learning

Arthur Charpentier
Université du Québec à Montréal
Montréal, Québec, Canada

Stéphane Tufféry
ENSAI & Université de Rennes 1
Rennes, France

CONTENTS

4.1 Introduction and Motivation

In this chapter, we will describe some techniques to learn from data, and to make a prediction based on a set of features. We will use a training set, where those features were observed, as well as our variable of interest, to build a predictive model. A good model will accurately predict the variable of interest. This is actually a standard procedure in actuarial science, where the variable of interest might be

- Whether an insured will buy additional (optional) coverage, or not
- Whether a claimant will be represented by an attorney, or not (see e.g. the automobile injury insurance claims in Frees (2009))
- Whether an insured will have some specific disease, or not
- Whether a loaner will be considered a good or a bad client (in this chapter)

All the techniques mentioned in this chapter will be used on a binary variable of interest (good or bad client), but one can easily extend most of them to an ordered discrete variable of interest.

4.1.1 The Dataset

The dataset is consumer credit files, called the *German Credit dataset* in Tufféry (2011) and Nisbet et al. (2011), with 1,000 instances, that can be found in the `CASdataset` package, under the name `credit`. New applicants for credit and loans can be evaluated using twenty explanatory variables:

Att. 1 (qualitative) `checking_status`, status of existing checking account, `A11` : less than 0 euro, `A12` : from 0 to 200 euros, `A13` : more than 200 euros, and `A14` : no checking account (or unknown).

Att. 2 (numeric) `duration`, credit duration in months.

Att. 3 (qualitative) `credit_history`, credit history A30 : delay in paying off in the past, A31 : critical account, A32 : no credits taken or all credits paid back duly, A33 : existing credits paid back duly till now, A34 : all credits at this bank paid back duly.

Att. 4 (qualitative) `purpose`, purpose of credit, A40 : car (new), A41 : car (used), A42 : furniture/equipment, A43 : radio/television, A44 : domestic appliances, A45 : repairs, A46 : education, A47 : (vacation—does not exist in the study sample), A48 : retraining, A49 : business, A410 : others.

Att. 5 (numerical) `credit_amount`, credit amount

Att. 6 (qualitative) `savings`, savings account A61 : less than 100 euros, A62 : from 100 to 500 euros, A63 : from 500 to 1,000, A64 : more than 1,000, A65 : no savings account (or unknown).

Att. 7 (qualitative) `employment`, Present employment since, A71 : unemployed, A72 : less than 1 year, A73 : from 1 to 4 years, A74 : from 4 to 7 years, A75 : more than 7 years.

Att. 8 (numerical) `installment_rate`, Installment rate (in percentage of disposable income)

Att. 9 (qualitative) `personal_status`, Personal status and sex, A91 : male : divorced/separated, A92 : female : divorced/separated/married, A93 : male : single, A94 : male : married/widowed, A95 : female : single.

Att. 10 (qualitative) `other_parties`, Other debtors or guarantors, A101 : none, A102 : co-applicant, A103 : guarantor.

Att. 11 (numerical) `residence_since`, Present residence since, 1 : $<$ 1 year, 2 : $\leq ... <$ 4 years, 3 : 4 $\leq ... <$ 7 years, 4 : $\geq$ 7 years

Att. 12 (qualitative) `property_magnitude`, Property (most valuable) , A121 : real estate, A122 : if not A121 : building society savings agreement/life insurance, A123 : if not A121/A122 : car or other, not in attribute 6, A124 : unknown / no property.

Att. 13 (numerical) `age`, Age (in years)

Att. 14 (qualitative) `other_payment_plans`, Other installment plans, A141 : bank, A142 : stores, A143 : none.

Att. 15 (qualitative) `housing`, Housing, A151 : rent, A152 : own, A153 : for free

Att. 16 (numerical) `existing_credits`, Number of existing credits at this bank (including the running one)

Att. 17 (qualitative) `job`, Job, A171 : unemployed/ unskilled - non-resident, A172 : unskilled - resident, A173 : skilled employee / official, A174 : management/ self-employed/highly qualified employee/officer.

Att. 18 (numerical) `num_dependents`, Number of people being liable to provide maintenance for.

Att. 19 (qualitative) `telephone`, Telephone, A191 : none, A192 : yes, registered under the customers name.

Att. 20 (qualitative) `foreign_worker`, foreign worker, A201 : yes, A202 : no.

We will remove the latter variable in the study.

```
> credit <- credit[,-which(myVariableNames == "foreign_worker")]
```

In the previous description, variables are characterized as either quantitative (numerical) or qualitative (called factors in the first chapter).

In this chapter, we will see how to develop a credit scoring rule that can be used to determine if a new applicant is more likely to be either a good credit risk, or a bad credit risk, based on values for one, or more, of the explanatory variables:

Att. 21 (qualitative) `class`, binary variable `0` stands for good and `1` bad (or credit-worthy against not credit-worthy, or no non-payments against existing non-payments)

In order to have a 0/1 variable, use

```
> credit$class <- credit$class-1
> table(credit$class)

  0   1
700 300
```

This dataset can be found in library `caret` using `data(GermanCredit)` (see Kuhn (2008)).

4.1.2 Description of the Data

Most of the variates are factor variables. It is also possible to cut continuous variates to define factor ones:

```
> credit.f <- credit
> credit.f$age <- cut(credit.f$age,c(0,25,Inf))
> credit.f$credit_amount <- cut(credit.f$credit_amount,c(0,4000,Inf))
> credit.f$duration <- cut(credit.f$duration,c(0,15,36,Inf))
```

A natural tool to quantify correlation between categorical variates is Cramér's V (from Cramér (1946)), defined as

$$V = \sqrt{\frac{\chi^2}{n(k-1)}},$$

where χ^2 is the statistics associated to Pearson's chi-squared test of independence, n is the number of observations, and k denotes the number of factors, for the two variates, whichever is less (here, $k-1$ will always be 1 because the variable of interest takes only two values).

```
> library(rgrs)
> k <- ncol(credit.f)-1
> cramer = function(i) cramer.v(table(credit.f[,i],credit.f$class))
```

It is also possible to use

```
> cramer = function(i) sqrt(chisq.test(
+ table(credit.f[,i],credit.f$class))$statistic/(length(credit.f[,i])))
```

Then,

```
> pv <- function(i) chisq.test(table(credit.f[,i],credit.f$class))$p.value
> CRAMER <- data.frame(variable=names(credit)[1:k],
> cramerv <- Vectorize(cramer)(1:k),
> p.value <- Vectorize(pv)(1:k))
> vCRAMER <- CRAMER[order(CRAMER[,2], decreasing=TRUE),]
```

It is possible to plot the variates, sorted by correlation level:

```
> par(mar=c(10,4,4,0))
> barplot(vCRAMER[,2],names.arg=vCRAMER[,1],las=3)
```

For continuous variates X (and categorical variable Y), it is possible to compare the conditional distribution of X given Y:

```
> aggregate(credit[,c("age","duration")],by=list(class=credit$class),mean)
class  Group.1      age duration
    1        0 36.22429 19.20714
    2        1 33.96333 24.86000
```

It is possible also to visualize the probability that $Y = 1$ for some value x of X, or for some partition of the X variable,

```
> Q <- quantile(credit$age,seq(0,1,by=.1))
> Q[1] <- Q[1]-1
> cut.age <- cut(credit$age,Q)
> (prop <- prop.table(table(cut.age,credit$class),1))
```

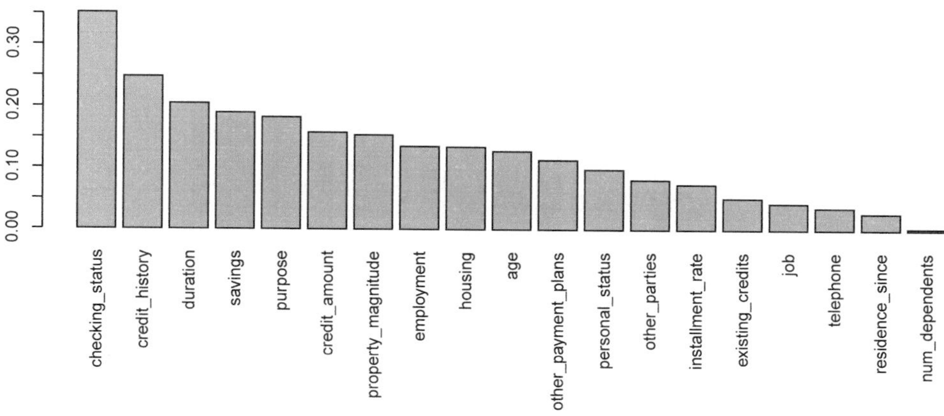

FIGURE 4.1
Evolution of Cramér's V, per variable.

```
cut.age            1         2
  (18,23] 0.6000000 0.4000000
  (23,26] 0.6148148 0.3851852
  (26,28] 0.7021277 0.2978723
  (28,30] 0.6623377 0.3376623
  (30,33] 0.6857143 0.3142857
  (33,36] 0.7927928 0.2072072
  (36,39] 0.7567568 0.2432432
  (39,45] 0.7256637 0.2743363
  (45,52] 0.8000000 0.2000000
  (52,75] 0.6979167 0.3020833
```

that can be visualized (Figure 4.2) using

```
> barplot(t(prop))
> abline(h=mean(credit$class==0),lty=2)
```

4.1.3 Scoring Tools

The variable of interest here is a 0/1 variate Y. The goal here is to use explanatory variables $\boldsymbol{X}$ to predict Y, based on a continuous score function $S(\boldsymbol{X}) \in [0, 1]$ (as introduced by Fisher (1940)). The prediction will then be

$$\widehat{Y} = \begin{cases} 1 \text{ if } S(\boldsymbol{X}) \geq s \\ 0 \text{ if } S(\boldsymbol{X}) < s \end{cases}$$

for some threshold $s \in (0, 1)$. From a decision theory perspective, it will be interesting to compare the true value, and the predicted one, as in Table 4.1.

Type I errors are also called *false positive*, while type II error are *false negative*. In a perfect world, we would like those two errors to be as small as possible. The choice of the

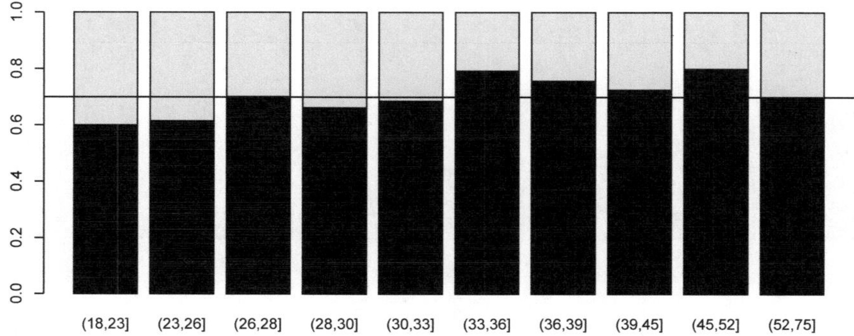

FIGURE 4.2
Proportion of `class` categories, per age category.

TABLE 4.1
Decision and errors in credit scoring.

		True value of Y	
		Y=0 (negative)	Y=1 (positive)
$\widehat{Y}=0$	(negative)	Correct decision	Type II error
$\widehat{Y}=1$	(positive)	Type I error	Correct decision

threshold will be used to make a tradeoff between the two kinds of errors, but we will have to spend some time to have both errors as small as possible.

A standard tool to visualize the fit is the ROC curve (*Receiver operating characteristic*, from communication theory) created by plotting the fraction of true positives out of the positives (the true positive rate) versus the fraction of false positives out of the negatives (false positive rate), at various threshold settings. Consider a simple latent-based model, where we assume that $Y = 1$ when some credit indicator (unobservable) $Y^\star$ is too high, where $Y^\star$ is assumed to be continuous and a function of some covariates. With a linear model, $Y^\star = \boldsymbol{X}^\top\boldsymbol{\beta} + \varepsilon$, where ε is a centered Gaussian noise. Thus, here

$$\widehat{Y} = \begin{cases} 1 \text{ if } S(\boldsymbol{X}) = \mathbb{P}(Y^\star < s^\star) = \Phi(\boldsymbol{X}^\top\boldsymbol{\beta}) \geq s \\ 0 \text{ if } S(\boldsymbol{X}) = \mathbb{P}(Y^\star < s^\star) = \Phi(\boldsymbol{X}^\top\boldsymbol{\beta}) < s \end{cases}$$

This is the so-called probit model. The code to construct the score function is

```
> Y <- credit$class
> reg <- glm(class~age+duration,data=credit,family=binomial(link="probit"))
> summary(reg)

Call:
glm(formula = class ~ age + duration, family = binomial(link = "probit"),
    data = credit)

Deviance Residuals:
```

```
    Min       1Q   Median       3Q      Max
-1.5364  -0.8400  -0.7084   1.2452   2.1074

Coefficients:
             Estimate Std. Error z value Pr(>|z|)
(Intercept) -0.645078   0.161121  -4.004 6.24e-05 ***
age         -0.010685   0.003888  -2.748  0.00599 **
duration     0.022887   0.003467   6.602 4.05e-11 ***
---
Signif. codes:  0 *** 0.001 ** 0.01 * 0.05 . 0.1   1

(Dispersion parameter for binomial family taken to be 1)

    Null deviance: 1221.7  on 999  degrees of freedom
Residual deviance: 1169.0  on 997  degrees of freedom
AIC: 1175

Number of Fisher Scoring iterations: 4

> S <- pnorm(predict(reg))
```

For that score, it is possible to plot the ROC curve using

```
> FP <- function(s) sum((S>s)*(Y==0))/sum(Y==0)*100
> TP <- function(s) sum((S>s)*(Y==1))/sum(Y==1)*100
> u <- seq(0,1,length=251)
> plot(Vectorize(FP)(u),Vectorize(TP)(u),type="s",
+ xlab="False Positive Rate (%)",ylab="True Positive Rate (%)")
> abline(a=0,b=1,col="grey")
```

This function can be obtained using functions `performance` and `prediction` from the `library(ROCR)`:

```
> library(ROCR)
> pred <- prediction(S,Y)
> perf <- performance(pred,"tpr", "fpr")
> plot(perf)
```

It is also possible to plot confidence bands, using either `library(verification)` or `library(pROC)`:

```
> library(verification)
> roc.plot(Y,S, xlab = "False Positive Rate",
+ ylab = "True Positive Rate", main = "", CI = TRUE,
+ n.boot = 100, plot = "both", binormal = TRUE)
> library(pROC)
> roc <- plot.roc(Y,S,main="", percent=TRUE, ci=TRUE)
> roc.se <- ci.se(roc,specificities=seq(0, 100, 5))
> plot(roc.se, type="shape", col="grey")
```

Those four graphs can be visualized on Figure 4.3.

The performance curve of the scoring function S is the graph $\{x(s), y(s)\}$, where

$$\left\{x(s) = \mathbb{P}(S \geq s); y(s) = \frac{\mathbb{P}(Y = 1|S \geq s)}{\mathbb{P}(Y = 1)}\right\}_{s\in(0,1)}$$

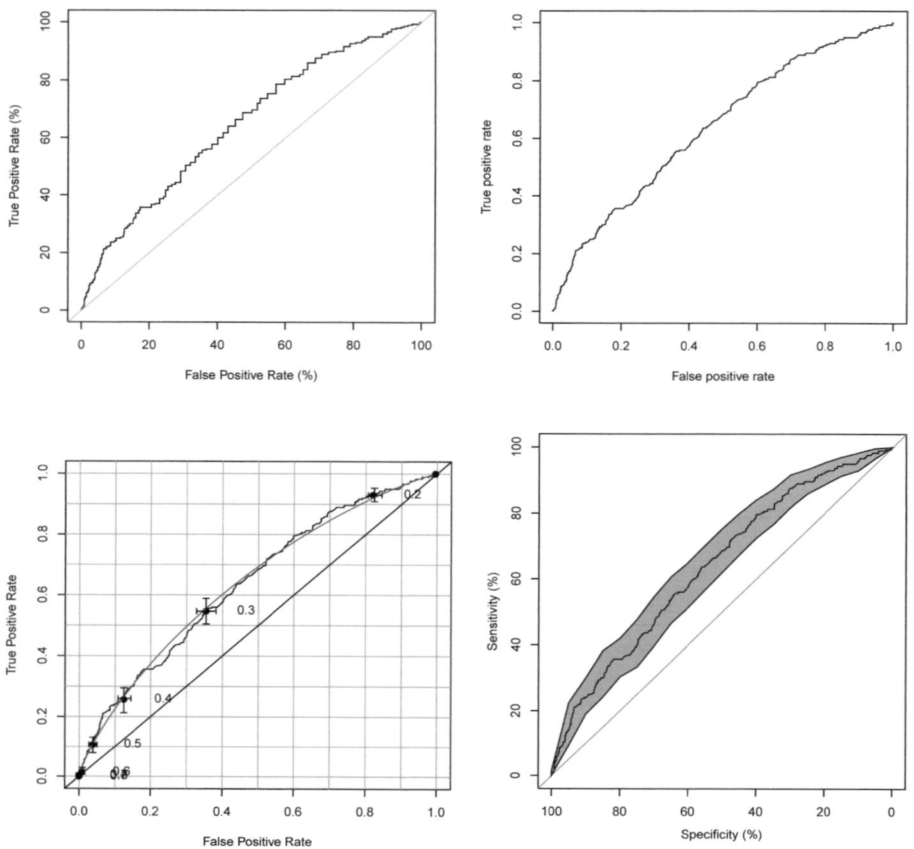

FIGURE 4.3
ROC curve (true positive rate versus false positive rate) from a probit model on the age and the duration of the loan, using our own codes, `ROCR`, `verification` and `pROC` packages.

The selection curve of the scoring function S is the function

$$\{x(s) = \mathbb{P}(S \geq s); y(s) = \mathbb{P}(S \geq s|Y = 1)\}_{s \in (0,1)}.$$

Practitioners also use those functionals to derive indicators (see Hand (2005) for a discussion). According to May (2004), Kolmogorov-Smirnov statistics (called the K-S statistic) "*is the most widely used statistic within the United States for measuring the predictive power of rating systems*," but "*this does not seem to be the case in other environments, where the Gini seem to be more prevalent*" (see also Anderson (2007)). The code to compute the K-S statistic, using outputs from the `ROCR` package, is

```
> max(attr(perf,'y.values')[[1]]-attr(perf,'x.values')[[1]])
[1] 0.1966667
```

which is the largest difference between the cumulative true positive and cumulative false positive rate. This statistic can be visualized on Figure 4.4.

```
> plot(ecdf(S[credit$class==0]),main="",xlab="", pch=19,cex=.2)
```

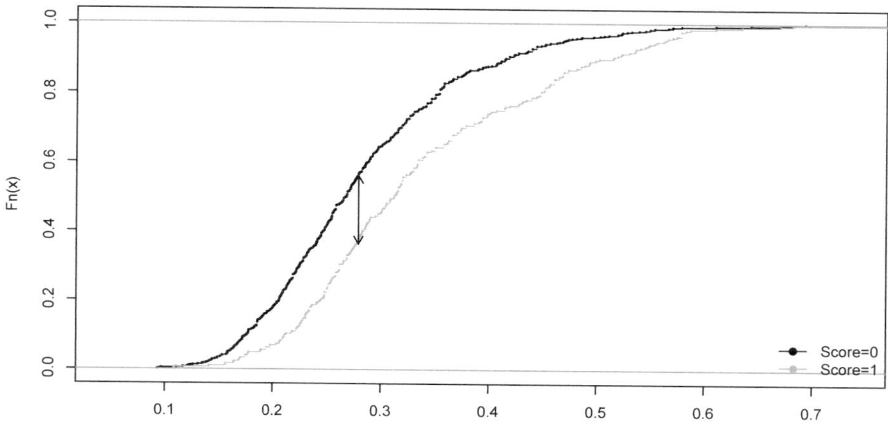

FIGURE 4.4
Kolmogorov–Smirnov statistic.

```
> plot(ecdf(S[credit$class==1]),, pch=19,cex=.2, col="grey",add=TRUE)
> legend("bottomright",c("Score=0","Score=1"), pch=19,col=c("black","grey"),lwd=1,bty="n")
> perf <- performance(pred,"tpr", "fpr")
> ks <- perf@y.values[[1]]-perf@x.values[[1]]
> (seuil <- pred@cutoffs[[1]][which.max(ks)])
     827
0.279034
> arrows(seuil,1-perf@y.values[[1]][which.max(ks)],seuil,
+ 1-perf@x.values[[1]][which.max(ks)],col='black',length = 0.1,code=3)
```

The AUC (area under the ROC curve) statistic is another way to measure predictive power of models, and the code is

```
> performance(pred,"auc")
```

Note that here we obtain 0.6405:

```
Slot "y.values":
[[1]]
[1] 0.6405286
```

This area is included between 0 and 1, and the more it is close to 1, the more predictive is the model. In that case, the ROC curve will hug the top left corner (indicating a high true positive rate and a low false positive rate).

4.1.4 Recoding the Variables

There is here a large number of factors, and all of them have a large number of modalities. Using the `recode` function, it is possible to recode the variables, and to merge modalities.

```
> library(car)
> credit.rcd <- credit.f
> credit.rcd$checking_status <- recode(credit.rcd$checking_status,
```

```
"'A14'='No checking account';'A11'='CA < 0 euros';'A12'=
+ 'CA in [0-200 euros[';'A13'='CA > 200 euros' ")
> credit.rcd$credit_history <- recode(
+ credit.rcd$credit_history,
+ "c('A30','A31')= 'critical account';
+ c('A32','A33')='existing credits paid back duly till now';
+ 'A34'='all credits paid back duly'")
> credit.rcd$purpose <- recode(
+ credit.rcd$purpose,"'A40'='Car (new)';
+ 'A41'='Car (used)';c('A42','A43','A44','A45')='Domestic equipment';c('A46','A48'
+ ,'A49')='Studies-Business';
+ 'A47'='Holidays';else='Else'")
> credit.rcd$savings <- recode(credit.rcd$savings,"c('A65','A63','A64')=
+ 'No savings or > 500 euros';c('A62','A61')='< 500 euros'")
> credit.rcd$employment <-
+ recode(credit.rcd$employment,"c('A71','A72')='unemployed or < 1 year';'A73'=
+ 'E [1-4[ years';c('A74','A75')='> 4 years'")
> credit.rcd$personal_status <-
+ recode(credit.rcd$personal_status ,
+ "'A91'=' male divorced/separated';'A92'='female divorced/separated/married';
+ c('A93','A94')='male single/married/widowed';'A95'='female : single'")
> credit.rcd$other_parties <- recode(credit.rcd$other_parties,"'A103'='guarantor';else='none'")
> credit.rcd$property_magnitude <-
+ recode(credit.rcd$property_magnitude,"'A121'='Real estate';'A124'='No property';
+ else='Else'")
> credit.rcd$other_payment_plans <-
+ recode(credit.rcd$other_payment_plans,"'A143'='None';else='Banks-Stores'")
> credit.rcd$housing <-
+ recode(credit.rcd$housing,"'A152'='Owner';else='Else'")
```

Using those labels will help in the interpretation of our models.

4.1.5 Training and Testing Samples

The natural technique to generate a sub-sample is to use the `sample` function:

```
> set.seed(123)
> index <- sort(sample(nrow(credit), 644, replace=FALSE))
> table(credit$class[index])

  1   2
447 197
```

In order to reproduce some outputs obtained in Tufféry (2011), using SAS®, we have to use another random number generator, based on Fishman & Moore (1982):

```
> library(randtoolbox)
> set.generator(name="congruRand", mod=2^(31)-1,mult=397204094, incr=0, seed=123)
> U=runif(1000)
> index <- sort(which(rank(U)<=644))
> table(credit$class[index])

  0   1
451 193
```

The training sample will be based on the 644 observations, and the remaining 356 will be used as the validation sample:

```
> train.db <- credit.rcd[index,]
> valid.db <- credit.rcd[-index,]
```

4.2 Logistic Regression

The first model that we will consider is logistic regression (see Hosmer & Lemeshow (2000) or Hilbe (2009)). As mentioned in Thomas (2000), logistic regression is now the most widely used method in credit scoring.

From a computational point of view, logistic regression is a Generalized Linear Model where the distribution of the endogenous variate is binary. Thus, in R, the code to run a logistic regression is

```
> reg <- glm(Y~X1+X2+X3,data=df,family=binomial(link='logit'))
```

4.2.1 Inference in the Logistic Model

Let Y_i denote random variables taking values $y_i \in \{0,1\}$, with probability $1-\pi_i$ and π_i (respectively), that is,

$$\mathbb{P}(Y_i = y_i) = \pi_i^{y_i}[1-\pi_i]^{1-y_i} = \begin{cases} \pi_i, \text{ if } y_i = 1 \\ 1-\pi_i, \text{ if } y_i = 0 \end{cases}, \quad \text{for } y_i \in \{0,1\},$$

with $\pi_i \in [0,1]$. Here $\mathbb{E}(Y_i) = \pi_i$ while $\text{Var}(Y_i) = \pi_i[1-\pi_i]$.

If we assume that π_i are identical (denoted π) and Y_i are independent, then—see Chapter 2—π can be estimated using maximum likelihood techniques. Here, the likelihood is

$$\mathcal{L}(\pi;\boldsymbol{y}) = \prod_{i=1}^{n} \mathbb{P}(Y_i = y_i) = \prod_{i=1}^{n} \pi^{y_i}[1-\pi]^{1-y_i}$$

and the log-likelihood becomes

$$\log \mathcal{L}(\pi;\boldsymbol{y}) = \sum_{i=1}^{n} y_i \log[\pi] + (1-y_i)\log[1-\pi]$$

The first-order condition is

$$\left.\frac{\partial \log \mathcal{L}(\pi;\boldsymbol{y})}{\partial \pi}\right|_{\pi=\widehat{\pi}} = \sum_{i=1}^{n} \frac{y_i}{\widehat{\pi}} - \frac{1-y_i}{1-\widehat{\pi}} = 0.$$

Assume now that π_i's are different, and a function of some covariates, $\boldsymbol{X}_i$, so that $\pi_i = \mathbb{E}(Y_i|\boldsymbol{X}_i)$. To be more specific, using the standard matrix notation,

```
> Y <- credit[index,"class"]
> X <- as.matrix(cbind(1,credit[index,c("age","duration")]))
```

A linear model, $\mathbb{E}(Y_i|\boldsymbol{X}_i) = \boldsymbol{X}_i'\boldsymbol{\beta}$, will not be suitable, because $\boldsymbol{X}_i'\boldsymbol{\beta}$ can be negative, or larger than 1. Parametrization with the probability yields too many constraints, so a natural idea is to model the odds, or the logarithm of the odds ratio:

$$\text{logit}(\pi_i) = \log\left(\frac{\pi_i}{1-\pi_i}\right) = \boldsymbol{X}_i'\boldsymbol{\beta}$$

or equivalently,

$$\pi_i = \text{logit}^{-1}(\boldsymbol{X}_i'\boldsymbol{\beta}) = \frac{\exp[\boldsymbol{X}_i'\boldsymbol{\beta}]}{1+\exp[\boldsymbol{X}_i'\boldsymbol{\beta}]}.$$

In that case, the log-likelihood becomes

$$\log \mathcal{L}(\boldsymbol{\beta}) = \sum_{i=1}^{n} y_i \log(\pi_i) + (1-y_i)\log(1-\pi_i) = \sum_{i=1}^{n} y_i \log(\pi_i(\boldsymbol{\beta})) + (1-y_i)\log(1-\pi_i(\boldsymbol{\beta})).$$

The gradient $\nabla \log \mathcal{L}(\boldsymbol{\beta}) = [\partial \log \mathcal{L}(\boldsymbol{\beta})/\partial \beta_k]$ is

$$\frac{\partial \log \mathcal{L}(\boldsymbol{\beta})}{\partial \beta_k} == \sum_{i=1}^{n} \frac{y_i}{\pi_i(\boldsymbol{\beta})}\frac{\partial \pi_i(\boldsymbol{\beta})}{\partial \beta_k} - \frac{1-y_i}{\pi_i(\boldsymbol{\beta})}\frac{\partial \pi_i(\boldsymbol{\beta})}{\partial \beta_k},$$

which can be written

$$\frac{\partial \log \mathcal{L}(\boldsymbol{\beta})}{\partial \beta_k} = \sum_{i=1}^{n} X_{k,i}[y_i - \pi_i(\boldsymbol{\beta})], \ \forall k$$

since, from the expression of $\pi_i(\boldsymbol{\beta})$,

$$\frac{\partial \pi_i(\boldsymbol{\beta})}{\partial \beta_k} = \pi_i(\boldsymbol{\beta})[1-\pi_i(\boldsymbol{\beta})]X_{k,i}.$$

Because there is no analytical solution for this expression, it is natural to use Newton–Raphson's algorithm:

1. Start from some initial value $\boldsymbol{\beta}_0$.
2. Set $\boldsymbol{\beta}_k = \boldsymbol{\beta}_{k-1} - H(\boldsymbol{\beta}_{k-1})^{-1}\nabla \log \mathcal{L}(\boldsymbol{\beta}_{k-1})$

a loop, where $\nabla \log \mathcal{L}(\boldsymbol{\beta})$ is the gradient of the log-likelihood, and $H(\boldsymbol{\beta})$ the Hessian matrix, with generic term

$$\frac{\partial^2 \log \mathcal{L}(\boldsymbol{\beta})}{\partial \beta_k \partial \beta_\ell} = \sum_{i=1}^{n} X_{k,i}X_{\ell,i}[y_i - \pi_i(\boldsymbol{\beta})].$$

Here,

$$\boldsymbol{\beta}_k = (\boldsymbol{X}^\top \boldsymbol{\Omega} \boldsymbol{X})^{-1}\boldsymbol{X}^\top \boldsymbol{\Omega} \boldsymbol{Z}, \text{ where } \boldsymbol{Z} = \boldsymbol{X}\boldsymbol{\beta}_{k-1} + \boldsymbol{\Omega}^{-1}(\boldsymbol{y}-\boldsymbol{\pi}),$$

and $\boldsymbol{\Omega} = [\omega_{i,j}] = \text{diag}(\widehat{\pi}_i(1-\widehat{\pi}_i))$. We do recognize a weighted regression, with weights $\boldsymbol{\Omega}$.

Our starting points can be estimators from a standard linear regression:

```
> beta <- as.vector(lm(Y~0+X[,1]+X[,2]+X[,3])$coefficients)
> BETA <- NULL
> for(s in 1:6){
+ pi <- exp(X%*%beta)/(1+exp(X%*%beta))
+ gradient <- t(X)%*%(Y-pi)
+ omega <- matrix(0,nrow(X),nrow(X));diag(omega)=(pi*(1-pi))
+ hessian <- -t(X)%*%omega%*%X
+ beta <- beta-solve(hessian)%*%gradient
+ BETA <- cbind(BETA,beta)
+ }
```

Observe that, actually, four iterations would be sufficient:

```
> BETA
                [,1]        [,2]        [,3]        [,4]        [,5]        [,6]
1        -1.08304581 -1.10983815 -1.10665019 -1.10663432 -1.10663432 -1.10663432
age      -0.01281867 -0.01754473 -0.01786453 -0.01786550 -0.01786550 -0.01786550
duration  0.03395857  0.03968785  0.03991297  0.03991347  0.03991347  0.03991347
```

Under mild regularity conditions

$$\sqrt{n}(\widehat{\boldsymbol{\beta}} - \boldsymbol{\beta}) \xrightarrow{\mathcal{L}} \mathcal{N}(\mathbf{0}, I(\boldsymbol{\beta})^{-1}), \text{ as } n \to \infty,$$

where $I(\boldsymbol{\beta}) = -H(\boldsymbol{\beta})$, thus the asymptotic variance-covariance matrix is `hessian`:

```
> (SD <- sqrt(diag(solve(-hessian))))
          1         age    duration
0.342223292 0.008673299 0.007171971
```

A coefficient is said to be significant if the p-value associated to the Student test is smaller than 5%. That p-value is computed as follows:

```
> cbind(BETA[,6],SD,BETA[,6]/SD,2*(1-pnorm(abs(BETA[,6]/SD))))
                                  SD
1         -1.10663432 0.342223292 -3.233662 1.222142e-03
age       -0.01786550 0.008673299 -2.059828 3.941499e-02
duration   0.03991347 0.007171971  5.565202 2.618487e-08
```

Those values can be obtained using `summary` of a `glm` object:

```
> reg <- glm(class~age+duration,data=credit[index,],
+ family=binomial(link="logit"))
> summary(reg)

Call:
glm(formula = class ~ age + duration, family = binomial(link = "logit"),
    data = credit[index, ])

Deviance Residuals:
    Min       1Q   Median       3Q      Max
-1.5814  -0.8382  -0.7007   1.2380   2.0974

Coefficients:
             Estimate Std. Error z value Pr(>|z|)
(Intercept) -1.106634   0.342223  -3.234  0.00122 **
age         -0.017866   0.008673  -2.060  0.03941 *
duration     0.039913   0.007172   5.565 2.62e-08 ***
---
Signif. codes:  0 *** 0.001 ** 0.01 * 0.05 . 0.1   1

(Dispersion parameter for binomial family taken to be 1)

    Null deviance: 786.45  on 643  degrees of freedom
Residual deviance: 750.43  on 641  degrees of freedom
AIC: 756.43

Number of Fisher Scoring iterations: 4
```

We do recognize the estimators $\widehat{\beta}_k$ and their standard errors obtained above (with only four iterations).

4.2.2 Logistic Regression on Categorical Variates

Consider, first, a model with *one* categorical regressor, for instance `credit_history`, with five modalities in the original `credit` database:

```
> reg <- glm(class~credit_history,data=credit,family=binomial(link="logit"))
> summary(reg)

Call:
glm(formula = class ~ credit_history, family = binomial(link = "logit"),
    data = credit)

Deviance Residuals:
    Min       1Q   Median       3Q      Max
-1.4006  -0.8764  -0.6117   1.0579   1.8805

Coefficients:
                  Estimate Std. Error z value Pr(>|z|)
(Intercept)         0.5108     0.3266   1.564 0.117799
credit_historyA31  -0.2231     0.4359  -0.512 0.608703
credit_historyA32  -1.2698     0.3396  -3.739 0.000185 ***
credit_historyA33  -1.2730     0.3988  -3.192 0.001413 **
credit_historyA34  -2.0919     0.3616  -5.784 7.28e-09 ***
---
Signif. codes:  0 *** 0.001 ** 0.01 * 0.05 . 0.1   1

(Dispersion parameter for binomial family taken to be 1)

    Null deviance: 1221.7  on 999  degrees of freedom
Residual deviance: 1161.3  on 995  degrees of freedom
AIC: 1171.3

Number of Fisher Scoring iterations: 4
```

In that case, the value predicted per category is exactly the empirical one:

```
> cbind(prop.table(table(credit$credit_history,credit$class),1),
+ logit=predict(reg,newdata=data.frame(credit_history=
+ levels(credit$credit_history)), type="response"))

            0         1     logit
A30 0.3750000 0.6250000 0.6250000
A31 0.4285714 0.5714286 0.5714286
A32 0.6811321 0.3188679 0.3188679
A33 0.6818182 0.3181818 0.3181818
A34 0.8293515 0.1706485 0.1706485
```

In the case of two categorical regressors, the interpretation is rather different. In the `credit.rcd` database (in order to have less modalities), consider the `credit_history` variable (three modalities) and the `purpose` variable (five modalities).

```
> reg <- glm(class~credit_history*purpose,data=credit.rcd,
+ family=binomial(link="logit"))
> p.class=matrix(predict(reg,newdata=data.frame(
```

```
+ credit_history=rep(levels(credit_history),each=length(levels(purpose))),
+ purpose=rep(levels(purpose),length(levels(credit_history)))),type="response"),
+ ncol=length(levels(credit_history)),
+ nrow=length(levels(purpose)))
> rownames(p.class) <- levels(purpose)
> colnames(p.class) <- levels(credit_history)
> p.class
                    all credits paid back duly critical account paid back duly
Car (new)                            0.2435897        0.7368421      0.4087591
Car (used)                           0.1111111        0.3750000      0.1694915
Domestic equipment                   0.1313869        0.5161290          0.2996942
Else                                 0.3333333        0.9999995      0.1666667
Studies-Business                     0.2051282        0.6071429        0.3595506
```

Here, a product of the two regressors was considered. But in the context of standard *linear* models, the prediction will not be equal to the empirical average value:

```
> reg <- lm(class~credit_history*purpose,data=credit.rcd)
> p.class.linear=matrix(predict(reg,newdata=data.frame(
+ credit_history=rep(levels(credit_history),each=length(levels(purpose))),
+ purpose=rep(levels(purpose),length(levels(credit_history)))),type="response"),
+ ncol=length(levels(credit_history)),
+ nrow=length(levels(purpose)))
> rownames(p.class.linear) <- levels(purpose)
> colnames(p.class.linear) <- levels(credit_history)
> p.class.linear
                    all credits paid back duly critical account paid back duly
Car (new)                           0.23626428        0.6872147        0.4198125
Car (used)                          0.08633173        0.4015839        0.1810066
Domestic equipment                  0.14817277        0.5526530        0.2891990
Else                                0.21699668        0.6631013        0.3932843
Studies-Business                    0.19263547        0.6288823      0.3581855
```

If instead of modelling probabilities, we compute odds ratios,

```
> p.class.linear/(1-p.class.linear)
                    all credits paid back duly critical account paid back duly
Car (new)                           0.30935346          2.197081      0.7235806
Car (used)                          0.09448914          0.671078      0.2210110
Domestic equipment                  0.17394698          1.235401    0.4068636
Else                                0.27713380          1.968251        0.6482185
Studies-Business                    0.23859789          1.694563      0.5580827
```

we observe that rows and columns, here, are proportionals, which is a consequence of the logistic regression, as odds ratios is a multiplicative model.

4.2.3 Step-by-Step Variable Selection

In stepwise selection, the choice of predictive variables is carried out by an automatic procedure, based on some specified criterion. As with the regression tree (and the CART algorithm), with a lot of covariates, the number of possible models can become too large extremely fast: with np covariates, there are $2^{\texttt{np}}$ possible models. With twenty covariates, there will be more than a million models to compare. Two approaches are usually considered:

- A forward selection, which involves starting with no variables in the model. Then the addition of each variable is tested (based on the criterion considered); we then add the variable (if any) that improves the model the most, and repeat this process until none improves the model significantly.

- A backward selection, which involves starting with all variables in the model. Then the deletion of each variable is tested (based on the criterion considered); we then remove the variable (if any) that improves the model the most by removing, and repeat this process until none improves the model significantly.

The most common criteria are based on penalization of the log-likelihood. One might think of the adjusted R^2 in the context of linear regression. For instance, if np denotes the number of parameters in the fitted model, $-\log \mathcal{L} + k \cdot \texttt{np}$ is the usual AIC (Akaike Information Criterion) when $k = 2$, while it is Schwarz's Bayesian Criterion (BIC) when $k = \log(n)$. The generic function to compute those values is

```
> AIC(model,k=2)
> AIC(model,k=log(nobs(model)))
```

for some fitted `model`. In the context of linear models, those criteria are equal (up to an additive constant) to n times the logarithm of the sum of squares of the residuals. For example, the AIC is

$$AIC = \text{ constant } + n \log(\text{sum of squares of the residuals}) + 2\texttt{np}.$$

Another popular criterion is Mallows' C_p, defined as

$$C_p = \frac{\text{sum of squares of the residuals}}{\widehat{\sigma}^2_\star} + 2\text{np} - n,$$

where $\widehat{\sigma}^2_\star$ is the estimated error variance for the largest model that can be considered. Again, this statistic adds a penalty to the (logarithm of the) sum of squares of the residuals. As earlier, the penalty increases as the number of predictors in the model increases. Interesting models should have a small C_p.

4.2.3.1 Forward Algorithm

The procedure here is rather simple. We start with no covariate, which is obtained using `glm(class~1)`. The variables are added, one by one, until we cannot decrease the AIC or BIC criterion by adding new variables:

```
> predictors <- names(credit.rcd) [-grep('class', names(credit.rcd))]
> formula <- as.formula(paste("y ~ ",
+ paste(names(credit.rcd[,predictors]),collapse="+")))
> logit <- glm(class~1,data=train.db,family=binomial)
> selection <- step(logit,direction='forward',trace=TRUE,k=log(nrow(train.db)),
               scope=list(upper=formula))
```

Variables are entered in the order in which they most improve the fit, viz they most decrease AIC (actually BIC here, with $k = \log(n)$). The first step was

```
Start:  AIC=792.92
class ~ 1


                    Df Deviance    AIC
+ checking_status    3   691.54 717.41
+ duration           2   748.47 767.88
```

```
+ credit_history        2    760.26 779.66
+ savings               1    767.59 780.53
+ housing               1    775.25 788.18
+ credit_amount         1    775.33 788.26
+ age                   1    775.39 788.33
+ other_payment_plans   1    778.54 791.48
+ purpose               4    759.81 792.15
<none>                       786.45 792.92
+ employment            2    773.94 793.34
+ installment_rate      1    780.41 793.34
+ other_parties         1    782.74 795.68
+ telephone             1    784.39 797.33
+ personal_status       2    779.58 798.99
+ num_dependents        1    786.39 799.32
+ residence_since       1    786.45 799.39
+ property_magnitude    2    780.23 799.63
+ existing_credits      3    784.70 810.57
+ job                   3    785.54 811.41
```

and we ended with

```
Step:  AIC=684.09
class ~ checking_status + duration + purpose + credit_history +
    savings + other_parties + age

                      Df Deviance    AIC
<none>                     587.08 684.09
+ other_payment_plans  1   582.23 685.71
+ installment_rate     1   582.61 686.09
+ housing              1   583.90 687.38
+ credit_amount        1   584.31 687.79
+ telephone            1   586.67 690.15
+ num_dependents       1   587.08 690.56
+ residence_since      1   587.08 690.56
+ employment           2   582.23 692.19
+ property_magnitude   2   584.12 694.07
+ personal_status      2   585.10 695.05
+ existing_credits     3   585.34 701.76
+ job                  3   586.84 703.26
```

Predictions for this model are the following, respectively on the training dataset, and also and the validation dataset:

```
> train.db$forward.bic <- predict(selection,newdata=train.db,type='response')
> valid.db$forward.bic <- predict(selection,newdata=valid.db,type='response')
```

4.2.3.2 Backward Algorithm

This time, the procedure is the exact opposite of the previous one. We start with all covariates, which is obtained using `glm(class~.)`. The variables are dropped, one by one, until we cannot decrease the AIC criterion by dropping variables:

```
> logit <- glm(class~.,data=train.db[,c("class",predictors)],family=binomial)
> selection <- step(logit,direction='backward',trace=TRUE,k=log(nrow(train.db)))
```

The first step is

```
Start:  AIC=775.85
class ~ checking_status + duration + credit_history + purpose +
    credit_amount + savings + employment + installment_rate +
    personal_status + other_parties + residence_since + property_magnitude +
    age + other_payment_plans + housing + existing_credits +
    job + num_dependents + telephone

                      Df Deviance    AIC
- job                  3   556.81 757.31
- existing_credits     3   557.88 758.38
- property_magnitude   2   557.24 764.20
- personal_status      2   558.69 765.66
- employment           2   559.56 766.53
- residence_since      1   555.96 769.39
- num_dependents       1   555.98 769.41
- telephone            1   556.59 770.02
- housing              1   557.23 770.66
- age                  1   559.36 772.79
- credit_history       2   566.59 773.55
- other_payment_plans  1   561.62 775.05
- credit_amount        1   562.30 775.73
<none>                     555.95 775.85
- other_parties        1   563.23 776.67
- savings              1   564.50 777.94
- installment_rate     1   565.29 778.72
- duration             2   575.42 782.39
- purpose              4   593.71 787.75
- checking_status      3   610.23 810.73
```

and the last step was (the same model as with forward selection)

```
Step:  AIC=684.09
class ~ checking_status + duration + credit_history + purpose +
    savings + other_parties + age

                  Df Deviance    AIC
<none>                 587.08 684.09
- age              1   593.74 684.29
- credit_history   2   601.55 685.63
- other_parties    1   595.16 685.71
- savings          1   596.18 686.73
- purpose          4   624.16 695.31
- duration         2   628.37 712.45
- checking_status  3   646.53 724.14
```

Forward selection here gives the same model as backward selection, but it is not always true. Again,

```
train.db$backward.bic <- predict(selection,newdata=train.db,type='response')
valid.db$backward.bic <- predict(selection,newdata=valid.db,type='response')
```

It is possible to visualize the accuracy of these two techniques in Figure 4.5,

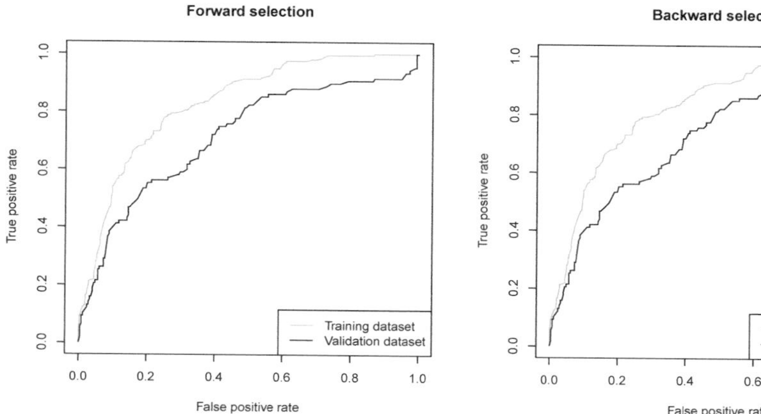

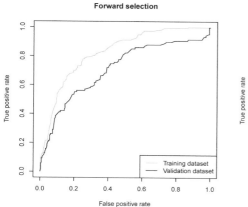

FIGURE 4.5
ROC curve (true positive rate versus false positive rate) from a logit model, using forward (on the left) and backward (on the right) stepwise selection.

```
> pred.train <- prediction(train.db$forward.bic,train.db$class)
> pred.valid <- prediction(valid.db$forward.bic,valid.db$class)
> perf.train <- performance(pred.train,"tpr", "fpr")
> perf.valid <- performance(pred.valid,"tpr", "fpr")
> plot(perf.train,col="grey",lty=2,main="Forward selection")
> plot(perf.valid,add=TRUE)
> legend("bottomright",c("Training dataset","Validation dataset"),
+ lty=c(2,1),col=c("grey","black"),lwd=1)
```

4.2.4 Leaps and Bounds

The `leaps()` function (from the eponym library) will search for the best subsets of our predictors using whichever criterion (e.g. Mallows' C_p), according to the leaps and bounds algorithm of Furnival & Wilson (1974). For each $p \geq 1$, we are looking for the `nbest` (here = 1) subset(s) of p predictors, giving the model(s) with the least Mallows' C_p (`method = "Cp"`). The difference with forward and backward selection is in the search without the constraint that the models are nested. Conversely, the limitation of the leaps() function is that it only applies on numeric variables, so that qualitative variables must be replaced by the indicators of their modalities.

```
> train.db <- credit.f[index,]
> y <- as.numeric(train.db[,"class"])
> x <- data.frame(model.matrix(~.,data=train.db[,-which(names(train.db)=="class")]))
> library(leaps)
> selec <- leaps(x,y,method="Cp",nbest=1,strictly.compatible=FALSE)
$label
 [1] "(Intercept)"              "X.Intercept."             "checking_statusA12"
 [4] "checking_statusA13"       "checking_statusA14"       "duration.15.36."
 [7] "duration.36.Inf."         "credit_historyA31"        "credit_historyA32"
[10] "credit_historyA33"        "credit_historyA34"        "purposeA41"
[13] "purposeA410"              "purposeA42"               "purposeA43"
```

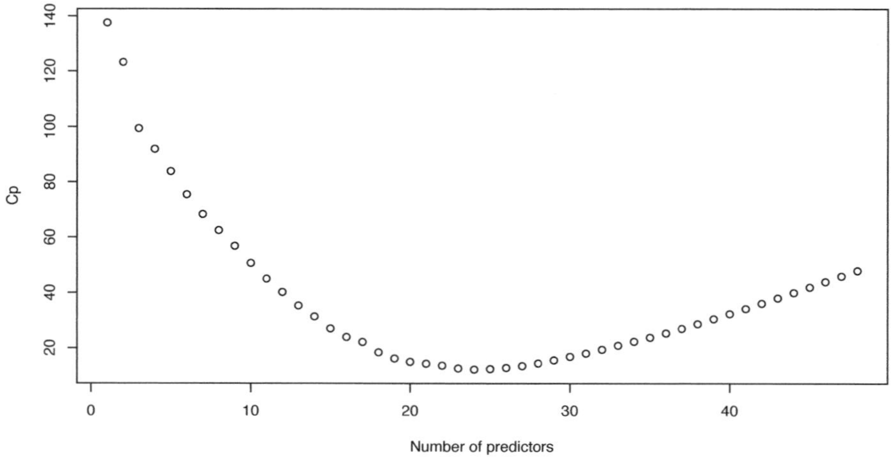

FIGURE 4.6
Evolution of Mallows' C_p as a function of the number of predictors.

```
[16] "purposeA44"                "purposeA45"                "purposeA46"
[19] "purposeA48"                "purposeA49"                "credit_amount.4e.03.Inf."
[22] "savingsA62"                "savingsA63"                "savingsA64"
[25] "savingsA65"                "employmentA72"             "employmentA73"
[28] "employmentA74"             "employmentA75"             "installment_rate"
[31] "personal_statusA92"        "personal_statusA93"        "personal_statusA94"
[34] "other_partiesA102"         "other_partiesA103"         "residence_since"
[37] "property_magnitudeA122"    "property_magnitudeA123"    "property_magnitudeA124"
[40] "age.25.Inf."               "other_payment_plansA142"   "other_payment_plansA143"
[43] "housingA152"               "housingA153"               "existing_credits"
[46] "jobA172"                   "jobA173"                   "jobA174"
[49] "num_dependents"            "telephoneA192"

$size
 [1]  2  3  4  5  6  7  8  9 10 11 12 13 14 15 16 17 18 19 20 21 22 23 24 25 26 27
[27]  28 29 30 31 32 33 34 35 36 37 38 39 40 41 42 43 44 45 46 47 48 49

$Cp
 [1] 137.62062 123.29162  99.42835  91.94227  83.88375  75.48833  68.37060  62.53615
 [9]  56.87359  50.73275  45.06511  40.25486  35.35416  31.38907  27.05092  23.97690
[17]  22.20806  18.48098 16.26368  15.07004  14.41763  13.70477  12.71125  12.30494
[25]  12.50403  12.90835  13.54865 14.52072  15.62970  16.92261  18.08451  19.45329
[33]  20.94871  22.35779  23.81829  25.37059 26.99681  28.71035  30.51637  32.34840
[41]  34.23233  36.14711  38.08677  40.03967  42.01328 44.00346  46.00022  48.00000

> plot(selec$size-1,selec$Cp, xlab="Number of predictors",ylab="Cp")
```

The 'best model' is the one that minimized Mallows's C_p (achieved with twenty-four predictors)

```
> best.model <- selec$which[which((selec$Cp == min(selec$Cp))),]
> z <- cbind(x,y)
```

```
> formula <- as.formula(paste("y ~",paste(colnames(x)[best.model],
   collapse="+")))
> logit <- glm(formula,data=z,family=binomial(link ="logit"))
> summary(logit)

Call:
glm(formula = formula, family = binomial(link = "logit"), data = z)

Deviance Residuals:
    Min       1Q   Median       3Q      Max
-2.3227  -0.6496  -0.3258   0.6392   2.7946

Coefficients:
                          Estimate Std. Error z value Pr(>|z|)
(Intercept)                 1.5531     0.6064   2.561 0.010425 *
checking_statusA12         -0.5222     0.2731  -1.912 0.055842 .
checking_statusA13         -0.8492     0.4417  -1.922 0.054572 .
checking_statusA14         -2.0301     0.2944  -6.896 5.35e-12 ***
duration.15.36.             1.1469     0.2579   4.447 8.70e-06 ***
duration.36.Inf.            1.2799     0.4501   2.843 0.004463 **
credit_historyA32          -0.5860     0.3772  -1.553 0.120308
credit_historyA33          -0.7710     0.5008  -1.540 0.123667
credit_historyA34          -1.3136     0.4092  -3.210 0.001327 **
purposeA41                 -2.1981     0.4774  -4.605 4.13e-06 ***
purposeA410               -16.7122   622.6691  -0.027 0.978588
purposeA42                 -1.2116     0.3175  -3.817 0.000135 ***
purposeA43                 -1.3125     0.3067  -4.280 1.87e-05 ***
purposeA48                 -2.0075     1.2321  -1.629 0.103236
purposeA49                 -1.0524     0.3815  -2.758 0.005808 **
credit_amount.4e.03.Inf.    0.9020     0.3153   2.861 0.004227 **
savingsA64                 -1.0801     0.5886  -1.835 0.066531 .
savingsA65                 -1.1594     0.3185  -3.640 0.000273 ***
employmentA72               0.4903     0.2718   1.804 0.071253 .
installment_rate            0.3088     0.1073   2.878 0.004008 **
other_partiesA103          -1.6873     0.6111  -2.761 0.005760 **
age.25.Inf.                -0.7495     0.2793  -2.683 0.007286 **
other_payment_plansA142    -1.1945     0.5500  -2.172 0.029868 *
other_payment_plansA143    -0.9798     0.2973  -3.296 0.000981 ***
housingA152                -0.3106     0.2462  -1.262 0.207072
---
Signif. codes:  0 *** 0.001 ** 0.01 * 0.05 . 0.1   1

(Dispersion parameter for binomial family taken to be 1)

    Null deviance: 786.45  on 643  degrees of freedom
Residual deviance: 546.44  on 619  degrees of freedom
AIC: 596.44

Number of Fisher Scoring iterations: 14
```

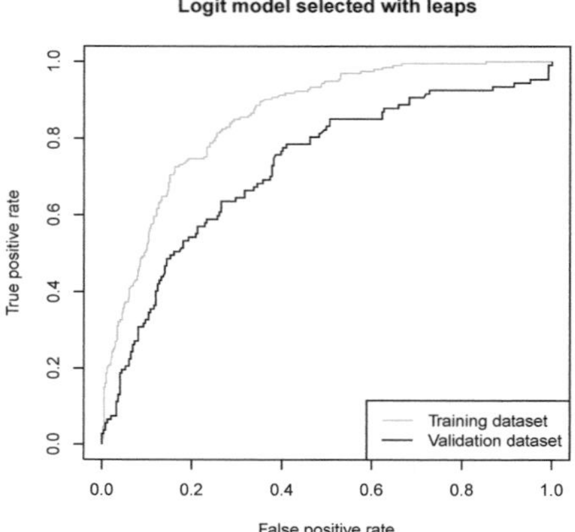

FIGURE 4.7
ROC curve obtained using the `leaps()` procedure, for the model that minimizes Mallows's C_p.

We can compute the performance of this model on the validation sample using the AUC criterion:

```
> xp <- data.frame(model.matrix(~.,data=valid.db[,-which(names(valid.db)=="class")]))
> predclass  <- predict(logit,newdata=xp,type="response")
> pred <- prediction(predclass,valid.db$class,label.ordering=c(0,1))
> performance(pred,"auc")@y.values[[1]]
[1] 0.7249934
```

or by plotting the ROC curve, in Figure 4.7,

```
> pred.train=prediction(predict(logit,newdata=x,type="response"),train.db$class)
> pred.valid=prediction(predict(logit,newdata=xp,type="response"),valid.db$class)
> perf.train=performance(pred.train,"tpr", "fpr")
> perf.valid=performance(pred.valid,"tpr", "fpr")
> plot(perf.train,col="grey",lty=2,main="Logit model selected with leaps")
> plot(perf.valid,add=TRUE)
> legend("bottomright",c("Training dataset","Validation dataset"),
+ lty=c(2,1),col=c("grey","black"),lwd=1)
```

We notice that the model minimizing Mallows' C_p on the training sample is not that one maximizing AUC on the validation sample.

4.2.5 Smoothing Continuous Covariates

So far, we have seen how to model credit applications using categorical variables. But instead of categorizing continuous variables, it could be interesting to keep them as continuous. Categorizing probably makes the model simpler to estimate, and interpret. But there might be serious drawbacks, as discussed in Royston et al. (2006), the main reason being that cutpoints are arbitrary and manipulable (and can result in both positive and negative associations, as mentioned in Wainer (2006)), unless we specify the cutpoints based on the proportion of data in each category of the predictor. As claimed in Harrell (2006), "*a better approach that maximizes power and that only assumes a smooth relationship is to use a restricted cubic spline function for predictors that are not known to predict linearly.*" Nevertheless, one has to admit that having, at the same time, a large number of categorical and continuous variables might yield complex and unstable models. So, smoothing continuous covariates will mainly be discussed to introduce some statistical techniques.

Consider a logistic regression on two continuous variates: the duration D and the age A. Then with a logistic regression,

$$\mathbb{P}(Y = 1|A = a, D = d) = \mathbb{E}(Y|A = a, D = d) = \frac{\exp(\beta_0 + \beta_1 a + \beta_2 d)}{1 + \exp(\beta_0 + \beta_1 a + \beta_2 d)}.$$

This is a *linear* model because a linear combination of explanatory variates is considered

```
> regglm <- glm(class~age+duration,data=credit[index,],family=binomial(link="logit"))
> summary(regglm)

Call:
glm(formula = class ~ age + duration, family = binomial(link = "logit"),
    data = credit[index, ])

Deviance Residuals:
    Min       1Q   Median       3Q      Max
-1.5814  -0.8382  -0.7007   1.2380   2.0974

Coefficients:
             Estimate Std. Error z value Pr(>|z|)
(Intercept) -1.106634   0.342223  -3.234  0.00122 **
age         -0.017866   0.008673  -2.060  0.03941 *
duration     0.039913   0.007172   5.565 2.62e-08 ***
---
Signif. codes:  0 *** 0.001 ** 0.01 * 0.05 . 0.1   1

(Dispersion parameter for binomial family taken to be 1)

    Null deviance: 786.45  on 643  degrees of freedom
Residual deviance: 750.43  on 641  degrees of freedom
AIC: 756.43

Number of Fisher Scoring iterations: 4
```

In those Generalized Linear Models, we considered the following (linear) relationship, based on link function g:

$$g(\mathbb{E}(Y|A = a, D = d)) = \beta_0 + \beta_1 a + \beta_2 d.$$

The idea now is to use non-parametric techniques to estimate $h : \mathbb{R}^2 \to \mathbb{R}$, where

$$g(\mathbb{E}(Y|A = a, D = d)) = \beta_0 + h(a, d).$$

Generalized Additive Models consider restricted a class of functions, namely those that have an additive form:

$$g(\mathbb{E}(Y|A = a, D = d)) = \beta_0 + h_1(a) + h_2(d).$$

As described in Wood (2006), the `gam` function of package `mgcv` can be used to estimate function h, using the Lanczos algorithm. The generic code is

```
> library(mgcv)
> reggam <- gam(Y~s(X1),data=db,family=binomial(link="logit"))
> reggam <- gam(Y~s(X1,by=F1),data=db,family=binomial(link="logit"))
> reggam <- gam(Y~s(X1)+s(X2),data=db,family=binomial(link="logit"))
> reggam <- gam(Y~s(X1,X2),data=db,family=binomial(link="logit"))
```

where X1 and X2 are continuous covariates, while F1 denotes some factor. Some arguments can be added to the s() function, but the default option is to use a cubic spline basis and to automatically choose the smoothing parameter via generalized cross validation. The idea of a cubic spline regression (see Wood (2006) for more details), with k knots $\{\xi_1,\dots,\xi_k\}$ is to start with a cubic polynomial regression, on X_1, X_1^2 and X_1^3, and then to add a truncated power basis function for each knot, $(X_1-\xi)_+^3$ where $(x)_+$ is equal to x if $x>0$, and 0 otherwise. A cubic spline regression is a standard regression on $k+3$ regressors (and the intercept).

Here, consider a bivariate spline smoothing function on the two (continuous) covariates,

```
> reggam <- gam(class~s(age,duration),data=credit[index,],family=binomial(link="logit"))
```

In order to compare the two models, regglm and reggam, it is possible to plot the predicted value of $\mathbb{E}(Y|A,D)$:

```
> pglm <- function(x1,x2){return(predict(regglm,newdata=
+ data.frame(duration=x1,age=x2),type="response"))}
> pgam <- function(x1,x2){return(predict(reggam,newdata=
+ data.frame(duration=x1,age=x2),type="response"))}
> M <- 31
> cx1 <- seq(from=min(credit$duration), to=max(credit$duration), length=M)
> cx2 <- seq(from=min(credit$age), to=max(credit$age), length=M)
> Pgam <- outer(cx1,cx2,pgam)
> persp(cx1,cx2,Pgam)
> contour(cx1,cx2,Pgam)
```

The output for these two models can be visualized in Figure 4.8.

4.2.6 Nearest-Neighbor Method

Another popular tool to classify with a non-parametric model is to use the k-th nearest-neighbor method. The use of this classifier in credit scoring was discussed, for instance, in Henley & Hand (1996).

Given our two covariates $\boldsymbol{X}=(A,D)$, the idea is to obtain a prediction for some $\boldsymbol{x}$ to consider the average of Y over the k-nearest neighbors $\boldsymbol{X}_i$ close to $\boldsymbol{x}$, for some distance d (such as the Euclidean distance). More precisely, if

$$I_k(\boldsymbol{x})=\left\{i\in\{1,\cdots,n\};\sum_{j=1}^n \mathbf{1}(d(\boldsymbol{X}_j,\boldsymbol{x})\leq d(\boldsymbol{X}_j,\boldsymbol{X}_i))=k\right\}$$

denote the k closest points $\boldsymbol{X}_i$ to $\boldsymbol{x}$ in the training sample, then

$$\widehat{Y}_k(\boldsymbol{x})=\frac{1}{k}\sum_{i\in I_k(\boldsymbol{x})} Y_i.$$

```
> pkmeans <- function(x1,x2,k=25){
```

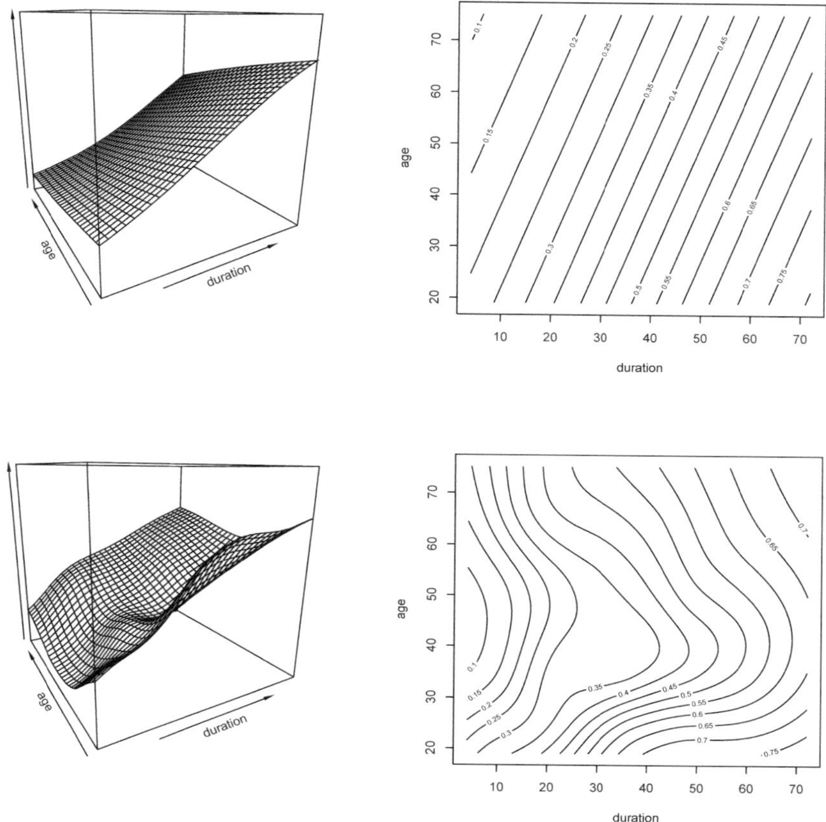

FIGURE 4.8
Linear (on top, with `glm`) versus nonlinear (below, with `gam`) logistic regression, on the duration and the age.

```
> D <- as.matrix(dist(rbind(credit[index,c("duration","age")],
+ c(x1,x2))))[length(index)+1,1:length(index)]
> i <- as.vector(which(D<=sort(D)[k]))
> return(mean((credit[index,"class"])[i])) }
```

4.3 Penalized Logistic Regression: From Ridge to Lasso

In the previous section, we have seen that $\boldsymbol{\beta}$'s in the logistic regression were obtained by maximizing the log-likelihood. But is it possible to add some penalty function, as discussed in Fu (1998) for instance,

$$\widehat{\boldsymbol{\beta}}^{(\lambda)} = \text{argmin}\left\{-\log \mathcal{L}(\boldsymbol{\beta};\boldsymbol{Y},\boldsymbol{X}) + \frac{\lambda}{n}p(\boldsymbol{\beta})\right\}$$

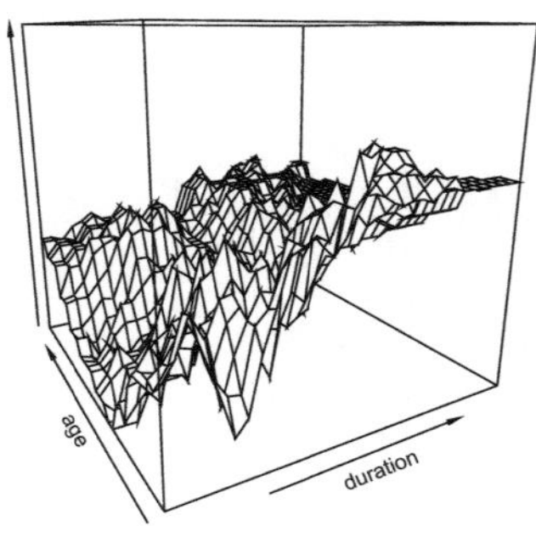

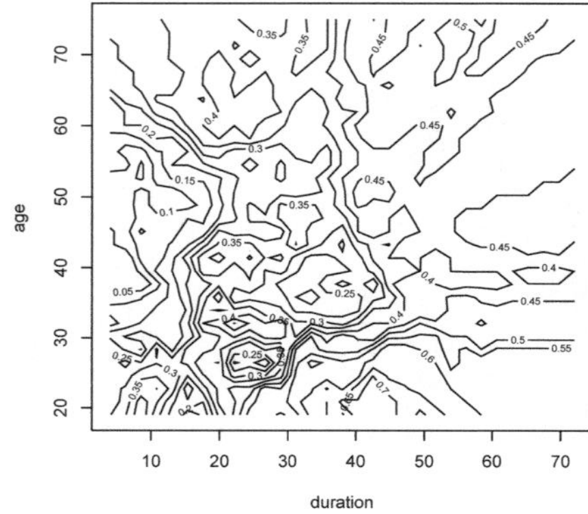

FIGURE 4.9
Nearest-neighbor method used to predict the probability that `class` will be equal to 1, on the duration and the age.

for some parameter $\lambda > 0$ and where p is the so-called penalty function. Penalizing the log-likelihood is common when the number of explanatory variable is large, or when those variables are correlated. The general form is, for $\alpha \in [0,1]$,

$$p(\boldsymbol{\beta}) = \left(\frac{1-\alpha}{2} \sum_k \beta_k^2 + \alpha \sum_k |\beta_k| \right),$$

where we obtain a *ridge* model when $\alpha = 0$ (L^2 norm) and a *lasso* model when $\alpha = 1$ (L^1 norm); see Fu (1998). This penalized logistic regression function can found in package `glmnet` (and also `penalized` , `lasso2` or `logistf`, even if those packages will not be used in this section).

4.3.1 Ridge Model

The generic code to run a ridge regression, introduced in Hoerl & Kennard (1970), is obtained when α is null, obtained setting `alpha=0`,

```
> library(glmnet)
> ridge <- glmnet(x,y,alpha=0,family ="binomial",lambda=c(0,1,2),standardize = TRUE)
```

Parameter λ can be estimated using cross-validation techniques, after having created indicators as *glmnet* function applies to numeric variables.

```
> train.db <- credit.rcd[index,]
> yt <- as.numeric(train.db[,"class"])
> xt <- model.matrix( ~ .,data=train.db[,-which(names(train.db)=="class")])
> set.seed(235)
> cvfit <- cv.glmnet(xt,yt,alpha=0, family = "binomial",type="auc",nlambda=100)
> cvfit$lambda.min
[1] 0.03983943
```

cvfit$lambda.min is the value of lambda giving the minimal error, viz here the maximal area below the ROC curve. On Figure 4.10 is represented the area below the ROC curve, for varying λ. The code is

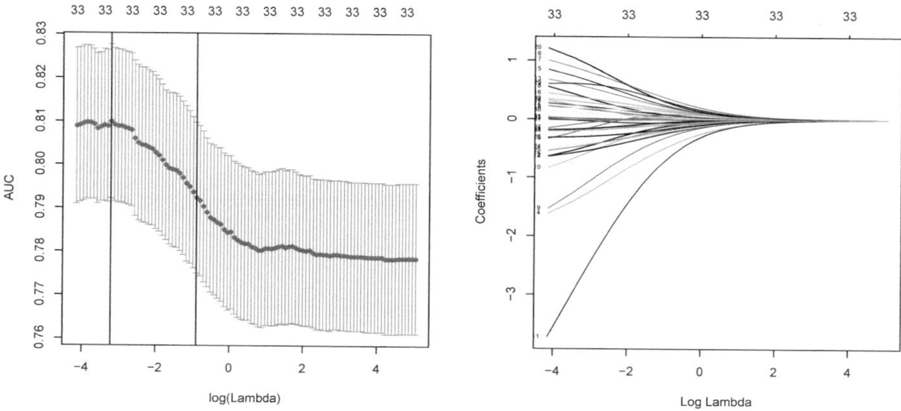

FIGURE 4.10
Optimal λ using cross validation, from a ridge regression ($\alpha = 0$).

```
> plot(cvfit)
> abline(v=log(cvfit$lambda.min),col='blue',lty=2)
```

It is also possible to visualize the evolution of parameters estimates $\widehat{\beta}_k^{(\lambda)}$ as a function of λ.

```
> fits <- glmnet(xt,yt,alpha=0,family = "binomial",lambda=seq(cvfit$lambda[1],
+ cvfit$lambda[100],length=10000),standardize = TRUE)
> plot(fits,xvar="lambda",label="T")
```

So far, we have use only the training sample. But it is also possible to see on the validation sample how those models actually perform:

```
> valid.db <- credit.rcd[-index,]
> yv <- as.numeric(valid.db[,"class"])
> xv <- model.matrix( ~ .,data=valid.db[,-which(names(train.db)=="class")])
> yvpred <- predict(fits,newx=xv,type="response")
> library(ROCR)
> roc <- function(x) {performance(prediction(yvpred[,x],yv),"auc")@y.values[[1]] }
> vauc <- Vectorize(roc)(1:ncol(yvpred))
> fits$lambda[which.max(vauc)]
[1] 2.137035
```

It is possible to visualize the AUC as a function of λ (Figure 4.11):

```
> plot(fit$lambda,vauc,type="l")
```

and we can get the maximal AUC obtained on the validation sample:

```
> vauc[which.max(vauc)]
```

4.3.2 Lasso Regression

The lasso regression, introduced in Tibshirani (1996), is obtained when α is one:

```
> lasso <- glmnet(x,y,alpha=1,family ="binomial",lambda=c(0,1,2),standardize = TRUE)
```

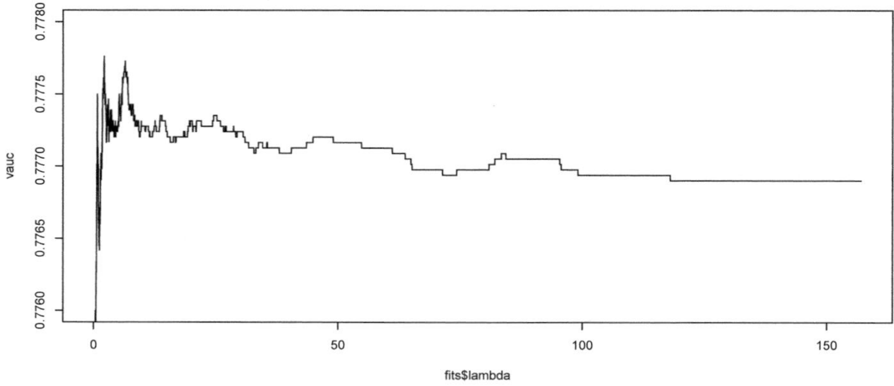

FIGURE 4.11
Evolution of AUC as a function of λ, with a ridge regression.

Again, parameter λ can be estimated using cross-validation techniques:

```
> set.seed(235)
> cvfit <- cv.glmnet(xt,yt,alpha=1, family = "binomial",type="auc",nlambda=100)
> cvfit$lambda.min
[1] 0.00217614
```

which can be visualized on Figure 4.12, representing the area below the ROC curve, for varying λ, using

```
> plot(cvfit)
> abline(v=log(cvfit$lambda.min),col=''grey'',lty=2)
```

It is also possible to visualize the evolution of parameters estimates $\widehat{\beta}_k^{(\lambda)}$ as a function of λ:

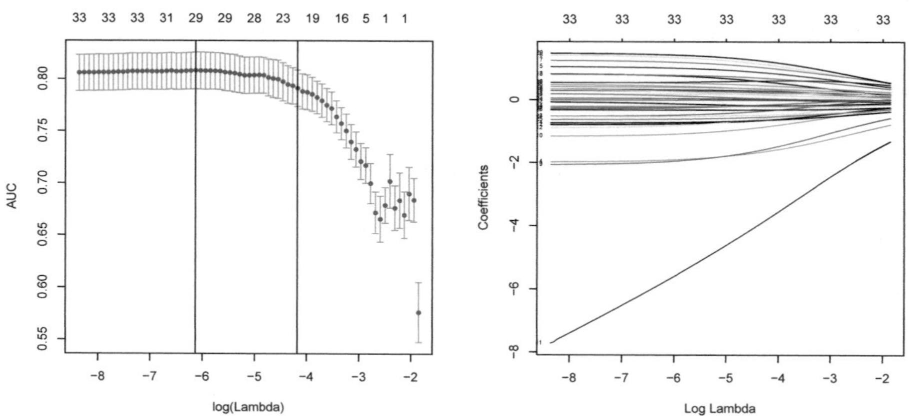

FIGURE 4.12
Optimal λ using cross validation, from a lasso regression ($\alpha = 1$).

```
> fits <- glmnet(xt,yt,alpha=0,family = "binomial",lambda=seq(cvfit$lambda[1],
+ cvfit$lambda[71],length=10000),standardize = TRUE)
> plot(fits,xvar="lambda",label="T")
```

Ridge regression has a major disadvantage: Unlike stepwise selection (which will generally suggest to consider only a subset of the covariables), ridge regression will include all predictors in the final model. Prediction will be good, but it might be a challenge to interpret values, and signs, when the number of regressors is large. Even if a model with three variables is relevant, with ridge regression, all variables will be considered. Increasing the value of λ will tend to reduce the magnitudes and the importance of the coefficients, but will not result in the exclusion of any of the variables. On the other hand, lasso performs variable selection, so models generated from the lasso are generally much easier to interpret than those produced by ridge regression. As claimed in Hastie et al. (2009), lasso yields sparse models.

4.4 Classification and Regression Trees

In the first part, we have used a logistic regression to compute the probability P(Y|X), given some covariates. Another class of popular predictive models is the class of regression trees and classification trees, introduced by Breiman et al. (1984). The latter technique will be described in this section (regression trees are used when the variable of interest is continuous, while classification trees are used for binary response); and in the next section, we will mention some so-called *ensemble* methods, that construct more than one decision tree:

- Bagged trees, which build multiple classification or regression trees, by repeatedly resampling training data with replacement;
- The random forest classifier, which uses multiple classification or regression trees, in order to minimize the error, by repeatedly resampling both training data (as in bagging) and predictors in building the trees,
- Boosted trees, which build multiple classification or regression trees on repeatedly modified versions of the data, before combining these trees through a weighted majority vote or weighted mean to produce the final prediction; the data modifications consist of applying weights to each of the training observations, with increased weights for misclassified observations and decreased weights otherwise.

4.4.1 Partitioning

In order to illustrate the use of trees, consider as in Sections 4.2.5 and 4.2.6, two continuous variates: the duration and the age, denoted (with very general notations) X_1 and X_2 respectively.

```
> X1 <- credit[,"age"]
> X2 <- credit[,"duration"]
> Y  <- credit[,"class"]
```

Starting with all the data, consider a splitting variable j (here in $\{1,2\}$), and a split point s (within the range of variable X_j). Define then the pair of half-planes

$$P_-(j,s) = \{i \in \{1,\cdots,n\}; X_{j,i} \leq s\} \text{ and } P_+(j,s) = \{i \in \{1,\cdots,n\}; X_{j,i} > s\},$$

where $X_{j,i}$ is the i-th value of variable X_j. If we adopt as a criterion minimization of the sum of squares (the choice of the criterion will be discussed in the next section), we seek *the* splitting variable j and split point s that solve

$$\min_{j,s}\left\{\min_{y\in\mathbb{R}}\left\{\sum_{i\in P_-(j,s)}(Y_i-y)^2\right\}+\min_{y\in\mathbb{R}}\left\{\sum_{i\in P_+(j,s)}(Y_i-y)^2\right\}\right\}.$$

Note that computation here is easy because for any choice j and s, the inner minimization is solved with

$$y_-=\frac{1}{n_-(j,s)}\sum_{i\in P_-(j,s)}Y_i \text{ and } y_+=\frac{1}{n_+(j,s)}\sum_{i\in P_+(j,s)}Y_i,$$

where $n_\pm(j,s)$ denotes the cardinal of partition $P_\pm(j,s)$. Thus, for each variable, the splitting point is obtained by scanning through possible outputs.

```
> criteria1 <- function(s){
+ sum((Y[X1<=s]-mean(Y[X1<=s]))^2)+sum((Y[X1>s]-mean(Y[X1>s]))^2)}
> criteria2<- function(s){
+ sum((Y[X2<=s]-mean(Y[X2<=s]))^2)+sum((Y[X2>s]-mean(Y[X2>s]))^2)}
```

It is possible to visualize those sums using

```
> S <- seq(0,100,length=501)
> plot(S,Vectorize(criteria2)(S),type="l",ylab="Sum of squares",xlab="Splitting point")
> lines(S,Vectorize(criteria1)(S),type="l",lty=2)
> legend(70,205,c("Variable X1","Variable X2"),lty=2:1)
```

Here, the best partition is obtained when the split is done according to the second variate (the duration of the credit), and where s is in $[33, 36)$ (here, the minimum is not unique as age and duration are discrete variables).

```
> S[which(Vectorize(criteria2)(S)==min(Vectorize(criteria2)(S)))]
 [1] 33.0 33.2 33.4 33.6 33.8 34.0 34.2 34.4
 [9] 34.6 34.8 35.0 35.2 35.4 35.6 35.8
```

Thus, it might be optimal to consider the midpoint of this interval, $s^\star = 34.5$, on variable $X2$.

Having found the best split (at least for this specific criterion, that yields the partition on the left of Figure 4.14), we partition the data into the two resulting regions, and repeat this splitting process on each of the two regions. And then, the process is repeated (again and again) on all of the resulting regions. For the second split, four cases should be considered, that can be visualized on the right of Figure 4.14

The following code can be used:

```
> s.star <- 34.5
> criteria1.lower <- function(s){
+ sum((Y[(X1<=s)&(X2<=s.star)]-meanY[((X1<=s)&(X2<=s.star)]))^2)+
+ sum((Y[(X1>s)&(X2<=s.star)]-meanY[((X1>s)&(X2<=s.star)]))^2)  +
+ sum((Y[(X2>s.star)]-mean Y[((X2>s.star)]))^2)}
> criteria1.upper <- function(s){
+ sum((Y[(X1<=s)&(X2>s.star)]-mean((Y[X1<=s)&(X2>s.star)]))^2)+
+ sum((Y[(X1>s)&(X2>s.star)]-mean((Y[X1>s)&(X2>s.star)]))^2) +
+ sum((Y[(X2<=s.star)]-mean(Y[(X2<=s.star)]))^2)}
```

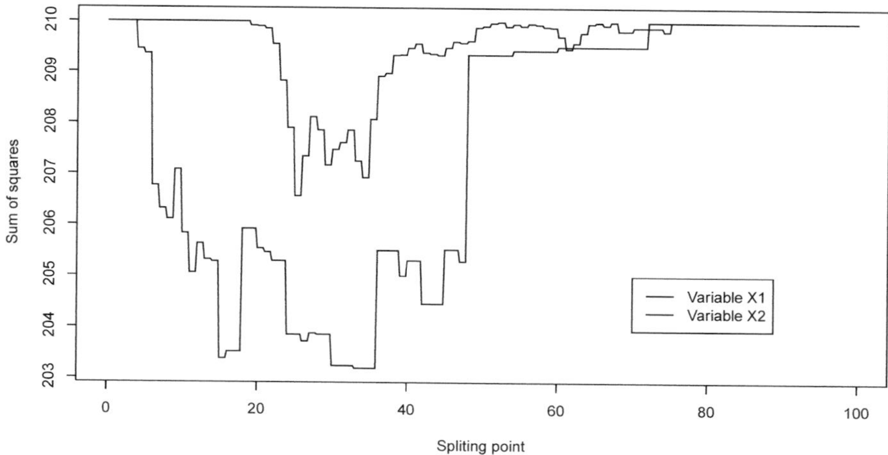

FIGURE 4.13
Evolution of the sum of squares, when the population is split into two groups, either `age<=s` and `age>s`, or `duration<=s` and `duration>s`.

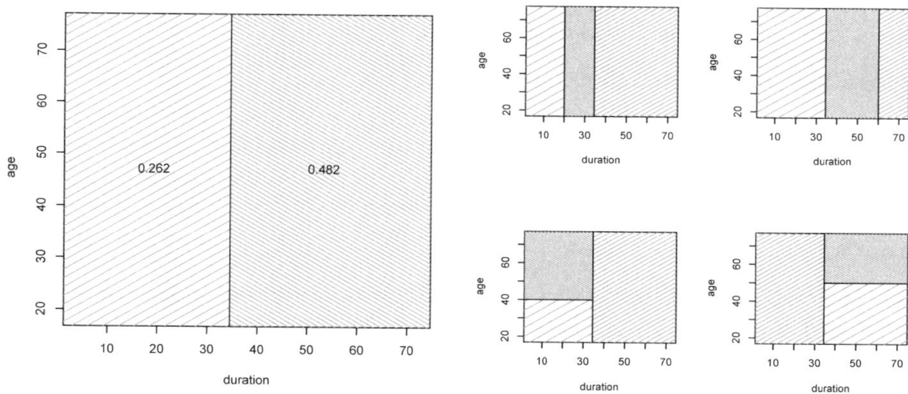

FIGURE 4.14
Regions obtained at the first splitting point, on the left, and possible regions obtained after the second splitting, on the right.

```
> criteria2 <- function(s){
+ sum((Y[(X2<=s)&(X2<=s.star)]-mean(Y[(X2<=s)&(X2<=s.star)]))^2)+
+ sum((Y[(X2>s)&(X2<=s.star)] -mean(Y[(X2>s)&(X2<=s.star)]))^2)+
+ sum((Y[(X2<=s)&(X2>s.star)]-mean(Y[(X2<=s)&(X2>s.star)]))^2)+
+ sum((Y[(X2>s)&(X2>s.star)]  -mean(Y[(X2>s)&(X2>s.star)]))^2)}
```

Here, the minimum of those three functions is again obtained when splitting is done according to the duration, and the splitting value is $s^\star \in [11, 12)$:

```
> S[which(Vectorize(criteria2)(S)==min(Vectorize(criteria2)(S)))]
[1] 11.0 11.2 11.4 11.6 11.8
```

Thus, it might be optimal to cut at the midpoint, $s^\star = 11.5$. Etc.

This splitting procedure can be done using function `tree` of library `tree`,

```
> library(tree)
> Tree <- tree(class~age+duration,data=credit)
> Tree
node), split, n, deviance, yval
      * denotes terminal node

1) root 1000 210.00 0.3000
  2) duration < 34.5 830 160.70 0.2627
    4) duration < 11.5 180  22.95 0.1500 *
    5) duration > 11.5 650 134.90 0.2938 *
  3) duration > 34.5 170  42.45 0.4824
    6) age < 29.5 58  12.78 0.6724 *
    7) age > 29.5 112  26.49 0.3839 *
```

The first node is here based on the duration, and the splitting value is 34.5. For the second one, we can either

- Consider the half-space with low durations, and then the optimal partition is based (again) on duration, with a splitting value equal to 11.5, or
- Consider the half-space with large durations, and then the optimal partition is based on age, with a splitting value equal to 29.5

This tree can be visualized on Figure 4.15 using

```
> plot(Tree)
> text(Tree)
```

while the partition (and the empirical probability that Y is 1) can be visualized using

```
> partition.tree(Tree)
```

Consider a loan of 20 months. On the right of Figure 4.15, we can see that the probability that `class` is equal to 1 is here 29.4%. On the tree (graph on the left of Figure 4.15) it is obtained going first on the left, as yes `duration<34.5`, and then on the right as no `duration<11.5`. The probability is then the second value at the bottom, 29.38%.

Those four regions (denoted `4`, `5`, `6` and `7` in the output) are called `terminal nodes`, or, if we keep a tree analogy, leaves of the tree. And decision trees are usually drawn upside down, with leaves at the bottom of the tree.

4.4.2 Criteria and Impurity

Assume that we have reached a node P, and we wonder if we should split it, or not. In a perfect world, all the population in the leaf will be identical, that is, either people where $Y = 0$ or $Y = 1$. In that case, $p = \mathbb{P}(Y = 1|P)$ will be either 0 or 1. The node will be be said to be *pure*. Conversely, the worst case scenario would be when half of the population is 0 and half is 1. In that case, $p = 1/2$ and the node would be said to be highly *impure*. Formally, impurity of a node is a function φ of $\mathbb{P}(Y = 1|P)$. Natural assumptions on φ are that φ is positive, symmetric, and minimal when the probability is either 0 or 1. Standard functions are

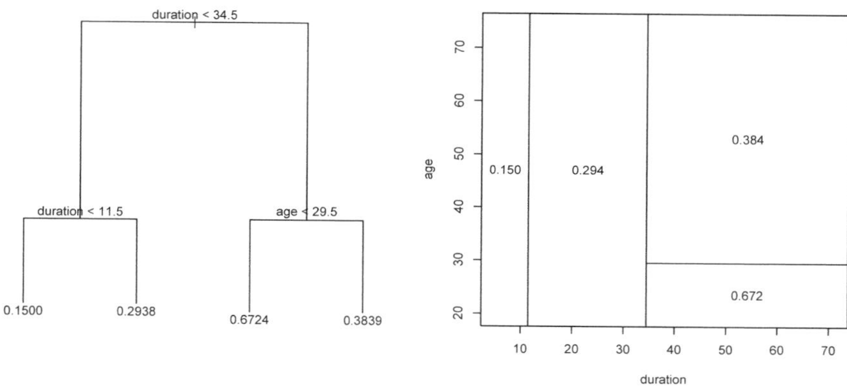

FIGURE 4.15
Classification tree with `tree()` on the left and `partition.tree()` on the right.

- $\varphi(p) = \min\{p, 1-p\}$, Bayes error
- $\varphi(p) = -p \log p - (1-p)\log(1-p)$, cross-entropy function
- $\varphi(p) = p \cdot (1-p)$, Gini index

These three functions are concave; they have minimums at 0 and 1, and maximum at 1/2.

As explained in Breiman et al. (1984), the Gini index is extremely popular to partition the data.

To illustrate the use of these functions, consider the very first partition. As mentioned earlier, if we split according to variable j, at splitting point s, two samples are obtained, in two half-spaces:

$$P_-(j,s) = \{i \in \{1, \cdots, n\}; X_{j,i} \leq s\} \text{ and } P_+(j,s) = \{i \in \{1, \cdots, n\}; X_{j,i} > s\}.$$

In each region, define $n_\pm(j,s) = n_\pm(j,s)^0 + n_\pm(j,s)^1$, where

$$n_\pm(j,s)^0 = \sum_{i \in P_\pm(j,s)} \mathbf{1}(Y_i = 0) \text{ and } n_\pm(j,s)^1 = \sum_{i \in P_\pm(j,s)} \mathbf{1}(Y_i = 1).$$

The entropy impurity is then the sum of

$$-\frac{n_\pm(j,s)^0}{n_\pm(j,s)} \log \frac{n_\pm(j,s)^0}{n_\pm(j,s)} - \frac{n_\pm(j,s)^1}{n_\pm(j,s)} \log \frac{n_\pm(j,s)^1}{n_\pm(j,s)},$$

while the Gini index is the sum of

$$\frac{n_\pm(j,s)^0}{n_\pm(j,s)} \cdot \frac{n_\pm(j,s)^1}{n_\pm(j,s)}.$$

These indices can be computed on contingency tables, as described in Table 4.2.

The code to produce this table (on the second variable, and with splitting value $s = 35$) is simply

TABLE 4.2
Contingency table and tree partitioning (first step), based on variable X_j.

	Value of Y	
	Y=0	Y=1
$\{X_j \leq s\}$	$n_-(j,s)^0$	$n_-(j,s)^1$
$\{X_j > s\}$	$n_+(j,s)^0$	$n_+(j,s)^1$

```
> s <- 35
> X0 <- (X2<=s)
> (T <- table(X0,Y))
       Y
X0        0   1
  FALSE  88  82
  TRUE  612 218
```

The Gini index is then computed using the following code:

```
> Nx <- apply(T,1,sum)
> (Pxy <- T/matrix(rep(Nx,2),2,nrow(T)))
       Y
X0              0         1
  FALSE 0.5176471 0.4823529
  TRUE  0.7373494 0.2626506

> Vxy <- Pxy*(1-Pxy)
> sum(Nx/sum(T)*apply(Vxy,1,sum))
[1] 0.4063785
```

This criterion can be used also to see how to split the population:

```
> gini <- function(X0){
+ T <- table(X0,Y)
+ Nx <- apply(T,1,sum)
+ Pxy <- T/matrix(rep(Nx,2),2,nrow(T))
+ Vxy <- Pxy*(1-Pxy)
+ return(sum(Nx/sum(T)*apply(Vxy,1,sum)))}
> criteria1=function(s){gini(X1<s)}
> criteria2=function(s){gini(X2<s)}
```

and as previously, it is possible to plot these indices for all s:

```
> S <- seq(0,80,length=501)
> plot(S,Vectorize(criteria2)(S),type="l",ylab="Gini index",xlab="Spliting point")
> lines(S,Vectorize(criteria1)(S),type="l",lty=2)
> legend(55,.408,c("Variable X1","Variable X2"),lty=2:1)
```

Again, we can see that the minimum is obtained when splitting variable $X2$ is at some splitting point $s \in [33, 36)$, as on Figure 4.13

Based on a set of predictors, for instance

```
> predictors <- names(credit)[-20]
```

run

```
> cart <- rpart(class ~ . ,data = credit[index,c("class",predictors)],
+ method="class",parms=list(split="gini"),cp=0)
> printcp(cart)

Classification tree:
rpart(formula = class ~ ., data = credit[index, c("class", predictors)],
    method = "class", parms = list(split = "gini"), cp = 0)

Variables actually used in tree construction:
 [1] age                checking_status     credit_amount
 [4] credit_history     duration            employment
 [7] job                other_payment_plans property_magnitude
[10] purpose            residence_since     savings

Root node error: 193/644 = 0.29969

n= 644

          CP nsplit rel error  xerror     xstd
1  0.0595855      0   1.00000 1.00000 0.060237
2  0.0518135      2   0.88083 1.00000 0.060237
3  0.0328152      3   0.82902 0.97409 0.059781
4  0.0310881      6   0.73057 0.96373 0.059592
5  0.0155440      8   0.66839 0.95337 0.059400
6  0.0138169     11   0.62176 0.99482 0.060148
7  0.0129534     14   0.58031 0.99482 0.060148
8  0.0103627     17   0.53886 0.99482 0.060148
9  0.0069085     18   0.52850 1.02591 0.060674
10 0.0051813     21   0.50777 1.02591 0.060674
11 0.0025907     26   0.48187 1.04663 0.061008
12 0.0000000     28   0.47668 1.05181 0.061090
```

It is possible to visualize the Complexity Parameter Table (Figure 4.16) using

```
> plotcp(cart)
```

Finally, in order to assess the quality of the model, it is possible to plot the ROC for the tree obtained using library `rpart` (on the left) and library `party` (on the right) on Figure 4.18:

```
> train.db$pred.tree <- predict(cart,type='prob',newdata=train.db)[,"1"]
> valid.db$pred.tree <- predict(cart,type='prob',newdata=valid.db)[,"1"]
> library(ROCR)
> pred.train=prediction(train.db$pred.tree,train.db$class)
> pred.valid=prediction(valid.db$pred.tree,valid.db$class)
> perf.train=performance(pred.train,"tpr", "fpr")
> perf.valid=performance(pred.valid,"tpr", "fpr")
> plot(perf.train,col="grey",lty=2,main="Tree (rpart)")
> plot(perf.valid,add=TRUE)

> library(party)
> ct.arbre <- ctree_control(mincriterion =0.95,
+ minbucket=10,minsplit=10*2,maxdepth=0)
```

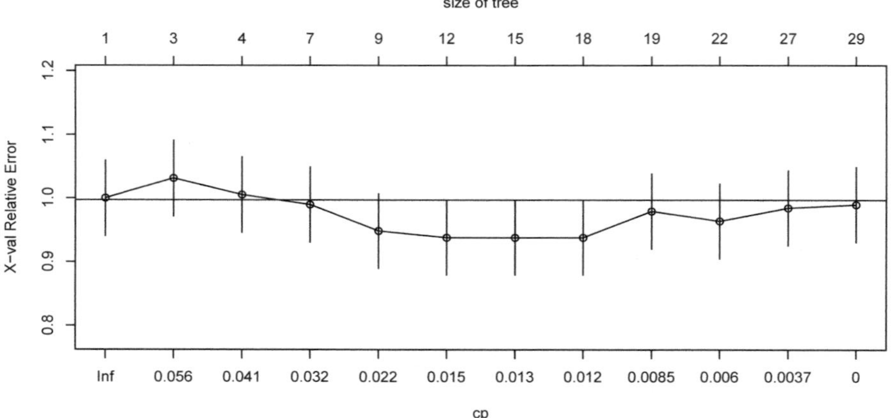

FIGURE 4.16
Complexity parameter graph using `printcp` on a tree.

```
> ctree <- ctree(class ~ .,data = credit[index,c("class",predictors)],control=ct.arbre)
> plot(ctree)
```

```
> train.db$pred.tree <- unlist(predict(ctree,type='prob',newdata=train.db))
> valid.db$pred.tree <- unlist(predict(ctree,type='prob',newdata=valid.db))
> pred.train=prediction(train.db$pred.tree,train.db$class)
> pred.valid=prediction(valid.db$pred.tree,valid.db$class)
> perf.train=performance(pred.train,"tpr", "fpr")
> perf.valid=performance(pred.valid,"tpr", "fpr")
> plot(perf.train,col="grey",lty=2,main="Tree (party)")
> plot(perf.valid,add=TRUE)
```

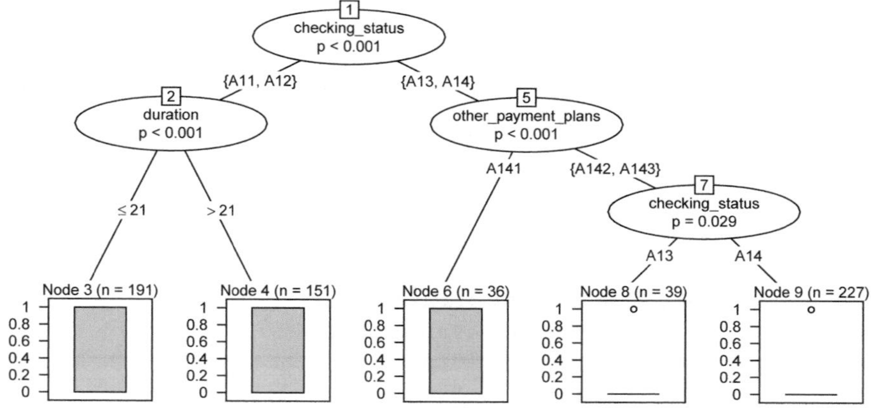

FIGURE 4.17
Output of the `ctree` function of library `party`.

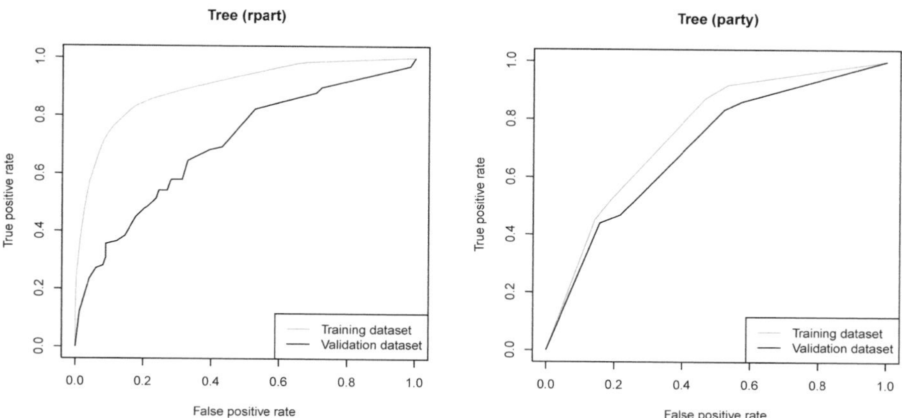

FIGURE 4.18
ROC curve (true positive rate versus false positive rate) from two classification trees, obtained using packages `rpart` and `party`.

Finally, note that it is also possible to use `prune` trees, to avoid overfitting the data, using a tree size that minimizes the cross-validated error (the `xerror` column printed by `printcp`), here

```
> prunecart <- prune(cart,cp=.015)
```

Nice graphs can be produced using functions of the `rpart.plot` package, such as `prp()`

```
> plot(prunecart,branch=.2, uniform=TRUE, compress=TRUE,margin=.1)
> text(prunecart, use.n=TRUE,pretty=0,all=TRUE,cex=.5)
> library(rpart.plot)
> prp(prunecart,type=2,extra=1)
```

Several other functions can be used on `rpart` objects; see Galimberti et al. (2012), Archer (2010).

4.5 From Classification Trees to Random Forests

Trees are nice, as they are extremely easy to understand. Unfortunately, trees are extremely volatile: If we split the training data into two parts (randomly) and fit decision trees to both halves, the output could be quite different (see Figure 4.20).

```
> library(tree)
> set.seed(1)
> indextree <- sample(1:1000,size=500,replace=FALSE)
> t1 <- tree(class~age+duration,data=credit[indextree,])
> t2 <- tree(class~age+duration,data=credit[-indextree,])
> partition.tree(t1)
> partition.tree(t2)
```

A natural idea is then to use bootstrap aggregation (called bagging) to reduce variance.

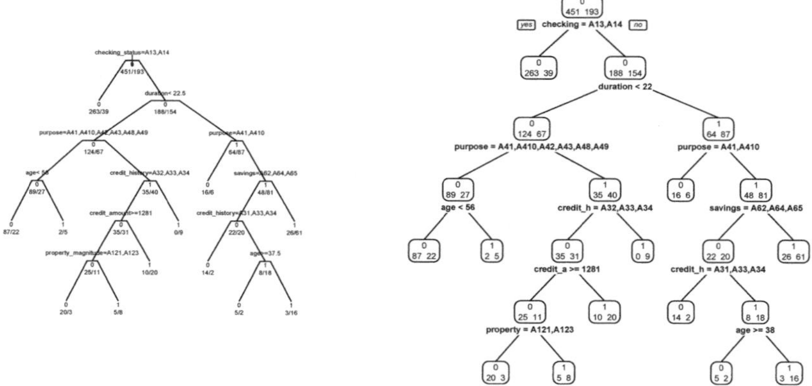

FIGURE 4.19
Pruned trees, obtained using `prune` function of `rpart` package.

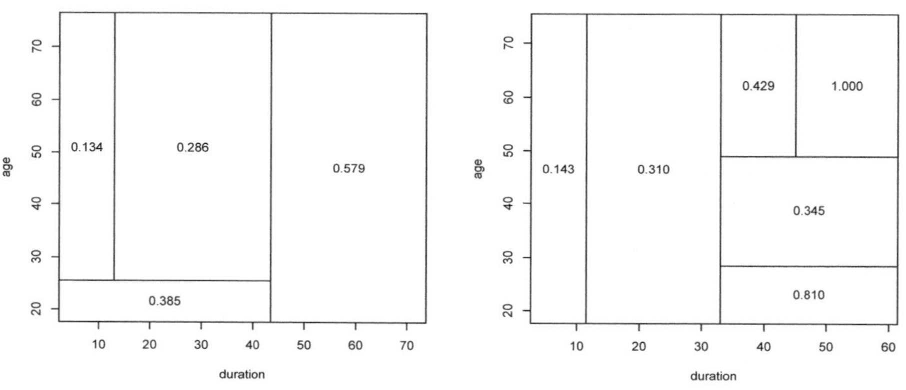

FIGURE 4.20
Partitions obtained from classification tree, on sub-samples of the dataset.

4.5.1 Bagging

The idea of bagging (from *bootstrap aggregating*) was proposed in Breiman (1996). This bagging procedure involves the following steps:

- Create multiple copies of the original training dataset using bootstrap procedures.
- Fit classification trees to each copy.
- Aggregate (or combine) all of the trees in order to create a single predictive model.

```
> library(ipred)
> set.seed(123)
> bag1 <- bagging(class ~ .,data=credit[index,c("class",predictors)],nbagg=200,coob=TRUE,
+ control= rpart.control(minbucket=5))
```

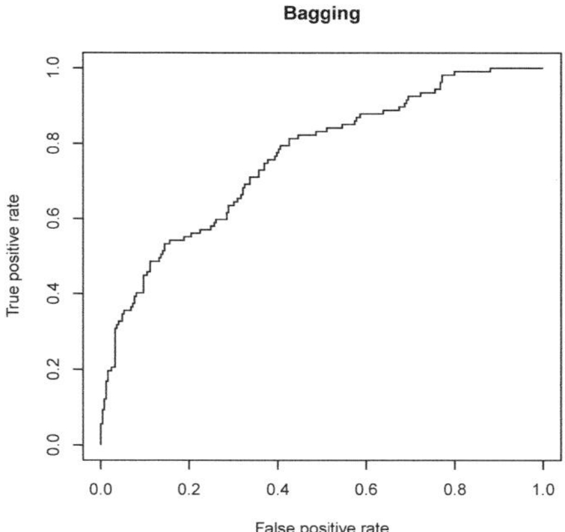

FIGURE 4.21
ROC curve obtained after bagging.

```
> pred=prediction(predict(bag1, test, type="prob",
+ aggregation="average"),credit[-index,"class"])
> perf=performance(pred,"tpr", "fpr")
```

4.5.2 Boosting

When bagging trees, boostrap techniques were considered, meaning that all copies were independent, so trees were built independently of the others. Boosting will work in a similar way, except that trees are grown sequentially: Each tree is grown using information obtained from previously grown trees. Thus, boosting does not involve bootstrap sampling, and each tree is obtained from a modified version of the original dataset.

Variable importance can be visualized using functions `ada` from the eponym library for boosting, and then `varplot` (see e.g. Culp et al. (2006) for more details). Boosting will be interesting only in the case where several predictors are important.

```
> library(ada)
> set.seed(123)
> boost <- ada(class ~ ., data=credit[index,c("class",predictors)],
+ type="discrete", loss="exponential", control =
+ rpart.control(cp=0), iter = 5000, nu = 1,
+ test.y=test[,"class"], test.x=test[,1:19])
> varplot(boost,type="scores")
      other_parties other_payment_plans      credit_history       credit_amount                 job
        0.011265924         0.010419375         0.009860286         0.009756094         0.009687740
            housing             savings            duration                 age
        0.009686706         0.009603497         0.009422427         0.009276642
            purpose      num_dependents    existing_credits     personal_status          employment
        0.009250920         0.009223420         0.009177000         0.009072014         0.008770979
 property_magnitude     residence_since    installment_rate     checking_status           telephone
        0.008407637         0.008280839         0.008194080         0.007814246         0.007088610
```

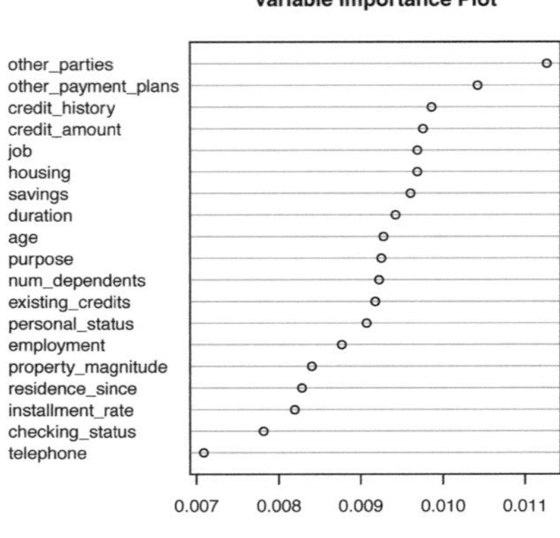

FIGURE 4.22
Variable importance and boosting.

4.5.3 Random Forests

The drawback of the bagging technique is that variance reduction is usually small. The intuitive idea behind this is simple: Assume that we have in our model one very strong predictor, and a couple of moderately strong ones. Using bagging technique will grow trees quite similar. Therefore, predictions will be extremely correlated. And averaging extremely correlated variables does not lead to a significant reduction in the variance.

The last method we will discuss provides an improvement over bagged trees that decorrelates the trees. In building a random forest, at each split in the tree, the algorithm is not allowed to consider all the available predictors.

In order to avoid this problem, random forest algorithms force us to consider only a subset of predictors. Then predictions will have more variability, and aggregation will then be extremely efficient. Resulting trees are then usually less variable, and hence more reliable.

The package `randomForest` (see Liaw & Wiener (2002)) can be used to build a predictive model.

```
> library(randomForest)
> set.seed(123)
> rf <- randomForest(class ~ ., data=credit[index,c("class",predictors)],
+ importance=TRUE, ntree=500, mtry=3, replace=TRUE, keep.forest=TRUE,
+ nodesize=5, ytest=test[,"class"], xtest=test[,1:19])
Message d'avis :
In randomForest.default(m, y, ...) :
  The response has five or fewer unique values.  Are you sure you want
  to do regression?
> rf

               Type of random forest: regression
                     Number of trees: 500
No. of variables tried at each split: 3

          Mean of squared residuals: 0.1638417
                    % Var explained: 21.93
```

```
                    Test set MSE: 0.17
                 % Var explained: 19.74
```

Then, the extractor function for variable importance measures as produced by the previous function can be obtained using either `type=1` for mean decrease in accuracy,

```
> importance(rf,type=1)
                       %IncMSE
checking_status     26.8158301
duration            17.2268269
credit_history       6.8096722
purpose              8.4606150
credit_amount        6.4029995
savings              6.7660725
employment           2.9760754
installment_rate     4.6122336
personal_status      2.2104102
other_parties        4.2947555
residence_since      1.5007471
property_magnitude   0.8281632
age                  5.5388709
other_payment_plans  8.0462958
housing              4.1083429
existing_credits     2.1720201
job                 -0.4039036
num_dependents      -0.1864291
telephone            1.0709050
```

or `type=2` for mean decrease in node impurity,

```
> importance(rf,type=2)
                    IncNodePurity
checking_status         14.712959
duration                11.359640
credit_history           6.876229
purpose                 11.722564
credit_amount           12.257516
savings                  6.135412
employment               6.780975
installment_rate         4.467483
personal_status          4.371698
other_parties            1.960203
residence_since          3.710073
property_magnitude       5.003570
age                     10.680848
other_payment_plans      3.843822
housing                  3.374496
existing_credits         2.307880
job                      3.392985
num_dependents           1.067552
telephone                1.798760
```

5

Spatial Analysis

Renato Assunção
Universidade Federal de Minas Gerais
Belo Horizonte, Brazil

Marcelo Azevedo Costa
Universidade Federal de Minas Gerais
Belo Horizonte, Brazil

Marcos Oliveira Prates
Universidade Federal de Minas Gerais
Belo Horizonte, Brazil

Luís Gustavo Silva e Silva
Universidade Federal de Minas Gerais
Belo Horizonte, Brazil

CONTENTS

5.1 Introduction

Statistical data often has associated a temporal and spatial coordinate reference system. Spatial statistics is the set of techniques to collect, visualize, and analyze statistical data taking into account their spatial coordinates. Therefore, spatial statistics is a natural tool when the interest lies on the geographical analysis of actuarial risk. To describe the spatial statistics methods, we need to consider first the spatial data nature. Spatial data can be classified into four types according to the stochastic nature of the random component in the statistical data: point pattern, random surface, spatial interaction and areal data.

5.1.1 Point Pattern Data

A point pattern dataset gives the locations $\mathbf{x}_1, \ldots, \mathbf{x}_n$ of n random events occurring in a continuous study region $\mathcal{R}$. The random locations of earthquake epicenters, burglarized households, or the coordinates of car accidents in an urban center are examples of point patterns. The random aspect of the data is its location $\mathbf{x}_i$ in the map. The interest is focused on describing and learning the mechanism that generates the spatial spread of the random events in the map. Many times, there will be additional information, called marks, associated with the events. In the previous examples, the marks could be the time of occurrence and the magnitude of each event.

Typical inference questions in this type of data are the following: Is the random spread of the events associated with some aspect of the region such as the presence of a river or an industrial center? Are the marks associated with the random location such as when the earthquake magnitude in a certain area tend to be larger than elsewhere? Do nearby cases tend to be also close in time?

5.1.2 Random Surface Data

The random aspect of this type of data is a potentially observable surface $Z(\mathbf{x})$ defined for every location $\mathbf{x}$ in a continuous study region $\mathcal{R}$. Examples for the surface height $Z(\mathbf{x})$ include the temperature at position $\mathbf{x}$ or the soil pH as a measure of the soil acidity viewed as a surface unfolded on $\mathcal{R}$. Although potentially observable at every $\mathbf{x}$, in practice the random surface is measured or observed only at a finite set of collection stations $\mathbf{x}_1, \ldots, \mathbf{x}_n$. In contrast with point pattern datasets, these stations are considered fixed and known previously to any measurement. The joint probability distribution of the random variables $(Z(\mathbf{x}_1), \ldots, Z(\mathbf{x}_n))$ is induced by the distribution of the random surface $Z(\mathbf{x})$ for $\mathbf{x} \in \mathcal{R}$. It is usual to observe the surface with some measurement error called the nugget effect.

Some of the typical inference problems in this type of spatial data are to predict the surface height in positions that are not monitored, to estimate the entire surface by interpolating between the monitoring stations and to select a position to receive a new monitoring station.

5.1.3 Spatial Interaction Data

In this type of data, we also have fixed, non-random stations or positions $\mathbf{x}_1, \ldots, \mathbf{x}_n$. The data, however, is not associated with the individual locations but rather to ordered pairs (i, j) of locations. The random data that require a statistical model are of the form $Z_{ij} = Z(\mathbf{x}_i, \mathbf{x}_j)$. Usually, we can see these locations as one origin station $\mathbf{x}_i$ flowing some random quantity Z_{ij} to a destination station $\mathbf{x}_j$. Typical examples include the airline traffic between airports in a region or the commuting patterns between neighborhoods in the morning.

In this type of spatial data, some of the typical inference questions are to describe quantitatively how the flow Z_{ij} is affected by the individual characteristics of stations i and j (such as their size, type, and location), and where a new station should be assigned to minimize the total flow cost.

5.1.4 Areal Data

This is the most common type of spatial data in insurance problems. Suppose a continuous study region $\mathcal{R}$ is partitioned into disjoint areas $\mathcal{R} = \mathcal{R}_1 \cup \ldots \cup \mathcal{R}_n$. In each area $\mathcal{R}_i$, we measure the random variable Y_i ending up with a random vector $\mathbf{Y} = (Y_1, \ldots, Y_n)$ whose joint distribution will reflect the spatial location of the areas.

The vector $\mathbf{Y}$ can be visualized in a thematic map, where each area is colored according to Y_i. Figure 5.1 shows a thematic map of lung cancer mortality among white males between 2006 and 2010, based on data from the Centers for Disease Control (CDC). The region is the continental United States partitioned into its counties. This thematic map was created using Rand following the design recommendations made by the CDC. These design rules were elaborated to maximize the atlas' effectiveness in conveying accurate mortality patterns to users and were obtained after extensive experimentation (see Pickle et al. (1999)).

To create the thematic map, we first read the shapefile file `uscancer` with the function `readShapeSpatial` of the `maptools` package. Further details of the `readShapeSpatial` function are provided later in this chapter.

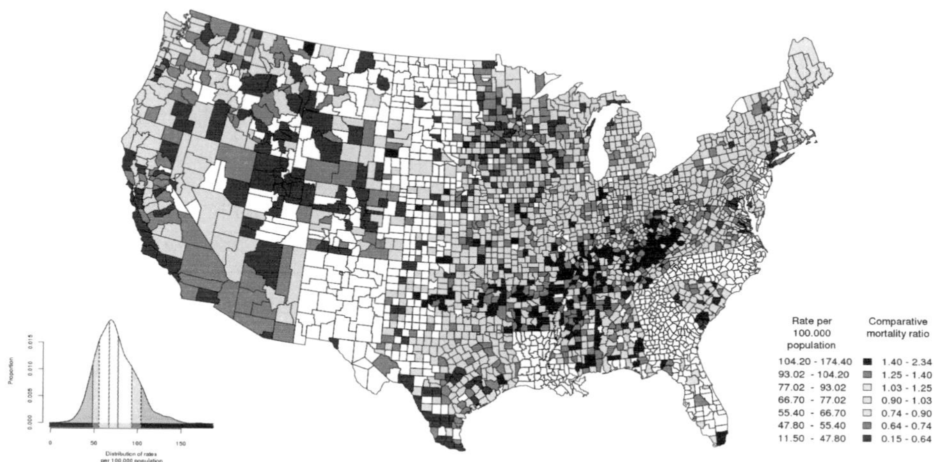

FIGURE 5.1
Map of lung cancer mortality among white males between 2006 and 2010 in the continental United States, by counties.

```
> library(maptools)
> source("AtlasMap.R")
> us.shp <- readShapeSpatial("uscancer")
```

The function `AtlasMap` creates the thematic map. The required arguments of the function are `shp`, the object of class `SpatialPolygonsDataFrame`, and `var.plot`, the variable name that is visualized on the map.

```
> AtlasMap(shp = us.shp, var.plot = "ageadj", mult=100000, size="full")
```

The remaining parameters are optional: the `mult` value is used as the multiplication factor of the rate; `rate.reg` is the reference rate. If not specified, the default value is the average of the `var.plot` variable; `r` is the rounding parameter; `colpal` is the palette of colors used for coloring the map; `ncls` is the number of color breaks of the `var.plot` variable; `brks` is the range of the data that defines the colors for each value of the `var.plot` variable. Default values are the quantiles 0.1, 0.2, 0.4, 0.6, 0.8 and 0.9. For more details, see `?classInt::classIntervals`; codesize is the size of the graphic window to be displayed.

The map shows that high rates are found in areas located in the Mississippi and Ohio River valleys. This spatial pattern reflects the high per-capita consumption of cigarettes in these areas.

The main spatial aspect of this type of data is that Y_i represents an aggregation of values dispersed in the i-th area. That is, Y_i is not associated to any specific location $\mathbf{x}$ within $\mathcal{R}_i$, but rather to the entire area $\mathcal{R}_i$.

Some of the main inference questions in this type of data are: to verify if the spatial pattern of the data is associated with attributes measured in the areas, to obtain a smooth version of the thematic map eliminating the random variation that can be assigned to random noise, and to detect sub-regions of higher values with respect to the rest of the map. We will study several other relevant questions after we specify models for this type of data.

5.1.5 Focus of This Chapter

The main difference between these spatial data types is the nature of their randomness. Therefore, it is natural that different statistical methods and models are adopted in each one of them. Indeed, they are so different as to prevent coverage of all of them due to lack of space in this chapter. Hence, we decided to focus on the type that is most common in actuarial studies, the areal data. We describe briefly the R capabilities for the other three types of data and point to relevant literature for the interested reader.

Presently, there are many R packages dedicated to spatial analysis. In the Task View in Analysis of Spatial Data (see http://cran.r-project.org/web/views/Spatial.html), there is a classification of these packages according to their main objectives: class definition, manipulation, reading, writing, and analysis of spatial data. As of May 2013, 108 packages that are involved with spatial analysis were listed. Because we cannot cover them all in this chapter, we focus on the most important and basic of them (the package `sp`) and present some of the packages that are associated with areal data, mixing the authors' preferences, experience and their sense of the packages' usefulness for actuarial work.

5.2 Spatial Analysis and GIS

Spatial statistical data can arise in many different formats and one needs a flexible medium for visualizing and interacting with them. This medium is a Geographic Information System

(GIS), a computer system integrating hardware, software, and data for storing, managing, analyzing and displaying geographically referenced information. Simply put, it is a combination of maps and a traditional database. Interacting with a GIS, a user can answer spatial questions by turning raw data into information. What most commercial GIS software calls a spatial analysis is a simple query in the spatial database. Typically, there is little sophisticated statistical analysis capability in the available GIS as compared with usual statistical software such as R.

In its core, every GIS establishes an architecture to combine the non-spatial attributes and the spatial information of geographical objets in a single database. Conceptually, the geographic world can be modeled according to two different and complementary approaches: using fields or geo-objects. In the geo-objects model, the aspect of reality under study is seen as a collection of distinct entities forming a collection of points, lines, or polygons corresponding to different objects in the real world. The computer implementation of these geo-objects uses vector data: one or more coordinate pairs rendering the visual aspect of points, lines or polygons. In the field model, the reality is seen as a continuous surface and the typical data model implementation is a grid image (or raster image).

The most simple geo-object is the point, a discrete location that is stored as a vector (s_1, s_2) giving its spatial coordinates. The geographical locations of crimes, traffic accidents, buildings, post offices, or gas stations can be represented by points. The choice of this data type depends on the map scale and the intended analysis. For example, the 1,000 largest American cities can be represented as points on a U.S. map if the aim is simply to visualize their geographical dispersion in the territory.

In addition to the spatial coordinates, attributes characterizing each point are stored in a table. In Figure 5.2, each crime in an urban area may be represented by a point with associated characteristics such as the type of crime, the closest address and the number of victims involved. There is 1-to-1 association between the rows of the attribute table and the set of points, with the linking provided by a unique id.

A line geo-object is an ordered sequence $(s_{11}, s_{21}), \ldots, (s_{1n}, s_{2n})$ of points together with the associated table attribute of this line. Connecting the points in the given order produces the visual appearance of a line. Typical entities represented by lines are roads, pipelines, rivers, bus routes and so on. As with point geo-objects, an attribute table associates statistical features to each line geo-object. Figure 5.3 shows a line representing the Amazon

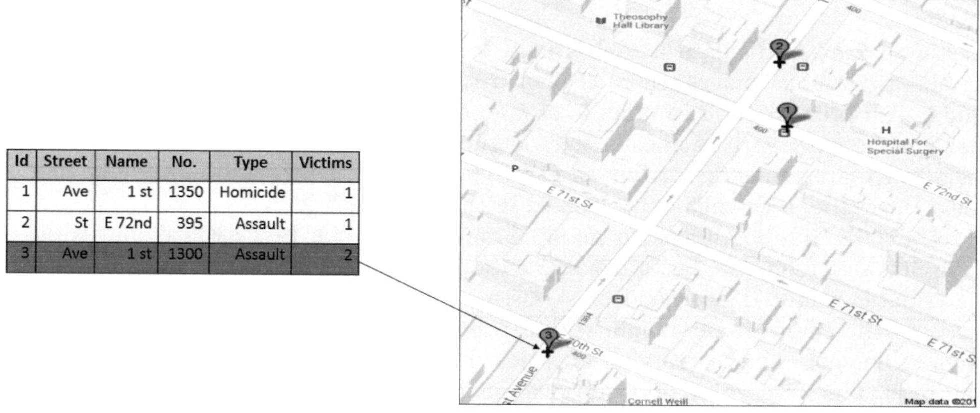

Id	Street	Name	No.	Type	Victims
1	Ave	1 st	1350	Homicide	1
2	St	E 72nd	395	Assault	1
3	Ave	1 st	1300	Assault	2

FIGURE 5.2
Locations of assaults in New York City, geo-points.

Id	Name	Length (km)	Drainage area	Average discharge
1	Tocantins	3,650	950,000	13,598
2	São Francisco	3,180	610,000	3,300
3	Amazonas	6,650	6,915,000	219,000

FIGURE 5.3
Amazon River in South America, a geo-line object.

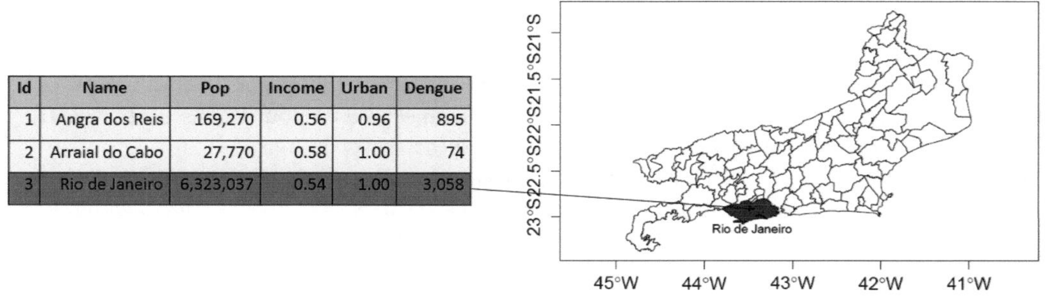

Id	Name	Pop	Income	Urban	Dengue
1	Angra dos Reis	169,270	0.56	0.96	895
2	Arraial do Cabo	27,770	0.58	1.00	74
3	Rio de Janeiro	6,323,037	0.54	1.00	3,058

FIGURE 5.4
Polygon of the city of Rio de Janeiro, Brazil.

River in South America with the associated attributes of length, drainage area and average discharge.

The geo-object polygon is the most important in this chapter and, as the geo-object line, it is represented by an ordered sequence of points. The difference with a line is that the first and the last points of the ordered sequence forming the polygon are the same. Hence, a polygon is simply a closed line. Figure 5.4 shows the municipalities of Rio de Janeiro, in Brazil. Each one is a polygon geo-object. Besides the geographical coordinates, each polygon geo-object has associated attributes stored in tables, such as population and average income.

The raster or image GIS data model is used to represent continuous fields. Satellite images or non-geographic images, such as photographic pictures, are examples of raster data. Regular grid cells are used as units to store raster data. It can be implemented as a rectangular matrix array with each cell corresponding to a geographical small area. Each cell has associated alphanumeric attributes. Figure 5.5 shows the region around Pampulha Lake in Belo Horizonte, a Brazilian city located in the Southeast.

In a typical GIS, the operations to manipulate the data can be classified into three groups. The first one refers to the traditional data summaries and calculations involving only the attributes values. The second one consists of operations involving only the geo-objects such as the count of geo-points within a polygon, the calculation of an area or perimeter of a polygon, or the creation of a new larger polygon as a result of merging two smaller ones. The third group of operations combines the attribute values and the geo-objects and could be represented, for example, by a function that returns the set of polygons that are within a certain distance from a given polygon and with yearly income per capita greater than a certain value.

FIGURE 5.5
Google® satellite image, raster data.

Commercial GIS are powerful environments where these operations are efficiently carried out. A large-scale project, involving extensive creation and manipulation of geo-objects, may require GIS software. Given its advanced analytical capability, R is an important option to those interested in spatial analysis. A complete book-length coverage of the spatial resources in R is available in Bivand et al. (2013). R is not a GIS but has a large number of features that allow it to be used as a spatial data analysis environment, mainly through contributed packages. These packages are mostly concerned with the input and output of spatial data and with spatial data analysis. There are some packages to interface between R and some GIS, such as `GRASS`, `pgrass6`, `StatConnector` (to link with ArcGIS), `geoR` and `ArcRstats`. However, the integration between R and GIS is not as developed as one might expect. Most spatial data analysis we carry out in this chapter is developed completely within R after inputting geo-objects prepared by GIS software. Although the analysis can be run in R, it is common to have the final results and presentation maps made using GIS software as it may have more visual tools for the specific task at hand. However, we stick with R, showing all our results using only its graphical capabilities.

5.3 Spatial Objects in R

One important development for the spatial analysis in R was the release of package `sp`; see Pebesma and Bivand (2005). They established a coherent set of classes and methods for the major spatial data types (points, lines, polygons and grids) following the same basic principles of GIS data models and following the principles of object-oriented programming. `sp` is a required package by most other R packages working with spatial data. Additionally, some spatial visualization and selection features are implemented for the classes defined by this package.

In `sp`, the basic building block is the `Spatial` class. Every object of this class has only two components, called slots in S4 terminology: a bounding box and a cartographic projection. The first one is named `bbox` and it gives the numeric limits of a given coordinate system within which lies the geographic object represented internally by an R object of `Spatial` class. The bounding box is a 2×2 matrix object with column names `c('min','max')`. The first row gives the bounds for the spatial objects coordinates in the East-West direction while the second row gives the bounds for the North-South direction.

The cartographic projection is specified by the second slot, named `proj4string`. It gives the reference system that needs to be used to make sense of the limits specified in the bounding box. The cartographic projections represent different mathematical attempts to represent the Earth's surface in a plane. It is impossible to avoid area, distance or angle distortions in this process. For instance, a point A could be equally distant from B and C on the Earth's surface but, in the two-dimensional map, it could be closer to B than to C.

Cartographic projection is an essential item when dealing with spatial data. Its relevance is even greater nowadays with GIS, whose main strength is its ability to aggregate disparate sources of information and data by overlaying different maps, delimitation of regions by different criteria, etc. Spatial information comes from different agencies and sources such as from the Census Bureau, from satellite images, police data, roads maps, etc. Different sources use different cartographic projections. We need to connect these different projections so the different spatial data can be overlaid properly.

The slot `proj4string` must contain an object of class `CRS`, which stands for coordinate reference system. Each `CRS` object has a single slot called `projargs`, and it is simply a character vector with values such as `"+proj=longlat"`, meaning that the longitude-latitude (longlat, from now on) coordinate system is to be considered. In fact, strictly speaking, the longlat coordinate system is not a cartographic projection (because it is spherical coordinate system, see Chapter 2 in Banerjee et al. (2004)).

To create a `Spatial` class object named `sobj`, one can type

```
> library(sp)
> bb <- matrix(c(-10,-10,10,10),ncol = 2,dimnames=list(NULL,c("min","max")))
> crs <- CRS(projargs = "+proj=longlat")
> sobj <- Spatial(bbox = bb, proj4string = crs)
```

We can represent the Earth locations more accurately if we approximate the globe by an ellipsoid model rather than a sphere. This can also be provided in the slot `projargs` as, for example, in the command `CRS("+proj=longlat +ellps=WGS84")` .

Two generic methods work on `Spatial` class objects, the `bbox` and `CRS`, which return the respective slots values:

```
> bbox(sobj)
     min max
[1,] -10  10
[2,] -10  10
> proj4string(sobj)
[1] "+proj=longlat"
```

5.3.1 `SpatialPoints` Subclass

There is little spatial content in an object such as `sobj` defined above. In fact, the geographical data are stored in `Spatial` subclasses objects rather than as `Spatial` objects directly. The subclass `SpatialPoints` extends the `Spatial` class and it can store the spatial coordinates of geographic points in an additional slot called `coords`. Usual plotting functions, such as `plot` and `points`, can be used with `SpatialPoints` objects.

The command to create an object of class `SpatialPoints` holding n points in a region is

```
> SpatialPoints(coords, proj4string=CRS(as.character(NA)), bbox = NULL)
```

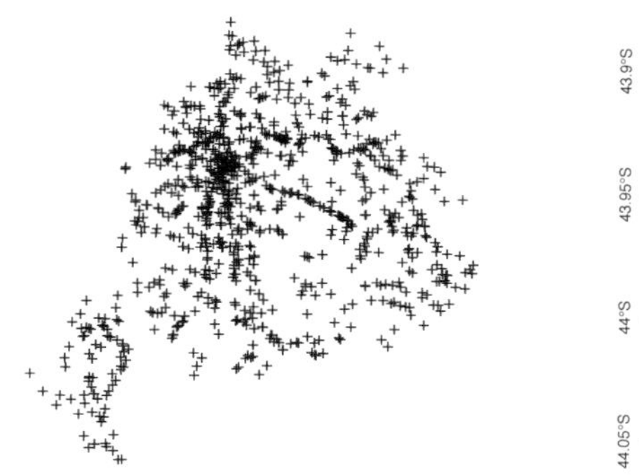

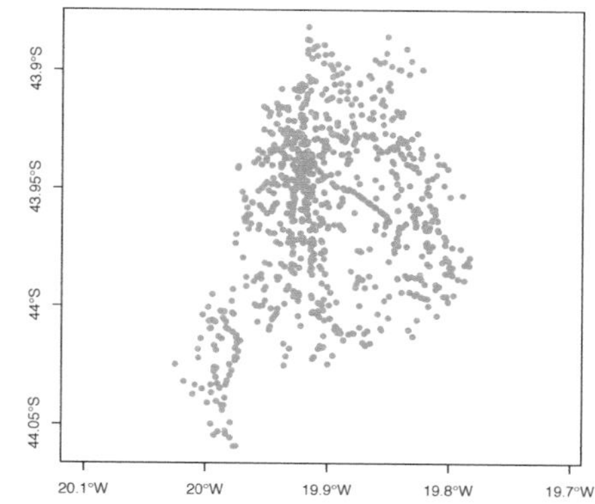

FIGURE 5.6
Use of SpatialPoints to visualize traffic accidents that occurred in February 2011, in Belo Horizonte, a Brazilian city.

where coords is a $n \times 2$ numeric matrix or dataframe. The cartographic projection default is missing and the bounding box matrix is built automatically from the data, if the default NULL value is used. As an example, we will read and plot the locations of traffic collisions that occurred in February 2011, in Belo Horizonte, a Brazilian city.

```
> mat <- read.table("accidents.txt",header=TRUE)
> crs <- CRS("+proj=longlat +ellps=WGS84")
> events <- SpatialPoints(mat, proj4string = crs)
> plot(events)
> plot(events, axes=T, asp=2, pch=19, cex=0.8, col="dark grey")
> summary(events)
Object of class SpatialPoints
Coordinates:
           min       max
lat  -20.02569 -19.78362
long -44.05925 -43.88172
Is projected: FALSE
proj4string : [+proj=longlat +ellps=WGS84]
Number of points: 1314
```

The left-hand side plot in Figure 5.6 shows the output of the simple plot command. Note that no axes are drawn when plotting SpatialPoints objects. If longlat or an NA coordinate reference system is used, one unit in the x direction equals one unit in the y direction. This default aspect ratio can be changed by the plot asp parameter, a positive value establishing how many units in the y direction is equal in length to one unit in the x direction. Axes can be added using axes=T in the plot command. For unprojected data (longlat or NA), the axis label marks will give units in decimal degrees as, for example, 80°W. The default symbol for the points is a cross but this can be changed using the parameter pch in the plot command (and using cex for the symbol size). An example of changing these default options is shown on the right-hand side of Figure 5.6, obtained with the plot command with additional parameters (pch=19 produces filled circles; see the help page of the points command).

The output of the summary command shows that the object does not have a cartographic projection (because longlat is not strictly a projection). A bounding box has been calculated automatically when the spatial object was created.

The bbox method returns the bounding box of the object (e.g., bbox(events)), and it is useful to set plotting axes and regions. A matrix with the points coordinates is the return value of coordinates(events), and proj4string(events) returns the cartographic projection adopted.

5.3.2 SpatialPointsDataFrame Subclass

In most spatial analysis, the points have nongeographic attributes. For example, besides its geographical locations, traffic accident locations can be classified according to hour of day, date, number of cars involved, either there were victims or not, among other additional information relevant for the statistical analysis. We now show how to add these attributes to a SpatialPoints object, creating a SpatialPointsDataFrame subclass object. The command to create such an object is as follows:

```
> SpatialPointsDataFrame(coords, data, coords.nrs = numeric(0),
+  proj4string = CRS(as.character(NA)), match.ID = TRUE, bbox = NULL)
```

where the arguments coords, proj4string, and bbox are the same as in the SpatialPoints command. The argument data is a dataframe with the same number of rows as coords and it holds the attributes of the points. When the row order of the points in the matrix data match the row order of the dataframe data, the user can set match.ID = FALSE, and the coordinates and data are bound together as in the cbind command. However, many times, attributes and geographical coordinates come from different sources and their row order do not match. In this case, assuming that coords and data have unique row names and there is an 1-to-1 correspondence between them, the argument match.ID = TRUE can be used. The dataframe rows are re-ordered to suit the coordinate points matrix. If any differences are found in the row names of the two objects, no SpatialPointsDataFrame object is created and an error message is issued. The argument coords.nrs is optional and is explained after a simple example.

We add hour, day, month, type, and severity of the accidents to their locations stored in the matrix mat. These attributes form the five columns of the dataframe x, and they are in the same row order as the events in mat. Hence, the SpatialPointsDataFrame object events2 is created and plotted in this way:

```
> x <- read.table("accidents_data.txt",header=TRUE)
> events2 <- SpatialPointsDataFrame(mat, x, proj4string = crs, match.ID = FALSE)
> summary(events2)
Object of class SpatialPointsDataFrame
Coordinates:
           min       max
lat  -20.02569 -19.78362
long -44.05925 -43.88172
Is projected: FALSE
proj4string : [+proj=longlat +ellps=WGS84]
Number of points: 1314
Data attributes:
      day              hour       weekDay        severity                type
 Min.   : 1.00   18:00  :  29   fri:218   Fatal   :  8   collision     :1040
```

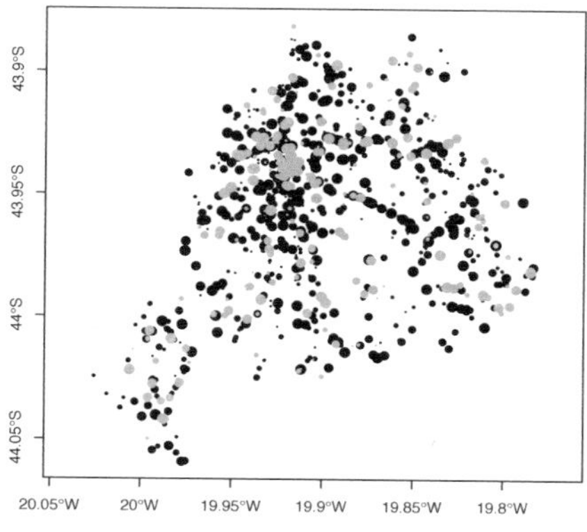

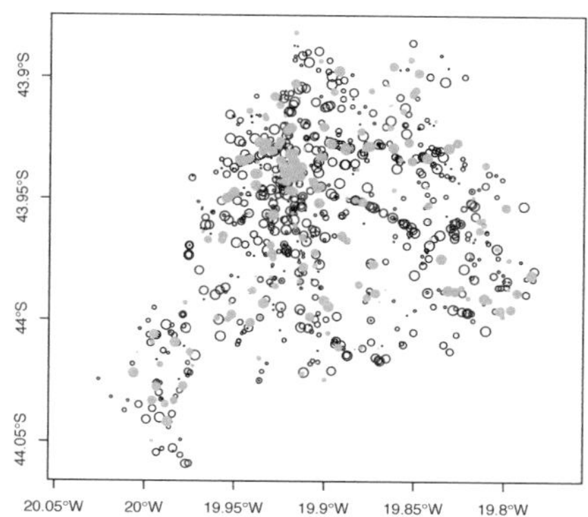

FIGURE 5.7
Use of `SpatialPointsDataFrame`.

```
 1st Qu.: 7.00   17:00   :  24   mon:196   Nonfatal:1306   running over: 274
 Median :14.00   15:30   :  23   sat:218
 Mean   :14.31   19:00   :  21   sun:152
 3rd Qu.:21.00   09:00   :  20   thr:155
 Max.   :28.00   14:00   :  20   tue:191
                 (Other):1177    wed:184
> events3 <- events2[order(events2$type),]
> plot(events3, axes=TRUE, pch = 19, cex = as.numeric(events3$weekDay)/5,
+ col=c(rep("black",sum(events3$type == "collision")),
+ rep("grey",sum(events3$type == "running over"))))
> plot(events3, axes=TRUE, pch = c(rep(1,sum(events3$type == "collision")),
+ rep(19,sum(events3$type == "running over"))), cex = as.numeric(events3$weekDay)/5,
+ col=c(rep("black",sum(events3$type == "collision")),
+ rep("grey",sum(events3$type == "running over"))))
```

The left-hand side plot of Figure 5.7 shows the result of the `plot` command. The circles are drawn with radii proportional to the week days (by means of the `cex` parameter) and with colors according to the type (by means of the `col` parameter, or if the dots are full, or not, by means of parameter `pch`).

The `SpatialPointsDataFrame` class behaves as a dataframe, both with respect to standard methods such as extraction operators and names, and to modeling functions such as formula. For example, `events2$day` extracts the column `day` from `events2`. The command `events2[13:25,]` returns the points (coordinates and attributes) corresponding to rows 13 to 25 and the command

```
> events2[events2$severity == "Fatal",c("weekDay","hour","type")]
              coordinates weekDay  hour         type
455 (-19.9281, -43.9828)      sat 00:30    collision
458 (-19.8356, -43.9397)      sat 01:20    collision
486 (-19.8122, -43.9586)      sat 15:01 running over
556 (-19.9602, -43.9992)      sun 01:00 running over
564 (-19.8618, -43.9323)      sun 04:00    collision
862 (-19.8904, -43.9283)      sat 15:30 running over
```

```
923 (-19.9934, -44.0258)      sun 22:50     collision
950  (-19.9209, -43.971)      mon 17:00     collision
```

returns the weekDay, hour, and type values of the accidents that had fatal victims in February of 2011. Typing names(events2) returns the character vector c("day","hour","weekDay","severity","type").

Another way to create a SpatialPointsDataFrame object is to add the attributes in a dataframe to an existing SpatialPoints object. For example, given the previously created events spatial object, the attributes in x can be added by issuing the command:

```
> events2.alt <- SpatialPointsDataFrame(events, x)
> all.equal(events2, events2.alt)
[1] TRUE
```

Still another way to create a SpatialPointsDataFrame object is to add the geographical coordinates to an existing dataframe. For example, the dataframe x can be transformed into a SpatialPointsDataFrame class object by adding as coordinates the numeric matrix mat.

```
> events2.alt2 <- x
> coordinates(events2.alt2) <- mat
> proj4string(events2.alt2) <- crs
> all.equal(events2, events2.alt2)
[1] TRUE
```

One last way to create a SpatialPointsDataFrame object is to assign two columns in a given dataframe as the points' coordinates. In this case, the coordinate columns are dropped from the attributes list. For example,

```
> aux0 <- cbind(x, mat)
```

creates a dataframe with seven columns. With

```
> coordinates(aux0) <- ~lat+long
```

aux0 is now a SpatialPointsDataFrame object, and has only five attributes columns. More information can be obtained using

```
> summary(aux0)
```

or

```
> proj4string(aux0) <- crs
```

to assign cartographic projection to aux0, or

```
> str(aux0)
```

to view the structure of aux0. The last command output will show that the coords.nrs slot, empty in all previous examples, has now the numeric vector c(6,7) giving the columns numbers of the original dataframe that were taken as the coordinates.

Spatial objects in the form of lines can model streets, rivers, and other linear features of the world. They are handled by the SpatialLines subclass in the sp package. Because most actuarial applications do not use these type of objects, we will not cover it in this chapter. Our main interest is the manipulation of polygons for areal data, a subject we turn to now.

5.3.3 SpatialPolygons Subclass

Basic R graphics has the function polygon to draw and shade one or more polygons. This is different from the subclasses Polygon and Polygons of the sp package. The definition of these sp subclasses is more complicated than those for point objects subclasses because, among other reasons, one spatial entity may be associated with more than one polygon. For example, a map of the fifty U.S. states needs more than fifty polygons. While most states are represented by a single polygon, Hawaii is composed of several islands, each one represented by a separate polygon. Furthermore, while attributes in a dataframe will be associated with a single polygon for most states, Hawaii attributes must be associated with all its constituent polygons. To deal with all the possible aspects of these geo-objects, the polygon subclasses are more complex than the others.

The basic class is Polygon, and an object of this class is created with the command Polygon(coords, hole=as.logical(NA)) where coords is an $n \times 2$ coordinate matrix, with no NA values, and with the last row equal to the first one. Successively connecting with line segments, the i-th and $i+1$-th rows of coords creates a closed line defining a polygon. The argument hole is explained later.

Given the possible need for representing a single geographical entity by more than one polygon (such as in the case of an administrative region with islands), sp also defines the Polygons class. Finally, to make a single bundle of all regions, we connect several Polygons objects in a SpatialPolygons object.

5.3.3.1 First Elementary Example

We will illustrate how to create a map with three administrative regions. The first region is composed of two polygons, one of them representing an island. The other two regions are composed of single polygons. Initially, let us define the four matrices containing the ordered vertices of each polygon and visualize them with the basic and nonspatial function polygon:

```
> p1 <- rbind(c(2,0), c(6,0), c(6,4), c(2,4), c(2,0)) # region 1, mainland
> p1i <- rbind(c(0,0), c(1,1), c(1,4), c(0,2), c(0,0)) # region 1, island
> p2 <- rbind(p1[2,], c(10,3), c(10,7), c(8,7), p1[3:2,]) # region 2
> p3 <- rbind(p1[4:3,], p2[4,], c(4,10), c(0,10), p1[4,]) # region 3
> plot(rbind(p1, p2, p3)); polygon(p1); polygon(p1i); polygon(p2); polygon(p3)
```

Note that some of the vertices are common to two or three areas due to their adjacency. To identify more precisely the areas, we can shade them:

```
> plot(rbind(p1, p2, p3))
> polygon(p1,density=20,angle=30)
> polygon(p1i,density=20,angle=30,col="grey")
> polygon(p2,density=10,angle=-30)
> polygon(p3,density=15,angle=-60,col="grey")
```

The class Polygon from sp is completely different from the function polygon from R, and they should not be confused. We will not make further use of the latter in this chapter. In fact, we use polygon simply to call attention to its difference to the Polygon class. Let us convert our objects to Polygon class objects.

```
> require(sp)
> pl1 <- Polygon(p1); pl1i <- Polygon(p1i); pl2 <- Polygon(p2); pl3 <- Polygon(p3)
> str(pl1)
Formal class 'Polygon' [package "sp"] with 5 slots
  ..@ labpt  : num [1:2] 4 2
  ..@ area   : num 16
```

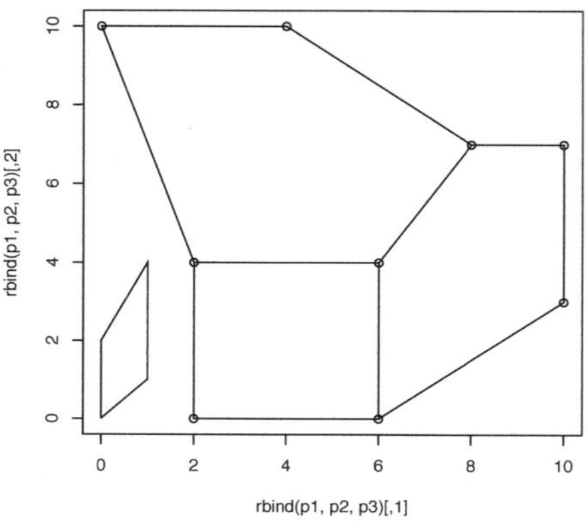

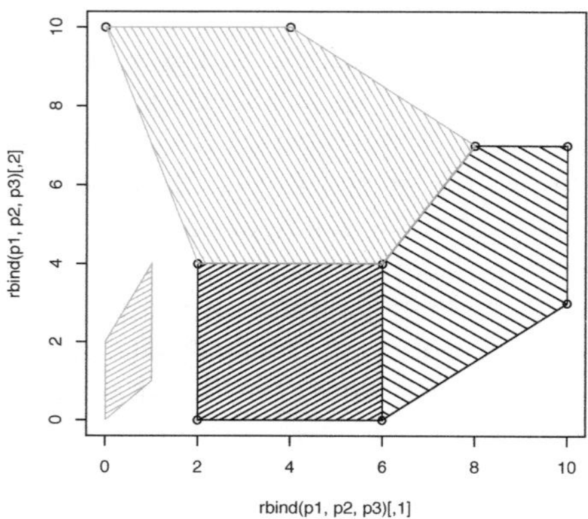

FIGURE 5.8
Use of polygon to visualize areas and regions.

```
  ..@ hole    : logi TRUE
  ..@ ringDir: int -1
  ..@ coords : num [1:5, 1:2] 2 6 6 2 2 0 0 4 4 0
  > names(attributes(pl1))
[1] "labpt"   "area"    "hole"    "ringDir" "coords"  "class"
```

A `Polygon` object has five slots, some of them automatically created when instantiating the object:

- `labpt` is a two-dimensional vector with the polygon centroid coordinates, the arithmetic average of the vertices coordinates.
- `area` is the polygon area.
- `hole` is a logical value indicating whether the polygon is a hole (see Section 5.3.3.2).
- `ringDir`: more about this in section 5.3.3.2.
- `coords` are the vertices coordinates.

Hence, it holds much more geographical information than a simple matrix with $x - y$ coordinates. We can access the attributes of a `Polygon` using and the slot name. For example,

```
> pl1@labpt
[1] 4 2
> pl1@area
[1] 16
```

There is no plot method for the `Polygon` class and therefore one cannot yet visualize the polygons stored in these objects.

We want to create one object in which the `Polygon` class `pl1` and `pl1i` objects are joined to compose a single administrative region. The way to do this is to create a `Polygons` class object (note the *s* at the end). It receives a list of `Polygon` objects and conforms them to a unique geographic unit. Its basic use is `Polygons(srl, ID)`, where `srl` is a list of `Polygon`

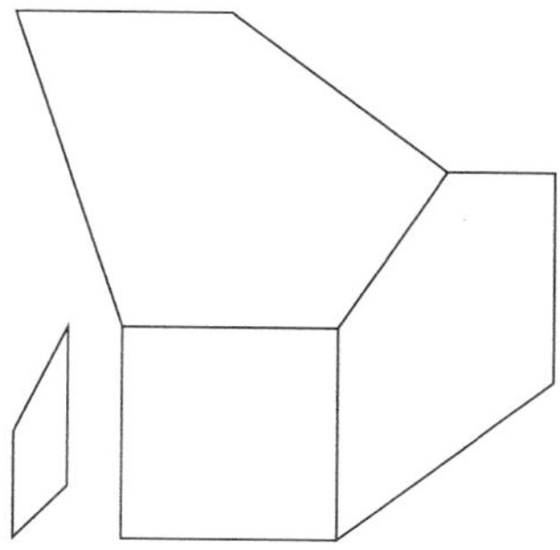

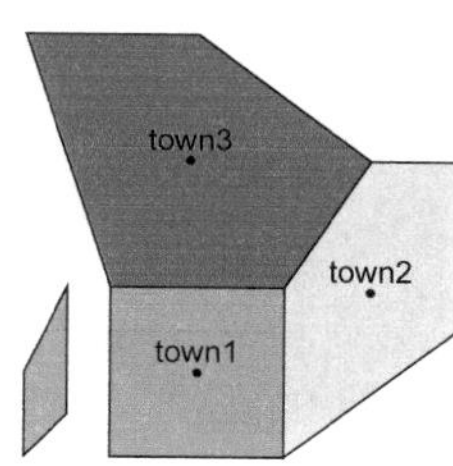

FIGURE 5.9
Use of `Polygon()` from library `sp` to visualize areas and regions.

objects and `ID` is a character vector of length one with a label identifier for the geographical entity created. For example,

```
> t1 <- Polygons(list(pl1,pl1i), "town1")
> t2 <- Polygons(list(pl2), "town2")
> t3 <- Polygons(list(pl3), "town3")
```

creates three objects of class `Polygons` in which `t1` is composed of two separate polygons and the other two are composed of single polygons. It seems redundant to define a `Polygons` object composed of a single `Polygon` object but we need some consistency when joining `Polygons` to create maps. The next step will clarify why we need this.

Finally, to create a geo-object holding the entire map of the region composed of these three towns, we use a `SpatialPolygons` class object, created with `SpatialPolygons(Srl, pO, crs)`, where `Srl` is a list with objects of class `Polygons` (note the plural, not `Polygon` class objects), `pO` is an optional parameter giving the plotting order of the polygons (if missing, they are plotted in reverse order of polygons area), and `crs`, also optional, is a string of class CRS with the cartographic projection:

```
> map3 <- SpatialPolygons(list(t1, t2, t3))
> plot(map3)
> plot(map3,col=grey(c(.7,.9,.5)))
> cents <- coordinates(map3)
> points(cents, pch=20)
> text(cents[,1], cents[,2]+0.5, c("town1","town2","town3"))
```

The left-hand side of Figure 5.10 shows the output of the first `plot` command, while the right-hand side shows that of the second.

5.3.3.2 Second Example

Let us modify the previous example to show more details of the `SpatialPolygons` class. Town `t1` will be the same. Within Town `t2` there is a lake of considerable area that cannot be ignored. This lake is considered a hole within Town `t2` polygon and it is represented

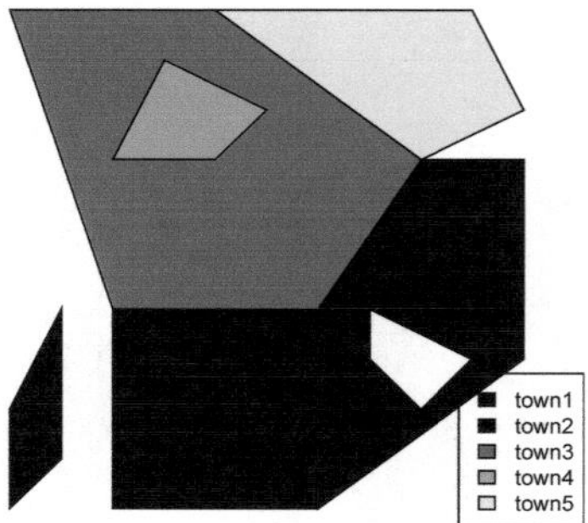

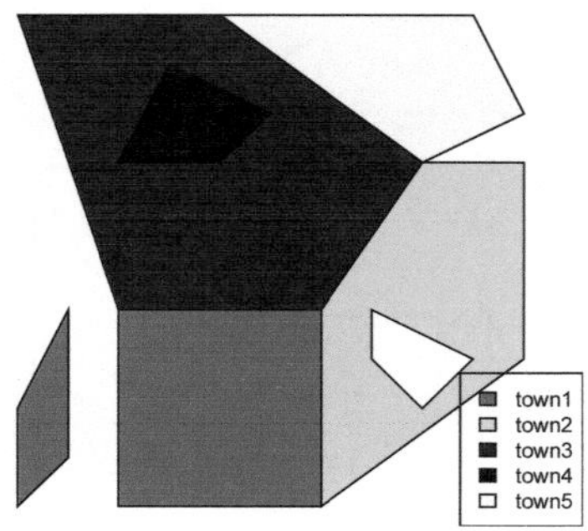

FIGURE 5.10
Use of `Polygon()` from library `sp` to visualize areas and regions, with holes.

by another polygon contained in the outer boundary of Town `t2`. A third situation is that Town `t3` also contains another smaller polygon that is also a hole. However, in this case, the hole represents Town4 `t4`, a fourth administrative region in our map. This is, for example, the situation of the Vatican, which is entirely contained within the boundaries of Rome. We finally add a fifth area to the set, ending with five towns in our schematic map.

For completeness, we will repeat some commands and create the matrices with the vertices coordinates, and define the Polygon objects:

```
> p1 <- rbind(c(2,0), c(6,0), c(6,4), c(2,4), c(2,0)) # region 1, mainland
> p1i <- rbind(c(0,0), c(1,1), c(1,4), c(0,2), c(0,0)) # region 1, island
> p2 <- rbind(p1[2,], c(10,3), c(10,7), c(8,7), p1[3:2,]) # region 2
> p2l <- rbind(c(8,2), c(9,3), c(7,4), c(7,3), c(8,2)) # region 2, lake
> p3 <- rbind(p1[4:3,], p2[4,], c(4,10), c(0,10), p1[4,]) # region 3
> p4 <- rbind(c(4,7), c(5,8), c(3,9), c(2,7), c(4,7)) # region 4, inside region 3
> p5 <- rbind(p3[4:3,], c(10,8), c(9,10), p3[4,]) # region 5
> pls5 <- list()
> pls5[[1]] <- Polygons(list(Polygon(p1, hole=FALSE),
+ Polygon(p1i, hole=FALSE)), "town1")
> pls5[[2]] <- Polygons(list(Polygon(p2, hole=FALSE),
+ Polygon(p2l, hole=TRUE)), "town2")
> pls5[[3]] <- Polygons(list(Polygon(p3, hole=FALSE),
+  Polygon(p4, hole=TRUE)), "town3")
> pls5[[4]] <- Polygons(list(Polygon(p4, hole=FALSE)), "town4")
> pls5[[5]] <- Polygons(list(Polygon(p5, hole=FALSE)), "town5")
> map5 <- SpatialPolygons(pls5)
> plot(map5)
> plot(map5, col=gray(c(.1,.3,.5,.7,.9)))
> legend("bottomright", c("town1", "town2", "town3", "town4", "town5"),
+ fill=gray(c(.1,.3,.5,.7,.9)))
> plot(map5, col=c("red", "green", "blue", "black", "yellow"))
> legend("bottomright", c("town1", "town2", "town3", "town4", "town5"),
+ fill=c("red", "green", "blue", "black", "yellow"))
```

Note that this time we avoid the intermediate creation of the `Polygon` and the `Polygons` objects by nesting the Polygon and Polygons commands when assigning the pls5 elements. The hole logical argument is optional and it establishes if the polygon is a hole. Note that p4 enters as a hole=TRUE in t3 (because it represents a hole within the limits of t3) but as a hole=FALSE in t4 (as p4 describes the limits of t4). The second plot command uses a sequential gray level for each town, the first town receiving the darkest tone. The third plot makes it more clear how towns and holes are considered when coloring.

If the hole argument is not given, the status of the polygon as a hole or an island will be taken from the ring direction, with clockwise meaning island, and counter-clockwise meaning hole.

5.3.4 SpatialPolygonsDataFrame Subclass

In addition to the polygons representing the geographical entities, we typically have attributes to be attached to them. In our schematic map with five administrative regions, we could be interested in analyzing their insured population size, income per capita, and incidence rate of a certain claim. Adding attributes to a SpatialPolygons object, we create a SpatialPolygonsDataFrame subclass object with the command

```
> SpatialPolygonsDataFrame(Sr, data, match.ID = TRUE)
```

The argument Sr is a SpatialPolygons class object, and data is a dataframe with number of rows equal to length(Sr). The third argument can be a logical flag indicating whether the row names should be used to match the Polygons ID slot values (match.ID = TRUE), re-ordering the dataframe rows if necessary. If match.ID = FALSE, the dataframe is merged with the Polygons assuming that they are in the same order. Compare the output of the first two commands SpatialPolygonsDataFrame below:

```
> x <- data.frame(x1 = c("F", "F", "T", "T", "T"), x2=1:5,
+ row.names = c("town4", "town5", "town1", "town2", "town3"))
> map5x <- SpatialPolygonsDataFrame(map5, x, match.ID = TRUE)
> map5x@data
> map5x <- SpatialPolygonsDataFrame(map5, x, match.ID = F)
> map5x@data
```

A third option is to pass a string to match.ID indicating the column name in the dataframe to match the Polygons IDs:

```
> x <- data.frame(x1 = c("F", "F", "T", "T", "T"), x2=1:5,
+ x3 = c("town4", "town5", "town1", "town2", "town3"))
> map5x <- SpatialPolygonsDataFrame(map5, x, match.ID = "x3")
```

5.4 Maps in R

The previous examples are schematic and purely illustrative of the type of spatial objects sp deals with. Most actual spatial analyses require maps built by official agencies or companies, maps of high quality, time consuming to produce, and for multiple purposes. These maps typically are acquired or downloaded and are composed of hundreds of areas and richly equipped with many attributes data. In this section we show how to read this type of map in R and how to add your own data to build a SpatialPolygonsDataFrame object.

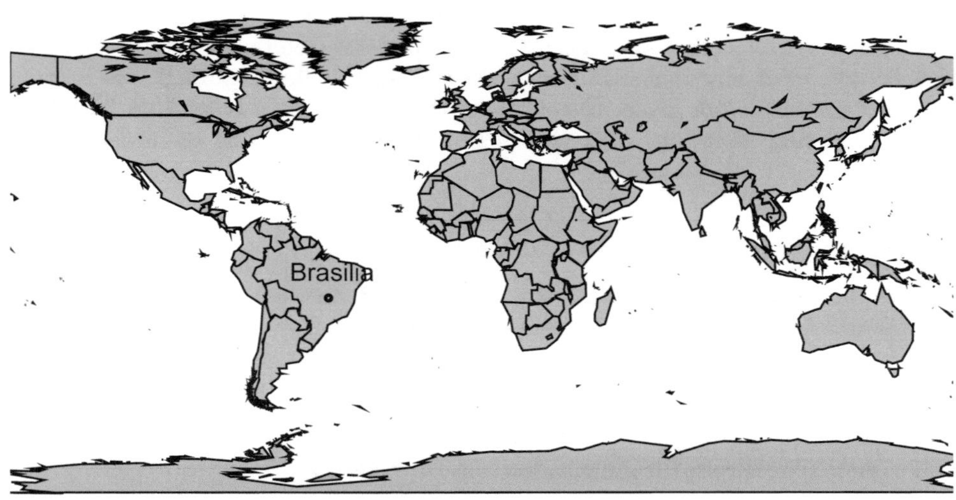

FIGURE 5.11
Use of `map()` to visualize the world map.

A first option is to use maps that come with the package `maps`. It makes available the world map, with the boundaries of the countries, as well as the United States map divided into states or counties. To visualize the world map, resulting in Figure 5.11, we can type

```
> require(maps)
> map("world", col = grey(0.8), fill=TRUE)
> map.cities(country = "Brazil", capitals = 1, cex=0.7)
```

The default projection is a rectangular one (with the aspect ratio chosen so that longitude and latitude scales are equivalent at the center of the picture), but it is possible to use another one using `mapproj` package. For instance, in Figure 5.12, several projections are considered for Canada.

```
> library(mapproj)
> map("world", "canada", proj="conic",  param=45, fill=TRUE, col=grey(.9))
> map("world", "canada", proj="bonne",  param=45, fill=TRUE, col=grey(.9))
> map("world", "canada", proj="albers", par=c(30,40), fill=TRUE, col=grey(.9))
> map("world", "canada", proj="lagrange", fill=TRUE, col=grey(.9))
```

However, most of the maps that users need in their analysis will be provided by third parties and in particular proprietary formats. The package `maptools` implements functions to read and write maps in the shapefile format, the most popular GIS format and specified by Environmental Systems Research Institute - ESRI. This package converts spatial classes defined by the package `sp` to classes defined in other R packages such as `PBSmapping`, `spatstat`, `maps`, `RArcInfo`, and others.

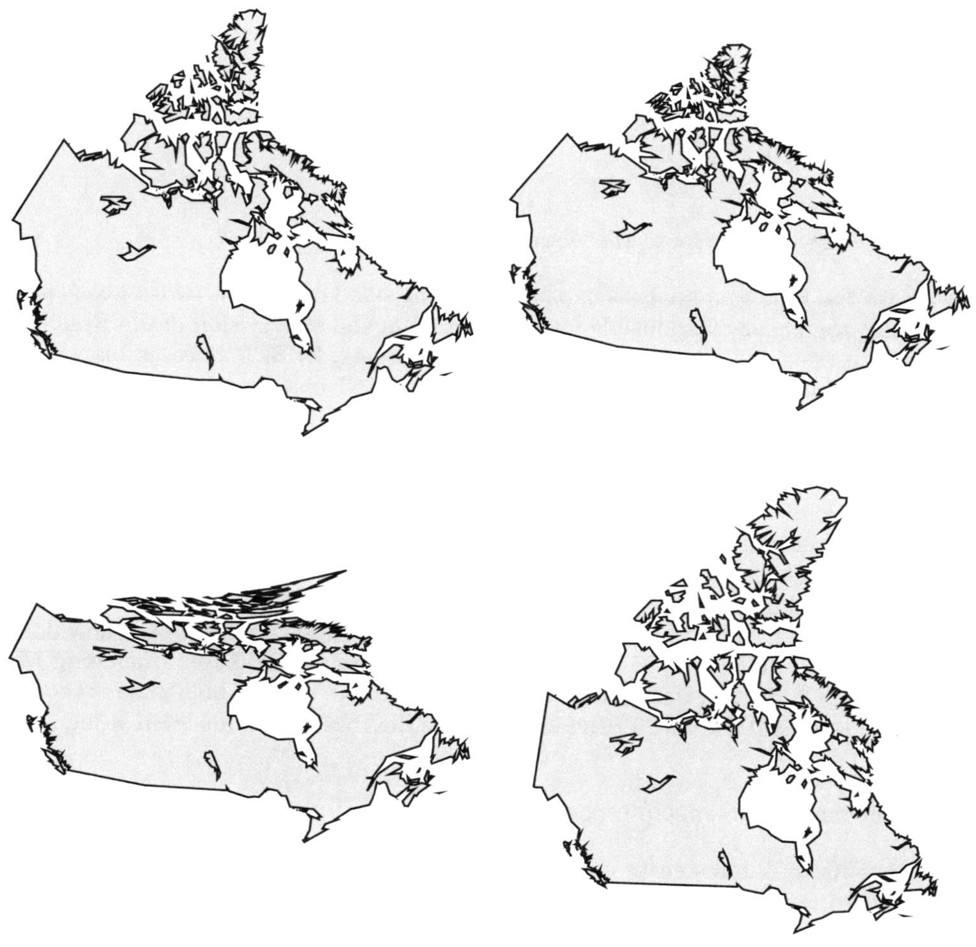

FIGURE 5.12
Use of `map("world", "canada")` and different projection techniques, using `proj`.

In the next section, we will show how to read a database with attribute data and to merge them with a spatial object created from an ESRI shapefile. The exploratory analysis of the spatial database formed is carried out in the remaining sections.

5.5 Reading Maps and Data in R

We read a spatial database provided by IBGE, the Brazilian governmental agency in charge of geographical issues and official statistics (ibge.gov.br, accessed in February, 2013). This database has a shapefile format with a set of at least three files: one containing the geographical coordinates of the polygons, lines or dots (extension `.shp`); another with attribute data (extension `.dbf`); and a third file with the index that allows the connection between the .shp and .dbf files (extension .shx). An additional file that can be part of the shapefile format is a file indicating the cartographic projection of the polygons, lines or points, with extension .prj.

We use the maptools functionalities to read the shapefile data organized by municipality as a SpatialPolygonsDataFrame, leaving for the next section a more detailed discussion of the object read in by R:

```
> library(maptools)
> shape.mun  <- readShapeSpatial("55mu2500gsd")
```

We illustrate the R spatial analysis capabilities using one large actuarial database provided by SUSEP, the agency responsible for the regulation and supervision of the Brazilian insurance, private pension, annuity, and reinsurance markets. SUSEP releases biannually a car insurance database composed of the aggregation of all insurance companies' information. Due to confidentiality concerns, there is no individual-level information, the data being aggregated into zip code areas.

The variables available are the number of vehicles-year exposed, the average premium, the average number of claims, and amount of damages, classified according to category, model and year of vehicle, region, working place ZIP code, and main driver age and sex. This database is known as AUTOSEG (an acronym for Statistical System for Automobiles) and can be accessed online (www2.susep.gov.br/menuestatistica/Autoseg, accessed February 2013). We used the 2011 AUTOSEG database in our analysis, available for download as an Access file, with 2GB. The package RODBC offers access to databases (including Microsoft Access and SQL Server) through an ODBC interface. We used its ODBConnectAccess function to fetch some data. First, a connection to the database file is achieved using

```
> library(RODBC)
> con <- odbcConnectAccess2007("baseAuto.mdb")
```

We are particularly interested in comparing premiums, claims, and reported damages for two specific groups: popular vehicles and luxury vehicles. The basic difference between the groups is the power of the engine and the materials and finishing quality. Popular cars have a power of 1,000 cc (cylinders), whereas luxury cars usually have a power of 2,000 cc or greater. In Brazil, popular cars also share a more simple internal structure and therefore their price is affordable to most car consumers. Selected popular vehicles in Brazil are Celta 1.0 (Chevrolet), Corsa 1.0 (Chevrolet), Prisma 1.0 (Chevrolet), Uno 1.0 (Fiat), Palio 1.0 (Fiat), Gol 1. (Volkswagen), Fox 1.0 (Volkswagen), Fiesta 1.0 (Ford), and Ka 1.0 (Ford). Similarly, selected luxury vehicles are Vectra (Chevrolet), Omega (Chevrolet), Linea (Fiat), Bravo (Fiat), Passat (Volkswagen), Polo (Volkswagen), Fusion (Ford), Focus (Ford), Corolla (Toyota), Civic (Honda), and Audi.

We can create a dataframe structure with detailed information of region, city code, yearly exposure, premium, and frequency of claims for the following categories: robbery or theft (RB), partial collision and total loss (COL), fire (FI), or others (OT). The transfer of the data stored in the Access database to R can be made using the function sqlFetch() from RODBC. We used the required parameters channel and sqtable with arguments con and susep:

```
> base <- sqlFetch(channel=con, sqtable="susep")
> odbcClose(con)
```

The final dataframe is available in this book' database package, and the different columns store information for popular and luxury groups. The frequency of claims for each category are found in columns CL_RB_LUX, CL_COL_LUX, CL_FI_LUX, CL_OT_LUX, CL_RB_POP, CL_COL_POP, CL_FI_POP, and CL_OT_POP. The name of the municipality is found in column NAME_MUN, and the numeric municipality code is found in column COD_MUN. The information on exposition and average premium for the popular and luxury groups is given in EXPO_POP and EXPO_LUX, respectively.

Originally, both SUSEP and IBGE databases did not present a unique identification column that provides a forward merge of the two databases. The joint information is the name and the state of each municipality. Due to discrepancies between the databases, some of the municipalities presented different names. An extensive code was written in order to properly adjust the names of the municipalities and merge the two different databases. We did not include this code in our example but we briefly describe how to merge the database generated from SUSEP and the database generated from IBGE. We make available the final database, which also includes the municipality population (POP_RES) based on the 2010 Census, and the 2000 municipality Human Development Index (HDI code HDIM_00). The Human Development Index (HDI) is a summary measure of long-term progress in three basic dimensions of human development: income, education, and health. The HDI provides a counterpoint to another widely used indicator, the Gross Domestic Product (GDP) per capita, which only considers economic dimensions. Both POP_RES and HDIM_00 columns were generated from the IBGE site (ibge.gov.br, accessed February 2013).

A new dataframe is created by merging the spatial dataframe and the SUSEP data frame:

```
> base.shp  <- merge(shape.mun@data, base, by="COD_MUN", all.x = TRUE)
```

The shape.mun object is of class SpatialPolygonsDataFrame, which stores polygons information and dataframe. To guarantee that the spatial polygon information available in shape.mun matches the order of municipalities in the new dataframe (base.shp), both databases are first sorted in ascending order of the municipality codes:

```
> base.shp  <- base.shp[order(base.shp$COD_MUN),]
> shape.mun <- shape.mun[order(shape.mun$COD_MUN),]
```

In sequence, the dataframe of the shape.mun object is updated with the new dataframe

```
> shape.mun@data <- base.shp
```

We restrict the size of the final database using only the municipalities from four Brazilian states. We use the state variable, ST, to select the following states: São Paulo (SP), Santa Catarina (SC), Paraná (PR), and Rio Grande do Sul (RS). These states are located in the southern region of Brazil and contain almost 70 million inhabitants (around 36% of the Brazilian population) and constitute one of the richest regions of the country (approximately 60% of the Brazilian gross product). It allows us to show how to select subregions in a map to carry out a spatial statistical analysis:

```
> sul_sp <- shape.mun[shape.mun$ST %in% c("SP", "SC", "PR", "RS"),]
> length(sul_sp@polygons)
[1] 1833
> dim(sul_sp@data)
[1] 1833   17
> names(sul_sp@data)
 [1] "COD_MUN"    "ST"         "NAME_MUN"   "EXPO_POP"   "PREMIO_POP"
 [6] "CL_RB_POP"  "CL_COL_POP" "CL_FI_POP"  "CL_OT_POP"  "EXPO_LUX"
[11] "PREMIO_LUX" "CL_RB_LUX"  "CL_COL_LUX" "CL_FI_LUX"  "CL_OT_LUX"
[16] "HDIM_00"    "POP_RES"
```

Therefore, our final `SpatialPolygonsDataFrame` has 1,833 administrative regions and 17 attributes associated to each of them. The names of the variables are displayed above and have been discussed previously in the text.

Out of the 1,833 available municipalities, approximately 20%, or 397 municipalities, have missing data. From the municipalities without missing values we adjusted a simple linear regression using the logarithm of the population as explanatory variable and the exposition as response. With the fitted regression, the logarithms of the municipalities with missing information were used to predict the exposition of these municipalities. Finally, the imputed value is the maximum between the exponential of the predicted value and zero because the exposition is a nonnegative variable. The claims in the municipalities with missing data are kept as `NA` and will be estimated in the model fitting. Some municipalities in the state of Rio Grande do Sul are recent and do not have any HDI information based on the most recent Brazilian Census. For these municipalities, the HDI information was set as the mean value of it neighbors.

The final dataset, with both spatial and insurance information, is saved as a shapefile for future use and is available as a separate file named `sul+sp_shape.shp`:

```
> writePolyShape(sul_sp, "sul+sp_shape")
```

5.6 Exploratory Spatial Data Analysis

Initially, load the package `maptools` and read the data from the shapefile `sul+sp_shape` as a `SpatialPolygonsDataFrame`:

```
> library(maptools)
> shape <- readShapeSpatial("sul+sp_shape")[,-1]
```

The `shape` object of class `SpatialPolygonsDataFrame` stores different information such as dataframe and polygons coordinates. The different structures are known as `slots` and their names can be accessed with

```
> slotNames(shape)
[1] "data"          "polygons"    "plotOrder"   "bbox"          "proj4string"
```

Further information about the `shape` object, such as the number of polygons and access to the coordinates of the polygons, is provided with

```
> class(shape)
[1] "SpatialPolygonsDataFrame"
attr(class(shape),"package")
[1] "sp"
> dim(shape)
[1] 1833   17
> str(shape@data)
> shape@polygons
> slotNames(shape@polygons[[1]])
> [1] "Polygons"  "plotOrder" "labpt"     "ID"         "area"
> names(shape)
 [1] "COD_MUN"     "ST"           "NAME_MUN"    "EXPO_POP"    "PREMIO_POP"
```

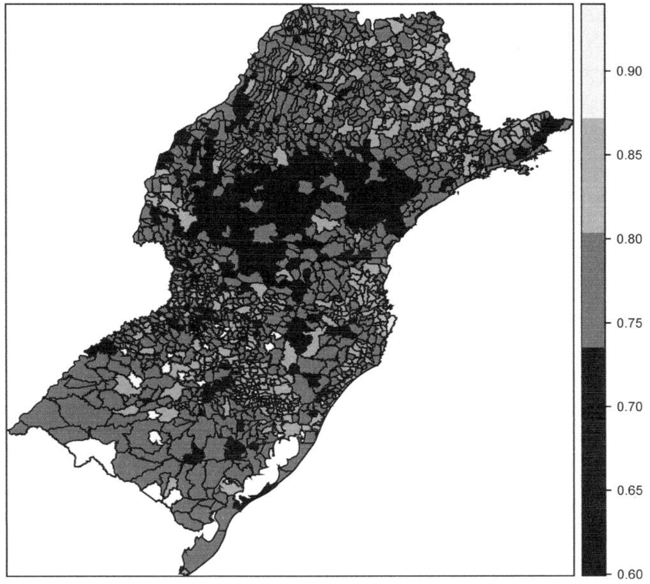

FIGURE 5.13
Visualizing shape maps with `spplot()`.

```
 [6] "CL_RB_POP"   "CL_COL_POP" "CL_FI_POP"  "CL_OT_POP"  "EXPO_LUX"
[11] "PREMIO_LUX" "CL_RB_LUX"  "CL_COL_LUX" "CL_FI_LUX"  "CL_OT_LUX"
[16] "HDIM_00"    "POP_RES"
```

5.6.1 Mapping a Variable

One advantage of working with `sp` class objects is that they can be passed directly as arguments to the function `plot()`, such as in `plot(shape)`. If `plot()` receives a `SpatialPolygonsDataFrame` class object, it returns a polygon map. Another useful way to visualize area data is to color each polygon according to an attribute. This is available through the function `spplot()` from the `sp` package. Taking the `HDIM_00` variable from the `shape` object as a theme, we create a five-dimensional vector `cols` with five gray shade values (more on colors soon) and draw a colored map with the following commands:

```
> cols <- rev(gray(seq(0.1, 0.9, length = 5)))
> cols
[1] "#E6E6E6" "#B3B3B3" "#808080" "#4D4D4D" "#1A1A1A"
> spplot(shape, "HDIM_00", col.regions = cols, cuts = length(cols) - 1)
```

The function `spplot()` automatically breaks the range of `HDIM_00` into equal length intervals. We need to provide the additional parameter `cuts` to force the number of colors to be equal to the number of cuts in the legend. Some municipalities are white painted due to their missing values in the `HDIM_00` variable. These are municipalities created after the 2000 Census and hence they had no `HDIM_00` value associated. We can input values to these municipalities using the average `HDIM_00` among their neighboring areas. We discuss how to perform this imputation in Section 5.7.3.

The automatic determination of the color classes by `spplot()` is an advantage when we want a quick spatial visualization of a certain variable. However, many times, the user will need customized breaks. This takes much longer than the simple use of `spplot()`. Suppose, for example, we wish gray shades associated with the intervals determined by the quantiles of `HDIM_00`. We need first to find the break points (in `brks`) and create a factor with levels indicating the class interval each area belongs to. As a more stringent need, this factor must be added to the `shape` object (as the variable `col_var`). This factor is now the variable to be mapped using the same gray shades as before:

```
> brks <- quantile(shape$HDIM_00, prob = c(0, .2, .4, .6, .8, 1), na.rm = TRUE)
> shape$col_var <- cut(shape$HDIM_00, brks)
> spplot(shape, "col_var", col.regions = cols, main = "Levels are intervals")
> levels(shape$col_var) <- c("Very Low", "Low", "Middle", "High", "Very High")
> spplot(shape, "col_var", col.regions = cols, main = "User defined levels")
```

Because we are mapping a factor (`col_var`) rather than the continuous variable `HDIM_00`, we are not required to use the parameter `cut` any more. The map is colored according to the values of the factor `col_var` and the legend uses its levels. The user can change these levels to obtain legends that could be more suitable. For example,

```
> levels(shape$col_var) <- c("Very Low", "Low", "Middle", "High", "Very High")
> spplot(shape, "col_var", col.regions = cols, main = "Levels are user defined")
```

5.6.2 Selecting Colors

Colors in R are represented by a vector of characters such as "#RRGGBB" where each of the pairs "RR", "GG", and "BB" consist of hexadecimal "digits" with a value ranging from "00" to "FF". Basic colors can also be obtained by typing their name between quotation marks, as for example, typing "black" or "red". The color names recognized by R can be obtained with the command `colors()`.

Typically, we prefer maps with a single color with different shades or two different colors as extreme values with gradual shading trends for the intermediate values. Some functions in R create shades between two colors as, for example, `heat.colors(n)` that creates an n-dimensional vector with the codes from intense red to white. Other color palettes are provided by R, such as `terrain.colors()`, `topo.colors()`, and `cm.colors`. The code below is a simple way to visualize some of the possibilities of color palettes in R:

```
> par(mfrow=c(2,2))
> pie(rep(1,10), col=heat.colors(10), main = "heat.colors()")
> pie(rep(1,10), col=topo.colors(10), main = "topo.colors()")
> pie(rep(1,10), col=terrain.colors(10), main = "terrain.colors()")
> pie(rep(1,10), col=cm.colors(10), main = "cm.colors()")
```

More options on colors are available using the package `RColorBrewer`, especially developed to generate pleasant color palettes to be used in thematic maps. For details, type `brewer.pal.info` to obtain a `data.frame` with the name of the palettes, their maximum number of colors, and category. To visualize all palettes in a single graphical window use the function `display.brewer.all()`. For more details, type `help(package="RColorBrewer")`.

We can customize the maps in R by using the many options on colors. The next set of commands illustrates how we can create the categories and colors to be used with `spplot()`.

```
> library(RColorBrewer)
> cols <- brewer.pal(5, "Reds")
```

```
> spplot(shape, "col_var", col.regions = cols,
+ main = "HDI by municipalities in South Brazil")
```

Instead of `spplot()`, we can use the usual function `plot()` with additional parameters to control colors and the legend position. We start defining the number of colors and categories as equal to five. To select the colors, we use the function `brewer.pal()` of package `RColorBrewer`. The first argument of the function is the number of colors, and the second is `"Greens"`, the name of the color palette.

```
> plotvar <- shape$HDIM_00
> ncls <- 5
> colpal <- brewer.pal(ncls,"Greens")
```

The next step is to define the interval categories to break down the range of the `HDIM_00` variable. From the package `classInt`, we use the function `classIntervals` requiring the parameter `style`, that specifies which method is used to create the class intervals. For example, `style="quantile"` indicates that the intervals are built using quantiles of the variable of interest, and `style="equal"` builds equal-amplitude intervals. Other options can be found by calling the help on `classIntervals`. Another parameter that must be specified is the number of classes to be created.

```
> library(classInt)
> classes <- classIntervals(plotvar, ncls, style = "equal")
> cols2 <- findColours(classes, colpal)
```

Rather than providing only the list of different colors of the categories as in `spplot`, the function `plot` needs to specify the color of each region in the map. In the last command above, we used the function `findColours`, specifying two arguments, the first one being a `classIntervals` class object, and the second is the color palette. The output returns the color of each polygon and two other attributes of help to draw a legend. More details can be found by typing `help(package="classInt")`.

In executing the above commands, a warning is issued because the variable `HDIM_M` has `NA`s. To highlight these polygons with missing values, we assign them the color `red`:

```
> cols2[is.na(shape$HDIM_00)] <- "red"
```

After defining classes and colors, the next step is to use the function `plot()` to generate the map. The first argument is `shape`, the `SpatialPolygonsDataFrame` class object, and the second one is the object `col2`, with the colors of each polygon. Finally, we build the legend through the function `legend()` passing the coordinates X and Y for the legend location in the graphical window, the character vector with the labels of each category in `legend`, and the parameter `fill` with the color of each category:

```
> plot(shape, col = cols2)
> legend(-47.85126, -29.96805, legend=c(names(attr(cols2, "table")), "NA"),
+ fill=c(attr(cols2, "palette"), "red"))
```

The results of these commands are shown in Figures 5.6.2 and 5.6.2.

5.6.3 Using the RgoogleMaps Package

We can use the `RgoogleMaps` package (Loecher et al. (2013)) to download into Ra static map from Google Maps™ (maps.google.com). This map is used as a canvas or background image where we overlay polygons, lines, and points as shown in Figure 5.15 where the

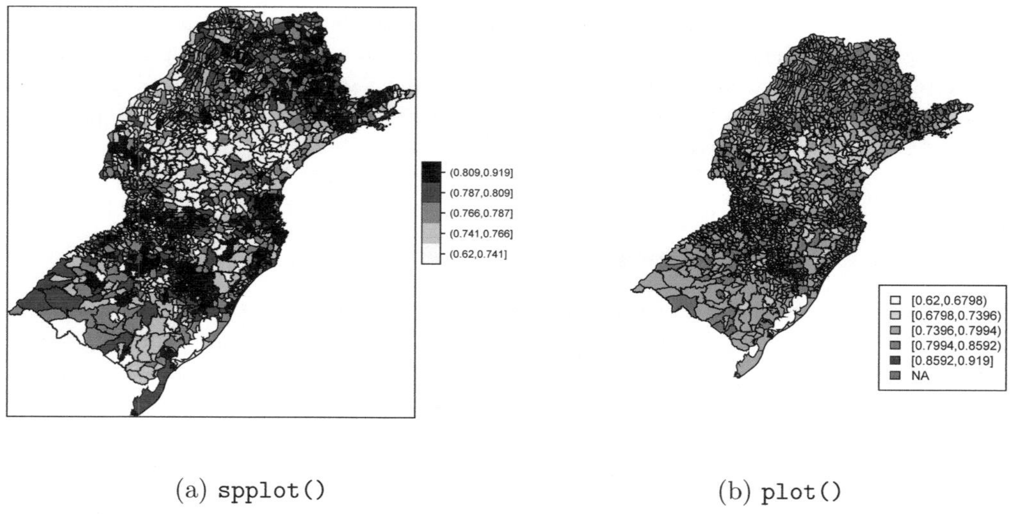

(a) `spplot()` (b) `plot()`

FIGURE 5.14
Example of maps using the functions `spplot` and `plot`.

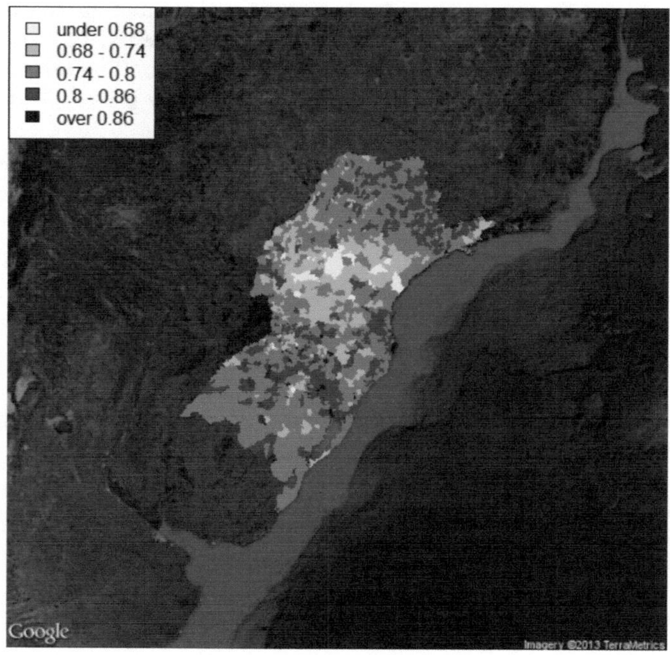

FIGURE 5.15
Visualizing shape maps with `RgoogleMaps` package. Variable `HDIM_00` overlaid on satellite map from Google Maps.

thematic map of the variable HDIM_00 is overlaid on a satellite image fetched from Google Maps.

The PBSmapping package was created by fisheries researchers to provide R with features similar to those available in a Geographic Information System (GIS). It is the standard package used by RgoogleMaps to plot a set of polygons in an image from Google Maps. The first step, using PBSmapping, is to convert shape, an object of class sp, into a PolySet dataframe.

```
> shape <- readShapeSpatial("sul+sp_shape")
> library(PBSmapping)
> map.susep <- SpatialPolygons2PolySet(shape)
> class(map.susep)
[1] "PolySet"    "data.frame"
> head(map.susep)
```

The PolySet dataframe is a standard R dataframe with five numerical columns: PID is the area ID, SID is a secondary id number when the area is composed of more than one polygon (such as when it has islands), POS is the index of the point forming the polygon contour, and finally the vertex coordinates X and Y.

The second step is to get the satellite image from Google Maps that contains the shape file with the South Brazilian states. For this, we need to specify a rectangle (or bounding box) enclosing the polygons we want to overlay on the Google Maps image. This bounding box can be obtained with the bbox() function from the sp package. RgoogleMaps requires this bounding box in lat-long coordinates. Because our shape spatial object has been specified with this coordinate system, it is the return value of bbox(). Next, we use the GetMap.bbox() function from the RgoogleMaps package for inputting the bounding box lat-long and some additional parameters. The resulting MyMap object is a list holding the image and can be inspected with str(MyMap). The Google Maps image can be visualized with R using the PlotOnStaticMap() command, the main RgoogleMaps plot function:

```
> bb <- bbox(shape) # getting the map bounding box
> bb
        min       max
x -57.64322 -44.16052
y -33.75158 -19.77919
> library(RgoogleMaps)
> MyMap <- GetMap.bbox(bb[1, ], bb[2, ],  # fetching the Google Maps image
+ maptype = "satellite",
+ destfile = "myMap.png",
+ GRAYSCALE = FALSE)
> str(MyMap)  # inspecting the MyMap object
> PlotOnStaticMap(MyMap) # plotting the image
```

The type of image can be controlled with the maptype parameter whose values include "roadmap", a map with the main roads shown as lines, and "terrain", an image showing the main relief aspects of the region (see ?GetMap() for further details). The destfile parameter identifies the name of the local png file holding the map image. The most important feature of RgoogleMaps is the overlay of the thematic maps on this image with the PlotOnStaticMap() command. Along with the map image MyMap returned from GetMap.bbox() and the spatial object map.susep with the polygons, we also need to pass a vector with the colors of each area in the map. Using the vector cols2 created for the Figure 5.6.2 map, we can overlay the polygons:

```
> PlotPolysOnStaticMap(MyMap, map.susep, col = cols2, lwd = 0.15,
+ border = NA, add = FALSE)
```

The other parameters are optional: `lwd` controls the line width; `border=NA` omits the polygon borders, while `border=NULL` is the default and plots the borders in black; and as usual, `add` starts a new plot or adds to an existing plot. To add a legend to the figure, we use the `classInt` package and `legend`. The `leglabs()` function makes the legend labels, whereas `cex`, `ncol`, `bg`, and `bty` parameters are related to character expansion (size) of the legend, number of columns, background color, and the type of box to be drawn around the legend (see `?legend()` for further details).

```
> legend("topleft", fill=attr(cols2, "palette"),
+        legend=leglabs( round(classes$brks, digits=2) ),
+        cex=1.0, ncol=1, bg="white", bty="o")
```

Finally, to save the map as a `png` file, or other formats, use `savePlot`:

```
> savePlot(filename="map.png", type="png")
```

It is worth mentioning that `plotGoogleMaps` Kilibarda (2013) and `ggmap` Kahle & Wickham (2013) packages also generate visualizations using Google Maps (see r-project.org). The R code of the previous examples using `RgoogleMaps` can be found in `plotmaps_RgoogleMaps.R`.

In data analysis of point processes, it is natural that the locations of the events are available as written addresses. For example, the number and the street name where an automobile accident occurred. However, to perform a spatial analysis of data points, we need the set of geographical coordinates that represent the exact locations of these events. We can obtain the coordinates from these addresses using the function `geocode()` in package `dismo`. The `dismo` package was originally written to implement models for species distributions and the function `geocode()` provides an interface to Google Maps API.

The `geocode )` function gets the geographical coordinates of a given written address through the Google Maps geocoding service. This service allows a maximum number of 2,500 queries per day. To use function `geocode()`, a vector with the addresses must be informed. Next, we create a sequence of addresses and apply the `geocode()` function to generate the coordinates.

```
> require(dismo)
> adress <- paste("Avenida Otacilio Negrao de Lima, ",
+ seq(1, 30000, by = 200),
+ " , Belo Horizonte - Minas Gerais",
+ sep = "")
> geo.pt <- geocode(adress)
> geo.pt <- rbind(geo.pt, geo.pt[1,])
```

The output of the `geocode()` function is a dataframe object with the following columns: the `originalPlace` column with the written address; the `interpretedPlace` column with the detected address by Google Maps API; the longitude column `lon`; the latitude column `lat`; the minimum longitude of the bounding box `monmin`; the maximum longitude of the bounding box `lonmax`; the minimum latitude of the bounding box, `latmin`; the maximum latitude of the bounding box, `latmax`; and the distance between the point and the farthest corner of the bounding box, `uncertainty`,

To visualize the points on the map, we use functions `GetMap()` and `PlotOnStaticMap()` from `RgoogleMaps` package to read a map from Google Maps, and use it as the background image.

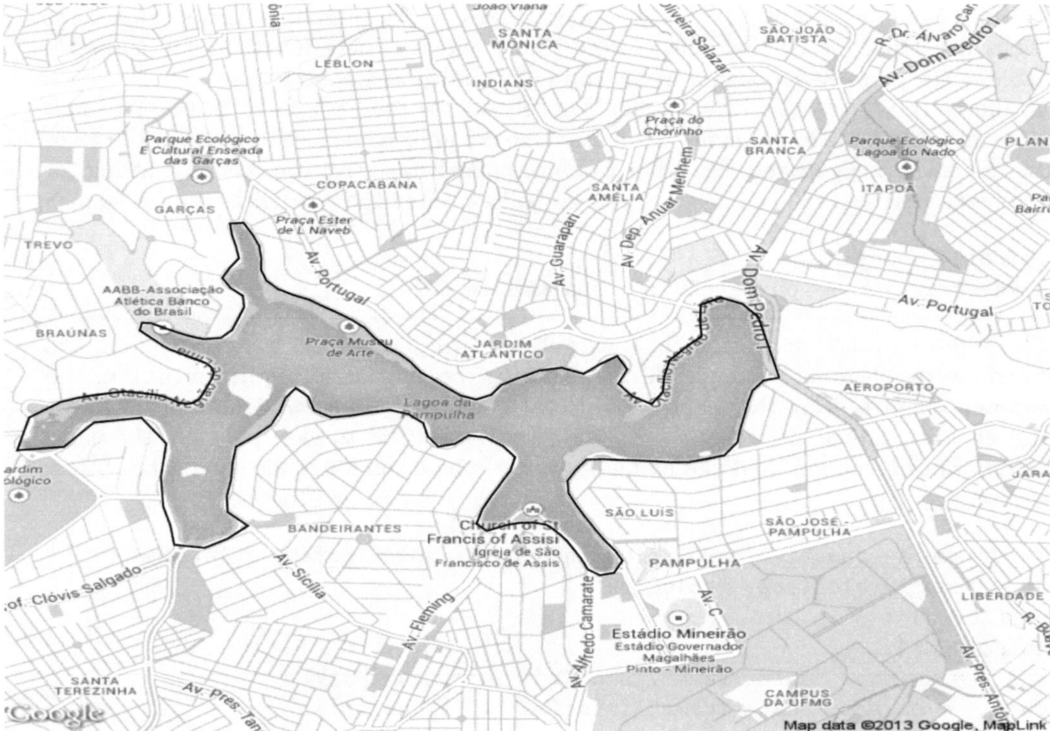

FIGURE 5.16
Example using the Google Maps API.

We use an optional argument, `FUN=lines`, from the `PlotOnStaticMap` function to connect the dots by lines, as shown in Figure 5.16:

```
> require(RgoogleMaps)
> center <- c(mean(geo.pt$lat), mean(geo.pt$lon))
> mymap  <- GetMap(center=center, zoom=14, GRAYSCALE = TRUE)
> map    <- PlotOnStaticMap(mymap, lat = geo.pt$latitude, lon = geo.pt$longitude,
+ lwd = 2.5, lty = 2, col="black", FUN = lines)
```

5.6.4 Generating KML Files

The `RgoogleMaps` package imports maps from Google Maps and use them as a canvas to overlay R spatial objects. This is a static view of the spatial information, with little user interaction. Another option is to visualize the spatial information generated by R in a dynamic environment such as Google Earth™. This allows for a rich interaction between the user and the Web interface, providing for use of zooming features and pop-up textual or graphical information to be added to the map. It is possible to visualize R-generated spatial information in Google Earth if we store this information in a `KML` file. `KML` (Keyhole Markup Language) is a file format developed by Google and based on the `XML` language. It is used to display geographic data in Google Earth, Google Maps, and Google Maps for mobile. Other applications displaying `KML` files include NASA WorldWind, Esri® ArcGIS Explorer,

Adobe PhotoShop®, AutoCAD®, and Yahoo! Pipes®. We present in this section some functions developed by us that link Google Earth and R. These functions are currently available in the package `spGoogle`, which can be uploaded at CRAN.

To illustrate the creation of a KML file from R, we use the variable `HDIM_00`. Two new columns are added to the shapefile database. One column, named `color`, stores the color scheme for each polygon and the other column, named `description`, stores a string for each polygon that is displayed inside a pop-up balloon when the user clicks on the map.

```
shape$color <- cols2
shape$description <- paste("HDI:", shape$HDIM_00)
```

The following function creates, for each polygon in the shapefile, a corresponding KML code and merges them into a single KML file. The `shp`, `color`, `namepoly`, `description`, and `file.name` parameters are the shape file object, the name of the color column in the shape file object, the name of the column with the name associated to each polygon, the name of the description column, and the name of the file where the KML code should be stored, respectively.

```
> KML.create <- function(shp, color, namepoly, description, file.name){
+ out <- sapply(slot(shp, "polygons"),
+ function(x) {
+  kmlPolygon(x,
+ name    = as(shp, "data.frame")[slot(x, "ID"), namepoly],
+  col     = as(shp, "data.frame")[slot(x, "ID"), color],
+ lwd     = 1,
+ border = "#C0C0C0",
+ description = as(shp, "data.frame")[slot(x, "ID"), description]
+ )
+ }
+ )
+ kmlFile <- file(file.name,"w")
+ cat(kmlPolygon(kmlname="KML", kmldescription="KML")$header, file=kmlFile,
+ sep="\n")
+ cat(unlist(out["style",]),   file=kmlFile, sep="\n")
+ cat(unlist(out["content",]), file=kmlFile, sep="\n")
+ cat(kmlPolygon()$footer,     file=kmlFile, sep="\n")
+ close(kmlFile)
+ }
```

The call of the `KML.create()` function is as follows

```
KML.create(shape, color="color", namepoly="NAME_MUN",
+ description="description", file.name="maps.kml")
```

5.6.4.1 Adding a Legend to a KML File

Creating a legend in a KML file is quite complex. One option is to draw colored polygons on the map, as if the boxes of the colors were regular polygons. Then add the text with legend information to the right of the boxes. By doing so, extra code lines are created in the KML file in order to account the legend boxes and texts. As a result, if the user zooms in on the map, then the legend is also expanded and vice versa. An alternative is to create a separate image, or a file, and, by using KML language, provide the link between the image

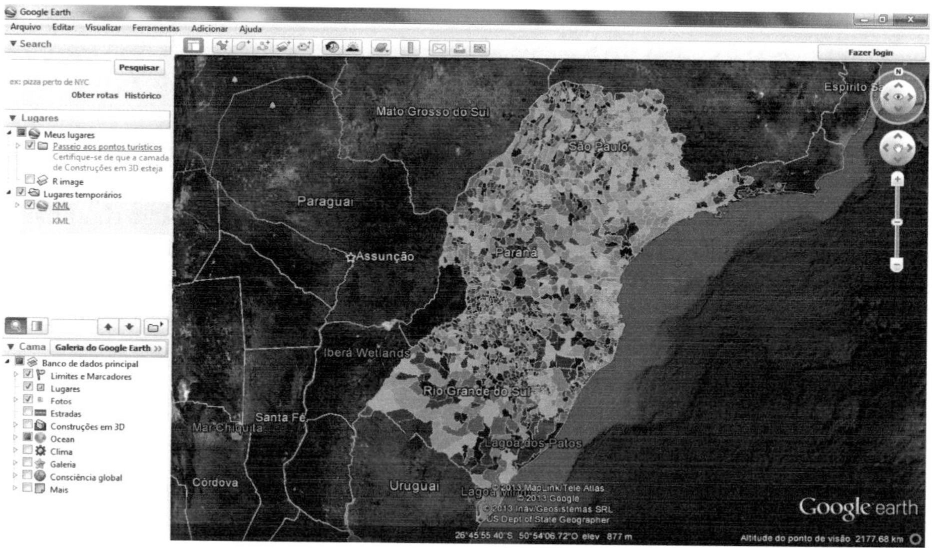

FIGURE 5.17
KML file generated using `maptools` and visualized in Google Earth.

and a specific location on the screen in the KML code. Thus, the size of the image of the legend is not changed if the user zooms in on or zooms out from the map. We created two functions named `generates_figure_legend()` and `generates_layer_legend()` that handle the image creation and the associated KML code. They can be used with

```
> source("kml_legend.R")
```

The following code creates the legend image and the KML code:

```
> library(Cairo)
> brks <- classes$brks
> dest.fig.attrs <-  generates_figure_legend(brks, colpal, 2, num.faixas = ncls)
> dest.fig       <-  dest.fig.attrs[1]
> fig.width      <-  dest.fig.attrs[2]
> fig.height     <-  dest.fig.attrs[3]
> legendkml      <-  generates_layer_legend(brks, colpal, dest.fig, fig.width,
+ fig.height, 2)
```

Responsible for creating the legend image in a png file, the function `generates_figure_legend()` uses the function `CairoPNG` from package `Cairo`. `Cairo` is a graphics device for R that provides high-quality output in various formats which includes png. The `legendkml` object has the KML code holding the information on the image filename, its location, and size.

```
> legendkml
 [1] "<ScreenOverlay>"
 [2] "<name>Legenda</name>"
 [3] "<color>ffffffff</color>"
 [4] "<visibility>1</visibility>"
 [5] "<Icon>"
 [6] "<href>legenda1584edf9b4.png</href>"
```

```
 [7] "</Icon>"
 [8] "<overlayXY x=\"0\" y=\"0\" xunits=\"fraction\" yunits=\"fraction\"/>"
 [9] "<screenXY x=\"148\" y=\"18\" xunits=\"insetPixels\" yunits=
      \"pixels\"/>"
[10] "<size x=\"-1\" y=\"-1\" xunits=\"pixels\" yunits=\"pixels\"/>"
[11] "</ScreenOverlay>"
```

This information is aggregated into the `KML.create()` function. Note that the function has a new parameter `legendkml` that must be specified with the object that we just created (also named `legendkml` in this example).

```
> KML.create <- function(shp, color, namepoly, description, file.name, legendkml){
+ out <- sapply(slot(shp, "polygons"),
+ function(x) {
+ kmlPolygon(x,
+ name   = as(shp, "data.frame")[slot(x, "ID"), namepoly],
+ col    = as(shp, "data.frame")[slot(x, "ID"), color],
+ lwd    = 1,
+ border = "#C0C0C0",
+ description = as(shp, "data.frame")[slot(x, "ID"), description]
+ )
+ }
+ )
+ kmlFile <- file(file.name,"w")
+ cat(kmlPolygon(kmlname="KML", kmldescription="KML")$header, file=kmlFile,
+ sep="\n")
+ cat(legendkml, file=kmlFile,sep="\n")
+ cat(unlist(out["style",]),   file=kmlFile, sep="\n")
+ cat(unlist(out["content",]), file=kmlFile, sep="\n")
+ cat(kmlPolygon()$footer,     file=kmlFile, sep="\n")
+ close(kmlFile)
+ }
```

Using the updated function `KML.create`, we created a new file `KML` that exhibits the legend in Google Earth, as shown in Figure 5.18.

```
> KML.create(shape, color="color", namepoly="NAME_MUN",
+ description="description", file.name="maps_lege.kml", legendkml=legendkml)
```

It is important to note that the `KML` code provides only the reference to the image. That is, the image is not stored in the `KML` code. An alternative is to encapsulate both `KML` and the image into a single file with a `KMZ` extension. A `KMZ` file is simply a standard zip file with both the `KML` code and the image. This file encapsulates all geographical information and image files and therefore it makes easy the transfer of information between users. Furthermore, `KMZ` files can also be visualized in Web browsers using *maps.google*, as long as the `KMZ` file is available online with public access.

```
> KMZ.create <- function(kml.name, legenda.name){
+ kmz.name <- gsub(".kml", ".kmz", kml.name)
+ zip(kmz.name, files = c(kml.name, legenda.name))
+ }

>  KMZ.create(kml.name = "maps_lege.kml", legenda.name = dest.fig)
  updating: maps_lege.kml (deflated 64%)
  adding: legenda158c4af43a70.png (deflated 2%)
```

The R code to create `KMZ` files can be found in `plotmaps_kml.R`.

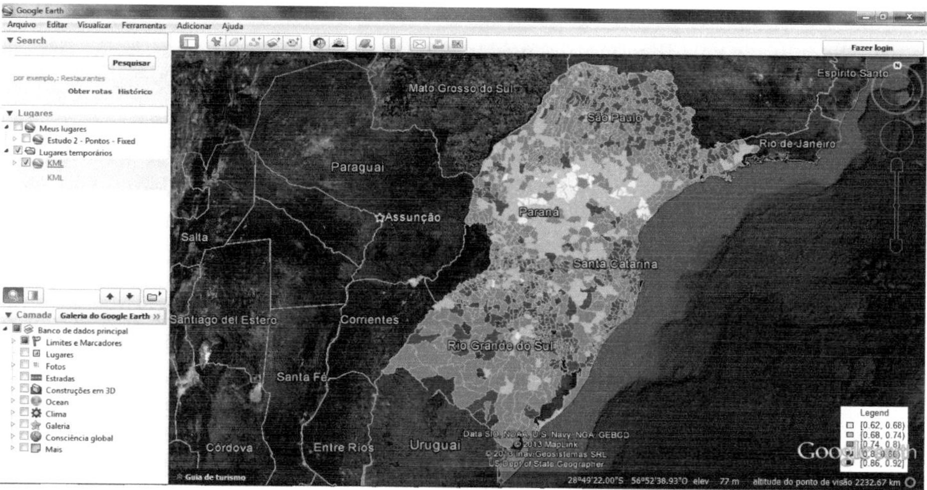

FIGURE 5.18
`KML` file with legend generated using `maptools` and visualized in Google Earth.

5.7 Testing for Spatial Correlation

One of the first tasks before embarking on a stochastic model for the spatial variation of a variable y is to test if there is evidence of a mechanism inducing spatial correlation between the areas. After all, by chance alone, we are almost certain to observe spatial clusters of high values or low values. These randomly formed clusters are likely to appear in any map, even if the values of y in each area are generated irrespective of their spatial location.

Spatial correlation tests depend on the definition of a neighborhood matrix. The `spdep` package provides basic functions for building and manipulating these neighborhood matrices as well as spatial inference tools in the form of correlation tests and spatial regression models.

5.7.1 Neighborhood Matrix

Given a map partitioned into n areas, we represent their degree of spatial association or proximity by an $n \times n$ matrix $\mathbf{W}$. The element w_{ij} represents the weight, the connectivity degree, or the spatial proximity intensity between areas i and j. The diagonal elements are null: $w_{ii} = 0$ for all $i = 1, \ldots, n$. The other elements can be selected rather arbitrarily but should be made considering specific aspects of the problem under analysis. However, in most applications, the matrix $\mathbf{W}$ is chosen in a relatively simple way. Some of the most common choices for the matrix $\mathbf{W}$ take $w_{ij} \geq 0$ and they are the following:

- $w_{ij} = 1$ if areas i and j share boundaries, and $w_{ij} = 0$ otherwise. This binary and symmetric matrix is called the *adjacency matrix.*
- $w_{ij} = 1$ if the distance d_{ij} between the centroids of areas i and j is less than a certain threshold d^*, and 0 otherwise.
- $w_{ij} = 1$ if the centroid of area j is one of the k nearest neighboring centroids of area i. Otherwise, $w_{ij} = 0$. Note that we can have $w_{ij} \neq w_{ji}$ in this case.
- $w_{ij} = 1/(1 + d_{ij}^{\gamma})$, where $\gamma > 0$.

- The influence of i intro j could depend on the relative size of i among j's neighbors. One way to capture this idea is to consider the common boundary length l_{ij} between areas i and j. Let l_i be the perimeter of area i. Then, $w_{ij} = l_{ij}/l_i$. Typically, we will have $w_{ij} \neq w_{ji}$.

In some statistical models, it is useful to standardize the rows of $\mathbf{W}$ so they are non-negative and sum to 1. Hence, with the above choices for $\mathbf{W}$, we can redefine the neighborhood matrix as $\boldsymbol{W}^*$ where $w^*_{ij} = w_{ij}/w_{i.}$ with $w_{i.} = \sum_j w_{ij}$. From now on, we will use only the notation $\mathbf{W}$ for the neighborhood matrix, making it clear from the context which matrix is under consideration.

Typically, $\mathbf{W}$ is sparse, with few non-zero elements in each row. Rather than storing a large $n \times n$ matrix with most elements equal to zero, it is more efficient to store a list with the few elements or neighboring areas that are non-zero, together with the indices of their areas. The package `spdep` defined the class `nb` to handle the sparse neighborhood matrix based on boundary adjacency as a neighborhood list. If a `SpatialPolygonsDataFrame` object has n `Polygons`, the neighborhood list will also have n elements. The i-th element is a vector with the integer indices of the adjacent regions of the i-th `Polygons`. To show how to create this list version of the adjacency matrix, we will use a subregion of our illustrative southern Brazil map, the Paraná state municipalities. The `spdep` function `poly2nb()` creates the neighborhood list:

```
> library(maptools)
> library(spdep)
> shape <- readShapeSpatial("sul+sp_shape")
> pos      <- which(shape@data$ST == "PR") # indices of selected rows
> prshape <- shape[pos,] # new SpatialPolygonsDataFrame Parana regions
> plot(prshape)  # plotting the map
> text(coordinates(prshape), label=prshape@data$NAME_MUN, cex=0.5) # adding areas names
> pr.nb    <- poly2nb(prshape) # Adjacency ngb list from SpatialPolygonsDataFrame
> is.list(pr.nb) # output is TRUE
> pr.nb[[1]] # neighbors of "ABATIA", the first data.frame region
[1]  30  86 181 306 321 336
> summary(pr.nb)
Neighbour list object:
Number of regions: 399
Number of nonzero links: 2226
Percentage nonzero weights: 1.398232
Average number of links: 5.578947
Link number distribution:

 2  3  4  5  6  7  8  9 10 11 12
 7 40 72 86 83 56 27 16  9  2  1
7 least connected regions:
32 50 176 257 277 333 361 with 2 links
1 most connected region:
69 with 12 links
```

The neighbors in the `pr.nb` list are identified by their order in relation to the list itself. Note how sparse the 399×399 adjacency neighborhood matrix is: only 1.4% of its elements are non-zero.

The `plot` command works with the neighborhood list and produces a graph where each centroid is connected by edges to its adjacent regions:

```
> plot(prshape)
> plot(pr.nb, coordinates(prshape), add=TRUE, col="blue")
```

The poly2nb function creates edges between regions that represent binary indicators of who is a neighbor of whom. It does not specify weights for these non-zero neighboring pairs. This is possible through the function nb2listw, whose most simple usage is

```
> pr.listw <- nb2listw(pr.nb, style="W") # weighted ngb list
> length(pr.listw); names(pr.listw);
> pr.listw$weights[[1]] # weights of the 1st region neighbors
[1] 0.1666667 0.1666667 0.1666667 0.1666667 0.1666667 0.1666667
> pr.listw$weights[[2]] # weights of the 2nd region neighbors
[1] 0.3333333 0.3333333 0.3333333
```

The neighborhood list pr.nb of class nb is transformed into a *weighted* neighborhood list. The output pr.listw is a list composed of three elements: "style", "neighbours", and "weights". The first one is simply a character indicating the style used to build the weighted list ("W" in this example). The second one is the list of neighbors, similar to pr.nb. The third element is the list pr.listw$weights whose element pr.listw$weights[[i]] is the vector with the *non-zero* weights w_{ij} assigned to the neighbors of region i. The argument to style determine the type of weights with "W" meaning that $w_{ij} = a_{ij}/sum_j a_{ij}$, where a_{ij} is the element (i, j) of the adjacency matrix.

Another option is to provide a list of weights directly. For example, for a map composed by seven polygons, the first municipality of Paraná plus its six adjacent neighbors, we could assign user-specified weights to neighbors in the following way. We start creating an indicator for region 1 and its six adjacent regions:

```
> pos <- c(1, pr.nb[[1]])
```

Then, we create a new SpatialPolygonsDataFrame with these seven regions:

```
> map4 <- prshape[pos,]
```

and we create the adjacency neighborhood list:

```
> plot(map4)
> map4.nb    <- poly2nb(map4)
```

We can count the number of neighbors in each area:

```
> sapply(map4.nb, length)
[1] 6 4 4 3 3 3 3
```

and finally assign weights:

```
> x <- rep(1/3,3) # auxiliary vector
> lweights <- list((1:6)/21, runif(4), 1/(1:4), x, x, x, x)
> map4.listw <- nb2listw(map4.nb, glist=lweights, style="W")
> map4.listw$weights
```

5.7.2 Other Neighborhood Options

Another way to build a neighborhood structure is to use the spdep function knearneigh which obtain the k nearest neighboring points of each point. We can use the centroids of the n regions or any other adequate point summary of the polygons. The return value of knearneigh is a list of class knn whose first element is $n \times k$ matrix holding the indices of the k nearest neighbors. This knn class object can be passed to knn2nb to generate the neighborhood list of class and then plotted on the map (see Figure :

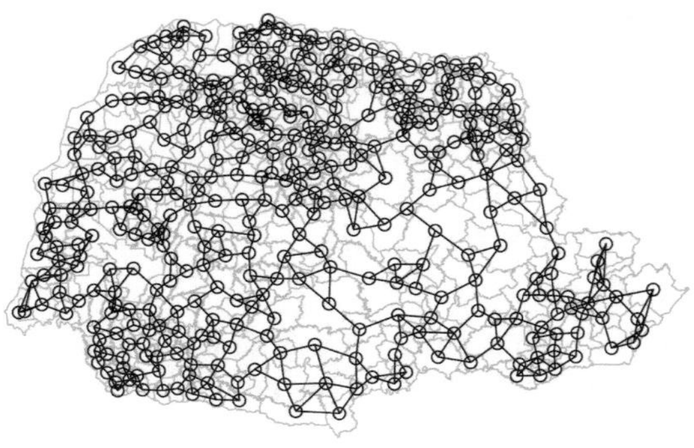

FIGURE 5.19
Neighborhood structure based on the three nearest neighbors of each area.

```
> coords <- coordinates(prshape)
> pr.knn <- knearneigh(coords, k=3, longlat = TRUE)
> pr.nbknn <- knn2nb(pr.knn) # ngb list
> plot(prshape, border="grey") # map of the Parana regions
> plot(pr.nbknn, coords, add=TRUE)
```

for plotting edges between neighbors

```
> title("K nearest neighbours, k = 3")
```

5.7.3 Moran's I Index

To evaluate the degree of spatial association of a variable y_i in a map, we use Moran's I index defined by

$$I = \frac{1}{\sum_{i \neq j} w_{ij}} \sum_{i \neq j} w_{ij} \left(\frac{y_i - \bar{y}}{s_y} \right) \left(\frac{y_j - \bar{y}}{s_y} \right). \tag{5.1}$$

This index is called Moran's I index in honor of P.A.P. Moran, an Australian statistician who studied their behavior (see Moran, 1950). Its intuition is clear. We calculate the standardized deviate of y_i with respect to the global variation of y in the map. Next, we calculate the correlation between pairs of areas weighting by the elements w_{ij} of the matrix $\mathbf{W}$.

As in the case of the usual correlation, if $I \approx 0$, there would not be evidence for spatial correlation. On the contrary, if I is much greater than zero, there is evidence of positive autocorrelation. That is, neighboring regions tend to deviate similarly from the mean $\bar{y}$. Finally, $I << 0$ indicates negative correlation, a situation rarely seen in practice. The statistical problem is to determine a critical value c_0 for I such that when $|I| > c_0$, we will admit that there is evidence for spatial autocorrelation.

Although it is possible to obtain the exact distribution of I under special situations, we prefer to evaluate the statistical significance of I using Monte Carlo methods. Assuming that $Y_1, \ldots, Y_n$ are i.i.d. random variables, the test based on Moran's index can be described as an algorithm:

- Calculate Moran's I index with the observed data obtaining $I^{(1)}$.

- For k varying from 1 up to a large N integer (say, $N = 999$), repeat independently the two next steps:

  - Generate a pseudo-map by randomly permuting the values y_i among the areas.
  - Calculate the Moran's index I for the pseudo-map obtaining $I^{(k+1)}$.

- Under H_0, all permutations are equally likely and the p-value is given by

$$p\text{-value} = \frac{\text{Number of } I^{(j)} \geq I^{(1)},\, j = 1, \ldots, N}{N+1}.$$

  Reject the null hypothesis at level α if p-value $< \alpha$.

The Moran index and the Monte Carlo test are easily obtained using the spdep package with the command moran.mc. For example, to run the test with $N = 999$ simulations and the variable HDIM_00 from prshape with the neighborhood structure pr.listw, we type

```
> imoran <- moran.mc(prshape$HDIM_00, pr.listw, nsim=999)
> par(mar=c(4,4,2,2))
> hist(imoran$res, xlab='Index', main='', col=gray(.5), border=gray(.7))
> arrows(imoran$stat,-2,imoran$stat,10,lwd=2,col=2,leng=.1,code=1)
> segments(imoran$stat, 3, 0.4, 120, lty=2)
> text(.4, 150, paste("Moran's I =", format(imoran$stat,dig=4)))
> text(.4, 130, paste("p-value =", format(imoran$p.val, dig=4)))
```

5.8 Spatial Car Accident Insurance Analysis

For the analysis of areal spatial site data, the Simultaneous Autoregressive (SAR), see Whittle (1954), and the Conditional Autoregressive (CAR), see Besag (1974), models are commonly applied. Both models account for spatial dependence by applying a spatial covariance structure. In this case, estimates for the parameters of the models are found by maximum likelihood or Bayesian methods. The package spdep provides both SAR and CAR maximum likelihood estimates through the spautolm() function. Other packages, like spbayes, perform Bayesian spatial estimates. Although maximum likelihood estimates (MLE) are possible for spatial random effects, it is extremely challenging because it requires numerical integration over all multivariate random effects. For this reason, for hierarchical spatial models, spatial data analysis is commonly performed in a Bayesian framework, as will be seen in our case.

Let ϕ be a CAR random vector. The CAR model is defined from the conditional distribution of each observation i given the neighboring sites (sites that share borders with site i) as

$$\phi_i|\phi_{-i} \sim \mathcal{N}(\rho\frac{\sum_{i\sim j}\phi_j}{n_i}, \frac{\sigma^2}{n_i}), \quad i = 1\ldots,n, \tag{5.2}$$

where $i \sim j$ denotes all neighboring sites of i, n_i is the number of neighbors of site i, $\boldsymbol{y}_{-i}$ is the observed responses for all sites but site i, σ^2 is the variance parameter, and ρ represents the strength of the spatial dependence. The precision of the conditional distribution (5.2) is directly proportional to the number of neighbors of site i, that is, the greater the number of neighbors of a particular site, the more local information is available for estimating the parameters of site i. If ρ is fixed as 1, then the CAR model is known as the Instrisic CAR (ICAR) model. The ICAR model is not a proper probability distribution. However, it can be used as a prior distribution in Bayesian hierarchical models because it usually leads to proper posterior distributions.

We revisit the data introduced in Section 5.5 to perform the analysis of the number of claims of popular and luxury vehicles by municipalities in the four selected Brazilian states. For each municipality i, $i = 1, \ldots, 1833$, let Y_i be the number of car claims. The hierarchical model is defined as

$$\begin{aligned} Y_i &\sim \mathcal{P}(E_i\theta_i) \quad i = 1, \ldots, 1833, \\ \log(\theta_i) &= \eta_i + \phi_i + e_i, \end{aligned} \tag{5.3}$$

where E_i is the expected number of car claims in each municipality i, $\eta_i = X_i'\beta$ is the fixed effect, ϕ_i is the spatial random effect parameter that captures the spatial variability, and e_i is the standard random noise.

We use the Human Development Index (HDI) and the logarithm of the population size (POP) of each municipality as possible explanatory variables. Thus, $\eta_i = \beta_0 + \beta_1 \text{HDI}_i + \beta_2 \text{POP}_i$, $i = 1, \ldots, 1833$.

In order to fit the model, the first step is to load the data into R and prepare the data structure for the analysis. First we define a function to calculate the neighbor mean given the map and the variable of interest:

```
> NbMean <- function(shp, vari){
+   library(spdep)
+   shpnb <- poly2nb(shp)
+   shpnb.mat <- nb2mat(shpnb, style="B",zero.policy=TRUE) #adjacency matrix
+   selNA <- which(is.na(shp@data[, vari]))
+   NAnb  <- shpnb.mat[selNA, ]
+   shp@data[selNA, vari] <- apply(NAnb, 1, FUN = function(x)
+                  mean(shp@data[which(x == 1), vari], na.rm = TRUE))
+   return(shp)
+ }
```

Now we can load the libraries

```
>  library(maptools)
>  library(INLA)
```

Then, we can read the shapefile:

```
>  shape <- readShapeSpatial("sul+sp_shape")
```

and calculate the number of accidents for popular and luxury cars

```
>  shape@data$SIN_LUX <- rowSums(shape@data[,20:23], na.rm=FALSE)
>  shape@data$SIN_POP  <- rowSums(shape@data[,14:17], na.rm=FALSE)
>  shape@data$POP  <- log(shape@data$POP_RES)
```

Some lines of codes are necessary to deal with the missing values for the exposition of the popular category:

```
> pos <- which(!is.na(shape@data$EXPO_POP)) #non missing positions
> pos.ms <- which(is.na(shape@data$EXPO_POP)) #missing positions
> expo <- log(shape@data$EXPO_POP[pos]+1)
```

```
> pop <- shape@data$POP[pos]
> reg <- lm(expo ~ pop)
> pred <- as.vector(predict(reg,data.frame(pop=shape@data$POP[pos.ms])))
> shape@data$EXPO_POP[pos.ms] <- sapply((exp(pred) - 1),
+ function(x) max(x,0))
```

A few more lines are used to input the exposition values for the luxury car:

```
> pos <- which(!is.na(shape@data$EXPO_LUX)) #non missing positions
> pos.ms <- which(is.na(shape@data$EXPO_LUX)) #missing positions
> expo <- log(shape@data$EXPO_LUX[pos]+1)
> pop <- shape@data$POP[pos]
> reg <- lm(expo ~ pop)
> pred <- as.vector(predict(reg,data.frame(pop=shape@data$POP[pos.ms])))
>         shape@data$EXPO_LUX[pos.ms] <- sapply((exp(pred) - 1),function(x) max(x,0))
```

To input the HDI missing values, we use the information of the neighboring areas:

```
> shape <- NbMean(shape,"HDIM_00")
```

We set the expected number of accidents based on the number of cars of each type:

```
> shape@data$E_POP <- shape@data$EXPO_POP *
+   sum(shape@data$SIN_POP,na.rm=TRUE)/sum(shape@data$EXPO_POP)
> shape@data$E_LUX <- shape@data$EXPO_LUX *
+   sum(shape@data$SIN_LUX,na.rm=TRUE)/sum(shape@data$EXPO_LUX)
```

We prepare the spatial structure for `inla()`:

```
>  shape@data$struct <- rep(1:dim(shape@data)[1])
>  shape@data$unstruct <- rep(1:dim(shape@data)[1])
```

Finally, we need to create the neighborhood structure for INLA

```
>   nb2INLA("ngbINLA.graph",poly2nb(shape))
```

where SIN_LUX and SIN_POP represents the number of accidents for the luxury and popular categories, respectively. The expected number of accidents E_LUX and E_POP were calculated as the number of cars exposed in municipality i times the overall accident risk (total number of accidents of the category divided by the total number of cars exposed).

R provides a variety of alternatives to adjust the spatial model. One option is to use the spbayes package. Another is to call OpenBUGS (Lunn et al. 2009) software from within R using one of the possible packages: rbugs (Yan & Prates 2012), R2OpenBugs (Sturtz et al. 2005), or BRugs (Thomas et al. 2006). The OpenBUGS is free software that performs Bayesian analysis in hierarchical models. However, both spbayes and OpenBUGS rely on Markov Chain Monte Carlo (MCMC) methods that can be computationally costly. For this reason, we use the R-INLA package (www.r-inla.org). The Integrated Nested Laplace Approximation (INLA) (Rue et al. 2009) is adequate for the analysis of Gaussian hierarchical models. INLA relies on directly approximating the posterior distributions of interest using numerical algorithms and, by doing so, it does not need to perform MCMC. Thus, it is extremely faster than the standard MCMC method. The R-INLA package is not directly available on CRAN and must be installed by the following command:

```
> source("http://www.math.ntnu.no/inla/givemeINLA.R")
```

The R-INLA function, `inla()`, is similar to `lm()` or `glm()` R functions. Briefly, to call `inla()` function, the user might provide a formula, the distribution of the response variable and a R dataframe, with the data:

```
> inla(formula, family = "gaussian", data = data.frame(),...)
```

More options are provided by the `inla()` function and are explored along the analysis.

To proceed with the analysis using `inla`, it is necessary to define the formulas that represent (11.4), with a formula for popular analysis:

```
> f.pop <- SIN_POP ~ HDIM_00 + POP + f(unstruct,model="iid")
+ + f(struct,model="besag",graph="ngbINLA.graph")
```

and a formula for luxury analysis:

```
> f.lux <- SIN_LUX ~ HDIM_00 + POP + f(unstruct,model="iid")
+ + f(struct,model="besag",graph="ngbINLA.graph")
```

The random effects in the `R-INLA` package are specified using the `f()` function. The generic syntax for `f()` function is

```
> f(name, model, ...)
```

where *name* is the name of the random effect variable, inside the dataframe, and *model* is the probability distribution of the random effects. `R-INLA` has several choices for random effect models, including, for example, *iid* for independent and identically distributed effects; *besagproper* for the CAR model; *besag* for the ICAR model; *besag2* for weighted ICAR model; *bym* which is a combination of ICAR and iid models; and many others. If the *besag* model is being used, then it is necessary to provide the neighboring structure of the map, for example, the *ngbINLA.graph* data. This variable contains the spatial structure of the municipalities and can be created using the function `nb2INLA()` from `spdep` package. The `struct` and `unstruct` elements in the `shape@data` dataframe represents the ϕ and e random effects, respectively.

To call the `inla()` function, use either; for popular analysis,

```
>  m.pop <- inla(f.pop, family="poisson", data=shape@data, E=E_POP,
+  control.compute=list(dic=TRUE,cpo=TRUE),
+  control.predictor= list(compute=TRUE,link=1))
```

or, for luxury analysis,

```
>  m.lux <- inla(f.lux, family="poisson", data=shape@data, E=E_LUX,
+  control.compute=list(dic=TRUE,cpo=TRUE),
+  control.predictor= list(compute=TRUE,link=1))
```

where `family` is set as Poisson because the response variable represents count data and the `offset` is the expected number of claims (see Chapter 13 for more details on the Poisson regression to count car accidents). `R-INLA` also provides model fit statistics such as DIC, see Spiegelhalter et al. (2002), and CPO, see Geisser & Eddy (1979) or Dey et al. (1997), using the command `control.compute`. Prediction estimates are provided by the command `control.prediction`. Further details of functionalities and options provided by the `R-INLA` package are found online. Results are stored in variables `m.pop` and `m.lux` for the popular and luxury analysis, respectively.

The `summary()` function summarizes the fitted model and shows estimated values and credible intervals:

```
> summary(m.pop)

Call:
```

```
c("inla(formula = f.pop, family = \"poisson\", data = shape@data,","E = E_POP,
control.compute = list(dic = TRUE, cpo = TRUE), ",  "control.predictor =
list(compute = TRUE, link = 1))")

Time used:
 Pre-processing    Running inla Post-processing            Total
        0.6833         10.2896          0.2356          11.2085

Fixed effects:
               mean     sd 0.025quant 0.5quant 0.975quant kld
(Intercept) -1.3909 0.4089    -2.1959  -1.3899    -0.5918   0
HDIM_00     -0.1796 0.5990    -1.3528  -0.1803     0.9977   0
POP          0.1109 0.0150     0.0816   0.1108     0.1404   0

Random effects:
Name          Model
 struct   Besags ICAR model
unstruct   IID model

Model hyperparameters:
                         mean   sd     0.025quant 0.5quant 0.975quant
Precision for struct    6.297  1.567  3.679      6.152    9.786
Precision for unstruct 28.220  8.367 15.825     26.848   48.352

Expected number of effective parameters(std dev): 368.41(18.30)
Number of equivalent replicates : 3.898

Deviance Information Criterion: 5669.35
Effective number of parameters: 366.75

Marginal Likelihood:  -4372.68
CPO and PIT are computed

Posterior marginals for linear predictor and fitted values computed
```

for popular cars, while for luxury cars,

```
> summary(m.lux)

Call:
c("inla(formula = f.lux, family = \"poisson\", data = shape@data,","E = E_LUX,
control.compute = list(dic = TRUE, cpo = TRUE), ",  "control.predictor =
list(compute = TRUE, link = 1))")

Time used:
 Pre-processing    Running inla Post-processing            Total
        1.4731          9.5868          0.4716          11.5315

Fixed effects:
               mean     sd 0.025quant 0.5quant 0.975quant kld
(Intercept) -1.0623 0.7300    -2.5038  -1.0590     0.3606   0
HDIM_00      0.1323 1.0576    -1.9333   0.1289     2.2181   0
```

```
POP           0.0506 0.0260    -0.0003   0.0506     0.1018   0

Random effects:
Name          Model
 struct   Besags ICAR model
unstruct   IID model

Model hyperparameters:
                        mean    sd     0.025quant 0.5quant 0.975quant
Precision for struct     4.827  1.743  2.214      4.567     8.970
Precision for unstruct 11.648  3.735  6.129     11.041    20.631

Expected number of effective parameters(std dev): 224.10(18.40)
Number of equivalent replicates : 6.408

Deviance Information Criterion: 3309.70
Effective number of parameters: 221.48

Marginal Likelihood:  -3115.58
CPO and PIT are computed

Posterior marginals for linear predictor and fitted values computed
```

Results show that the population variable has a positive effect on the number of claims. For popular cars, an increase of 1,000 people in the municipality causes an increase in accident risk of 115%; while for the luxury cars, the increase risk factor is 42%. Thus, larger cities tend to increase the number of accidents as compared to the expected value. The HDI index has no significant effect for either popular cars or luxury cars. Thus, the HDI index has no direct connection with the number of claims in each municipality. Results also show that the spatial effects in both models have a smaller precision value, which suggests that most variability is captured in the spatial trend.

`R-INLA` also provides the predicted number of accidents, as required by the `control.prediction` command. Next, we update the shapefile with the estimated posterior mean information. First, we save in the `shape` variable the posterior mean of the popular vehicles' number of accidents fitted values:

```
> shape@data$POP_SPAT <- m.pop$summary.random$struct$mean
> shape@data$POP_PRED <- m.pop$summary.fitted.values$mean * shape@data$E_POP
> shape@data$POP_RR   <- m.pop$summary.linear.predictor$mean
```

And then we save in the `shape` variable the posterior mean of the luxury vehicles' number of accidents fitted values:

```
> shape@data$LUX_SPAT <- m.lux$summary.random$struct$mean
> shape@data$LUX_PRED <- m.lux$summary.fitted.values$mean * shape@data$E_LUX
> shape@data$LUX_RR   <- m.lux$summary.linear.predictor$mean
```

Thus, the `spplot()` function can be used to plot the posterior mean of the spatial random effects, ϕ_i (e.g., `m.pop$summary.random$struct$mean`). Figure 5.20 shows an increasing spatial trend from the south to north.

```
> spplot(shape,c("POP_SPAT","LUX_SPAT"), layout = c(2,1),
+ main = "Spatial Dependence", cuts=5, col.regions=grey.colors(50,1,0))
```

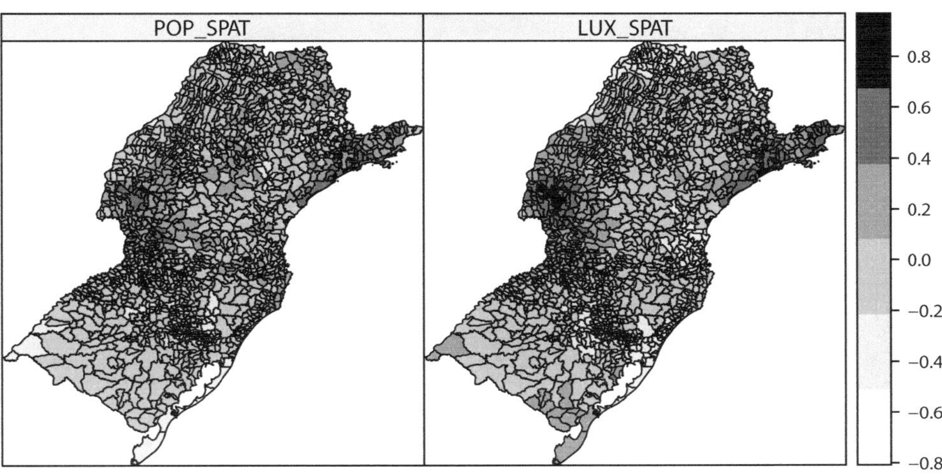

FIGURE 5.20
Posterior mean spatial pattern by category.

where `layout` is the parameter responsible for the number of graphical displays, `main` is the parameter of the figure title, `cuts` is the number of categories in the scale, and `col.regions` is the color scheme.

It is also possible to update the shapefile with the estimated prediction of the number of accidents for each category. Figure 5.21 shows that the model prediction is quite accurate for the number of accidents in each scenario.

```
> par(mfrow=c(1,2))
> plot(log(shape@data$POP_PRED+1),log(shape@data$SIN_POP+1),
+       xlab="log pred #s of popular cars accidents",
+       ylab="log original #s of popular cars accidents",
+       main="log-predicted x log-accidents")
> abline(a=0,b=1)
> plot(log(shape@data$LUX_PRED+1),log(shape@data$SIN_LUX+1),
+       xlab="log pred #s of luxury cars accidents",
+       ylab="log original #s of luxury cars accidents",
+       main="log-predicted x log-accidents")
> abline(a=0,b=1)
```

Another interesting result is the logarithm of the accident risk rate, returned by the model. The relative accident risk is defined as

$$\log(\theta_i) = \eta_i + \phi_i + e_i, \quad i = 1, \ldots, 1833.$$

To plot the logarithm of the estimated posterior mean relative risk, use

```
> spplot(shape,c("POP_RR","LUX_RR"), layout = c(2,1),
+ main="Log relative risk by category", col.regions=grey.colors(50,1,0))
```

Figure 5.22 shows that the relative risk is higher for some specific areas. The higher risk municipality is the municipality of Cascavel, which has that have one of the most violent transits in Brazil with high rates of car robbery and accidents.

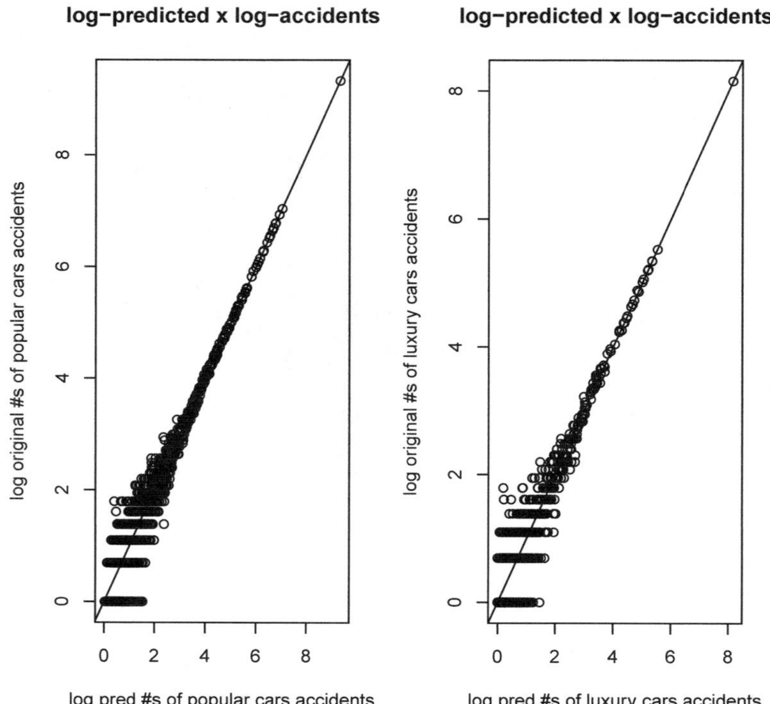

FIGURE 5.21
Predicted Values × Number of Accidents for the popular and luxury analysis.

5.9 Spatial Car Accident Insurance Shared Analysis

Previous results relied on separate analysis for each car category. Figure 5.20 shows that there is evidence of a common spatial trend in both categories. In this section we present an advanced spatial analysis, in which both categories are analyzed simultaneously. In this case, the statistical model assumes a joint spatial effect that is shared between the two car categories. Following the idea proposed by Knorr-Held & Best (2001), we propose the following model:

$$\begin{aligned} Y_{ji} &\sim \mathcal{P}(E_{ji}\theta_{ji}) \quad j=\{\text{popular}, \text{luxury}\}, \quad i=1,\ldots,1833, \\ \log(\theta_{ji}) &= \eta_{ji} + \gamma(j)\psi_i + \phi_{ji} + e_{ji}, \end{aligned} \tag{5.4}$$

where $\eta_{ji} = \beta_0 + \beta_1^{(j)}\text{HDI}_i^{(j)} + \beta_2^{(j)}\text{POP}_i^{(j)}$, ψ_i is the joint spatial component and

$$\gamma(j) = \begin{cases} 1 & \text{if} \quad j = \text{popular} \\ \delta & \text{if} \quad j = \text{luxury} \end{cases}.$$

The other elements of the equation are similar to the ones presented in Section 11.4. Notice that there is one joint spatial random effect parameter, ψ, in model (5.4), and a second spatial effect parameter, ϕ, that captures spatial singularities of each category. All spatial

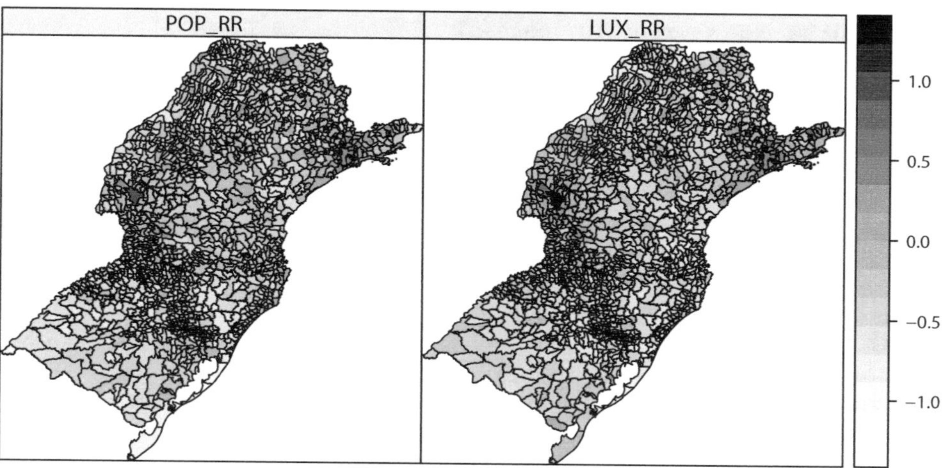

FIGURE 5.22
Log relative risk of the number of accidents for the popular and luxury vehicles per county.

effects are modeled using an ICAR structure. Due to the fact that the numbers of claims of popular and luxury cars are not on the same scale, there are many more claims for popular cars, it is necessary to include a parameter related to the variability in each category. Therefore, although both models share one spatial structure, the scale structure is different. Thus, the parameter δ is responsible for re-scaling the variability between the car categories.

To perform the analysis using `R-INLA`, it is necessary to define properly the model, and the joint and individual random effects. Next, we define the model in which `shared` is the joint spatial component and `shared.copy` is the joint component of one car category:

```
> k  <- 2
> n  <- dim(shape@data)[1]
> Y  <- matrix(NA, n, k)
> Y[1:n, 1] <- shape@data$SIN_POP
> Y[1:n, 2] <- shape@data$SIN_LUX
```

Create the shared list to store the data:

```
> share.dat <- list(Y=matrix(NA, nrow=n*2, ncol=2))
> share.dat$Y[1:n, 1]     <- Y[,1]
> share.dat$Y[n+(1:n), 2] <- Y[,2]
```

Set the expected values:

```
> share.dat$E <- c(shape@data$E_POP,shape@data$E_LUX)
```

Now some code is necessary to prepare random effects:

```
> share.dat$shared           <- c(1:n, rep(NA,n))
> share.dat$shared.copy <- c(rep(NA,n), 1:n)
> share.dat$spat.pop    <- c(1:n, rep(NA,n))
> share.dat$spat.lux <- c(rep(NA,n), 1:n)
> share.dat$random.pop    <- c(1:n, rep(NA,n))
> share.dat$random.lux <- c(rep(NA,n), 1:n)
```

Finally, include intercept and attributes information in the shared model:

```
> share.dat$alpha_POP <- rep(1:0, each=n)
> share.dat$alpha_LUX <- rep(0:1, each=n)
> share.dat$POP_POP <- c(shape@data$POP,rep(0,n))
> share.dat$POP_LUX <- c(rep(0,n),shape@data$POP)
> share.dat$HDI_POP <- c(shape@data$HDIM_00 ,rep(0,n))
> share.dat$HDI_LUX <- c(rep(0,n),shape@data$HDIM_00)
```

The `share.dat` variable contains all the required elements of the model. The formula of the model (5.4) is now written as

```
> f.shared <- Y ~ 0 + alpha_POP + POP_POP + HDI_POP +
+ alpha_LUX + POP_LUX + HDI_LUX +
+ f(shared, model="besag", graph="ngbINLA.graph") +
+ f(shared.copy, copy="shared", hyper=list(theta=list(fixed=FALSE,
+ param=c(1,1), range=c(0,Inf)))) +
+ f(spat.pop, model="besag", graph="ngbINLA.graph",
+ hyper=list(theta=list(initial=log(6.30),
+ param=c(0.5,0.0005)))) +
+ f(random.pop, model="iid", hyper=list(theta=list(initial=log(28.22),
+ param=c(0.5,0.0005)))) +
+ f(spat.lux, model="besag", graph="ngbINLA.graph",
+ hyper=list(theta=list(initial=log(4.83),
+ param=c(0.5,0.0005)))) +
+ f(random.lux, model="iid", hyper=list(theta=list(initial=log(11.65),
+ param=c(0.5,0.0005))))
```

where the `f()` function defines the random effects and the `copy` parameter provides the `inla()` function the information that there is one joint component that is shared by both categories. Initial values were set based on previous analysis, presented in Section 5.8. Then we can proceed calling the `inla()` function and run the analysis. This model takes a longer time to run due to its complex structure. The following command is used to run the model:

```
> m.shared <- inla(f.shared, family=rep("poisson", 2), data=share.dat, E=E,
+ control.inla=list(h=0.005),
+ control.compute=list(dic=TRUE, cpo=TRUE),
+ control.predictor= list(compute=TRUE,link=c(rep(1,n),rep(2,n))) )
```

Next, we generate a summary of the results:

```
> summary(m.shared)

Call:
c("inla(formula = f.shared, family = rep(\"poisson\", 2), data = share.dat,",
"E = E, control.predictor = list(compute = TRUE, link = c(rep(1, ", "      n),
rep(2, n))), control.inla = list(h = 0.005))")

Time used:
 Pre-processing    Running inla Post-processing           Total
         1.9053       1167.6059          0.8915       1170.4027

Fixed effects:
              mean     sd 0.025quant 0.5quant 0.975quant kld
```

```
alpha_POP -1.3064 0.3957    -2.0855  -1.3053    -0.5326  0
POP_POP    0.1057 0.0142     0.0780   0.1057     0.1338  0
HDI_POP   -0.2142 0.5831    -1.3577  -0.2145     0.9311  0
alpha_LUX -1.2520 0.6334    -2.5008  -1.2498    -0.0156  0
POP_LUX    0.0637 0.0215     0.0217   0.0637     0.1059  0
HDI_LUX    0.2014 0.9212    -1.6006   0.1994     2.0147  0

Random effects:
Name          Model
shared    Besags ICAR model
spat.pop    Besags ICAR model
random.pop    IID model
spat.lux    Besags ICAR model
random.lux    IID model
shared.copy    Copy

Model hyperparameters:
                              mean       sd        0.025quant 0.5quant  0.975quant
Precision for shared          3.4007     0.5303    2.5158       3.3393    4.6404
Precision for spat.pop      103.1491    41.9399   40.4108      96.5179  205.4203
Precision for random.pop     97.3670    34.6254   43.2435      93.0040  179.4224
Precision for spat.lux       27.9985    13.9083    9.1927      25.3512   62.7788
Precision for random.lux    786.1263   986.4267   13.8756     441.9154 3452.4998
Beta for shared.copy          1.3596     0.0182    1.3248       1.3589    1.3981

Expected number of effective parameters(std dev): 476.52(27.80)
Number of equivalent replicates : 6.027

Marginal Likelihood:  -26114.53
CPO and PIT are computed

Posterior marginals for linear predictor and fitted values computed
```

Results show that the population variable has a positive effect on the number of car claims for both categories in the same scale presented in Section 5.8. Similarly, the HDI variable has no significant effect on either category. These results are similar to previous results presented in Section 5.8, and possibly share the same interpretation. The precision estimate of the joint spatial effects is much higher than the precision estimate of the individual spatial effects. Therefore, it can be said that the joint spatial effects have captured the spatial variability of the data. Moreover, the precision estimate of the random effects for both popular and luxury cars is large, which indicates that there is no other important source of variability in the dataset.

One of the main interests in our proposed analysis is to observe the joint spatial pattern and to understand the effect of the individual spatial pattern of each category. Previous results show that the joint spatial effect is very strong; therefore, it can be concluded that both vehicle types share the same spatial structure with a slight variation difference of $\delta \approx 1.36$. To plot the shared spatial dependence first, obtain the spatial shared components from inla output:

```
> shape@data$SHARED_SPAT <- m.shared$summary.random$shared$mean
```

And now we can use `spplot()` to plot it:

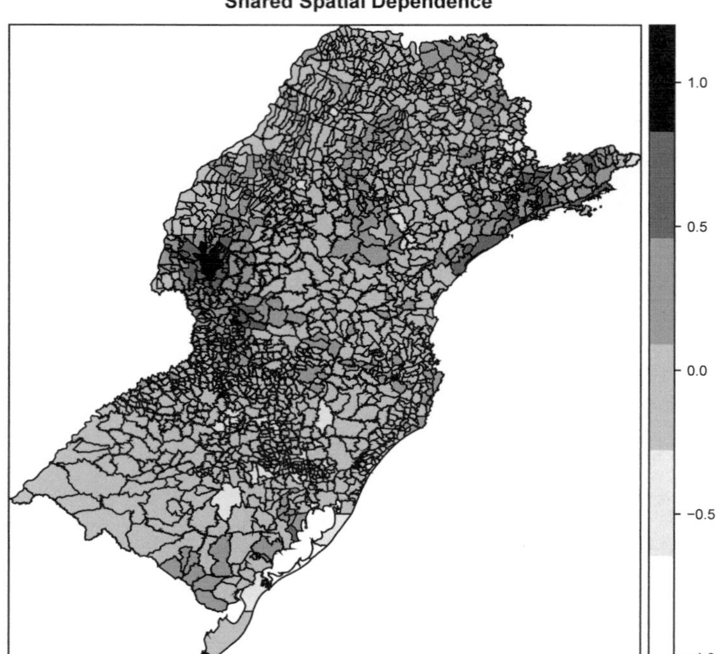

FIGURE 5.23
Posterior mean spatial pattern of the shared structure and by category.

```
> spplot(shape,"SHARED_SPAT", main = "Shared Spatial Dependence", cuts=5,
+ col.regions=grey.colors(50,1,0))
```

Figure 5.23 shows that Cascavel municipality and its surroundings is the region with higher risk. This results agrees with the analysis obtained in Figure 5.22 in the separate analysis.

Nevertheless, we can fit separately the spatial models for each car category and the joint spatial model, and compare them in terms of prediction. Although the joint model is more complex and provides prediction for both categories, Table 5.1 shows that the joint model does not improve prediction. The separate models for each category present a smaller estimated mean absolute prediction, and therefore the separate analysis is more effective.

The relative risk estimates of the joint model is very close to the estimates presented in Figure 5.22, and therefore were not presented. Finally, we can conclude that, for the

TABLE 5.1
Mean estimated prediction error, $\hat{E}_j = \frac{1}{1833}\sum_{i=1}^{1833}(|\hat{Y}_{ji} - Y_{ji}|)$, for $j = \text{popular}, \text{luxury}$.

Model	Category	
	$\hat{E}_{\text{popular}}$	$\hat{E}_{\text{luxury}}$
Shared	1.38	0.75
Popular	1.25	–
Luxury	–	0.62

analysis of the number of car claims, the joint analysis and the analysis with separate models achieved very similar results. The joint spatial model has the advantage of being a more complete model, and it presents the separate models as a special case. However, the model fit of the joint spatial model requires more expertise in terms of R computing language, and demands more computing time. Moreover, although there were no significant differences between the joint and the separate models using our dataset, results may be completely different for different datasets.

5.10 Conclusion

R provides a variety of packages to access, visualize, and fit spatial models. Most of these packages rely on the **sp** and **maptools** packages. These two packages can be considered as the standard packages in which geographical data bases can be read and manipulated. Recently, new R packages have been created providing new visualization tools and new models for spatial and spatial temporal analysis. In particular, the following packages: `RgoogleMaps`, `plotKML`, `googleVis`, among others, have integrated Rwith open resources from maps.google. As a result, the user can interact with the map in a friendly way, either by zooming in, zooming out, using the Google Street View^{TM} application, or overlaying the map with Google online geographical databases such as roads and locations of hospitals. The R environment provides statistical maps but it does not provide interactive maps such as those provided by Google. Thus, we predict that new spatial packages will appear, further connecting R to external visualization resources. For example, we are currently investigating the `SVG` package as a potential language for building Web interactive maps.

Different from visualization tools, there have been an increasing number of new models for statistical analysis of spatial data. Packages such as `rbugs` and `INLA` provide a model building environment to run very sophisticated Bayesian spatial analysis using MCMC and Integrated Laplacian Approximations. We could only touch on the possibilities for spatial analysis in this chapter. We emphasized in the chapter the description of the `sp` package, which is the building block for all the others. We also showed some of the R capabilities for spatial analysis in actuarial science, focusing on Bayesian methods. Hopefully, the readers will find this useful as a first step in the widely varied and beautiful world of spatial analysis in R.

6

Reinsurance and Extremal Events

Eric Gilleland
Research Application Laboratory, National Center for Atmospheric Research
Boulder, Colorado, USA

Mathieu Ribatet
Department of Mathematics, University of Montpellier
Montpellier, France

CONTENTS

6.1 Introduction

In risk analysis and especially for insurance applications, it is important to anticipate the losses that a given company might face in the near future. From a statistical point of view, the observed losses are assumed to be independent realizations from a non-negative random variable X whose distribution function is F.

In this chapter we will focus only on the largest losses, that is, the ones impacting the company's solvency. That is, we restrict our attention to the (right) tail of X. The theory related to the modeling of the tail of a distribution is known as extreme value theory and is widely applied in several domains such as finance, insurance, or risk analysis of natural hazards.

Contrary to inferences pertaining to averages of a random variable where larger values may be considered as outliers, and are often removed from the data, when considering extremes, it is the bulk of the data that are discarded, and the outliers that are of interest. Typically, therefore, one is left with a paucity of data for analyzing the extremes.

6.2 Univariate Extremes

Let $X_1, X_2, \ldots$ be independent copies of a random variable X with distribution F. If one is interested in modeling the (right) tail of X, different modeling strategies are possible:

- Analyzing block maxima, that is, characterizing the distribution of
$$M_n = \max\{X_1, \ldots, X_n\}, \tag{6.1}$$
for some suitable large $n \in \mathbb{N}$;
- Looking at exceedances, that is, characterizing the distribution of
$$X \mid \{X > u\}, \tag{6.2}$$
for some suitable high threshold $u \in \mathbb{R}$.

Remark 6.1 *When interest is in the minima or deficits below a low threshold, the same theory applies after making a suitable transformation (e.g.,* $\min\{X_1, \ldots, X_n\} = -\max\{-X_1, \ldots, -X_n\}$*).*

The left panel of Figure 6.1 shows the 2,167 fire losses collected by Copenhagen Reinsurance between 1980 and 1990. It is clear that the exceedances approach makes more efficient

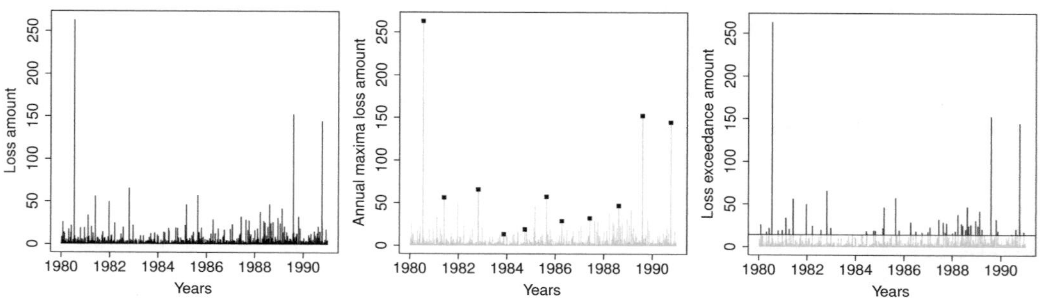

FIGURE 6.1
The left panel shows the time series of the Danish reinsurance claim dataset `danishuni` from the `CASdatasets` package. The two right-most panels illustrate the block maxima and exceedances approaches for modeling large losses. The claims are in millions of Danish Kroner and cover the time period 1980–1990.

use of the data because, contrary to the annual maxima, several exceedances might occur in a given year. However, this approach may induce serial dependence between excesses if the initial assumption of mutual independence of losses is wrong. This is usually not a problem with the block maxima methodology provided that the starting and closing times for the blocks have been defined with caution.

6.2.1 Block Maxima

Because the X_i are independent and identically distributed, it is clear that $\mathbb{P}(M_n \leq x) = F(x)^n$. Unfortunately, for practical purposes, this result is of little interest because typically F has to be estimated, and a small error made on the estimation of F can have catastrophic consequences on the estimation of large quantiles of M_n. Similar to the use of the normal distribution appearing in the central limit theorem, the strategy consists of using asymptotic results and we are therefore interested in characterizing the limiting distribution G such that

$$\frac{M_n - b_n}{a_n} \xrightarrow{\text{d}} G, \qquad n \to \infty, \tag{6.3}$$

for some sequences of constants $a_n > 0$ and $b_n \in \mathbb{R}$.

The Fisher–Tippett theorem (see Embrechts et al. (1997), Section 3.3; Coles (2001), Theorem 3.3; or Beirlant et al. (2004), Theorem 2.1) states that, provided G is nondegenerate, there exist constants $a > 0$ and $b \in \mathbb{R}$ such that $G(ax + b) = G_\xi(x)$, where

$$G_\xi(x) = \exp\left\{-\left(1 + \xi x\right)^{-1/\xi}\right\}, \qquad 1 + \xi x > 0, \tag{6.4}$$

where $\xi \neq 0$. As $\xi \to 0$, G_ξ is prolonged by continuity to give

$$G_0(x) = \exp\left\{-\exp(-x)\right\}, \qquad x \in \mathbb{R}.$$

The parameter ξ is known as the extreme value index and plays a major role in the behavior of the tail of X. When $\xi < 0$, G is upper bounded by $-1/\xi$ and is said to be short tailed. When $\xi = 0$, G corresponds to the standard Gumbel distribution and is said to be light tailed (i.e., decays exponentially). When $\xi > 0$, H is lower bounded by $-1/\xi$ and is said to be heavy tailed (i.e., decays geometrically). The limiting distribution is the Fréchet distribution with parameter $1/\xi$.

For statistical purposes, (6.3) justifies the use of the class of distributions generated by (6.4) as a suitable class of distributions for modeling sample maxima. More precisely, we suppose that the asymptotic result in (6.3) is (approximately) met for finite but large enough $n \in \mathbb{N}$, that is,

$$\mathbb{P}(M_n \leq x) \approx \exp\left\{-\left(1 + \xi\frac{x-\mu}{\sigma}\right)^{-1/\xi}\right\}, \qquad 1 + \xi\frac{x-\mu}{\sigma} > 0, \tag{6.5}$$

for some unknown location $\mu \in \mathbb{R}$, scale $\sigma > 0$ and shape $\xi \neq 0$ parameters.

The distribution (6.5) is known as the generalized extreme value distribution and takes the Gumbel distribution as a special case when $\xi \to 0$, giving

$$\mathbb{P}(M_n \leq x) \approx \exp\left\{-\exp\left(-\frac{x-\mu}{\sigma}\right)\right\}, \qquad x \in \mathbb{R}. \tag{6.6}$$

Figure 6.2 plots the probability density function of a generalized extreme value distribution with different shape parameters ξ.

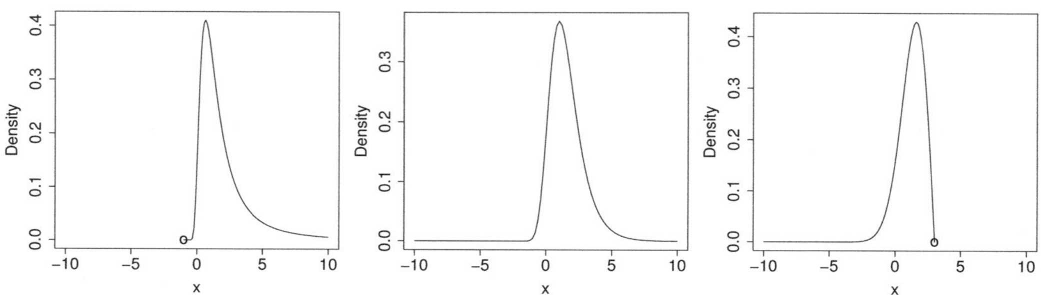

FIGURE 6.2
Plot of generalized extreme value probability density functions with $\mu = \sigma = 1$ and, from left to right, $\xi = 0.5, 0, -0.5$.

The parameters μ and σ are required to embed both the unknown normalizing constants a_n and b_n in (6.3) and the scaling constants a and b in the expression $G(ax+b) = G_\xi(x)$.

A fundamental property of the block maxima approach is the max-stable property. A random variable Y is said to be max-stable if there exist constants $c_n > 0$ and $d_n \in \mathbb{R}$ such that

$$\frac{\max\{Y_1, \dots, Y_n\} - d_n}{c_n} \stackrel{\mathrm{d}}{=} Y, \tag{6.7}$$

for all $n \in \mathbb{N}$ and where $Y_1, \dots, Y_n$ are independent copies of Y. Loosely speaking, the max-stability property states that, up to a shift and scale transformation, the random variable Y is stable by the maximum operator and is similar to what the α-stability is for the addition operator (see Section 3.3. in Embrechts et al. (1997)).

It is straightforward to show that the generalized extreme value distribution at the righthand side of (6.5) is max-stable with

$$c_n = n^\xi, \qquad d_n = (1 - c_n)\left(\mu - \frac{\sigma}{\xi}\right).$$

Less formally, the max-stability property is desirable because for all $n, k \in \mathbb{N}$, we have

$$\max\{X_1, \dots, X_{nk}\} = \max\left\{\max(X_1, \dots, X_n), \dots, \max\left(X_{n(k-1)+1}, \dots, X_{nk}\right)\right\},$$

that is, the "overall maximum" can be thought of as the maxima of several "sub-maxima." Now as $n \to \infty$, both the overall maximum and these sub-maxima converge to (different) generalized extreme value distributions—under linear normalizations. The difference between the two limiting generalized extreme value distributions coming from the normalizing constants $c_n > 0$ and $d_n \in \mathbb{R}$ in (6.7).

The following R code show how it is possible to retrieve the annual maxima from the Danish claim dataset.

```
> danish.claim <- danishuni[,2]
> years <- as.numeric(substr(danishuni[,1], 1, 4))
> danish.max <- aggregate(danish.claim, by=list(years), max,
+ na.rm=TRUE)[,2]
```

6.2.2 Exceedances above a Threshold

As illustrated by Figure 6.1, the threshold excess approach makes better, more efficient use of the available data by including all of the extreme values regardless of whether they happen within some arbitrarily chosen block or not; with the potential issue of introducing dependence among those exceedances.

Because our interest is in the tail of X and we aim to derive asymptotic distributions for extreme events, it makes sense to characterize the limiting distribution H,

$$\left.\frac{X-u}{a(u)}\,\right|\{X>u\} \xrightarrow{\text{d}} H, \qquad u \to x_+, \tag{6.8}$$

for some positive function $a(\cdot)$ and where $x_+ = \sup\{x \in \mathbb{R}\colon F(x) < 1\}$.

Provided the limiting distribution G in (6.3) is nondegenerate, the Pickands–Balkema–de Haan theorem (see Embrechts et al. (1997), Section 3.4; Coles (2001), Theorem 4.2; or Beirlant et al. (2004), Section 4.5) states that there exists $a > 0$ such that $H(ax) = H_\xi(x)$, where

$$H_\xi(x) = 1 - (1+\xi x)^{-1/\xi}, \qquad 1+\xi x > 0, \tag{6.9}$$

where $\xi \neq 0$. Again, as $\xi \to 0$, H_ξ is prolonged by continuity to give the exponential distribution

$$H_0(x) = 1 - \exp(-x), \qquad x > 0.$$

The shape parameter ξ appearing in H_ξ is the same as the one arising in the generalized extreme value distribution (6.4).

For statistical purposes, (6.8) justifies the use of the distributions generated by (6.9) to model excesses above a threshold and similarly to the reasoning made for block maxima, we assume that for $u > 0$ large enough, we have

$$\mathbb{P}\left(X-u \le x \mid X>u\right) \approx 1 - \left(1+\xi\frac{x}{\tau}\right)^{-1/\xi}, \qquad 1+\xi\frac{x}{\tau} > 0, \tag{6.10}$$

for some unknown scale $\tau > 0$ and shape parameter $\xi \in \mathbb{R}$ parameters.

The distribution (6.10) is known as the generalized Pareto distribution and takes as a special case the exponential distribution when $\xi \to 0$

$$\mathbb{P}\left(X-u < x \mid X>u\right) \approx 1 - \exp\left(-\frac{x}{\tau}\right), \qquad x > 0. \tag{6.11}$$

The generalized extreme value and the generalized Pareto distributions share strong connections because $1 - H_\xi(x) = -\log G_\xi(x)$, which makes sense because when u is large enough, a first-order approximation gives

$$\begin{aligned}
\mathbb{P}(X-u > x \mid X>u) &\approx \frac{n\log \mathbb{P}(X-u<x)}{n\log\mathbb{P}(X<u)}\\
&\approx \frac{\log\mathbb{P}(M_n < u+x)}{\log\mathbb{P}(M_n<u)}\\
&\approx \left(\frac{1+\xi\frac{u+x-\mu}{\sigma}}{1+\xi\frac{u-\mu}{\sigma}}\right)^{-1/\xi}\\
&\approx \left(1+\xi\frac{x}{\tau}\right)^{-1/\xi},
\end{aligned}$$

with $\tau = \sigma + \xi(u-\mu)$ and provided n is large enough to ensure a good approximation.

Up to a transformation, the generalized Pareto distribution is stable by "thresholding." Let Y be a generalized Pareto random variable with scale $\tau > 0$ and shape $\xi \in \mathbb{R}$; then for all $u > 0$, we have

$$\mathbb{P}(Y > x+u \mid Y>u) = \frac{\{1+\xi(x+u)/\tau\}^{-1/\xi}}{(1+\xi u/\tau)^{-1/\xi}} = \left\{1+\xi\frac{x}{\tau(u)}\right\}^{-1/\xi},$$

where $\tau(u) = \tau + \xi u$ and $x > 0$. In particular as we increase the threshold u, the scale parameter τ is a linear function of u while the shape parameter ξ is constant. As expected, we recover the memorylessness property of the exponential distribution, as in that case $\xi = 0$ and hence $\tau(u) \equiv \tau$ for all $u > 0$.

The following R code shows how it is possible to retrieve exceedances above the threshold $u = 10$ for the Danish claim dataset.

```
> u <- 10
> danish.exc <- danishuni[danishuni[,2] > u, 2]
```

From a technical point of view (see discussion in Beirlant et al. (2004) or Embrechts et al. (1997)), Y is said to have Pareto tails, with tail index $\xi > 0$ if

$$\mathbb{P}(Y > x) = x^{-1/\xi}\mathcal{L}(x),$$

where $\mathcal{L}$ is some slowly varying function,

$$\lim_{t\to\infty} \frac{\mathcal{L}(tx)}{\mathcal{L}(x)} = 0.$$

Equivalently, if F^{-1} denotes the quantile function, then

$$F^{-1}(1-u) = u^{-\xi}\mathcal{L}^\star(1/u)$$

where $\mathcal{L}^\star$ is some slowly varying function (see Figure 6.1) (the de Bruyn conjugate of the slowly varying function $\mathcal{L}$ associated with F, from Proposition 2.5 in Beirlant et al. (2004)).

6.2.3 Point Process

A point process is a collection of some arbitrary objects, called the points, that appear randomly in some state space E, where typically $E \subset \mathbb{R}^d$. For example, these points can be the times of arrival in a queue or even positions of stars in the sky. Any point process is completely characterized by its counting measure $N\colon A \mapsto N(A)$ defined for any subset A of E and where $N(A)$ is the random number of points lying in A.

With a slight abuse of notation, it is often said that N is a point process where N might denote the points or the counting measure, depending on the context. Note that $N(A)$ is a random variable whose mean $\mathbb{E}\{N(A)\} = \Lambda(A)$ is called the intensity measure. Under some mild regularity conditions, the intensity measure has an intensity function, that is, there exists a non-negative function λ such that

$$\Lambda(A) = \int_A \lambda(x)\mathrm{dx},$$

for all $A \subset E$.

Poisson point processes are probably the most simple and useful point processes. A point process N is a Poisson point process if

- $N(A)$ is Poisson distributed with mean $\Lambda(A)$ for any $A \subset E$;
- $N(A_1), \ldots, N(A_k)$ are mutually independent for any $k \in \mathbb{N}$ and disjoint subsets $A_1, \ldots, A_k$ of E.

For our purposes of exceedances, we will consider the points living in $E = (0,1) \times \mathbb{R}$ defined by

$$\left\{\left(\frac{i}{n+1}, \frac{X_i - b_n}{a_n}\right) : i = 1, \ldots, n\right\},$$

where $a_n > 0$ and $b_n \in \mathbb{R}$ are the normalizing constants appearing in (6.3).

Provided the limiting distribution G in (6.3) is nondegenerate, it can be shown that this point process converges as $n \to \infty$ to a Poisson point process with intensity

$$\Lambda\{(t_1, t_2) \times (x, \infty)\} = (t_2 - t_1)\,(1 + \xi x)^{-1/\xi}\,,$$

where $x \in \mathbb{R}$ and $0 < t_1 < t_2 < 1$.

Because most of the points $(X_i - b_n)/a_n$ will move toward $-\infty$ as n gets large, we will assume that the ones lying in $(0, 1) \times (u, \infty)$ are distributed according to the above Poisson process provided that the threshold $u \in \mathbb{R}$ is large enough. Again, for practical purposes, it is convenient to embed the normalizing constants a_n and b_n into the intensity measure to give

$$\Lambda\{(t_1, t_2) \times (x, \infty)\} = (t_2 - t_1)\left(1 + \xi\frac{x - \mu}{\sigma}\right)^{-1/\xi}, \tag{6.12}$$

where $x > u$ and $0 < t_1 < t_2 < 1$.

Note how the intensity measure (6.12) resembles the generalized extreme value distribution (6.6). This is not a coincidence as, provided n is large enough, we have

$$\begin{aligned}\mathbb{P}(M_n \le x) &= \mathbb{P}\{\text{no points in } (0,1) \times (x, \infty)\}\\ &\approx \exp\left[-\Lambda\{(0,1) \times (x, \infty)\}\right],\end{aligned}$$

where we used the fact that if N is a Poisson process with intensity measure Λ, we have $\mathbb{P}\{N(A) = 0\} = \exp\{-\Lambda(A)\}$. It follows that

$$\Lambda\{(t_1, t_2) \times (x, \infty)\} = (t_2 - t_1)\left(1 + \xi\frac{x - \mu}{\sigma}\right)^{-1/\xi},$$

where the term $(t_2 - t_1)$ comes from a time homogeneity property of the exceedances because the X_i were supposed to be independent and identically distributed, that is, there is no reason to assume that the exceedances are more likely to occur in a given time period.

Compared to the generalized Pareto approach, the point process representation of exceedances is advantageous in that it is parametrized according to the generalized extreme value distribution, and as such, the parameters μ, σ, and ξ do not depend on the threshold u.

6.3 Inference

In Section 6.2 we introduced asymptotic results for the extremes of random variables. This section uses these results for the statistical modeling of high quantiles. Modeling extreme events is challenging and typically is a trade-off between being as close as possible to the "asymptotic regime" so that the use of the limiting distributions G and H in (6.3) and (6.8) is sensible, and simultaneously having as much data as possible so that the estimates are reliable.

Taking n or u too small will induce a bias, but small standard errors while taking n or u too large yields a smaller bias but larger standard errors and the trade-off is therefore a compromise between bias and variance. Before introducing some tools to choose a suitable block size n or threshold value u, we will first focus on how it is possible to fit extreme value distributions to a sample of extreme values.

6.3.1 Visualizing Tails

In Section 6.2.2, we have seen that the generalized Pareto distribution should provide a good approximation in tails because

$$\mathbb{P}(Y > x + u \mid Y > u) \approx \left\{1 + \xi \frac{x}{\tau(u)}\right\}^{-1/\xi}.$$

If we use a standard Pareto, above u, it can be written that

$$\mathbb{P}(Y > x \mid Y > u) \approx \left(\frac{x}{u}\right)^{-\alpha}, \qquad x > u.$$

Equivalently, it means that

$$\log \mathbb{P}(Y > x \mid Y > u) \approx \alpha \left(\log u - \log x\right).$$

The empirical version of the exceedance probability from sample x_i is the proportion of observations exceeding x, among those exceeding u (usually +1 is we want to avoid 0's and 1's). Let $x_{i:n}$ denote the order statistic, in the sense that

$$x_{1:n} \leq x_{2:n} \leq \cdots \leq x_{i:n} \leq \cdots \leq x_{n-1:n} \leq x_{n:n}.$$

Given u, let n_u be the number of observations exceeding u; then for $i \in \{1, \ldots, n_u\}$, using $x = x_{n-i:n}$, we have

$$\log \frac{i}{n_u} \approx \alpha \left(\log u - \log x_{n-i:n}\right),$$

which is a linear relationship. With Pareto tails, the scatterplot on a log–log scale of the points

$$\left\{\left(x_{n-i:n}, \frac{i}{n_u + 1}\right) : i = 1, \ldots, n_u\right\}$$

should be linear, and the slope should be the opposite of α.

```
> n.u <- length(danish.exc)
> surv.prob <- 1 - rank(danish.exc)
> plot(danish.exc, surv.prob, log = "xy", xlab = "Exceedances",
+        ylab = "Survival probability")
```

This scatterplot can be visualized in Figure 6.3. It is possible to add the (theoretical) survival probability from the Pareto distribution if the tail index α is estimated using least squares techniques:

```
> alpha <- - cov(log(danish.exc), log(surv.prob)) / var(log(danish.exc))
> alpha
[1] 1.582581
> x = seq(u, max(danish.exc), length = 100)
> y = (x / u)^(-alpha)
> lines(x, y)
```

It is also possible to plot the cumulative distribution function, given that u is exceeded:

```
> prob <- rank(danish.exc) / (n.u + 1)
> plot(danish.exc, prob, log = "x", xlab= "Exceedances",
+       ylab = "Probability of non exceedance")
> y = 1 - (x / u)^(-alpha)
> lines(x, y)
```

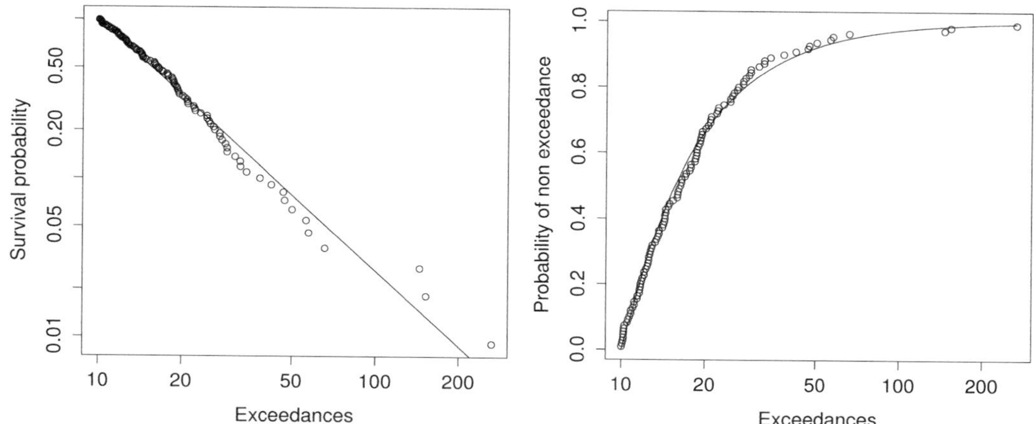

FIGURE 6.3
Plotting Pareto tails. The left panel plots the empirical and fitted survival function. The right panel plots the empirical and fitted cumulative distribution function.

6.3.2 Estimation

Although other approaches are possible, in this section we restrict our attention to the maximum likelihood estimator as this approach can easily handle more complex models, for example, the addition of covariates. Maximum likelihood estimation is carried out by optimizing the likelihood of observing a given sample according to the appropriate distribution. For the extreme value distributions discussed here, the optimization must be performed numerically as no analytic solution exists. Further, the case where the shape parameter is zero requires special attention. It is a single point in a continuous parameter space, and as such will not be estimated with probability 1 when optimizing over the generalized extreme value or generalized Pareto likelihoods. The most common approach is to optimize both the generalized extreme value/generalized Pareto likelihoods and the Gumbel/Exponential likelihoods, and apply a significance test (e.g., using the likelihood ratio test, via confidence intervals for the shape parameter) to test whether or not the parameter should be zero.

6.3.2.1 Generalized Extreme Value Distribution

Letting $z_i = \max\{x_{1i}, \ldots, x_{ki}\}$, the i-th block maximum of an observed time series $x_1, x_2, \ldots$, the generalized extreme value log-likelihood for $\boldsymbol{z} = (z_1, \ldots, z_n)$ with $\xi \neq 0$ is given by

$$\ell(\mu, \sigma, \xi; \boldsymbol{z}) = -n \log \sigma - \left(1 + \frac{1}{\xi}\right) \sum_{i=1}^{n} \log\left(1 + \xi \frac{z_i - \mu}{\sigma}\right) - \sum_{i=1}^{n} \left(1 + \xi \frac{z_i - \mu}{\sigma}\right)^{-1/\xi},$$

provided that $1 + \xi(z_i - \mu)/\sigma > 0$ for $i = 1, \ldots, n$, and in the case of $\xi = 0$, is

$$\ell(\mu, \sigma; \boldsymbol{z}) = -n \log \sigma - \sum_{i=1}^{n} \frac{z_i - \mu}{\sigma} - \sum_{i=1}^{n} \exp\left(-\frac{z_i - \mu}{\sigma}\right).$$

Although when $\xi < -1$ the likelihood is not upper bounded anymore because

$$\ell(\mu, \sigma, \xi; \mathbf{z}) \longrightarrow \infty, \qquad \mu - \frac{\sigma}{\xi} \downarrow \max\{z_1, \ldots, z_n\},$$

this is usually not a problem because such small values for ξ correspond to very short-tailed distributions that are unlikely to occur in concrete applications. When $\xi \geq -1$, the maximum likelihood estimator is convergent and asymptotically normal when $\xi > -0.5$.

The following R code computes the negative log-likelihood for the generalized extreme value distribution:

```
> nllik.gev <- function(par, data){
+    mu <- par[1]
+    sigma <- par[2]
+    xi <- par[3]
+  if ((sigma <= 0) | (xi <= -1))
+        return(1e6)
+    n <- length(data)
+    if (xi == 0)
+       n * log(sigma) + sum((data - mu) / sigma) +
+           sum(exp(-(data - mu) / sigma))
+    else {
+        if (any((1 + xi * (data - mu) / sigma) <= 0))
+            return(1e6)
+        n * log(sigma) + (1 + 1 / xi) *
+           sum(log(1 + xi * (data - mu) / sigma)) +
+           sum((1 + xi * (data - mu) / sigma)^(-1/xi))
+    }
+}
```

Some care is needed when trying to optimize the generalized extreme value likelihood because the likelihood is typically erratic in regions far away from its global maximum and hence numerical optimization might fail. A reasonable strategy is to provide sensible starting values such as the moment estimates for the Gumbel distribution. The following R code does this job for the Danish claim dataset:

```
> sigma.start <- sqrt(6) * sd(danish.max) / pi
> mu.start <- mean(danish.max) + digamma(1) * sigma.start
> fit.gev <- nlm(nllik.gev, c(mu.start, sigma.start, 0),
+ hessian = TRUE, data = danish.max)
> fit.gev
$minimum
[1] 58.2333

$estimate
[1] 37.7934096 28.9358639  0.6384013

$gradient
[1] -4.878783e-07  2.600457e-07 -7.716494e-06

$hessian
            [,1]        [,2]       [,3]
[1,]  0.03796779 -0.02924794  0.3404126
[2,] -0.02924794  0.03073232 -0.2542531
[3,]  0.34041257 -0.25425309  8.8896897

$code
```

```
[1] 1

$iterations
[1] 37
```

The maximum likelihood estimates can be obtained from `fit$estimate` and the associated standard errors from

```
> sqrt(diag(solve(fit.gev$hessian))) #$
[1] 10.7183026 11.0494255  0.4141631
```

In particular, the maximum likelihood estimates (and the associated standard errors) are $\hat{\mu} = 38$ (11), $\hat{\sigma} = 29$ (11), and $\hat{\xi} = 0.64$ (0.41).

6.3.2.2 Poisson-Generalized Pareto Model

Similar to the generalized extreme value distribution, letting $y_i = x_i - u$ for only those $x_i > u$, then for the generalized Pareto distribution, the log-likelihood is given by

$$\ell(\tau, \xi; \boldsymbol{y}) = -m \log \tau - \left(1 + \frac{1}{\xi}\right) \sum_{i=1}^{m} \log \left(1 + \xi \frac{y_i}{\tau}\right),$$

where m is the number of exceedances above the threshold u, and in the case of $\xi = 0$,

$$\ell(\tau; \boldsymbol{y}) = -m \log \tau - \frac{1}{\tau} \sum_{i=1}^{m} y_i.$$

As for the generalized extreme value distribution, the log-likelihood can be made arbitrarily large when $\xi < -1$ because

$$\ell(\tau, \xi; \mathbf{y}) \longrightarrow \infty, \qquad \frac{\tau}{\xi} \downarrow - \max\{y_1, \ldots, y_m\}.$$

Similar to the generalized extreme value distribution, the maximum likelihood estimator is consistent when $\xi > -1$ and asymptotically normal when $\xi > -0.5$.

The following R code computes the negative log-likelihood for the generalized Pareto distribution:

```
> nllik.gp <- function(par, u, data){
+     tau <- par[1]
+     xi <- par[2]
+
+     if ((tau <= 0) | (xi < -1))
+         return(1e6)
+
+     m <- length(data)
+
+     if (xi == 0)
+         m * log(tau) + sum(data - u) / tau
+
+     else {
+         if (any((1 + xi * (data - u) / tau) <= 0))
+             return(1e6)
+
```

```
+         m * log(tau) + (1 + 1 / xi) *
+             sum(log(1 + xi * (data - u) / tau))
+     }
+ }
```

The use of the generalized Pareto model for modeling exceedances is a two-step procedure. First, one must optimize the above likelihood and then estimate the rate of exceedances ζ_u, that is, $\zeta_u = \mathbb{P}(X > u)$.

For the Danish claim dataset, the likelihood is maximized by invoking

```
> u <- 10
> tau.start <- mean(danish.exc) - u
> fit.gp <- nlm(nllik.gp, c(tau.start, 0), u = u, hessian = TRUE,
+ data = danish.exc)
> fit.gp
$minimum
[1] 374.893

$estimate
[1] 6.9754659 0.4969865

$gradient
[1] 4.783492e-06 7.958079e-05

$hessian
         [,1]      [,2]
[1,] 1.138184  5.021902
[2,] 5.021902 75.965994

$code
[1] 1

$iterations
[1] 21
```

where `tau.start` is the moment estimator of the exponential distribution. Independently, the rate parameter ζ_u is easily estimated by $\hat{\zeta}_u = m/n$ with associated standard error $\{\hat{\zeta}_u(1-\hat{\zeta}_u)/n\}^{1/2}$. The estimates and related standard errors for the generalized Pareto can be obtained similarly to the generalized extreme value case, and we found $\hat{\zeta}_u = 0.050\ (0.004), \hat{\tau} = 7\ (1)$ and $\hat{\xi} = 0.50\ (0.14)$.

Although from a theoretical point of view the shape parameters of the generalized extreme value and generalized Pareto distributions are the same, the estimates for ξ differ.

It is also possible to use a profile likelihood technique because the main parameter of interest is ξ (see Venzon & Moolgavkar (1988)). Recall that if we consider a parametric model with parameter $\boldsymbol{\theta} = (\boldsymbol{\theta}_1, \boldsymbol{\theta}_2)$, define

$$\ell_P(\boldsymbol{\theta}_1) = \max_{\boldsymbol{\theta}_2} \ell(\boldsymbol{\theta}_1, \boldsymbol{\theta}_2).$$

We can then compute the maximum of the profile log-likelihood

$$\widehat{\boldsymbol{\theta}}_1 = \underset{\boldsymbol{\theta}_1}{\text{argmax}}\ \ell_P(\boldsymbol{\theta}_1) = \underset{\boldsymbol{\theta}_1}{\text{argmax}}\ \max_{\boldsymbol{\theta}_2}\ \ell(\boldsymbol{\theta}_1, \boldsymbol{\theta}_2).$$

This maximum is not necessarily the same as the (global) maximum obtained by maximizing the likelihood, on a finite sample. Under standard suitable conditions,

$$2\left\{\ell_P(\widehat{\boldsymbol{\theta}}_1) - \ell_P(\boldsymbol{\theta}_1)\right\} \xrightarrow{\text{d}} \chi^2(\dim(\boldsymbol{\theta}_1)),$$

so it is possible to derive confidence intervals for $\widehat{\boldsymbol{\theta}}_1$. The code will be

```
> prof.nllik.gp <- function(par,xi, u, data)
+ nllik.gp(c(par,xi), u, data)
> prof.fit.gp <- function(x)
+ -nlm(prof.nllik.gp, tau.start, xi = x, u = u, hessian = TRUE,
+      data = danish.exc)$minimum
> vxi = seq(0,1.8,by=.025)
> prof.lik <- Vectorize(prof.fit.gp)(vxi)
> plot(vxi, prof.lik, type="l", xlab = expression(xi),
+      ylab = "Profile log-likelihood")
> opt <- optimize(f = prof.fit.gp, interval=c(0,3), maximum=TRUE)
> opt
$maximum
[1] 0.496993

$objective
[1] -374.893

> up <- opt$objective
> abline(h = up, lty=2)
> abline(h = up-qchisq(p = 0.95, df = 1), col = "grey")
> I <- which(prof.lik >= up-qchisq(p = 0.95, df = 1))
> lines(vxi[I], rep(up-qchisq(p = 0.95, df = 1), length(I)),
+       lwd = 5, col = "grey")
> abline(v = range(vxi[I]), col = "grey", lty = 2)
> abline(v = opt$maximum, col="grey")
```

6.3.2.3 Point Process

As alluded to previously, fitting the two-dimensional model better accounts for the uncertainty in estimating these parameters simultaneously. In this case, the log-likelihood to optimize is, using $x_1, \dots, x_k$ as for the generalized Pareto likelihood (but now with m exceedances from n total observations), given by

$$\ell(\mu,\sigma,\xi;\boldsymbol{x}) = -m\log\sigma - \left(1+\frac{1}{\xi}\right)\sum_{i=1}^{m}\log\left(1+\xi\frac{x_i-\mu}{\sigma}\right) - n_b\left(1+\xi\frac{u-\mu}{\sigma}\right)^{-1/\xi},$$

if $\xi \neq 0$ and

$$\ell(\mu,\sigma;\boldsymbol{x}) = -m\log\sigma - \sum_{i=1}^{m}\frac{x_i-\mu}{\sigma} - n_b\exp\left(-\frac{u-\mu}{\sigma}\right),$$

if $\xi = 0$ and where n_b is the number of blocks appearing in the time series.

The following R code computes the negative log-likelihood for the Poisson point process model:

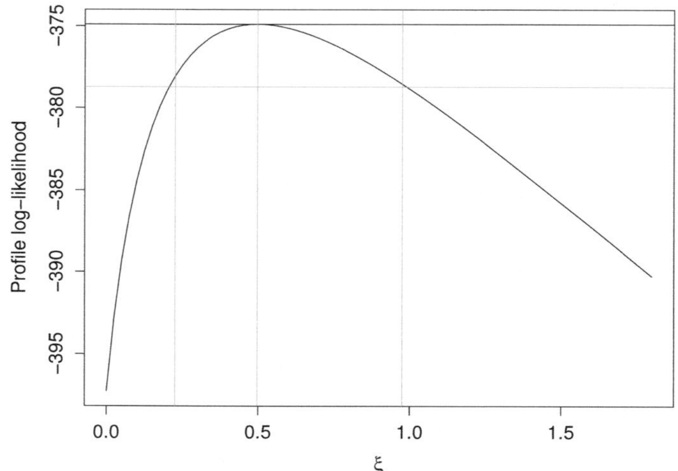

FIGURE 6.4
Profile likelihood for ξ in the generalized Pareto distribution above $u = 10$.

```
> nllik.pp <- function(par, u, data, n.b){
+     mu <- par[1]
+     sigma <- par[2]
+     xi <- par[3]
+     if ((sigma <= 0) | (xi <= -1))
+         return(1e6)
+     if (xi == 0)
+         poiss.meas <- n.b * exp(-(u - mu) / sigma)
+     else
+         poiss.meas <- n.b * max(0, 1 + xi * (u - mu) / sigma)^(-1/xi)
+   exc <- data[data > u]
+     m <- length(exc)
+     if (xi == 0)
+         poiss.meas + m * log(sigma) + sum((exc - mu) / sigma)
+     else {
+         if (any((1 + xi * (exc - mu) / sigma) <= 0))
+             return(1e6)
+         poiss.meas + m * log(sigma) + (1 + 1 / xi) *
+             sum(log(1 + xi * (exc - mu) / sigma))
+     }
+ }
```

As previously, it is desirable to set suitable starting values. For the Poisson point process approach, the starting values are the same as for the generalized extreme value model. The Poisson point process model is fitted by invoking the following code:

```
> n.b <- 1991 - 1980
> u <- 10
> sigma.start <- sqrt(6) * sd(danish.exc) / pi
> mu.start <- mean(danish.exc) + (log(n.b) + digamma(1)) *
+     sigma.start
```

```
> fit.pp <- nlm(nllik.pp, c(mu.start, sigma.start, 0), u = u,
+              hessian = TRUE, data = danishuni[,2], n.b = n.b)
> fit.pp
$minimum
[1] 233.9067

$estimate
[1] 39.8421585 21.8065885  0.4969855

$gradient
[1]  4.518800e-07 -6.474347e-07  1.358003e-05

$hessian
          [,1]        [,2]       [,3]
[1,]  2.524825   -3.631807   76.93413
[2,] -3.631807    5.325742 -114.35596
[3,] 76.934131 -114.355962 2532.88928

$code
[1] 1

$iterations
[1] 43
```

The maximum likelihood estimates for the Danish claim dataset are $\hat{\mu} = 40$ (5), $\hat{\sigma} = 22$ (6), and $\hat{\xi} = 0.50$ (0.14). Note that the shape parameter estimate $\hat{\xi}$ is exactly the same as the one we get for the generalized Pareto approach. Further, $\hat{\sigma}$ and the generalized Pareto scale parameter estimate $\hat{\tau}$ are connected by the relation $\hat{\tau} = \hat{\sigma} + \hat{\xi}(u - \hat{\mu})$.

6.3.2.4 Other Tail Index Estimates

From Section 6.2.2 (see also Section 6.3.1), if Y has Pareto tails, with tail index ξ, then

$$F^{-1}(1-u) = u^{-\xi}\mathcal{L}^\star(1/u)$$

or equivalently,

$$\log F^{-1}(1-u) = -\xi \log u + \log \mathcal{L}^\star(1/u).$$

If $p = i/(n+1)$, then ξ should be the opposite of the slope of the linear regression of $\log x_{n-i:n}$ against $\log\{i/(n+1)\}$ (on Figure 6.3, we did plot $\log\{i/(n+1)\}$ against $\log x_{n-i:n}$, so that the slope was $-\alpha = -1/\xi$).

If we consider only the k largest observations (as again, we focus on tails only here), the estimator of ξ is

$$\widehat{\xi} = \frac{\displaystyle\frac{1}{k}\sum_{i=1}^{k} \log x_{n-i+1:n} - \log x_{n-k+1:n}}{\displaystyle\frac{1}{k}\sum_{i=1}^{k} \log \frac{i}{n+1} - \frac{k}{n+1}}.$$

Hill's estimator (introduced in Hill (1975)) is obtained by assuming that the denominator is almost 1 (which means that k/n is small and n is large):

$$\widehat{\xi}_k = \frac{1}{k}\sum_{i=1}^{k} \log x_{n-i+1:n} - \log x_{n-k+1:n}.$$

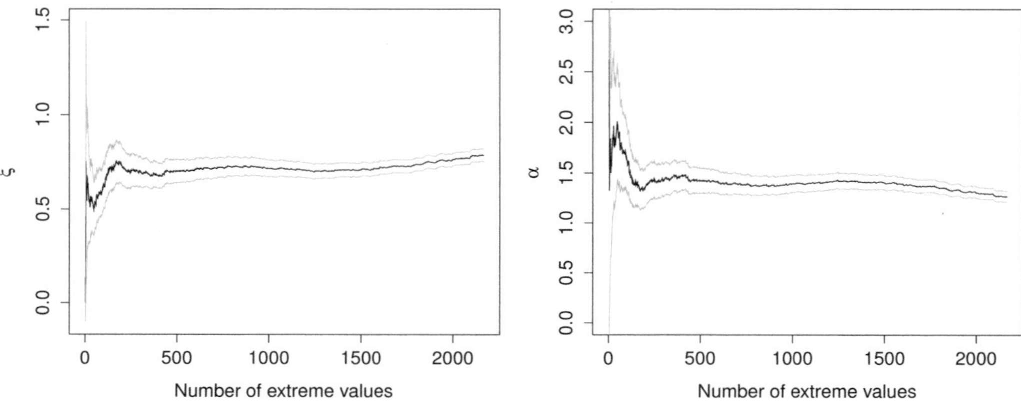

FIGURE 6.5
Hill plots for ξ and α.

From asymptotic theory (k should be large, but still, k/n should tend to 0), $\widehat{\xi}_k$ will be converging to ξ. Further (see Beirlant et al. (2004)),

$$\sqrt{k}\left(\widehat{\xi}_k - \xi\right) \xrightarrow{\text{d}} \mathcal{N}(0, \xi^2).$$

It is possible to obtain (and plot) an asymptotic confidence interval for ξ:

```
> logXs <- log(sort(danishuni[,2], decreasing=TRUE))
> n <- length(logXs)
> xi <- 1/1:n * cumsum(logXs) - logXs
> ci.up <- xi + qnorm(0.975) * xi / sqrt(1:n)
> ci.low <- xi - qnorm(0.975) * xi / sqrt(1:n)
> matplot(1:n, cbind(ci.low, xi, ci.up),lty = 1, type = "l",
+         col = c("grey", "black", "grey"), ylab = expression(xi),
+         xlab = "Number of extreme values")
```

Using the delta-method, it is also possible to prove that $\widehat{\alpha}_k = 1/\widehat{\xi}_k$ satisfies

$$\sqrt{k}(\widehat{\alpha}_k - \alpha) \xrightarrow{\text{d}} \mathcal{N}(0, \alpha^2).$$

Then it is possible to estimate $\alpha = 1/\xi$, and to plot a confidence interval:

```
> alpha <- 1 / xi
> alpha.std.err <- alpha / sqrt(1:n)
> ci.up <- alpha + qnorm(0.975) * alpha / sqrt(1:n)
> ci.low <- alpha - qnorm(0.975) * alpha / sqrt(1:n)
> matplot(1:n, cbind(ci.low, alpha, ci.up), lty = 1, type = "l",
+         col = c("grey", "black", "grey"), ylab = expression(alpha),
+         xlab = "Number of extreme values")
```

6.3.3 Checking for the Asymptotic Regime Assumption

There exist some theoretical results about the rate of convergence to the generalized extreme value or Pareto distributions. Unfortunately, they are usually of limited interest because they rely on the knowledge of the distribution of X in $M_n = \max\{X_1, \dots, X_n\}$, which is typically unknown for concrete applications.

If one decides to opt for the block maxima framework, one usually chooses convenient block sizes $n \in \mathbb{N}$, for example, $n = 365$ for daily values and annual maxima, and checks the goodness of fit using standard tools such as quantile-quantile plots. It is more difficult to choose a suitable threshold level as one decides to model threshold exceedances using the generalized Pareto approach or the point process representation. One common way is to check if some specific properties of exceedances above a candidate threshold u are met.

6.3.3.1 Mean Excess Plot

A first strategy relies on the behavior of the expectation of a random variable $X - u \mid X > u$ as u increases. Suppose $X - u \mid X > u$ is generalized Pareto distributed with scale $\tau > 0$ and shape $\xi < 1$. Then it is not difficult to show that

$$\mathbb{E}\left(X - \tilde{u} \mid X > \tilde{u}\right) = \frac{\tau + \xi \tilde{u}}{1 - \xi}, \tag{6.13}$$

for all $\tilde{u} > u$.

Equation (6.13) serves as a basis for choosing a suitable threshold $u \in \mathbb{R}$. If for a given $u \in \mathbb{R}$, $X - u$ is (approximately) generalized Pareto distributed, then the function

$$\begin{aligned} [u, \infty) &\longrightarrow (0, \infty) \\ \tilde{u} &\longmapsto \mathbb{E}(X - \tilde{u} \mid X > \tilde{u}) \end{aligned}$$

is a linear function of $\tilde{u}$ provided that $\xi < 1$ to ensure that $\mathbb{E}(X - \tilde{u} \mid X > \tilde{u})$ is finite.

The conditional expectation $\mathbb{E}(X - u \mid X > u)$ is called the mean excess and can be estimated by its empirical version

$$\frac{1}{N(u)} \sum_{i=1}^{n} (X_i - u) 1_{\{X_i > u\}}, \qquad N(u) = \sum_{i=1}^{n} 1_{\{X_i > u\}},$$

where $1_{\{\cdot\}}$ is the indicator function and $X_1, \dots, X_n$, $n \in \mathbb{N}$, are i.i.d. random variables. Confidence intervals based on the central limit theorem can also be derived.

The following R function creates the mean excess plot

```
> meanExcessPlot <- function(data, u.range = range(data),
+                            n.u = 100){
+     mean.excess <- ci.up <- ci.low <- rep(NA, n.u)
+     all.u <- seq(u.range[1], u.range[2], length = n.u)
+     for (i in 1:n.u){
+         u <- all.u[i]
+         excess <- data[data > u] - u
+         n.u <- length(excess)
+         mean.excess[i] <- mean(excess)
+         var.mean.excess <- var(excess)
+         ci.up[i] <- mean.excess[i] + qnorm(0.975) *
+             sqrt(var.mean.excess / n.u)
+         ci.low[i] <- mean.excess[i] - qnorm(0.975) *
+             sqrt(var.mean.excess / n.u)
+     }
+
+     matplot(all.u, cbind(ci.low, mean.excess, ci.up), col = 1,
```

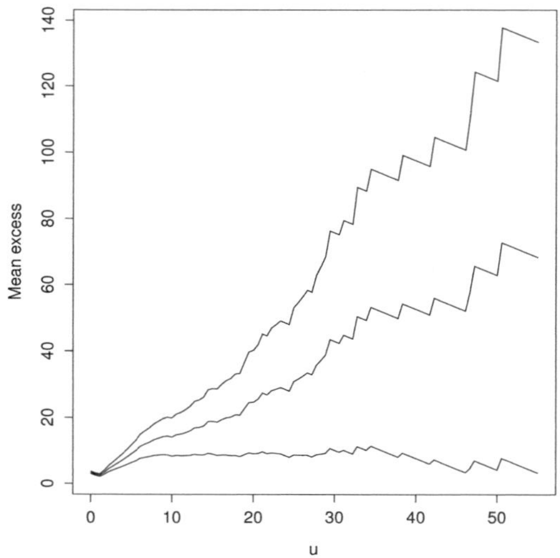

FIGURE 6.6
Mean excess plot for the Danish claim data set. The plot seems to be linear above the threshold $u = 10$.

```
+                lty = c(2, 1, 2), type = "l", xlab = "u",
+                ylab = "Mean excess")
+ }
```

If we use this function on the complete dataset, we obtain the graph of Figure 6.6.

```
> meanExcessPlot(danish.exc)
```

Recall that the aim is to detect a suitable threshold u above which the plot appears to be linear—taking into account the estimation uncertainties with the help of the pointwise confidence intervals. For the Danish claim dataset, it seems that there are two possible candidates for u, $u = 10$ and $u = 20$. It is more sensible to choose the former because more data will be kept in our analysis.

Remark 6.2 *Identifying a suitable threshold using the mean excess plot is usually tricky and highly subjective. Often we will fit extreme value models using several threshold values and check goodness of fit.*

6.3.3.2 Parameter Stability

More precisely, let $X - u_0 \mid X > u_0$ be a generalized Pareto distributed with scale $\tau > 0$ and shape $\xi \in \mathbb{R}$. As stated in Section 6.2.2, because the generalized Pareto distribution is stable by "thresholding" (see Section 6.2.2), a second strategy relies on the behavior of the parameters of a generalized Pareto as the threshold increases. More precisely, if $X - u_0 \mid X > u_0$ is generalized Pareto distributed with scale $\tau > 0$ and shape $\xi \in \mathbb{R}$, then $X - u \mid X > u$ is generalized Pareto distributed with scale $\tau(u) = \tau + \xi(u - u_0)$ and shape $\xi(u) = \xi$ for all $u \geq u_0$.

The parameter stability procedure consists of plotting

$$\left\{\left(u, \hat{\tau}(u) - \hat{\xi}(u)u\right) : u \in \mathbb{R}\right\} \qquad \text{and} \qquad \left\{\left(u, \hat{\xi}(u)\right) : u \in \mathbb{R}\right\},$$

where $\hat{\tau}(u)$ and $\hat{\xi}(u)$ are the estimates of the generalized Pareto distribution related to the threshold u, for example, maximum likelihood estimates, and identify regions where the plot remains constant. Usually it is desirable to add confidence intervals on the plot. The parameter stability procedure can be implemented in R with the following code:

```
> parStabilityPlot <- function(data, u.range = range(data),
+ n.thresh = 50){
+  modified.scales <- rep(NA, n.thresh)
+  shapes <- rep(NA, n.thresh)
+  ci.scales <- ci.shapes <- matrix(NA, 2, n.thresh)
+  all.u <- seq(u.range[1], u.range[2], length = n.thresh)
+  for (i in 1:n.thresh){
+      u <- all.u[i]
+       excess <- data[data > u]
+       tau.start <- mean(excess) - u
+       fit <- nlm(nllik.gp, c(tau.start, 0), u = u, data = excess,
+                   hessian = TRUE)
+       mle <- fit$estimate
+       var.cov <- solve(fit$hessian)
+       modified.scales[i] <- mle[1] - mle[2] * u
+       gradient <- c(1, -u)
+       sd.mod.scale <- sqrt(t(gradient) %*% var.cov %*% gradient)
+       ci.scales[,i] <- modified.scales[i] + c(-1, 1) *
+           qnorm(0.975) * sd.mod.scale
+
+       shapes[i] <- mle[2]
+       ci.shapes[,i] <- mle[2] + c(-1, 1) *
+          qnorm(0.975) * sqrt(var.cov[2,2])
+   }
+
+   par(mfrow = c(1, 2), mar = c(4, 5, 1, 0.5))
+   plot(all.u, modified.scales, xlab = "u", ylab = "Modified scale",
+        ylim = range(ci.scales))
+   segments(all.u, ci.scales[1,], all.u, ci.scales[2,])
+
+   plot(all.u, shapes, xlab = "u", ylab = expression(xi(u)),
+        ylim = range(ci.shapes))
+   segments(all.u, ci.shapes[1,], all.u, ci.shapes[2,])
+ }
```

Again, if we use this function on the complete dataset, we obtain the graph of Figure 6.7.

```
> parStabilityPlot(danishuni[,2], c(0, 50))
```

6.3.4 Quantile Estimation

For the extreme value distributions, it is easy to obtain quantiles by inverting the distributions. For the GEV distribution, the quantiles are given by

$$z_p = \begin{cases} \mu + \frac{\sigma}{\xi}\left\{(-\log p)^{-\xi} - 1\right\}, & \xi \neq 0 \\ \mu - \sigma \log(-\log p), & \xi = 0. \end{cases} \tag{6.14}$$

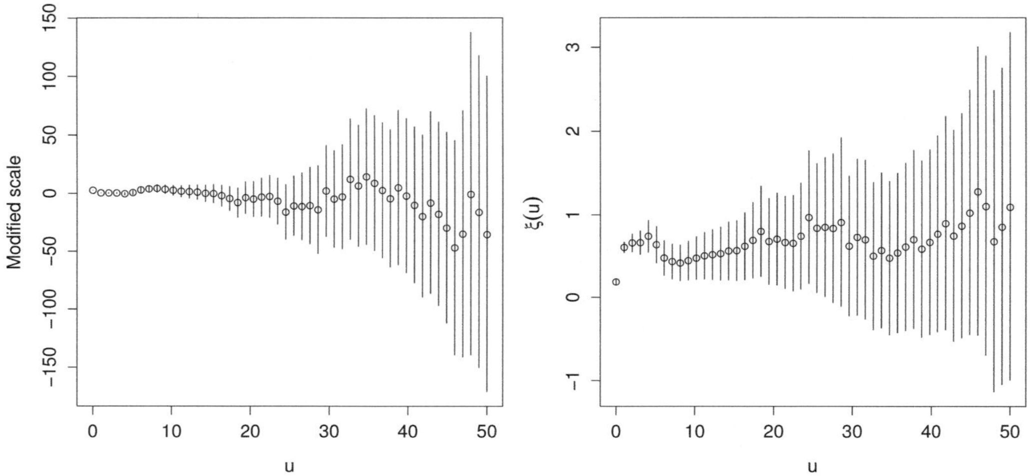

FIGURE 6.7
Parameter stability plot.

Using the extreme value theory terminology, quantiles are often called return levels—or value at risk for financial applications. For the generalized extreme value distribution, z_p is the return level associated with the $T = 1/(1-p)$ return period. Return levels are widely used in concrete applications because they have a convenient interpretation.

To see this, let $X_1, X_2, \ldots$ be independent copies of a random variable X and z_p a quantile of X such that $\mathbb{P}(X < z_p) = p$. Consider the first occurrence where the X_i exceed the level z_p, that is,

$$I = \min\{i \in \mathbb{N}\colon X_i > z_p\}.$$

Clearly, the random variable I follows a geometric distribution where the probability to have success at each trial is $\mathbb{P}(X > z_p) = 1 - p$. This implies that $\mathbb{E}(I) = 1/(1-p)$, or in other words, that the quantile z_p is expected to be exceeded once every $T = 1/(1-p)$ trials. For the generalized extreme value distribution, assuming that the X_i are annual maxima, the return level z_p would be expected to be exceeded once every $T = 1/(1-p)$ years.

It is important to mention that the notion of return period does not imply any periodicity in the occurrence of extreme events because within each block (typically years), there is the same chance of exceeding the level z_p. Furthermore, the notion of a return period does not make sense in the case of non-stationarity; for example, the random variables $X_1, X_2, \ldots$ are still independent but do not share the same distribution.

For the generalized Pareto distribution, one must account for the frequency of occurrence of exceedances in order to define return levels because this distribution is usually used for modeling values given that we exceed a threshold u. We then have to solve

$$\begin{aligned}\mathbb{P}(X \leq x_p) = p &\iff \mathbb{P}(X > x_p, X > u) = 1 - p \\ &\iff \mathbb{P}(X > u)\mathbb{P}(X > x_p \mid X > u) = 1 - p,\end{aligned}$$

where $x_p > u$.

It is not difficult to show that the return levels are

$$x_p = \begin{cases} u + \frac{\tau}{\xi}\left\{\left(\frac{1-p}{\zeta_u}\right)^{-\xi} - 1\right\}, & \xi \neq 0 \\ u - \tau \log\left(\frac{1-p}{\zeta_u}\right), & \xi = 0, \end{cases} \tag{6.15}$$

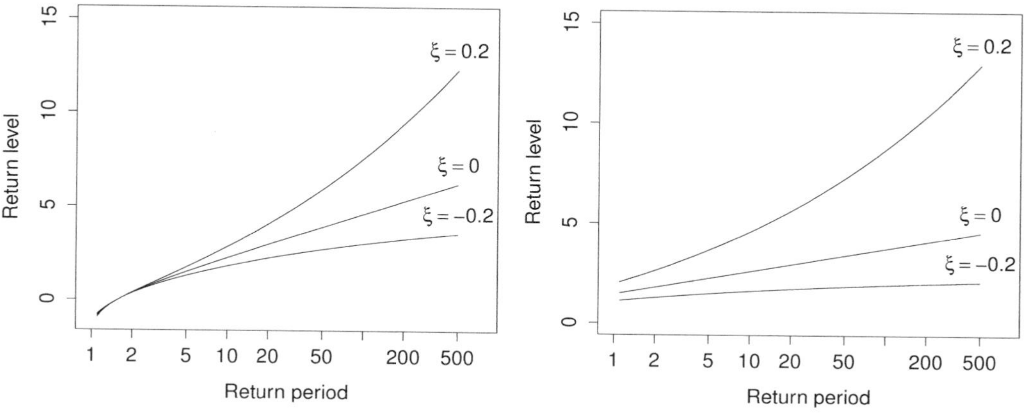

FIGURE 6.8
Return level plots for different values of the shape parameter ξ. The left panel corresponds to the generalized extreme value distributions with $\mu = 1, \sigma = 1$. The right panel corresponds to the generalized Pareto distribution with $u = 0, \tau = 0.5, \zeta_u = 0.05$ and $n_b = 365$.

where $\zeta_u = \mathbb{P}(X > u)$.

Using similar arguments as those used for the generalized extreme value distribution, the return level x_p is expected to be exceeded once every $1/(1-p)$ observations. For practical reasons, it is often more convenient to modify x_p to a suitable scale. For example if we observed daily values, then the return levels x_p would be expected to be exceeded once every $T = 1/\{n_b(1-p)\}$ years with $n_b = 365$.

It is common in extreme value analysis to plot the return levels as the return period increases. Such a plot is called a return level plot and consists of plotting the function

$$\begin{aligned} (1, \infty) &\longmapsto \mathbb{R} \\ T &\longrightarrow z_{1-1/T} \end{aligned} \tag{6.16}$$

for the generalized extreme value distribution and

$$\begin{aligned} (n_b^{-1}, \infty) &\longmapsto (u, \infty) \\ T &\longrightarrow x_{1-1/(n_b T)} \end{aligned} \tag{6.17}$$

for the generalized Pareto distribution.

Figure 6.8 shows return levels plots for both the generalized extreme value and the generalized Pareto distributions with different values for the shape parameter ξ. Typically, return level plots use a logarithmic scale for the x-axis so that the Gumbel/Exponential cases, that is, $\xi = 0$, appear as a straight line—at least for large return periods. As expected, the cases $\xi > 0$ yield the largest return levels because, in such a situation, the distribution is not upper bounded.

To estimate quantiles, plug in the maximum likelihood estimates in (6.14) or (6.15) depending on the model considered. For the point process characterization (6.12), quantile estimates are given by that of the generalized Pareto distribution with scale $\hat{\sigma} + \hat{\xi}(u - \hat{\mu})$ and shape $\hat{\xi}$.

6.4 Model Checking

Although our statistical models are based on asymptotic theory, that is, (6.3) and (6.8), it is wise (as for any statistical model) to check whether our modeling is accurate. There is nothing specific to extreme values here, and standard graphical checks such as quantile-quantile or probability-probability plots are widely used. We will see how return levels can be used to assess goodness of fit. Whatever graphical tool is used, they all rely on the same idea: comparing what we get (the data) to what would expect from the model (the fitted values).

6.4.1 Quantile Quantile Plot

A first strategy is to compare quantiles. The idea is the following. For each ordered observation $z_{1:n} < z_{2:n} < \cdots < z_{n:n}$, we associate the corresponding empirical non-exceedance probability

$$\frac{1}{n+1}, \frac{2}{n+1}, \ldots, \frac{n}{n+1}, \tag{6.18}$$

that is, we treat the ordered observation $z_{i:n}$ as a sample quantile with non exceedance probability $i/(n+1)$. The model counterpart of each $z_{i:n}$ is the estimated quantiles associated with the non-exceedance probabilities $i/(n+1)$, i.e., $\hat{z}_{i/(n+1)}$ in (6.14) or $\hat{x}_{i/(m+1)}$ in (6.15). Note that for the generalized Pareto case, we have to set $\zeta_u = 1$ in (6.15) because the quantile-quantile plot focus is only on exceedances and not all of the data.

The quantile quantile plots for the generalized extreme value and generalized Pareto distributions plot, respectively, the points

$$\left\{(z_{i:n}, \hat{z}_{i/(n+1)}) \colon i = 1, \ldots, n\right\}, \qquad \left\{(x_{i:m}, \hat{x}_{i/(m+1)}) \colon i = 1, \ldots, m\right\}.$$

If the model is correct, then these points should lie approximately around the $y = x$ line.

The following R code produces a quantile-quantile plot for the generalized Pareto distribution:

```
> qqgpd <- function(data, u, tau, xi){
+   excess <- data[data > u]
+   m <- length(excess)
+   prob <- 1:m / (m + 1)
+   x.hat <- u + tau / xi * ((1 - prob)^-xi - 1)
+
+   ylim <- xlim <- range(x.hat, excess)
+   plot(sort(excess), x.hat, xlab = "Sample quantiles",
+        ylab = "Fitted quantiles", xlim = xlim, ylim = ylim)
+   abline(0, 1, col = "grey")
+ }

> qqgpd(danishuni[,2], 10, 7, 0.5)
```

The left panel of Figure 6.9 shows the quantile-quantile plot obtained from the fitted generalized Pareto distribution. We can see that most of the points lie around the $y = x$ line, indicating a good fit. Although the three largest deviate from the $y = x$ line, it is not worrying because these points are associated to the largest plotting position uncertainties.

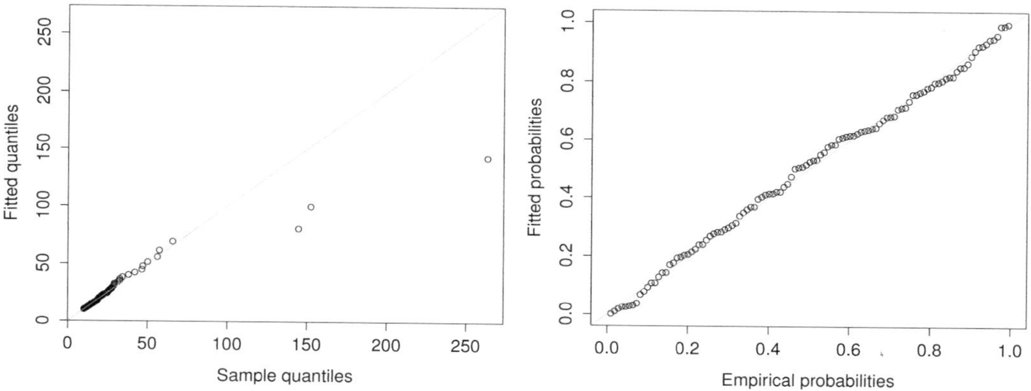

FIGURE 6.9
Model checking for the fitted generalized Pareto distribution to the Danish claim dataset.

6.4.2 Probability–Probability Plot

The idea behind the probability-probability plot is almost the same but compares non-exceedance probabilities instead of quantiles. Given the empirical non-exceedance probabilities (6.18), we compute for each observation the estimated non-exceedance probability from the fitted model. For the generalized extreme value distribution, we have

$$\hat{p}_{i:n} = \exp\left\{-\left(1+\hat{\xi}\frac{z_{i:n}-\hat{\mu}}{\hat{\sigma}}\right)^{-1/\hat{\xi}}\right\}, \qquad i = 1,\dots,n,$$

while for the generalized Pareto distribution, it is given by

$$\tilde{p}_{i:m} = 1 - \left(1+\hat{\xi}\frac{x_{i:m}-u}{\hat{\tau}}\right)^{-1/\hat{\xi}}, \qquad i = 1,\dots,m.$$

The probability-probability plots for the generalized extreme value and generalized Pareto distribution plot, respectively, the points

$$\left\{\left(\frac{i}{n+1}, \hat{p}_{i:n}\right) : i = 1,\dots,n\right\}, \qquad \left\{\left(\frac{i}{m+1}, \tilde{p}_{i:m}\right) : i = 1,\dots,m\right\},$$

and, similar to the quantile-quantile plot, points lying around the $y = x$ line indicate a good fit.

The following R code produces a probability-probability plot for the generalized Pareto distribution:

```
> ppgpd <- function(data, u, tau, xi){
+ excess <- data[data > u]
+ m <- length(excess)
+ emp.prob <- 1:m / (m + 1)
+ prob.hat <- 1 - (1 + xi * (sort(excess) - u) / tau)^(-1/xi)
+ plot(emp.prob, prob.hat, xlab = "Empirical probabilities",
+      ylab = "Fitted probabilities", xlim = c(0, 1),
+      ylim = c(0, 1))
+ abline(0, 1, col = "grey")
```

```
+ }

> ppgpd(danishuni[,2], 10, 7, 0.5)
```

The right panel of Figure 6.9 shows the probability-probability plot for the fitted generalized Pareto distribution. Similar to the quantile-quantile plot, most of the points lie around the $y = x$ line, indicating a good fit.

6.4.3 Return Level Plot

In Section 6.3.4 we introduced the notion of return levels and return periods. These two concepts can be used to check the goodness of fit of our model. Note that return level plots are not only restricted to model checking.

The use of return level plots for model checking is to plot the fitted return level curve as in Figure 6.8 and compare it to the data. In other words, it is a kind of mix between the two previous tools because the x-axis plots (some kind) of probability, as in the probability-probability plot while the y-axis plots quantiles as in the quantile-quantile plot.

First, based on our fitted model, we produce a return level plot using (6.14) or (6.15). The second step involves adding the observations to this plot. As in Section 6.4.1, for each ordered observation $z_{1:n} < z_{2:n} < \cdots < z_{n:n}$, we associate the corresponding empirical return period based on (6.18). For the generalized extreme value distribution, and because $T = 1/(1-p)$, the empirical return periods are

$$\frac{n+1}{n}, \frac{n+1}{n-1}, \ldots, \frac{n+1}{1},$$

and the following points are added to the return level plot,

$$\left\{ \left(\frac{n+1}{n+1-i}, z_{i:n} \right) : i = 1, \ldots, n \right\}.$$

For the generalized Pareto model, it is a little bit more complicated because the return period depends on the probability of observing an exceedance $\zeta_u = \mathbb{P}(X > u)$ and the number of observation per block, for example, year, n_b. Because, in that case, $T = 1/\{n_b\zeta_u(1-p)\}$, the empirical return periods are

$$\frac{1}{n_b\zeta_u}\frac{m+1}{m}, \frac{1}{n_b\zeta_u}\frac{m+1}{(m-1)}, \ldots, \frac{1}{n_b\zeta_u}\frac{m+1}{1},$$

and the following points are added to the return level plot,

$$\left\{ \left(\frac{1}{n_b\zeta_u}\frac{m+1}{m+1-i}, x_{i:m} \right) : i = 1, \ldots, m \right\}.$$

Remark 6.3 *There is no contradiction in saying that in such cases, $T = 1/\{n_b\zeta_u(1-p)\}$ instead of $T = 1/\{n_b(1-p)\}$ as in Section 6.3.4. While the former uses the conditional non-exceedance probability $p = \mathbb{P}(X < x_p \mid X > u)$, the latter is unconditional, that is, $p = \mathbb{P}(X < x_p)$.*

The following R code produces return level plots for the generalized Pareto model:

```
> rlgpd <- function(data, u, tau, xi, nb){
+ excess <- data[data > u]
+ n <- length(data)
```

```
+ m <- length(excess)
+ zeta.u <- m / n
+
+ rl <- function(T){
+   prob <- 1 - 1 / (nb * T)
+   u + tau / xi * (((1 - prob) / zeta.u)^-xi - 1)
+ }
+
+ plot(rl, from = 1 / nb + 0.1, to = 100, log = "x",
+      xlab = "Return period", ylab = "Return level")
+
+ points((m+1) / (nb * zeta.u * m:1), sort(excess))
+ }
```

Because the Danish dataset has irregularly spaced observations, it is necessary to appropriately set n_b. We use $n_b = 365 \times 2,167/4,015 = 197$ as there are 2,167 claims, 4,015 consecutive days of observations, and we want return levels on an annual scale. The corresponding return level plot is obtained by invoking

```
> rlgpd(danishuni[,2], 10, 7, 0.5, 197)
```

Figure 6.10 plots two return level plots for the Danish claim dataset based on exceedances. Apart from the three largest observations, all the points lie around the fitted return level curve. For pedagogical reasons, we artificially set the largest claim to 500 million Danish Kroner in the right panel. As expected, the largest claim lies far away from the fitted return level curve. But because it is based on ranks, the associated empirical return period remains the same. Although extreme value theory aims at modeling the most severe events, this example is illuminating because it says that one should not focus too much on the largest observations in a return level plot because these observations are associated with the largest uncertainties. Further, it is heavily recommended to add confidence intervals to return level plots either using bootstrap or the delta method on the fitted quantiles.

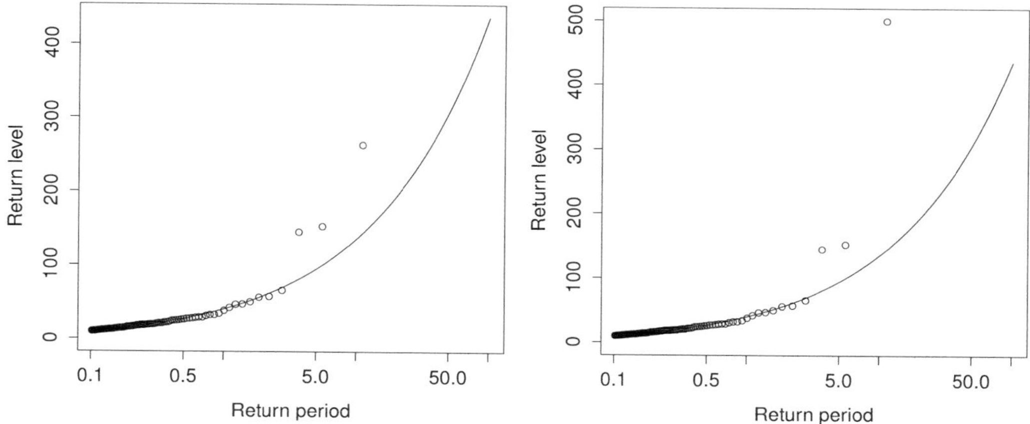

FIGURE 6.10
Checking the goodness of fit of the generalized Pareto distribution to the Danish claim dataset using return level plots. The left panel plots the points defined by (6.18) while the right panel still uses (6.18) but the largest claim has been artificially set to 500 millions Danish Kroner.

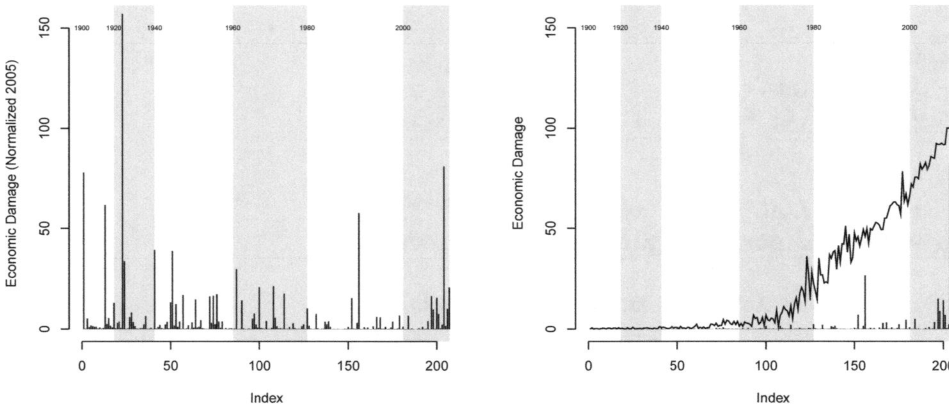

FIGURE 6.11
Economic losses on U.S. mainland, due to tropical cyclones, actual versus normalized losses.

6.5 Reinsurance Pricing

Consider observations, over k years. In year i, n_i losses were observed. For convenience, let $X_{i,1}, \cdots, X_{i,n_i}$ denote reported losses, for year (of occurrence) i. One can imagine some inflation index I_i, so that we have to replace original observations with adjusted (or normalized) losses $I_i^{-1} X_{i,k}$'s. For instance, in Pielke et al. (2008), to adjust losses related to tropical cyclones (U.S. mainland damage), consider the I_i^{-1} function of the GNP inflation index, a wealth factor index, and an index to take into account coastal county population change. This index I_i can be visualized on Figure 6.11, with actual economic losses, on the left, and normalized losses, on the right.

```
> plot(base$Base.Economic.Damage/1e9,type="h",
+ ylab="Economic Damage",ylim=c(0,155))
> lines(base$Base.Economic.Damage/base$Normalized.PL05*100,lwd=2)

> plot(base$Normalized.PL05/1e9,type="h",
+ ylab="Economic Damage (Normalized 2005)",ylim=c(0,155))
```

For damages due to tropic cyclones, we consider an adjustment factor function of inflation, wealth, and population increase. For aviation losses, it is common to take into account fleet value, and passenger per kilometers flown.

Consider here a standard nonproportional reinsurance contract, namely an excess of loss treaty C xs D with deductible D and an upper limit C, in excess of the deductible. The upper limit is then $C + D$. For the reinsurance company, the cost of a claim with economic loss X will be $X_R = (X - D)_+ + (X - [C + D])_+$, while for the (direct) insurance company, the cost is $X_D = X - X_R$. For convenience, let $L_{D,C}(x)$ denote the loss function, from the reinsurer's point of view,

$$L_{D,C}(x) = (x - D)_+ + (x - [C + D])_+.$$

Some specific contracts can be found to deal with aggregate losses, the so-called stop-loss treaties. For year i, one can consider either

$$S_{D,C\ (XL)} = \sum_{k=1}^{N_i} L_{D,C}(X_{i,k}),$$

where reinsurance is used for all contracts, or

$$S_{AD,AC\ (AXS)} = L_{AD,AC}\left(\sum_{k=1}^{N_i} X_{i,k}\right),$$

where an excess of loss treaty is considered, on the overall loss, with aggregate deductible AD and aggregate cover AC. A reinsurance treaty is said to have k reinstatements if $AC = (k+1)\cdot C$.

6.5.1 Modeling Occurence and Frequency

In this section and the next, we will use an approach similar to the one used in Schmock (1999). Consider the number of tropical storms per year. Using `table()` , we will get counts only for years that appear in the dataset (which experienced a tropical cyclone). So we need to add a `0` to the years that did not appear in the `names` of the `table` object:

```
> TB <- table(base$Year)
> years <- as.numeric(names(TB))
> counts <- as.numeric(TB)
> years0=(1900:2005)[which(!(1900:2005)%in%years)]
> db <- data.frame(years=c(years,years0),
+ counts=c(counts,rep(0,length(years0))))
> db[88:93,]
   years counts
88  2003      3
89  2004      6
90  2005      6
91  1902      0
92  1905      0
93  1907      0
```

A natural idea can be to assume that $N_i \sim \mathcal{P}(\lambda)$, with a constant intensity λ,

```
> mean(db$counts)
[1] 1.95283
```

One can also consider that the intensity cannot be considered constant with time if we look at Figure 6.12. Two models will be considered here. A first one will be obtained when λ is assumed to be affine in i:

```
> reg0 <- glm(counts~years,data=db,family=poisson(link="identity"),
+ start=lm(counts~years,data=db)$coefficients)
```

(Observe that we have to specify starting values for the algorithm to ensure that the numerical algorithms converge; see Chapter 14 for a discussion). An alternative can be to consider a standard Poisson regression, with a logarithm link function, so that λ will be exponentially growing (or decaying) with i:

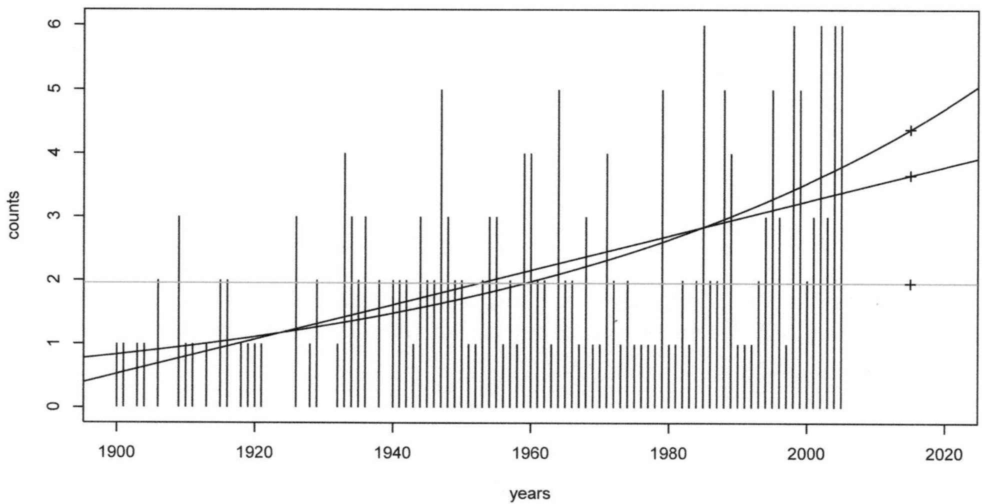

FIGURE 6.12
Number of tropical cyclones per year with three models.

```
> reg1 <- glm(counts~years,data=db,family=poisson(link="log"))
```

Those three models can be compared on Figure 6.12:

```
> plot(years,counts,type='h',ylim=c(0,6),xlim=c(1900,2020))
> cpred1=predict(reg1,newdata=data.frame(years=1890:2030),type="response")
> lines(1890:2030,cpred1,lwd=2)
> cpred0=predict(reg0,newdata=data.frame(years=1890:2030),type="response")
> lines(1890:2030,cpred0,lty=2)
> abline(h=mean(db$counts),col="grey")
```

If we want to price a reinsurance contract for 2015, we need a prediction of that specific year:

```
> cbind(constant=mean(db$counts),linear=cpred0[126],exponential=cpred1[126])
    constant   linear exponential
126  1.95283 3.573999    4.379822
```

Observe that other datasets can be used to model the number of landfalling tropical storms, such as MacAdie et al. (2009).

6.5.2 Modeling Individual Losses

Two adjustments were considered in Pielke et al. (2008):

```
> hill(base$Normalized.PL05)
> hill(base$Normalized.CL05)
```

Assume that a company has a 5% market share, so that we can assume that if a loss is X, the insurance company will pay $X/20$.

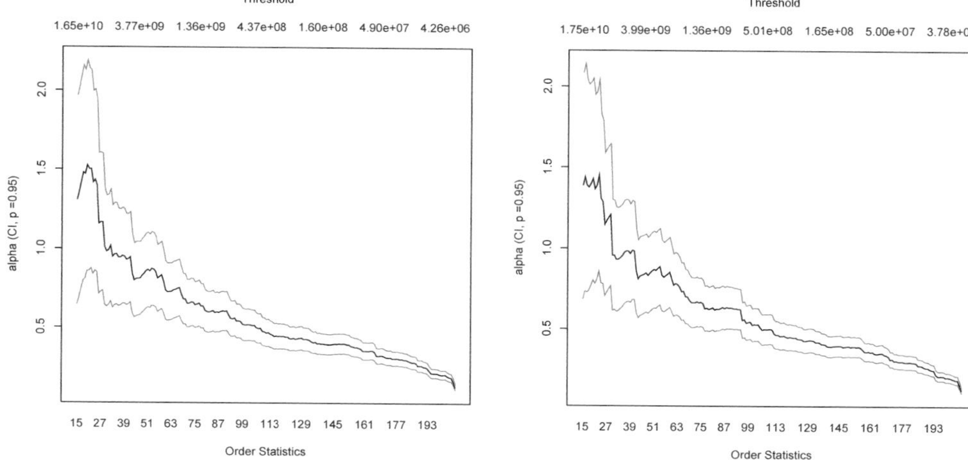

FIGURE 6.13
Hill plot of Economic losses on U.S. mainland, due to tropical cyclones.

Consider a threshold of $c_0 = 0.5$, so that above that threshold, we assume that losses have a GP distribution. Note that 12.5% of the storms exceed that threshold (the same percentage using `Normalized.CL05` and `Normalized.PL05`):

```
> threshold=.5
> mean(base$Normalized.CL05/1e9/20>.5)
[1] 0.1256039
```

Above c_0, parameters of the GP distribution are

```
> (gpd.CL <- gpd(base$Normalized.CL05/1e9/20,threshold)$par.ests)
       xi      beta
0.3634010 0.7122717
> (gpd.PL <- gpd(base$Normalized.PL05/1e9/20,threshold)$par.ests)
       xi      beta
0.4424669 0.6705315
```

Then we should compute

$$\mathbb{E}(L_{D,C}(X)|X > c_0) = \int_D^{D+C} L_{D,C}(x)dG(x) + (D+C)(1 - G(D+C)),$$

where G denotes the Pareto distribution.

```
> E <- function(yinf,ysup,xi,beta){
+  as.numeric(integrate(function(x) (x-yinf)*dgpd(x,xi,mu=threshold,beta),
+  lowe=yinf,upper=ysup)$value+
+  (1-pgpd(ysup,xi,mu=threshold,beta))*(ysup-yinf))
+  }
> E(2,6,gpd.PL[1],gpd.PL[2])
[1] 0.3309865
> E(2,6,gpd.CL[1],gpd.CL[2])
[1] 0.3058124
```

The pure premium is then

$$\mathbb{E}(N) \cdot \mathbb{P}(X > c_0) \cdot \mathbb{E}(L_{D,C}(X)|X > c_0),$$

where the first term is the expected number of landfall storms, the second term is the probability that the loss exceeds the threshold, and the third term is the expected reinbursment of a reinsurance treaty, given that threshold c_0 has been exceeded. Here, if we consider a linear model for frequency, the pure premium of a nonproportional contract, with deductible 2 and cover 4, will be

```
> cpred1[126]*mean(base$Normalized.CL05/1e9/20>.5)*
+ E(2,6,gpd.CL[1],gpd.CL[2])
       126
0.05789891
```

(in USD billions). A company with 5% market share should be ready to pay a 5.789 million USD premium to have a reinsurance cover, in excess of 2 billion, with a limited cover of 4 billion.

Part II

Life Insurance

7

Life Contingencies

Giorgio Spedicato
UnipolSai Assicurazioni
Bologna, Italy

CONTENTS

7.1 Introduction

This chapter aims to show how standard financial and actuarial mathematics calculations for life-contingent risks can be easily performed within the R statistical software and `lifecontingencies` package, Spedicato (2013a). While theoretical concepts will be briefly recalled during the exposition, the interested reader could find deeper details within standard actuarial references such as Finan (n.d.), Dickson et al. (2009), Bowers et al. (1997), and Ruckman & Francis (n.d.).

The chapter will be organized as follows: Section 7.2 will deal with financial mathematics applications, Section 7.3 will show life table analysis and applications, Section 7.4 will discuss pricing life insurance with R, while Section 7.5 will deal with life insurance reserves. More advanced topics will be discussed in Section 7.6. Finally, Section 7.8 will propose selected exercises for each previous section, whose solutions are reported in the demo files of the `lifecontingencies` package.

7.2 Financial Mathematics Review

Financial mathematics deals with the time value of the money. An amount X after t years will become $A(t) = X \cdot (1+i)^t$ if it follows the law of compound interest. i represents the effective rate of interest, that is, the amount of money that one unit invested at the beginning of a period will earn during that period, with interest being paid at the end of the period. Similarly, if the amount i had been earned at the beginning of the period, it would have been valued at $\frac{i}{1+i} = d$, d being the discount rate. A naive example follows: $100 grows to

```
> 100 * ( 1 + 0.05 ) ^ 2
[1] 110.25
```

After 2 years, using a 5% interest rate, the discount rate would be

```
> 0.05 / ( 1 + 0.05 )
[1] 0.04761905
> interest2Discount( i = 0.05 )
[1] 0.04761905
```

which yields (for this 2-year contract)

```
> 110.25 * ( 1 - interest2Discount(i=0.05) ) ^ 2
[1] 100
```

Interest can be paid (reinvested) more frequently than once per period, being the fractions of a period called interest conversion periods. Often, the interest rate is expressed in nominal (also named "convertible") terms, $i^{(m)}$, that is, the effective interest rate earned in the conversion period multiplied by m. A similar relationship exists for the discount rate; thus Equation (7.1) holds.

$$A(t) = (1+i)^t = \left(1 + \frac{i^{(m)}}{m}\right)^t = \left(1 - \frac{d^{(m)}}{m}\right)^{-t} = (1-d)^{-t} = v^{-t} \tag{7.1}$$

The `nominal2Real()` function computes the real interest rate given the nominal one:

```
> nominal2Real(i=0.06, k=12)
[1] 0.06167781
```

Rates of discount, for example, $d^{(m)} \to d$, can be given as input too, as the following code displays

```
> real2Nominal(i=0.04, k=12, type="discount")
[1] 0.04075264
```

Equation (7.1) can be used to perform basic financial mathematics calculations, for which functions within the `lifecontingencies` package help.

Example 7.1 *Find the present value of $1,000 to be paid at the end of 6 years at 6% per year payable in advance and convertible semiannually.*

```
> 1000 * (1 - 0.06 / 2) ^ 12
[1] 693.8424
```

or equivalently,

```
> annualDiscount = nominal2Real(i=0.06, k=2, type="discount")
> i = discount2Interest(annualDiscount)
> presentValue(cashFlows=1000, timeIds=6, interestRates=i, probabilities=1)
[1] 693.8424
```

The Net Present Value (NPV) is a milestone concept within financial mathematics. It is the value in current money of a series of cash flows that are available at different points in time. It allows to value on a common basis different streams of cash flows. This means, in other words, ranking different investments according to their expected profit.

$$NPV = \sum_{j=1}^{K} CF_j (1 + i_j)^{-t_j} \tag{7.2}$$

Equation (7.2) shows its definition, while examples that follow display how the `lifecontingencies` package helps in calculating NPVs.

Example 7.2 *Which investment is preferred between:*

1. *Invest \$500 now and receive \$100, 200, 300, 250 at the end of the following four years.*
2. *Invest \$700 now and receive \$1000 at the end of time 2.*

```
> # example 2.1
> cfs1 = c(-500,100,200,300,250)
> times1 = 0:4
> NPV1 = presentValue(cashFlows=cfs1, t=times1, i=0.05)
> NPV1
[1] 241.4709
> # example 2.2
> cfs2 = c(-700,1000)
> times2 = c(0,2)
> NPV2 = presentValue(cashFlows=cfs2, t=times2, i=0.05)
> NPV2
[1] 207.0295
```

Thus, the first one is preferred here.

The inverse function of NPV is the Internal Rate of Return (IRR). The following example shows how to calculate IRR using R and the `lifecontingencies` package's functions.

Consider the following function whose minimum represents IRR:

```
> f<-function(irr)  (presentValue(cashFlows=CF,
+         timeIds=TI,
+         interestRates=irr))^2

> CF = cfs1
> TI = times1
> IRR1 = nlm(f, 0.01)$estimate
> IRR1
[1] 0.2147816
```

This means that with a 21.478% rate of return, the discounted value of the cash flow is null.

```
> presentValue(cashFlows=cfs1, t=times1, i=IRR1)
[1] 0.0005410981
```

For the second investment, IRR is

```
> CF <- cfs2
> TI <- times2
> IRR2 <- nlm(f,0.01)$estimate
> IRR2
[1] 0.1952281
```

Financial mathematics developed formulas to price annuities that are a series of payments made at equal intervals of time. If payments are guaranteed to occur for a fixed period of time, the annuity is called "annuity-certain." If payments are made at the end of each period for n periods, we have an "annuity-immediate or ordinary annuity. Otherwise, if payments are performed at the beginning of each period, we have a an annuity-due. In addition, fractional payments can be performed within each period or for an infinite term (see Bowers et al. (1997) or Dickson et al. (2009) for further details).

As a remark, Equation (7.3) of an annuity-immediate of n-years term is

$$a_{\overline{n}|} = \frac{1-(1+i)^{-n}}{i} \tag{7.3}$$

Example 7.3 *Price the following annuities:*

> 1. *Calculate the present value of an annuity-immediate of amount $100 paid annually for 5 years at a rate of interest of 9%.*
>
> 2. *Suppose a company issues a stock that pays a dividend at the end of each year of 10 indefinitively, and the company cost of capital is 6%. What would be the value of the stock at the beginning of the year?*
>
> 3. *What amount must you invest today at 6% interest rate compounded annually so that you can withdraw $5,000 at the beginning of each year for the next 5 years?*

For Example 3.1., the present value is

```
> i <- 0.09
> n <- 5
> 100*(1-(1+i)^(-n))/i
[1] 388.9651
```

which can be obtained using the `annuity` function,

```
> #example 3.1
> 100 * annuity(i=0.09, n=5, type="immediate")
[1] 388.9651
> #example 3.2
> 10 * annuity(i=0.06, n=Inf)
[1] 166.6667
> #example 3.3
> 5000 * annuity(i=0.06, n=5, type="due")
[1] 22325.53
```

Similarly, the accumulated value is computed by capitalizing the corresponding annuity. Equation (7.4) gives the accumulated value equation and symbol.

$$s_{\overline{n}|} = (1+i)^n a_{\overline{n}|} = \frac{(1+i)^n - 1}{i} \tag{7.4}$$

Example 7.4 *What amount will accumulate if we deposit \$5,000 at the beginning of each year for the next 5 years? Assume an interest of 6% compounded annually.*

From the relationship given above, one gets

```
> annuity(i=0.06, n=5, type="due")*5000*1.06^5
[1] 29876.59
```

or we can use the `accumulatedValue` function

```
> 5000 * accumulatedValue(i=0.06, n=5,type="due")
[1] 29876.59
```

Accumulated values are often used to perform financial planning for individuals, as the following example shows.

Example 7.5 *A man wants to save \$100,000 to pay for the education of his son in 10 years time. An education fund requires investors to deposit equal installments annually at the end of each year. If the guaranteed interest rate is 5%, how much does the man need to save each year in order to meet his target?*

```
> C <- 100000
> R <- C/accumulatedValue(i=0.05,n=10)
> R
[1] 7950.457
```

It is possible to define series of payments where amounts paid in each period are not constant. All such cash flows can be evaluated using the NPV Equation (7.2). However, closed formulas have been derived for some such sequences. The `lifecontingencies` package contains functions to evaluate the present value of arithmetically increasing, $(IA)_n$, or decreasing, $(DA)_n$, annuities, as shown by the following examples.

Example 7.6

1. *The following payments are to be received: \$500 at the end of the first year, \$520 at the end of the second year, \$540 at the end of the third years and so on, until the final payment is \$800. Using an annual effective interest rate of 2%, determine the present value of these payments at time 0.*
2. *Determine the accumulated value of these payments at the time of the last payment.*

In this example, we can either consider a fixed annuity of \$480, plus an arithmetical increase of \$20 (since the first payment of the increasing component will start with \$20, and not \$0),

```
> # example 6.1
> 480*annuity(i=0.02,n=16)+ 20*increasingAnnuity(i=0.02,n=16)
[1] 8711.431
```

For the accumulated value, one can use

```
> # example 6.2
> (480*annuity(i=0.02,n=16)+ 20*increasingAnnuity(i=0.02,n=16))*1.02^16
[1] 11958.93
```

or

```
> 480*accumulatedValue(i=0.02,n=16)+ 20*Isn(i=0.02,n=16)
[1] 11958.93
```

Similarly, the present value of annuities that grow following a geometric progression can be easily computed. If the payment grows by a rate of r and the interest valuation rate is i, the annuity can be calculated using a synthetic rate $j = \left(\frac{1+i}{1+r}\right) - 1$.

Example 7.7 *An annual annuity due pays \$1 at the beginning of the first year. Each subsequent payment is 5% greater than the preceding payment. The last payment is at the beginning of the tenth year. Calculate the present value using an annual effective interest rate of 4%.*

```
> j=(1+0.04)/(1+0.05)-1
> 1 * annuity(i=j,n=10,type="due")
[1] 10.44398
```

Annuities are said to be deferred if their inception date is different from the moment when they are being evaluated. The `lifecontingencies` package easily handles deferment.

Example 7.8 *Calculate the present value of an annuity due that pays annual amounts of \$1200 for 12 years with the first payment 2 years from now. The annual effective interest rate is 6%.*

```
> 1200*annuity(i=0.06,n=12,m=1,type="immediate")
[1] 9491.144
> 1200*annuity(i=0.06,n=12,m=2,type="due")
[1] 9491.144
```

The basic concepts presented up to now are used to perform financial calculations used in real business environments. One example is represented by bond pricing, which the following example shows.

Example 7.9 *A 5-year bond offers 3% coupons paid twice a year. The par value is \$100. Find its price using a 5% annual effective interest rate.*

A first strategy is to compute

```
> isemestral <- ( 1 + 0.05 )^0.5 - 1
> 100 * (1+0.05)^-5 + (100*0.03/2) * annuity(i=isemestral, n=10)
[1] 91.50142
```

but it is also possible to use

```
> 100 * (1+0.05)^-5 + 3 * annuity(i=0.05, n=5, k=2)
[1] 91.50142
```

A loan amortization program can be easily set up with `lifecontingencies` package help, as the following example shows.

Example 7.10 *A \$10,000 loan is stipulated with 7% APR and monthly installments for repayment. Find the montly installment if the loan is planned to be repaid in 5 years.*

```
> R <- 10000/annuity(i=0.07,n=5,k=12)/12
> instalments <- rep(R,5*12)
> balance <- numeric(60)
> balance[1] <- 10000*(1+0.07)^(1/12)-R
> for(i in 2:60) balance[i] <- balance[i-1]*(1+0.07)^(1/12)-R
> amortization <- data.frame(month=1:60, rate=round(R,1),
+ balance <- round(balance,1))
> head(amortization)
  month rate balance
1     1  197  9859.5
2     2  197  9718.3
3     3  197  9576.2
4     4  197  9433.4
5     5  197  9289.7
6     6  197  9145.3
> tail(amortization)
   month rate balance
55    55  197   968.5
56    56  197   777.0
57    57  197   584.4
58    58  197   390.7
59    59  197   195.9
60    60  197     0.0
```

Finally, the `duration` and `convexity` functions can be used to calculate the duration and convexity of a series of cash flows. We will exemplify the use of such functions with a fictional 5-year term bond paying 3% coupon assuming a 5% yield rate. We will test Equation (7.5), D^* being the effective duration, $\Delta y = y_1 - y_0$ the variation in market yield, and P being the original price of the bond at initial yield y_0.

$$\frac{\Delta P}{P} \approx -D^* \Delta y \tag{7.5}$$

```
> CFs = c(3,3,3,3,103)
> TIs = seq(1,5,1)
> bondPrice = presentValue(cashFlows = CFs,
+ timeIds = TIs,interestRates = 0.05)
> duration(cashFlows = CFs, timeIds=TIs, i=0.05,
+                 macaulay=FALSE)
[1] 4.701744
> bondPrice2=presentValue(cashFlows = CFs,
+ timeIds = TIs, interestRates = 0.04)
> bondPrice2
[1] 95.54818
> bondPrice * (1 + duration(cashFlows = CFs,timeIds=TIs,
+ i=0.05,macaulay=FALSE) * 0.01)
[1] 95.63567
```

7.3 Working with Life Tables

Life insurance and annuities can be seen as the expected value of future cash flows currently valued, that is, actuarial present values (APV). In fact, we use the term *expectation* because APV is the expected value of a random variable that is itself a function of the future lifetime of the policyholder. While the present value of future cash flows is evaluated using the tools presented in Section 7.2, this section discusses how to handle life tables, from which the probabilities underlying actuarial mathematics calculations come.

Life tables consist of a nonincreasing sequence of l_x, $x = 0, 1, \ldots, \omega$ that represents the number of subjects alive at the beginning of age x, ω being the terminal age. The number of subjects at time 0, l_0, is the radix of the table.

Because a table counts survivors, one can compute a survival probability from l_x's. If T_x denotes the (stochastic) remaining lifetime of some individual aged x, the probability that (x) reaches age $x + t$ is

$$ {}_t p_x = \mathbb{P}(T_x > t) = \frac{l_{x+t}}{l_x}, $$

and the probability that (x) does not reach age $x + t$ is

$$ {}_t q_x = 1 - {}_t p_x = \mathbb{P}(T_x \leq t) = \frac{l_x - l_{x+t}}{l_x}. $$

As standard convention, ${}_1 p_x$ will be denoted p_x.

Different ways can be approached to create a life table with the `lifecontingencies` package.

A direct approach is to impute directly x and l_x, as the code below shows:

```
> newLifeTable1 <- new("lifetable", x=seq(0,10,1),
+ lx=seq(from=1000,to=0,by=-100),name="Sample life table 1")
removing NA and 0s
```

or, one can provide 1-year survival/mortality rates (p_x, q_x) and build a life table using convenience functions as the code below shows:

```
> newLifeTable2 <- probs2lifetable(probs=seq(from=0.1,to=1,by=0.1),
+ radix=100000,type="qx",name="Sample life table 2")
```

S3 methods are available, like `head`, `tail`, `plot`, and `summary`.

```
> head(newLifeTable1)
  x   lx
1 0 1000
2 1  900
3 2  800
4 3  700
5 4  600
6 5  500
> summary(newLifeTable1)
This is lifetable:  Sample life table 1
 Omega age is:  10
 Expected curtated lifetime at birth is:  4.5
```

Moreover it is possible to export `lifetable` objects into `data.frame` objects as the code below shows:

```
> lifeTableDf <- as(newLifeTable1, "data.frame")
> class(lifeTableDf)
[1] "data.frame"
```

Demographic analysis can be performed using several functions within the package. The basic probabilities underlying life contingencies insurance are the probability that an insured aged x survives until age $x+t$, that is ${}_tp_x$, and its complement to one, ${}_tq_x$.

Example 7.11 *Using the Society of Actuaries illustrative life table, evaluate*

1. *The probability that an insured aged 65 will die before reaching 85.*
2. *The probability that a policyholder aged 25 will survive until 65.*

The Society of Actuaries (SoA) illustrative life table can be loaded using

```
> data(soa08Act)
```

Because it is a S4 class object (see Chapter 1), elements are obtained using the `@` symbol. For instance, the number of survivors aged 65 will be

```
> soa08Act@lx[soa08Act@x==65]
[1] 7533964
```

To compute the first probability, we can use either

```
> #example 11.1
> (soa08Act@lx[soa08Act@x==65]-soa08Act@lx[soa08Act@x==85])/
+  soa08Act@lx[soa08Act@x==65]
[1] 0.6869847
```

or, more conveniently, functions `pxt` and `qxt`

```
> qxt(soa08Act, 65,20)
[1] 0.6869847
> #example 11.2
> pxt(soa08Act, 25,40)
[1] 0.7876582
```

Similarly, the curtate expectation of life at age x, define as

$$e_x = \sum_{t=1}^{\infty} {}_tp_x \text{ and } e_{\overline{x:n}|} = \sum_{t=1}^{n} {}_tp_x$$

(expected life span between ages x and $x+n$) can be easily computed as shown in the code below.

Example 7.12 *Using SoA illustrative life table, calculate*

1. *The curtate expectation of birth embedded in the life table.*
2. *The curtate expectation of life between 50 and 60 on the same life table.*

For the (complete) curtate expectation of life, at birth, we just have to sum all the ${}_tp_0$'s, or equivalently, all the l_x's except the first one, and divide it by l_0

```
> #example 12.1
> sum(soa08Act@lx[soa08Act@x%in%(1:110)]/soa08Act@lx[soa08Act@x==0])
[1] 71.34692
> sum(soa08Act@lx/soa08Act@lx[soa08Act@x==0])-1
[1] 71.34692
```

or use the `exn` function

```
> exn(object = soa08Act)
[1] 71.34692
```

For the expected life span between ages 50 and 60, we can obtain it also by summing some l_x's:

```
> #example 12.2
> sum(soa08Act@lx[soa08Act@x%in%(51:60)]/soa08Act@lx[soa08Act@x==50])
[1] 9.583979
```

or using

```
> exn(object = soa08Act, x=50,n=60-50,type="curtate")
[1] 9.583979
```

The complete expectation of life between age x and $x+n$ can be expressed as

$$\mathring{e}_x = \int_{t=0}^{\infty} {}_tp_x \text{ and } \mathring{e}_{\overline{x:n}|} = \int_{t=0}^{n} {}_tp_x$$

or

$$\mathring{e}_{\overline{x:n}|} = \frac{{}_tL_x}{l_x}, \text{ where } {}_tL_x = \int_x^{x+t} l_s ds.$$

Example 7.13 *Evaluate the complete expectation of life between ages 80 and 90.*

The natural idea is to assume that when someone dies at age x, the death will occur at "age" $x+1/2$. Thus, the complete expectation of life will be, here

```
> (sum(soa08Act@lx[soa08Act@x%in%(81:90)]/soa08Act@lx[soa08Act@x==80])+
+ sum(soa08Act@lx[soa08Act@x%in%(80:89)]/soa08Act@lx[soa08Act@x==80]))/2
[1] 6.136299
```

or using the `exn` function

```
> exn(soa08Act,80,10,"complete")
[1] 6.136299
```

Another quantity relevant for demographers is the central mortality rate,

$${}_tm_x = \frac{{}_td_x}{{}_tL_x} \text{ where } {}_td_x = {}_tL_x - {}_tL_{x+1}.$$

Example 7.14 *Evaluate ${}_2m_8$ on the SoA illustrative life table.*

```
> mxt(soa08Act,80,2)
[1] 0.0874816
```

`lifecontingencies` package allows the analyst to evaluate survival probabilities at fractional intervals, using different kinds of interpolation. For instance, with a linear interpolation, ${}_{[h]}p_x$ et ${}_{[h]+1}p_x$ (where $[h]$ is the integer part of $h \geq 0$),

$$ {}_h\tilde{p}_x = (1 - h + [h])\ {}_{[h]}p_x + (h - [h])\ {}_{[h]+1}p_x . $$

For a constant force of mortality, recall that

$$ {}_hp_x = \exp\left(- \int_0^h \mu_{x+s} ds \right), \text{ where } \mu_x \text{ is the force of mortality.} $$

Assume that $h \in [0,1)$, and that $s \mapsto \mu_{x+s}$ is constant on $[0,1)$; then

$$ {}_h\tilde{p}_x = \exp\left(- \int_0^h \mu_{x+s} ds \right) = \exp[-\mu_x \cdot h] = (p_x)^h . $$

Finally, if we still assume that $h \in [0,1))$, another alternative is to assume that

$$ \frac{1}{{}_hp_x} = \frac{1 - h + [h]}{{}_{[h]}p_x} + \frac{h - [h]}{{}_{[h]+1}p_x}. $$

or equivalently, we define

$$ {}_h\tilde{p}_x = \frac{{}_{[h]+1}p_x}{1 - (1 - h + [h])\ {}_{[h+1]h}q_x}. $$

Example 7.15 *Evaluate the probability that a policyholder aged 80 1/4 will die within half a year,assuming:*

1. *Linear mortality interpolation.*
2. *Constant force of mortality behavior.*

```
> #example 15.1
> pxt(object=soa08Act,x=80.25,t=0.5, fractional="linear")
[1] 0.959027
> #example 15.2
> pxt(object=soa08Act,x=80.25,t=0.5, fractional="constant force")
[1] 0.959027
```

In addition, the `lifecontingencies` package allows one to work with multiple lifes. Consider two policyholders and their future lifetime random variables T_x, T_y. The joint life status is defined as $T_{xy} = \min(T_x, T_y)$, while the last survivor life status is defined as $T_{\overline{xy}} = \max(T_x, T_y)$. The joint survival probability with time horizon t is

$$ \mathbb{P}(T_x > t \text{ and } T_y > t) = \mathbb{P}(T_{xy} > t) \text{ denoted } {}_tp_{xy} $$

while the last survivor probability with time horizon t is

$$ \mathbb{P}(T_x > t \text{ or } T_y > t) = \mathbb{P}(T_{\overline{xy}}) > t) \text{ denoted } {}_tp_{\overline{xy}}. $$

Assuming independent lifetime yields, ${}_tp_{xy} = {}_tp_x \cdot {}_tp_y$ and ${}_tp_{\overline{xy}} = {}_tp_x + {}_tp_y - {}_tp_{xy}$.

The example shown below displays how to perform demographic analysis with multiple heads.

Example 7.16 *Assume the SoA illustrative life table applies to two policyholders aged 65 and 65, respectively. Find*

1. *The probability that the two policyholders will be both alive after 20 years.*
2. *The probability that at least one of them will be alive after 20 years.*
3. *The expected joint life period.*

Consider here the same life table, for the husband and his wife (but different life tables can be considered).

```
> maleTable <- soa08Act; femaleTable <- soa08Act;
> tables <- list(male=maleTable,female=femaleTable)
```

For the joint probability, if we assume independent lifetimes, then the probability is

```
> #example 16.1
> pxt(tables$male,x=65,t=20)*pxt(tables$female,x=60,t=20)
[1] 0.1496391
```

but one can use the pxyzt function

```
> pxyzt(tablesList=tables, x=c(65,60),t=20,status="joint")
[1] 0.1496391
> #example 16.2
> pxyzt(tablesList=tables, x=c(65,60),t=20,status="last")
[1] 0.6414331
> #example 16.3
> exyzt(tablesList=tables, x=c(65,60),t=20,status="joint")
[1] 11.08887
```

For these two individuals, it is possible to plot survival probabilities; see Figure 7.1.

```
> probjoint <- function(t){pxyzt(tablesList=tables, x=c(65,60),t,status="joint")}
> problast <- function(t){pxyzt(tablesList=tables, x=c(65,60),t=20,status="last")}
> vecT <- seq(0,45)
> plot(vecT,Vectorize(probjoint)(vecT),type="l",lty=2,ylab="Survival probability",xlab="")
> lines(vecT,Vectorize(problast)(vecT),col="grey",lwd=2)
```

7.4 Pricing Life Insurance

Life-contingent insurances are contracts that promise one or more payments upon occurrence of a life-contingent event. For example,

1. Life insurance contracts promise payments of a lump sum in case the insured dies within the period defined in the contract.
2. Annuity contracts pay a sum of money at the beginning (or the end) of each period, until the the term of the contract or the insured's death, whichever occurs first.
3. Endowment contracts pay a sum at the minimum between insured's death or the term of the contract.

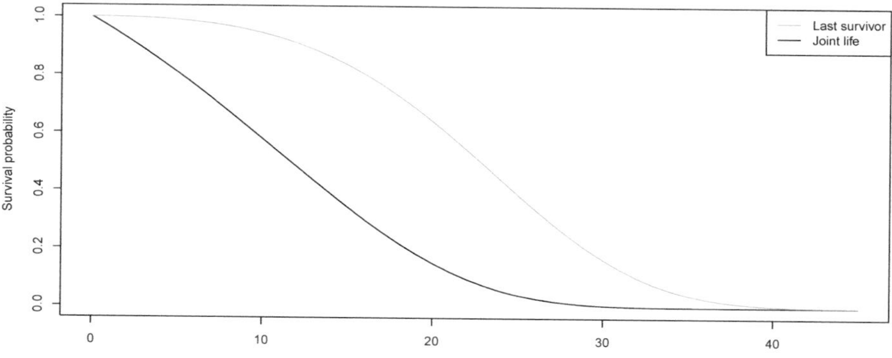

FIGURE 7.1
Last survivor and joint life survival probability, when $x = 65$ and $y = 60$.

Pricing life insurance contracts is performed in several steps:

1. Define the financial and demographical assumptions, that is, the interest rate that will be used to discount future cash flows as well as the life table to estimate survival probabilities.
2. Determine the actuarial present value (APV) of life-contingent cash flows, that is, the actual price of the contractually promised benefits and eventual expense and profit loads.
3. Determine the premium(s), taking into account that it can be paid as a lump sum or within periodic installments.

Several examples will better explain such concepts. From now on, the following notation will be used in this chapter as well as the `lifecontingencies` package:

- x will represent insured's age at the inception of the contract.
- n will represent the term of the contract.
- m will represent the deferring period, that is, broadly speaking, the period since the benefits coverage would start (default value would be $m = 0$).
- i will represent the interest rate that discounts the contractually granted cash flows.
- k will represent the fractional payments frequency (default value would be $k = 1$).

All examples in this section will be based on the illustrative Society of Actuaries life table at an interest valuation rate of 6%. In addition, we will assume that benefits are paid at the end of each period. We remind interested readers of the cited references for deepening and additional examples.

The standard insurance contract that will be considered here is the n-year term life insurance policy that will pay \$1 at the death of the insured (x) if the death occurs before n years. The expected value of this contract is

$$ {}_nA_x = \sum_{k=1}^{n} \frac{\mathbb{P}(T_x \in [k-1,k))}{(1+i)^k} = \sum_{k=1}^{n} \frac{1}{(1+i)^k} \cdot {}_{k-1}p_x \cdot {}_1q_{x+k-1}. \tag{7.6} $$

Recall that the whole life contract is obtained when n is equal to ω, and the deferred contract when the sum is no longer from 1 to n, but from m to $n+m-1$.

Before computer era, life insurance was evaluated using functions of the insureds' age tabulated in sheets called actuarial tables, the so-called "commutation functions." The `lifecontingencies` package offers functions to generate tables of commutation function.

Example 7.17 *From the SoA illustrative life table,*

1. *Generate the actuarial table object using an interest rate of 6%.*
2. *Export such an object in a dataframe showing corresponding commutation functions.*

```
> data(soaLt)
> #example 1
> soaAct <- new("actuarialtable", x=soaLt$x,
+ lx=soaLt$Ix, interest=0.06)
> #example 2
> soaActDf <- as(soaAct, "data.frame")
> head(soaActDf)

  x       lx       Dx        Nx        Cx       Mx       Rx
1 0 10000000 10000000 168358017 47263.585 470300.9 12487975
2 1  9949901  9386699 158358017 44588.288 423037.4 12017674
3 2  9899801  8810788 148971318 42064.422 378449.1 11594637
4 3  9849702  8270000 140160530 39683.417 336384.6 11216188
5 4  9799602  7762203 131890531 37437.186 296701.2 10879803
6 5  9749503  7285396 124128328  6191.668 259264.0 10583102
```

The first example shows how to price a simple 3-year term life insurance policy. Assume the policyholder's age to be 36 and insured sum to be $100,000. Using `lifecontingencies`' package functions, we can directly compute the APV of the selected insurance, $100{,}000 \cdot A^{1}_{36:\overline{3}|}$.

The probability to actually pay $100,000 is, for each age,

```
> (probdeath <- -diff(soaActDf$lx)[soaActDf$x%in%36:38]/soaActDf$lx[soaActDf$x==36])
[1] 0.002140254 0.002274272 0.002420523
```

As the discount factors are, respectively,

```
> disc <- (1+0.06)^(-(1:3))
```

then the actuarial present value is

```
> sum(disc*probdeath) * 100000
[1] 607.5519
```

using Equation (7.6). Note that probabilities can also be obtained using (e.g., for the second one)

```
> k <- 2
> qxt(soaAct,x=36+k-1,t=1)*pxt(soaAct,x=36,t=k-1)
[1] 0.002274272
```

But all those computations can be done also using the `Axn` function:

```
> P <- 100000 * Axn(actuarialtable=soaAct, x=36, n=3)
> P
[1] 607.5519
```

Otherwise, using the "old" commutation functions, we would have written

$$A^{1}_{36:\overline{3}|} = \frac{M_{36} - M_{36+3}}{D_{36}}.$$

```
> 100000 * with(soaActDf, (Mx[37]-Mx[40])/Dx[37])
[1] 607.5519
```

Another type of insurances consists of increasing or decreasing insurances with benefits payable at the end of the year of death. An n-year increasing term life insurance pays $k+1$ if the curtate future lifetime of the insured, $\tilde{K}_x$, is between $0, 1, \ldots, n-1$. n-year term insurance, n-year increasing and decreasing term insurances are bounded by the relationship expressed in Equation (7.7).

$$(n+1)\, A^{1}_{x:\overline{n}|} = (DA)^{1}_{x:\overline{n}|} + (IA)^{1}_{x:\overline{n}|} \tag{7.7}$$

```
> (10+1) * Axn(soaAct,60,10)
[1] 1.504674
> IAxn(soaAct,60,10) + DAxn(soaAct,60,10)
[1] 1.504674
```

Moreover, it can be assumed that the year can be divided into m subperiods and benefits to be paid on an m^{tly} basis, as the following examples show.

Example 7.18 *Verify (numerically)* ${A^{1}_{x:\overline{n}|}}^{m} = \frac{i}{i^m} A^{1}_{x:\overline{n}|}$.

```
> Axn(actuarialtable=soaAct,x=30,k=12)
[1] 0.1052721
> 0.06/real2Nominal(0.06,12)*Axn(actuarialtable=soaAct,x=30)
[1] 0.1052721
```

A whole life annuity-due is a series of payments made at the beginning of the year while the annuitant is alive. The whole life annuity for insured (x) has the following actuarial present value:

$$a_x = \sum_{k=1}^{\infty} \frac{\mathbb{P}(T_x > k)}{(1+i)^k} = \sum_{k=1}^{\infty} \frac{1}{(1+i)^k} \cdot {}_kp_x$$

The APV of $100 payable at the beginning of each period to a 65-year-old policyholder until death would be

```
> sum((soaActDf$lx)[soaActDf$x%in%66:111]/soaActDf$lx[soaActDf$x==65]*
+ 1.06^(-(1:45)))
[1] 8.896928
```

which can be obtained using the function `axn`:

```
> axn(actuarialtable=soaAct,x=65)
[1] 9.896928
```

while if it were paid at the end of each period, it would be an annuity-immediate.

```
> axn(actuarialtable=soaAct,x=65, m=1)
[1] 8.896928
```

Another annuity example is the n-year temporary life annuity-due, where amounts of \$1 are paid at the end of each year as long as the policyholder survives, for up to a total of n years, or n payments otherwise.

Example 7.19 *Find the APV of \$1000 paid by a policyholder between 25 and 45 years old at the beginning of each period.*

```
> 1000 * axn(actuarialtable=soaAct, x=25, n=45-25)
[1] 12008.43
```

Sometimes annuities are deferred from the inception of the contract. In other words, the payment of the promised amount is deferred to the future and is contingent upon the survival of the subject.

Example 7.20 *Find the APV of a whole life insurance on a policyholder aged 30, if it were deferred 10 years.*

```
> Axn(actuarialtable=soaAct, x=30, m=10)
[1] 0.0882981
```

Handling geometrically increasing benefits is not complicated, as shown by the example that follows.

Example 7.21 *A policyholder aged 50 wants to invest now a sum to purchase a deferred annuity-due with annual payment geometrically increasing by a ratio of 2% per annum. If the initial term of the annuity is 10,000, what is the amount the policyholder will be asked to pay, beyond any expense consideration?*

Initially, we have to find the synthetic interest rate

```
> irate = 1.06 / 1.02 - 1
```

Then we can evaluate the annuity

```
> P=10000 * axn(actuarialtable=soaAct, x=50, i=irate)
> P
[1] 164275.2
```

Often, life annuities are paid m times a year. For example, a m-thly life annuity-due makes a payment of $\frac{1}{m}$ at the beginning of every m-thly period so that in 1 year the total payment is \$1.

Example 7.22 *Find the APV of an annuity due if the annuitant's age is 60, assuming a fractional payment at the beginning of each month.*

```
> axn(actuarialtable=soaAct, x=60, k=12)
[1] 10.68036
```

The `lifecontingencies` package easily handles fractional payments by determining the fractional payments cash flow pattern, their probabilities, and discounting them. Approximation formulas have been developed, for example, $\ddot{a}_x^{(m)} \approx \ddot{a}_x - \frac{m-1}{2m}$, but the algorithm within the package does not make use of them.

We have assumed, until now, the policyholder to want a policy with a single payment at the time the policy begins, being charged therefore the APV (also known as the net single premium) at the time of issuance. But, more frequently, the premium is paid in several installments, as following examples show.

Example 7.23 *Find the benefit premium for a whole life insurance of a policyholder aged 25 paid at the beginning of the first 10 years the insured is alive,* ${}_{10}P(A_{25}) = \frac{A_{25}}{\ddot{a}_{25:\overline{10|}}}$.

```
> P <- Axn(soa08Act,25)/axn(soa08Act,25,10)
> P
[1] 0.01052354
```

Modify the policy terms to take into account that deferring is easy.

Example 7.24 *Calculate* $P({}_{20|}\ddot{A}_{50}) = \dfrac{{}_{20|}\ddot{A}_{50}}{\ddot{a}_{50}}$.

```
> P <- Axn(soa08Act,x=50,m=20)/axn(soa08Act,x=50)
> P
[1] 0.008945766
```

A more complex benefit premium computation is shown by the following example.

Example 7.25 *For a special, fully discrete 5-year deferred whole life insurance of $100,000 on a policyholder aged 40, it is assumed that the death benefit during the 5-year deferral period is the return of benefit premiums paid without interest and that the annual benefit premiums are payable only during the deferral period. Calculate the 5-year benefit premium. As a hint, the benefit premium, π, solves*

$$\pi \ddot{a}_{40:\overline{5|}} = \pi \cdot (IA)^{\,1}_{40:\overline{5|}} + 100000 \cdot {}_{5|}A_{40} \qquad (7.8)$$

```
> pi <- 100000 * Axn(actuarialtable=soa08Act,x=40,m=5) /
+ (axn(actuarialtable=soa08Act,x=40,n=5)-
+ IAxn(actuarialtable=soa08Act,x=40,m=5))
```

We remind the interested reader of the cited literature to deepen the topic.

7.5 Reserving Life Insurances

The "benefit reserve" of an insurance policy consists of the amount of money the insurance company must have saved up to be able to provide for the future benefits of the policy, Finan (n.d.). Benefit reserves at time t are estimated using two methods, both assuming that the policyholder has survived at time t:

1. Under the prospective method, the benefit reserve is computed as the difference between the APV of future coverage benefits and the APV of future benefit premiums.

2. Under the retrospective method, the benefit reserve is computed as the difference between the accumulated value of past benefits paid less the accumulated value of past premiums received.

This chapter will only show examples regarding prospective benefit reserve calculation as well as fully discrete life-contingent policies. Equation (7.9) shows the general relationship that holds for benefit reserve. This is the so-called *prospective method.*

$$V_t = APV_t\,(\text{Future Benefits}) - APV_t\,(\text{Future Premiums}) \tag{7.9}$$

Examples will better exemplify such concepts. The benefit reserve for a whole life insurance is expressed by the following equation

$${}_tV_x = A_{x+t} - P \cdot \ddot{a}_{x+t}, \tag{7.10}$$

where P is the yearly benefit premium (fixed at time $t = 0$), $P = \frac{A_x}{\ddot{a}_x}$.

Example 7.26 *Calculate the reserve for a (whole) life insurance policy on a 60-year-old policyholder paid by annual premiums at year 10.*

First, we get the yearly benefit premium using

```
> P <- Axn(soa08Act,60)/axn(soa08Act,60)
```

and then we get the benefit reserve at year 10

```
> V <- Axn(soa08Act,60+10)-P*axn(soa08Act,60+10)
> V
[1] 0.2311365
```

The benefit reserve for an n-year temporary life insurance is expressed by the following equation

$${}_tV_x = {}_{n-k}A_{x+t} - P \cdot {}_{n-k}\ddot{a}_{x+t}, \tag{7.11}$$

where P is the yearly benefit premium (fixed at time $t = 0$), $P = \frac{{}_nA_x}{{}_n\ddot{a}_x}$.

Example 7.27 *Calculate the reserve for a temporary life insurance policy on a 60-year-old policyholder paid by annual premiums for 30 years at year 10.*

The yearly benefit premium will be

```
> P <- Axn(soa08Act,60,30)/axn(soa08Act,60,30)
```

and then, we get the benefit reserve at year 10

```
> V <- Axn(soa08Act,60+10,30-10)-P*axn(soa08Act,60+10,30-10)
> V
[1] 0.209061
```

If we want to visualize the evolution of the reserve, with time, the code will be

```
> V <- function(t) Axn(soa08Act,60+t,30-t)-P*axn(soa08Act,60+t,30-t)
> VecT  <- seq(0,30)
> plot(VecT,Vectorize(V)(VecT),type="b")
```

Note there is also a *retrospective method* and an *iterative method* (the latter will be discussed at the end of this section). Because the APV of benefits and premiums are equal when the insured signs the contract (this is the principle of actuarial valuation), then

$$V_t < -APV_t\,(\text{Past Premiums}) - APV_t\,(\text{Past Benefits})\,. \tag{7.12}$$

This is the *retrospective method.*

In case the premium is paid for a limited number of periods, the following equation applies.

$$ {}_k^hV^{\,1}_{x:\overline{n}|} = \begin{cases} A_{x+k} - {}_hP^{1}_{x:\overline{n}|} * \ddot{a}_{x+h:\overline{n-h}|}, k < h \\ A_{x+k}, k \geq h \end{cases} \tag{7.13}$$

Example 7.28 *Evaluate* ${}_3^5V^{\,1}_{25:\overline{40}|}$ *and* ${}_{10}^{5}V^{\,1}_{25:\overline{40}|}$.

```
> #lump sum premium
> U <- Axn(soa08Act, x=25,n=40)
> #5-year benefit premium
> P <- U/axn(soa08Act, x=25,n=5)
> #Benefit reserves
> V3 <- Axn(soa08Act, x=25+3,n=40-3)-P*axn(soa08Act, x=25+3,n=5-3)
> ##at t=10
> V10 <- Axn(soa08Act, x=25+10,n=40-10) #benefit reserve at 10
> V3
[1] 0.03237489
> V10
[1] 0.06748181
```

Reserves for other life contingencies insurance follow a similar approach, that is, subtracting from the APV of prospective benefits the APV of prospective premiums.

Example 7.29 *Calculate the reserve for a 20-year endowment at* $t = 10$, *for a policyholder aged 60.*

```
> #Endowment insurance
> #full value premium
> U=AExn(soa08Act, 60,20)
> #benefit yearly premium
> P <- U/axn(soa08Act, 60,20)
> #reserve
> V <- AExn(soa08Act, 60+10,20-10)-P*axn(soa08Act, 60+10,20-10)
> V
[1] 0.355253
```

Example 7.30 *Calculate the following reserves on annuities:*

1. ${}_{10}V({}_{20|}\ddot{a}_{55})$.
2. ${}_{30}V({}_{20|}\ddot{a}_{55})$.

```
> #full value premium
> U <- axn(soa08Act,x=55,m=20)
> #yearly benefit premium
```

```
> P <- U/axn(soa08Act,x=55,n=20)
> #reserve 1
> V10 <- axn(soa08Act,x=55+10,m=20-10)-P*axn(soa08Act,x=55+10,n=20-10)
> V10
[1] 1.980102
> #reserve 2
> V30 <- axn(soa08Act,x=55+30)
> V30
[1] 4.698031
```

The idea of the *iterative method* is that

$$V_t = V_{t+1} - APV_t\,(\text{Future Premiums on}(t, t+1)) + APV_t\,(\text{Future Benefits on}(t, t+1)) \tag{7.14}$$

Thus, it is possible to use recursive techniques to compute V_t because at time $t = 0$, there should be no reserve. Giles (1993) suggested a simple algorithm to solve recursive equations.

Remark 7.1 *We do discuss here iterative equations for the reserves, but most actuarial quantities can actually be seen as solutions of such equations. For instance, A_x satisfies*

$$A_x = \frac{1}{1+i}\left(q_x + p_x \cdot A_{x+1}\right).$$

Similarly, $\ddot{a}_x$ is the solution of the following recursive equation,

$$\ddot{a}_x = 1 + \frac{1}{1+i} p_x \cdot \ddot{a}_{x+1},$$

and even e_x satisfies $e_x = p_x + p_x \cdot e_{x+1}$.

Consider a sequence $\boldsymbol{u} = (u_n)$ satisfying

$$u_n = a_n + b_n u_{n+1},$$

for all $n = 1, 2, \cdots, m$ such that u_{m+1} is known, as well as both $\boldsymbol{a} = (a_n)$ and $\boldsymbol{b} = (b_n)$. The general solution is

$$u_n = \frac{u_{m+1}\prod_{i=0}^{m} b_i + \sum_{j=n}^{m} a_j \prod_{i=0}^{j-1} b_i}{\prod_{i=0}^{n-1} b_i} \tag{7.15}$$

with convention $b_0 = 1$. Then the following code can be used to solve this recursive equation:

```
> recurrent <- function(a,b,ufinal){
+ s <- rev(cumprod(c(1, b)))
+ return( (rev(cumsum(s[-1] * rev(a))) + s[1] * ufinal)/rev(s[-1]) )
}
```

Consider the temporary life insurance of Example 7.27. Then one can get that

$${}_tV_x = \frac{p_{x+t}}{1+i} \cdot {}_{t+1}V_x - P + {}_1A_{x+t}.$$

Then, based on notations of Equation (7.15), $u_{m+1} = 0$ with here $m = n$, and if the index of the recursive equation is t, then

$$a_t = {}_1A_{x+t} - P$$

and

$$b_t = \frac{p_{x+t}}{1+i}.$$

The code is then simply

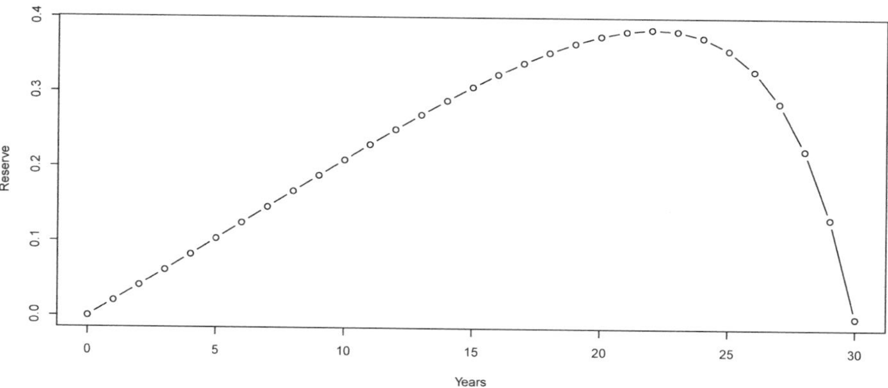

FIGURE 7.2
Evolution of $t \mapsto {}_tV_x$ for a temporary life insurance over $n = 30$ years.

```
> P <- Axn(soa08Act,60,30)/axn(soa08Act,60,30)
> Vecta <- Vectorize(function(t) Axn(soa08Act,t,1))(60+0:29) - P
> Vectb <- Vectorize(function(t) pxt(soa08Act,t,1))(60+0:29)/1.06
> Vectv <- c(recurrent(a=Vecta,b=Vectb,ufinal=0),0)
> plot(0:30,Vectv,type="b")
```

And the graph is exactly the same as obtained on Figure 7.2.

7.6 More Advanced Topics

This section discusses, and exemplifies with R code, special and more advanced topics like contingent policies for multiple lives, consideration of expenses, and stochastic analysis.

The general theory about insurance and annuities can be applied to multiple lives. A key relationship regarding joint and last survival status is expressed in Equation (7.16), which holds also for APVs:

$$T_{xy} = T_x + T_y - T_{\overline{xy}}. \tag{7.16}$$

Example 7.31 *Evaluate* $A_{\overline{60:70}}$.

```
> #define the table
> coupleLifeTables=list(soa08Act,soa08Act)
> #direct approach
> Axyzn(tablesList=coupleLifeTables,x=c(60,70),status="joint")
[1] 0.5722832
> #indirect approach
> Axn(soa08Act,60)+Axn(soa08Act,70)-
+ Axyzn(tablesList=coupleLifeTables, x=c(60,70),
```

```
+ status="last")
[1] 0.5722827
```

Multiple life insurance theory is applicable in the evaluation of reversionary annuities, a special type of two-life annuities. A reversionary annuity pays benefits only when one of the lives has failed and then for as long as the other continues to survive. For example, the APV of an annuity that pays lifetime \$1 to y starting from the death of x is $a_{x|y} = a_y - a_{xy}$.

Example 7.32 *Compute the reversionary annuity on a 60-year-old policyholder, assuming the other insured is 70 years old.*

```
> revAnn <- axn(soa08Act,60)-axyzn(tablesList=coupleLifeTables,
+ x=c(60,70),status="joint")
> revAnn
[1] 3.589022
```

The second topic this section will discuss is the incorporation of expenses in calculating premiums and reserves. In practice, expenses are treated as if they were benefits. The gross premium G is determined as the sum of the APV of benefits and expenses. Similarly, the expense-augmented reserve is determined as the difference between the sum of prospective benefits and expenses and the APV of gross premiums.

Example 7.33 *For a fully discrete whole life insurance of \$100,000 on a policy holder aged 35 years, it is given that*

1. *Percent of premium expenses are 10% per year.*
2. *Per-policy expenses are \$25 per year.*
3. *Annual maintenance expense of \$2.50 for each \$1,000 of face value.*
4. *All expenses are paid at the beginning of the year.*

Fing G. As a hint, Equation (7.17) shows the expression for G.

$$G\ddot{a}_{35} = 100{,}000A_{35} + \left(25 + \frac{2.5}{1000}100{,}000 + 0.1G\right)\ddot{a}_{35} \tag{7.17}$$

```
> G <- (100000*Axn(actuarialtable=soa08Act,x=35) +
+ (25 + 250)*axn(actuarialtable=soa08Act,x=35)) /
+ (0.9*axn(actuarialtable=soa08Act,x=35))
> G
[1] 1234.712
```

Example 7.34 *For a fully discrete whole life insurance of \$1,000 on a policyholder aged 45 years, it is given that:*

- *Expenses include 10% per year of gross premium.*
- *Additional expenses of \$3 per year.*
- *All expenses are paid at the beginning of the year.*

Calculate:

1. *The annual benefit premium.*

2. *The annual gross premium.*
3. *The annual expense premium.*
4. *The benefit reserves at the end of year 1.*
5. *The total reserves at the end of year 1.*

```
> #example 34.1
> U <- 1000*Axn(soa08Act,x=45); P=U/axn(soa08Act,x=45)
> P
[1] 14.25744
> #example 34.2
> G <- (U+3*axn(soa08Act,x=45))/((1-.1)*axn(soa08Act,x=45))
> G
[1] 19.17493
> #example 34.3
> E <- G-P
> E
[1] 4.917493
> #example 34.4
> V <- 1000*Axn(soa08Act,x=45+1)-P*axn(soa08Act,x=45+1)
> V
[1] 11.16089
> #example 34.5
> Vt <- 1000*Axn(soa08Act,x=45+1)+(0.1*G+3)*axn(soa08Act,x=45+1)
> -G*axn(soa08Act,x=45+1)
[1] -267.5782
> Vt
[1] 278.7391
```

The final topic covered in this section is the use of stochastic simulation of life-contingent insurances. It is worth remembering that life contingencies are themselves stochastic variables, as they are functions of the discount factor c and the future lifetime random variable. The `lifecontingencies` package contains functions that allow the analyst to generate variates from the expected future lifetime and life contingencies.

Example 7.35 *Generate* 10^3 *variates of the curtate future lifetime at birth from the SoA illustrative life table.*

Let us generate the variates using `rLife`:

```
> K0 <- rLife(n=10^3,object=soa08Act,x=0,type="Kx")
\begin{rcode}
From the law of large number, the average of those life times should not be
far away from the expected lifetime,\index{com}{t.test@\code{t.test}}
\begin{rcode}
> exn(soa08Act)
[1] 71.34692
> mean(K0)
[1] 73.129
> t.test(x=K0, mu=exn(soa08Act))

        One Sample t-test
```

```
data:  K0
t = 3.2432, df = 999, p-value = 0.001221
alternative hypothesis: true mean is not equal to 71.34692
95 percent confidence interval:
 72.05071 74.20729
sample estimates:
mean of x
   73.129
```

Generating variates from life contingencies can be used in pricing using the percentile premium principle approach. This approach prices the life contingency insurance finding the minimum premium that makes the insurer reduce the loss probability (the present value of prospective costs less the present value of prospective premiums) to not exceed a selected threshold.

Example 7.36 *A policyholder aged 25 purchases a 40-year term life insurance. Find the smallest premium the insurer could charge to set the probability of loss no greater than 5%.*

```
> samples <- rLifeContingencies(n=10^4,lifecontingency="Axn",
+ object=soa08Act,x=25,t=40,parallel=TRUE)
```

We can compute the classical benefit premium using

```
> APV <- Axn(soa08Act,x=25,n=40)
```

and percentile premium using

```
> percentilePremium <- quantile(x=samples, p=1-0.05)
```

We can also compare the two

```
> APV
[1] 0.0479709
> percentilePremium
      95%
0.2775051
```

Example 7.37 *A life insurance company is going to underwrite a pool of 1,000 retirees, all aged 65. Find the amount to be charged to each insured to make the insurance company not to lose with 99% probability.*

As a hint, remember that if $\tilde{Z}$ is the random variable representing the present value of the annuity, the present value of the total cost of the pool will be distributed as a normal distribution with parameters expressed by Equation (7.18)

$$\begin{aligned} \mu &= 1000 * \mathbb{E}(\tilde{Z}) \\ \sigma &= \sqrt{1000 * var(\tilde{Z})} \end{aligned} . \tag{7.18}$$

The Monte Carlo sample of A_{65} is obtained using

```
> ax65 <- rLifeContingencies(n=10^5, lifecontingency = "axn",
+ object = soa08Act,x=65,parallel=TRUE)
```

and the first two empirical moments are

```
> muax65 <- mean(ax65)
> sdax65 <- sd(ax65)
> qnorm(p=0.99,mean=1000*muax65,sd=(1000)^0.5*sdax65)/1000
[1] 10.16101
```

Finally, the `lifecontingencies` package can evaluate the APV of higher moments, by the use of the "power" parameter, as the following example shows.

Example 7.38 *Show that the variance of* $(IA)^{1}_{45:\overline{20|}}$ *as sample estimate is very close to its estimation by formula.*

```
> sampleIAxn <- rLifeContingencies(n=50000,lifecontingency="IAxn",object=soa08Act,
+ x=45, t=20,parallel=TRUE)
```

The sample variance is here

```
> var(sampleIAxn)
[1] 4.722363
```

which is comparable with the analytical one,

```
> IAxn(actuarialtable=soa08Act, x=45,n=20,power=2)-(IAxn(actuarialtable=soa08Act,
+ x=45,n=20,power=1))^2
[1] 4.719925
```

7.7 Health Insurance and Markov Chains

This section deals with the application of discrete Markov chains to Health Insurance. Discrete Markov chains represent a class of stochastic processes defined on a discrete set of possible states characterized by Equation (7.19). A classical application of Markov chains in insurance lies in Health and Disability coverage pricing and reserving. Another notable actuarial application of Markov chains is no-claim discount and bonus-malus policyholders' evolution within a portfolio. Daniel (n.d.) provides an introduction to Markov chains application in the insurance business, while Haberman & Pitacco (1999) and Deshmukh (2012) provide a broader overview of actuarial topics, like health insurance, on which Markov chains can be effectively applied.

$$\mathbb{P}\left(X_{n+1}=x|X_1=x_1,X_2=x_2,\ldots,X_n=x_n\right)=\Pr\left(X_{n+1}=x|X_n=x_n\right) \tag{7.19}$$

For a broader mathematical introduction, Ross (2009), Norris (1997), and Ching & Ng (2006) provide a good introduction to the topic.

7.7.1 Markov Chain with R

The R package `markovchain`, see (Spedicato 2013*b*), will be used throughout the chapter. The package vignettes provide much deeper detail on the package's capatibilites to handle and manipulate discrete Markov chains within an R environment, as well as how to perform probabilitistic and statistical analyses on it.

```
> library(markovchain)
```

To show how to define a Markov chain within the package, let's consider a critical illness model with three states: healthy (H), critically ill (I), and dead (D). A Markov chain with associated transition probabilities can be defined as follows:

```
> stateNames <- c("H","I","D")
> cimMc <- new("markovchain",states=stateNames,
+ transitionMatrix=matrix(c(0.92,0.05,0.03,
+ 0.00,0.76,0.24,0.00,0.00,1.00),nrow=3, byrow=TRUE))
> cimMc
A Markov chain
 A  3 - dimensional discrete Markov Chain with following states
 H I D
 The transition matrix   (by rows)  is defined as follows
     H    I    D
H 0.92 0.05 0.03
I 0.00 0.76 0.24
D 0.00 0.00 1.00
```

Various methods have been defined within the package to make probabilistic analysis easier, as the package vignettes reveal. For example, the probability that a subject initially healthy is dead in the fourth step is

```
> transitionProbability(cimMc^4, t0="H","D")
[1] 0.163991
```

Identifying steady probabilities vectors and absorbing states is easy as well:

```
> steadyStates(cimMc)
> absorbingStates(cimMc)
```

To perform such calculations, the power method, defined for the `markovchain` `S4` (see (Chambers & Hastie 1991) and Chapter 1) class has been used. Similarly, a plotMc method is available for the class, as Figure 7.3 displays, based on Csardi & Nepusz (2006).

```
> plotMc(cimMc)
```

If the transition probabilities between states vary by time, they can be modeled by non-homogeneous Markov chains, as the following example displays.

Example 7.39 *The status of residents in a Continuing Care Retirement Community (CCRC) is modeled by a nonhomogeneous Markov chain with three states: Independent Living ("H"), Health Center ("I"), and Gone ("D"). The transition probabilities are modeled in the following R code by* $Q_0, Q_1, \ldots, Q_3$.

```
> Q0 <- new("markovchain", states=stateNames,
+ transitionMatrix=matrix(c(0.7, 0.2, 0.1,0.1, 0.6, 0.3,0, 0, 1),
+ byrow=TRUE, nrow=3))
> Q1 <- new("markovchain", states=stateNames,
+ transitionMatrix=matrix(c(0.5, 0.3, 0.2,0, 0.4, 0.6,0, 0, 1),
+ byrow=TRUE, nrow=3))
> Q2 <- new("markovchain", states=stateNames,
+ transitionMatrix=matrix(c(0.3, 0.2, 0.5,0, 0.2, 0.8,0, 0, 1),
```

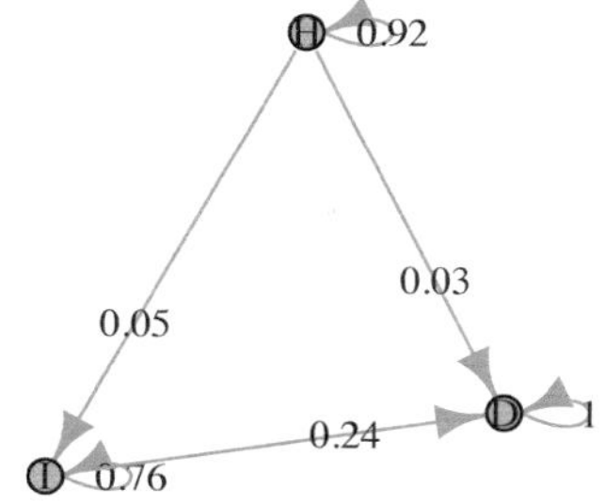

FIGURE 7.3
Graph of the transition matrix `cimMc`.

```
+ byrow=TRUE,nrow=3))
> Q3 <- new("markovchain", states=stateNames, transitionMatrix=
+ matrix(c(0, 0, 1,0, 0, 1,0, 0, 1),byrow=TRUE, nrow=3))
> mcCCRC<-new("markovchainList",markovchains=list(Q0,Q1,Q2,Q3))
```

7.7.2 Valuation of Cash Flows

Two kinds of events result in payments: cash flow upon transitions, when payments are made upon transition from one state to another; and cash flows while in states represent payments made due to being in a certain state for a particular time period.

Example 7.40 *Suppose a 2-year Accident and Death policy pays \$250 per year in case of illness and \$1000 in case of death. Suppose transition probabilities to be defined by* ***cimMc*** *matrix, the interest rate to be 5%. Suppose premiums are paid only when the policyholder is healthy, when no benefit is paid. Compute the APV.*

Here, the subject is `H`

```
initialState<-c(1,0,0);attr(initialState, which="name")<-stateNames
v=1.03^-1
```

The possible transitions are the following:

```
> 250*transitionProbability(cimMc^2, "H","I")*v^2+
+ 1000*transitionProbability(cimMc^2, "H","D")*v^2+
+ 250*transitionProbability(cimMc, "H","I")*v+
+ 250*transitionProbability(cimMc, "H","I")*v+
+   250*transitionProbability(cimMc, "I","I")*v^2+
+ 250*transitionProbability(cimMc, "H","I")*v+
+  1000*transitionProbability(cimMc, "I","D")*v^2+
+ 1000*transitionProbability(cimMc, "H","D")*v
[1] 556.2494
```

Example 7.41 *Each patient of the CCRC costs \$50 when healthy, \$200 when ill, \$10 when dead. Suppose the interest rate is 3%. Compute the APV of inpatient support using Monte Carlo simulation.*

Consider the following function:

```
> getAPV<-function(mcList,t1="H"){
+   lifeStates<-character()
+   t2<-markovchainSequence(n=1,markovchain=mcList@markovchains[[1]],t0=t1) #state during second year
+   t3<-markovchainSequence(n=1,markovchain=mcList@markovchains[[2]],t0=t2) #state during third year
+   t4<-markovchainSequence(n=1,markovchain=mcList@markovchains[[3]],t0=t3) #state during fourth year
+   t5<-markovchainSequence(n=1,markovchain=mcList@markovchains[[4]],t0=t4) #state during fifht year
+    lifeStates<-c(t1,t2,t3,t4,t5)
+   APV<-0
+   v<-1.03^-1
+   for(i in 1:5){
+     value<-ifelse(lifeStates[i]=="H",50, ifelse(lifeStates[i]=="I",200,10))*v^((i-0.5))
+     APV<-APV+value
+     if(lifeStates[i]=="D") break}
+   return(APV)}
```

We can use that function , with 1,000 Monte Carlo simulations,

```
> simulations<-numeric(1000)
> set.seed(1)
> for(i in 1:1000) simulations[i]<-getAPV(mcCCRC)
```

and the average of those simulations is

```
> mean(simulations)
[1] 221.178
```

7.7.3 APV of Benefits and Reserves

Benefit premiums and reserves can be computed by applying the Equivalence Principle to any contingent payment situation, that is, equating the Actuarial Present Value of Premiums (APVP) to the Actuarial Present Value of the Benefits (APVB).

Example 7.42 *Suppose a four-state homogeneous Markov model represents the joint mortality of a married couple: a husband and a wife. The states are 1 = husband alive, wife alive; 2 = husband dead, wife alive; 3 = husband alive, wife dead, and 4 = both husband and wife dead. A life insurer sells a 2-year term insurance contract to a married couple who are both age 60. The death benefit of \$100 is payable at the end of the year in which the second life dies, if both die within 2 years. Premiums are payable as long as at least one of them is alive and annually in advance. Interest rate i = 5%. Calculate the annual benefit premium.*

```
> mc2Lifes<-new("markovchain", states<-c("1","2","3","4"),transitionMatrix=matrix(
+ c(0.95, 0.02, 0.02, 0.01,
+ 0.00, 0.90, 0.0, 0.10,
+ 0.00, 0.00, 0.85, 0.15,
+ 0.00, 0.00, 0.00, 1.00),byrow=TRUE,nrow=4))
```

The APVP and APVB are computed as follows, as premiums are paid at the beginning of period unless both subjects are dead.

```
> APVP<-1+(1-transitionProbability(mc2Lifes,"1","4"))*1.05^-1
> APVB<-100*(transitionProbability(mc2Lifes,"1","4")*1.05^-1
    +(transitionProbability(mc2Lifes^2,"1","4")
    -transitionProbability(mc2Lifes,"1","4"))*1.05^-2)
> P<-APVB/APVP
> P
[1] 1.167134
```

7.8 Exercises

The solutions to the following exercises are stored in demo files of the `lifecontingencies` package. For example, to load pricing section exercises, send the following code to R console:

```
> demo("pricing")
```

7.8.1 Financial Mathematics

7.1. Find the compound interest rate equivalent to a rate of compound interest of 8% payable semiannually.

7.2. In return for a payment of $1,000 at the end of 10 years, a lender agrees to pay $200 immediately, $500 at the end of 6 years, and a final amount at the end of 15 years. Find the amount of the final payment at the end of 15 years if the nominal rate of interest is 5% converted semiannually.

7.3. Find the IRR of an investment that requires investing $1,000 for 5 years and returning $500 for the following 15 years.

7.4. What would you be willing to pay for an infnite stream of $37 annual payments (cash inflows) beginning now if the interest rate is 8% per annum?

7.5. John receives $400 at the end of the frst year, $350 at the end of the second year, $300 at the end of the third year, and so on; the final payment is $50. Using an annual effective rate of 3.5%, calculate the present value of these payments at time 0.

7.6. What is the difference between an annuity due with a 5-year term, interest rate 3% 12 times?

7.7. A bond whose term is 5 years, par value $100, and yearly coupon rate is 5% prices at $95.45. Find its yield.

7.8. Redo the duration example taking into account the convexity.

7.8.2 Demography

7.9. Create one lifetable for tables CL1 through CL3 shown in the `demoChina` dataset and compare the expected residual lifetime.

7.10. Find the probability of a life aged 2 to survive to age 4.

7.11. Find the number of people dying between 35 and 45.

7.12. Calculate the probability that three brothers aged 14, 15, 16 will all be alive after 60 years.

7.8.3 Pricing Life Insurance

Use `soa08Act` lifetable unless otherwise stated.

7.13. Create the Italian 92 male actuarial table at 3% from the `demoIta` dataset and name the table SIM92.

7.14. How much should a 25-year-old policyholder pay to receive \$100,000 when he turns 65, assuming he is alive?

7.15. How much should a 25-year-old policyholder pay to receive \$100,000 when he turns 65 or when he dies?

7.16. On the SOA actuarial table, calculate the APV whole life insurance for a policyholder aged 30 with benefit payable at the end of month of death at a 4% interest rate, $A_{30}^{(12)}$.

7.17. For the APV calculated in Exercise 7.16, determine the level benefit premium assuming it will be paid until the policyholder's death.

7.18. Calculate the quarterly premium that a policyholder aged 50 will pay until the year of death to insure a face value of \$100,000. The face value will be paid at the end of the quarter of death.

7.19. Compute the 15-year benefit premium for 35-year term insurance with benefit payable at the end of month of death. Assume the interest rate is 3% and the premium is to be paid at the beginning of each month.

7.8.4 Reserving Life Insurances

7.20. Calculate the reserve at time 4 for life insurance on a policyholder aged 60 paid with five yearly premiums as long as the policyholder is alive from year.

7.21. Calculate ${}_5V_{\overset{1}{75:\overline{20|}}}$.

7.22. Calculate the benefit reserve for a 20-year endowment on a policyholder aged 60 at time 10 assuming level benefit premiums to be paid for 15 years.

7.23. Calculate the benefit reserve at time 10 for whole life insurance on 60 payable at the end of year of death with benefit premiums paid quarterly.

7.8.5 More Advanced Topics

7.24. Calculate a 10-year last survivor annuity-due on a policyholder aged 30 and on a policyholder aged 40.

7.25. Consider a 20-year term insurance on a policyholder aged 30, face value equal to \$1000. Calculate the gross level yearly premium, G, allowing for \$30 flat expenses and yearly maintenance expenses equal to 15% of G.

7.26. Sample a vector of size 10^4 from K_{25} and compute summary statistics.

7.27. Find the smallest premium to charge a policyholder aged 65 who wants to purchase an annuity-due in order to limit the probability of loss to 25%.

8

Prospective Life Tables

Heather Booth
Australian National University
Canberra, Australia

Rob J. Hyndman
Monash University
Melbourne, Australia

Leonie Tickle
Macquarie University
Sydney, Australia

CONTENTS

8.1 Introduction

Prospective life tables depend on forecasting age-specific mortality. Considerable attention has been paid to methods for forecasting mortality in recent years. Much of this work has grown out of the seminal Lee-Carter method (Lee & Carter 1992). Other extrapolative approaches use Bayesian modelling, generalized linear modelling and state-space approaches. Methods for forecasting mortality have been extensively reviewed by Booth (2006) and Booth & Tickle (2008). This chapter covers various extrapolative methods for forecasting age-specific central death rates (also referred to as mortality rates). Also covered is the derivation of stochastic life expectancy forecasts based on mortality forecasts.

The main packages on CRAN for implementing life tables and mortality modelling are **demography** (see Hyndman 2012) and **MortalitySmooth** (see Camarda 2012), and we will concentrate on the methods implemented in those packages. However, mention is also made of other extrapolative approaches, and related R packages where these exist.

We will use, as a vehicle of illustration, U.S. mortality data. This can be extracted from the Human Mortality Database (2013) using the **demography** package.

```
> library(demography)
> usa <- hmd.mx("USA", "username", "password", "USA")
```

The `username` and `password` are for the Human Mortality Database.

The `hmd.mx` function downloads all available annual data by single years of age. Note that it is also possible to download data in 5-year age groups and that the methods described in this chapter can accommodate 5-year age groups. We use single years of age and annual years over time.

The object `usa` contains all the necessary data to forecast mortality:

```
> names(usa)
[1] "type"    "label"   "lambda" "year"    "age"     "pop"     "rate"
```

A useful summary is given by

```
> usa
Mortality data for USA
    Series: female male total
    Years: 1933 - 2010
    Ages:  0 - 110
```

`usa` is a '`demogdata`' object of type '`mortality`'. The functions in the **demography** package are designed to use `demogdata` objects.

In order to stabilize the high variance associated with high age-specific rates, it is necessary to transform the raw data by taking logarithms. Consequently, the mortality models considered in this chapter are all in log scale.

8.2 Smoothing Mortality Data

Suppose $D_{x,t}$ is the number of deaths in calendar year t of people aged x, and $E^c_{x,t}$ is the total years of life lived between ages x and $x+1$ in calendar year t, which can be approximated by the mid-year (central) population at age x in year t.

Mid-year populations can be obtained from the `usa$pop` list:

```
> names(usa$pop)
[1] "female" "male"   "total"
```

For example, the matrix `usa$pop$total` contains the total population at age x (per row) at mid-year t (per column):

```
> usa$pop$total[1:5,1:6]
     1933    1934    1935    1936    1937    1938
0 1963038 1920566 1940669 1953470 1967325 2012973
1 2034701 1927298 1896434 1924241 1945812 1965169
```

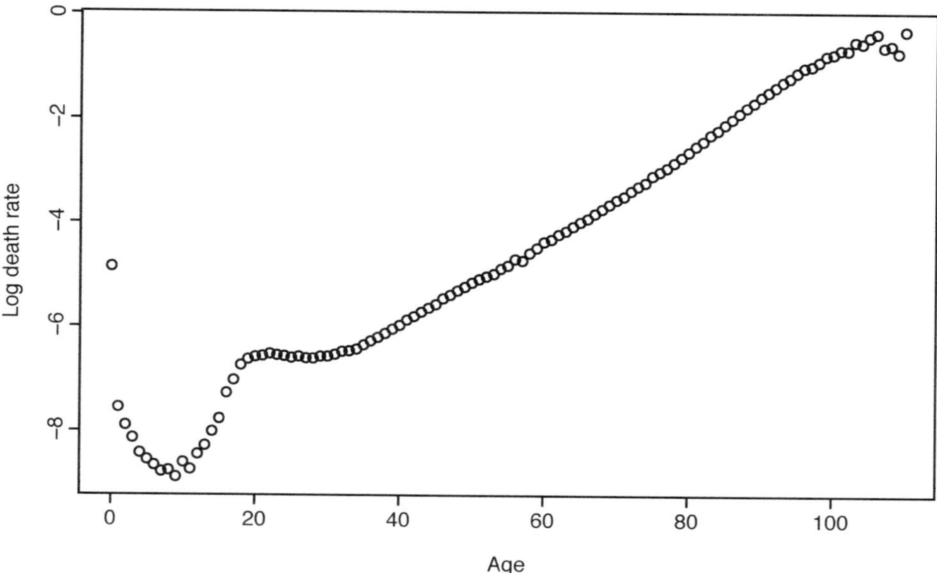

FIGURE 8.1
Male mortality rates for single years of age, United States 2003.

```
2 2178069 2106591 2037580 2020441 2063568 2102951
3 2229949 2191411 2126538 2056968 2040318 2086125
4 2274434 2226345 2187956 2122040 2053175 2038711
```

Similarly, the observed mortality rates, defined as

$$m_{x,t} = D_{x,t}/E^c_{x,t},$$

are obtained from the list of matrices `usa$rate`. For example, rates for the total population are obtained by

```
> usa$rate$total[1:5,1:6]
      1933     1934     1935     1936     1937     1938
0 0.061667 0.067866 0.061964 0.062786 0.061008 0.058018
1 0.009459 0.010622 0.008831 0.008895 0.008366 0.007793
2 0.004351 0.004852 0.004171 0.004199 0.004000 0.003563
3 0.003104 0.003230 0.002980 0.002870 0.002681 0.002463
4 0.002386 0.002451 0.002404 0.002272 0.002151 0.001912
```

Figure 8.1 shows mortality rates for U.S. males in 2003, based on the following R code:

```
> plot(usa$age, log(usa$rate$male[,"2003"]), xlab= "Age", ylab="Log death rate")
```

Because `usa` is a '`demogdata`' object, this graph can also be obtained by

```
> plot(usa, series="male", years=2003, type="p", pch=1)
```

This example shows that the mortality rates follow a smooth function with some observational error. The observational error has higher variance at very old ages (because the populations are small) and at childhood ages (because the mortality rates are low). (Note that observed mortality rates at very old ages can exceed 1 because the number of deaths at age x may exceed the mid-year population aged x.)

Thus, we observe $\{x_i, m_{x_i,t}\}$, $t = 1, \ldots, n$, $i = 1, \ldots, p$, where

$$\log m_{x_i,t} = f_t(x_i^*) + \sigma_t(x_i^*)\varepsilon_{t,i} ,$$

log denotes the natural logarithm, $f_t(x)$ is a smooth function of x, x_i^* is the mid-point of age interval x_i, n is the number of years and p is the number of ages in the observed data set, $\varepsilon_{t,i}$ is an iid random variable and $\sigma_t(x)$ allows the amount of noise to vary with x.

Then the observational variance, $\sigma_t^2(x)$, can be estimated assuming deaths are Poisson distributed (Brillinger 1986). Thus, $m_{x,t}$ has approximate variance $D_{x,t}/(E_{x,t}^c)^2$, and the variance of $\log m_{x,t}$ (via a Taylor approximation) is

$$\sigma_t^2(x) \approx 1/D_{x,t}.$$

Life tables constructed from the smoothed $f_t(x)$ data have lower variance than tables constructed from the original $m_{t,x}$ data, and thus provide better estimates of life expectancy. To estimate f, we can use a nonparametric smoothing method such as kernel smoothing, loess, or splines. Two smoothing methods for estimating $f_t(x)$ have been widely used, and both involve regression splines. We will briefly describe them here.

8.2.1 Weighted Constrained Penalized Regression Splines

Hyndman & Ullah (2007) proposed using constrained and weighted penalized regression splines for estimating $f_t(x)$. The weighting takes care of the heterogeneity due to $\sigma_t(x)$, and a monotonic constraint for upper ages can lead to better estimates.

Following Hyndman & Booth (2008), we define weights proportional to the approximate inverse variances, $w_{x,t} \propto D_{x,t}$, and use weighted penalized regression splines (Wood 2003, He & Ng 1999) to estimate the curve $f_t(x)$ in each year. Weighted penalized regression splines are preferred because they can be computed quickly and allow monotonicity constraints to be imposed relatively easily.

We apply a qualitative constraint to obtain better estimates of $f_t(x)$, especially when $\sigma_t(x)$ is large. We assume that $f_t(x)$ is monotonically increasing for $x > b$ for some b (say, $b = 65$ years). This monotonicity constraint allows us to avoid some of the noise in the estimated curves for high ages, and is not unreasonable for this application (after middle age, the older you are, the more likely you are to die). We use a modified version of the approach described in Wood (1994) to implement the monotonicity constraint.

Figure 8.2 shows the estimated smooth curve, $f_t(x)$, for the U.S. male mortality data plotted in Figure 8.1. This is easily implemented in the `demography` package using the following code:

```
> smus <- smooth.demogdata(usa)
> plot(usa, years=2003, series="male", type="p", pch=1, col="gray")
> lines(smus, years=2003, series="male")
```

8.2.2 Two-Dimensional P-Splines

The above approach assumes $f_t(x)$ is a smooth function of x, but not of t. Hyndman & Ullah (2007) argued that the occurrence of wars and epidemics meant that $f_t(x)$ should not be assumed to be smooth over time. In the absence of wars and epidemics, it is reasonable to assume smoothness in both the time and age dimensions. For this reason, we use data from 1950 in this subsection.

```
> usa1950 <- extract.years(usa, years=1950:2010)
```

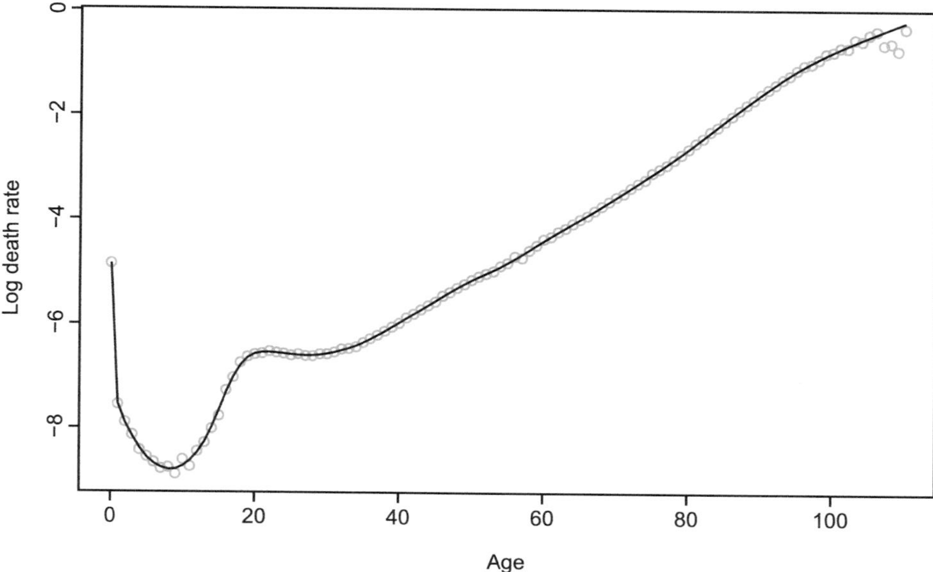

FIGURE 8.2
Smoothed male mortality rates for single years of age, United States 2003. The smooth curve, $f_t(x)$, is estimated using weighted penalized regression splines with a monotonicity constraint for ages greater than 65.

Currie et al. (2004) proposed using two-dimensional splines. We will call this approach the Currie-Durban-Eilers or CDE method.

The CDE method adopts the generalized linear modelling (GLM) framework for the Poisson deaths $D_{x,t}$ with two-dimensional P-splines. Here

$$D_{x,t} \sim \mathcal{P}(E^c_{x,t} \cdot \exp[s(x,t)])$$

where the number of deaths, $D_{x,t}$, will be proportional to the exposure, $E^c_{x,t}$, and $f_t(x) = \exp[s(x,t)]$ will denote the mortality rate.

This is implemented in the `MortalitySmooth` package in R (Camarda 2012) and compared with the Hyndman & Ullah (2007) approach using the following code:

```
> library(MortalitySmooth)
> Ext <- usa1950$pop$male
> Dxt <- usa1950$rate$male * Ext
> fitBIC <- Mort2Dsmooth(x=usa1950$age, y=usa1950$year, Z=Dxt, offset=log(Ext))

> par(mfrow=c(1,2))
> plot(fitBIC$x, log(usa1950$rate$male[,"2003"]), xlab="Age", ylab="Log death rate",
+ main="USA: male death rates 2003", col="gray")
> lines(fitBIC$x, log(fitBIC$fitted.values[,"2003"]/Ext[,"2003"]))
> lines(smus,year=2003, series="male", lty=2)
> legend("topleft",lty=1:2, legend=c("CDE smoothing", "HU smoothing"))

> plot(fitBIC$y, log(Dxt["65",]/Ext["65",]), xlab="Year", ylab="Log death rate",
+ main="USA: male death rates age 65", col="gray")
> lines(fitBIC$y, log(fitBIC$fitted.values["65",]/Ext["65",]))
```

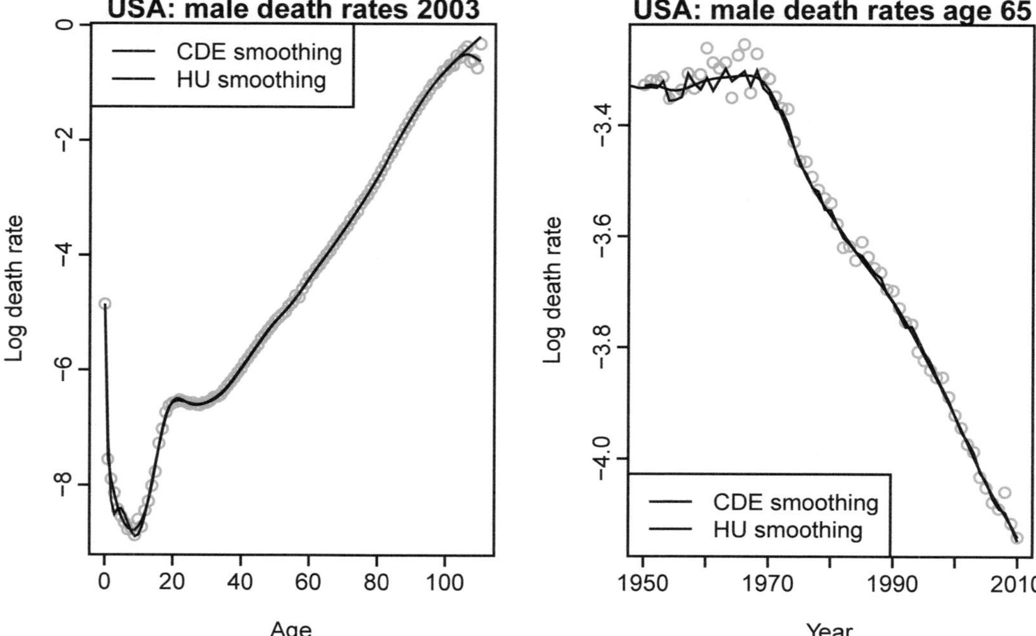

FIGURE 8.3
Smoothed male mortality rates using bivariate P-splines, United States.

```
> lines(smus$year, log(smus$rate$male["65",]), lty=2)
> legend("bottomleft", lty=1:2, legend=c("CDE smoothing", "HU smoothing"))
```

Figure 8.3 shows the estimated smooth curve, $f_t(x)$, for the U.S. male mortality data using the bivariate P-spline (CDE) method (Currie et al. 2004) and the univariate penalized regression spline method of Hyndman & Ullah (2007). Note that the univariate method is not smooth in the time dimension (right panel), but gives a better estimate for the oldest ages due to the monotonic constraint.

```
> par(mfrow=c(1,2))
> year=usa1950$year[seq(1,62,by=3)]
> age=usa1950$age[seq(1,111,by=3)]
> persp(age, year, log(usa1950$rate$male[seq(1,111,by=3), seq(1,62,by=3)]), theta=-30,
+ main="Observed death rates", col=grey(.93), shade=TRUE, xlab="Age", ylab="",
+ zlab="Log death rate", ticktype="detailed",cex.axis=0.6, cex.lab=0.6)
> persp(age, year, predict(fitBIC)[seq(1,111,by=3),seq(1,62,by=3)], theta=-30,
+ main="Smoothed death rates", col=grey(.93), shade=TRUE, xlab="Age",  ylab="",
+ zlab="Log death rate", ticktype="detailed", cex.axis=0.6, cex.lab=0.6)
```

Figure 8.4 shows smoothing of the 1950–2010 dataset. On the left are observed death rates, $m_{x,t}$, and on the right are the estimated smoothed rates, $f_t(x)$.

8.3 Lee–Carter and Related Forecasting Methods

The Lee–Carter (LC) method introduced in Lee & Carter (1992) for forecasting mortality rates uses principal components analysis to decompose the age-time matrix of log central

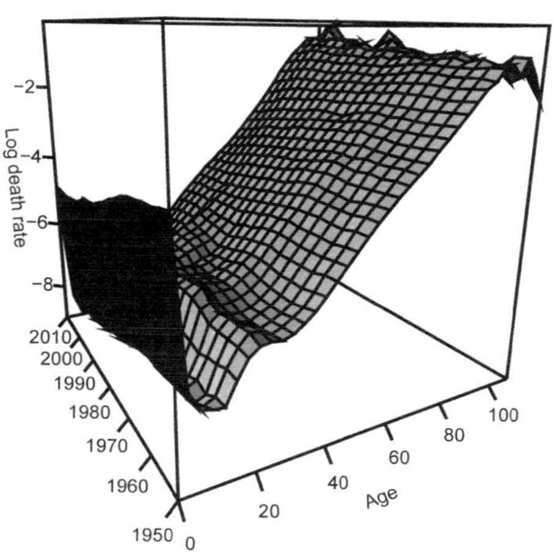

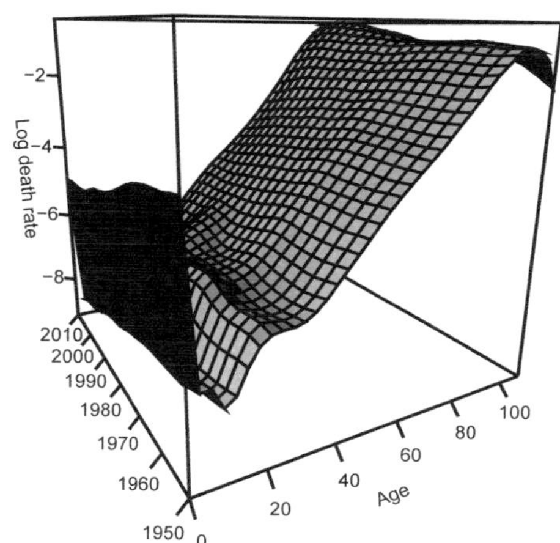

FIGURE 8.4
Smoothed male mortality rates using bivariate P-splines, United States, 1950–2010.

death rates into a linear combination of age and time parameters. The time parameter is used in forecasting.

LC has spawned numerous variants and extensions. The two main variants of LC are Lee-Miller (LM) (Lee & Miller 2001) and Booth-Maindonald-Smith (BMS) (Booth et al. 2002). Others result from different combinations of possible options. These variants are collectively referred to as "LC methods". A major extension of this approach uses functional data models (FDM); first proposed by Hyndman & Ullah (2007), it was further developed by Hyndman & Booth (2008) and Hyndman & Shang (2009). Again, various combinations of options produce variations within the collectively labeled "HU methods".

We identify six methods by their proponents; these are listed in Table 8.1 where the defining features of the models are shown. Most authors referring to the "Lee-Carter method" actually refer to the generic model in which all available data are used, there is no adjustment of the time parameter prior to forecasting, and fitted rates are used as jump-off rates; Booth et al. (2006) labeled this "LCnone". Note that within the options listed in Table 8.1 there are twenty-four possible combinations (4 adjustment options × 3 data period options × 2 jump-off options) for the LC methods. For the HU methods, additional options have been defined by varying the data period option to include 1950 (Shang et al. 2011). Clearly, any date can be used for the start of the data period.

In this section, we begin with the complete available dataset `usa`, for which ages range from 0 to 110+ and years range from 1933 to 2010.

```
> usa
Mortality data for USA
    Series: female male total
    Years: 1933 - 2010
    Ages:  0 - 110
```

Appropriate age ranges and fitting periods are defined for each method.

As the Lee-Carter methods do not incorporate smoothing, it is advisable to avoid erroneous rates at the oldest ages when using these methods.

TABLE 8.1
Lee–Carter and Hyndman–Ullah methods by defining features.

Method	Data Period	Smoothing	Adjustment to Match	Jump-off Rates	Reference
			Lee–Carter Methods		
LC	All	No	D_t	Fitted	Lee & Carter (1992)
LM	1950	No	$e(0)$	Observed	Lee & Miller (2001)
BMS	Linear	No	$D_{x,t}$	Fitted	Booth et al. (2002)
LCnone	All	No	–	Fitted	–
			Hyndman–Ullah Methods		
HU	–	Yes	–	Fitted	Hyndman & Ullah (2007)
HUrob	–	Yes	–	Fitted	Hyndman & Ullah (2007)
HUw	–	Yes	–	Fitted	Hyndman & Shang (2009)

```
> usa.90 <- extract.ages(usa, ages=0:90)
```

In this example, the upper age group is then `90+`.

8.3.1 Lee–Carter (LC) Method

The model structure proposed by Lee & Carter (1992) is given by

$$\log(m_{x,t}) = a_x + b_x k_t + \varepsilon_{x,t}, \tag{8.1}$$

where a_x is the age pattern of the log mortality rates averaged across years, b_x is the first principal component reflecting relative change in the log mortality rate at each age, k_t is the first set of principal component scores by year t and measures the general level of the log mortality rates, and $\varepsilon_{x,t}$ is the residual at age x and year t. The model assumes homoskedastic error and is estimated using a singular value decomposition.

The LC model in Equation (8.1) is over-parameterized in the sense that the model structure is invariant under the following transformations:

$$\{a_x, b_x, k_t\} \mapsto \{a_x, b_x/c, ck_t\},$$
$$\{a_x, b_x, k_t\} \mapsto \{a_x - cb_x, b_x, k_t + c\}.$$

In order to ensure the model's identifiability, Lee & Carter (1992) imposed two constraints, given as

$$\sum_{t=1}^{n} k_t = 0, \qquad \sum_{x=x_1}^{x_p} b_x = 1.$$

In addition, the LC method adjusts k_t by refitting to the total number of deaths. This adjustment gives more weight to high rates, thus roughly counterbalancing the effect of using a log transformation of the mortality rates. The adjusted k_t is then extrapolated using ARIMA models. Lee & Carter (1992) used a random walk with drift (RWD) model, which can be expressed as

$$k_t = k_{t-1} + d + e_t,$$

where d is known as the drift parameter and measures the average annual change in the series, and e_t is an uncorrelated error. It is notable that the RWD model provides satisfactory results in many cases (Tuljapurkar, Li & Boe 2000, Lee & Miller 2001, Lazar & Denuit

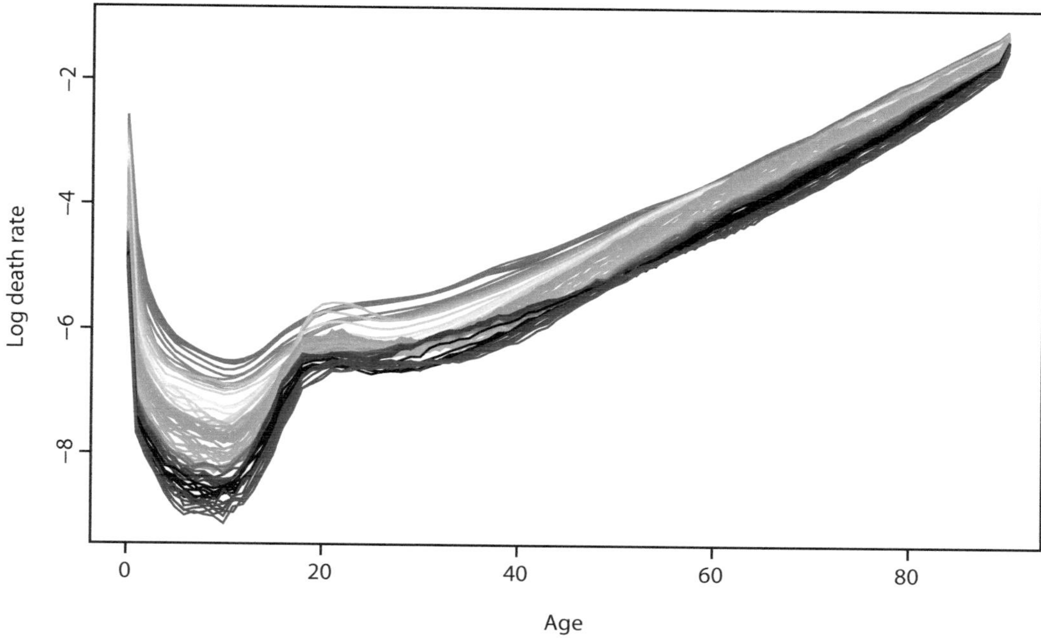

FIGURE 8.5
U.S. male mortality rates, 1933–2010. Note the emergence of the "accident hump" at about age 20 and the effect of deaths due to AIDS at about ages 25–44 in 1985–1995.

2009). From this forecast of the principal component scores, the forecast age-specific log mortality rates are obtained using the estimated age effects a_x and b_x, and setting $\varepsilon_{x,t} = 0$, in Equation (8.1).

The LC method is implemented in the R `demography` package as follows:

```
> lc.male <- lca(usa.90, series="male")
> forecast.lc.male <- forecast(lc.male, h=20)
```

The estimated age parameters a_x and b_x, and the estimated time parameter k_t are respectively obtained using `lc.male$ax`, `lc.male$bx` and `lc.male$kt`. Similarly, the forecast time parameter (rescaled to zero in the jump-off year 2010) is obtained using `forecast.lc.male$kt`.

The data (Figure 8.5), model parameters (Figure 8.6) and forecasts can be viewed via

```
> plot(usa.90, series="male")
> plot(lc.male)
> plot(forecast.lc.male, plot.type="component")
> plot(usa.90, series="male", ylim=c(-10,0), lty=2)
> lines(forecast.lc.male)
```

The LC method without adjustment of k_t (LCnone) is achieved by choosing the adjustment option `adjust="none"`.

```
> lcnone.male <- lca(usa.90, series="male", adjust="none")
```

The effect of the LC adjustment of k_t is seen in Figure 8.7 via

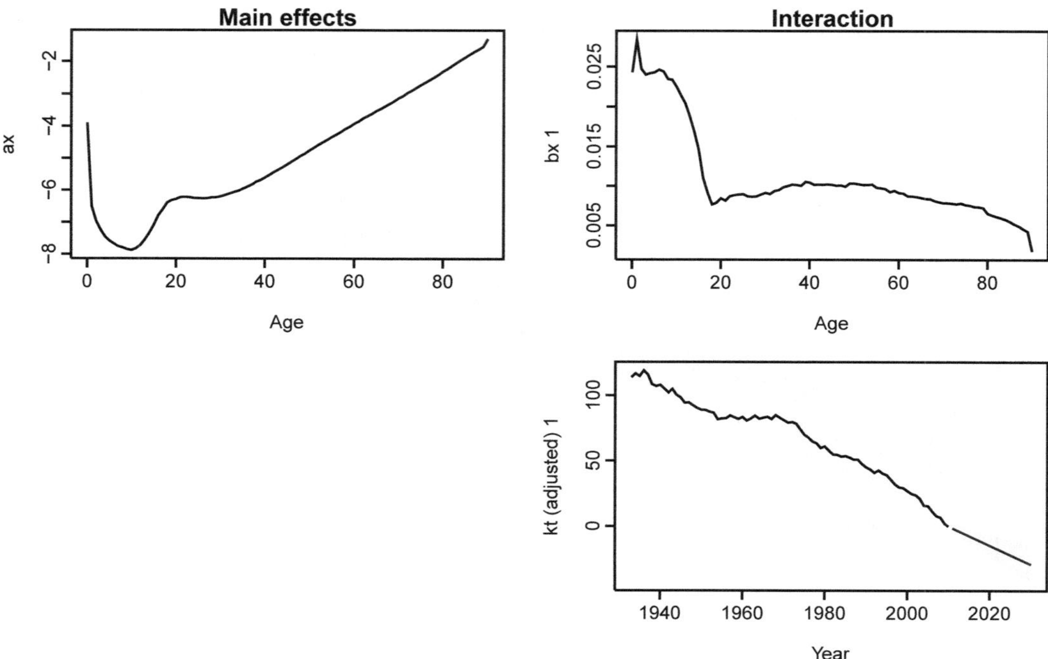

FIGURE 8.6
LC model and forecast, U.S. male mortality. Fitting period = 1933–2010; forecasting horizon = 20 years.

```
> plot(lcnone.male$kt, ylab="kt",ylim=c(-70,90), xlab="")
> lines(lc.male$kt, lty=2)
> legend("topright", lty=1:2, legend=c("LCnone","LC"))
```

Differences between these two lines in the first and last years of the fitting period can have a substantial effect on the forecast.

An alternative, and more efficient, approach to estimating a Lee–Carter model was described by Brouhns et al. (2002); it involves embedding the method in a Poisson regression model, and using maximum likelihood estimation. This can be achieved in R using, for example,

```
> lca(usa, series="male", adjust="dxt")
```

8.3.2 Lee–Miller (LM) Method

The LM method is a variant of the LC method. It differs from the LC method in three ways:

- The fitting period begins in 1950;
- The adjustment of k_t involves fitting to the life expectancy $e(0)$ in year t;
- The jump-off rates are the observed rates in the jump-off year instead of the fitted rates.

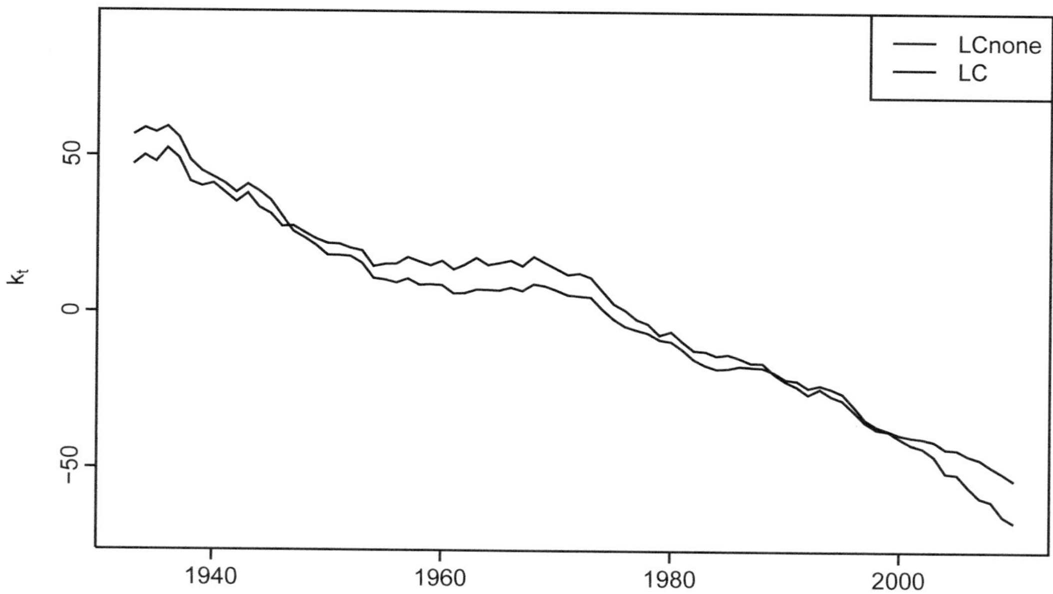

FIGURE 8.7
LC k_t with and without adjustment, U.S. male mortality, 1933–2010.

In their evaluation of the LC method, Lee & Miller (2001) found that the pattern of change in mortality rates was not constant over time, which is a strong assumption of the LC method. Consequently, the adjustment of historical principal component scores resulted in a large estimation error. To overcome this, Lee & Miller (2001) adopted 1950 as the commencing year of the fitting period due to different age patterns of change for 1900–1949 and 1950–1995. This fitting period had previously been used by Tuljapurkar et al. (2000).

In addition, the adjustment of k_t was done by fitting to observed life expectancy in year t, rather than by fitting to total deaths in year t. This has the advantage of eliminating the need for population data. Further, Lee & Miller (2001) found a mismatch between fitted rates for the final year of the fitting period and observed rates in that year. This jump-off error was eliminated by using observed rates in the jump-off year.

The LM method is implemented as follows:

```
> lm.male <- lca(usa.90, series="male", adjust="e0", years=1950:max(usa$year))
> forecast.lm.male <- forecast(lm.male, h=20, jumpchoice = "actual")
```

The LM method has been found to produce more accurate forecasts than the original LC method (Booth et al. 2005, 2006).

8.3.3 Booth–Maindonald–Smith (BMS) Method

The BMS method is another variant of the LC method. The BMS method differs from the LC method in three ways:

- The fitting period is determined on the basis of a statistical ‘goodness of fit’ criterion, under the assumption that the principal component score k_1 is linear;
- The adjustment of k_t involves fitting to the age distribution of deaths rather than to the total number of deaths;

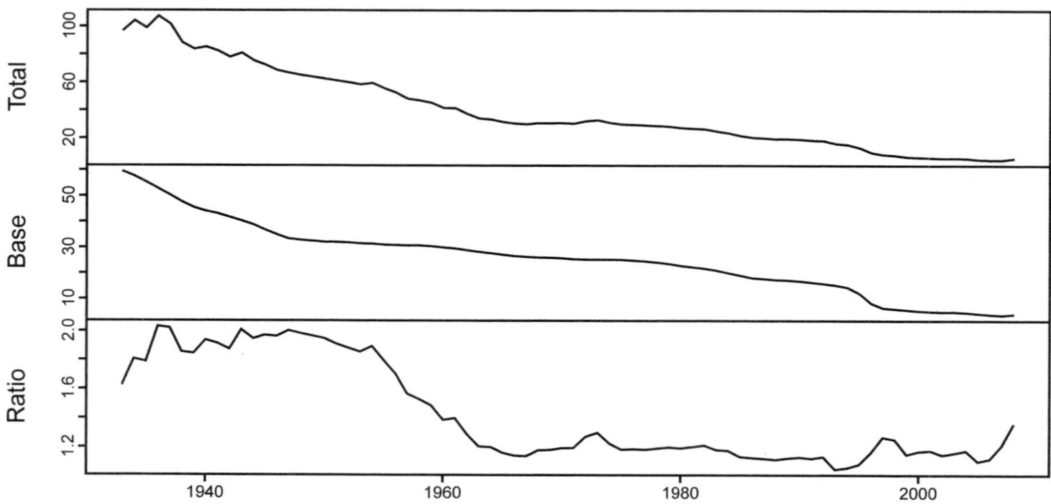

FIGURE 8.8
Mean deviances for base and total models and their ratio, U.S. male mortality, 1933–2010.

- The jump-off rates are the fitted rates under this fitting regime.

A common feature of the LC method is the linearity of the best fitting time series model of the first principal component score, but Booth, Maindonald & Smith (2002) found the linear time series to be compromised by structural change. By first assuming the linearity of the first principal component score, the BMS method seeks to achieve the optimal 'goodness of fit' by selecting the optimal fitting period from all possible fitting periods ending in year n. The optimal fitting period is determined based on the smallest ratio of the mean deviances of the fit of the underlying LC model to the overall linear fit.

Instead of fitting to the total number of deaths, the BMS method uses a quasi-maximum likelihood approach by fitting the Poisson distribution to model age-specific deaths, and using deviance statistics to measure the 'goodness of fit' (Booth, Maindonald & Smith 2002). The jump-off rates are taken to be the fitted rates under this adjustment.

The BMS method is implemented thus:

```
> bms.male <- bms(usa.90 series="male", minperiod = 30, breakmethod = "bms")
> forecast.bms.male <- forecast(bms.male, h=20)
```

It is advisable to review the deviances (Figure 8.8) and the automatically chosen fitting period, which is based on a local minimum of the ratio of total to base mean deviances.

```
> plot(bms.male$mdevs, main="Mean deviances for base and total models", xlab="")
> bms.male$year[1]
[1] 1967
```

Estimated k_t and forecast rates (Figure 8.9) are obtained via

```
> plot(bms.male$kt)
> plot(usa.90, series="male", years=bms.male$year[1]:max(usa$year),
+ ylim=c(-10,0), lty=2, main="BMS method: observed (1979-2010)
     and forecast (2011-2030) rates")
> lines(forecast.bms.male)
```

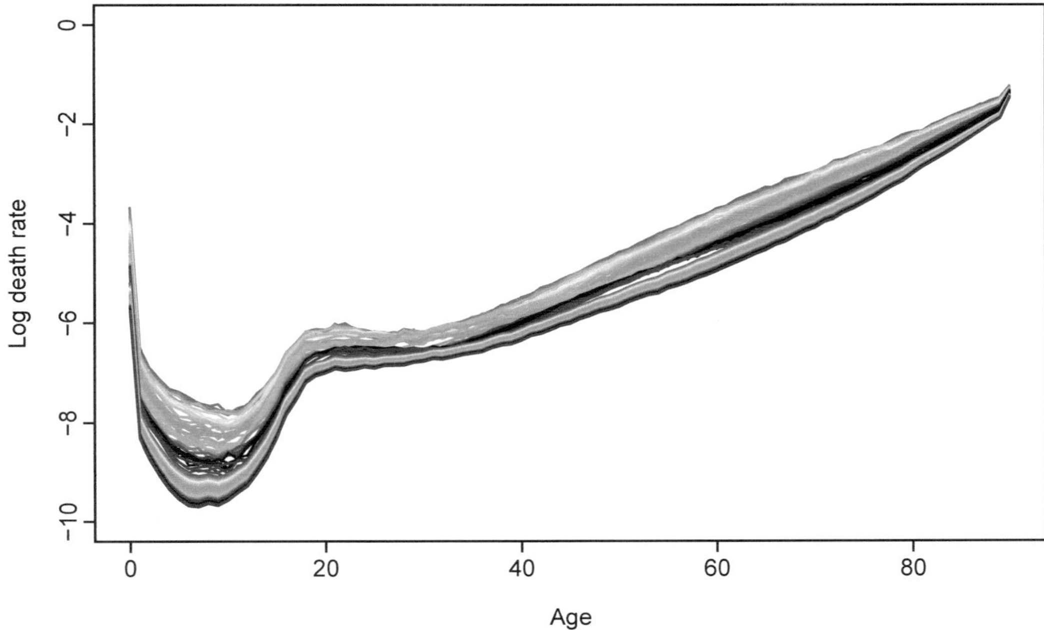

FIGURE 8.9
Observed (1979–2010) and forecast (2011–2030) mortality rates using the BMS method for U.S. males.

An alternative implementation using the `lca()` function, which permits all possible variants to be produced, is:

```
> bms.male <- lca(usa.90, series="male", adjust="dxt", chooseperiod=TRUE,
+ minperiod = 30, breakmethod = "bms")
> forecast.bms.male <- forecast(bms.male, h=20)
```

Forecasts from the BMS method have been found to be more accurate than those from the original LC method and of similar accuracy as those from the LM method (Booth et al. 2005, 2006).

8.3.4 Hyndman–Ullah (HU) Method

Using the functional data analysis paradigm of Ramsay & Silverman (2005), Hyndman & Ullah (2007) proposed a nonparametric method for modeling and forecasting log mortality rates. This approach extends the LC method in four ways:

- The log mortality rates are smoothed prior to modeling;
- Functional principal components analysis is used;
- More than one principal component is used in forecasting;
- The forecasting models for the principal component scores are typically more complex than the RWD model.

The log mortality rates are smoothed using penalized regression splines as described in Section 8.2. To emphasize that age, x, is now considered a continuous variable, we write $m_t(x)$ to represent mortality rates for age $x \in [x_1, x_p]$ in year t. We then define $z_t(x) = \log m_t(x)$ and write

$$z_t(x_i) = f_t(x_i) + \sigma_t(x_i)\varepsilon_{t,i}, \quad i = 1, \dots, p, \; t = 1, \dots, n, \tag{8.2}$$

where $f_t(x_i)$ denotes a smooth function of x as before; $\sigma_t(x_i)$ allows the amount of noise to vary with x_i in year t, thus rectifying the assumption of homoskedastic error in the LC model; and $\varepsilon_{t,i}$ is an independent and identically distributed standard normal random variable.

Given continuous age x, functional principal components analysis (FPCA) is used in the decomposition. The set of age-specific mortality curves is decomposed into orthogonal functional principal components and their uncorrelated principal component scores. That is,

$$f_t(x) = a(x) + \sum_{j=1}^{J} b_j(x) k_{t,j} + e_t(x), \tag{8.3}$$

where $a(x)$ is the mean function estimated by $\hat{a}(x) = \frac{1}{n}\sum_{t=1}^{n} f_t(x)$; $\{b_1(x), \dots, b_J(x)\}$ is a set of the first J functional principal components; $\{k_{t,1}, \dots, k_{t,J}\}$ is a set of uncorrelated principal component scores; $e_t(x)$ is the residual function with mean zero; and $J < n$ is the number of principal components used. Note that we use $a(x)$ rather than a_x to emphasise that x is not treated as a continuous variable.

Multiple principal components are used because the additional components capture non-random patterns that are not explained by the first principal component (Booth, Maindonald & Smith 2002, Renshaw & Haberman 2003, Koissi, Shapiro & Högnäs 2006). Hyndman & Ullah (2007) found $J = 6$ to be larger than the number of components actually required to produce white noise residuals, and this is the default value. The conditions for the existence and uniqueness of $k_{t,j}$ are discussed by Cardot, Ferraty & Sarda (2003).

Although Lee & Carter (1992) did not rule out the possibility of a more complex time series model for the k_t series, in practice an RWD model has typically been employed in the LC method. For higher order principal components, which are orthogonal by definition to the first component, other time series models arise for the principal component scores. For all components, the HU method selects the optimal time series model using standard model-selection procedures (e.g. AIC). By conditioning on the observed data $\boldsymbol{\mathcal{I}} = \{z_1(x), \dots, z_n(x)\}$ and the set of functional principal components $\boldsymbol{B} = \{b_1(x), \dots, b_J(x)\}$, the h-step-ahead forecast of $z_{n+h}(x)$ can be obtained by

$$\hat{z}_{n+h|n}(x) = \mathbb{E}[z_{n+h}(x)|\boldsymbol{\mathcal{I}}, \boldsymbol{B}] = \hat{a}(x) + \sum_{j=1}^{J} b_j(x)\hat{k}_{n+h|n,j},$$

where $\hat{k}_{n+h|n,j}$ denotes the h-step-ahead forecast of $k_{n+h,j}$ using a univariate time series model, such as the optimal ARIMA model selected by the automatic algorithm of Hyndman & Khandakar (2008), or an exponential smoothing state space model (Hyndman et al. 2008).

Because of the orthogonality of all components, it is easy to derive the forecast variance as

$$\hat{v}_{n+h|n}(x) = \text{var}[z_{n+h}(x)|\boldsymbol{\mathcal{I}}, \boldsymbol{B}] = \sigma_a^2(x) + \sum_{j=1}^{J} b_j^2(x) u_{n+h|n,j} + v(x) + \sigma_t^2(x),$$

where σ_a^2 is the variance of $\hat{a}(x)$; $u_{n+h,n,j}$ is the variance of $k_{n+h,j} \mid k_{1,j}, \dots, k_{n,j}$ (obtained from the time series model); $v(x)$ is the variance of $e_t(x)$ and $\sigma_t(x)$ is defined in (8.2). This expression is used to construct prediction intervals for future mortality rates in R.

For the Hyndman–Ullah methods, we return to the complete dataset usa.

```
> usa
Mortality data for USA
    Series: female male total
    Years: 1933 - 2010
    Ages:  0 - 110
```

In order to avoid the war years, the fitting period commences in 1950. Given that smoothing is employed, the age range is 0 to 100+.

```
> usa1950.100 <- extract.years(extract.ages(usa, 0:100), years=1950:max(usa$year))
> usa1950.100
Mortality data for USA
    Series: female male total
    Years: 1950 - 2010
    Ages:  0 - 100
```

The HU method is implemented as below. The model and forecast are seen in Figures 8.10 and 8.11.

```
> smus1950.100 <- smooth.demogdata(usa1950.100)
> fdm.male <- fdm(smus1950.100, series="male", order=3)
> forecast.fdm.male <- forecast.fdm(fdm.male, h=20)
> plot(forecast.fdm.male, plot.type="component")
> plot(smus1950.100, series="male", ylim=c(-10,0), lty=2, main ="HU method:
+ observed (1950-2010) and forecast (2011-2030) rates")
> lines(forecast.fdm.male)
```

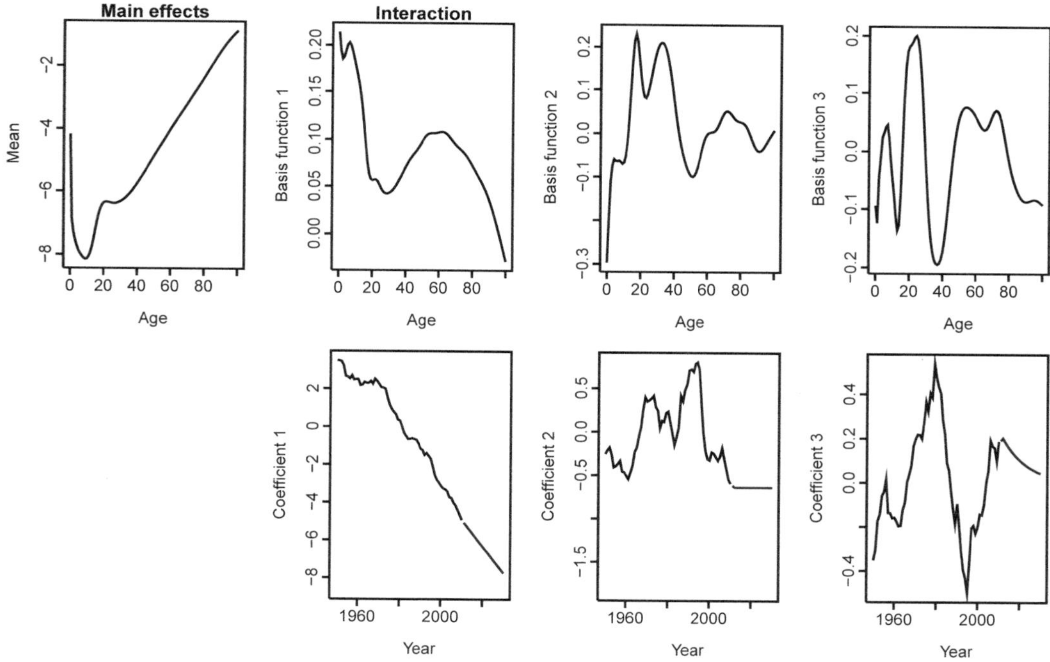

FIGURE 8.10
HU model and forecast, U.S. male mortality. Fitting period = 1950–2010; forecasting horizon = 20 years.

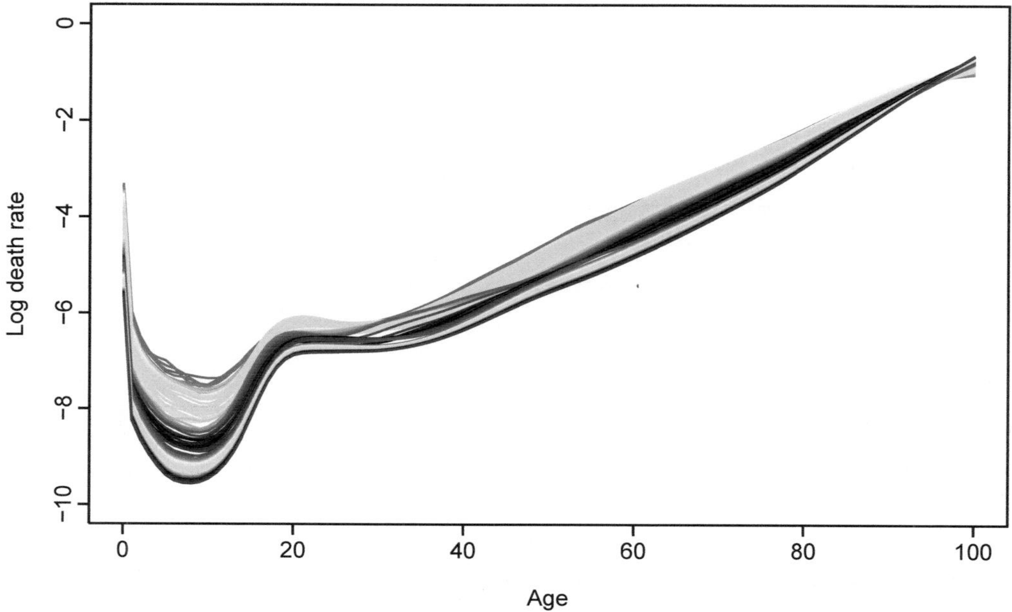

FIGURE 8.11
Observed (1950-2000) and forecast (2011-2030) mortality rates using the HU method for U.S. males.

8.3.5 Robust Hyndman–Ullah (HUrob) Method

The presence of outliers can seriously affect the performance of modelling and forecasting. The HUrob method is designed to eliminate their effect. This method utilizes the reflection-based principal component analysis (RAPCA) algorithm of Hubert, Rousseeuw & Verboven (2002) to obtain projection-pursuit estimates of principal components and their associated scores. The integrated squared error provides a measure of the accuracy of the principal component approximation for each year (Hyndman & Ullah 2007). Outlying years would result in a larger integrated squared error than the critical value obtained by assuming normality of $e_t(x)$ (see Hyndman & Ullah 2007, for details). By assigning zero weight to outliers, the HU method can then be used to model and forecast mortality rates without the possible influence of outliers.

The HUrob method is implemented as follows:

```
> fdm.male <- fdm(smus1950.100, series="male", method="rapca")
> forecast.fdm.male <- forecast.fdm(fdm.male, h=20)
> plot(smus1950.100, series="male", ylim=c(-10,0), lty=2, main ="HUrob method:
+ observed (1950-2010) and forecast (2011-2030) rates")
> lines(forecast.fdm.male)
```

8.3.6 Weighted Hyndman–Ullah (HUw) Method

The HU method does not weight annual mortality curves in the functional principal components analysis. However, it might be argued that more recent experience has greater relevance to the future than more distant experience. The HUw method uses geometrically decaying weights in the estimation of the functional principal components, thus allowing these quantities to be based more on recent data than on data from the distant past.

The weighted functional mean $a^*(x)$ is estimated by the weighted average

$$\hat{a}^*(x) = \sum_{t=1}^{n} w_t f_t(x), \tag{8.4}$$

where $\{w_t = \beta(1-\beta)^{n-t}, t = 1, \ldots, n\}$ denotes a set of weights, and $0 < \beta < 1$ denotes the weight parameter. Hyndman & Shang (2009) describe how to estimate β from the data. The set of weighted curves $\{w_t[f_t(x) - \hat{a}^*(x)]; t = 1, \ldots, n\}$ is decomposed using FPCA:

$$f_t(x) = \hat{a}^*(x) + \sum_{j=1}^{J} b_j^*(x) k_{t,j} + e_t(x), \tag{8.5}$$

where $\{b_1^*(x), \ldots, b_J^*(x)\}$ is a set of weighted functional principal components. By conditioning on the observed data $\boldsymbol{\mathcal{I}} = \{z_1(x), \ldots, z_n(x)\}$ and the set of weighted functional principal components $\boldsymbol{B}^*$, the h-step-ahead forecast of $z_{n+h}(x)$ can be obtained by

$$\hat{z}_{n+h|n}(x) = \mathbb{E}[z_{n+h}(x)|\boldsymbol{\mathcal{I}}, \boldsymbol{B}^*] = \hat{a}^*(x) + \sum_{j=1}^{J} b_j^*(x)\hat{k}_{n+h|n,j}.$$

The HUw method is implemented as follows:

```
> fdm.male <- fdm(smus1950.100, series="male", method="classical", weight=TRUE, beta=0.1)
> forecast.fdm.male <- forecast.fdm(fdm.male, h=20)
> plot(smus1950.100, series="male", ylim=c(-10,0), lty=2, main ="HUw method:
    + observed (1950-2010) and forecast (2011-2030) rates")
> lines(forecast.fdm.male)
```

8.4 Other Mortality Forecasting Methods

Other extrapolative mortality forecasting methods are included here for completeness, but are not considered in detail as the methods are not fully implemented in packages available on CRAN.

A number of methods have been developed to account for the significant impact of cohort (year of birth) in some countries. In the United Kingdom, males born around 1931 have experienced higher rates of mortality improvement than earlier or later cohorts (Willets 2004); less marked effects have also been observed elsewhere (Cairns 2009).

The Renshaw and Haberman (RH) (2006) extension to Lee-Carter to include cohort effects can be written as[1]

$$\log(m_{x,t}) = a_x + b_x^1 k_t + b_x^2 \gamma_{t-x} + \varepsilon_{x,t}, \tag{8.6}$$

[1]For clarity, models have been written in a standardised format which may in some cases differ from the form used by the authors originally. a_x and b_x terms are used for age-related effects, k_t terms for period-related effects, and γ_{t-x} terms for cohort-related effects. Models originally expressed in terms of the force of mortality are expressed in terms of the central death rate; these are equivalent under the assumption of a constant force of mortality over each year of age.

where a_x is the age pattern of the log mortality rates averaged across years, k_t represents the general level of mortality in year t, γ_{t-x} represents the general level of mortality for the cohort born in year $(t-x)$, b_x^1 and b_x^2 measure the relative response at age x to changes in k_t and γ_{t-x}, respectively, and $\varepsilon_{x,t}$ is the residual at age x. The fitted k_t and γ_{t-x} parameters are forecast using univariate time series models. The model can be implemented using the **ilc** functions (Butt & Haberman 2009).

A subsequent related model in which b_x^2 is set equal to 1 at all ages (Haberman & Renshaw 2011) was found to resolve some forecasting issues associated with the original. The Age-Period-Cohort (APC) model (Currie 2006) incorporates age, time and cohort effects that are independent in their effects on mortality,

$$\log(m_{x,t}) = a_x + k_t + \gamma_{t-x} + \varepsilon_{x,t}. \tag{8.7}$$

The two-dimensional P-spline method of Currie et al. (2004) has already been described in Section 8.2. Forecast rates are estimated simultaneously with fitting the mortality surface. Implementation of the two-dimensional P-spline method to produce mortality forecasts uses the `MortalitySmooth` package. Forecasts of U.S. male mortality rates and plots of age 65 and age 85 forecast rates with prediction intervals can be produced as follows, following the commands already shown in Section 8.2:

```
> library(MortalitySmooth)
> forecastyears <- 2011:2030
> forecastdata <- list(x=usa1950$age, y=forecastyears)
> CDEpredict <- predict(fitBIC, newdata=forecastdata, se.fit=TRUE)
> whiA <- c(66,86)
> plot(usa1950, series="male", age=whiA-1, plot.type="time",
+ xlim=c(1950,2030), ylim=c(-6.2,-1), xlab="years",
+ main="USA: male projected death rates using 2-dimensional CDE method", col=c(1,2))
> matlines(forecastyears, t(CDEpredict$fit[whiA,]), lty=1, lwd=2)
> matlines(forecastyears, t(CDEpredict$fit[whiA,]+2*CDEpredict$se.fit[whiA,]), lty=2)
> matlines(forecastyears, t(CDEpredict$fit[whiA,]-2*CDEpredict$se.fit[whiA,]), lty=2)
> legend("bottomleft", lty=1, col=1:2, legend=c("Age 65", "Age 85"))
```

In addition to being applied in the age and period dimensions, the two-dimensional P-spline method can incorporate cohort effects by instead being applied to age-cohort data.

Cairns et al. (2006*a*) have forecast mortality at older ages using a number of models for $\text{logit}(q_{x,t}) = \log[q_{x,t}/(1-q_{x,t})]$, where $q_{x,t}$ is the probability that an individual aged x at time t will die before time $t+1$. The original CBD model (Cairns et al. 2006*a*) is

$$\text{logit}(q_{x,t}) = k_t^1 + (x-\bar{x})k_t^2 + \varepsilon_{x,t}, \tag{8.8}$$

where $\bar{x}$ is the mean age in the sample range. Later models (Cairns 2009) incorporate a combination of cohort effects and a quadratic term for age:

$$\text{logit}(q_{x,t}) = k_t^1 + (x-\bar{x})k_t^2 + \gamma_{t-x} + \varepsilon_{x,t}, \tag{8.9}$$

$$\text{logit}(q_{x,t}) = k_t^1 + (x-\bar{x})k_t^2 + ((x-\bar{x})^2 - \hat{\sigma}_x^2) + \gamma_{t-x} + \varepsilon_{x,t}, \tag{8.10}$$

$$\text{logit}(q_{x,t}) = k_t^1 + (x-\bar{x})k_t^2 + (x_c - x)\gamma_{t-x} + \varepsilon_{x,t}, \tag{8.11}$$

where the constant parameter x_c is to be estimated and the constant $\hat{\sigma}_x^2$ is the mean of $(x-\bar{x})^2$. Other authors (e.g., Plat 2009) have proposed related models.

The LifeMetrics R software package implements the Lee-Carter method (using maximum likelihood estimation and a Poisson distribution for deaths) along with RH, APC, P-splines and the four CBD methods. The software, which is not part of CRAN, is available from www.

jpmorgan.com/pages/jpmorgan/investbk/solutions/lifemetrics/software. The software and the methods it implements is described in detail in Coughlan et al. (2007).

De Jong & Tickle (2006) (DJT) tailor the state space framework to create a method that integrates model estimation and forecasting, while using B-splines to reduce dimensionality and build in the expected smooth behaviour of mortality over age. Compared with Lee-Carter, the method uses fewer parameters, produces smooth forecast rates and offers the advantages of integrated estimation and forecasting. A multi-country evaluation of out-of-sample forecast performance found that LM, BMS, HU and DJT gave significantly more accurate forecast log mortality rates relative to the original LC, with no one method significantly more accurate than the others (Booth et al. 2006).

8.5 Coherent Mortality Forecasting

In modelling mortality for two or more sub-populations of a larger population simultaneously, it is usually desirable that the forecasts are non-divergent or "coherent". The Product-Ratio method (Hyndman et al. 2013) achieves coherence through the convergence to a set of appropriate constants of forecast age-specific ratios of death rates for any two sub-populations. The method makes use of functional forecasting (HU methods).

The method is presented here in terms of forecasting male and female age-specific death rates; extension to more than two sub-populations is straightforward (Hyndman et al. 2013). Let $s_{t,F}(x) = \exp[f_{t,F}(x)]$ denote the smoothed female death rate for age x and year t, $t = 1, \ldots, n$. Similar notation applies for males.

Let the square roots of the products and ratios of the smoothed rates for each sex be

$$p_t(x) = \sqrt{s_{t,M}(x)s_{t,F}(x)} \quad \text{and} \quad r_t(x) = \sqrt{s_{t,M}(x)/s_{t,F}(x)}.$$

These are modeled by functional time series models:

$$\log[p_t(x)] = \mu_p(x) + \sum_{k=1}^{K} \beta_{t,k}\phi_k(x) + e_t(x), \tag{8.12a}$$

$$\log[r_t(x)] = \mu_r(x) + \sum_{\ell=1}^{L} \gamma_{t,\ell}\psi_\ell(x) + w_t(x), \tag{8.12b}$$

where the functions $\{\phi_k(x)\}$ and $\{\psi_\ell(x)\}$ are the principal components obtained from decomposing $\{p_t(x)\}$ and $\{r_t(x)\}$, respectively, and $\beta_{t,k}$ and $\gamma_{t,\ell}$ are the corresponding principal component scores. The function $\mu_p(x)$ is the mean of the set of curves $\{p_t(x)\}$, and $\mu_r(x)$ is the mean of $\{r_t(x)\}$. The error terms, given by $e_t(x)$ and $w_t(x)$, have zero mean and are serially uncorrelated.

The coefficients $\{\beta_{t,1}, \ldots, \beta_{t,K}\}$ and $\{\gamma_{t,1}, \ldots, \gamma_{t,L}\}$ are forecast using time series models, as detailed in Section 8.3.4. To ensure the forecasts are coherent, the coefficients $\{\gamma_{t,\ell}\}$ are constrained to be stationary processes. The forecast coefficients are then multiplied by the basis functions, resulting in forecasts of the curves $p_t(x)$ and $r_t(x)$ for future t. If $p_{n+h|n}(x)$ and $r_{n+h|n}(x)$ are h-step forecasts of the product and ratio functions, respectively, then forecasts of the sex-specific death rates are obtained using $s_{n+h|n,M}(x) = p_{n+h|n}(x)r_{n+h|n}(x)$ and $s_{n+h|n,F}(x) = p_{n+h|n}(x)/r_{n+h|n}(x)$.

The method makes use of the fact that the product and ratio behave roughly independently of each other provided the sub-populations have approximately equal variances. (If

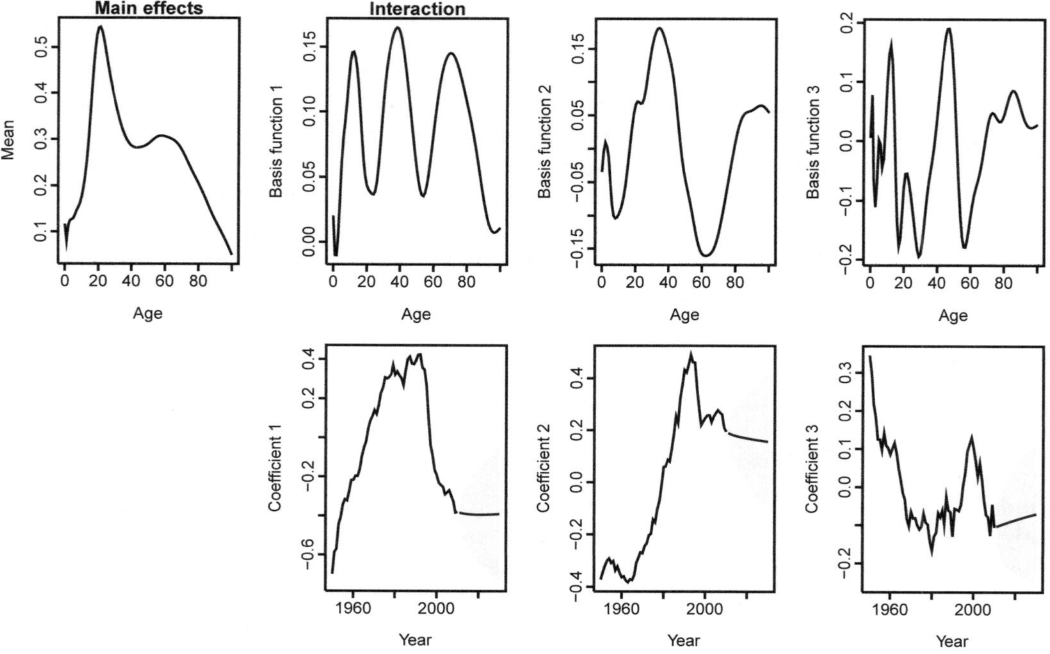

FIGURE 8.12
Ratio function decomposition (1950–2010) with forecast (2011–2030), U.S. mortality.

there are substantial differences in the variances, the forecasts remain unbiased but less efficient.)

The Product-Ratio method is illustrated in Figures 8.12 to 8.14 and implemented as follows:

```
> usa.pr <- coherentfdm(smus1950.100, weight=TRUE, beta=0.05)
> usa.pr.f <- forecast(usa.pr, h=20)

> plot(usa.pr.f$product, plot.type="component", components=3)
> plot(usa.pr.f$ratio$male, plot.type="component", components=3)

> par(mfrow=c(1,2))
> plot(usa.pr$product$y, ylab="Log of geometric mean death rate", font.lab=2,
+ lty=2, las=1, ylim=c(-10,-1), main="Product function")
> lines(usa.pr.f$product)
> plot(sex.ratio(smus1950.100), ylab="Sex ratio of rates: M/F", ylim=c(0.7,3.5),
+ lty=2, las=1, font.lab=2, main="Ratio function")
> lines(sex.ratio(usa.pr.f))

> plot(smus1950.100, series="male", lty=2, ylim=c(-11,-1), main="Males")
> lines(usa.pr.f$male)
> plot(smus1950.100, series="female", lty=2, ylim=c(-11,-1), main="Females")
> lines(usa.pr.f$female)
```

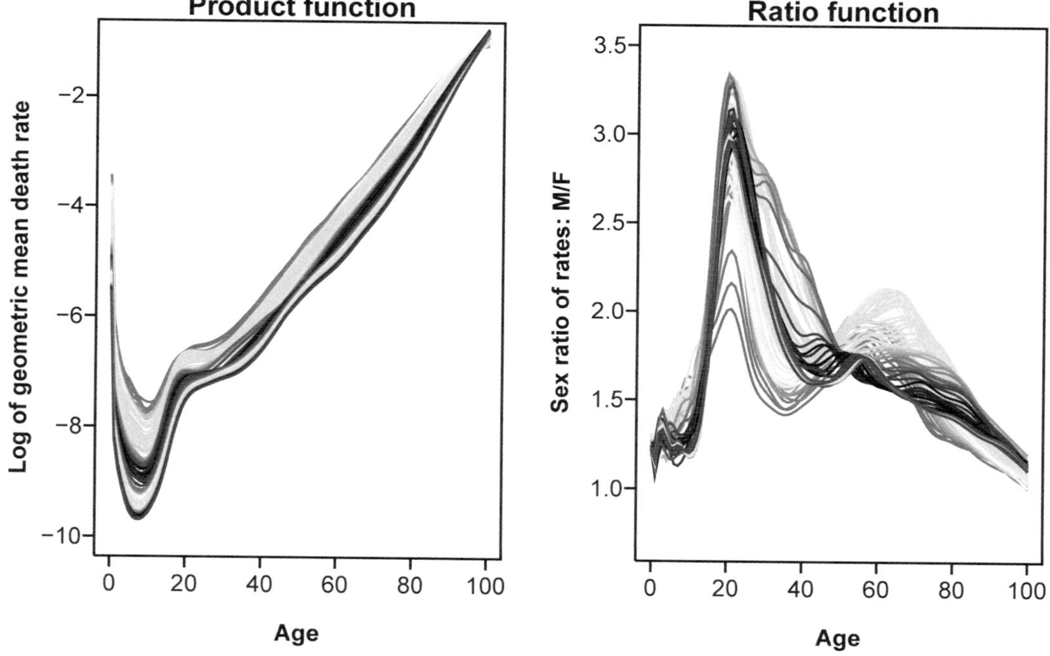

FIGURE 8.13
Observed(1950–2010) and forecast(2011–2030) product and ratio functions, U.S. mortality.

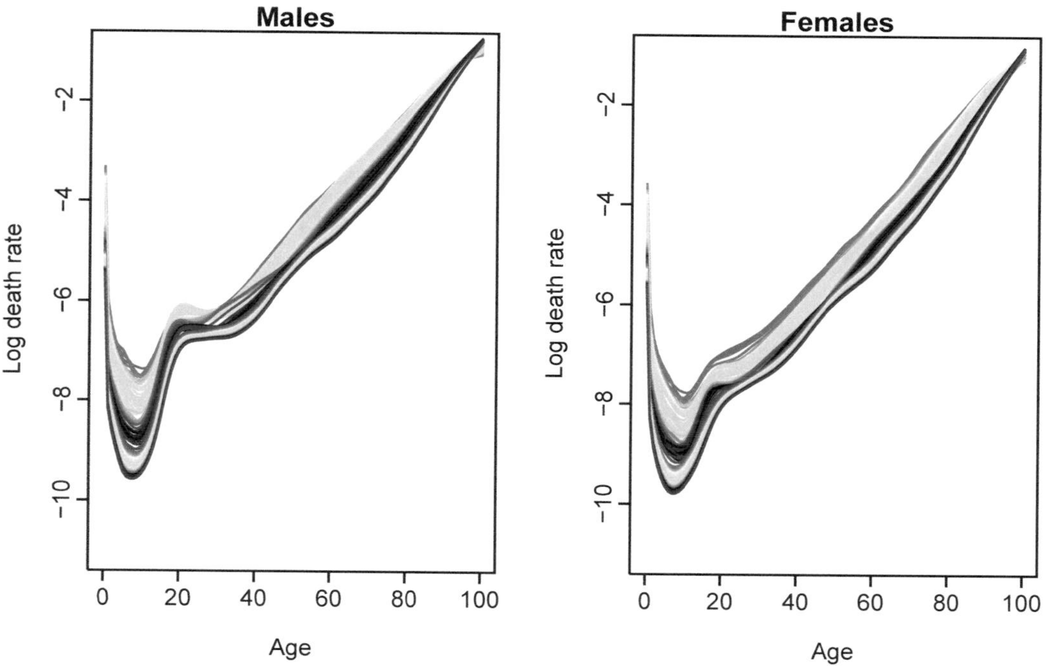

FIGURE 8.14
Observed(1950–2010) and forecast(2011–2030) male and female mortality rates using the product-ratio method with functional data models, United States.

8.6 Life Table Forecasting

The methods described in this chapter generate forecast m_x rates, which can then be used to produce forecast life table functions using standard methods (e.g. Chiang 1984). Assuming that m_x rates are available for ages $0, 1, \ldots, \omega - 1, \omega+$, the **lifetable** function in **demography** generates life table functions from a radix of $l_0 = 1$ as follows for single years of age to $\omega - 1$:

$$
\begin{aligned}
q_x &= m_x/(1 + (1 - a_x)m_x), &(8.13)\\
d_x &= l_x q_x, &(8.14)\\
l_{x+1} &= l_x - d_x, &(8.15)\\
L_x &= l_x - d_x(1 - a_x), &(8.16)\\
T_x &= L_x + L_{x+1} + \cdots + L_{\omega-1} + L_{\omega+}, &(8.17)\\
e_x &= T_x/l_x, &(8.18)
\end{aligned}
$$

where $a_x = 0.5$ for $x = 1, \ldots, \omega - 1$, and a_0 values (which allow for the fact that deaths in this age group occur earlier than midway through the year of age on average) are from Coale et al. (1983). For the final age group, $q_{\omega+} = 1$, $L_{\omega+} = l_x/m_x$, and $T_{\omega+} = L_{\omega+}$. For life tables commencing at an age other than zero, the same formulae apply, generated from a radix of 1 at the commencing age.

The **demography** package produces life tables using `lifetable`, and life expectancies using the function `life.expectancy`. For forecast life expectancies, `flife.expectancy` is used to produce the point forecast and the prediction interval. Additionally, `e0` is a shorthand wrapper for `flife.expectancy` with `age=0`. The m_x rates on which the life table is based can be the rates applying in a future forecast year t, in which case a period or cross-sectional life table is generated, or can be rates that are forecast to apply to a certain cohort, in which case a cohort life table is generated. All functions use the `cohort` argument to give cohort rather than period life tables and life expectancies.

For example, to generate the cross-sectional life table for males in 1980, we use:

```
> lifetable(usa, series="male", year=1980, type="period")
Period lifetable for USA : male

Year: 1980
        mx     qx     lx     dx     Lx      Tx      ex
0   0.0142 0.0140 1.0000 0.0140 0.9871 69.9883 69.9883
1   0.0011 0.0011 0.9860 0.0011 0.9854 69.0012 69.9835
2   0.0007 0.0007 0.9849 0.0007 0.9845 68.0157 69.0584
3   0.0006 0.0006 0.9842 0.0006 0.9839 67.0312 68.1089
4   0.0005 0.0005 0.9836 0.0004 0.9834 66.0473 67.1491
:
99  0.4066 0.3379 0.0051 0.0017 0.0043  0.0123  2.3891
100 0.4249 1.0000 0.0034 0.0034 0.0080  0.0080  2.3533
```

To generate the cohort life expectancy for males aged 65 by year in which aged 65 (using the coherent functional model obtained in Section 8.5), we use:

```
> usa.pr <- coherentfdm(smus1950.100, weight=TRUE, beta=0.05)
> usa.pr.f40 <- forecast(usa.pr,h=40)
> flife.expectancy(usa.pr.f40$male, age=65, type="cohort")
```

```
     Point Forecast
1976       14.45048
1977       14.54158
1978       14.63950
1979       14.75728
1980       14.88563
:
2013       19.30161
2014       19.39335
2015       19.48484
```

To obtain prediction intervals for future life expectancies, we simulate the forecast log mortality rates as described in Hyndman & Booth (2008). Briefly, the simulated forecasts of log mortality rates are obtained by adding disturbances to the forecast basis function coefficients $k_{t,j}$ which are then multiplied by the fixed basis functions, $b_j(x)$ (assuming Equation 8.3). Then we calculate the life expectancy for each set of simulated log mortality rates. Prediction intervals are constructed from percentiles of the simulated life expectancies. This is all implemented in the `demography` package.

Using the coherent functional model obtained in Section 8.5, we can forecast period life expectancies (Figure 8.15) as follows:

```
> e0.fcast.m <- e0(usa.pr.f, PI=TRUE, series="male")
> e0.fcast.f <- e0(usa.pr.f, PI=TRUE, series="female")
> plot(e0.fcast.m, ylim=c(65,85), col="blue", fcol="blue", ylab="Years",
+ main="Product-Ratio method: coherent life expectancy forecasts")
> par(new=TRUE)
> plot(e0.fcast.f, ylim=c(65,85), col="red", fcol="red",main="")
> legend("topleft", lty=c(1,1),lwd=c(2,2), col=c("red", "blue"),
+ legend=c("female","male"))
```

An alternative approach to life expectancy forecasting is direct modeling, rather than via mortality forecasts. This is the approach taken by Raftery et al. (2013) who use a Bayesian hierarchical model for life expectancy, and pool information across countries in order to improve estimates. Their model is implemented in the `bayesLife` package, available on CRAN.

8.7 Life Insurance Products

Connections between demographic functions, described in this chapter, and life contingencies (described in Chapter 7) can be made using the function `probs2lifetable()` in library `lifecontingencies`. The input used in this function is a vector of probabilities, either p_x or q_x.

To generate period q_x rates that can be input into the `lifecontingencies` package, we simply use the `lifetable` function for either observed (past) or forecast rates:

```
> lifetable1980 <- lifetable(smus1950.100, series="male", year=1980, type="period")
> qx1980 <- lifetable1980$qx
> lifetable2025 <- lifetable(usa.pr.f$male, series="male", year=2025, type="period")
> qx2025 <- lifetable2025$qx
```

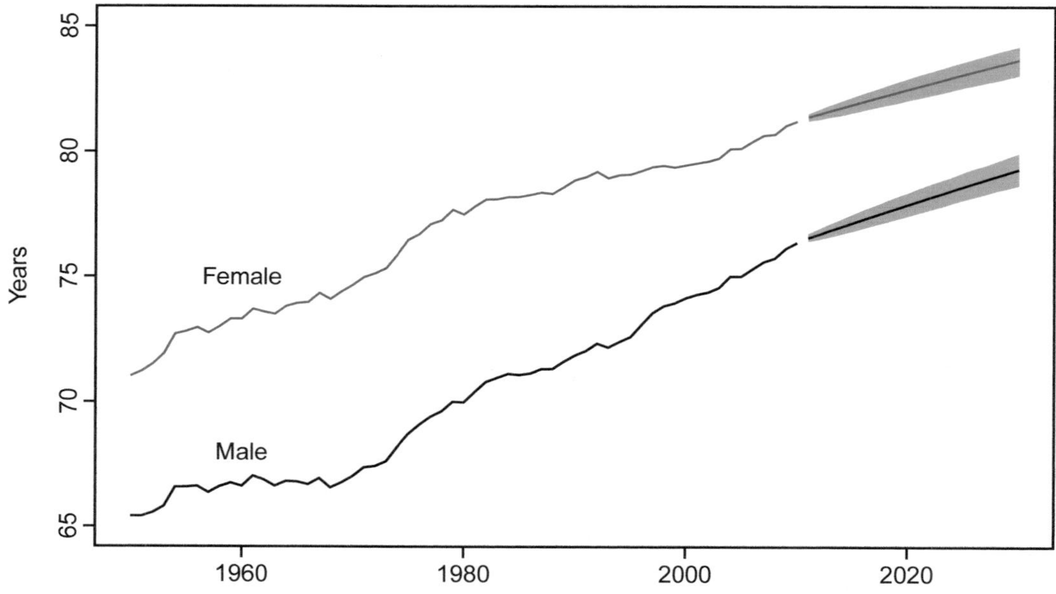

FIGURE 8.15
Observed(1950–2010) and forecast(2011–2030) male and female period life expectancy using the product-ratio method with functional data models, United States.

If we want to use cohort probabilities, we will in most cases need to use both observed $m_{x,t}$ and forecast $\widehat{m}_{x,t}$. Using the computations described above, and smoothed U.S. male rates at ages 0 to 100+ for 1950 to 2010, we obtain

```
> usa.pr <- coherentfdm(smus1950.100, weight=TRUE, beta=0.05)
> usa.pr.f150 <- forecast(usa.pr, h=150)
```

The matrix `smus1950.100$rate$male` contains past rates, $m_{x,t}$, and `usa.pr.f150$male$rate$male` contains forecast rates, $\widehat{m}_{x,t}$. We define the combined matrix, `com.mx`:

```
> com.mx <- cbind(smus1950.100$rate$male[1:nrow(usa.pr.f150$male$rate$male),],
+ usa.pr.f150$male$rate$male)
```

The cohort mortality rate for males born in 1950 is obtained using

```
> birthyear <- 1950
> mx <- com.mx[1:nrow(com.mx),(birthyear-1950+1):ncol(com.mx)]
> cohort.mx <- diag(mx)
> plot(log(cohort.mx), main="Cohort mortality rates, US males born in 1950",
+ ylab="Log death rate", xlab="Age")
```

Figure 8.16 shows the mortality rate surface as a function of x and t. The diagonal line is the cohort rate of mortality for males born in 1950.

```
> mat3d<-persp(seq(0,99,by=3),seq(1950,1950+209,by=5),
+ log(com.mx[seq(1,100,by=3),seq(1,210,by=5)]), theta=-30,
+ xlab="Age", ylab="Year", zlab="Log death rate", ticktype="detailed",
+ col=grey(.93), shade=TRUE)
> xyz3d <- trans3d(0:100,birthyear+0:100, log(cohort.mx), mat3d)
> lines(xyz3d, col="red", lwd=2)
```

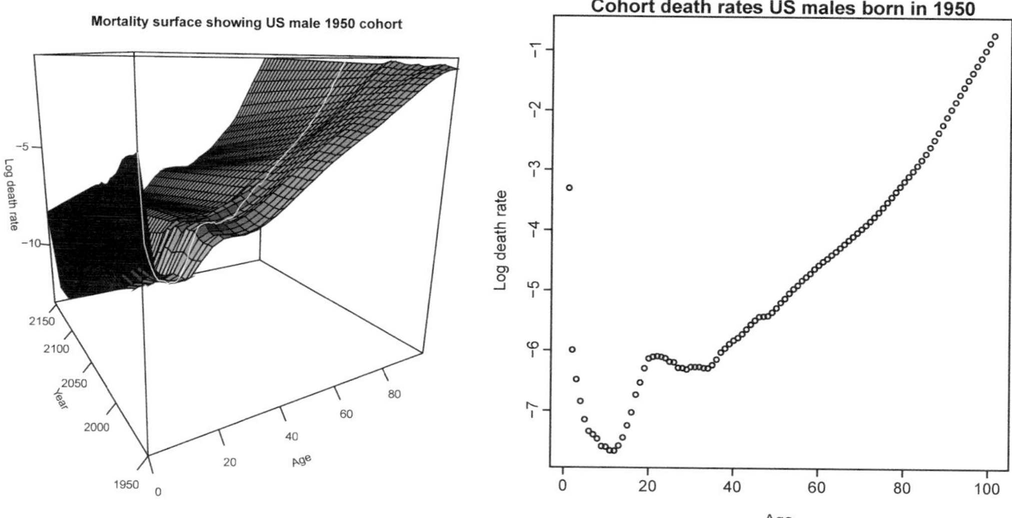

FIGURE 8.16
Observed (1950-2010) and forecast (2011-2150) rates, showing cohort rates for U.S. males born in 1950.

Based on vector `cohort.mx`, we compute q_x using the equations given in Section 8.6. Note that Chapters 7 and 8 use slightly different assumptions for the construction of the life table, and therefore calculated life expectancy and other life table quantities will differ between the two for the same set of input q_x rates.

8.8 Exercises

8.1 Download the Human Mortality Database (HMD) mortality data for Denmark and plot male mortality rates at single ages 0 to 95+ for the 20th century.

8.2 Using data from 1950 for Danish females aged 0–100+, smooth the data by the Currie–Durban–Eilers and Hyndman–Ullah methods. Plot the two smoothed curves and the observed data for 1950 and 2000.

8.3 Download HMD data for Canada. Using data for the total population, compare forecast life expectancy for the next 20 years from the Lee–Carter and Lee–Miller methods.

8.4 Apply the Booth–Maindonald–Smith method to total mortality data for Canada. What is the fitting period? How does this forecast compare with the Lee–Miller forecast in terms of life expectancy after 20 years?

8.5 Using female data for Japan (from HMD), apply the Hyndman–Ullah method and plot the first three components. Plot forecast mortality rates for the next 20 years. How does the forecast differ from a forecast of the same data using the Lee–Carter method without adjustment?

8.6 Using male data for Japan, apply the Hyndman–Ullah method to forecast 20 years ahead, and plot male and female observed and forecast life expectancies on the same graph.

8.7 Apply the product-ratio method of coherent forecasting to data by sex for Japan. Plot past and future product and ratio functions. Add coherent male and female forecast life expectancies to the previous life expectancy graph.

8.8 Plot the sex difference over time in observed life expectancy, in independently forecast life expectancy and in coherently forecast life expectancy.

9

Prospective Mortality Tables and Portfolio Experience

Julien Tomas
Université de Lyon 1
Lyon, France

Frédéric Planchet
Université de Lyon 1 - Prim'Act
Lyon, France

CONTENTS

9.1 Introduction and Motivation

In this chapter[1], we will describe an operational framework for constructing and validating prospective mortality tables specific to an insurer. The material is based on studies[2] carried out by the Institut des Actuaries. This research has been conducted with the aim of providing to French insurance companies methodologies to take into account their own mortality experience for the computation of their best estimate reserves.

We will present several methodologies and the process of validation allowing an organism to adjust a mortality reference to get closer to a best estimate adjustment of its mortality and longevity risks. The techniques proposed are based only on the two following elements:

- A reference of mortality
- Data, line by line originating from a portfolio provided by the insurer

Various methods of increasing complexity will be presented. They allow the organism some latitude of choice while preserving simplicity of implementation for the basic methodology. In addition, they provide a simple adjustment without the intervention of an expert.

The simplest approach is the application of a single factor of reduction / increase to the probabilities of death of the reference. In practice, this coefficient is the Standardized Mortality Ratio (SMR) of the population considered. The second method is a semiparametric Brass-type relational model. This model implies that the differences between the observed mortality, and the reference can be represented linearly with two parameters. For the third method, we will consider a Poisson generalized linear model including the baseline mortality of the reference as a covariate and allowing interactions with age and calendar year. The fourth method includes, in a first step, a nonparametric smoothing of the periodic table and, in a second step, the application of the rates of mortality improvement derived from the reference.

The validation will be assessed on three levels. The first level concerns the proximity between the observations and the model. It is assessed by the likelihood ratio test, the SMR test, and the Wilcoxon Matched-Pairs Signed-Ranks test. As a complement, the validation of the fit involves graphical diagnostics such as the analysis of the response, Pearson and deviance residuals, as well as the comparison between the predicted and observed mortality by attained age and calendar year. The second level involves the regularity of the fit. It is assessed by the runs test and signs test. Finally, the third level covers the plausibility and consistency of the mortality trends. It is evaluated by single indices summarizing the lifetime probability distribution for different cohorts at several ages. Moreover, it involves graphical diagnostics assessing the consistency of the observed and forecasted life expectancy. Additionally, if we have at our disposal the male and female mortality, we can compare the improvement and judge the plausibility of the common evolution of the mortality of the two genders. We ask the question where are the data originating from and based on this knowledge, what mixture of biological factors, medical advances and environmental changes would have to happen to cause this particular set of forecasts?

---

[1]This chapter has benefitted substantially from the constructive comments of the working group *Mortalité* of the Institut des Actuaries. This research was supported by grants from BNP Paribas Cardif Chair *Management de la modélisation* and the Institut des Actuaries.

[2]Available publicly here: www.ressources-actuarielles.net/gtmortalite.

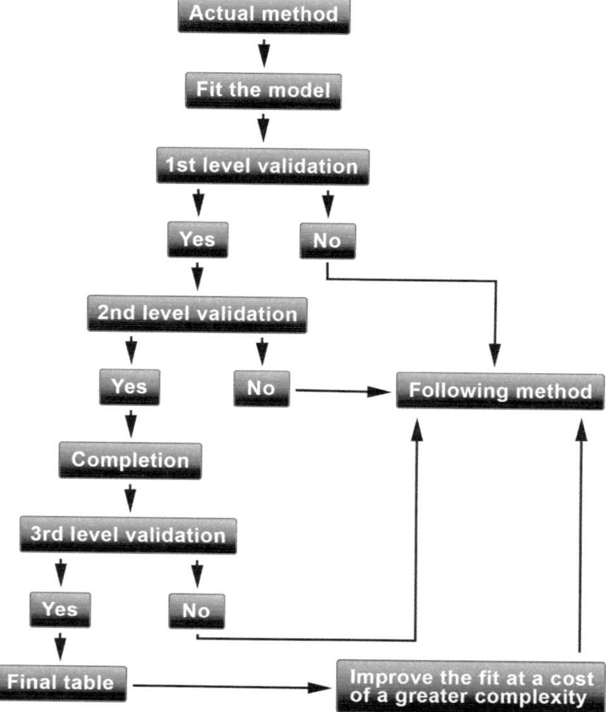

FIGURE 9.1
Process of validation.

The methods articulate around an iterative procedure allowing one to compare them and to choose parsimoniously the most satisfying one. We will present an operational framework based on the software R and on the package ELT available on CRAN. Briefly, the procedure is as follows. We will start with the first method; if the criteria corresponding to the first level of validation are not satisfied, we then switch to the second method as it is useless to continue the validation with the first method. If the criteria corresponding to the first level are satisfied, we continue the validation with the criteria of the second level. We can also turn to the following method to improve the fit at a cost of somewhat greater complexity without degrading the results of the criteria of the first level. The process of validation is summarized in Figure 9.1.

9.2 Notation, Data, and Assumption

We analyze the mortality as a function of both the attained age x and the calendar year t. The number of individuals at attained age x during calendar year t is denoted by $L_{x,t}$, and $D_{x,t}$ represents the number of deaths recorded from an exposure-to-risk $E_{x,t}$ that measures the time during which individuals are exposed to the risk of dying. It is the total time lived by these individuals during the period of observation. The probability of death at attained

TABLE 9.1
Observed characteristics of the male and female population of the portfolio.

	Mean Age In	Mean Age Out	Average Exposition	Mean Age at Death	Period of Observation Beginning	End
Male pop.	42.27	49.69	7.42	70.31	1996/01/01	2007/12/31
Female pop.	45.09	52.91	7.81	81.01	1996/01/01	2007/12/31

age x for the calendar year t, is denoted by $q_x(t)$ and computed according to the Hoem estimator, $q_x(t) = \frac{D_{x,t}}{E_{x,t}}$. We suppose that we have data line by line, originating from a portfolio, with a comma-separated variable in a `csv` file. The variables

- `Gender` with modalities `Male` and/or `Female`,
- `DateOfBirth`, `DateIn`, and `DateOut`,
- `Status` with modalities `other` and `deceased`,

should imperatively appear as displayed below:

```
Id     Gender DateOfBirth DateIn     DateOut    Status
100001 Female 1973/10/10  1995/12/01 2003/12/01 other
100002 Male   1901/05/12  1996/01/01 2001/04/21 deceased
100003 Female 1970/07/10  1995/11/01 2000/02/01 other
100004 Male   1916/07/07  1996/01/01 2002/01/28 deceased
100005 Female 1950/10/31  1995/11/01 2003/11/01 other
100006 Male   1918/04/06  1996/01/01 2002/06/01 deceased
```

The data in input should ideally be validated by the organism and it is supposed that they can be used without re-treatment.

Table 9.1 presents the observed characteristics of the male and female population of the data `MyPortfolio` used for our application. These are real data that, for confidentiality reasons, are composed by a mix of portfolios and, in consequence, can present atypical behaviors. We refer to Tomas & Planchet (2013) for additional illustrations of the methodologies proposed.

To each of the observations i, we associate the dummy variable δ_i indicating if the individual i dies or not,

$$\delta_i = \begin{cases} 1 & \text{if individual } i \text{ dies,} \\ 0 & \text{otherwise,} \end{cases}$$

for $i = 1, \ldots, L_{x,t}$. We define the time lived by individual i before $(x+1)th$ birthday by τ_i. We assume that we have at our disposal i.i.d. observations (δ_i, τ_i) for each of the $L_{x,t}$ individuals. Then,

$$\sum_{i=1}^{L_{x,t}} \tau_i = E_{x,t} \text{ et } \sum_{i=1}^{L_{x,t}} \delta_i = D_{x,t}.$$

Figures 9.2 and 9.3 display the observed statistics of the male and female population respectively. We refer to Section 10.2 for an example of R codes used to produce such statistics.

In this chapter, we will use the national demographic projections for the French population over the period 2007–2060, provided by the French National Office for Statistics,

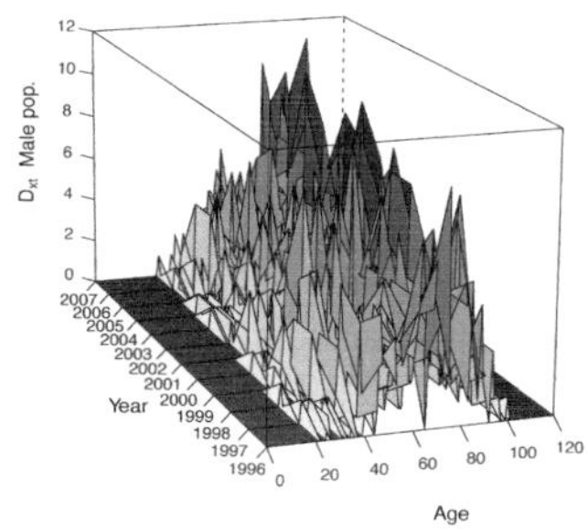

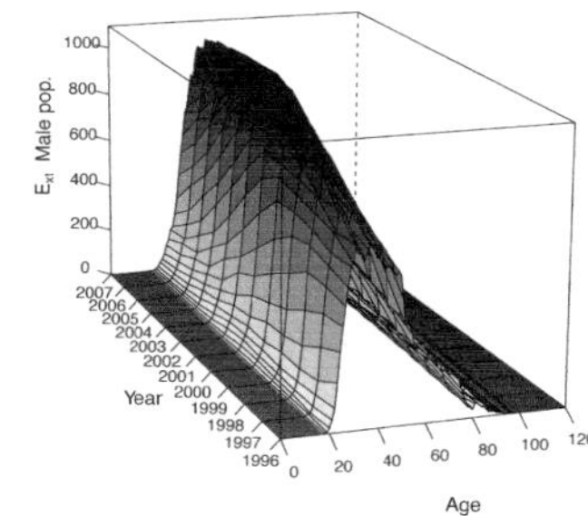

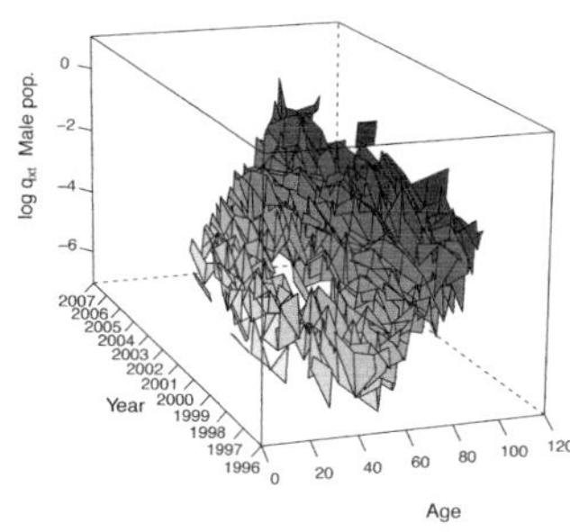

FIGURE 9.2
Observed statistics, male population.

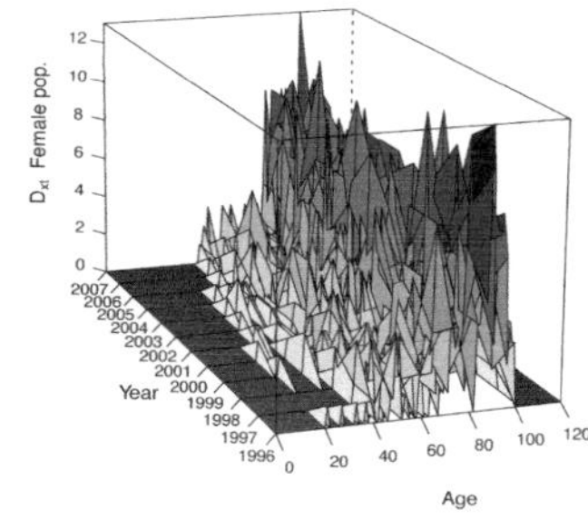

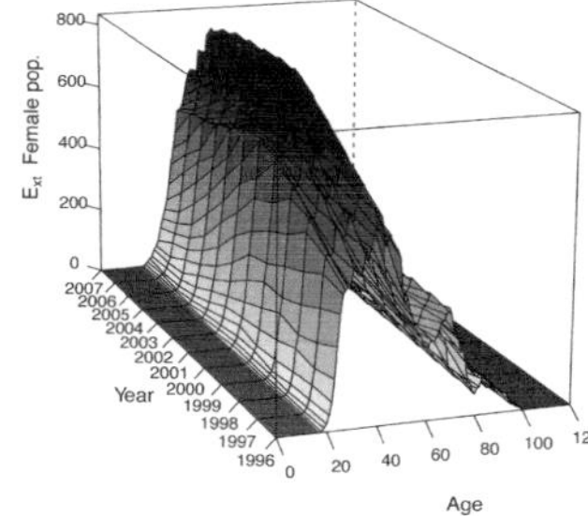

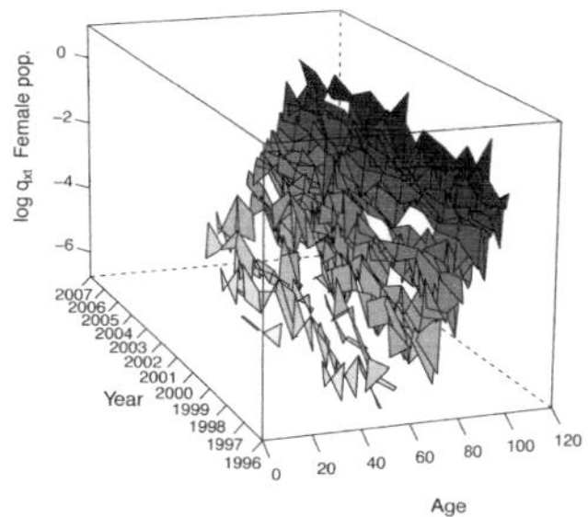

FIGURE 9.3
Observed statistics, female population.

INSEE, Blanpain & Chardon (2010) as a reference to adjust the mortality experience of our portfolio. These projections are based on assumptions concerning fertility, mortality, and migrations and we chose the baseline scenario for our application.

9.3 The Methods

In the following, we present the methodological aspects of the four methods considered, as well as the core of the functions contained in the package ELT.

9.3.1 Method 1: Approach Involving One Parameter with the SMR

The approach involving one parameter is the simplest methodology considered. It consists of applying a single factor of reduction / increase to the probability of death of the reference, denoted $q_x^{\text{ref}}(t)$. In practice, this coefficient is the Standardized Mortality Ratio (SMR) of the population considered, see Liddell (1984). Then we obtain the probabilities of death of the organism, denoted $\widetilde{q}_x(t)$, for $x \in [\underline{x}, \overline{x}]$ et $t \in [\underline{t}, \overline{t}]$ by

$$\widetilde{q}_x(t) = \text{SMR} \times q_x^{\text{ref}}(t) \quad \text{with} \quad \text{SMR} = \frac{\sum_{(x^*,t^*)} D_{x,t}}{\sum_{(x^*,t^*)} E_{x,t}\, q_x^{\text{ref}}(t)},$$

where x^* and t^* correspond to the age range and to the period of observation in common with the reference of mortality, respectively. The method is implemented in R as follows:

```
> FctMethod1 = function(d, e, qref, x1, x2, t1, t2){
+     SMR <- sum(d[x1 - min(as.numeric(rownames(d))) + 1,]/ sum(e[x1 -
min(as.numeric(rownames(l))) + 1,] * log(1 - qref[x1 - min(x2) + 1,
as.character(t1)])
+     SMR <- mean(SMRxt[SMRxt != 0 & is.na(SMRxt) == F & SMRxt != Inf])
+     QxtFitted <- SMR * qref[, as.character(min(t1) : max(t2))]
+     colnames(QxtFitted) <- min(t1) : max(t2); rownames(QxtFitted) <- x2
+     return(list(SMR = SMR, QxtFitted = QxtFitted, NameMethod = "Method1"))
+     }
```

In consequence, we adjust the mortality of the organism only with one parameter, the SMR. It represents the observed deviation between the deaths recorded by the organism and the ones predicted by the reference.

The choice of x^* is of great importance because the table constructed is only valid on the age range considered and its choice may be complicated. For the male population, after setting the age range, we obtain,

```
> AgeRange <- 30 : 90
> M1Male <- FctMethod1(MyData$Male$Dxt, MyData$Male$Ext, MyData$Male$QxtRef,
AgeRange, MyData$Male$AgeRef, MyData$Male$YearCom, MyData$Male$YearRef)
> print(paste("QxtFittedMale = ",M1Male$SMR," * QxtRefMale"))
[1] "QxtFittedMale = 0.614721542286672 * QxtRefMale"
```

9.3.2 Method 2: Approach Involving Two Parameters with a Semiparametric Relational Model

The second approach is a semiparametric Brass-type relational model. The fit is performed using the logistic function,

$$\text{logit } \hat{q}_{x^*}(t^*) = \alpha + \beta \text{ logit } q_{x^*}^{\text{ref}}(t^*),$$

where x^* and t^* correspond to the age range and to the period of observation in common with the reference, respectively, and $q_{x^*}^{\text{ref}}(t^*)$ is the reference of mortality.

This model implies that the differences between the observed mortality and the reference can be represented linearly with two parameters. The parameter α is an indicator of mortality affecting all ages identically while the parameter β modifies this effect with age. The estimation is done by minimizing a weighted distance between the estimated and observed mortality,

$$\sum \left| E_{x^*,t^*} \times \left(\hat{q}_{x^*}(t^*) - \widehat{q}_{x^*}(t^*) \right) \right|.$$

This model has the advantage of integrated estimation and forecasting, as the parameters α and β are constant. We refer to Section 10.4.2 and Planchet & Thérond (2011, Chapter 7) for more details.

We obtain the probabilities of death of the organism $\widetilde{q}_x(t)$, for $x \in [\underline{x}, \overline{x}]$ et $t \in [\underline{t}, \overline{t}]$ by

$$\widetilde{q}_x(t) = \frac{\exp\big(\widehat{\alpha} + \widehat{\beta}\,\text{logit}\; q_x^{\text{ref}}(t)\big)}{1 + \exp\big(\widehat{\alpha} + \widehat{\beta}\,\text{logit}\; q_x^{\text{ref}}(t)\big)}.$$

The semiparametric relational model is defined in R with the following functions:

```
> FctLogit = function(q) { log(q / (1 - q)) }

> FctMethod2 = function(d, e, qref, x1, x2, t1, t2){
+       Qxt <- (d[x1 - min(as.numeric(rownames(d))) + 1, ] /
e[x1 - min(as.numeric(rownames(l))) + 1, ])
+       LogitQxt <- FctLogit(Qxt)
+       LogitQxt[LogitQxt == -Inf] <- 0
+       LogitQxtRef <- FctLogit(qref[x1 - min(x2) + 1, as.character(t1)])
+       LogitQxtRef[LogitQxtRef == -Inf] <- 0
+       Distance = function(p){
+          LogitQxtFit <- (p[1] + p[2] * LogitQxtRef)
+          QxtFit <- exp(LogitQxtFit) / (1 + exp(LogitQxtFit))
+          sum(abs(e[x1 - min(as.numeric(rownames(e))) + 1, ] * (Qxt - QxtFit)))
+          }
+       ModPar <- constrOptim(c(0, 1), Distance, ui = c(0, 1), ci = 0,
control = list(maxit = 10^3), method = "Nelder-Mead")$par
+       LogitQxtRef <- FctLogit(qref[, as.character(min(t1) : max(t2))])
+       QxtFitted <- as.matrix(exp(ModPar[1] + ModPar[2] * LogitQxtRef) /
(1 + exp(ModPar[1] + ModPar[2] * LogitQxtRef)))
+       colnames(QxtFitted) <- as.character(min(t1) : max(t2))
+       rownames(QxtFitted) <- x2
+       return(list(ModPar = ModPar, QxtFitted = QxtFitted,
NameMethod = "Method2"))
+       }
```

After setting the age range, we obtain for the male population:

```
> AgeRange <- 30 : 90
> M2Male <- FctMethod2(MyData$Male$Dxt, MyData$Male$Ext, MyData$Male$QxtRef,
AgeRange, MyData$Male$AgeRef, MyData$Male$YearCom, MyData$Male$YearRef)
> print(paste("logit (QxtFittedMale) = ",M2Male$ModPar[1],"+",M2Male$ModPar[2],"
* logit (QxtRefMale)"))
[1] "logit (QxtFittedMale) = 0.247405963276836 + 1.18826425350273 *
logit (QxtRefMale)"
```

9.3.3 Method 3: Poisson GLM Including Interactions with Age and Calendar Year

The third approach is a Poisson Generalized Linear Model (GLM) including the baseline mortality of the reference as a covariate and allowing interactions with age and calendar year.

With the notation of Section 9.2 and under the assumption of a piecewise constant force of mortality, the likelihood becomes

$$\mathcal{L}\big(q_x(t)\big) = \exp\big(-E_{x,t}\, q_x(t)\big)\big(q_x(t)\big)^{D_{x,t}}.$$

The associated log-likelihood is

$$\ell\big(q_x(t)\big) = \log \mathcal{L}\big(q_x(t)\big) = -E_{x,t}\, q_x(t) + D_{x,t} \log q_x(t).$$

Maximizing the log-likelihood $\ell\big(q_x(t)\big)$ gives $\widehat{q}_x(t) = D_{x,t}/E_{x,t}$, which coincides with the central death rates $\widehat{m}_x(t)$.

It is then apparent that the likelihood $\ell\big(q_x(t)\big)$ is proportional to the Poisson likelihood based on

$$D_{x,t} \sim \mathcal{P}\big(E_{x,t}\, q_x(t)\big). \tag{9.1}$$

Thus, it is equivalent to work on the basis of the *true* likelihood or on the basis of the Poisson likelihood, as recalled in Delwarde, A. & Denuit, M. (2005). In consequence, under the assumption of constant forces of mortality between non-integer values of x and t, we consider (9.1) to take advantage of the GLMs framework.

We suppose that the number of deaths of the organism at attained age x^* and calendar year t^* is determined by

$$\begin{aligned} &D_{x^*,t^*} \sim \mathcal{P}\big(E_{x^*,t^*}\, \mu_{x^*}(t^*)\big),\\ \text{with}\quad &\mu_{x^*}(t^*) = \beta_0 + \beta_1 \log q_{x^*}^{\text{ref}}(t^*) + \beta_2\, x^* + \beta_3\, t^* + \beta_4\, x^*\, t^*, \end{aligned}$$

where x^* and t^* correspond to the age range and to the period of observation in common with the reference, respectively, and $\mu_{x^*}^{\text{ref}}(t^*)$ are the baseline forces of mortality.

If we do not allow for interactions, we will observe parallel shifts of the mortality according to the baseline mortality for each dimension. This view is certainly unrealistic, and interactions need to be incorporated. However, they can only be reasonably taken into account if we have at our disposal a sufficient historic in common with the reference.

We obtain the probabilities of death of the organism $\widetilde{q}_x(t)$, for $x \in \big[\,\underline{x}\,,\overline{x}\,\big]$ et $t \in \big[\,\underline{t}\,,\overline{t}\,\big]$ by

$$\widetilde{q}_x(t) = 1 - \exp\left(\widehat{\beta}_0 + \widehat{\beta}_1 \log q_x^{\text{ref}}(t) + \widehat{\beta}_2\, x + \widehat{\beta}_3\, t + \widehat{\beta}_4\, x\, t\right).$$

The method is implemented in R with the following function.

```
> FctMethod3 = function(d, e, qref, x1, x2, t1, t2){
+      DB <- cbind(expand.grid(x1, t1), c(d[x1 - min(as.numeric(rownames(d))) +
1, ]), c(e[x1 - min(as.numeric(rownames(e))) + 1, ]), c(qref[x1 - min(x2)+
1, as.character(t1)])); colnames(DB) <-c("Age", "Year", "D_i", "E_i", "mu_i")
+      DimMat <- dim(qref[, as.character(min(t1) : max(t2))])
+      if(length(t1) < 10){
+      PoisMod <- glm(D_i ~ as.numeric(log(mu_i)) + as.numeric(Age),
family = poisson, data = data.frame(DB), offset = log(E_i))
+      QxtFitted <- matrix(exp(as.numeric(coef(PoisMod)[1]) +
as.numeric(coef(PoisMod)[2]) * as.numeric(log(qref[, as.character(min(t1):
max(t2))])) + (x2) * as.numeric(coef(PoisMod)[3])), DimMat[1], DimMat[2])
+           }
+      if(length(t1) >= 10){
+           PoisMod <- glm(D_i ~  as.numeric(log(mu_i)) + as.numeric(Age) *
as.numeric(Year), family = poisson, data = data.frame(DB), offset = log(E_i))
+           DataGrid <- expand.grid(x2, min(t1) : max(t2))
+           IntGrid <- matrix(DataGrid[, 1] * DataGrid[, 2], length(x2),
length(min(t1) : max(t2)))
+          QxtFitted <- matrix(exp(as.numeric(coef(PoisMod)[1]) +
as.numeric(coef(PoisMod)[2]) * as.numeric(log(qref[, as.character(min(t1):
max(t2))])) + (DataGrid[,1]) * as.numeric(coef(PoisMod)[3]) + (DataGrid[,2]) *
```

```
as.numeric(coef(PoisMod)[4]) + IntGrid * as.numeric(coef(PoisMod)[5])), DimMat[1],
DimMat[2])
+         }
+       as.character(min(t1) : max(t2))
+       colnames(QxtFitted) <- as.character(min(t1) : max(t2))
+       rownames(QxtFitted) <- x2
+       return(list(PoisMod = PoisMod, QxtFitted = QxtFitted,
NameMethod = "Method3"))
+       }
```

For our application, as we only have one year in common with the reference, the interactions with the calendar year are not considered.

The estimated parameters of the Poisson model for the male population are displayed below.

```
> AgeRange <- 30 : 90
> M3Male <- FctMethod3(MyData$Male$Dxt, MyData$Male$Ext, MyData$Male$QxtRef,
AgeRange, MyData$Male$AgeRef, MyData$Male$YearCom, MyData$Male$YearRef)
> M3Male$PoisMod
Call: glm(formula = D_i ~ as.numeric(log(mu_i)) + as.numeric(Age),
     family = poisson, data = data.frame(DB), offset = log(E_i))
Coefficients:
          (Intercept) as.numeric(log(mu_i)) as.numeric(Age)
             -3.37679           0.80810          0.03165
Degrees of Freedom: 60 Total (i.e. Null); 58 Residual
Null Deviance: 390.9; Residual Deviance: 51.3; AIC: 200.6
```

It should be noticed that this method is not flexible when we forecast the mortality trends. In addition, it is also relatively unstable if we do not have a sufficient historic at our disposal. This method should be used with care, especially when data have a high underlying heterogeneity, as is observed in our data `MyPortfolio`.

9.3.4 Method 4: Nonparametric Smoothing and Application of the Improvement Rates

The fourth approach consists of, in a first step, smoothing the periodic table computed from the portfolio provided by the organism and, in a second step, the application of the rates of mortality improvement derived from the reference.

We consider the following nonparametric relational model applied to the periodic table of the organism,

$$D_x \sim \mathcal{P}\left(E_x\, q_x^{\text{ref}}\, \exp(f(x))\right),$$

including the expected number of deaths $E_x\, q_x^{\text{ref}}$ according to the reference and where f is an unspecified smooth function of attained age x.

Similar to Method 3, see Section 9.3.3, we are taking advantage of the GLMs framework. However, here, the role of the GLMs is of a background model that is fitted locally.

We consider the local kernel-weighted log-likelihood method to estimate the smooth function $f(x)$. Statistical aspects of local likelihood techniques have been discussed extensively in Tomas (2013). These methods have been used in a mortality context by Delwarde et al. (2004), Debón et al. (2006), and Tomas (2011), to graduate life tables with attained age. More recently, Tomas & Planchet (2013) have covered smoothing in two dimensions and introduced adaptive parameters choice with an application to long-term care insurance.

Local likelihood methods have the ability to model very well the mortality patterns even in presence of complex structures and avoid to rely on experts opinion. These techniques have been implemented in R with the package `locfit` by Loader (1999).

The selection of the smoothing parameters, that is, the window width λ, the polynomial degree p, and the weight function, is an effective compromise between two objectives: the elimination of irregularities and the achievement of a desired mathematical shape to the progression of the mortality rates. This underlines the importance of thorough investigation of data as the prerequisites of reliable judgment, as we must first inspect the data and take the decision as the type of irregularity we wish to retain. The strategy is to evaluate a number of candidates and to use criteria to select among the fits the one with the lowest score. Here, we use the Akaike information criterion (AIC).

It is well known that between the three smoothing parameters, the weight function has much less influence on the bias and variance trade-off. The choice is not too crucial; at best; it changes the visual quality of the regression curve. In the following, the weight function is set to be Epanechnikov kernel.

We obtain the values of the criterion for a range of window width $\lambda = \left[3, 41\right]$ observations and polynomial degrees $p = \left[0, 3\right]$ as follows:

```
> GetCrit = function(z, x, hh, pp, k, fam, lk, ww) {
+       AICMAT <- V2MAT <- SpanMat <- matrix(,length(hh),length(pp))
+       colnames(AICMAT) <- colnames(V2MAT) <- colnames(SpanMat) <- c("Degree 0",
"Degree 1", "Degree 2", "Degree 3")
+       rownames(AICMAT) <- rownames(V2MAT) <- rownames(SpanMat) <- c(1 : 20)
+       for ( h in hh ) {
+               for( p in (pp + 1) ) {
+                       SpanMat[h, p] <- (2 * h + 1) / length(z)
+                       AICMAT[h, p] <- aic(c(z) ~ lp(c(x), deg = p - 1,
nn = SpanMat[h, p],  scale = 1), ev = dat(), family = fam, link = lk,
 kern = k, weights = ww)[4]
+                       V2MAT[h, p] <-  locfit(c(z) ~ lp(c(x), deg = p - 1,
nn = SpanMat[h, p],  scale = 1), ev = dat(), family = fam, link = lk,
kern = k, weights = ww)$dp["df2"] } }
+       return(list(AICMAT = AICMAT, V2MAT = V2MAT, SpanMat = SpanMat ))
+       }

> FctMethod4_1stPart = function(d, e, qref, x1, x2, t1){
+     Dx <- apply(as.matrix(d[x1 - min(as.numeric(rownames(d)))+1, ]), 1, sum)
+     DxRef <- apply(as.matrix(as.matrix(e)[x1 - min(as.numeric(rownames(e)))+
1, ] * qref[x1 - min(x2) + 1, as.character(t1)]), 1, sum)
+     ModCrit <- GetCrit(Dx, x1, 1 : 20, 0 : 3, c("epan"), "poisson", "log",
c(DxRef))
+     return(ModCrit)
+     }

> AgeRange <- 30 : 90
> M4Male <- FctMethod4_1stPart(MyData$Male$Dxt, MyData$Male$Ext, MyData$Male$QxtRef,
AgeRange, MyData$Male$AgeRef, MyData$Male$YearCom)
```

In practice, the smoothing parameters are selected using graphical diagnostics. The graphic displays the AIC scores against the fitted degrees of freedom. This aids interpretation: 1 degree of freedom represents a smooth model with very little flexibility while 10 degrees of freedom represents a noisy model showing many features. It also aids comparability as we can compute criteria scores for other polynomial degrees or for other smoothing methods and add them to the plot.

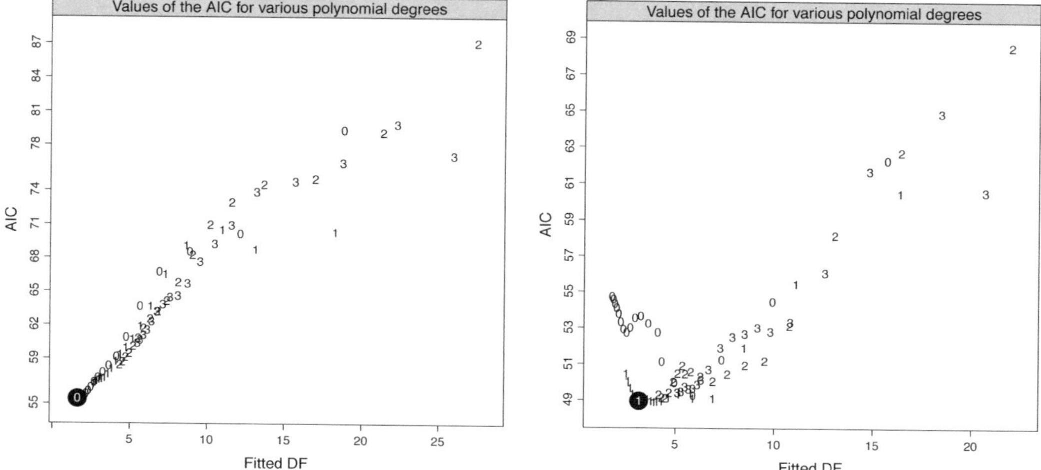

FIGURE 9.4
Values of the AIC criterion for various polynomial degrees, male population (left panel) and female population (right panel).

We select the smoothing parameters at the point when the criterion reaches a minimum or a plateau after a steep descent. To display the AIC scores, we define the following function:

```
> PlotCrit = function(crit, v2, LimCrit, LimV2) {
+        FigName <- paste("Values of the AIC for various polynomial degrees")
+        VECA <- c(crit); VECB <- c(v2)
+        xyplot(VECA ~ VECB|as.factor(FigName),xlab = "Fitted DF", ylab = "AIC",
type = "n", par.settings = list(fontsize = list(text = 15, points = 12)),
par.strip.text = list(cex = 1, lines = 1.1), strip = strip.custom(bg =
"lightgrey"), scales = list(y = list(relation = "free", limits = list(LimCrit),
at = list(c(seq(min(LimCrit), max(LimCrit), by = (max(LimCrit) - min(LimCrit)) /
10)))), x = list(relation = "free", limits = LimV2, tck = c(1, 1))),
panel = function(...) {
+                llines(VECA[1:20] ~ VECB[1:20],type = "p", pch = "0", col = 1)
+                llines(VECA[21:40] ~ VECB[21:40],type = "p", pch = "1", col = 1)
+                llines(VECA[41:60] ~ VECB[41:60],type = "p", pch = "2", col = 1)
+                llines(VECA[61:80] ~ VECB[61:80],type = "p", pch = "3", col = 1)
+                } )
+        }
```

With aid of this graphical diagnostic, we select the smoothing parameter:

```
> PlotCrit(M4Male$AICMAT, M4Male$V2MAT, XLim, YLim)
```

Figure 9.4 presents the values of the AIC criterion according the fitted degrees of freedom v_2 for the male and female populations. We select a constant and a linear fit with $v_2 = 1.63$ and 3.15 for the male and female population, corresponding to a window width of 41 and 31 observations respectively.

We select a constant and a linear fit with $v_2 = 1.63$ and 3.15 for the male and female population, corresponding to a window width of 41 and 31 observations respectively.

Having selected the optimal window width and polynomial degree, denoted `h.Opt` and `P.Opt` in the function below, we obtain the fitted probabilities of death by applying the improvement rates $q_x(t+1)/q_x(t)$ derived from the reference with the following function:

```
> FctMethod4_2ndPart = function(d, e, qref, x1, x2, t1, t2, P.Opt, h.Opt){
+       Dx <- apply(as.matrix(d[x1 - min(as.numeric(rownames(d)))+1, ]), 1, sum)
+       DxRef <- apply(as.matrix(as.matrix(e[x1 - min(as.numeric(rownames(e))) +
1, ]) * qref[x1 - min(x2) + 1, as.character(t1)]), 1, sum)
+       QxtFitted <- predict(locfit(Dx ~ lp(x1, deg = P.Opt, nn = (h.Opt *
2 + 1) / length(x1), scale=1), ev = dat(), family = "poisson", link = "log",
kern = c("epan"), weights = c(DxRef))) * qref[x1 - min(x2) + 1,
as.character(min(t1):max(t2))]
+       colnames(QxtFitted) <- as.character(min(t1) : max(t2))
+       rownames(QxtFitted) <- x1
+       return(list(QxtFitted = QxtFitted, NameMethod = "Method4"))
+       }
```

9.3.5 Completion of the Tables: The Approach of Denuit and Goderniaux

Finally, we need to complete the tables. Due to the probable lack of data beyond a certain age, we do not have valid information to derive mortality at older ages. Actuaries and demographers have developed various techniques for the completion of the tables at older ages. In this chapter, we use a simple and efficient method proposed by Denuit & Goderniaux (2005). This method relies on the fitted one-year probabilities of death and introduces two constraints about the completion of the mortality table. It consists of fitting, by ordinary least squares, the following log-quadratic model:

$$\log\, \widehat{q}_x(t) = a_t + b_t\, x + c_t\, x^2 + \epsilon_x(t), \tag{9.2}$$

where $\epsilon_x(t) \sim$ iid $\mathcal{N}(0, \sigma^2)$, separately for each calendar year t at attained ages $x^\star$. Two restrictives conditions are imposed:

- First a completion constraint,

  $$q_{130}(t) = 1\ , \quad \text{for all } t.$$

  Even though human lifetime does not seem to approach any fixed limit imposed by biological factors or other, it seems reasonable to accept the hypothesis that the age limit of end of life 130 will not be exceeded.

- Second an inflexion constraint,

  $$\frac{\partial}{\partial x}\, q_x(t)|_{x=130} = 0\ , \quad \text{for all } t.$$

These constraints impose concavity at older ages in addition to the existence of a tangent at the point $x = 130$. They lead to the following relation between the parameters a_t, b_t and c_t for each calendar year t:

$$a_t + b_t\, x + c_t\, x^2 = c_t\, (130 - x)^2,$$

for $x = x_t^\star, x_t^\star + 1, \ldots$. The parameters c_t are estimated from the series $\{\widehat{q}_x(t), x = x_t^\star, x_t^\star + 1, \ldots\}$ of calendar year t with equation (9.2) and the constraints imposed. We implement the completion method in R as follows:

```
> CompletionDG2005 = function(q, x, y, RangeStart, RangeCompletion, NameMethod) {
+      x <- min(x) : pmin(max(x),100); RangeStart <- RangeStart - min(x)
+      RangeCompletion <- RangeCompletion - min(x)
+      CompletionValues <- matrix(, length(y), 3)
+      rownames(CompletionValues) <- y
+      colnames(CompletionValues) <- c("OptAge", "OptR2", "OptCt")
+      CompletionMat <- matrix(, RangeCompletion[2] - RangeCompletion[1] + 1,
length(y))
+      QxtFinal <- matrix(, RangeCompletion[2] + 1, length(y))
+      colnames(QxtFinal) <- y; rownames(QxtFinal) <- min(x) : 130
+      for(j in 1 : length(y)) {
+           R2Mat <- matrix(,RangeStart[2] - RangeStart[1] + 1, 2)
+           colnames(R2Mat) <- c("Age", "R2")
+           quadratic.q <- vector("list", RangeStart[2] - RangeStart[1] + 1)
+           for (i in 0 : (RangeStart[2] - RangeStart[1])) {
+                AgeVec <- ((RangeStart[1] + i) : (max(x) - min(x)))
+                R2Mat[i + 1, 1] <- AgeVec[1]
+                FitLM <- lm(log(q[AgeVec + 1, j])  AgeVec + I(AgeVec^2))
+                R2Mat[i + 1, 2] <- summary(FitLM)$adj.r.squared
+                quadratic.q[[i + 1]] <- as.vector(exp(fitted(FitLM)))
+                }
+           if(any(is.na(R2Mat[,2])) == F){
+                OptR2 <- max(R2Mat[, 2], na.rm = T)
+                OptAge <- as.numeric(R2Mat[which(R2Mat[, 2] == OptR2), 1])
+                quadratic.q.opt <- quadratic.q[[which(R2Mat[, 2]==OptR2)]]
+                }
+           if(any(is.na(R2Mat[,2]))){
+                OptR2 <- NA; OptAge <- AgeVec[1]
+                quadratic.q.opt <- quadratic.q[[i + 1]]
+                }
+           OptCt <- coef(lm(log(quadratic.q.opt)  I((RangeCompletion[2] -
(OptAge : (max(x) - min(x))))^2) - 1))
+           CompletionVal <- vector(,RangeCompletion[2]-RangeCompletion[1]+1)
+           for (i in 0 : (RangeCompletion[2] - RangeCompletion[1])) {
+                CompletionVal[i + 1] <- exp(OptCt * (RangeCompletion[2] -
(RangeCompletion[1] + i))^2)
+                }
+           CompletionValues[j, 1] <- OptAge + min(x)
+           CompletionValues[j, 2] <- OptR2; CompletionValues[j, 3] <- OptCt
+           CompletionMat[, j] <- CompletionVal
+           QxtFinal[, j] <- c(q[1 : RangeCompletion[1], j], CompletionVal)
+           k <- 5; QxtSmooth <- vector(,2 * k + 1)
+           for(i in (RangeCompletion[1] - k) : (RangeCompletion[1] + k)){
+                QxtSmooth[1 + i - (RangeCompletion[1] - k)] <- prod(
QxtFinal[(i - k) : (i + k), j])^(1 / (2 * k + 1))
+                }
+           QxtFinal[(RangeCompletion[1] - k) : (RangeCompletion[1] + k), j]
<- QxtSmooth
+           }
+      return(list(CompletionValues = CompletionValues, CompletionMat =
CompletionMat, QxtFinal = QxtFinal, NameMethod = NameMethod))
+      }
```

As an illustration, the R^2 and corresponding estimated regression parameters c_t for the male and female population are displayed in Figure 9.5, left panel.

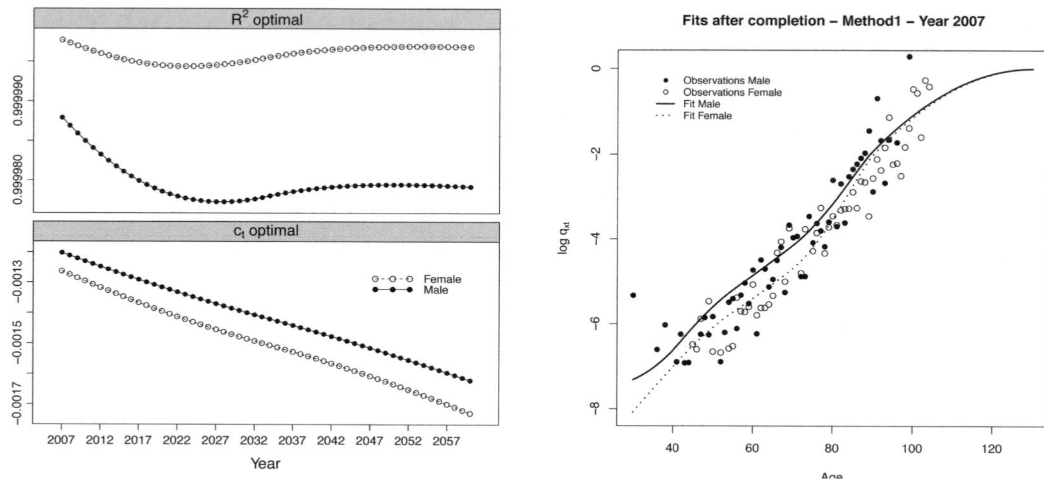

FIGURE 9.5
Regression parameters and fits obtained after the completion for the year 2007 with method 1.

We produced the figure with the following function, where the object `CompletionMethod1` has been obtained when fitting the completion procedure, see Section 9.5.5.

```
> PlotParamCompletion(CompletionMethod1, MyData$Param$Color)
```

The models capture more than 99.9% of the variance of the probabilities of death at high ages for both populations. The regression parameter $\widehat{c}_t$ represents the evolution of the mortality trends at high ages. We observe that the mortality at high ages decreases and the speed of improvement is relatively similar for both populations.

We keep the original $\widehat{q}_x(t)$ for ages below 90 years old for both populations, and replace the annual probabilities of death beyond this age by the values obtained from the quadratic regression. The results obtained with method 1 for the calendar year 2007 are presented in Figure 9.5, right panel, for both populations.

The figure has been obtained with the function:

```
> FitPopsAfterCompletionLog(CompletionMethod1, MyData, 30 : 130,
MyData$Param$Color, "2007")
```

It should be noted that the completed part of the table is rather formal and is not validated.

9.4 Validation

The validation is assessed on three levels. This concerns the proximity between the observations and the model, the regularity of the fit, and the consistency and plausibility of the mortality trends.

9.4.1 First Level: Proximity between the Observations and the Model

The first level of the validation assesses the overall deviation with the observed mortality. It involves the SMR test proposed by Liddell (1984), the likelihood ratio test, and the

Wilcoxon Matched-Pairs Signed-Ranks test. In addition, we find it useful to compare criteria measuring the distance between the observations and the models with the χ^2 applied by Forfar et al. (1988), the mean average percentage error (MAPE), and R^2 applied by Felipe et al. (2002), as well as the deviance, the SMR, and the number of standardized residuals larger than 2 and 3.

The tests and quantities summarizing the overall deviation between the observations and the model are described in the following:

- χ^2. This indicator allows one to measure the quality of the fit of the model. It is defined as

$$\chi^2 = \sum_{(x,t)} \frac{\left(D_{x,t} - E_{x,t}\,\widehat{q}_x(t)\right)^2}{E_{x,t}\,\widehat{q}_x(t)\left(1 - \widehat{q}_x(t)\right)}$$

  We will privilege the model having the lowest χ^2.

- **MAPE.** This is a measure of accuracy of the fit to the observations. This indicator is the average of the absolute values of the deviations from the observations,

$$\text{MAPE} = \frac{\sum_{(x,t)} \left|(D_{x,t}/E_{x,t} - \widehat{q}_x(t))/(D_{x,t}/E_{x,t})\right|}{\sum_{(x,t)} D_{x,t}} \times 100.$$

  It is a percentage and thus a practical indicator for the comparison. However, in the presence of zero observations, there will be divisions by zero, and these observations must be removed.

- R^2. The coefficient of determination measures the adequacy between the model and the observation. It is defined as the part of variance explained with respect to the total variance,

$$R^2 = 1 - \left(\frac{\sum_{(x,t)} \left(D_{x,t}/E_{x,t} - \widehat{q}_x(t)\right)^2}{\sum_{(x,t)} \left(D_{x,t}/E_{x,t} - (\sum_{(x,t)}(D_{x,t}/E_{x,t})/n)\right)^2}\right),$$

  where n is the number of observations.

- **The deviance.** This a measure of the quality of the fit. Under the hypothesis of the number of deaths following a poisson law $D_{x,t} \sim P\left(E_{x,t}\; q_x(t)\right)$, the deviance is defined as

$$\text{If } D_{x,t} > 1 \text{ and } 0 < D_{x,t} < E_{x,t},\ \text{Deviance}_{x,t} =$$
$$2\left(D_{x,t}\ln\left(\frac{D_{x,t}}{E_{x,t}\,\widehat{q}_x(t)}\right) + (E_{x,t} - D_{x,t})\ln\left(\frac{E_{x,t} - D_{x,t}}{E_{x,t} - L_{x,t}\,\widehat{q}_x(t)}\right)\right).$$

$$\text{If } D_{x,t} > 0,\ \text{Deviance}_{x,t} = 2\left(D_{x,t}\ln\left(\frac{D_{x,t}}{E_{x,t}\,\tilde{q}_x(t)}\right) - (D_{x,t} - E_{x,t}\tilde{q}_x(t))\right).$$

$$\text{If } D_{x,t} = 0,\ \text{Deviance}_{x,t} = 2\,E_{x,t}\tilde{q}_x(t).$$

  And $\text{Global deviance} = \sum_{(x,t)} \text{Deviance}_{x,t}$.

- **The likelihood ratio test.** We can also consider the p-value associated to the likelihood ratio test (or drop-in deviance test). We seek, here, to determine if the fit corresponds to the underlying mortality law (null hypothesis $\mathcal{H}_0$). The likelihood ratio test statistic, ξ^{LR}, is defined as

$$\xi^{\text{LR}} = \sum_{(x,t)} \left(D_{x,t} \ln \left(\frac{D_{x,t}}{E_{x,t}\, \widehat{q}_x(t)} \right) - (D_{x,t} - E_{x,t}\tilde{q_x}(t)) \right).$$

If $\mathcal{H}_0$ is true, this statistic follows a χ^2 law with the number of degrees of freedom equal to the number of observations n:

$$\xi^{\text{LR}} \sim \chi^2(n).$$

Hence, the null hypothesis $\mathcal{H}_0$ is rejected if

$$\xi^{\text{LR}} > \chi^2_{1-\alpha}(n),$$

where $\chi^2_{1-\alpha}(n)$ is the $(1-\alpha)$ quantile of the χ^2 distribution with n degrees of freedom. The p-value is the lowest value of the type I error (α) for which we reject the test. We will privilege the model having the closest p-value to 1,

$$p\text{-value} = \mathbb{P}\big[\chi^2_{1-\alpha}(n) > \xi^{\text{LR}}\big] = 1 - \text{F}_{\chi^2(n)}(\xi^{\text{LR}}).$$

- **The SMR.** This is the ratio between the observed and fitted number of deaths. If we consider that the number of deaths follows a poisson law $D_{x,t} \sim \mathcal{P}\big(E_{x,t}\, q_x(t)\big)$,

$$\text{SMR} = \frac{\sum_{(x,t)} D_{x,t}}{\sum_{(x,t)} E_{x},+ \tilde{q}_x(t)}.$$

Hence, if SMR > 1, the fitted deaths are underestimated, and vice versa if SMR < 1.

- **The SMR test.** We can also apply a test to determine if the SMR is significatively different from 1, see Liddell (1984). We compute the following statistic:

$$\begin{aligned} \text{If SMR } &> 1, \xi^{\text{SMR}} = 3 \times D^{\frac{1}{2}}\big(1 - (9D)^{-1} - (D/E)^{\frac{1}{3}}\big). \\ \text{If SMR } &< 1, \xi^{\text{SMR}} = 3 \times D^{*\,\frac{1}{2}}\big((D^*/E)^{\frac{1}{3}} + (9D^*)^{-1} - 1\big), \end{aligned}$$

where $D = \sum_{(x,t)} D_{x,t}$, $D^* = \sum_{(x,t)} D_{x,t} + 1$ and $E = \sum_{(x,t)} E_{(x,t)}\, \tilde{q}_x(t)$.
If the SMR is not significatively different from 1 (null hypothesis $\mathcal{H}_0$), this statistic follows a standard Normal law,

$$\xi^{\text{SMR}} \sim \mathcal{N}(0,1).$$

Thus, the null hypothesis $\mathcal{H}_0$ is rejected if

$$\xi^{\text{SMR}} > \mathcal{N}_{1-\alpha}(0,1),$$

where $\mathcal{N}_{1-\alpha}(0,1)$ is the $(1-\alpha)$ quantile of the standard Normal distribution.
The p-value is given by

$$p\text{-value} = 1 - \text{F}_{\mathcal{N}(0,1)}(\xi^{\text{SMR}}).$$

We will seek to obtain the closest p-value to 1.

- **The Wilcoxon Matched-Pairs Signed-Ranks test.** The framework of this test is very similar to the signs test. While the signs test only uses the information on the direction of the differences between pairs composed of the observed and fitted probabilities of death, the Wilcoxon test also takes into account the magnitude of these differences.

  We test the null hypothesis $\mathcal{H}_0$ that the median between the difference of each pairs is null. We compute the difference between the observed and fitted probabilities of death, and rank them by increasing order of the absolute values, omitting the null differences. We assign to each non-zero difference its rank. We note w_+ et w_- the sum of the ranks of the differences strictly positive and negative, respectively. Finally, we note w, the maximum between the two numbers: $w = \max\{w_+, w_-\}$.

  If the observed and fitted probabilities of death are equal, that is if $\mathcal{H}_0$ is true, the sum of the rank having a positive and negative sign should be approximatively equal. But if the sum of the ranks of positive signs differs largely from the sum of the ranks of the negative signs, we will deduce that the observed probabilities of death differ from the fitted ones, and we will reject the null hypothesis.

  We can compute the statistic ξ^{WIL},

$$\xi^{\text{WIL}} = \frac{\left(w - 1/2 - n\,(n+1)\right)/4}{\sqrt{n\,(n+1)\,(2\,n+1)/24}}.$$

  If $\mathcal{H}_0$ is true, this statistic follows a standard Normal law,

$$\xi^{\text{WIL}} \sim \mathcal{N}(0,1).$$

  Thus, the null hypothesis $\mathcal{H}_0$ is rejected if

$$|\xi^{\text{WIL}}| > \mathcal{N}_{1-\alpha/2}(0,1),$$

  where $\mathcal{N}_{1-\alpha/2}(0,1)$ is the $(1-\alpha/2)$ quantile of the standard Normal distribution. The p-value is given by:

$$p\text{-value} = \mathbb{P}\big[\mathcal{N}_{1-\alpha/2}(0,1) > |\xi^{\text{WIL}}|\big] = 2 \times \left(1 - \mathrm{F}_{\mathcal{N}(0,1)}(|\xi^{\text{WIL}}|)\right).$$

  We will seek to obtain the closest p-value to 1.

  The Wilcoxon test uses the magnitude of the differences. The result may be different from the signs test that uses the number of positive and negative signs of the difference.

We implement the criteria and tests assessing the first level of validation in R as the following:

```
> TestsLevel1 = function(x, y, z, vv, AgeVec, AgeRef, YearVec, NameMethod){
+      quantities <- LRTEST <- matrix(, 4, 1)
+      SMRTEST <- WMPSRTEST <- matrix(, 5, 1)
+      RESID <- matrix(, 2, 1)
+      colnames(LRTEST) <- colnames(SMRTEST) <- colnames(RESID) <-
colnames(WMPSRTEST) <- colnames(quantities) <- NameMethod
+      rownames(LRTEST) <- c("Xi", "Threshold", "Hyp", "p.val")
+      rownames(SMRTEST) <- c("SMR", "Xi", "Threshold", "Hyp", "p.val")
+      rownames(RESID) <- c("Std. Res. > 2", "Std. Res. > 3")
+      rownames(quantities) <- c("Chi2", "R2", "MAPE", "Deviance")
+      rownames(WMPSRTEST) <- c("W", "Xi", "Threshold", "Hyp", "p.val")
+      xx <- x[AgeVec - min(as.numeric(rownames(x))) + 1, YearVec]
```

```
+       yy <- y[AgeVec - min(as.numeric(rownames(y))) + 1, YearVec]
+       zz <- z[AgeVec - min(as.numeric(rownames(z))) + 1, YearVec]
+ ## ---------- LR test
+       val.all <- matrix(, length(AgeVec), length(YearVec))
+       colnames(val.all) <- as.character(YearVec)
+       val.all[(yy > 0)] <- yy[(yy > 0)] * log(yy[(yy > 0)] / (xx[(yy > 0)] *
zz[(yy > 0)])) - (yy[(yy > 0)] - xx[(yy > 0)]) * (zz[(yy > 0)]
+       val.all[(yy == 0)] <- 2 * xx[(yy == zz)] * (zz[(yy == zz)] /
+       xi.LR <- sum(val.all)
+       Threshold.LR <- qchisq(1 - vv, df = length(xx))
+       pval.LR <- 1 - pchisq(xi.LR, df = length(xx))
+       LRTEST[1, 1] <- round(xi.LR, 2)
+       LRTEST[2, 1] <- round(Threshold.LR,2)
+       if(xi.LR <= Threshold.LR){ LRTEST[3, 1] <- "H0" }
else { LRTEST[3, 1] <- "H1" }
+       LRTEST[4, 1] <- round(pval.LR, 4)
+ ## ---------- SMR test
+       d <- sum(yy); e <- sum( zz * xx); val.SMR <- d / e
+       if(val.SMR > -1){
+       xi.SMR <- 3 * d^(1/2) * (1 - (9 * d)^(- 1) - (d / e)^(- 1 / 3)) }
+       if(val.SMR < -1){
+       xi.SMR <- 3 * (d + 1)^(1 / 2) * (((d + 1)/e)^(- 1 / 3) + (9 *
(d + 1))^(- 1) - 1) }
+       Threshold.SMR <- qnorm(1 - vv)
+       p.val.SMR <- 1 - pnorm(xi.SMR)
+       if(xi.SMR <= Threshold.SMR){ SMRTEST[4, 1] <- "H0" }
else { SMRTEST[4,1] <- "H1" }
+       SMRTEST[1, 1] <- round(val.SMR, 4)
+       SMRTEST[2, 1] <- round(xi.SMR, 4);
+       SMRTEST[3, 1] <- round(Threshold.SMR, 4)
+       SMRTEST[5, 1] <- round(p.val.SMR, 4);
+ ## ---------- Wilcoxon Matched-Paris Signed-Ranks test
+       tab.temp <- matrix(, length(xx), 3)
+       tab.temp[, 1] <- yy / zz - xx
+       tab.temp[, 2] <- abs(tab.temp[, 1])
+       tab.temp[, 3] <- sign(tab.temp[, 1])
+       tab2.temp <- cbind(tab.temp[order(tab.temp[, 2]), ], 1 : length(xx))
+       W.pos <- sum(tab2.temp[, 4][tab2.temp[, 3] > 0])
+       W.neg <- sum(tab2.temp[, 4][tab2.temp[, 3] < 0])
+       WW <- pmax(W.pos, W.neg)
+       Xi.WMPSRTEST <- (WW - .5 - length(xx) * (length(xx) + 1) / 4) /
sqrt(length(xx) * (length(xx) + 1) * (2 * length(xx) + 1) / 24)
+       Threshold.WMPSRTEST <- qnorm(1 - vv / 2)
+       p.val.WMPSRTEST <- 2 * (1 - pnorm(abs(Xi.WMPSRTEST)))
+       if(Xi.WMPSRTEST <= Threshold.WMPSRTEST){ WMPSRTEST[4, 1] <- "H0" }
else { WMPSRTEST[4, 1] <- "H1" }
+       WMPSRTEST[1, 1] <- WW
+       WMPSRTEST[2, 1] <- round(Xi.WMPSRTEST, 4)
+       WMPSRTEST[3, 1] <- round(Threshold.WMPSRTEST, 4)
+       WMPSRTEST[5, 1] <- round(p.val.WMPSRTEST, 4)
+ ## ---------- Standardized residuals
+       val.RESID <- (xx - yy / zz) / sqrt(xx / zz)
+       RESID[1, 1] <- length(val.RESID[(abs(val.RESID)) > 2])
+       RESID[2, 1] <- length(val.RESID[(abs(val.RESID)) > 3])
+ ## ---------- chi^2, R2, MAPE
```

TABLE 9.2
First level of validation, male population.

		Method 1	Method 2	Method 3	Method 4
Standardized residuals	> 2	2	2	2	2
	> 3	0	1	1	1
χ^2		53.13	57.79	55.82	51.81
R^2		0.7644	0.7312	0.7654	0 : 7554
MAPE (%)		65.27	50.84	50.88	50.29
Deviance		96.43	68.54	68.50	70.98
Likelihood ratio test	ξ^{LR}	48.22	34.27	34.25	35.49
	p-value	0.8826	0.9978	0.9978	0.9963
SMR test	SMR	0.7913	0.934	0.9681	0.9786
	ξ^{SMR}	3.0098	0.8165	0.3573	0.2197
	p-value	0.0013	0.2071	0.3604	0.4131
Wilcoxon test	w	1154	1004	1014	1004
	ξ^{WIL}	1.494	0.4166	0.4884	0.4166
	p-value	0.1352	0.6770	0.6252	0.6770

```
+       quantities[1, 1] <- round(sum(((yy - zz * xx)^2) / (zz * xx *
(1 - xx))), 2)
+       quantities[2, 1] <- round(1 - (sum((yy / zz - xx)^2) / sum((yy / zz -
mean(yy / zz))^2)), 4)
+       quantities[3, 1] <- round(sum(abs((yy / zz - xx) / (yy / zz))[yy > 0])
/ sum(yy > 0) * 100, 2)
+       quantities[4, 1] <- round(2*sum(val.all), 2)
+       RSLTS <- vector("list", 6)
+       RSLTS[[1]] <- LRTEST; RSLTS[[2]] <- SMRTEST; RSLTS[[3]] <- WMPSRTEST
+       RSLTS[[4]] <- RESID; RSLTS[[5]] <- quantities; RSLTS[[6]] <- NameMethod
+       names(RSLTS) <- c("Likelihood ratio test", "SMR test",
"Wilcoxon Matched-Pairs Signed-Ranks test", "Standardized residuals",
"Quantities", "NameMethod")
+       Return(RSLTS)
+       }
```

We carry out the proposed tests and quantities. Tables 9.2 presents the results for the four methods and for the male population.

In a general manner, the four methods give acceptable results.

Besides the tests and quantities, the process of validation for the first level involves graphical analysis. It consists of representing graphically the fitted values against the observations for a given attained age or calendar year.

Figures 9.6 and 9.7 present the fitted probabilities of death in original and log scale obtained by the four methods for the age range $[30, 90]$ and the calendar year 2007 for the male and female population, respectively. These graphics give a first indication about the quality of the fits. They have been produced using the following functions,

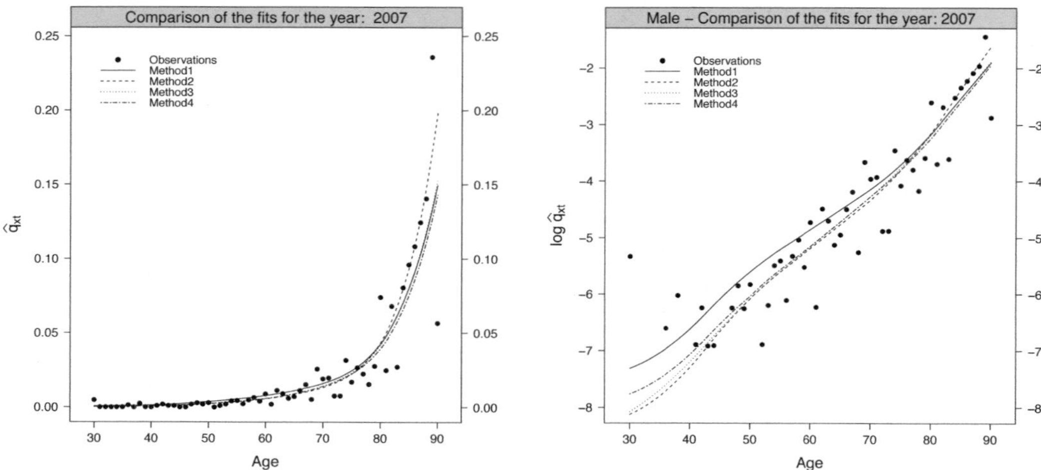

FIGURE 9.6
Fitted probabilities of death against the observations in original (left panel) and log scale (right panel), male population.

```
> ComparisonFitsMethods(ListOutputs, 2007, MyData$Male, AgeCrit, "Male",
ColorComp, LtyComp)
> ComparisonFitsMethodsLog(ListOutputs, 2007, MyData$Male, AgeCrit, "Male",
ColorComp, LtyComp)
```

where `ListOutputs` is the list of the objects `OutputMethod1` to `OutputMethod4`, see Section 9.5.3.

In conjunction with looking to the plots of the fits, we should study the residuals plots. We determine

- **Response residuals.** $r_{x,t} = q_x(t) - \widehat{q}_x(t)$;
- **Pearson residuals.** $r_{x,t} = \big(D_{x,t} - L_{x,t}\, \widehat{q}_x(t)\big) / \sqrt{\mathrm{var}\big[L_{x,t}\, \widehat{q}_x(t)\big]}$; and
- **Deviance residuals.** $r_{x,t} = \mathrm{sign}\big(D_{x,t} - L_{x,t}\, \widehat{q}_x(t)\big) \times \sqrt{\mathrm{Deviance}_{x,t}}$.

Such residual plots provide a powerful diagnostic that nicely complements the criteria. The diagnostic plots can show lack of fit locally, and we have the opportunity to judge the lack of fit based on our knowledge of both the mechanism generating the data and of the performance of the model.

We compute the residuals in R as follows:

```
> ResFct = function(x, y, z, AgeVec, YearVec, DevMat, NameMethod){
+      xx <- as.matrix(x)[AgeVec - min(as.numeric(rownames(x))) + 1, YearVec]
+      yy <- as.matrix(y)[AgeVec - min(as.numeric(rownames(y))) + 1, YearVec]
+      zz <- as.matrix(z)[AgeVec - min(as.numeric(rownames(z))) + 1, YearVec]
+      RespRes <- as.matrix(yy / zz - xx)
+      PearRes <- as.matrix((yy - xx * zz) / sqrt(xx * zz))
+      DevRes <- as.matrix(sign(yy / zz - xx) * sqrt(DevMat))
+      colnames(RespRes) <- colnames(PearRes) <- colnames(DevRes) <- YearVec
+      rownames(RespRes) <- rownames(PearRes) <- rownames(DevRes) <- AgeVec
```

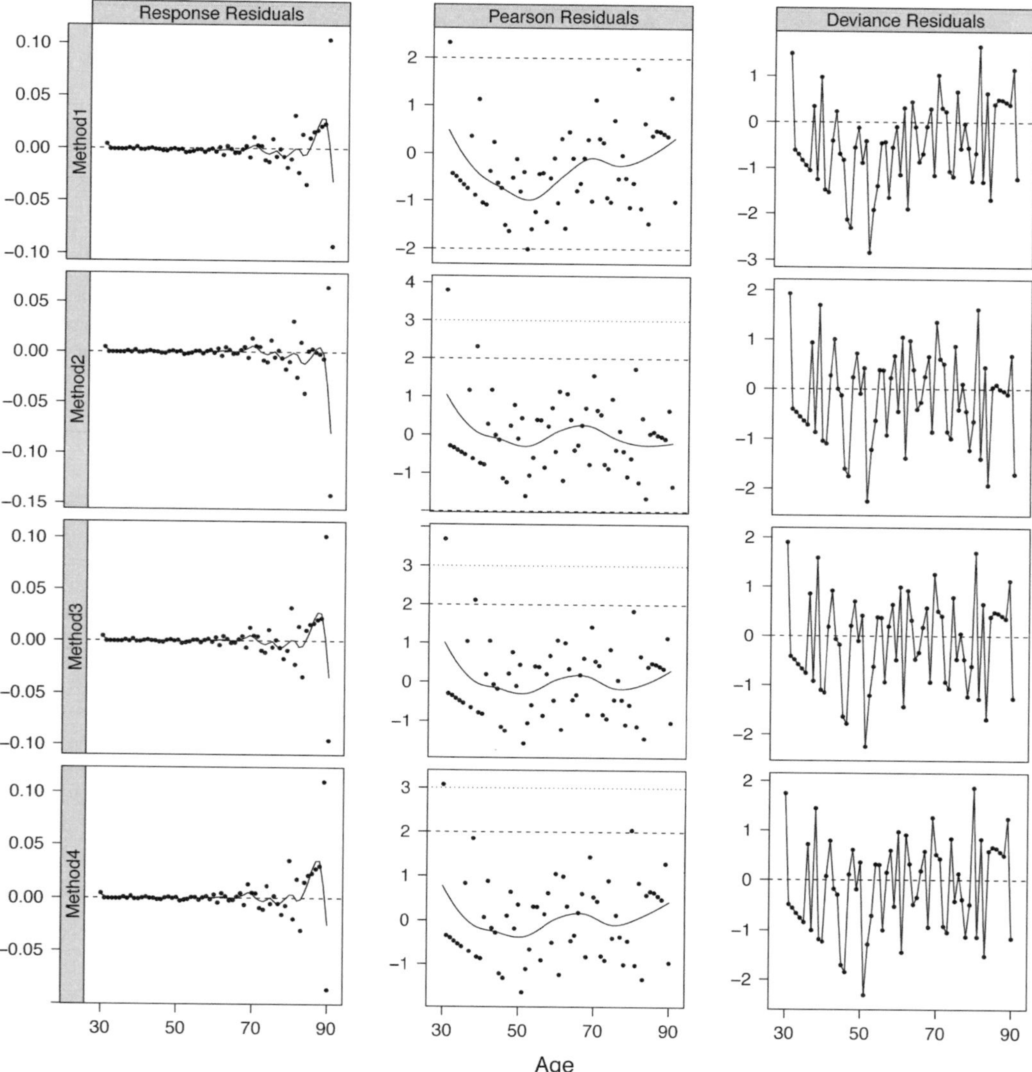

FIGURE 9.7
Comparison of the residuals for calendar year 2007, male population.

```
+       Residuals <- list(RespRes, PearRes, DevRes, NameMethod)
+       names(Residuals) <- c("Response Residuals", "Pearson Residuals",
"Deviance Residuals", "NameMethod")
+       return(Residuals)
+       }
```

where the deviance is defined as

```
> DevFct = function(x, y, z, AgeVec, YearVec){
+      DevMat <- matrix(, length(AgeVec), length(YearVec))
```

```
+       colnames(DevMat) <- YearVec
+       xx <- as.matrix(x)[AgeVec - min(as.numeric(rownames(x))) + 1, YearVec]
+       yy <- as.matrix(y)[AgeVec - min(as.numeric(rownames(y))) + 1, YearVec]
+       zz <- as.matrix(z)[AgeVec - min(as.numeric(rownames(z))) + 1, YearVec]
+       DevMat[(yy > 0)] <- 2 * (yy[(yy > 0)] * log(yy[(yy > 0)] / (xx[(yy > 0)]
* zz[(yy > 0)] ) ) - (yy[(yy > 0)] - xx[(yy > 0)]) * ((zz[(yy > 0)]
yy[(yy > 0)]) / (zz[(yy > 0)] - xx[(yy > 0)] * zz[(yy > 0)] )))
+       DevMat[(yy == 0)] <- 2 * xx[(yy == 0)] * zz[(yy == 0)]
+       return(DevMat)
+       }
```

We superimposed a loess smooth curve on the response and Pearson residuals. These smooths help search for clusters of residuals that may indicate lack of fit. If the approach correctly models the data, no strong patterns should appear in the response and Pearson residuals. The plots of the residuals obtained by the four methods for the calendar year 2007 are displayed in Figure 9.7 for the male population. The figure has been obtained with:

```
> ComparisonResidualsMethods(ListValidationLevel1, 2007, AgeCrit, "Male",
ColorComp)
```

where `ListValidationLevel1` is the list of the objects `ValidationLevel1Method1` to `ValidationLevel1Method4`, see Section 9.5.4.

The Pearson residuals are mainly in the interval $[-2, 2]$, indicating that the models adequately capture the variability of the dataset. The Pearson and response residuals obtained by fitting method 1 displays a more pronounced trend on the age range $\left[30, 60\right]$ than the other ages. This indicates an inappropriate fit in this region. We can visualize this lack of fit in the deviance residuals where several successive residuals exhibit a negative sign. It illustrates that the data have been over-smoothed locally.

Finally, a classical step of the validation consists of comparing the observed and fitted deaths for a given attained age or calendar year over the common period of observation. Using the usual Normal approximation of a Poisson law, we can confront graphically the observed and fitted deaths as well as the lower and upper bounds of the pointwise confidence intervals.

Let suppose the following relation:

$$D_{x,t} \sim \mathcal{N}\left(E_{x,t}\, q_x(t), E_{x,t}\, q_x(t)\,(1 - q_x(t))\right).$$

An approximation of the $100\,\% \times (1 - \alpha)$ pointwise confidence intervals of $D_{x,t}$ is

$$\left(E_{x,t}\, \widehat{q}_x(t) - z_{1-\alpha/2}\sqrt{E_{x,t}\, \widehat{q}_x(t)\left(1 - \widehat{q}_x(t)\right)},\ E_{x,t}\, \widehat{q}_x(t) + z_{1-\alpha/2}\sqrt{E_{x,t}\, \widehat{q}_x(t)\left(1 - \widehat{q}_x(t)\right)}\right),$$

where $z_{1-\alpha/2}$ is the $(1 - \alpha/2)$ quantile of the Normal distribution.

```
> FittedDxtAndConfInt = function(q, e, x1, t1, ValCrit, NameMethod){
+       DxtFitted <- as.matrix(q[x1 - min(as.numeric(rownames(q))) + 1,
as.character(t1)] * e[x1 - min(as.numeric(rownames(l))) + 1, as.character(t1)])
+       DIntUp <- as.matrix(DxtFitted + qnorm(1 - ValCrit / 2) * sqrt(DxtFitted *
(1 - q[x1 - min(as.numeric(rownames(q))) + 1, as.character(t1)])))
+       DIntLow <- as.matrix(DxtFitted - qnorm(1 - ValCrit / 2) * sqrt(DxtFitted
* (1 - q[x1 - min(as.numeric(rownames(q))) + 1, as.character(t1)])))
+       DIntLow[DIntLow < 0 ] <- 0
+       colnames(DIntUp) <- colnames(DIntLow) <- colnames(DxtFitted) <-
as.character(t1)
+       rownames(DIntUp) <- rownames(DIntLow) <- rownames(DxtFitted) <- x1
```

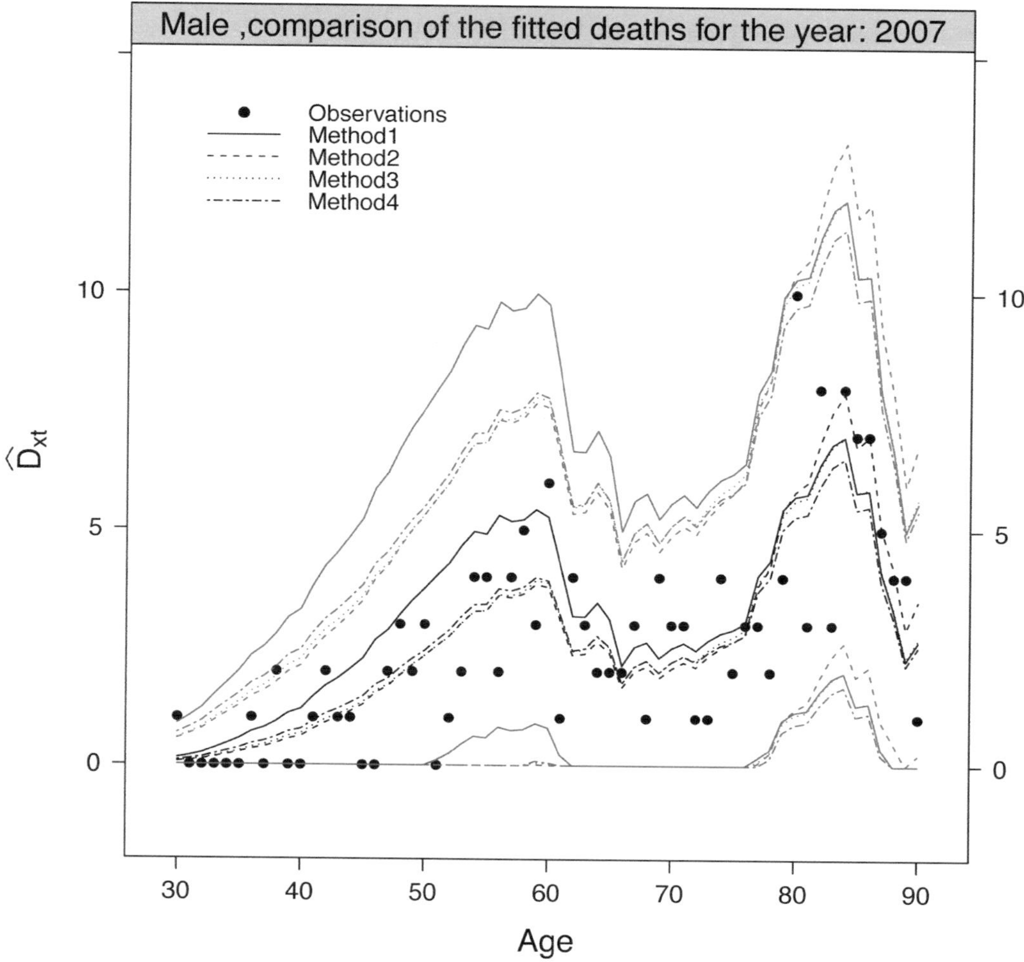

FIGURE 9.8
Comparison of the observed and fitted deaths for calendar year 2007, male population.

```
+       return(list(DxtFitted = DxtFitted, DIntUp = DIntUp, DIntLow = DIntLow,
NameMethod = NameMethod))
+       }
```

Figure 9.8 compares the observed and fitted deaths for the range [30, 90] in 2007 for the male population. The pointwise confidence intervals at 95% of the fitted deaths have been added to the plots. The figure have been produced with:

```
> ComparisonFittedDeathsMethods(ListValidationLevel1, 2007, MyData$Male, AgeCrit,
"Male", ColorComp, LtyComp)
```

The observed deaths are within the bands of the theoretical pointwise confidence intervals at 95% over the age range considered. This indicates a correct representation of the reality by the male and female mortality tables.

9.4.2 Second Level: Regularity of the Fit

The second level of the validation assesses the regularity of the fit. It examines if the data have been over- or under-smoothed. It involves the number of positive and negative signs of the response residuals and the corresponding signs test, the number of runs, and the corresponding runs test used in Forfar et al. (1988) and Debón et al. (2006).

The tests and quantities summarizing the regularity of the fit are described in the following:

- **Signs test.** This is a nonparametric test that examines the frequencies of the signs changes of the difference between the observed and fitted probabilities of death. Under the null hypothesis $\mathcal{H}_0$, the median between the positive and negative signs of this difference is null.

  Let n_+ and n_- be the numbers of positive and negative signs, respectively, with $n = n_+ + n_-$; the signs test statistic, ξ^{SIG}, is defined by:

  $$\xi^{\text{SIG}} = \frac{|n_+ - n_-| - 1}{\sqrt{n}}.$$

  If $\mathcal{H}_0$ is true, this statistic follows a standard Normal law,

  $$\xi^{\text{SIG}} \sim \mathcal{N}(0,1).$$

  Thus, the null hypothesis $\mathcal{H}_0$ is rejected if

  $$|\xi^{\text{SIG}}| > \mathcal{N}_{1-\alpha/2}(0,1),$$

  where $\mathcal{N}_{1-\alpha/2}(0,1)$ is the $(1-\alpha/2)$ quantile of the standard Normal distribution.

  The p-value is given by

  $$p\text{-value} = \mathbb{P}\big[\mathcal{N}_{1-\alpha/2}(0,1) > |\xi^{\text{SIG}}|\big] = 2 \times \big(1 - \text{F}_{\mathcal{N}(0,1)}(|\xi^{\text{SIG}}|)\big).$$

  We will seek to obtain the p-value closest to 1.

- **Runs test.** This is a nonparametric test that determines if the elements of a sequence are mutually independent. A run is the maximal non-empty segment of the sequence consisting of adjacent equal elements. For instance, the following sequence composed of twenty elements,

  $$\{+ + - - - - + + + + - - + + + + - - - + +\},$$

  consists of seven *runs* with four composed of $+$ and three of $-$.

  Under the null hypothesis $\mathcal{H}_0$, the number of runs of a sequence of n elements is a random variable whose conditional distribution given the numbers n_+ and n_- of positive and negative signs, with $n = n_+ + n_-$ is approximatively Normal, with:

  $$\mu = \frac{2\,n_+\,n_-}{n_+ + n_-} + 1 \qquad \text{and} \qquad \sigma^2 = \frac{2\,n_+\,n_-\,\big(2\,n_+\,n_- - (n_+ + n_-)\big)}{(n_+ + n_-)^2(n_+ + n_- - 1)}.$$

  The runs test statistic, ξ^{RUN}, is defined as

  $$\xi^{\text{RUN}} = \frac{\text{Number of runs} - \mu}{\sigma}.$$

If $\mathcal{H}_0$ is true, this statistic follows a standard Normal law,

$$\xi^{\text{RUN}} \sim \mathcal{N}(0,1).$$

Hence, the null hypothesis $\mathcal{H}_0$ is rejected if

$$|\xi^{\text{RUN}}| > \mathcal{N}_{1-\alpha/2}(0,1),$$

where $\mathcal{N}_{1-\alpha/2}(0,1)$ is the $(1-\alpha/2)$ quantile of the standard Normal distribution. The p-value is given by

$$p\text{-value} = \mathbb{P}\big[\mathcal{N}_{1-\alpha/2}(0,1) > |\xi^{\text{RUN}}|\big] = 2 \times \big(1 - \text{F}_{\mathcal{N}(0,1)}(|\xi^{\text{RUN}}|)\big).$$

We will seek to obtain the p-value closest to 1. This test is also called the Wald–Wolfowitz test.

```
> TestsLevel2 = function(x, y, z, vv, AgeVec, AgeRef, YearVec, NameMethod){
+      RUNSTEST <- matrix(, 7, 1)
+      SIGNTEST <- matrix(, 6, 1)
colnames(RUNSTEST) <- colnames(SIGNTEST) <- NameMethod
+      rownames(RUNSTEST) <- c("Nber of runs", "Signs (-)", "Signs (+)",
"Xi (abs)", "Threshold", "Hyp", "p.val")
+      rownames(SIGNTEST) <- c("Signs (+)", "Signs (-)", "Xi", "Threshold",
"Hyp", "p.val")
+      xx <- x[AgeVec - min(AgeVec) + 1, YearVec]
+      yy <- y[AgeVec - min(AgeRef) + 1, YearVec]
+      zz <- z[AgeVec - min(AgeRef) + 1, YearVec]
+ ## ---------- Runs test
+      val.SIGN <- sign(yy / zz - xx)
+      fac.SIGN <- factor(as.matrix(val.SIGN))
+      sign.neg <- sum(levels(fac.SIGN)[1] == fac.SIGN)
+      sign.pos <- sum(levels(fac.SIGN)[2] == fac.SIGN)
+      val.RUNS <- 1 + sum(as.numeric(fac.SIGN[-1] !=
fac.SIGN[-length(fac.SIGN)]))
+      mean.run <- 1 + 2 * sign.neg*sign.pos / (sign.neg+sign.pos)
+      var.run <- 2 * sign.neg * sign.pos * (2 * sign.neg * sign.pos -
sign.neg - sign.pos) / ((sign.neg + sign.pos)^2 * (sign.neg + sign.pos - 1))
+      Threshold.RUNS <- qnorm(1 - vv/2)
+      xi.RUNS <- (val.RUNS - mean.run) / sqrt(var.run)
+      p.val.RUNS <- 2*(1-pnorm(abs(xi.RUNS)))
+      if(abs(xi.RUNS) <= Threshold.RUNS){ RUNSTEST[6, 1] <- "H0" }
else { RUNSTEST[6,1] <- "H1" }
+      RUNSTEST[1, 1] <- val.RUNS
+      RUNSTEST[2, 1] <- sign.neg
+      RUNSTEST[3, 1] <- sign.pos
+      RUNSTEST[4, 1] <- round(abs(xi.RUNS), 4)
+      RUNSTEST[5, 1] <- round(Threshold.RUNS, 4);
+      RUNSTEST[7, 1] <- round(p.val.RUNS, 4)
+ ## ---------- Sign test
+      SIGNTEST[1, 1] <- sign.pos
+      SIGNTEST[2, 1] <- sign.neg
+      xi.SIGN <- (abs(sign.pos - sign.neg) - 1) / sqrt(sign.pos + sign.neg)
```

TABLE 9.3
Second level of validation, male population.

		Method 1	Method 2	Method 3	Method 4
Signs test	+(−)	25(36)	29(32)	30(31)	30(31)
	ξ^{SIG}	2.2804	0.2561	0	0
	p-value	0.2004	0.7979	1	1
Runs test	Nber of *runs*	32	36	34	34
	$\lvert\xi^{\text{RUN}}\rvert$	0.3984	1.2233	0.6479	0.6479
	p-value	0.6903	0.2212	0.5171	0.5171

```
+       Threshold.SIGN <- qnorm(1 - vv/2)
+       p.val.SIGN <- 2 * (1 - pnorm(abs(xi.SIGN)))
+       if(xi.SIGN <= Threshold.SIGN){ SIGNTEST[5, 1] <- "H0" }
else { SIGNTEST[5, 1] <- "H1" }
+       SIGNTEST[3, 1] <- round(xi.SIGN, 4)
+       SIGNTEST[4, 1] <- round(Threshold.SIGN, 4)
+       SIGNTEST[6, 1] <- round(p.val.SIGN, 4)
+       RSLTS <- vector("list", 2)
+       RSLTS[[1]] <- RUNSTEST; RSLTS[[2]] <- SIGNTEST;
+       RSLTS[[3]] <- NameMethod
+       names(RSLTS) <- c("Runs test", "Signs test", "NameMethod")
+       return(RSLTS)
+       }
```

The results of tests carried out are presented in Table 9.3 for the four methods, for the male population.

From Table 9.3 we observe that, the regularity of the fit increases with the complexity of the models.

The two first levels of validation evaluate the fit according its regularity and the overall deviation from the past mortality. A satisfying fit, characterized by a homogeneous repartition of positive and negative signs of the response residuals and a high number of runs, should not lead to a significant gap with the past mortality, or vice versa. Accordingly, the two first levels of validation balance these two complementary aspects.

9.4.3 Third Level: Consistency and Plausibility of the Mortality Trends

The third level of the validation covers the plausibility and consistency of the mortality trends. It is evaluated by singles indices summarizing the lifetime probability distribution for different cohorts at several ages, such as the cohort life expectancies ${}_{\omega}\overset{\nearrow}{e}_{\tilde{x}}$, median age at death $\text{Med}\big[{}_{\omega}T_{\tilde{x}}\big]$ and the entropy $\text{H}\big[{}_{\omega}T_{\tilde{x}}\big]$.

It also involves graphical diagnostics assessing the consistency of the historical and forecasted periodic life expectancy ${}_{\omega}\overset{\uparrow}{e}_{x}(t)$.

In addition, if we have at our disposal the male and female mortality, we can compare the trends of improvement and judge the plausibility of the common evolution of the mortality of the two genders.

We refer to the concept of *biological reasonableness* proposed by Cairns et al. (2006*b*) as a means to assess the coherence of the extrapolated mortality trends. We ask the question:

where are the data originating from, and based on this knowledge, what mixture of biological factors, medical advances, and environmental changes would have to happen to cause this particular set of forecasts?

It consists, initially, of obtaining the survival function calculated from the completed tables, see Section 9.3.5, resulting from the different approaches. From the survival function, we can derive a series of markers summarizing the lifetime probability distribution. We are interested in the survival distribution of cohorts for a given age $\widetilde{x}$ at time t. Hence, we are working along the diagonal of the Lexis diagram. As a result, we can determine the mortality trends and compare the level and speed of improvement between the models.

We expose the indices summarizing the lifetime probability distribution in the following.

- **Survival function.** The survival function of a cohort aged $\widetilde{x}$ at time t measures the proportion of individuals of the cohort aged $\widetilde{x}$ at time t being alive at age $\widetilde{x}+x$ (or equivalently at time $t+x$). Under the condition $\overset{\nearrow}{S}_{\widetilde{x}}(0)=1$, it writes

$$\overset{\nearrow}{S}_{\widetilde{x}}(x)=\prod_{j=0}^{x-1}\left(1-q_{\widetilde{x}+j}(t+j)\right),$$

  where the upper indices $\nearrow$ recalls that we are working along a diagonal of the Lexis diagram.

- **Cohort life expectancy.** This is the partial life expectancy (over ω years) of an individual of a cohort aged $\widetilde{x}$ at time t. It is defined as

$${}_{\omega}\overset{\nearrow}{e}_{\widetilde{x}}=\int_{1}^{\omega}\overset{\nearrow}{S}_{\widetilde{x}}(u)\,du\,.$$

  we obtain

$${}_{\omega}\overset{\nearrow}{e}_{\widetilde{x}}=\sum_{x=1}^{\omega}\prod_{j=0}^{x-1}\left(1-q_{\widetilde{x}+j}(t+j)\right).$$

- **Periodic life expectancy.** This is the residual life expectancy (over ω years) of an individual aged x at time t. It is defined as

$${}_{\omega}\overset{\uparrow}{e}_{x}(t)=\sum_{\delta=1}^{\omega}\prod_{j=0}^{\delta-1}\left(1-q_{x+j}(t)\right),$$

  where the upper indices $\uparrow$ recalls that we are working along a vertical of the Lexis diagram.

- **Median age at death.** The median age at death of an individual of a cohort aged $\widetilde{x}$ at time t (over ω years), denoted $\mathrm{Med}\left[{}_{\omega}T_{\widetilde{x}}\right]$, is the median of the lifetime probability distribution $T_{\widetilde{x}}$,

$$\overset{\nearrow}{S}_{\widetilde{x}}\left(\mathrm{Med}\left[{}_{\omega}T_{\widetilde{x}}\right]\right)=0.5\,.$$

- **Entropy.** The entropy, $\mathrm{H}\left[{}_{\omega}T_{\widetilde{x}}\right]$, is the mean of $\ln\overset{\nearrow}{S}_{\widetilde{x}}(x)$ weighted by $\overset{\nearrow}{S}_{\widetilde{x}}(x)$ (over ω years),

$$\mathrm{H}\left[{}_{\omega}T_{\widetilde{x}}\right]=-\frac{\int_{1}^{\omega}\overset{\nearrow}{S}_{\widetilde{x}}(u)\,\ln\overset{\nearrow}{S}_{\widetilde{x}}(u)\,du}{\int_{1}^{\omega}\overset{\nearrow}{S}_{\widetilde{x}}(u)\,du}\,.$$

  When the deaths become more concentrated, $\mathrm{H}\left[{}_{\omega}T_{\widetilde{x}}\right]$ decreases. In particular, $\mathrm{H}\left[{}_{\omega}T_{\widetilde{x}}\right]=0$ if the survival function has a perfect rectangular shape.

The single indices summarizing the lifetime probability distribution, presented above, are implemented in R as follows:

```
> FctSingleIndices = function(q , Age, t1, t2, NameMethod){
+ ## ---------- Survival functions
+       LSF <- floor(length(min(t1) : max(t2)) / 10) * 10 - 1
+       AgeSFMat <- min(Age) : (min(Age) + LSF)
+       NAgeSF <- AgeSFMat + 10
+       repeat{
+               if(max(NAgeSF) > 130){ break }
+               AgeSFMat <- rbind(AgeSFMat, NAgeSF)
+               NAgeSF <- NAgeSF + 10
+               }
+       Start  <- AgeSFMat[, 1]
+       NberLignes <- nrow(AgeSFMat) - length(Start[Start >= min(Age)])
+       AgeSFMat <- AgeSFMat[(NberLignes + 1) : nrow(AgeSFMat), ]
+       Sx <- matrix(, ncol(AgeSFMat), nrow(AgeSFMat))
+       for(i in 1 : nrow(AgeSFMat)){
+               Sx[,i] <- SurvivalFct(q[AgeSFMat[i, ] - min(Age) + 1, ], 100,
AgeSFMat[i, ])
+               }
+ ## ---------- Median age at death
+       print("Median age at death ...")
+       NameMedian <- vector(, nrow(AgeSFMat))
+       RangeIndices <- t(apply(AgeSFMat, 1, range))
+       for(i in 1 : nrow(AgeSFMat)){
+               NameMedian[i] <- paste("Med[", LSF+1, "_T_",RangeIndices[i,1],
"]", sep = "")
+               }
+       Sxx <- vector("list", nrow(AgeSFMat))
+       for(i in 1 : nrow(AgeSFMat)){
+               Sxx[[i]] <- vector("list",        2)
+               Sxx[[i]][[1]] <- Sxx[[i]][[2]] <- vector(, LSF * 100)
+               Temp <- GetSxx(t(Sx[, i]), c(1, LSF))
+               Sxx[[i]][[1]] <- Temp$a.xx
+               Sxx[[i]][[2]] <- Temp$Sxx
+               }
+       MedianAge <- matrix(, nrow(AgeSFMat), 1)
+       rownames(MedianAge) <- NameMedian
+       colnames(MedianAge) <- NameMethod
+       for (i in 1 : nrow(AgeSFMat)) {
+               MedianAge[i] <- Sxx[[i]][[1]][max(which(Sxx[[i]][[2]] >= 0.5)) +
1]
+               }
+       print(MedianAge)
+ ## ---------- Entropy
+       print("Entropy ...")
+       NameEntropy <- vector(, nrow(AgeSFMat))
+       for(i in 1 : nrow(AgeSFMat)){
+               NameEntropy[i] <- paste("H[", LSF + 1, "_T_", RangeIndices[i, 1],
"]", sep = "")
+               }
+       Entropy <- matrix(, nrow(AgeSFMat), 1)
+       colnames(Entropy) <- NameMethod
+       rownames(Entropy) <- NameEntropy
+       for(i in 1 : nrow(AgeSFMat)){
```

```
+                Entropy[i] <- round(- mean(log(Sx[,i]))/ sum(Sx[,i]), 4)
+                }
+        print(Entropy)
+ ## ---------- Cohort life expectancy for cohort in min(t1)
+        print(paste("Cohort life expectancy for cohort in",min(t1),"over",LSF+1,
years ..."))
+        NameEspGen <- vector(, nrow(AgeSFMat))
+        for(i in 1 : nrow(AgeSFMat)){
+                NameEspGen[i] <- paste(LSF+1,"_e_",RangeIndices[i,1],sep="")
+                }
+        CohortLifeExp <- matrix(, nrow(AgeSFMat), 1)
+        for(i in 1 : nrow(AgeSFMat)){
+                AgeVec <- AgeSFMat[i, ]
+                CohortLifeExp[i] <- sum(cumprod(1 - diag(q[AgeVec + 1 - min(Age),
 ])))
+                }
+        colnames(CohortLifeExp) <- NameMethod
+        rownames(CohortLifeExp) <- NameEspGen
+        print(CohortLifeExpMale)
+        return(list(MedianAge = MedianAge, Entropy = Entropy, CohortLifeExp =
CohortLifeExp, NameMethod = NameMethod))
+ }
```

From the survival function, we derive a series of markers summarizing the lifetime probability distribution. We are interested in the survival distribution of cohorts aged $\widetilde{x} = 30$ to 80 years old in 2007 over 50 years.

As an illustration, Table 9.4 presents the cohort life expectancies, the median age at death, and the entropy obtained with method 1 for the male population. Method 1 leads to the largest life expectancy. In consequence, the median age at death is the highest and the deaths are more concentrated.

In addition, if we have at our disposal the male and female mortalities, we can compare the trends of improvement and judge the plausibility of the common evolution of mortality of the two genders. For this purpose, we compare the male and female cohort life expectancies over 5 years.

```
> FctCohortLifeExp5 = function(q, Age, t1, t2, NameMethod){
+      AgeExp <- min(Age) : (min(Age)+4)
+      AgeMat <- matrix(, length(Age) - length(AgeExp) + 1, length(AgeExp))
+      AgeMat[1, ] <- AgeExp
+      for (i in 2 : (length(Age) - length(AgeExp) + 1)){
+           AgeMat[i, ] <- AgeMat[i - 1, ] + 1
+           }
+      NameCohortLifeExp5 <- vector(, nrow(AgeMat))
+      for(i in 1 : nrow(AgeMat)){
+           NameCohortLifeExp5[i] <- paste(5, "_e_", AgeMat[i, 1], sep = "")
+           }
+      CohortLifeExp5 <- matrix(,length(Age) - length(AgeExp) + 1, length(t2) -
length(AgeExp) + 1)
+      colnames(CohortLifeExp5) <- as.character(min(t1) : (max(t2) - 4))
+      rownames(CohortLifeExp5) <- NameCohortLifeExp5
+      for (i in 1 : (length(t2) - length(AgeExp) + 1)){
+           for (j in 1 : nrow(AgeMat)){
+                AgeVec <- AgeMat[j, ] + 1 - min(Age)
+                CohortLifeExp5[j, i] <- sum(cumprod(1 - diag(q[AgeVec,
i : (i + 4)])))
```

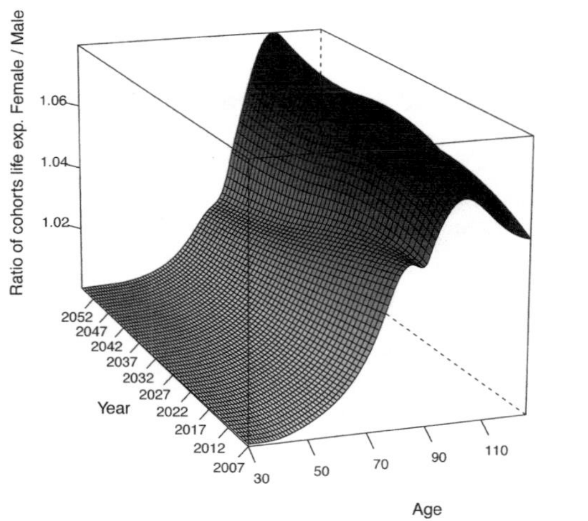

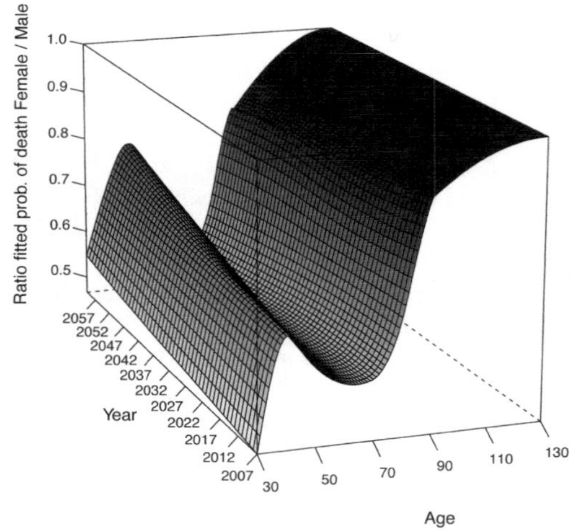

FIGURE 9.9
Ratio between the cohorts life expectancies (right panel) and the fitted probabilities of death (left panel) of the two genders obtained with method 1.

```
+          } }
+       return(list(CohortLifeExp5 = CohortLifeExp5, NameMethod = NameMethod))
+       }
```

Figure 9.9 displays the ratio between the cohorts life expectancies (right panel) and between the fitted probabilities of death (left panel) of the two genders obtained with method 1.

The cohorts life expectancies of the female population are 8% larger than the male ones at 100 years old and 5% larger at 90 years old. The ratio between the cohorts life expectancies as well as the one between the fitted probabilities of death tend to get closer to 1 with the calendar year, indicating that the male mortality is improving more rapidly than the female, which seems plausible. However, after this age, the female mortality improvement increases more rapidly than the male mortality which does not sound coherent.

Finally, the third level of validation also involves graphical diagnostics assessing the consistency of the historical and forecasted periodic life expectancy.

```
> FctPerLifeExp = function(q, Age, a, t1, t2, NameMethod){
+       AgeComp <- min(Age) : pmin(a, max(Age))
+       PerLifeExp <- matrix(, length(AgeComp), length(min(t1) : max(t2)))
+       colnames(PerLifeExp) <- as.character(min(t1) : max(t2))
+       rownames(PerLifeExp) <- AgeComp
+       for (i in 1 : (length(min(t1) : max(t2)))){
+            for (j in 1 : length(AgeComp)){
+                 AgeVec <- ((AgeComp[j]) : max(AgeComp)) + 1 - min(Age)
+                 PerLifeExp[j,i] <- sum(cumprod(1 - q[AgeVec, i]))
+                 }
+            }
+       return(list(PerLifeExp = PerLifeExp, NameMethod = NameMethod))
+       }
```

To illustrate the consistency of the historical and forecasted periodic life expectancy,

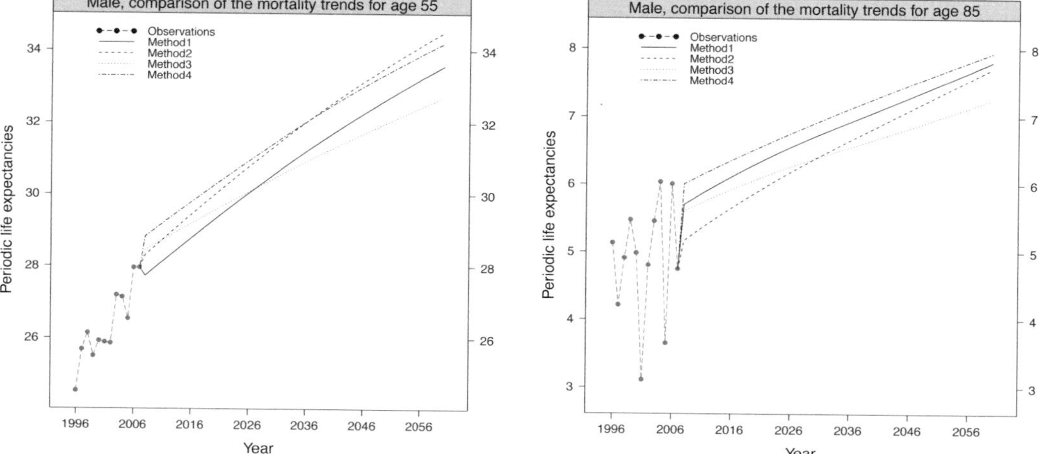

FIGURE 9.10
Comparison of the trends in periodic life expectancies for ages 55 and 70, male population.

Figure 9.10 compares the trends for the ages 55 and 70 obtained by the four methods for the male population.

The plots have been produced using:

```
> ComparisonTrendsMethods(ListValidationLevel3, 55, "Male", ColorComp, LtyComp)
> ComparisonTrendsMethods(ListValidationLevel3, 85, "Male", ColorComp, LtyComp)
```

The predicted periodic life expectancies obtained by the four methods adequately follow the observed trend. The trend derived from method 3 seems to deviate from the ones obtained by the other methods, leading to the slowest mortality improvement.

9.5 Operational Framework

In the following, we present an operational framework to implement the methodologies developed previously. The procedure is based on the software R and on package ELT (Experience Life Table) available on CRAN. Briefly, the procedure is as follows:

- **i.** We start by computing the number of deaths and exposition from the data line by line.
- **ii.** We import the reference table by gender.
- **iii.** We execute method 1.
- **iv.** We validate the resulting table with the first level of criteria. If the criteria assessing the proximity between the observation and the model are not satisfied, we turn to the following method of adjustment as it is useless to continue the process of validation.
- **v.** If the criteria assessing the proximity are satisfied, we continue the validation with the criteria of the second level assessing the regularity of the fit.

vi. If the criteria assessing the regularity of the fit are satisfied, we complete the table at high ages using the procedure of Denuit & Goderniaux (2005).

vii. We then pursue the validation with the criteria of the third level assessing the plausibility and coherence of the forecasted mortality trends. We can also turn to the next method to improve the fit at a cost of a somewhat greater complexity without degrading the results of the criteria corresponding to the first level. We repeat the step **iii.** to **vii.** with the following method.

9.5.1 The Package ELT

The package ELT allows one to implement the proposed methodology by the following functions.

- `ReadHistory()` to compute of the number of deaths, number of individuals, and exposure.
- `AddReference()` to import the reference of mortality.
- `Method1()` to execute the method 1, that is, applying the SMR to the probabilities of death of the reference.
- `Method2()` to execute the method 2, that is, a semi-parametric Brass-type relational model.
- `Method3()` to execute the method 3, that is, a Poisson GLM including the baseline mortality of the reference as a covariate and allowing interactions with age and calendar year.
- `Method4A()` to execute the first step of the method 4, the choice of the parameters for the local likelihood smoothing.
- `Method4B()` to execute the second step of the method 4, the application of the improvement rates derived from the reference.
- `CompletionA()` to execute the first step of the completion procedure, the choice of the age range.
- `CompletionB()` to execute the second step of the completion procedure.
- `ValidationLevel1()` to validate the methods according to the first level of criteria, that is, the proximity between the observations and the model.
- `ValidationLevel2()` to validate the methods according to the second level of criteria, that is, the regularity of the fit.
- `ValidationLevel3()` to validate the methods according to the third level of criteria, that is, the plausibility and coherence of the mortality trends.

In addition, the functions create the folder `Results`, which will be located in the current working directory, with the following subfolders:

- `Excel` where the files `History.xlsx` containing the number of deaths, number of individuals; and exposure of the population(s); `Validation.xlsx` containing the results of the level(s) of the validation, and `FinalTables.xlsx` storing the final mortality tables, will appear.

- Graphics, also decomposed according to
  - graphics of the observed statistics in the folder Data
  - graphics of the final tables in the folder FinalTables
  - graphics allowing one to judge the coherence of the mortality tables in the folder Validation
  - graphics allowing one to judge the regularity of the completed tables in the folder Completion

Finally, the package ELT uses the following dependencies, which need to be installed in advance:

- xlsx to export tables in Excel®
- locfit for the local likelihood smoothing

as well as the graphical packages

- lattice, grid, and latticeExtra

9.5.2 Computation of the Observed Statistics and Importation of the Reference

To obtain the observed number of deaths, number of individuals and exposure, we first need to read the data from the .csv file. For example:

```
> MyPortfolio = read.table(".../MyPortfolio.csv", header = TRUE, sep = ",",
colClasses = "character")
```

We then execute the function ReadHistory() with specifying the date of beginning and end of the period of observation as well as the format of the dates contained in the MyPortfolio.csv file.

```
> History <- ReadHistory(MyPortfolio = MyPortfolio, DateBegObs = "1996/01/01",
DateEndObs = "2007/12/31", DateFormat = "%Y/%m/%d", Plot = TRUE, Excel = TRUE)
```

By setting Excel = TRUE, we obtain an Excel file History.xlsx containing the observed statistics of the male and female population in the folder Results/Excel/ located in the current working directory.
In addition, by setting Plot = TRUE, we obtain graphics similar than Figures 9.2 and 9.3 in the folder Results/Graphics/Data.
Finally, we execute the function AddReference() to import the references of mortality. In this chapter, we use the national demographic projections for the French population over the period 2007-2060. Beforehand, we have extracted the references from the .csv files.

```
> ReferenceMale = read.table(".../QxtRefInseeMale.csv", sep = ",", header = TRUE)
> ReferenceMale <- ReferenceMale[,2:ncol(ReferenceMale)]
> ReferenceFemale = read.table(".../QxtRefInseeFemale.csv", sep = ",",
header = TRUE)
> ReferenceFemale <- ReferenceFemale[,2:ncol(ReferenceFemale)]
> colnames(ReferenceMale) <- colnames(ReferenceFemale) <- 2007 : 2060
> rownames(ReferenceMale) <- rownames(ReferenceFemale) <- 30 : 95
> MyData <- AddReference(History = History, ReferenceMale = ReferenceMale,
ReferenceFemale = ReferenceFemale)
```

9.5.3 Execution of the Methods

The approach involving one parameter is the simplest method considered, see Section 9.3.1. We specify the age range used to compute the SMR and execute the function `Method1()` to perform an adjustment with method 1.

```
> OutputMethod1 <- Method1(MyData = MyData, AgeRange = 30 : 90, Plot = TRUE)
```

The second approach implies that the differences between the observed mortality and the reference can be represented linearly with two parameters through a semi-parametric Brass-type relational model, see Section 9.3.2. We specify the age range used to compute the parameters and execute the function `Method2()` to perform an adjustment with method 2.

```
> OutputMethod2 <- Method2(MyData = MyData, AgeRange = 30 : 90, Plot = TRUE)
```

The third approach is a Poisson GLM including the baseline mortality of the reference as a covariate and allowing interactions with age and calendar year, see Section 9.3.3. We specify the age range used to compute the parameters of the Poisson GLMs and execute the function `Method3()` to perform an adjustment with method 3.

```
> OutputMethod3 <- Method3(MyData = MyData, AgeRange = AgeRange, Plot = TRUE)
```

The fourth approach involves a non-parametric smoothing of the periodic table by local likelihood and the application of the rates of mortality improvement derived from the reference. Its execution is in two steps, see Section 9.3.4. We specify the age range used to construct the periodic table and execute the function `Method4A()` to smooth the table with local likelihood technics.

```
> OutputMethod4PartOne <- Method4A(MyData = MyData, AgeRange = 30 : 90,
AgeCrit = AgeCrit, ShowPlot = TRUE)
```

By setting `ShowPlot = TRUE`, we obtain the graphical diagnostics used to select the smoothing parameters similar than Figure 9.4. The point corresponding to the minimum or a plateau after a steep descent of the criterion determines the optimal window width and polynomial degree, denoted by `OptMale` and `OptFemale` respectively for the male and female population. Finally, we execute the function `Method4B()` corresponding to the second step of the method to apply, to the smoothed periodic tables, the rates of mortality improvement derived from the references.

```
> OutputMethod4 <- Method4B(PartOne = OutputMethod4PartOne, MyData = MyData,
OptMale = c(0, 20), OptFemale = c(1, 15), Plot = TRUE)
```

We obtain after the execution of each method, the fitted and extrapolated surfaces of mortality before the completion in the folder `Results/Graphics/Completion` by setting `Plot = TRUE`.

9.5.4 Process of Validation

We execute the function `ValidationLevel1()` to perform the validation with the criteria corresponding to the first level, see Section 9.4.1. We specify the age range used for the computation of the quantities and tests of the first and second levels as well as the critical value used by these tests.

```
> ValidationLevel1Method1 <- ValidationLevel1(OutputMethod = OutputMethod1,
MyData = MyData, AgeCrit = 30 : 90, ValCrit = 0.05, Plot = TRUE, Excel = TRUE)
```

The execution of the function produces in the `R` console the values of the quantities and proposed tests used to assess the proximity between the observations and the model, see Table 9.2. These values are also recorded and stored in the Excel file `Validation.xlsx` in the folder `Results/Excel/ if Excel = TRUE`. As an example, the values obtained with method 1 for the male population are displayed below.

```
[1] "Validation: Level 1"
[1] "Male population: "
$`Likelihood ratio test`
          Method1
Xi        "69.71"
Threshold "80.23"
Hyp       "H0"
p.val     "0.2078"
$`SMR test`
          Method1
SMR       "1"
Xi        "0.0265"
Threshold "1.6449"
Hyp       "H0"
p.val     "0.4894"
$`Wilcoxon Matched-Pairs Signed-Ranks test`
          Method1
W         "969"
Xi        "0.1652"
Threshold "1.96"
Hyp       "H0"
p.val     "8688"
$`Standardized residuals`
               Method1
Std. Res.  > 2       2
Std. Res.  > 3       0
$Quantities
          Method1
Chi2      55.7900
R2         0.7201
MAPE      50.3200
Deviance 139.4300
```

We note that the null hypothesis $\mathcal{H}_0$ of the likelihood ratio test is verified, the fit corresponds to the underlying mortality law as well as the one of the Wilcoxon test, that is, the median of the difference of each pair composed of the observed and fitted probabilities of death is null.

The SMR is exactly equal to 1 as the model has the capacity to reproduce the observed number of deaths. The number of standardized residuals larger than 2 or 3, as well as the quantities, are indicative and provide a complement to the comparison of fits between the different methods.

In addition to the results of the tests of the first level, the validation involves graphical diagnostics such as the figures presented in Section 9.4.1. Those graphics are available in the folder `Results/Graphics/Validation if Plot = TRUE`.

If the criteria of the first level are not satisfied, we turn to the following method as it is useless to continue the validation. If the criteria are satisfied, we pursue the second level of validation. As an example, we continue the validation with the criteria assessing the regularity of the fit; see Section 9.4.2. We execute the function `ValidationLevel2()`:

```
> ValidationLevel2Method1 <- ValidationLevel2(OutputMethod = OutputMethod1,
MyData = MyData, AgeCrit = AgeCrit, ValCrit = 0.05, Excel = TRUE)
```

The execution of the script produces in the R console the values of the quantities and tests assessing the second level of validation; see Table 9.2. These values are also recorded and stored in the Excel file `Validation.xlsx` in the folder `Results/Excel/` if `Excel = TRUE` is specified. In the following, we display the R output obtained for the male population after execution of method 1.

```
[1] "Validation: Level 2"
[1] "Male population: "
$`Runs test`
              Method1
Nber of runs "32"
Signs (-)    "36"
Signs (+)    "25"
Xi (abs)     "0.3984"
Threshold    "1.96"
Hyp          "H0"
p.val        "0.6903"
$`Signs test`
          Method1
Signs (+) "25"
Signs (-) "36"
Xi        "1.2804"
Threshold "1.96"
Hyp          "H0"
p.val        "0.2004"
```

We note that the null hypothesis $\mathcal{H}_0$ of the runs test as well as the signs tests are verified.

If the criteria of the second level are not satisfied, we turn to the following method. If they are satisfied, we continue with the completion of the table at high ages. Having applied the completion method, see Sections 9.3.5 and 9.5.5, we have at our disposal completed tables until 130 years old. We execute the function `ValidationLevel3()` to assess the plausibility and coherence of the forecasted mortality trends.

```
> ValidationLevel3Method1 <- ValidationLevel3(FinalMethod = FinalMethod1,
MyData = MyData, Plot = TRUE, Excel = TRUE)
```

The execution of the function produces singles indices summarizing the probability lifetime distribution for several cohorts; see Section 9.4.3 and Table 9.4. These values are also recorded and stored in the Excel file `Validation.xlsx` in the folder `Results/Excel/` if `Excel = TRUE` is specified.

In addition, the validation involves graphical diagnostics such as the figures presented in Section 9.4.3, available in the folder `Results/Graphics/Validation` by setting `Plot = TRUE`.

TABLE 9.4
Single indices summarizing the lifetime probability distribution for cohorts of several ages in 2007 over 50 years, male population.

	Method 1	Method 2	Method 3	Method 4
${}_{50}\overset{\nearrow}{e}_{30}$	47.41	47.98	47.56	47.67
${}_{50}\overset{\nearrow}{e}_{40}$	44.21	44.61	43.76	44.29
${}_{50}\overset{\nearrow}{e}_{50}$	37.09	36.47	35.64	36.72
${}_{50}\overset{\nearrow}{e}_{60}$	27.43	26.18	25.77	26.76
${}_{50}\overset{\nearrow}{e}_{70}$	18.35	16.71	16.73	17.54
${}_{50}\overset{\nearrow}{e}_{80}$	10.42	8.61	9.06	9.78
$\text{Med}\left[{}_{50}T_{50}\right]$	41.64	40.39	39.44	40.83
$\text{Med}\left[{}_{50}T_{60}\right]$	30.75	29.23	28.73	29.80
$\text{Med}\left[{}_{50}T_{70}\right]$	20.56	18.75	18.73	19.56
$\text{Med}\left[{}_{50}T_{80}\right]$	11.63	9.69	10.15	10.85
$\text{H}\left[{}_{50}T_{30}\right]$	0.0011	0.0008	0.0010	0.0009
$\text{H}\left[{}_{50}T_{40}\right]$	0.0027	0.0025	0.0030	0.0027
$\text{H}\left[{}_{50}T_{50}\right]$	0.0093	0.0108	0.0121	0.0100
$\text{H}\left[{}_{50}T_{60}\right]$	0.0400	0.0522	0.0539	0.0439
$\text{H}\left[{}_{50}T_{70}\right]$	0.1782	0.2400	0.2356	0.1963
$\text{H}\left[{}_{50}T_{80}\right]$	0.8511	1.1915	1.1038	0.9349

9.5.5 Completion of the Tables

The completion of the tables is performed in two steps We first define the age range `AgeRangeOpt` in which the optimal starting age is determined and used to estimate the parameters c_t; see Section 9.3.5. In addition, we have to initialize the age `BegAgeComp` from which we replace the fitted probabilities of death with the values obtained by the completion method. Then we execute the function `CompletionA()`:

```
> CompletionMethod1 <- CompletionA(OutputMethod = OutputMethod1, MyData = MyData,
AgeRangeOptMale = c(80, 80), AgeRangeOptFemale = c(80, 80), BegAgeCompMale = 85,
BegAgeCompFemale = 85, ShowPlot = TRUE)
```

By setting `ShowPlot = TRUE`,
the execution of this function produces the completed tables for the population(s) as well as graphics comparing the fit before and after the completion procedure; see Figure 9.5.

We then verify visually, in the second step, if the completed surface at high ages does not present breaks with the initial fit. We note, in Figure 9.5, that the junction between the fitted probabilities of death and the values obtained after the completion method appears relatively smooth.

If the completion is not satisfying, we modify the values `AgeRangeOpt` and `BegAgeComp` in the object `ParamMortExp` and repeat the execution of the function `CompletionA()`. If the completion is satisfying, we execute the function `CompletionB()`.

```
> FinalMethod1 <- CompletionB(ModCompletion = CompletionMethod1,
OutputMethod = OutputMethod1, MyData = MyData, Plot = TRUE, Excel = TRUE)
```

We obtain, after the execution of the function, the completed tables for the populations in the folder `Results/Graphics/FinalTables`, as well as the graphics comparing the fit before and after the completion procedure by calendar year in the folder `Results/Graphics/Completion`if `Plot = TRUE`. Finally, the final tables are stored in the Excel file `FinalTables.xlsx` in the folder `Results/Excel/` if `Excel = TRUE` is specified.

10

Survival Analysis

Frédéric Planchet
ISFA - Université de Lyon 1, France — Prim'Act
Lyon, France

Pierre-E. Thérond
ISFA - Université de Lyon 1, Galea & Associés
Lyon, France

CONTENTS

10.1 Introduction

The purpose of this chapter dedicated to duration models is to provide operational tools with R to implement them effectively in the context of their most common use in insurance. After a brief reminder of the theoretical context, examples that illustrate the practical implementation of the R functions concerned are provided. Each example begins with a reminder of the underlying theory and the associated code is commented. The reader interested in a complete presentation of theoretical concepts may consult for example, Planchet & Thérond (2011) or Martinussen & Scheike (2006).

We will here consider the following framework, which is very often used in actuarial problems:

- From the data, we must compute crude estimators $\hat{q}_x$ of conditional exit (death for example) probabilities q_x;
- The crude rates must then be graduated;
- Finally, it is essential to validate that the model and the data are consistent.

These three steps are described later in this chapter.

Let us consider a positive random variable T. In the following, we denote

$$F(t) = \mathbb{P}(T \leq t),$$

its cumulative distribution function. When F is differentiable, we will denote f, the density function of T defined by

$$f(t) = \frac{d}{dt}F(t).$$

The survival function of T is

$$S(t) = 1 - F(t).$$

We can define the `sfd` R function in order to plot survival distribution functions:

```
> sdf <- function (x)
+ {
+   x <- sort(x)
+   n <- length(x)
+   if (n < 1)
+     stop("'x' must have 1 or more non-missing values")
+   vals <- unique(x)
+   rval <- approxfun(vals, 1-cumsum(tabulate(match(x, vals)))/n,
    + method = "constant", yleft = 1, yright = 0,
    + f = 0, ties = "ordered")
+   class(rval) <- c("ecdf", "stepfun", class(rval))
+   assign("nobs", n, envir = environment(rval))
+   attr(rval, "call") <- sys.call()
+   rval
+ }
```

The survival function S is non-increasing such that $S(0) = 1$ and $\lim_{t\to\infty} S(t) = 0$. The survival expectation is easily expressed with S by

$$\mathbb{E}(T) = \int_0^\infty t dF(t) = -\int_0^\infty t dS(t) = \int_0^\infty S(t)dt.$$

The variance of T may be expressed as

$$\text{var}(T) = 2\int_0^\infty tS(t)dt - \mathbb{E}T^2.$$

In the following, we will consider the conditional survival function defined by

$$S_u(t) = \mathbb{P}\left(T > u + t | T > u\right) = \frac{\mathbb{P}\left(T > t + u\right)}{\mathbb{P}\left(T > u\right)} = \frac{S(u+t)}{S(u)}.$$

If we consider that T is the lifetime of a policyholder, $S_u(t)$ represents the probability that he survives until time $t + u$ considering that he is alive at time t.

When F is differentiable, we define the hazard function by

$$h(t) = \frac{f(t)}{S(t)} = -\frac{S'(t)}{S(t)} = -\frac{d}{dt}\log S(t).$$

As a consequence of the definition, S can be expressed in h terms as

$$S(t) = \exp - \int_0^t h(s)ds.$$

The cumulative hazard function of T is defined by

$$H(t) = \int_0^t h(s)ds.$$

We immediately have $S(t) = \exp(-H(t))$. We can observe that $H(T)$ is a random variable exponentially distributed with parameter $\lambda = 1$ because

$$\begin{aligned}\mathbb{P}(H(T) > x) &= \mathbb{P}\left(T > H^{-1}(x)\right) = S\left(H^{-1}(x)\right)\\ &= \exp\left\{-H\left(H^{-1}(x)\right)\right\} = e^{-x}, x \geq 0.\end{aligned}$$

In this chapter, two sets of data are used:

- Mortality portfolio (Sections 10.2 and 10.3);
- Disability portfolio (Section 10.4).

.

But before that, let us mention the specificities of incomplete data.

10.2 Working with Incomplete Data

The aim of this section is to examine how to import and which data manipulations to operate in order to build a survival model. The basic data of this kind of model are incomplete data (i.e., truncated or censored observations).

We assume in the remainder of this chapter that we have individual data on a duration phenomenon. Thus, there is statistical material for estimating the different laws of interest: mortality, time in disability, survival of an annuitant, etc. Because of the conditions for collecting data, right-censored observations are available (due to the presence of individuals at risk at the time of extraction) and left-truncated (because tracking individuals starts generally not from the beginning of the modeled state). From a formal point of view, we have a sample of survival times $(X_1, \ldots, X_n)$ and a second independent sample of variables of censorship $(C_1, \ldots, C_n)$. Instead of observing directly $(X_1, \ldots, X_n)$, we observe

$$Y_1 = (T_1, D_1), \ldots, Y_n = (T_n, D_n),$$

where $T_i = X_i \wedge C_i$ and

$$D_i = \begin{cases} 1, & \text{if} X_i \leq C_i, \\ 0, & \text{if} X_i > C_i. \end{cases}$$

Consider, for instance, exponential durations X, and deterministic (and constant) censorship:

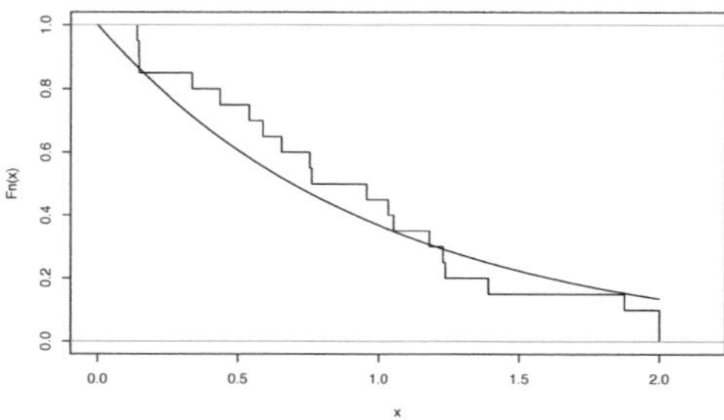

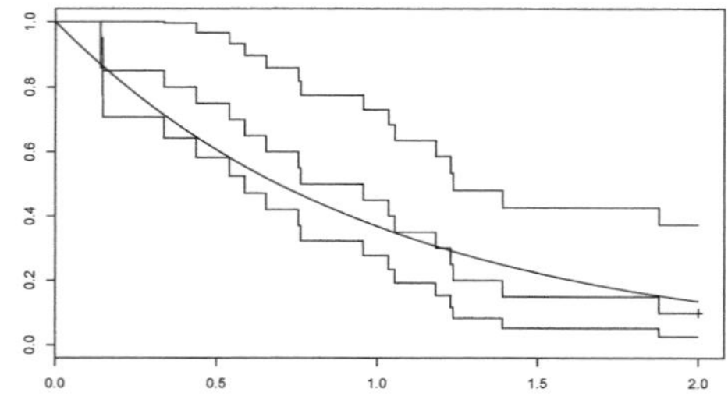

FIGURE 10.1
Survival distribution function: empirical distribution from censored survival times, and Kaplan–Meier's correction.

```
> set.seed(1)
> n <- 20; X <- rexp(n); C <- rep(2,n)
> T <- apply(data.frame(X,C),1,min)
> D <- (T==X)*1
> SampleData <- Surv(T,D)
```

If we plot the empirical distribution function on (censored) survival times, we get the graphic on the left of of Figure 10.1.

```
> t <- seq(0,2,by=.01)
> plot(sdf(T),verticals=TRUE,do.points = FALSE,main="")
> lines(t,exp(-t),lwd=2)
```

The estimation of the survival function is obtained using

```
> plot(survfit(Surv(T,D)~1))
> lines(t,exp(-t),lwd=2)
```

See graphic on the right of Figure 10.1

In addition, the data are left-truncated, that is to say, we did observe individuals whose time of entry into the observation is $T_i \geq E_i$.

10.2.1 Data Importation and Some Statistics

In this subsection, we consider a .csv file containing the incomplete data. We are going to import it; to build the variables, we will have to use for constructing a survival model, and to identify and delete any anomalies in the database.

A key point here is to define the so-called observation period: we have to know the earliest and the latest dates for which the information is complete.

At first we import the data:

```
> data=read.table("data\\DataMortality.csv",header=TRUE,sep=";")
```

Then we convert the dates that are in the database.

```
> data$BirthDate <- as.Date(data$BirthDate,"%d/%m/%Y")
> data$BegDate <- as.Date(data$BegDate,"%d/%m/%Y")
> data$EndDate <- as.Date(data$EndDate,"%d/%m/%Y")
```

Let us observe some preliminary statistics using the summary function:

```
> summary(data)
```

```
      Id              BirthDate             Sex            BegDate
 Min.   :     1   Min.   :1911-10-13   F: 43755   Min.   :1935-05-01
 1st Qu.: 40238   1st Qu.:1947-05-15   M:117194   1st Qu.:1976-03-01
 Median : 80475   Median :1961-03-19              Median :1991-09-01
 Mean   : 80475   Mean   :1960-02-27              Mean   :1988-11-15
 3rd Qu.:120712   3rd Qu.:1975-08-01              3rd Qu.:2003-02-01
 Max.   :160949   Max.   :1994-12-07              Max.   :2012-03-01
    EndDate                   Status          GuarCode      GuarOption
 Min.   :1995-05-02   autre    :   407   DCC1 :10023   A01:141042
 1st Qu.:2007-12-31   encours :152610   DCE1 :     4   B01:  9880
 Median :2009-11-30   deceased:  7932   DCP65:    12   C01:  1561
 Mean   :2011-03-24                       DCPA1: 9868   C02:  8462
 3rd Qu.:2011-10-31                       DCSTA:85101   E02:     4
 Max.   :2057-12-31                       DCSTR:55941
```

Before building the appropriate database, it is useful to test the consistency of the database with some tests such as

- Number of rows with the birth date posterior to the inception date:

```
> nrow(subset(data,data$BirthDate>data$BegDate))
[1] 0
```

- Number of rows with a `deceased` status and without any exit date:

```
> nrow(subset(data, data$Status=="deceased" & is.na(data$EndDate)))
[1] 0
```

10.2.2 Building the Appropriate Database

Let us suppose that we choose the following observation period for the study: from 2009 to 2011. Let us define the following parameters:

```
> BegObsDate <- as.Date("01/01/2009","%d/%m/%Y")
> EndObsDate <- as.Date("31/12/2011","%d/%m/%Y")
```

We are going to select the data observed during the observation period. In the first time, we define the boundaries of the observation period:

```
> BegObsDate <- as.Date("01/01/2009","%d/%m/%Y")
> EndObsDate <- as.Date("31/12/2011","%d/%m/%Y")
```

Then we compute for each observation the beginning and ending dates of observation:

```
> data$BegObs <- pmax(BegObsDate, data$BegDate)
> data$EndObs <- pmin(EndObsDate, data$EndDate)
```

The true exits are those for which the death time is within the observation period:

```
> data$non_censored[(data$Status=="deceased")&
+   (data$EndDate<=EndObsDate)] <- TRUE
> data$non_censored[(data$Status!="deceased")|
+   (data$EndDate>EndObsDate)] <- FALSE
```

Finally, we do not keep in the database the policyholder with

- An exit date prior to the first observation date, or
- An inception date posterior to the last observation date.

```
> data_obs <- subset(data,
+   (data$EndObs>=BegObsDate) & (data$BegObs<=EndObsDate))
```

For each policyholder, we calculate the ages at the observation beginning and ending:

```
> data_obs$BegAge <- as.double(
+   difftime(data_obs$BegObs,data_obs$BirthDate))/365.25
> data_obs$EndAge <- as.double(
+   difftime(data_obs$EndObs,data_obs$BirthDate))/365.25
```

Now we are going to build two subsets for male and female observations and determine a (nonweighted) sex-ratio:

```
> data_obsH=subset(data_obs,(data_obs$Sex=="M"))
> data_obsF=subset(data_obs,(data_obs$Sex=="F"))
> print(paste("Sex-ratio : ",nrow(data_obsH)/nrow(data_obs)))
[1] "Sex-ratio :  0.512871549893843"
```

Now we have in our hands a dataframe object with all the variables we need to build a survival model.

10.2.3 Some Descriptive Statistics

We are now going to build some variables and compute some statistics we will later use in order to check the survival model.

We begin by computing the number of true exits (e.g., deaths for a mortality risk) by age and year during the observation period:

```
> BegDateYear <- as.Date(
+   c("01/01/2009","01/01/2010","01/01/2011","01/01/2012"),"%d/%m/%Y")
```

In order to do that, we create a `data.frame` with the only observations of true exit (just a matter of computing time):

```
> data_TrueExit <- subset(data_obs, data_obs$non_censored==TRUE)
```

Then we build a dataframe TrueExitNumber of size

- Rows: the number of considered ages
- Columns: one for the total on the observation period plus one per year in this period (4 in our example)

```
> TrueExitNumber <- as.data.frame(matrix(nrow = 121, ncol = 5))
> colnames(TrueExitNumber) <-
+   c("Age","Total","2009","2010","2011")
```

```
> for (x in 0:120){
  +   TrueExitNumber[x+1,1] <- x
  +   TrueExitNumber[x+1,2] <- sum(floor(data_TrueExit$EndAge)==x)
  +   TrueExitNumber[x+1,3] <- sum((floor(data_TrueExit$EndAge)==x) &
      + (data_TrueExit$EndDate < BegDateYear[2]))
  +   TrueExitNumber[x+1,4] <- sum((floor(data_TrueExit$EndAge)==x) &
      + (data_TrueExit$EndDate < BegDateYear[3]))
  +   TrueExitNumber[x+1,5] <- sum((floor(data_TrueExit$EndAge)==x) &
      + (data_TrueExit$EndDate < BegDateYear[4]))
  + }
> TrueExitNumber[,5]=TrueExitNumber[,5]-TrueExitNumber[,4]
```

We control the number of true exits

```
> sum(TrueExitNumber[,2])
[1] 4458
```

Now we want to build some exposures for each observation year and each age. We first have to determine the age of each policyholder, both at the beginning and the ending of each observation year (only the code for the first year is disclosed below).

```
> data_obs$BegDate2009 <- BegDateYear[1]

> data_obs$BegDate2009[
+       (data_obs$BegDate>=BegDateYear[1])&(data_obs$BegDate<BegDateYear[2])] <-
+   data_obs$BegDate[
+       (data_obs$BegDate>=BegDateYear[1])&(data_obs$BegDate<BegDateYear[2])]
> is.na(data_obs$BegDate2009) <-
+   (data_obs$BegDate>=BegDateYear[2])|(data_obs$EndDate<BegDateYear[1])
> data_obs$BegAge2009 <-
+   as.double(data_obs$BegDate2009 - data_obs$BirthDate)/365.25

> summary(data_obs$BegAge2009)
   Min. 1st Qu.  Median    Mean 3rd Qu.    Max.    NA's
  18.17   36.08   56.23   54.14   65.74   96.97     777
```

We follow the same way for the end age:

```
> data_obs$EndDate2009 <- BegDateYear[2]-1
> data_obs$EndDate2009[(data_obs$EndDate<BegDateYear[2]) & data_obs$non_censored] <-
   + data_obs$EndDate[(data_obs$EndDate<BegDateYear[2]) & data_obs$non_censored]
> is.na(data_obs$EndDate2009) <-
   + (data_obs$BegDate>=BegDateYear[2])|(data_obs$EndDate<BegDateYear[1])
> data_obs$EndAge2009 <-
   + as.double(data_obs$EndDate2009 - data_obs$BirthDate)/365.25

> summary(data_obs$EndAge2009)
   Min. 1st Qu.  Median    Mean 3rd Qu.    Max.    NA's
  18.52   37.06   57.20   55.09   66.73   97.97     777
```

Now we have to determine, for each observation year and each age, the observation time of each policyholder. In a first step, we create an `ExpoNumber` variable and three subsets of `data_obs` (one for each observation year):

```
> ExpoNumber <- as.data.frame(matrix(nrow = 121, ncol = 5))
> colnames(ExpoNumber) <- c("Age","Total","2009","2010","2011")

> data_obs2009 <- subset(data_obs,!(is.na(data_obs$BegDate2009)))
> data_obs2010 <- subset(data_obs,!(is.na(data_obs$BegDate2010)))
> data_obs2011 <- subset(data_obs,!(is.na(data_obs$BegDate2011)))
```

```
> for (x in 0:120){
+   ExpoNumber[x,1] <- x
+   ExpoNumber[x,3] <- sum(data_obs2009$EndAge2009[
+   (floor(data_obs2009$BegAge2009)==x)&(floor(data_obs2009$EndAge2009)==x)]
+   - data_obs2009$BegAge2009[
+   (floor(data_obs2009$BegAge2009)==x)&(floor(data_obs2009$EndAge2009)==x)])
+   ExpoNumber[x,3] <- ExpoNumber[x,3]
    +sum(x+1-data_obs2009$BegAge2009[(floor(data_obs2009$BegAge2009)==x)&
    +(floor(data_obs2009$EndAge2009)==x+1)])
+   ExpoNumber[x,3] <- ExpoNumber[x,3] + sum(data_obs2009$EndAge2009[
+   (floor(data_obs2009$BegAge2009)==x-1)&(floor(data_obs2009$EndAge2009)==x)]-x)

+   ExpoNumber[x,4] <-sum(data_obs2010$EndAge2010[
+   (floor(data_obs2010$BegAge2010)==x)&(floor(data_obs2010$EndAge2010)==x)]-
+   data_obs2010$BegAge2010[
+   (floor(data_obs2010$BegAge2010)==x)&(floor(data_obs2010$EndAge2010)==x)])
+   ExpoNumber[x,4] <- ExpoNumber[x,4] +
+   sum(x+1-data_obs2010$BegAge2010[
+   (floor(data_obs2010$BegAge2010)==x)&(floor(data_obs2010$EndAge2010)==x+1)])
+   ExpoNumber[x,4] <- ExpoNumber[x,4]+ sum(data_obs2010$EndAge2010[
+   (floor(data_obs2010$BegAge2010)==x-1)&(floor(data_obs2010$EndAge2010)==x)]-x)

+   ExpoNumber[x,5] <- sum(data_obs2011$EndAge2011[
+   (floor(data_obs2011$BegAge2011)==x)&(floor(data_obs2011$EndAge2011)==x)] -
+   data_obs2011$BegAge2011[
+   (floor(data_obs2011$BegAge2011)==x)&(floor(data_obs2011$EndAge2011)==x)])
+   ExpoNumber[x,5] <- ExpoNumber[x,5] +
    + sum(x+1-data_obs2011$BegAge2011[(floor(data_obs2011$BegAge2011)==x)&
    + (floor(data_obs2011$EndAge2011)==x+1)])
+   ExpoNumber[x,5] <- ExpoNumber[x,5]+sum(data_obs2011$EndAge2011[
+   (floor(data_obs2011$BegAge2011)==x-1)&(floor(data_obs2011$EndAge2011)==x)]-x)
+ }
> ExpoNumber[,2] <- ExpoNumber[,3]+ExpoNumber[,4]+ExpoNumber[,5]
```

Now we can create some graphics (see Figures 10.2, 10.3 and 10.4):

```
> xx <- 18:65
> plot(xx,ExpoNumber[xx+1,2],type="h",lwd=2,xlab="Age",
+   ylab="Exposure")
> grid(NA,8,lwd=1, col="black",lty=2)

> barplot(TrueExitNumber[xx+1,2],names.arg=xx,xlab="Age",
+   ylim=c(0,100), ylab="Number of deaths")
> grid(NA,10,lwd=1, col="black",lty=2)
```

We can also use some advanced graphic functions such as barplot2 (from the `gplots` library):

```
> TH0002 <- readNamedRegionFromFile("data/TH0002.xlsx",
+   "TH_0002" , header = TRUE )
> ci.l <- pmax(TrueExitNumber[xx+1,2]
+   -1.96*sqrt(TrueExitNumber[xx+1,2]),0)
> ci.u <- TrueExitNumber[xx+1,2]
+   +1.96*sqrt(TrueExitNumber[xx+1,2])
> barplot2(3*TH0002[xx+1,2]*ExpoNumber[xx+1,2],
+          names.arg=xx,col="red",xlab="Age",ylim=c(0,120),
+          plot.ci=TRUE,ci.lwd=2,ci.color="blue",
+          ci.l = ci.l, ci.u = ci.u,
+          plot.grid=TRUE)
```

We can observe that the underlying population has significantly higher mortality rates than the mortality table.

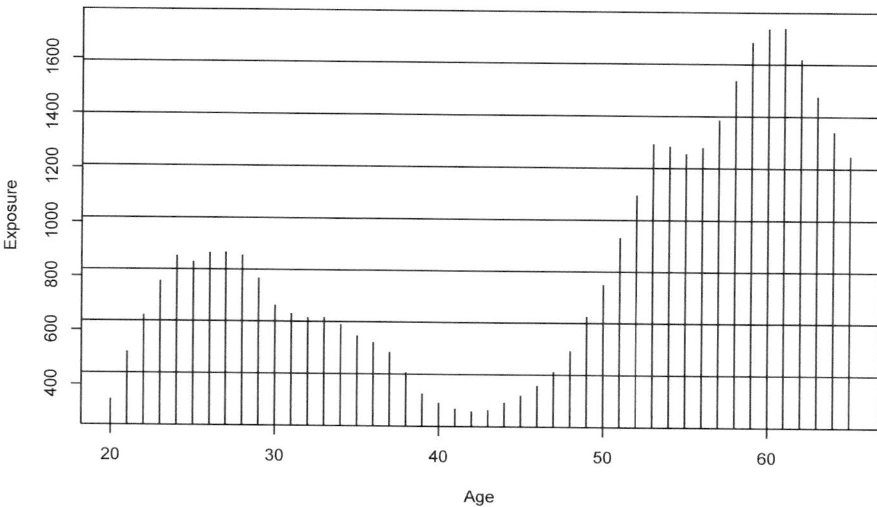

FIGURE 10.2
Exposure.

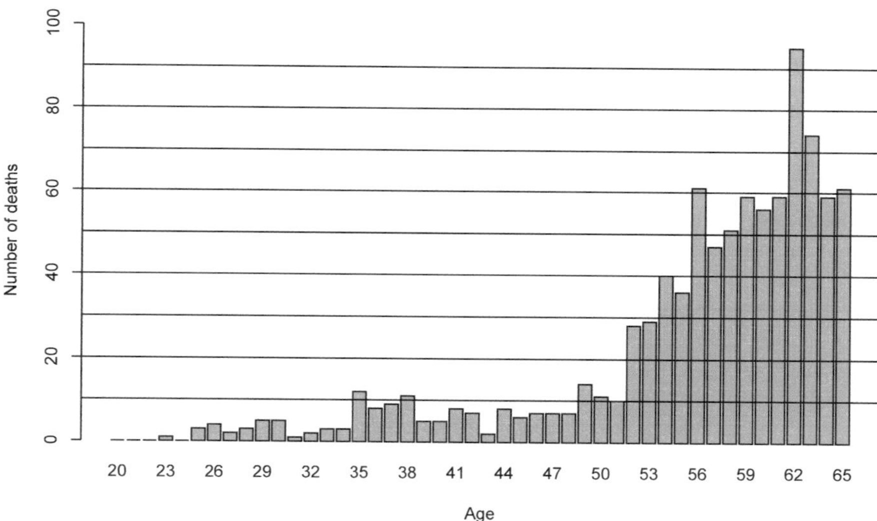

FIGURE 10.3
Number of deaths.

10.3 Survival Distribution Estimation

The aim of this section is to estimate some survival distributions from data. There are numerous estimators of survival distribution, such as Nelson–Aalen and Hoem estimators

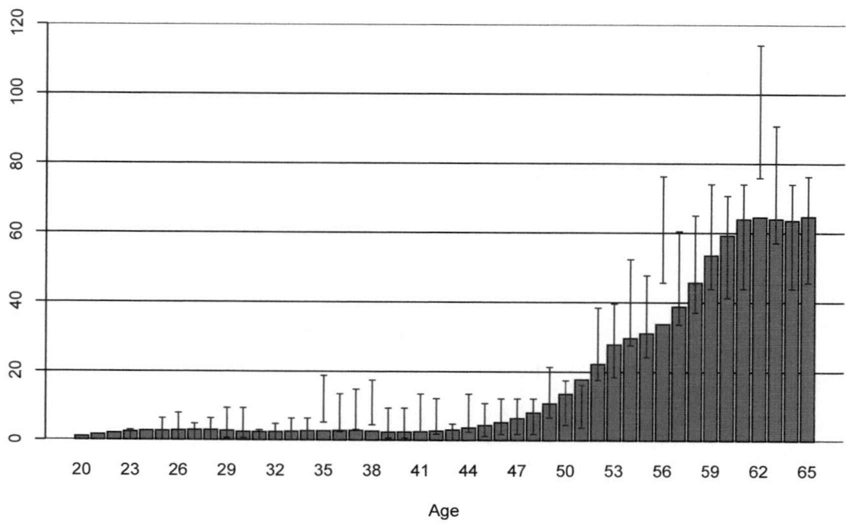

FIGURE 10.4
Number of predicted deaths versus CI.

(see Planchet & Thérond (2011)). After introducing the Hoem estimator (also known as actuarial estimator), we will mainly focus on the Kaplan–Meier estimator of the survival function.

10.3.1 Hoem Estimator of the Conditional Rates

The Hoem estimator of the conditional mortality rates consists of dividing the number of deaths at age x by a measure of the exposure at the same age (see Planchet & Thérond (2011)).

With the previously introduced R code, it consists of

```
Hoem <- TrueExitNumber[,2]/ExpoNumber[,2]
```

10.3.2 Kaplan–Meier Estimator of the Survival Function

Let us denote $\hat{S}$ the Kaplan–Meier estimator of the (unknown) survival function S which is defined by

$$\hat{S}_t = \prod_{t_i < t} \left(1 - \frac{d_i}{r_i}\right),$$

where

- The time serial $(t_i)_{i \geq 0}$ represents the time (or ages) at which a true exit (e.g., a death) is observed,
- d_i represents the number of true exits (e.g., deaths) at time t_i, and
- r_i represents the exposure just before t_i.

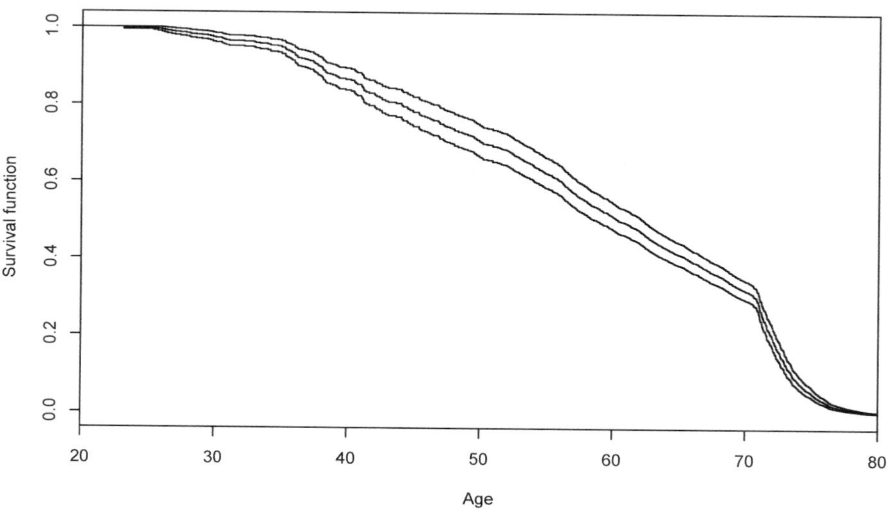

FIGURE 10.5
Kaplan–Meier estimator of the survival function.

The properties of the Kaplan–Meier estimator have been well-studied and it is the unique estimator of the survival function which is consistent.

We will use the `survfit` function of the `survival` library to proceed to this estimation:

```
> library(survival)
> data_km <- subset(data_obs, data_obs$BegAge<data_obs$EndAge)
> w <- survfit(Surv(BegAge,EndAge,non_censored,type="counting")~1,
+    data=data_km, type="kaplan-meier")
```

`w` is a `surv` object. Its main characteristics are obtained by

```
> print(w)
Call: survfit(formula = Surv(BegAge, EndAge, non_censored,
     type = "counting")~1, data = data_km, type = "kaplan-meier")

records   n.max n.start  events  median 0.95LCL 0.95UCL
22598.0  1719.0     0.0  4450.0    60.6    58.9    62.4

> summary(w)
```

The last instruction enables as to get the values of the estimator at each death time.

We can make a graphical representation (see Figure 10.5) of `w` using the `plot` function:

```
> plot(w, mark.time=FALSE, lty=1, xlim=c(20,80),
+     xlab = "Age", ylab = "Survival function", conf.int=TRUE)
```

From the `summary` instruction, we can obtain the value of the estimator at desired times (or ages). For example, if we want to compute some conditional mortality rates (q_x), we are interested in the value of the Kaplan–Meier estimator at each integer age.

```
> summary(w, times = c(40:60))
Call: survfit(formula = Surv(BegAge, EndAge, non_censored,
    type = "counting")~1, data = data_km, type = "kaplan-meier")

 time n.risk n.event survival std.err lower 95% CI upper 95% CI
   40    349      77    0.866  0.0144        0.839        0.895
   41    308       5    0.853  0.0153        0.824        0.884
   42    290       8    0.831  0.0169        0.798        0.864
   43    266       7    0.810  0.0182        0.775        0.846
   44    273       2    0.804  0.0185        0.769        0.841
   45    301       8    0.782  0.0196        0.744        0.821
   46    331       6    0.767  0.0201        0.729        0.808
   47    351       7    0.751  0.0206        0.712        0.793
   48    401       7    0.738  0.0209        0.698        0.780
   49    454       7    0.725  0.0210        0.685        0.768
   50    550      14    0.706  0.0211        0.666        0.748
   51    671      11    0.693  0.0211        0.653        0.736
   52    822      10    0.684  0.0210        0.644        0.726
   53    960      28    0.663  0.0207        0.623        0.705
   54   1167      29    0.645  0.0204        0.606        0.686
   55   1267      40    0.624  0.0200        0.586        0.665
   56   1156      36    0.605  0.0197        0.567        0.644
   57   1224      61    0.575  0.0191        0.538        0.613
   58   1269      47    0.553  0.0186        0.518        0.591
   59   1435      51    0.532  0.0181        0.498        0.569
   60   1573      58    0.512  0.0176        0.479        0.548
```

Now we are able to determine the conditional mortality rates.

```
> sum_w <- summary(w, times = c(20:66))
> q_x <- data.frame(nrow = max(sum_w$time), ncol = 2)
> colnames(q_x) <- c("Age","Rate")

> for (x in min(sum_w$time):(max(sum_w$time)-1)){
+   q_x[x,1] <- x
+   q_x[x,2] <- 1-
+   sum_w$surv[sum_w$time==x+1]/sum_w$surv[sum_w$time==x]
+ }
```

And to plot them (see Figure 10.6):

```
> plot(q_x, xlim=c(20,65), ylim=c(0,.06), pch=3)#,
+   xlab = "Age", ylab = "Rates")
> grid(nx=0, ny=6)
```

We may finally compare them with those obtained by the Hoem estimators (see Figure 10.7).

```
> plot(q_x, xlim=c(20,65), ylim=c(0,.06), pch=3)
> grid(nx=0, ny=6)
> points(Hoem,pch=1)
```

10.4 Regularization Techniques

When building a table of experience, the first step is the estimation of conditional probabilities of death (exit rate) and this step is essential, both for parametric and nonparametric

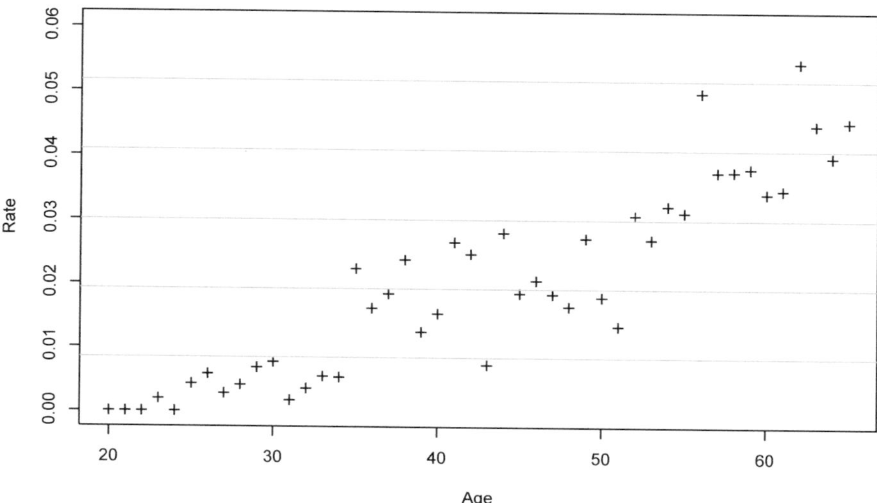

FIGURE 10.6
Crude conditional mortality rates (deducted from the Kaplan–Meier estimator).

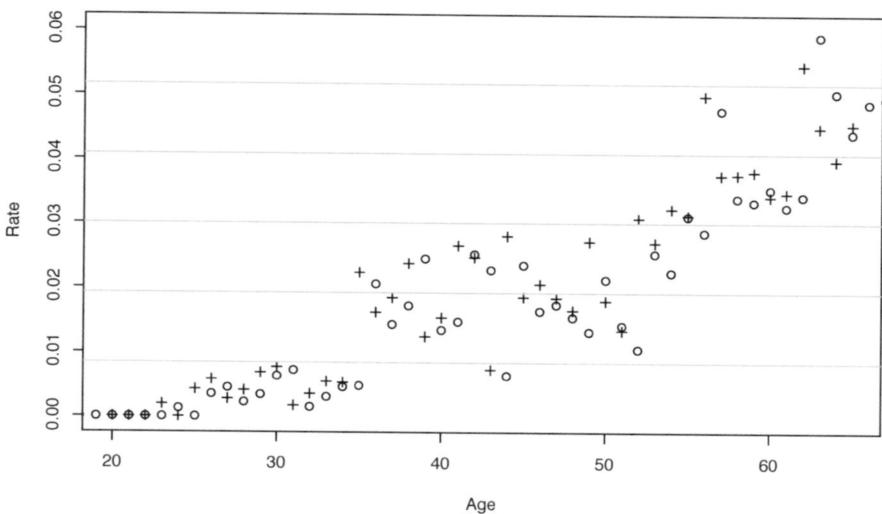

FIGURE 10.7
Crude conditional mortality rates: Kaplan–Meier versus Hoem.

approaches. The estimator of the exit rate shows some irregularities, which is legitimate to think that it does not reflect the underlying phenomenon that seeks to measure, but it is the result of imperfect conditions of the experience. The sampling fluctuations thus induce parasite variability in the estimated values. We then want to "adjust" or "smooth" the raw empirical values to represent a more accurate law (unknown) that is to be estimated. The process of revising the original estimate can be conducted in three ways:

- We can fix an a priori form for the underlying law, assuming for example that the hazard function is a Makeham function. This is a fitting approach defined by a parameter distribution.
- We may not find a parametric representation, but simply define a number of treatments to be applied to the original raw data to make them more "smooth." This is typically the nonparametric methods, for example, Whittaker–Henderson smoothing and its extension in a more general Bayesian framework.
- And we may apply both techniques together, leading to semiparametric models. The Brass model is one of them.

We now assume that we have estimators of the underlying, unknown, conditional exit probabilities (*crude rates*) and we have to graduate them.

10.4.1 Parametric Adjustment

A very simple approach to address a parametric graduation is to observe that one can often assume that the crude rate estimator at time x, $\hat{q}_x$ is normally distributed, that is,

$$\hat{q}_x \approx N\left(q_x(\theta); \sigma^2(\theta) = \frac{q_x(\theta)(1 - q_x(\theta))}{N_x}\right).$$

It is then easy to show that when this assumption is true, the optimal parameter θ is the one that minimizes the function

$$\phi(\theta) = \sum_x \frac{E_x}{\hat{q}_x(1 - \hat{q}_x)} (q_x(\theta) - \hat{q}_x)^2,$$

where E_x denotes the exposure at time x (that is, between x and $x + 1$). The key point is to specify the form of the function $\theta \to q_x(\theta)$.

In the framework of duration models, the most natural way to do that is to make a parametric assumption for the hazard function h_θ and use the link between the hazard function and the conditional exit probabilities

$$q_x(\theta) = 1 - \exp\left(-\int_x^{x+1} h_\theta(u)\, du\right).$$

For example, for human mortality, it is well known that the Gompertz–Makeham model

$$h_\theta(t) = \alpha + \beta \cdot \gamma^t,$$

where $\theta = (\alpha, \beta, \gamma)$, fits the data relatively well (for not so-aged populations). In this special case, we have

$$q_x(\theta) = \exp(-\alpha) \exp\left[-\frac{\beta}{\log(\gamma)} \gamma^x (\gamma - 1)\right].$$

We will now assume that we have the input data $(E_x, \hat{q}_x)$ for a range $x = x_m, \ldots, x_M$. The numerical computation $\theta^* = \arg\min\{\phi(\theta)\}$ will be done using numerical algorithms like Newton–Raphson or a gradient algorithm. Such an algorithm requires the use of *well-chosen* initial value for θ, not too far from the optimal one.

In the Gompertz-Makeham case, one can observe that

$$\log(q_{x+1} - q_x) = x \log(\gamma) + \log\left((\gamma - 1)^2 \log(\delta)\right),$$

with $\delta = \exp\left(-\frac{\beta}{\log(\gamma)}\right)$.

Initial values for β and γ can then be computed using a simple linear regression. The programming of this method with R software is based on the `qxAdjust` function defined as follows:

```
> qAdjustH_M <- qxAdjust(Q_H,rowSums(expoVentilesH),xMin,xMax,TRUE,"Makeham")
```

with

```
> qxAdjust <- function(qxBrut,expo,xMin,xMax,pond,model){
> pInitial <- getValeursInitiales(qxBrut,xMin,xMax)
> f <- function(p){
+   getEcart(qxBrut,expo,xMin,xMax,p[1],p[2],p[3],pond,model)}
> argmin <- constrOptim(pInitial,f,ui=rbind(c(1,0,0),c(0,1,0),c(0,0,1)),
+   ci=c(0,0,0),control=list(maxit=10^3),method="Nelder-Mead")$par

> q <- as.vector(0:120)
> for (x in 0:120){
+         q[x+1]=qxModel(model,argmin[1],argmin[2],argmin[3],x)
+}
> list(argmin,q)
+}
```

As can be seen, this function uses three auxiliary functions:

```
> qxModel <- function(model,a,b,c,x){
+         if (model=="Thatcher"){q=qxThatcher(b,log(c),a,x)}
+         else if (model=="Makeham") {q=qxMakeham(a,b,c,x)}
+         else {q=qxLogit(a,b,c,x)}
+         if (is.na(q)){0}
+         else {q}
+ }

> getValeursInitiales <- function(qxBrut,xMin,xMax){
+         q <- pmax(qxBrut,10^-10)
+         x <- as.vector(xMin:xMax)
+         y <- log(pmax(abs(q[(xMin+1):(xMax+1)]
+        -q[(xMin+2):(xMax+2)]),10^-10))
+         t <- data.frame(cbind(x,y))
+         d <- lm(y~x,data=t)
+         p <- as.vector(1:3)
+         p[1] <- 2*10^-4
+         p[3] <- exp(d$coefficients[2])
+         p[2] <- max(10^-10,-log(p[3])*exp(d$coefficients[1])/(p[3]-1)^2)
+         return(p)
+ }
```

```
>getEcart <- function(qxBrut,expo,xMin,xMax,a,b,c,pond,model){
+        w <- pmin(1,as.vector(1:120))
+        if (pond & !(model=="Logit")){
+                        expo <- expo[1:120]
+                         w <- expo/(qxBrut*(1-qxBrut))
+                }
+        else if (pond & (model=="Logit")){
+                        expo <- expo[1:120]
+                         w <- expo
+                }
+        w[(is.na(w))|(is.infinite(w))]<-0
+        s <- 0
+        for (x in xMin:xMax){
+                s <- s+w[x+1]*(qxModel(model,a,b,c,x)-qxBrut[x+1])^2
+         }
+        s
+ }
```

At least we need the last simple function:

```
> qxMakeham <- function(a,b,c,x){
+        1-exp(-a)*exp(-b/log(c)*(c-1)*c^x)
+ }
```

The above graphics show that the Makeham model may not be the best one. We will now illustrate the use of the Brass relational model.

10.4.2 Semiparametric Adjustment: Brass Relational Model

Modeling the function $x \to q_x(\theta)$ with a*small* number of parameters may be too restrictive. A simple solution is to take a larger parameter. We then have to choose a smart way to choose this less-restrictive parameter. It seems quite natural to use another mortality table (for our sample) and to describe the link between this table and the (unknown) mortality table we want to build. The Brass model relies on the assumption that

$$\log \frac{q_x(\theta)}{1-q_x(\theta)} = a \cdot \log \frac{q_x^{ref}}{1-q_x^{ref}} + b,$$

where the parameter $\theta = (a, b)$ must be fitted. We use the same criterion as previously, that is, we are looking for the value of the parameter $\theta^* = \arg\min \phi(\theta)$, where we denote

$$\phi(\theta) = \sum_x \frac{E_x}{\hat{q}_x(1-\hat{q}_x)} (q_x(\theta) - \hat{q}_x)^2.$$

It is sufficient to slightly change the code used in the previous section:

```
> qxAjusteLogit <- function(qxBrut,expo,xMin,xMax,pond,qxRef){
+        modelLogit <- "Logit"
+        f <- function(p){
+    getEcart(qxBrut,expo,xMin,xMax,p[1],p[2],qxRef,pond,modelLogit)}
+        argmin <- nlm(f,c(1,0))$estimate
```

```
+        q <- as.vector(0:120)
+        for (x in 0:120){
+                q[x+1] <- qxModel(modelLogit,argmin[1],argmin[2],qxRef,x)
+        }
+        list(argmin,q)
+ }
```

with

```
> qxLogit <- function(a,b,q,x){
+        z=exp(b)*(q[x+1]/(1-q[x+1]))^a
+        z/(1+z)
+ }
```

10.4.3 Nonparametric Techniques: Whittaker–Henderson Smoother

The idea of the method of Whittaker–Henderson is to combine a fidelity criterion and a regularity criterion and find the fitted values that minimize the sum of the two criteria.

We fixed weights (ω_i) and we set the following criteria:

- Fidelity: $F = \sum_{i=1}^{p} \omega_i \left(q_i - \hat{q}_i\right)^2$,
- Regularity: $S = \sum_{i=1}^{p-z} \left(\Delta^z(q_i)\right)^2$,

where z is a parameter of the model. The criterion is to minimize a linear combination of fidelity and regularity functions; the weight of each of the two terms is controlled by a second parameter h:

$$M = F + h \cdot S.$$

One can show that

$$M = (q - \hat{q})'\omega(q - \hat{q}) + hq'{K_z}'K_z q,$$

with K_z the $(p - z, p)$ matrix whose terms are the binomial coefficients of order z whose sign alternates and began positively to peer z. For example, with $p = 5$ and $z = 2$, we have

$$K_2 = \begin{pmatrix} 1 & -2 & 1 & 0 & 0 \\ 0 & 1 & -2 & 1 & 0 \\ 0 & 0 & 1 & -2 & 1 \end{pmatrix}$$

If $p = 3$ and $z = 1$, we have $K_1 = \begin{pmatrix} -1 & 1 & 0 \\ 0 & -1 & 1 \end{pmatrix}$. One can easily show that $\Delta^Z q = K_z q$. This leads to

$$M = (q - \hat{q})'\omega(q - \hat{q}) + hq'{K_z}'K_z q.$$

We then compute

$$\frac{\partial M}{\partial q} = 2\omega q - 2\omega\hat{q} + 2h{K_z}'K_z q,$$

and solve the equation

$$\frac{\partial M}{\partial q} = 0,$$

to get the following result:

$$q^* = \left(\omega + h{K_z}'K_z\right)^{-1} \omega\hat{q}.$$

The method of Whittaker–Henderson is very interesting because its extension in dimension two (or more) is simple. First we distinguish the vertical regularity via the operator Δ_ν^z (which acts on q_{ij}, j set seen as a series indexed by i) to calculate an index of vertical regularity:

$$S_v = \sum_{j=1}^{q}\sum_{i=1}^{p-z}(\Delta_\nu^z q_{ij})^2.$$

In the same way, we calculate an index of horizontal regularity S_h , and we denote

$$M = F + \alpha S_v + \beta s_h,$$

which must be minimized. The resolution of the optimization problem is made by rearranging the elements to be reduced to a one-dimensional case. We construct the vector u, of size $p \times q : u_{q(i-1)+j} = \hat{q}_{ij}$. Similarly, we construct a weight matrix by copying the diagonal lines of the matrix (ω_{ij}). We denote $\omega^*_{q(i-1)+j,q(i-1)+j} = \omega_{ij}$. We proceed in the same way to define the matrices K_z^v and K_y^h. The smoothed values are then obtained by

$$q^* = \left(w^* + \alpha K_z^{v\prime} K_z^v + \beta K_y^{h\prime} K_y^h\right)^{-1} w^* u.$$

10.4.3.1 Application

Here is a simple concrete example that illustrates this method. Crude rates are put in a $p \times q$ matrix with $p = 3$ and $q = 4$.

```
> TCrude <- read.table("data/Q_Moment_h.txt", sep=";")
> TCrude <- as.numeric(TCrude)
> TCrude <- as.matrix(TCrude)

> q <- ncol(TCrude)
> p <- nrow(TCrude)

> Weights <- matrix(1,p*q,p*q)
```

We choose $z = 2$ (vertical regularity index) and $y = 1$ (horizontal regularity index), so that we get K_v^z , a $(q(p-z), pq) = (6, 12)$ matrix and K_h^y , a $(p(q-y), pq) = (8, 12)$ matrix:

$$K_h^2 = \begin{pmatrix} 1 & 0 & 0 & -2 & 0 & 0 & 1 & 0 & 0 & 0 & 0 & 0 \\ 0 & 0 & 0 & 1 & 0 & 0 & -2 & 0 & 0 & 1 & 0 & 0 \\ 0 & 1 & 0 & 0 & -2 & 0 & 0 & 1 & 0 & 0 & 0 & 0 \\ 0 & 0 & 0 & 0 & 1 & 0 & 0 & -2 & 0 & 0 & 1 & 0 \\ 0 & 0 & 1 & 0 & 0 & -2 & 0 & 0 & 1 & 0 & 0 & 0 \\ 0 & 0 & 0 & 0 & 0 & 1 & 0 & 0 & -2 & 0 & 0 & 1 \end{pmatrix}$$

$$K_h^1 = \begin{pmatrix} -1 & 1 & 0 & 0 & 0 & 0 & 0 & 0 & 0 & 0 & 0 & 0 \\ 0 & -1 & 1 & 0 & 0 & 0 & 0 & 0 & 0 & 0 & 0 & 0 \\ 0 & 0 & 0 & -1 & 1 & 0 & 0 & 0 & 0 & 0 & 0 & 0 \\ 0 & 0 & 0 & 0 & -1 & 1 & 0 & 0 & 0 & 0 & 0 & 0 \\ 0 & 0 & 0 & 0 & 0 & 0 & -1 & 1 & 0 & 0 & 0 & 0 \\ 0 & 0 & 0 & 0 & 0 & 0 & 0 & -1 & 1 & 0 & 0 & 0 \\ 0 & 0 & 0 & 0 & 0 & 0 & 0 & 0 & 0 & -1 & 1 & 0 \\ 0 & 0 & 0 & 0 & 0 & 0 & 0 & 0 & 0 & 0 & -1 & 1 \end{pmatrix}.$$

```
> U <- matrix(0,q*p,1)
> v <- matrix(0,VertOrder+1)
> h <- matrix(0,HorOrder+1)
> Kv <- matrix(0,(p-VertOrder)*q,q*p)
> Kh <- matrix(0,p*(q-HorOrder),q*p)
> M <- (matrix(0,q*p,q*p))
> W <- matrix(0,q*p,q*p)
> Qsmooth <- matrix(0,p,q)
> Tsmooth <- matrix(0,p,q)

> for (j in 1:q) {
+   for (i in 1:p) {
+     U[(i-1)*q+j,1] <- as.numeric(TCrude[i,j])
+     W[q*(i-1)+j,q*(i-1)+j] <-Weights[i,j]
+   }
+ }

> for (k in 0:VertOrder) {
+   v[(k+1),1]=(-1)^(VertOrder-k)*
+   factorial(VertOrder)/(factorial(k)*factorial(VertOrder-k))
+ }

> for (j in 1:q) {
+   for (z in 1:(p-VertOrder)) {
+     for (i in 1:(VertOrder+1)) {
+       Kv[z+(j-1)*(p-VertOrder),j+(q)*(i-1)+(z-1)*(q)]=v[i,1]
+     }
+   }
+ }

> for (k in 0:HorOrder) {
+   h[(k+1),1]=(-1)^(HorOrder-k)*factorial(HorOrder)
+       /(factorial(HorOrder-k)*factorial(k))
+ }

> M <- W+alpha*(t(Kv))%*%Kv+beta*(t(Kh))%*%Kh
> Qsmooth <- solve(M)%*%W%*%U

> for (j in 1:q) {
+   for (i in 1:p) {
+     Tsmooth[i,j]=Qsmooth[(i-1)*q+j,1]
+   }
+ }
```

With the sample data, we get Figure 10.8.

```
> q <- ncol(TCrude)
> p <- nrow(TCrude)
> x <- matrix(1:p,p,1)
> y <- matrix(1:q,q,1)
> par(mfrow=c(1,2))
```

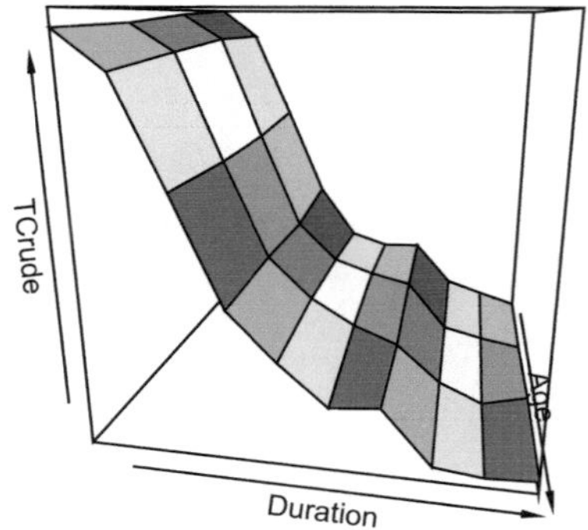

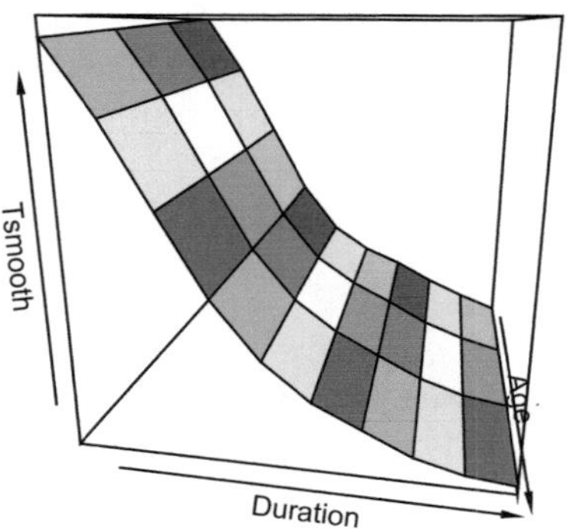

FIGURE 10.8
Crude and smoothed rates (Whittaker–Henderson).

```
> persp(x,y,TCrude,theta=+100,col=brewer.pal(9,"RdYlGn"),
+       xlab="Age",ylab="Duration",main="Crude rates",adj=0.5,font=2)
> persp(x,y,Tsmooth,theta=+100,col=brewer.pal(9,"RdYlGn"),
+       xlab="Age",ylab="Duration",main="Smooth rates",adj=0.5,font=2)
```

10.5 Modeling Heterogeneity

Heterogeneity is a mixture of populations with different characteristics. It has a significant impact on the perception of the hazard function of the model. For example, a mixture of exponential population (thus constant hazard) leads to a decreasing aggregate hazard function. This phenomenon is called "heterogeneity bias."

Many models take into account this mixture of laws: frailty models (Vaupel et al. (1979), combined fragility (Barbi (1999)), common shocks models, Cox model (Cox (1972)), additive hazard decomposition (Aalen (1978)), combinations of both, etc. In the context of actuarial models, the Cox model and, more recently, the Aalens one, are widely used, in particular because of their ease of implementation and interpretation, and also because the presence of censorship (right) and truncation (left) are taken into account.

The reference model for proportional hazard is the model of Cox (1972), specified as follows from the hazard function:

$$\log h\,(t|Z=z) = \log h_0(t) + \sum_{i=1}^{p} z_i\beta_i.$$

So this is a regression model whose coefficients can be estimated by (partial) maximum likelihood. The link with the discrete exit rate is immediate:

$$q\,(x|z,\theta) = 1-(1-q_0(x))^{\exp\left(-z'\theta\right)} \approx q_0(x)\cdot\exp\left(-z'\theta\right).$$

The approximation is valid when the exit rate is low. This framework may actually be declined in several ways, depending on how you consider the baseline hazard function:

- If a parameter is specified, one obtains in practice a complete parametric model that one can try to estimate by maximum likelihood.
- If the baseline hazard function is known, one can use least squares methods.
- If one does not specify a priori any particular form for the baseline hazard function, the model is a semiparametric one, which is the framework chosen by Cox.

The hazard function is then of the form

$$h\left(x|z;\alpha,\beta\right) = h_\alpha(x)\cdot\exp\left(-z'\beta\right),$$

and we can calculate the log-likelihood in the presence of right censoring and left truncation (see Planchet & Thérond (2011)):

$$\log L(\theta) = k + \sum_{i=1}^{n} d_i \log\left(h_{\theta|Z}(t_i)\right) + \log S_{\theta|Z}(t_i) - \log S_{\theta|Z}(e_i),$$

with k a constant and $\theta = (\alpha,\beta)$. For example, if we choose a hazard function based on a Weibull distribution,

$$h\left(x|z;\alpha,\beta\right) = \exp\left(-z'\beta\right)\alpha x^{\alpha-1}.$$

This case has, in practice, been discussed above and requires no special comment.

10.5.1 Semiparametric Framework: Cox Model

The idea here is to separate the estimation of the heterogeneity parameter on one hand and the baseline hazard function on the other. Cox showed in 1972 that this could be done using the following partial likelihood:

$$L_{cox} = \prod_{i=1}^{n}\left(\frac{\exp -\beta' z_i}{\sum_{j=1}^{n}\exp -\beta' z_j \mathbf{1}_{T_i\leq T_j}}\right)^{d_i}.$$

```
> cox <- coxph(Surv(AncEntree,AncSortie,
+   non_censored,"counting")~Sexe+CSP,data=t)
> summary(cox)

Call:
coxph(formula = Surv(AncEntree, AncSortie,
    non_censored, "counting") ~ Sexe + CSP, data = t)


              coef exp(coef) se(coef)      z       p
SexeH       0.1512     1.163  0.00298  50.83 0.0e+00
CSPENSPER -0.0186      0.982  0.00290  -6.41 1.4e-10
CSPCADRES -0.1868      0.830  0.01005 -18.59 0.0e+00

Likelihood ratio test=2864  on 3 df,
    p=0  n= 542619, number of events= 520238
```

There is a test to validate the proportionality assumption underlying the model, based on the Schoenfeld residuals:

```
> cox.zph(cox)

              rho chisq p
SexeH     -0.0765  3031 0
CSPENSPER  0.0488  1238 0
CSPCADRES  0.0302   476 0
GLOBAL         NA  4784 0
```

10.5.2 Additive Models

When the proportional hazard hypothesis is not satisfied, we can turn to the additive models (Aalen (1978)), whose general shape is based on the following specification of the hazard function:

$$h(t) = Z^T(t)\beta(t).$$

So, for an individual i, we have

$$h\left(t|z_i(t)\right) = \beta_0(t) + \sum_{j=1}^{p} \beta_j(t) \times z_{ij}(t).$$

Depending on the assumptions on the model, coefficients can be parametric, semiparametric, or nonparametric. For example, in the nonparametric framework, by introducing the matrix $n \times p$:

$$X(t) = \left(R_1(t)Z^1(t), ..., R_n(t)Z^n(t)\right),$$

and its generalized inverse (see Ben-Israel & Greville (2003) of size $p \times n$:

$$X^-(t) = \left(X^T(t)X(t)\right)^{-1} X^T(t),$$

we know how to estimate the cumulative regression coefficients $B(t) = \int_0^t \beta(u)du$ by

$$\hat{B}_j(t) = \sum_{T_i \leq t} X_{ji}^-(T_i) \cdot D_i.$$

However, we can observe that the volume of calculations to be performed is important, since it is necessary to calculate $X^-(t)$ at every time we observe an uncensored exit. This involves to calculate the inverse $(X^T(t)X(t))^{-1}$ (size $p \times p$) at every time of an uncensored exit, then the matrix products for $X^-(t)$ and finally $\hat{B}_j(t)$.

```
> aa <- aareg(Surv(AncEntree,AncSortie,non_censored,
+    "counting")~CSP,data=t[1:5000,])

Call:
aareg(formula = Surv(AncEntree, AncSortie, non_censored,
    "counting") ~ CSP, data = t[1:5000, ])

  n= 5000
    1434 out of 1440 unique event times used
```

```
             slope      coef se(coef)     z       p
Intercept  0.06310  0.001440 2.73e-05 52.50 0.00000
CSPENSPER -0.00540 -0.000145 4.46e-05 -3.24 0.00118
CSPCADRES -0.00824 -0.000200 6.99e-05 -2.86 0.00420

Chisq=14.73 on 2 df, p=0.00063; test weights=aalen
```

10.6 Validation of a Survival Model

To validate a model, the most typical approach is to compare the number of observed exits and the number of exits from the model. This comparison is performed over the period of observation. To determine the number of exits estimated by the model, a one-per-capita calculation is performed. The number of predicted exits at time (resp. age) x is equal to the exposure multiplied by the conditional probability of exiting at this time (resp. age).

The calculation conventions may have a significant effect on the output. They are specified here to avoid ambiguity.

The integer age of death is equal to the integer part of the exact age, the latter being calculated by a difference of days and then divided by 365.25.

The exposures are therefore determined on a daily basis by observing, for each insured, the time spent between two ages. The calculation of the contribution ν of the insured age x exposure period 01/01/N–31/12/N is performed as follows:

$$\nu([x]) = \frac{\min\{([x]+1)\cdot 365.25, \overline{x}\} - \underline{x}}{365.25},$$

and

$$\nu([x]+1) = \frac{\max\{([x]+1)\cdot 365.25, \overline{x}\} - \underline{x}}{365.25},$$

with

- $\overline{x} = 31/12/N - BirthDate$ refers to the age in days at the end of the observation year N;
- $\underline{x} = 01/01/N - BirthDate$ refers to the age in days at the beginning of the observation year N;
- $[x]$ denotes the integer part of x.

To underestimate the bias associated with the above formulas, the age at the end of observation of a person released uncensored (deceased) is taken as equal to the smallest integer greater than the exit age.

Denote by N_x the exposure at age x, D_x the number of deaths in the year of age x people. q_x is estimated by $\hat{q}_x$. According to the Central Limit theorem,

$$\sqrt{N_x}\frac{q_x - \hat{q}_x}{\sqrt{\hat{q}_x(1-\hat{q}_x)}} \to_{N\to\infty} \mathcal{N}(0,1).$$

The asymptotic confidence interval of level α is given by

$$I_\alpha = \left[\hat{q}_x - u_{\alpha/2}\sqrt{\frac{\hat{q}_x(1-\hat{q}_x)}{N_x}}, \hat{q}_x + u_{\alpha/2}\sqrt{\frac{\hat{q}_x(1-\hat{q}_x)}{N_x}}\right],$$

where u_α denotes the percentile function of the standard normal distribution.

Part III

Finance

11

Stock Prices and Time Series

Yohan Chalabi
ETH Zürich
Zürich, Switzerland

Diethelm Würtz
ETH Zürich
Zürich, Switzerland

CONTENTS

11.1 Introduction

The analysis and modeling of financial time series have been, and are continuing to be, actively developed as to ways to identify and model statistical properties that appear to remain consistent over a period of time. With regard to financial data, such consistent properties are known as *stylized facts*, and they are often used to support investment decisions. For example, an investor might be interested in building a risk measure based on the statistical distribution of asset returns. Another example is the modeling of time-dependent structures, such as conditional volatility.

In this chapter, we give an overview of established models used for time series analysis and show how to implement them in R from scratch. To demonstrate the practical limitations of the models, we give simple examples of their implementation in their simplest forms. This will provide a good grounding before using full-featured R packages. These simplified demonstrations retain the core ideas of the models, and so should aid understanding of their basic principles. The snippets are constructed such that the interested reader can extend them to formulate more complex approaches.

The first model we consider is the *Generalized Autoregressive Conditional Heteroskedastic* (GARCH) model with Student-t innovation distribution as introduced in Bollerslev & Ghysels (1996). This is an important econometric model because it can model the conditional volatility of financial returns. The second model considered is the Peak-over-Threshold (POT) method from Extreme Value Theory (EVT); see Chapter 6. We will combine both methodologies to estimate tail risk measures, as presented in McNeil & Frey (2000).

This chapter also demonstrates how the computation of a time-consuming R code can be made significantly quicker by calling an external C/C++ routine. Embarrassingly parallel computation is common in time series models; a typical example is the replication of a model on a basket of time series. Several parallel computation approaches have been developed in R; an example of one that can be used in either Windows or Unix-like operating systems is presented in this chapter.

The remainder of this chapter is organized as follows. Section 11.2 introduces time series analysis in R. We discuss the properties of financial data, including stylized facts. We also show how to import, manipulate, and display time series in R. Section 11.3 presents the GARCH model with Student-t innovation distribution as described in Bollerslev & Ghysels (1996). Its R implementation is explained in detail, along with how to reduce the computation time by calling an external routine implemented in C/C++. The method is applied to past foreign exchange data for the U.S. dollar against the British pound (DEXUSUK) in order to reproduce parts of the results in Bollerslev & Ghysels (1996)). Section 11.4 demonstrates an application of combining EVT and the GARCH model to build a dynamic VaR estimator, as done in McNeil & Frey (2000). The method is backtested with the `DEXUSUK` dataset and shown to be performable as an embarrassingly parallel application. The chapter ends with concluding remarks in Section 13.7 and references to third-party packages that can be used to implement and extend the presented procedures.

11.2 Financial Time Series

11.2.1 Introduction

Financial data range from macroeconomic values such as the gross domestic product of a country to the tick-by-tick data used in high-frequency trading. However, the concepts underlying the analysis of these time series remain the same. The analysis aim is to identify statistical properties that remain constant over time and that can be reliably estimated given historical data. In stochastic jargon, time series analysis is the search for stationary ergodic processes.

A stationary process is a stochastic process for which the joint probability distribution does not change when shifted in time. Let X_t be a stochastic process with joint probability distribution F. (X_t) is a stationary process when the joint probability of the time-shifted process is the same for any time shift τ and position t_k,

$$F(x_{t_1+\tau}, x_{t_2+\tau}, \dots, x_{t_k+\tau}) = F(x_{t_1}, x_{t_2}, \dots, x_{t_k}).$$

The above condition, often referred to as the strong stationarity condition, is rarely encountered in practice. A relaxed version, the weak stationarity condition, is used where only the first two moments of the joint distribution should remain constant over shifts of time and position.

Given a set of financial data, part of the analytic process is to transform it to a stationary ergodic process. For example, to describe the daily closing prices of a stock as a stationary process, it can be transformed to daily returns. The challenges of time series modeling, therefore, lie in constructing and applying the appropriate model and data transformation

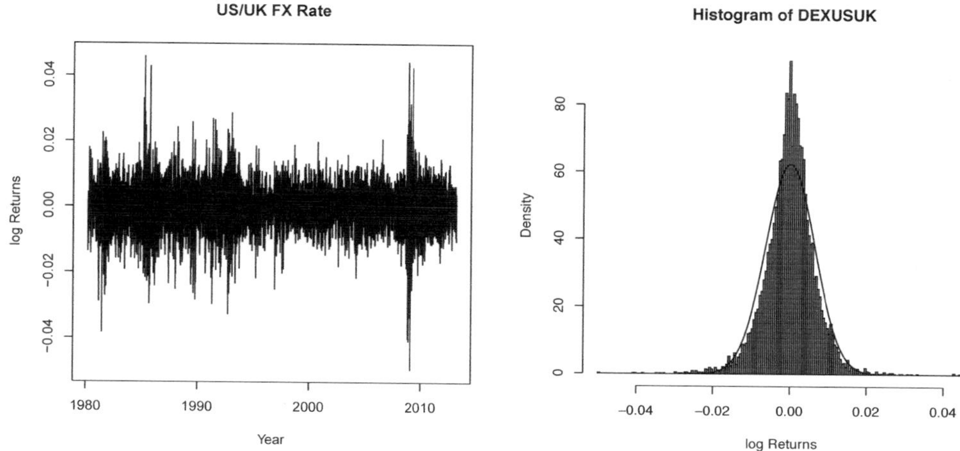

FIGURE 11.1
Daily logarithm returns of the U.S./U.K. foreign exchange rate on the left, and histogram of the daily logarithm returns of the U.S./U.K. foreign exchange rate on the right.

given the data at hand. In this regard, there exists a wide range of support tools to aid the formulation of the proper approach.

11.2.2 Data Used in This Chapter

All the analyses in this chapter use the same dataset. The set appears later when we implement the $GARCH(1,1)$ model. We will reproduce some of the results reported by Bollerslev (1987), and therefore use the DEXUSUK dataset closest to the one used in the original paper. The dataset is the daily buying rates in New York City for cable transfers payable in foreign currencies from January 4, 1971 to March 1, 2013. The data can be downloaded from the FRED website[1]. We recommend downloading the data in comma-delimited format, as this format is used in the following demonstrations.

The following code snippet demonstrates the loading of the DEXUSUK dataset in R. We use a simple `data.frame` object that will hold the dates and prices. When the datafile is located at "data/DEXUSUK.csv" relative to the working directory of the running R process, we can calculate the logarithm returns ($r_t = \log(P_t/P_{t-1})$):

```
> DEXUSUK <- read.csv("data/DEXUSUK.csv",
+ colClasses = c("Date", "numeric"))
> DEXUSUK$RETURN <- c(NA, diff(log(DEXUSUK$VALUE)))
```

Then we extract the subset of dataset, starting March 1, 1980:

```
> DEXUSUK <- DEXUSUK[DEXUSUK$DATE >= "1980-03-01", ]
```

Now that we have loaded our dataset into R and extracted entries from March 3, 1980, we can plot the time series with the `plot()` function as follows. The graphics are displayed in Figure 11.1 on the left.

[1]Data Source: FRED, Federal Reserve Economic Data, Federal Reserve Bank of St. Louis: U.S. / U.K. Foreign Exchange Rate (DEXUSUK); http://research.stlouisfed.org/fred2/series/DEXUSUK; accessed March 6, 2012.

```
> plot(DEXUSUK$DATE, DEXUSUK$RETURN, main = "US/UK FX Rate",
+ xlab = "Year", ylab = "log Returns", type = "l")
```

11.2.3 Stylized Facts

At the heart of time series analysis is the identification of patterns within the stochastic processes that influence data, as seen in the previous section. These patterns are often referred for financial time series to as the *stylized facts* of the dataset. In this section, we illustrate some of these stylized facts using the DEXUSUK dataset.

Financial returns often exhibit distributions with fatter tails than the normal distribution. To illustrate this, Figure 11.1 shows a histogram of the DEXUSUK log returns overlain with a normal density line fitted to the mean and standard deviation of the empirical returns. The normal distribution is clearly not adequate to describe the data. The next code snippet illustrates how to create the histogram plot:

```
> hist(DEXUSUK$RETURN, freq = FALSE, breaks = "FD", col = "gray",
+ main = "Histogram of DEXUSUK", xlab = "log Returns")
> v <- seq(min(DEXUSUK$RETURN), max(DEXUSUK$RETURN),
+ length.out = 100)
> lines(v, dnorm(v, mean(DEXUSUK$RETURN),
+ sd = sd(DEXUSUK$RETURN)),
+ lty = 2, lwd = 2)
```

The quantile–quantile (QQ) plot, which relates empirical and theoretical quantiles, confirms that the tails of the empirical distribution are not well described by the normal distribution. The following snippet shows how to create Figure 11.2, on the left:

```
> qqnorm(DEXUSUK$RETURN)
> qqline(DEXUSUK$RETURN)
```

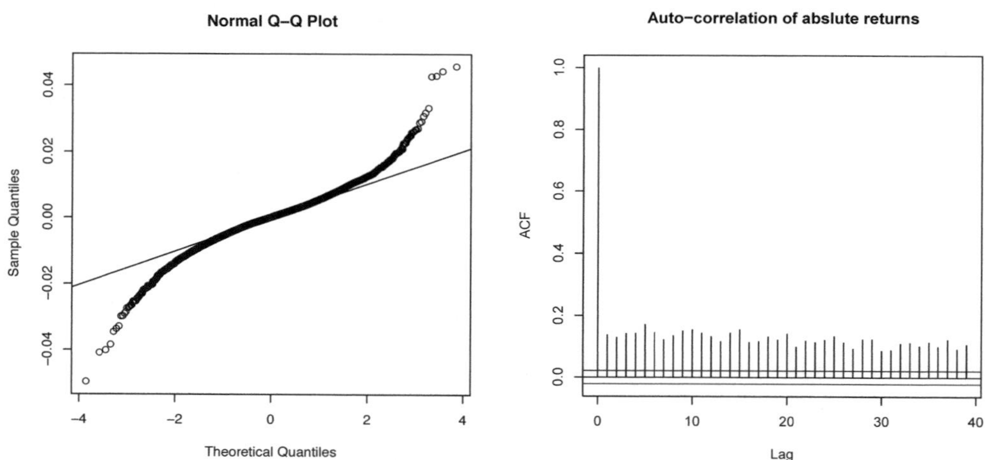

FIGURE 11.2
Q–Q plot of the daily logarithm returns of the U.S./U.K. foreign exchange rate on the left, and auto-correlation plot of the daily logarithm returns of the U.S./U.K. foreign exchange rate on the right.

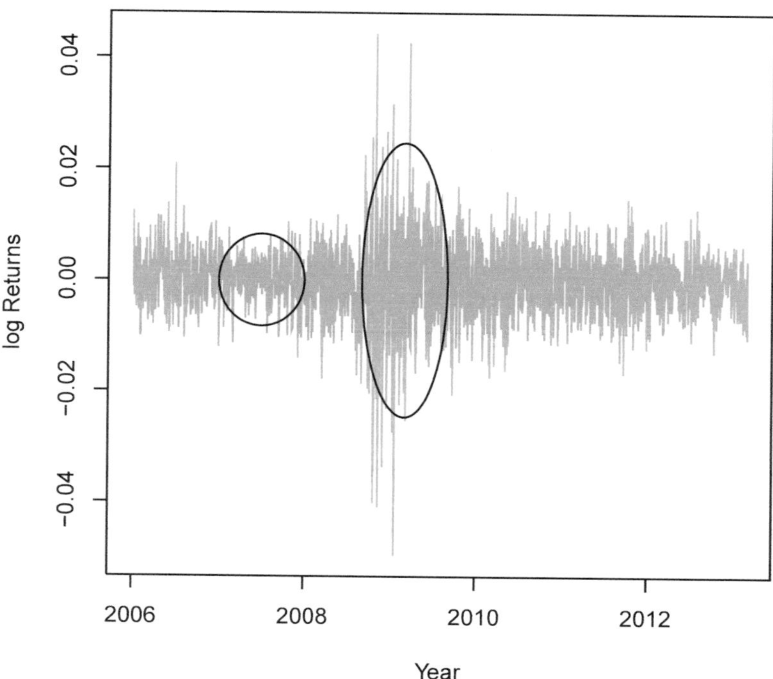

FIGURE 11.3
Illustration of conditional volatility of financial returns. The ellipse major axis corresponds to the time window being studied, and the semi-minor axis to 2 standard deviations of the considered time window data.

Another typical stylized fact is the auto-correlation of the absolute returns. This is illustrated by the auto-correlation in Figure 11.2 on the right. Given that the absolute values of returns can be used as a proxy for volatility, the auto-correlation of the absolute returns indicates the time dependency of the volatility.

```
> acf(abs(DEXUSUK$RETURN),
+ main = "Auto-correlation of abslute returns")
```

Figure 11.3 also illustrates the time-conditional structure of the volatility. That is, a previous period of low volatility denotes a high probability that volatility the following day will also be low. Conversely, when the volatility is high, there is a high probability that it will also be high on the following days. The ellipse added to the time series plot has its major axis corresponding to the time window being studied, and its semi-minor axis corresponding to two standard deviations of the considered time window data.

11.3 Heteroskedastic Models

11.3.1 Introduction

GARCH models are important in time series analysis, particularly in financial applications when the goal is to analyze and forecast volatility. In this section, we describe the estimation and forecasting of the univariate *GARCH*-type time series models. We present a numerical implementation of the maximum log-likelihood approach with Student-t innovation distribution. Although we consider Student-t innovation distribution, the code is constructed such that it would be straightforward to extend it to other distribution functions.

The number of heteroskedastic models is immense, but the earliest models remain the most influential: the standard ARCH model introduced by Engle (1982) and the GARCH model introduced by Bollerslev & Ghysels (1996). We describe the mean equation of a univariate time series (x_t) as follows:

$$x_t = \mathbb{E}\left[x_t|\Omega_{t-1}\right] + \varepsilon_t,$$

where $\mathbb{E}\left[x_t|\Omega_{t-1}\right]$ denotes the conditional expectation operator, Ω_{t-1} the information set at time $t-1$, and ε_t the residuals of the time series. ε_t describes uncorrelated disturbances with zero mean and represents the unpredictable part of the time series. For example, the mean equation might be represented as an *auto-regressive moving average* (*ARMA*) process of autoregressive order m and moving average order n:

$$x_t = \mu + \sum_{i=1}^{m} a_i x_{t-i} + \sum_{j=1}^{n} b_j \varepsilon_{t-j} + \varepsilon_t \ , \tag{11.1}$$

with mean μ, autoregressive coefficients a_i, and moving average coefficients b_i. If $n = 0$, we have a pure autoregressive process; while if $m = 0$, we have a pure moving average process.

The mean equation cannot consider heteroskedastic stylized facts, as shown in the previous section. In this context, Engle (1982) introduced the *Autoregressive Conditional Heteroskedastic* (*ARCH*) model, later generalized by Bollerslev (1986) to the *GARCH* model. The ε_t terms in the mean equation (Equation (11.1)) are the residuals of the time series process. Engle (1982) defined them as an autoregressive conditional heteroskedastic process where all the ε_t terms are of the following form:

$$\varepsilon_t \ = \ z_t \, \sigma_t \quad \text{where} \quad z_t \sim \mathcal{D}_\vartheta(0,1)$$

The innovation (z_t) is an independent and identically distributed process with zero mean and unit variance with distribution $\mathcal{D}_\vartheta$, $z_t \sim \mathcal{D}_\vartheta(0,1)$. ϑ represents additional distributional parameters. For example, for the Student-t ditribution, the additional distriubutional parameter would be the degrees of freedom.

The variance equation of the GARCH(p,q) model can then be expressed as follows:

$$\sigma_t^2 = \omega + \sum_{i=1}^{p} \alpha_i \varepsilon_{t-i}^2 + \sum_{j=1}^{q} \beta_j \sigma_{t-j}^2 \ .$$

If all the coefficients β_j are zero, the *GARCH* model is reduced to the *ARCH* model. As for ARMA models, a GARCH specification often leads to a more parsimonious representation of the temporal dependencies, and thus provides a similar added flexibility over the linear ARCH model when parameterizing the conditional variance.

11.3.2 Standard GARCH(1,1) Model

Bollerslev (1987) was the first to model financial time series for foreign exchange rates and stock indexes using $GARCH(1,1)$ models extended by the use of standardized Student-t distributions. When compared with conditionally normal errors, he found that t-$GARCH(1,1)$ errors much better captured the leptokurtosis seen in the data. As the benchmark dataset, we use the daily DEXUSUK exchange rate as presented in Section 11.2.2. This set spans the same time period considered by Bollerslev. The series contains a total of 1231 daily observations sampled from March 1, 1980 to January 28, 1985.

```
> pos <- (DEXUSUK$DATE >= "1980-03-01" &
+ DEXUSUK$DATE <= "1985-01-28")
> x <- DEXUSUK$RETURN[pos]
```

Previous studies have shown that the Student-t distribution performs well in capturing the observed kurtosis in empirical log-return time series (Bollerslev (1987), Baillie & Bollerslev (1989), Hsieh (1989), Bollerslev et al. (1992), Pagan (1996), Palm (1996)). The density $f^\star(z|\nu)$ of the standardized Student-t distribution can be expressed as follows:

$$
\begin{aligned}
f^\star(z|\nu) &= \frac{\Gamma(\frac{\nu+1}{2})}{\sqrt{\pi(\nu-2)}\Gamma(\frac{\nu}{2})} \frac{1}{\left(1+\frac{z^2}{\nu-2}\right)^{\frac{\nu+1}{2}}} \qquad (11.2) \\
&= \frac{1}{\sqrt{\nu-2}\,\mathrm{B}(\frac{1}{2},\frac{\nu}{2})} \frac{1}{\left(1+\frac{z^2}{\nu-2}\right)^{\frac{\nu+1}{2}}},
\end{aligned}
$$

where $\nu > 2$ is the shape parameter and $\mathrm{B}(a,b) = \Gamma(a)\Gamma(b)/\Gamma(a+b)$ is the Beta function. Note that setting $\mu = 0$ and $\sigma^2 = \nu/(\nu-2)$ leads Equation (11.2) to results in the usual one-parameter expression for the Student-t distribution as implemented in the R function `dt()`. We implement in the following code snippet the density and quantile function of the standardized Student-t distribution based on the `dt()` and `qt()` functions.

```
> dst <- function (x, mean = 0, sd = 1, nu = 5) {
+     stopifnot(nu > 2)
+     s <- sqrt(nu / (nu - 2))
+     z = (x - mean) / sd
+     s * dt(x = z * s, df = nu) / sd
+ }
> qst <- function (p, mean = 0, sd = 1, nu = 5) {
+     stopifnot(p >= 0, p <= 1, nu > 2)
+     s <- sqrt(nu/(nu-2))
+     mean + sd * qt(p = p, df = nu) / s
+ }
```

Given the model for the conditional mean and variance and an observed univariate return series, we can use the maximum likelihood estimation (MLE) approach to fit the parameters for the specified model of the return series. The procedure infers the process innovations or residuals by inverse filtering. This filtering transforms the observed process ε_t into an uncorrelated white noise process z_t. The log-likelihood function then uses the inferred innovations z_t to infer the corresponding conditional variances ${\sigma_t}^2$ via recursive substitution into the model-dependent conditional variance equations. Finally, the procedure uses the inferred innovations and conditional variances to evaluate the appropriate log-likelihood objective function. The MLE concept interprets the density as a function of the parameter set, conditional on a set of sample outcomes. Using $\varepsilon_t = z_t\sigma_t$, the log-likelihood function of the distribution $\mathcal{D}_\vartheta$ is given by

$$\mathcal{L}_N(\theta) = \log \prod_t \mathcal{D}_\vartheta(x_t, E(x_t|\Omega_{t-1}), \sigma_t),$$

where $\mathcal{D}_\vartheta$ is the conditional distribution function. The second argument of $\mathcal{D}_\vartheta$ denotes the mean, and the third argument the standard deviation. The full set of parameters θ includes the parameters from the mean equation μ, from the variance equation $(\omega, \alpha_p, \gamma_p, \beta_q, \delta)$, and the distributional parameters (ϑ). For Bollerslev's Student-t $GARCH(1,1)$ model, the parameter set reduces to $\theta = \{\mu, \omega, \alpha_1, \beta_1, \nu\}$. In the following, we will suppress the index on the parameters α and β when considering the $GARCH(1,1)$ model.

We first implement the calculation of the conditional variance in the form of a filtering function:

```
> garch11filter <- function(x, mu, omega, alpha, beta) {
+
+     # extract error terms
+     e <- x - mu
+     e2 <- e^2
+
+     # estimate the conditional variance
+     s2 <- numeric(length(x))
+
+     # set first value of the conditional variance
+     init <- mean(e2)
+     s2[1] <- omega + alpha * init + beta * init
+
+     # calculate the conditional variance
+     for (t in 2:length(x))
+         s2[t] <- omega + alpha * e2[t-1] + beta * s2[t-1]
+
+     list(residuals = e, sigma2 = s2)
+ }
```

The following code snippet implements the log-likelihood of the $GARCH(1,1)$ given the previous R function. Note the way in which the function is implemented for any standardized distribution function `ddist()`.

```
> garch11nlogl <- function(par, x) {
+
+     filter <- garch11filter(x, mu = par[1], omega = par[2],
+                             alpha = par[3], beta = par[4])
+     e <- filter$residuals
+     sigma2 <- filter$sigma2
+     if (any(sigma2 < 0)) return(NA)
+     sd <- sqrt(sigma2)
+
+     - sum(log(ddist(e, sd = sd, theta = par[-(1:4)])))
+ }
```

In our example, we consider innovations modeled by Student-t distribution. Therefore, for the `ddist()` function, we have

```
> ddist <- function(x, sd, theta) dst(x, sd = sd, nu = theta)
```

The best-fit parameters θ are obtained by minimizing the "negative" log-likelihood function. Some of the values of this parameter set are constrained to a finite or semi-finite range. Note the positive-value constraint of $\omega > 0$; also, α and β must be constrained in a finite interval [0,1]. This requires a solver for constrained numerical optimization problems. R offers the solvers `nlminb()` and `optim(method="L-BFGS-B")` for constrained optimization. These solvers are R interfaces to the underlying Fortran routines from the PORT Mathematical Subroutine Library (Blue et al. (1978)) and TOMS Algorithm 778 (Zhu et al. (1997)), respectively. We will use the `nlminb()` routine in our implementation.

The optimizer requires a vector of initial parameters for the mean μ, as well as for the $GARCH$ coefficients ω, α, and β and the degrees of freedom ν of the Student-t distribution. The initial value for the mean is estimated from the mean of the time series observations $\mu = \text{mean}(x)$. For the $GARCH(1,1)$ model, we start with $\alpha = 0.1$ and $\beta = 0.8$ as typical values of financial time series, and set ω as the variance of the series adjusted by the persistence $\omega = \text{Var}(x) * (1 - \alpha - \beta)$. Further arguments to the $GARCH$ fitting function are the upper and lower box bounds, in addition to optional control parameters.

We can now define the objective function and fit the $GARCH(1,1)$ parameters to the dataset x as follows:

```
> objective <- function(par) garch11nlogl(par, x = x)
> # set initial values and lower and upper bounds
> start <- c(mu = mean(x), omega = 0.1 * var(x), alpha = 0.1,
+            beta = 0.8, theta = 8)
> lb <- c(-Inf, 0, 0, 0, 3)
> # Estimate Parameters and Compute Numerically Hessian:
> nlminb(start, objective, lower = lb, scale = 1/abs(start),
+        control = list(trace = 5))
  0:     -4514.5174: -0.000581511 4.14852e-06 0.100000 0.800000  8.00000
  5:     -4521.9518: -0.000555754 1.06559e-06 0.0646560 0.910217  9.32111
 10:     -4523.4039: -0.000476499 4.99108e-07 0.0432756 0.945164  8.76353
 15:     -4523.4138: -0.000488913 5.18420e-07 0.0433510 0.944838  8.64621
$par
           mu         omega         alpha          beta
-4.889221e-04  5.183968e-07  4.335198e-02  9.448386e-01
        theta
 8.646856e+00


$objective
[1] -4523.414


$convergence
[1] 0


$iterations
[1] 16


$evaluations
function gradient
      33      111
```

```
$message
[1] "relative convergence (4)"
```

Note the use of the `scale` argument in the call of the `nlminb()` function. Scaling parameters are important in optimization problems to make the unknown variables of the same scale order. In the previous snippet, the omission of the `scale` would result in the failure of the optimization routine.

Everything can now be combined to obtain a function to estimate the GARCH(1,1) parameters for a time series and wrap the results as an S3 class object `mygarch11`.

```
> garch11Fit <- function(x, hessian = TRUE, ...) {
+
+     call <- match.call()
+
+     ddist <- function(x, sd, theta) dst(x, sd = sd, nu = theta)
+     nloglik <- function(par) {
+         filter <- garch11filter(x, mu = par[1], omega = par[2],
+                                 alpha = par[3], beta = par[4])
+         e <- filter$residuals
+         sigma2 <- filter$sigma2
+         if (any(sigma2 < 0)) return(NA)
+         sd <- sqrt(sigma2)
+         - sum(log(ddist(e, sd = sd, theta = par[5])))
+     }
+
+     # set initial values and lower and upper bounds
+     start <- c(mu = mean(x), omega = 0.1 * var(x),
+                alpha = 0.1, beta = 0.8, theta = 8)
+     lb <- c(-Inf, 0, 0, 0, 3)
+
+     # Estimate Parameters
+     fit <- nlminb(start, nloglik, lower = lb,
+                   scale = 1/abs(start), control = list(...))
+     coef <- fit$par
+     npar <- length(fit$par)
+
+     # Compute Numerically Hessian:
+     hess <-
+         if (hessian)
+             optimHess(coef, function(par) {
+                 ans <- nloglik(par)
+                 if (is.na(ans)) 1e7 else ans
+                 }, control = list(parscale = start))
+         else
+             NA
+
+     filter <- garch11filter(x, mu = coef[1], omega = coef[2],
+                             alpha = coef[3], beta = coef[4])
+
+     # build our own mygarch11 class object
+     ans <- list(call = call, series = x, residuals = filter$residuals,
+                 sigma2 = filter$sigma2, details = fit,
```

```
+                   loglik = -fit$objective, coef = coef,
+                   vcov = solve(hess), details = fit)
+     class(ans) <- "mygarch11"
+     ans
+ }
```

We also add printing and other common methods for the newly created class:

```
> coef.mygarch11 <- function(object, ...) object$coef
> print.mygarch11 <- function(x, ...) {
+     cat("\nCall:\n")
+     print(x$call)
+     cat("\nCoefficients:\n")
+     print(coef(x))
+ }
> summary.mygarch11 <- function(object, digits = 3, ...) {
+
+     mat <- cbind("Estimate" = object$coef,
+                  "Std. Error" = sqrt(abs(diag(object$vcov))))
+
+     cat("My GARCH(1,1) Model:\n")
+     cat("\nCall:\n")
+     print(object$call)
+     cat("\nCoefficients:\n")
+     print(mat, digits = 3)
+ }
> logLik.mygarch11 <- function(object, ...) object$loglik
> residuals.mygarch11 <- function(object, standardize = TRUE, ...) {
+     if (standardize)
+         object$residuals / sqrt(object$sigma2)
+     else
+         object$residuals
+ }
```

This gives, for the dataset in use,

```
> fit <- garch11Fit(x)
> summary(fit)
My GARCH(1,1) Model:

Call:
garch11Fit(x = x)

Coefficients:
       Estimate Std. Error
mu    -4.89e-04   1.65e-04
omega  5.18e-07   1.86e-07
alpha  4.34e-02   1.11e-02
beta   9.45e-01   9.44e-03
theta  8.65e+00   1.99e+00
```

The fitted values are close to the estimates obtained in Bollerslev's paper (1987) $\{\mu =$

$-4.56 \cdot 10^{-4}$, $\omega = 0.96 \cdot 10^{-6}$, $\alpha = .057$, $\beta = .921$, $\nu = 8.13\}$. The differences can be explained by the fact that the dataset used in this chapter is not exactly the same but is a rather good proxy to the one used in his work.

11.3.3 Forecasting Heteroskedastic Model

A major objective of the investigation of heteroskedastic time series is forecasting. Expressions for forecasts of both the conditional mean and the conditional variance can be derived, with the two properties capable of being forecast independently of each other. For a $GARCH(p,q)$ process, the n-step-ahead forecast of the conditional variance $\hat{\sigma}^2_{t+n|t}$ is computed recursively from

$$\hat{\sigma}^2_{t+n|t} = \hat{\omega} + \sum_{i=1}^{q} \hat{\alpha}_i \varepsilon^2_{t+n-i|t} + \sum_{j=1}^{p} \hat{\beta}_j \sigma^2_{t+n-j|t} ,$$

where $\varepsilon^2_{t+i|t} = \sigma^2_{t+i|t}$ for $i > 0$ while $\varepsilon^2_{t+i|t} = \varepsilon^2_{t+i}$ and $\sigma^2_{t+i|t} = \sigma^2_{t+i}$ for $i \leq 0$.

The following code snippet implements the n-step ahead forecast for the $GARCH(1,1)$ model

```
> predict.mygarch11 <- function(object, n.ahead = 1, ...) {
+
+     par <- coef(object)
+     mu <- par[1]; omega <- par[2];
+     alpha <- par[3]; beta <- par[4]
+     len <- length(object$series)
+
+     mean <- rep(as.vector(mu), n.ahead)
+
+     e <- object$residuals[len]
+     sigma2 <- object$sigma2[len]
+     s2 <- numeric(n.ahead)
+     s2[1] <- omega + alpha * e^2 + beta * sigma2
+     if (n.ahead > 1)
+         for (i in seq(2, n.ahead))
+             s2[i] <- (omega +
+                       alpha * s2[i - 1] +
+                       beta * s2[i - 1])
+     sigma <- sqrt(s2)
+
+     list(mu = mean, sigma = sigma)
+ }
```

The 5-steps ahead forecast for the `DEXUSUK` exchange returns becomes

```
> fit <- garch11Fit(x)
> predict(fit, n.ahead = 5)
$mu
[1] -0.0004889221 -0.0004889221 -0.0004889221 -0.0004889221
[5] -0.0004889221

$sigma
[1] 0.006486122 0.006487785 0.006489428 0.006491051 0.006492655
```

11.3.4 More Efficient Implementation

In this section, we profile the code so far implemented to identify the most time-consuming part to assess whether quicker computation could be achieved by implementing the identified bottleneck as an external routine.

We use the base `Rprof()` function to profile the code. This function writes the code performance profile to an external file, which can be displayed by the `summaryRprof()` function as an R object, as illustrated by the following code snippet:

```
> Rprof(filename = "Rprof.out")
> fit <- garch11Fit(x)
> Rprof(NULL)
> out <- summaryRprof("Rprof.out")
> head(out$by.self)
               self.time self.pct total.time total.pct
"garch11filter"     1.32    62.86       1.98     94.29
"*"                 0.28    13.33       0.28     13.33
"+"                 0.18     8.57       0.18      8.57
"-"                 0.14     6.67       0.14      6.67
"dt"                0.08     3.81       0.08      3.81
"mean"              0.04     1.90       0.04      1.90
```

When the functions are written entirely in R, the bottleneck appears in the computation of the log-likelihood function, more specifically in the `garch11filter()` function for the recursive estimation of the conditional variance. This can be alleviated by implementing this time-consuming function in an equivalent C/C++ routine. The following code snippet is a direct translation of the R function `garch11filter` to C/C++ code using the `Rcpp` package (see Chapter 1, and Eddelbuettel & François (2011)).

```
#include <Rcpp.h>
using namespace Rcpp;

RcppExport SEXP garch11filterCpp(SEXP xs, SEXP mus,  SEXP omegas,
                                 SEXP alphas, SEXP betas)
{

    NumericVector x(xs);
    double mu = as<double>(mus);
    double omega = as<double>(omegas);
    double alpha = as<double>(alphas);
    double beta = as<double>(betas);

    NumericVector e = x - mu;
    NumericVector e2 = e * e;

    NumericVector s2(x.size());

    double init = mean(e2);
    s2[0] = omega + alpha * init + beta * init;

    for (int i = 1; i < s2.size(); ++i)
        {
```

```
            s2[i] = omega + alpha * e2[i-1] + beta * s2[i-1];
        }

    return wrap(List::create(Named("residuals") = e,
                             Named("sigma2") = s2));
}
```

What remains to be done is to compile and load the dynamic library into the running R process, which is demonstrated in the following code snippet:

```
> library(Rcpp)
> Rcpp:::SHLIB("garch11filterCpp.cpp")
> dyn.load("garch11filterCpp.so")
> garch11filterCpp <- function (x, mu, omega, alpha, beta)
+     .Call("garch11filterCpp", x, mu, omega, alpha, beta)
```

Given the new garch11filterCpp() function, it is straightforward to modify the garch11Fit() as follows:

```
> garch11FitCpp <- function(x, hessian = TRUE, ...) {
+     call <- match.call()
+     ddist <- function(x, sd, theta) dst(x, sd = sd, nu = theta)
+     nloglik <- function(par) {
+         filter <- garch11filterCpp(x, mu = par[1], omega = par[2],
+                                    alpha = par[3], beta = par[4])
+         e <- filter$residuals
+         sigma2 <- filter$sigma2
+         if (any(sigma2 < 0)) return(NA)
+         sd <- sqrt(sigma2)
+         - sum(log(ddist(e, sd = sd, theta = par[5])), na.rm = TRUE)
+     }
+
+     # set initial values and lower and upper bounds
+     start <- c(mu = mean(x), omega = 0.1 * var(x),
+                alpha = 0.1, beta = 0.8, theta = 8)
+     lb <- c(-Inf, 0, 0, 0, 3)
+
+     # Estimate Parameters
+     fit <- nlminb(start, nloglik, lower = lb, scale = 1/abs(start),
+                   control = list(...))
+     coef <- fit$par
+     npar <- length(fit$par)
+
+     # Compute Numerically Hessian:
+     hess <-
+     if (hessian)
+         optimHess(coef, function(par) {
+             ans <- nloglik(par)
+             if (is.na(ans)) 1e7 else ans
+             }, control = list(parscale = start))
+     else
+         NA
```

```
+
+     filter <- garch11filterCpp(x, mu = coef[1], omega = coef[2],
+                                alpha = coef[3], beta = coef[4])
+
+     # build our own mygarch11 class object
+     ans <- list(call = call, series = x, residuals = filter$residuals,
+                 sigma2 = filter$sigma2, details = fit,
+                 loglik = -fit$objective, coef = coef,
+                 vcov = solve(hess), details = fit)
+     class(ans) <- "mygarch11"
+     ans
+ }
```

A naive speed comparison between `garch11Fit()` and `garch11FitCpp()` where we take the ratio of the median of time run for ten runs of each routines gives the following:

```
> mySystemTime <- function (expr, gcFirst = TRUE, n = 10) {
+     time <- sapply(integer(n), eval.parent(substitute(
+         function(...) system.time(expr, gcFirst = gcFirst))))
+     structure(apply(time, 1, median), class = "proc_time")
+ }
> rate <- (mySystemTime(garch11Fit(x))[3] /
+          mySystemTime(garch11FitCpp(x))[3])
```

The `garch11FitCpp()` function is approximatively 11 times faster with the C/C++ implementation of the filtering function than the R implementation.

11.4 Application: Estimation of the VaR Based on the POT and GARCH Model

In the context of the time series of financial returns, the modeling of the lower tail of their distributions is the primary focus of risk measures. A good example is the value-at-risk (VaR_α), that is, the value that one might lose with a given level of probability α. In this context, the extreme value theory (EVT) becomes of very high interest. It consists of modeling the tail of the distributions. As introduced in Chapter 6, one of the approaches used in the EVT is the peak-over-threshold (POT) method. This approach finds its root in the EVT theorem, which states that when one selects a threshold high enough, the distribution of the values exceeding the threshold converges in distribution to the generalized Pareto distribution (GPD, see Chapter 6). The probability density function of the GPD is defined as

$$g_{\xi,\beta}(x) = \begin{cases} \frac{1}{\beta}\left(1+\xi x/\beta\right)^{-\frac{1}{\xi}-1} & \text{if } \xi \neq 0, \\ \frac{1}{\beta}\exp\left(-\frac{x}{\beta}\right) & \text{if } \xi = 0, \end{cases}$$

where β is a scale parameter and ξ a shape parameter. The support of the distribution is, when $\xi \leq 0$, $x \leq 0$, and when $\xi < 0$, it is $0 \geq x \geq -\beta/\xi$. The following code snippet implements the density function of the GPD:

```
> dgp <- function(x, beta = 1, xi = 0) {
```

```
+
+     if (is.na(beta) || beta <= 0 || is.na(xi))
+         return(rep(NA, length(x)))
+
+     den <- numeric(length(x))
+
+     # look for variates that are in the support range
+     idx <-
+         if (xi >= 0)
+             x >= 0
+         else
+             (0 <= x & x <= - beta / xi)
+
+     den[idx] <-
+         if (xi != 0)
+             ((1 + xi * x[idx] / beta)^(-1 / xi - 1)) / beta
+         else
+             exp(-x[idx] / beta) / beta
+     den
+ }
```

The VaR_α can then be expressed in terms of the distribution of the exceeding points as

$$\text{VaR}_\alpha = u + \frac{\hat{\beta}}{\hat{\xi}}\left[\left(\frac{n}{N_u}(1-\alpha)\right)^{-\hat{\xi}} - 1\right], \tag{11.3}$$

where $\hat{\beta}$ and $\hat{\xi}$ are the parameters of the GPD, u is the threshold, N_u is the number of points exceeding the threshold and, n is the number of total values of the dataset.

The estimation of the VaR based on the POT method can be implemented as follows. First, we count the number of points that exceed the threshold u. By default, we keep the default threshold at 5%. Second, we estimate the parameters of the GPD by means of the MLE. The VaR is then calculated based on Equation (11.3).

```
> evtVaR <- function(loss, alpha = 0.9, u = NULL) {
+     if (is.null(u))
+         u <- quantile(loss, .95)
+     y <- loss[loss > u] - u
+     Nu <- length(y)
+     n <- length(loss)
+     fit <- nlminb(c(beta = 1, xi = 0),
+                   function(par)
+                     -sum(log(dgp(y, par[1], par[2]))),
+                   lower = c(1e-8, -Inf))
+     beta <- fit$par[1]
+     xi <- fit$par[2]
+     VaR <- u + beta / xi * (((n / Nu) * (1 - alpha))^(-xi) - 1)
+     names(VaR) <- NULL
+     VaR
+ }
```

We want now to combine the POT method from EVT together with the GARCH(1,1) model as done by McNeil & Frey (2000). We consider the simple time series model where the events at time t are composed of a mean term and some stochastic process where $z_t \sim D(0,1)$ and with some time dependent volatility σ_t:

$$x_t = \mu_t + \sigma_t z_t, \tag{11.4}$$

where σ_t is modeled by a GARCH(1,1) process. The one-step VaR forecast can take the form:

$$\text{VaR}_\alpha = \mu_{t+1} + \sigma_t \text{VaR}_\alpha(Z).$$

This is implemented in the next code snippet:

```
> dynEvtVaR <- function(loss, alpha = 0.95,
+     fit = garch11FitCpp(loss)) {
+     pred <- predict(fit)
+     z <- residuals(fit)
+     pred$mu + pred$sigma * evtVaR(z, alpha)
+ }
```

The same approach based only on the GARCH(1,1) model and not on the POT would give

```
> dynVaR <- function(loss, alpha = 0.95,
+     fit = garch11FitCpp(loss)) {
+     pred <- predict(fit)
+     qst(alpha, pred$mu, pred$sigma, coef(fit)["theta"])
+ }
```

Now that we have implemented the routines, we would like to compare the different approaches. Given the length of a times series, we know that, by definition, for the VaR at level α, the number of values exceeding the VaR is approximately $\alpha \cdots$ (time series length). We can then compare the methods by counting the number of points exceeding the VaR to assess the efficiency of the estimator. In this regard, we use the `DEXUSUK` time series as done in the previous section. We apply the different VaR estimators based on a range of 1,000 historical points, make the one-step forecast, and count the number of events that exceed the forecast VaR. Such calculations can be performed in the form of an embarrassingly parallel problem.

In recent years, several contributed packages to perform parallel computation in R have appeared. Functionalities from the packages `snow` and `multicore` have been merged into the recommended R package `parallel`. The package `parallel` offers two approaches to perform parallel computation in R; see also Chapter 1. One can either create a set of R processes running in parallel and use sockets communication or use function based on the fork mechanism of the operating system. The latter is not supported on a Windows platform. In this chapter, we use the approach based on the creation of R processes that communicate via sockets. In this regard, one first creates a cluster of R processes with the `makeCluster()` function. Second, one exports the functions that have been defined in the current R R session as well as the shared library we have implemented in C/C++ with the functions `clusterExport()` and `clusterEvalQ()`. The remaining code is straightforward. The parallel calculation is done via the `parLapply()` function, which mimics the standard `lapply()` function. When the calculation is done, the cluster of processes is stopped with `stopCluster()`.

```
> alphas <- c(0.95, 0.99, 0.995)
> loss <- - DEXUSUK$RETURN[1:2500]
> len <- length(loss)
> range <- 1000
> library(parallel)
> cl <- makeCluster(4) # To be adapted to your system
> clusterExport(cl, ls())
> clusterEvalQ(cl, dyn.load("garch11filterCpp.so"))
> ans <- parLapply(cl, seq(len - range), function(i) {
+
+     y <- loss[i:(i + range - 1)]
+     fit <- garch11FitCpp(y, hessian = FALSE)
+
+     VaR <- sapply(alphas, function(alpha) {
+         c(dynEvtVaR(y, alpha = alpha, fit = fit),
+           dynVaR(y, alpha = alpha, fit = fit),
+           evtVaR(y, alpha = alpha),
+           quantile(y, alpha))
+     })
+
+     loss[i + range] > VaR
+ })
> stopCluster(cl)
> # Add expected number of violation
> count <- matrix(0, nrow = 4, ncol = 3)
> for (i in seq_along(ans))
+     count <- count + ans[[i]]
> colnames(count) <- paste(100 * alphas, "%", sep = "")
> rownames(count) <- c("dynEvtVaR", "dynVaR", "evtVaR",
+                      "empVaR")
> count <- rbind(expected = round((1 - alphas) * (len - range)),
+                count)
```

The results of the comparison are reported in Table 11.1. The expected number of points exceed VaR three levels of confidence levels, that is, 95%, 99%, 99.5%. One can clearly notice that the combination of the GARCH(1,1) and the VaR estimator based on the EVT gives the most encouraging results.

TABLE 11.1
Comparions of the number of violations of the VaR estimator based on the GARCH(1,1)-EVT, GARCH(1,1), EVT and empirical approaches.

	$\text{VaR}_{0.95}$	$\text{VaR}_{0.99}$	$\text{VaR}_{0.995}$
Expected violations	75	15	8
GARCH(1,1)-EVT	79	17	11
GARCH(1,1)	80	19	11
EVT	87	20	15
Empirical	87	22	13

11.5 Conclusion

In this chapter we have presented and discussed the implementation of R functions for modeling univariate time series processes from the GARCH(p,q) family allowing for arbitrary conditional distributions. Through the modular concept of the estimation, the software can easily be extended to other GARCH and GARCH-related models. Moreover, we have presented an R implementation of the estimation of the VaR based on the POT and GARCH(1,1) models.

12

Yield Curves and Interest Rates Models

Sergio S. Guirreri
Accenture S.p.A.
Milan, Italy

CONTENTS

12.1 A Brief Overview of the Yield Curve and Scenario Simulation

In recent years, the simulation of scenarios of macroeconomic variables has became a crucial issue. Institutional operators, such as national banks, insurance companies, and investment banks must evaluate different scenarios of financial and macroeconomic variables before adopting a monetary policy or promoting a new pension plan. Economic variables such as interest rates and monetary aggregates are often studied by economists to forecast the path of the economy or to predict an ongoing recession. The relationship among interest rates at different maturities contains relevant information about future economic activities. A research of the Federal Reserve Bank (Estrella & Mishkin (1996)) emphasizes the role of the yield curve as a predictor of an ongoing recession. There is a strong relationship among the yield curve, monetary policy, and investor expectations. The monetary policy and the changes in investor expectations affect the slope of the term structure of interest rates; therefore it has been shown that the role of the yield curve is as an indicator of an ongoing recession, or to forecast the future inflation (Estrella & Trubin (2006)). The interest rates are also used for other purposes such as pricing derivatives, evaluating investment projects, or computing risk measures; thus, a good representation of the evolution of the yield curve in the long horizon is a key risk factor for decision makers.

Scenarios of the yield curve represent different states of the world economy; therefore a simulation model of the yield curve is a useful tool, both for a decision makers and for a forecaster. In fact, the more trustworthy are the scenarios simulated and the more accurate will be the risk assessment of a future choice for a decision maker, while for a forecaster the more accurate will be the forecasting model.

In continuous-time classical actuarial pricing, the value (at time t) of a future payoff (as time $\tau > t$) denoted F_τ is $V_t(F_\tau) = e^{-r(\tau-t)}F_\tau$, where r denotes the discount factor, or interest rate. In Chapter 7, a discrete interest rate was considered, and was denoted i. In that context, if t and Y are discrete, $V_t(F_\tau) = (1+i)^{-(\tau-t)}F_T$. But in both cases, r and i were considered as constant, which might not be realistic. Thus, more generally, one can assume that $V_t(F_\tau) = B(t,\tau)F_\tau$ where $B(t,\tau)$ is the value of a risk-free asset, at time t, that will pay 1 at time τ (with $\tau > t$). This product is usually called a *zero-coupon* bond. If such products were exchanged on markets, for all possible maturities τ, and all times t, then prices of those bounds could be used as discount factors. Unfortunately, only some of them can be observed, and therefore it is necessary to extrapolate $B(t,\tau)$'s based on some observed values.

But more than zero-coupon prices, it may be interesting to model instantaneous (stochastic) rates r_s, such that

$$V_t(F_\tau) = \mathbb{E}_{\mathbb{Q}}\left(\exp\left(\int_t^\tau r_s ds\right) F_\tau\right),$$

where $\mathbb{Q}$ denotes the risk-neutral measure (assumed here to be unique; see Privault (2008) for more details).

At time t, the dynamic of forward rates is given either by the yield to maturity curve, defined as

$$\tau \mapsto y(t,\tau) = \frac{1}{\tau - t}\int_t^\tau r_s ds$$

or instantaneous forward curve, defined as

$$\tau \mapsto y_f(t,\tau) = -\frac{\partial}{\partial\tau}\log B(t,\tau) = -\frac{\partial}{\partial\tau}\log\left(\exp\left(\int_t^\tau r_s ds\right)\right).$$

For convenience, in this chapter, we will consider the case where $t=0$, and denote $y_f(\tau)$ and $y(t,\tau)$, respectively, the instantaneous forward rate, and the yield to maturity.

The financial and economic literature have proposed two different frameworks to fit the evolution of the yield curve. The first one is based on term-structure models designed for pricing purposes (Vasicek (1977), Cox et al. (1985), Duffie & Kan (1996)). In Vasicek (1977), short rate is supposed to satisfy

$$dr_t = a(b - r_t)\,dt + \sigma\,dW_t,$$

while in Cox et al. (1985),

$$dr_t = a(b - r_t)\,dt + \sqrt{r_t}\,\sigma\,dW_t.$$

The second one is based models that attempt to provide a statistical description of the evolution of rates in the objective measure (Nelson & Siegel (1987), Svensson (1994)). Recently, it has been shown (Diebold & Li (2006)) that a dynamic approach based on the factor of Nelson-Siegel's model can reproduce key features of U.S. interest rate curves with good forecasting performances.

Some key factors of the yield curve are that the linear correlation between the short and long term rates, and correlations at different lags, are very high; the marginal distributions of the yield curve differ from the Normal distribution; and the time series of each interest rate are not stationary. Some statistical methodologies that have been recently applied to estimate the previous features of the term structure of interest rates are the vector autoregressive model (VAR) and the copula approach.

In particular the VAR model is able to capture most of the key factors of the yield curve such as the cross and lagged correlation, but it is limited to simulating the interest rates because the marginal distributions are assumed Gaussian (Ang & Piazzesi (2003)). The marginal distributions of interest rates are significantly different from the Gaussian distribution, especially when, by the first differences, the interest rates are transformed into stationary time series. For this reason, the Gaussian vector auto-regressive model is not able to capture the entire feature of the term structure interest rates.

The copula method (Cherubini et al. (2004)) allows us to model the dependence structure independently of the marginal distributions. One may construct a multivariate distribution with different marginals with the copula function that describes the dependence structure. The main limit of the model based on this approach is that they focused on estimating copula, when data are dependent, and did not supply the VAR dynamics.

Recently, a new methodology has been proposed to generate the scenario of the term structure of interest rates (Consiglio & Guirreri (2011)) using the VARTA (Vector Autoregressive To Anything) approach of Biller & Nelson (2003). In this framework, the two authors are able to simulate a scenario of the yield curve in the long horizon replicating: the correlation among interest rates, the autocorrelation at different lags of each rate time series, and the marginals distribution, using the Johnson' distributions (Johnson (1949)). The non-stationarity of the interest rates time series is one limit of the IRTA (Interest Rates To Anything) methodology, whereas the second one is the number of interest rates that the IRTA is able to simulate. The latter issue can be solved by simulating the coefficients of Nelson–Siegel's model, using the VARTA approach, and then build the term structure of the interest rates using the number of pillars as input (Guirreri (2010)).

Rebonato et al. (2005) emphasize the role of the scenarios simulation of the yield curve in the real world. The yield curve is a key factor in numerous applications; some examples are the following:

- *Evaluation of potential future exposure (PFE) for counter party credit risk assessment*: In order to evaluate the PFE, the relevant yield curve must be simulated, typically using a Monte Carlo procedure, and the conditional exposure in the various states of the world computed.

- *Assessment of the hedging performance of interest rate option models*: In this context, one wants to judge the quality of a model from its assumptions. One method is to shock the real-world yield curve by a few eigenvalues obtained from the orthogonalization of the covariance matrix. The hedging strategy suggested by a simple diffusive pricing model with deterministic volatilities will produce a good replication of the payoff of a plain-vanilla option. This is more likely to be due, however, to the very simplified nature of the yield curve evolution rather than to any intrinsic virtues of the modeling approach.

- *Assessment of different investment strategies in interest rate-sensitive investment portfolios (Asset/Liability management)*: Investment portfolios are usually evaluated with reference to the net interest income (NII) that they generate. It is not a complex issue to generate a favorable NII over a short period of time, but it is essential to compare the NII performance of competing interest rate-based portfolios, evaluating over a suitably long time horizon.

- *Economic capital calculations*: These techniques estimate the marginal contribution of a new investment or a business activity to a total loss profile. In these applications, the evolution of the yield curve plays an important role because the loss profile is affected by the interest rates not only directly, but also indirectly via the depositors' behavior or mortgage prepayment patterns.

There are three main reasons why scenario analysis is often preferred over alternative approaches (Steehouwer (2005)). The first reason is the flexibility it offers to model the often complex interactions and relations within and between the components of an ALM problem.

The second reason for the popularity of scenario analysis is that it offers great possibilities for learning about the problem under consideration in addition to just obtaining some "optimal" solution.

The third reason is that the strong visual aspects of scenario analysis cause the models and the solutions obtained to be more easily accepted by decision makers. In conclusion, the scenarios play an important role in strategic policy-making processes of institutions such as pension funds, insurance companies, and banks.

Macroeconomic scenarios are fundamental inputs for Asset and Liability Management (ALM) models. The output of ALM models strongly depends on the input scenarios. For this reason, it is of paramount importance to have a scenario generator as reliable as possible. Reliability, in this respect, means that the distribution of the set of scenarios resembles as much as possible the empirical one.

The concepts of *scenario analysis*, also called *Monte Carlo simulation* or *stochastic simulation*, are often applied to study financial tools and modeling the economic risk and return factors. An economic scenario is a possible future evolution of all relevant (uncertain) macroeconomic (and other) variables. Usually, a large number of scenarios of economic variables is generated, for example, several hundreds with a horizon of 15 years. Together with the strategic policy under consideration, these are fed into a model that states all relations between policy instruments, scenario variables, and relevant output measures with respect to the objectives of the stakeholders[1]. With the system of the above relations at hand, the model simulates what would happen to the objectives of the stakeholders if the policy under consideration were applied during the simulation period. For example, the output of the model could be the future evolution of the solvency ratio of an insurance company for each of the economic scenarios.

It is important to note that the scenarios should be neutral with respect to the objectives and constraints of the various stakeholders. The scenarios represent one and the same independent, macroeconomic world in which the economic entity under investigation and its stakeholders need to operate and which they cannot change by themselves. Based on the scenarios, different risk and return measures can be calculated. Examples of such measures are the probability of the solvency ratio falling below the legally required 100% (a risk measure) or the expected return on equity during the next 15 years (a return measure). In the next step, decision makers have to evaluate these risk and return numbers of the policy under consideration and decide whether, for example, the expected return on equity is above or below the expectations.

The following sections will focus on the Nelson–Siegel's model and Svensson's model using the R package `YieldCurve`; see Guirreri (2012)

12.2 Yield Curves

12.2.1 Description of the Datasets

Interest rates of the central bank of the United States, the Federal Reserve, can be downloaded from the Fed website [2] selecting the variables in scope. We will use, for different dates t and different maturities τ, interest rates $y(t,\tau)$. Usually, maturities are given (1 year, 2

[1] A stakeholder is a third party who temporarily holds money or property while its owner is still being determined. In a business context, a stakeholder is a person or organization that has a legitimate interest in a project or entity.

[2] http://www.federalreserve.gov/datadownload/Build.aspx?rel=H15.

years, 10 years, etc.), and thus, for all maturities, we do have a time series $t \mapsto y(t, \tau)$. Interest rates are a collection of time series $(y(t, \tau))$. One can either use functional time series from the `ftsa` library (as in Chapter 9), or extensible time series from the `xts` library. The package `YieldCurve` involves the following datasets:

1. `FedYieldCurve`: an `xts` object that contains the interest rates of the Federal Reserve, from January 1982 to December 2012. The interest rates are Market yield on U.S. Treasury securities constant maturity (CMT)[3] at different maturities (3 months, 6 months, 1 year, ..., 10 years), quoted on investment basis and have been gathered with monthly frequency.
2. `ECBYieldCurve`: an `xts` object that consists of a collection of interest rates of the European Central Bank [4] at different maturities.

We visualized the Fed yield curve at the several selected dates. Indeed, the first 3 and the last 3 months were selected from the dateset `FedYieldCurve`.

```
> require(xts)
> require(YieldCurve)
> data(FedYieldCurve)

> first(FedYieldCurve,'3 month')
            R_3M  R_6M  R_1Y  R_2Y  R_3Y  R_5Y  R_7Y R_10Y
1982-01-01 12.92 13.90 14.32 14.57 14.64 14.65 14.67 14.59
1982-02-01 14.28 14.81 14.73 14.82 14.73 14.54 14.46 14.43
1982-03-01 13.31 13.83 13.95 14.19 14.13 13.98 13.93 13.86
> last(FedYieldCurve,'3 month')
           R_3M R_6M R_1Y R_2Y R_3Y R_5Y R_7Y R_10Y
2012-10-01 0.10 0.15 0.18 0.28 0.37 0.71 1.15  1.75
2012-11-01 0.09 0.14 0.18 0.27 0.36 0.67 1.08  1.65
2012-12-01 0.07 0.12 0.16 0.26 0.35 0.70 1.13  1.72
> mat.Fed<-c(3/12, 0.5, 1,2,3,5,7,10)
> par(mfrow=c(2,3))
> for( i in c(1,2,3,370,371,372) ){
+ plot(mat.Fed, FedYieldCurve[i,], type="o",
  xlab="Maturities structure in years", ylab="Interest rates values")
+ title(main=paste("Federal Reserve yield curve observed at",
  time(FedYieldCurve[i], sep=" ") ))
+ grid()
}
```

The interest rate surface can also be visualized in Figure 12.2.

```
> persp(1982:2012,maturity,FedYieldCurve[seq(2,nrow(FedYieldCurve),by=12),],
+ theta=30,xlab="Year",ylab="Maturity (in years)",
+ zlab="Interest rates (in %)",ticktype = "detailed",shade=.2,expand=.3)
```

[3]More information on the Treasury yield curve can be found at the following website, http://www.treasury.gov/resource-center/data-chart-center/interest-rates/Pages/yieldmethod.aspx.

[4]The interest rate can be downloaded from the ECB's website, http://www.ecb.europa.eu/stats/money/yc/html/index.en.html.

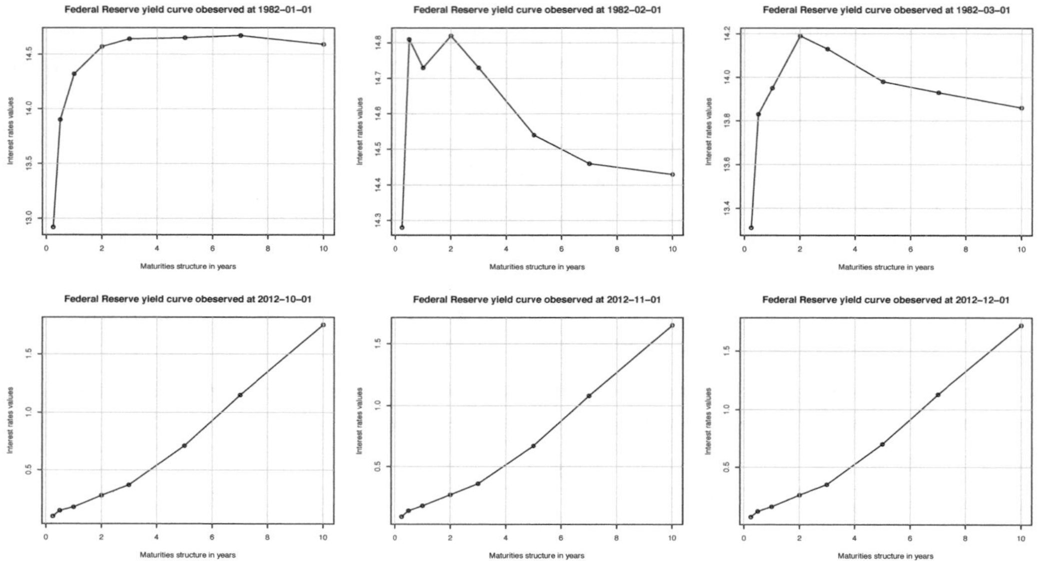

FIGURE 12.1
Observed Federal Reserve yield curves at different dates.

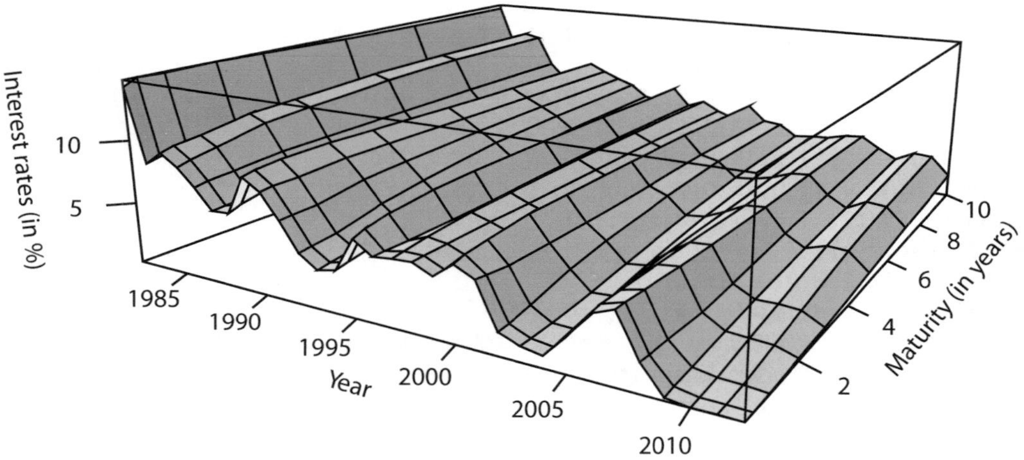

FIGURE 12.2
Observed Federal Reserve yield surface.

12.2.2 Principal Component Analysis

As mentioned previously in Vasicek (1977), the short rate is modeled, assuming that $dr_t = a(b - r_t)\,dt + \sigma\,dW_t$, and the the rest of the curve is derived from it. This can be seen as a *one-factor* model. But it is clearly not enough to obtain stylized facts discussed in the previous section. Litterman & Scheikman (1991) looked at the treasury yield curve, and obtained that only a few factors were necessary because only a three eigenvalues were significant. Those three factors could actually be used to model the level of the yield curve, the slope of the curve, and the curvature of the curve.

To perform a principal component analysis, let $\boldsymbol{Y} = [\boldsymbol{Y}_1, \cdots, \boldsymbol{Y}_m]$ be the collection of all time series, in matrix form. Vector $\boldsymbol{Y}_j$ is the time series associated to jth maturity. Classically, we should normalize the data by subtracting the mean from each of the data dimensions. The goal is to find the linear combination of the $\boldsymbol{Y}_j$'s with maximum variance. Thus, we have to solve

$$\max_{\|\boldsymbol{u}\|=1} \{\boldsymbol{u}'\boldsymbol{X}'\boldsymbol{X}\boldsymbol{u}\}.$$

Because $\boldsymbol{X}'\boldsymbol{X}$ is a positive definite matrix, it admits a decomposition $\boldsymbol{P\Delta P}'$, where $\boldsymbol{P}$ is an orthogonal matrix, and $\boldsymbol{\Delta}$ is a diagonal matrix, containing all eigenvalues. Assume that eigenvalues are ordered in decreasing order, $\boldsymbol{\Delta} = \text{diag}(\lambda_1, \cdots, \lambda_m)$, then $\lambda_1 \geq \cdots \geq \lambda_n > 0$ (because $\boldsymbol{X}'\boldsymbol{X}$ is positive definite). Let $\boldsymbol{F} = \boldsymbol{XP}$. Columns of $\boldsymbol{F}$ are called factors. By construction, note that those factors are orthogonals. And the $\boldsymbol{X}_t$'s are linear combinations of the factors because $\boldsymbol{X} = \boldsymbol{FP}'$:

$$\boldsymbol{X}_t = \sum_{j=1}^{m} \boldsymbol{P}_{j,t}\boldsymbol{F}_j.$$

To perform such a decomposition, use the `princomp` function:

```
> M <- as.matrix(FedYieldCurve)
> pca.rates <- princomp(M, scale=TRUE)
```

As mentioned previously, three factors are clearly sufficient, as those first three explain 99.9% of the total variance,

```
> summary(pca.rates)
> summary(pca.rates)
Importance of components:
                          Comp.1     Comp.2       Comp.3       Comp.4
Standard deviation     8.5598756 1.16055974 0.2557040238 0.1246520910
Proportion of Variance 0.9808032 0.01802943 0.0008752298 0.0002079918
Cumulative Proportion  0.9808032 0.99883266 0.9997078851 0.9999158769
                             Comp.5       Comp.6       Comp.7       Comp.8
Standard deviation     5.544781e-02 3.941926e-02 0.0341450611 2.214145e-02
Proportion of Variance 4.115435e-05 2.080003e-05 0.0000156064 6.562348e-06
Cumulative Proportion  9.999570e-01 9.999778e-01 0.9999934377 1.000000e+00
```

The three factors can be visualized in Figure 12.3.

```
> factor.loadings <- pca.rates$loadings[,1:3]
> matplot(maturity,factor.loadings,type="l", lwd=c(2,1,1),
+ lty=c(1,1,2),xlab = "Maturity (in years)", ylab = "Factor loadings")
```

The first one is (almost) flat and corresponds to the level of the yield curve. The second one is a monotone function, which will explain the slope of the curve. And the third one is a convex function that will explain the curvature of the curve.

The evolution of weights $\boldsymbol{P}_{j,t}$ for $j = 1, 2, 3$ can be visualized in Figure 12.4.

```
> vtime=seq(1981+11/12,length=nrow(M),by=1/12)
> for(j in 1:3) plot(vtime,pca.rates$scores[,j],type="l",xlab="Year")
```

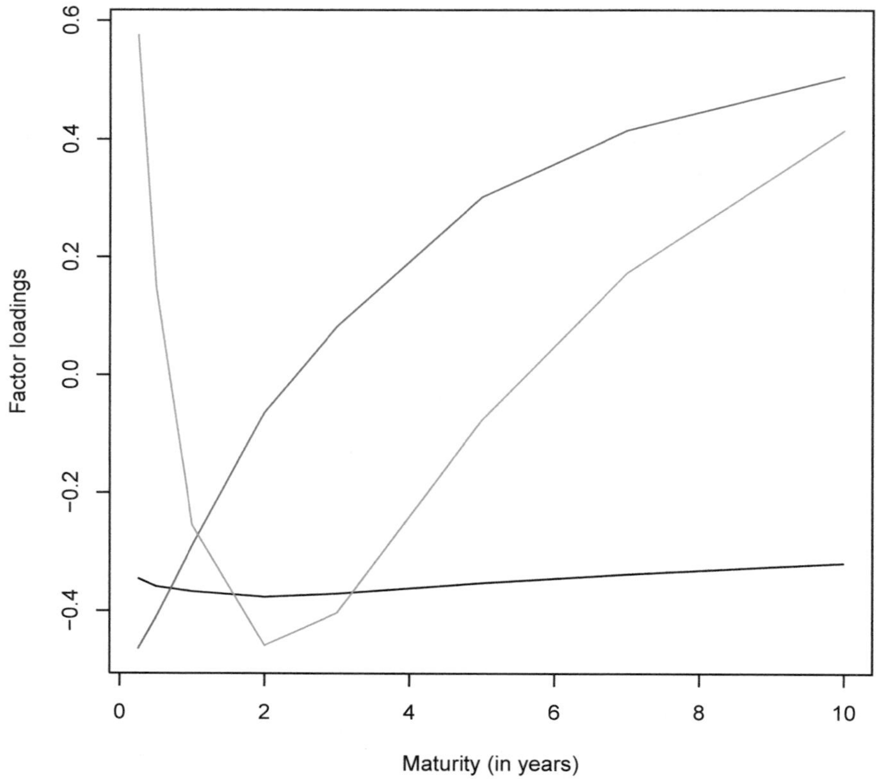

FIGURE 12.3
Three principal components from the Federal Reserve yield surface.

12.3 Nelson–Siegel Model

The core of this section is the Nelson–Siegel model and the dynamic approach used by Diebold & Li (2006) to model and forecast the yield curve. The Nelson–Siegel model is widely used among central banks and financial analysts, and its popularity is due to its flexible and parsimonious form. In Diebold & Li (2006), it is observed that the Nelson–Siegel model fits both the cross section and time series of yields remarkably well, in many countries and periods. They described the yield curve evolution using the Nelson–Siegel model, giving an interpretation to the three factors as level, slope, and curvature of the term structure interest rates. To model the dynamics of the whole yield curve, Diebold and Li estimate an auto-regressive model for each of the three coefficients.

The aim of Nelson & Siegel (1987) was to define a parsimonious model to be flexible enough to describe the various shapes and features of the yield curve. The most frequent shapes of the yield curve are the humped and $\mathcal{S}$ forms. Nelson and Siegel focused on a class of functions that could be able to generate the typical form of the yield curve. This class of

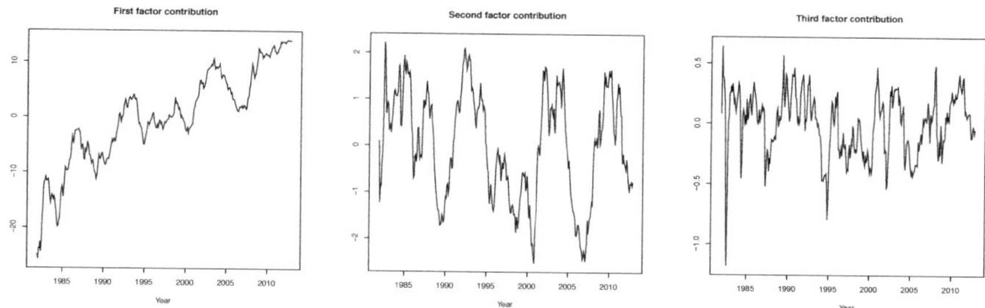

FIGURE 12.4
Three principal components scores from the Federal Reserve yield surface.

functions is related to the Laguerre functions, which consist of a polynomial term and an exponential decay term. The instantaneous forward rate, y_f, at maturity τ is described by the following relation:

$$y_f(\tau) = \beta_0 + \beta_1 \exp\left(-\frac{\tau}{\lambda}\right) + \beta_2 \left[\frac{\tau}{\lambda} \exp\left(-\frac{\tau}{\lambda}\right)\right], \tag{12.1}$$

where λ is a time constant related to the maturity τ and $(\beta_0, \beta_1, \beta_2)$ are unknown parameters. Equation (12.1) can be viewed as a constant plus a Laguerre function (see furthers details in Heiberger & Neuwirth (1953)). The yield to maturity can be obtained from the instantaneous forward rate:

$$y(\tau) = \frac{1}{\tau} \int_0^{\tau} y_f(x)\, dx. \tag{12.2}$$

Substituting Equation (12.1) in (12.2) and solving the integral, we obtain the Nelson–Siegel yield curve:

$$y(\tau) = \beta_0 + \beta_1 \left(\frac{1 - \exp(-\frac{\tau}{\lambda})}{\frac{\tau}{\lambda}}\right) + \beta_2 \left(\frac{1 - \exp(-\frac{\tau}{\lambda})}{\frac{\tau}{\lambda}} - \exp(-\frac{\tau}{\lambda})\right), \tag{12.3}$$

where β_0 represents the long-term component that does not decay to zero in the limit. The coefficient β_1 has the loading factor

$$\frac{1 - \exp(-\frac{\tau}{\lambda})}{\frac{\tau}{\lambda}}$$

that approaches 1 as $\tau \to 0$ and decays quickly to 0 as $\tau \to +\infty$; hence β_1 identifies the short-term component.

The factor loading of β_2 models a hump shape, starting at zero and reaching its maximum and then decreasing to zero; thus the coefficient β_2 can be viewed as the medium-term factor. It is interesting to note that the instantaneous yield is governed by the long and short term, $y(0) = \beta_0 + \beta_1$.

The parameter λ controls the decay of the regressors and determines at which maturity the medium-term factor reaches its maximum, namely the position of the hump. In particular, large values of λ produce slow decay in the regressors and better fitting on the longer maturity, while small values of λ produce a quick decay in the regressors and a better fit on the short maturity.

The following R code will help us to understand the role of each component of Equation (12.3):

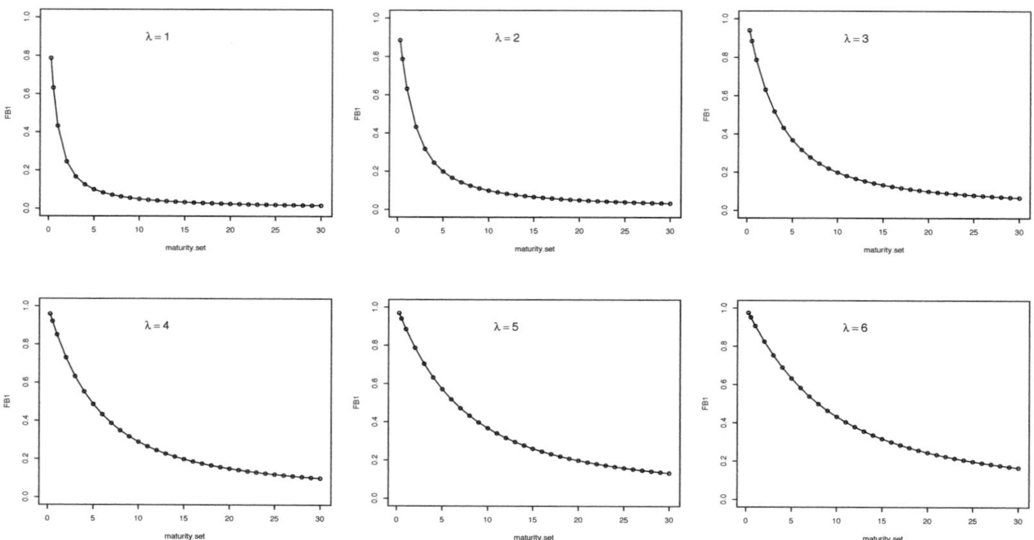

FIGURE 12.5
β_1 factor loading for different $\dot{\lambda}$ values.

```
> factorBeta1 <- function(lambda, Tau)
+   {
+     (1-exp(-Tau/lambda)) / (Tau/lambda)
+   }
> maturity.set<-c(3/12,6/12,seq(1:30))
> lambda.set <- c(0.5,1,2,3,4,5)
> par(mfrow=c(2,3))
> for(i in 1:length(lambda.set)){
+   FB1 <- factorBeta1(lambda.set[i],maturity.set)
+   plot(maturity.set, FB1, type="o", ylim=c(0,1))
+   text(12,0.9, substitute(list(lambda) == group("",list(x),""),list(x=i)),cex=1.5)
+   }

> factorBeta2 <- function(lambda, Tau)
+   {
+     (1-exp(-Tau/lambda)) / (Tau/lambda) - exp(-Tau/lambda)
+   }
>   par(mfrow=c(2,3))
> for(i in 1:length(lambda.set)){
+   FB2 <- factorBeta2(lambda.set[i],maturity.set)
+   plot(maturity.set, FB2, type="o", ylim=c(0,0.4))
+   text(i+2,0.35, substitute(list(lambda) == group("",list(x),""),list(x=i)),
      cex=1.5)
+   abline(v=i, lty=2)
+   }
```

Therefore, the loading factors of β_1 and β_2 depend on τ, that is, a set of maturities for which we want to represent the rate curve, and $\dot{\lambda}$, that is, an unknown key factor for the Nelson–Siegel model. More generally, to fit, Equation (12.3) should use nonlinear methods to estimate the set of unknown parameters $(\beta_0, \beta_1, \beta_2, \dot{\lambda})$, when facing some problem related to this method, such as time computing, performance, constrains verification.

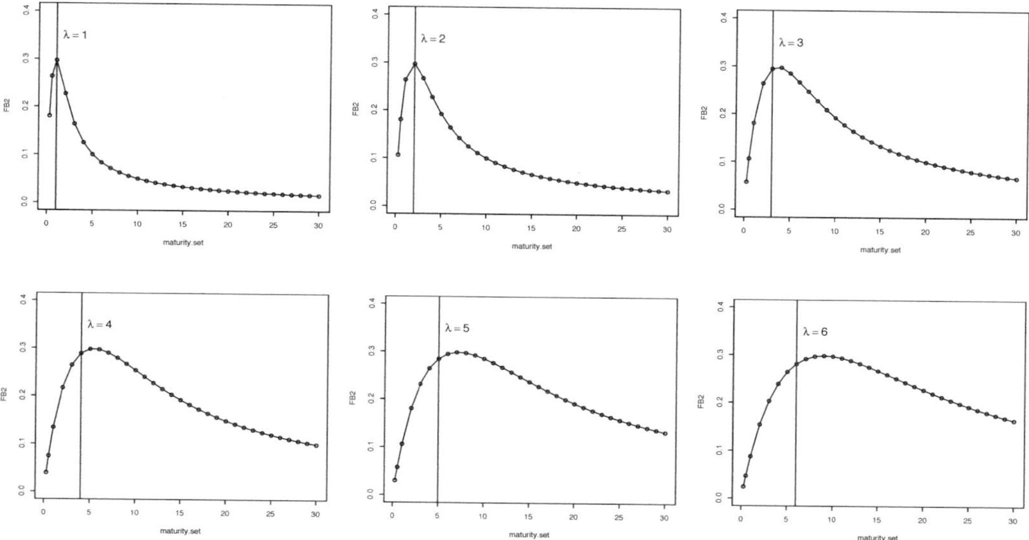

FIGURE 12.6
β_2 factor loading for different $\dot{\lambda}$ values.

To avoid the implementation of some nonlinear optimization algorithms, the R package `YieldCurve` has been used as one of the ideas behind the paper of Diebol–Li. The two main ideas of Diebold & Li (2006) were

1. Estimating the Nelson–Siegel model as a linear model, fixing a $\dot{\lambda}$ value for each period;
2. Forecasting the term structure of interest rates based on the parameters of the Nelson–Siegel model, instead of each single interest rate used in the structure of the rate curve.

The first point was developed in the function `NelsonSiegel()` of the R package `YieldCurve` fixing a grid of $\dot{\lambda}$ and for each value an estimate of the Nelson–Siegel model, by ordinary least square (OLS), has been obtained. The best estimate of $\hat{\lambda}$ is the value that maximizes the medium-term factor loading:

$$\max_{\dot{\lambda}_j}\left[\frac{1-\exp(-\frac{\tau_j}{\dot{\lambda}_j})}{\frac{\tau_j}{\dot{\lambda}_j}}-\exp(-\frac{\tau_j}{\dot{\lambda}_j})\right], \qquad j=1,\ldots,n. \tag{12.4}$$

In correspondence with each $\dot{\lambda}_j$, I is estimated by OLS parameters $(\hat{\beta}_{0,j}, \hat{\beta}_{1,j}, \hat{\beta}_{2,j})$. The best fitting of the Equation (12.3) is obtained for parameters $(\hat{\beta}_0, \hat{\beta}_1, \hat{\beta}_2, \hat{\lambda})$ that minimize the sum of the square of residuals:

$$\min_{\hat{\beta}_{0,j},\hat{\beta}_{1,j},\hat{\beta}_{2,j},\hat{\lambda}_j}\sum_j\left[y_{obs}-\hat{y}_j(\tau)\right]^2, \qquad j=1,\ldots,n, \tag{12.5}$$

where y_{obs} is the yield curve observed and $\hat{y}_j(\tau)$ corresponds to the yield curve fitted by the Nelson-Siegel model with the parameters $(\hat{\beta}_{0,j}, \hat{\beta}_{1,j}, \hat{\beta}_{2,j}, \hat{\lambda}_j)$. The above estimation method avoids a non-trivial problem due to the presence of multiple local optima in the cross section of coupon bond prices, as the function to optimize is not guaranteed to be convex.

12.3.1 Estimating the Nelson–Siegel Model with R

The estimation of the yield curve shown in the graph (12.1) can be done using the Nelson–Siegel model (12.3) implemented in the function `Nelson.Siegel()`, where the input arguments are

- `rate`: It can be a vector or a matrix with interest rates observed at several periods, one row for each day. The default class of the `rate` object is `xts`.
- `maturity`: It is a vector that represents the structure of the interest rate and it must have the same length as the number of columns of the `rate` object.

Recall that at time t, Equation (12.3) is

$$y(t,\tau) = \beta_0(t) + \beta_1(t)\left(\frac{1-\exp(-\frac{\tau}{\lambda(t)})}{\frac{\tau}{\lambda(t)}}\right) + \beta_2(t)\left(\frac{1-\exp(-\frac{\tau}{\lambda(t)})}{\frac{\tau}{\lambda(t)}} - \exp(-\frac{\tau}{\lambda(t)})\right). \quad (12.6)$$

Thus, at each time t, four coefficients are estimated:

```
> Fed.Rate1 <- Nelson.Siegel(first(FedYieldCurve, '3 month'), mat.Fed)

> Fed.Rate1
              beta_0       beta_1   beta_2    lambda
1982-01-01 14.34594 -1.76249751 3.650061 0.9999507
1982-02-01 14.14681  0.05426534 2.219142 0.9999507
1982-03-01 13.61065 -0.54316951 2.708078 0.9999507

> Fed.Rate2 <- Nelson.Siegel(last(FedYieldCurve, '3 month'), mat.Fed)
> Fed.Rate2
              beta_0   beta_1    beta_2   lambda
2012-10-01 6.752246 -6.62480 -6.609233 0.183919
2012-11-01 6.292996 -6.17330 -6.111852 0.183919
2012-12-01 6.590762 -6.49626 -6.364102 0.183919
```

The output of the function is the estimation of the Nelson-Siegel coefficients and the λ parameter for each period. Comparing the coefficients of the two outputs, `Fed.Rate1` and `Fed.Rate2`, can be noted as the long-run component $\hat{\beta}_0$, has d significantly change; and, in the same way, the slope and the curvature of the yield curve.

Proceeding with the estimation of the yield curve for each period in the dataset `FedYieldCurve`, one can observe the behavior of the multivariate time series of the Nelson-Siegel coefficient to obtain some information on the trend of the term structure of interest rates.

```
> system.time(Nelson.Siegel(FedYieldCurve, mat.Fed))
   user  system elapsed
  18.93    0.08   19.30
> Fed.Rates <- Nelson.Siegel(FedYieldCurve, mat.Fed)

> first(Fed.Rates,'year')
              beta_0       beta_1     beta_2    lambda
1982-01-01 14.34594 -1.76249751 3.65006071 0.9999507
```

```
1982-02-01 14.14681  0.05426534 2.21914158 0.9999507
1982-03-01 13.61065 -0.54316951 2.70807842 0.9999507
1982-04-01 13.61517 -0.51818755 2.75748648 0.9999507
1982-05-01 13.52630 -1.16619236 2.39279341 0.9999340
1982-06-01 14.13378 -1.43747932 3.23769741 0.9999507
1982-07-01 13.84696 -2.49133196 3.69440615 0.9999507
1982-08-01 13.02162 -4.77732118 4.90917816 0.9999507
1982-09-01 12.10749 -4.77732737 6.11249115 0.9999507
1982-10-01 11.02616 -3.75388964 2.73096552 0.9999340
1982-11-01 10.80609 -2.77592647 0.51759030 0.9999507
1982-12-01 10.84695 -2.96572285 0.05023213 0.9999507
> last(Fed.Rates,'year')
             beta_0    beta_1    beta_2    lambda
2012-01-01 5.067016 -5.002900 -5.291108 0.2869276
2012-02-01 5.415061 -5.293744 -5.645563 0.2656801
2012-03-01 5.266186 -5.166166 -5.092655 0.2869276
2012-04-01 5.520023 -5.401772 -5.608334 0.2656801
2012-05-01 6.755632 -6.641670 -6.408147 0.1839190
2012-06-01 5.829454 -5.717205 -5.360561 0.1839190
2012-07-01 5.942878 -5.805087 -5.918784 0.1839190
2012-08-01 6.007142 -5.890642 -5.770203 0.1938867
2012-09-01 6.877328 -6.739946 -6.945611 0.1839190
2012-10-01 6.752246 -6.624800 -6.609233 0.1839190
2012-11-01 6.292996 -6.173300 -6.111852 0.1839190
2012-12-01 6.590762 -6.496260 -6.364102 0.1839190

> par(mfrow=c(2,2))

> plot(Fed.Rates$beta_0, main='Beta_0 coefficient', ylab='Values')

> plot(Fed.Rates$beta_1, main='Beta_1 coefficient', ylab='Values')

> plot(Fed.Rates$beta_2, main='Beta_2 coefficient', ylab='Values')

> plot(Fed.Rates$lambda, main='Lambda coefficient', ylab='Values')
```

From the trend of the β_0 it is evident that the long-term interest rate decreases during the last years, the time series of the slope, β_1, increases over the crisis period, while the hump of the term structure is located in the long period [5] and tends to be an inverted hump.
To obtain the interest rates from the Nelson–Siegel estimates, they have developed the function `NSrates()`, where the input variables are

1. `Coeff`: A vector or matrix of the Nelson–Siegel function, such as returned from the `Nelson.Siegel()`, indeed the `Fed.Rates` object.
2. `maturity`: A vector that represents the structure of the yield curve, indeed the `mat.Fed` object.

[5]The λ parameter in the `YieldCurve` package is represented as the inverse of the Nelson-Siegel representation, such as $\lambda = \frac{1}{\lambda}$

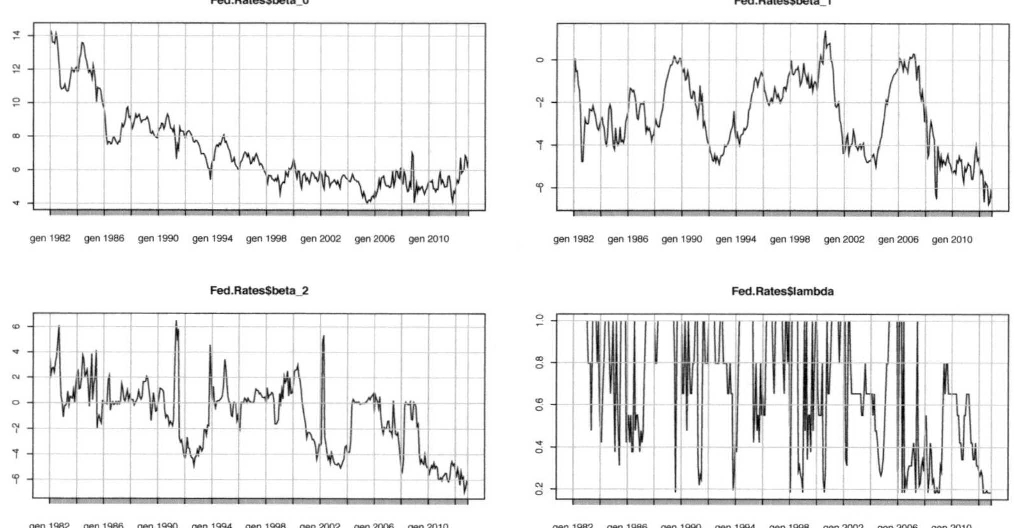

FIGURE 12.7
Time series of the Nelson-Siegel coefficients.

For example, if we want to obtain the estimated interest rates, we can run the following code:

```
> Fed.yield.curve <- NSrates(Fed.Rates, mat.Fed)
> par(mfrow=c(2,2))
> plot(mat.Fed, as.numeric(FedYieldCurve[1,]), type="o", col=2,
+  ylab="Interest rates", xlab="Maturity in years", ylim=c(0,15))
> lines(mat.Fed, as.numeric(Fed.yield.curve[1,]), type="o", col=3)
> title(main="Observed vs Fitted yield curve")
> legend('bottomright', legend=c("Observed","Fitted"),col=c(2,3), lty=1, bg='gray90')
> grid()
> plot(mat.Fed, as.numeric(FedYieldCurve[120,]), type="o", col=2,
+  ylab="Interest rates", xlab="Maturity in years", ylim=c(0,15))
> lines(mat.Fed, as.numeric(Fed.yield.curve[120,]), type="o", col=3)
> title(main="Observed vs Fitted yield curve after 10 years")
> legend('bottomright', legend=c("Observed","Fitted"),col=c(2,3), lty=1, bg='gray90')
> grid()
> plot(mat.Fed, as.numeric(FedYieldCurve[240,]), type="o", col=2,
+  ylab="Interest rates", xlab="Maturity in years", ylim=c(0,15))
> lines(mat.Fed, as.numeric(Fed.yield.curve[240,]), type="o", col=3)
> title(main="Observed vs Fitted yield curve after 20 years")
> legend('topright', legend=c("Observed","Fitted"),col=c(2,3), lty=1, bg='gray90')
> grid()
> plot(mat.Fed, as.numeric(FedYieldCurve[360,]), type="o", col=2,
+  ylab="Interest rates", xlab="Maturity in years", ylim=c(0,15))
> lines(mat.Fed, as.numeric(Fed.yield.curve[360,]), type="o", col=3)
> title(main="Observed vs Fitted yield curve after 30 years")
> legend('topright', legend=c("Observed","Fitted"),col=c(2,3),
+ lty=1, bg='gray90')
> grid()
```

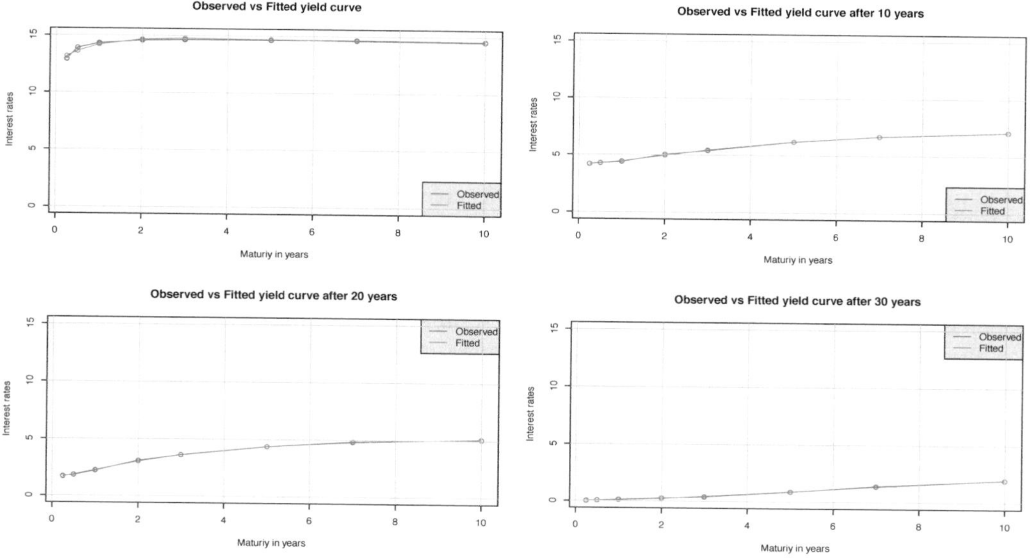

FIGURE 12.8
Observed versus fitted yield curves at different periods.

The graphic in Figure 12.8 shows the evolution of the yield curve since 1982 in steps of 10 years. [It is evident as the curve is shifted down, β_0 coefficient, whereas the curvature, β_2 coefficient, is changed from a "positive" to "negative" hump.] To better understand the role of each coefficient of the Nelson–Siegel model we can run the following code for the curvature using the estimated value of the last period and keeping fixed all other coefficients:

```
> last(Fed.Rates,'month')
              beta_0    beta_1    beta_2   lambda
2012-12-01 6.590762 -6.49626 -6.364102 0.183919

> b0<-rep(6.590762,21)
> b1<-rep(-6.496260,21)
> b2<-seq(-10, 10,by=1) # create a sequence of fictive beta_2 coeff
> lambda<-rep(0.1839190,21)
> B<-ts(cbind(b0,b1,b2,lambda), start=c(2000,1,1),
+ frequency=12)#create a time series object
> B<-as.xts(B,RECLASS=FALSE)# trasform the time series object in xts
> A<- NSrates(B ,mat.Fed)#create the fictive yield curves
# create an interactive plot shows the movement of the yield curve
# at different values of beta_2
> for(i in 1:nrow(A))
+ {
+ plot(mat.Fed, A[i,],type="l", ylim=c(-1,6))
+ title(main=paste("beta_2",B[i,3], sep="=") )
+ par(ask=TRUE)
+ }
```

12.4 Svensson Model

12.4.1 Estimating the Svensson Model with R

The Nelson–Siegel model has optimal performance and returns good results if the maturities are not longer than 15 years. When the term structure is more complex, fitting by the Nelson–Siegel model is unsatisfactory. This is due to the convexity effect that tends to pull down the yields on longer-term securities, giving the yield curve a concave shape at longer maturities. Svensson proposed an extension of the Nelson–Siegel approach by increasing its flexibility and improving the fitting on a richer term structure. Svensson methodology assumes that the spot rates are governed by six parameters according to the following functional form:

$$y(\tau) = \beta_0 + \beta_1 \left(\frac{1-\exp(-\frac{\tau}{\lambda_1})}{\frac{\tau}{\lambda_1}} \right) + \beta_2 \left(\frac{1-\exp(-\frac{\tau}{\lambda_1})}{\frac{\tau}{\lambda_1}} - \exp(-\frac{\tau}{\lambda_1}) \right) + \\ + \beta_3 \left(\frac{1-\exp(-\frac{\tau}{\lambda_2})}{\frac{\tau}{\lambda_2}} - \exp(-\frac{\tau}{\lambda_2}) \right) \quad (12.7)$$

Equation (12.7) is equal to Equation (12.3) when β_3 is set to zero. The factor loading of the coefficient β_3 captures a second "hump" of the yield curve at longer maturities, and λ_2 indicates the position of the second hump. A survey of the Bank for International Settlements (2005) showed that many central banks use the Nelson-Siegel-Svensson model to describe the evolution of the yield curve.

The estimation methodology implemented in the R package `YieldCurve` is similar to that used for the Nelson-Siegel model:

1. Two vectors of maturities, in the medium- and long-term ranges, have been selected: $\boldsymbol{\omega}_1 = (1, \ldots, n_1)'$ and $\boldsymbol{\omega}_2 = (1, \ldots, n_2)'$.
2. A grid of estimations of λ_1 and λ_2 is obtained, respectively, by maximizing the β_2 and β_3 factor loadings:

$$\hat{\lambda}_{1j} = \max_{\lambda_1} \left[\frac{1-\exp(-\frac{\omega_{1j}}{\lambda_1})}{\frac{\omega_{1j}}{\lambda_1}} - \exp(-\frac{\omega_{1j}}{\lambda_1}) \right], \qquad j = 1, \ldots, n_1 \quad (12.8)$$

$$\hat{\lambda}_{2i} = \max_{\lambda_2} \left[\frac{1-\exp(-\frac{\omega_{2j}}{\lambda_2})}{\frac{\omega_{2j}}{\lambda_2}} - \exp(-\frac{\omega_{2j}}{\lambda_2}) \right], \qquad i = 1, \ldots, n_2 \quad (12.9)$$

3. Substituting $\hat{\lambda}_{1j}$ and $\hat{\lambda}_{2i}$ into Equation (12.7), a linear function is obtained that can be rewritten as

$$\mathbf{y}_{obs} = \mathbf{X}_{(\hat{\lambda}_{1j};\hat{\lambda}_{2i})} \boldsymbol{\beta} + \boldsymbol{\epsilon}, \quad (12.10)$$

where

$$\boldsymbol{\beta} = (\beta_0, \beta_1, \beta_2, \beta_3)',$$

$$\mathbf{y}_{obs} = (y_{obs}(\tau_1), \ldots, y_{obs}(\tau_n))'$$

are, respectively, the vector of the parameters and the observed interest rates at different maturities, and

$$\mathbf{X}_{(\hat{\lambda}_{1j};\hat{\lambda}_{2i})} = \begin{pmatrix} 1 & f_1(\tau_1;\hat{\lambda}_{1j}) & f_2(\tau_1;\hat{\lambda}_{1j}) & f_3(\tau_1;\hat{\lambda}_{2i}) \\ \vdots & \vdots & \vdots & \vdots \\ 1 & f_1(\tau_n;\hat{\lambda}_{1j}) & f_2(\tau_n;\hat{\lambda}_{1j}) & f_3(\tau_n;\hat{\lambda}_{2i}) \end{pmatrix}_{(n\times 4)},$$

where $\left[f_1(\boldsymbol{\tau};\hat{\boldsymbol{\lambda}}_1), f_2(\boldsymbol{\tau};\hat{\boldsymbol{\lambda}}_1), f_3(\boldsymbol{\tau};\hat{\boldsymbol{\lambda}}_2)\right]$ are the factor loadings of $(\beta_1, \beta_2, \beta_3)$[6] and $\boldsymbol{\epsilon}$ is a white noise vector.

We obtain a grid of estimators for the parameters of Equation (12.7) using the ordinary least squares method:

$$\hat{\boldsymbol{\beta}}_{ij} = \left(\mathbf{X}_{ij}{}^T \boldsymbol{\Sigma}_{\boldsymbol{\epsilon}}{}^{-1} \mathbf{X}_{ij}\right)^{-1} \left(\mathbf{X}_{ij}{}^T \boldsymbol{\Sigma}_{\boldsymbol{\epsilon}}{}^{-1} \mathbf{y}_{obs}\right) \qquad \forall\, i,j \tag{12.11}$$

The optimal solution $\hat{\boldsymbol{\Theta}}^* = (\hat{\boldsymbol{\beta}}_0^*, \hat{\boldsymbol{\beta}}_1^*, \hat{\boldsymbol{\beta}}_2^*, \hat{\boldsymbol{\beta}}_3^*, \hat{\lambda}_1^*; \hat{\lambda}_2^*)$ corresponds to those $\hat{\boldsymbol{\Theta}}_{ij}$ that minimize the sum of the squares of the residuals:

$$\hat{\boldsymbol{\Theta}}^* = \text{minimize}_{\hat{\boldsymbol{\Theta}}_{ij}} \left[\hat{\mathbf{y}}(\boldsymbol{\tau};\hat{\boldsymbol{\Theta}}_{ij}) - \mathbf{y}_{obs}(\boldsymbol{\tau})\right]^2 \qquad \forall\ i, j, \tag{12.12}$$

where $\hat{\mathbf{y}}(\boldsymbol{\tau};\hat{\boldsymbol{\Theta}}_{ij})$ is the term structure fitted with the parameters $\hat{\boldsymbol{\Theta}}_{ij}$.

[6]For simplicity we will refer to the matrix $\mathbf{X}_{(\hat{\lambda}_{1j};\hat{\lambda}_{2i})}$ indicating only the indexes i and j.

13

Portfolio Allocation

Yohan Chalabi
ETH Zürich
Zürich, Switzerland

Diethelm Würtz
ETH Zürich
Zürich, Switzerland

CONTENTS

13.1 Introduction

Nobel Laureate Harry H. Markowitz provided one of the first formulations of portfolio allocation as an optimization problem (Markowitz (1952)); since then, portfolio allocation has been widely studied and numerous models have been introduced, although the underlying concepts have remained the same. As summarized by Meucci (2009), portfolio allocation can be viewed as a method of maximizing the degree of satisfaction of the investor. For example, one investor might seek a portfolio that minimizes risk represented by a covariance estimator of the daily returns on assets, whereas another might consider risk in terms of the draw-down of wealth over a given time period.

This chapter takes a step-by-step approach to portfolio allocation by introducing the reader to portfolio optimization using R, thus allowing the reader to implement his or her own routines in the process. We deliberately do not use third-party packages so that users can more readily grasp the principles behind portfolio optimization using R. The examples are chosen to be sufficiently brief to be represented by a few lines of code, although general enough to be extended to more complex situations. We take care to introduce problems that require different types of solvers, so that the reader can extend the code snippets to meet their own needs. The packages used in this chapter are **Rglpk**, **quadprog**, **Rsolnp**, **DEoptiom**, and **robustbase**, which can be installed as follows:

```
> library(Rglpk)
> library(quadprog)
> library(Rsolnp)
> library(DEoptim)
> library(robustbase)
```

Before describing the details of R implementation, we first review in Section 13.2 the portfolio optimization problems that we will be using in this chapter. In Section 13.3, we introduce the dataset that is used in the R code snippets. Section 13.5 introduces typical portfolio problems, and Section 13.6 introduces two graphical approaches for comparing sets of feasible portfolios, which include the efficient frontier and the weighted return plot. Section 13.7 concludes the chapter and includes recommendations for third-party R packages that can be used to extend the portfolio problems presented in this chapter.

13.2 Optimization Problems in R

13.2.1 Introduction

We first review the method for solving optimization problems using R. The optimization field is wide, and optimization problems can be classified on the basis of whether or not a dedicated algorithm exists to solve the problem, and if so, on the type of algorithm that is used in the solution. In this section, we review the optimization problems required for solving the examples in this chapter.

Several algorithms and packages are available for solving a given optimization problem. For simplicity, in this chapter we consider one optimization solver for each type of optimization problem considered. Our selection criteria are that the package is readily available in R, it can model the constraint used in the examples, it is open source, and it is actively maintained. However, readers should bear in mind that many solver routines are available, and the selection of a routine for a given optimization problem should be carefully investigated. Types of solvers in R include (1) optimization routines that are available by default in the base environment. The general-purpose nonlinear optimization routines in R are `optim()` and `nlminb()`; however, these routines only support simple bound constraints and do not provide sufficient solvers for our portfolio examples. (2) Several third-party packages are available for implementation of different optimization algorithms. The packages may be available either in the R environment or in external libraries (the libraries may not be shipped with the R package for licensing reasons, or because they require external installation). (3) Optimization solvers in R may interface with a dedicated optimization platform; such platforms typically have their own modeling languages, such as the AMPL modeling language. Although an interface to an external platform does not provide a pure R approach, it offers access to a large set of both open-source and commercial solvers. Indeed, the main drawback of using an optimization routine is that each routine expresses the programming problem in terms of different input arguments, which requires the user to first understand how each interface functions.

The remainder of this section reviews linear, quadratic, and nonlinear optimization problems. The optimization R packages are presented, and wrapper functions are implemented so that each function has a common interface, thus easing implementation of the ensuing portfolio optimization problems.

13.2.2 Linear Programming

For $\mathbf{x} \in \mathbb{R}^n$, a set of vector variables subject to linear equality and inequality constraints, the linear programming problem (LP) can be formulated as

$$\begin{aligned} \underset{\mathbf{x}}{\text{minimize}} \quad & \mathbf{c}^\top \mathbf{x} \\ \text{subject to} \quad & \mathbf{A}_{\text{eq}}\,\mathbf{x} = \mathbf{a}_{\text{eq}}, \\ & \mathbf{A}\,\mathbf{x} \geq \mathbf{a}, \end{aligned} \tag{13.1}$$

where $\mathbf{A}_{\text{eq}}$ and $\mathbf{a}_{\text{eq}}$ are the matrix and vector coefficients, respectively, describing the equality linear constraints; $\mathbf{A}$ and $\mathbf{a}$ are the matrix and vector coefficients, respectively, describing the inequality linear constraints; and $\mathbf{c}$ is the vector of coefficients of the objective function.

At the time of writing, the R packages that provide linear programming solvers are, as reported in the Optimization and Mathematical Programming CRAN Task View Theussl (2013): `boot`, `clpAPI`, `cplexAPI`, `glpkAPI`, `limSolve`, `linprog`, `lpSolve`, `lpSolveAPI`, `quantreg`, `rcdd`, `Rcplex`, `Rglpk`, `Rmosek`, and `Rsymphony`. In this chapter, we use the `Rglpk` package; `Rglpk`, which was developed and is maintained by Kurt Hornik and Stefan Theussl (Hornik & Theussl (2012)), is a high-level interface of the GNU Linear Programming Kit (GLPK), which solves linear as well as mixed-integer linear programming (MILP) problems.

The arguments for the R function in the GLPK routine, `Rglpk_solve_LP()`, are

```
> args(Rglpk_solve_LP)
function (obj, mat, dir, rhs, bounds = NULL, types = NULL, max = FALSE,
    control = list(), ...)
NULL
```

where `obj` is the vector holding the linear coefficients of the objective function, `mat` is the general constraint matrix, `dir` describes the direction and types of inequalities, and `rhs` is the right-hand side vector of the constraints. The remaining arguments are not required for our application.

Following the formulation of the linear programming problem in Equation 13.1, our wrapper function becomes

```
> LP_solver <- function(c, cstr = list(), trace = FALSE) {
+
+     Aeq <- Reduce(rbind, cstr[names(cstr) %in% "Aeq"])
+     aeq <- Reduce(c, cstr[names(cstr) %in% "aeq"])
+     A <- Reduce(rbind, cstr[names(cstr) %in% "A"])
+     a <- Reduce(c, cstr[names(cstr) %in% "a"])
+
+     sol <- Rglpk_solve_LP(obj = c,
+                           mat = rbind(Aeq, A),
```

```
+                              dir = c(rep("==", nrow(Aeq)),
+                                      rep(">=", nrow(A))),
+                              rhs = c(aeq, a),
+                              verbose = trace)
+
+       status <- sol$status
+       solution <- if (status) rep(NA, length(c)) else sol$solution
+       list(solution = solution, status = status)
+ }
```

Here, the constraints are provided as a list object, where the components of the linear constraints are provided as the named entries, Aeq, A, aeq, and a. Note that the list can have several entries with the same name, a feature that is useful when implementing the portfolio constraints. The Reduce function merges all the entries of the cstr list that have the same name. After the constraints have been constructed, the optimization routine is called. The returned object of LP_solver() is a list with two elements: the first is the optimized solution, and the second is the status of the optimization routine if it has completed successfully. We use the same calling convention for the other optimization routines in this chapter.

13.2.3 Quadratic Programming

Compared to linear programming problems, quadratic programs (QP) contain a quadratic term ($\mathbf{x}^\top \mathbf{Q}\,\mathbf{x}$) in the objective function. The linear constraints $\mathbf{A}_{\mathrm{eq}}$, $\mathbf{A}$, $\mathbf{a}_{\mathrm{eq}}$, and $\mathbf{a}$ remain similar. The quadratic formulation is presented in a canonical form as

$$\begin{aligned} \underset{\mathbf{x}}{\text{minimize}} \quad & \mathbf{c}^\top \mathbf{x} + \mathbf{x}^\top \mathbf{Q}\,\mathbf{x} \\ \text{subject to} \quad & \mathbf{A}_{\mathrm{eq}}\,\mathbf{x} = \mathbf{a}_{\mathrm{eq}}, \\ & \mathbf{A}\,\mathbf{x} \geq \mathbf{a}. \end{aligned} \tag{13.2}$$

The R packages providing quadratic solvers are cplexAPI, kernlab, limSolve, LowRankQP, quadprog, Rcplex, and Rmosek. The package selected for use in this chapter is quadprog port (2013). The solve.QP() function implements the dual method of Goldfarb & Idnani (1982, 1983) for solving quadratic programming problems of the form $\min_{\mathbf{x}} -\mathbf{c}^\top \mathbf{x} + 1/2\,\mathbf{x}^\top \mathbf{Q}\,\mathbf{x}$ with the constraints $\mathbf{A}\,\mathbf{x} \geq \mathbf{a}$, where the arguments of solve.QP are

```
> args(solve.QP)
function (Dmat, dvec, Amat, bvec, meq = 0, factorized = FALSE)
NULL
```

Dmat is the quadratic matrix $\mathbf{Q}$, dvec is the linear part of $\mathbf{c}$ in the objective function, Amat defines the constraints matrix $\mathbf{A}$, and bvec is the vector holding the values of $\mathbf{a}$. The argument meq is used to specify how many of the first linear constraints should be considered equality constraints.

The canonical form used in the quadprog package is slightly different from that used in the linear approach. Therefore, a small wrapper is used to maintain a similar canonical form:

```
> QP_solver <- function(c, Q, cstr = list(), trace = FALSE) {
+
+     Aeq <- Reduce(rbind, cstr[names(cstr) %in% "Aeq"])
+     aeq <- Reduce(c, cstr[names(cstr) %in% "aeq"])
```

```
+     A <- Reduce(rbind, cstr[names(cstr) %in% "A"])
+     a <- Reduce(c, cstr[names(cstr) %in% "a"])
+
+     sol <- try(solve.QP(Dmat = Q,
+                         dvec = -2 * c,
+                         Amat = t(rbind(Aeq, A)),
+                         bvec = c(aeq, a),
+                         meq = nrow(Aeq)),
+               silent = TRUE)
+     if (trace) cat(sol)
+     if (inherits(sol, "try-error"))
+         list(solution = rep(NA, length(c)), status = 1)
+     else
+         list(solution = sol$solution, status = 0)
+ }
```

Note that the objective function defined in the previous snippet is equal to the canonical form in Equation (13.2) times a factor of 2. However, the solution remains the same as in the canonical formulation as the minimum of both objective functions is attained using the same set of parameter values.

13.2.4 Nonlinear Programming

The canonical form of the nonlinear programming (NLP) model is characterized by a nonlinear objective function, represented by the function f, which has as its argument the vectors of unknown variables $\mathbf{x}$:

$$\begin{aligned} \underset{\mathbf{x}}{\text{minimize}} \quad & f(\mathbf{x}) \\ \text{subject to} \quad & \mathbf{A}_{\text{eq}}\,\mathbf{x} = \mathbf{a}_{\text{eq}}, \\ & \mathbf{A}\,\mathbf{x} \geq \mathbf{a}, \\ & h_i^{\text{eq}}(\mathbf{x}) = 0, \\ & h_i(\mathbf{x}) \geq 0. \end{aligned} \tag{13.3}$$

The model contains both the linear ($\mathbf{A}_{\text{eq}}$, $\mathbf{a}_{\text{eq}}$) and ($\mathbf{A}$, $\mathbf{a}$), and nonlinear constraints h_i^{eq} and h_i, which are the equality and inequality constraints, respectively. At the time of writing, two R packages are available that can solve NLP problems: `Rdonlp2` and `Rsolnp`. In this chapter, we have selected the open-source package `Rsolnp`, developed by Ghalanos Ghalanos & Theussl (2012); `Rsolnp` is based on the SOLNP routine of Ye (1987). SOLNP implements the augmented Lagrange multiplier method with a sequential quadratic programming interior algorithm.

As before, we implement a small wrapper function to maintain a common interface:

```
> NLP_solver <- function(par, f, cstr = list(), trace = FALSE) {
+
+     Aeq <- Reduce(rbind, cstr[names(cstr) %in% "Aeq"])
+     aeq <- Reduce(c, cstr[names(cstr) %in% "aeq"])
+     A <- Reduce(rbind, cstr[names(cstr) %in% "A"])
+     a <- Reduce(c, cstr[names(cstr) %in% "a"])
+     heq <- Reduce(c, cstr[names(cstr) %in% "heq"])
+     h <- Reduce(c, cstr[names(cstr) %in% "h"])
+
```

```
+     leqfun <- c(function(par) c(Aeq %*% par), heq)
+     eqfun <- function(par)
+         unlist(lapply(leqfun, do.call, args = list(par)))
+     eqB <- c(aeq, rep(0, length(heq)))
+
+     lineqfun <- c(function(par) c(A %*% par), h)
+     ineqfun <- function(par)
+         unlist(lapply(lineqfun, do.call, args = list(par)))
+     ineqLB <- c(a, rep(0, length(h)))
+     ineqUB <- rep(Inf, length(ineqLB))
+
+     sol <- solnp(par = par,
+                  fun = f,
+                  eqfun = eqfun,
+                  eqB = eqB,
+                  ineqfun = ineqfun,
+                  ineqLB = ineqLB,
+                  ineqUB = ineqUB,
+                  control = list(trace = trace))
+
+     status <- sol$convergence
+     solution <- if (status) rep(NA, length(par)) else sol$pars
+     list(solution = solution, status = status)
+ }
```

Note that implementation of the canonical form used in Equation (13.3) requires more work than is required in the previous cases, as the constraints in `solnp()` must be directly expressed in terms of functions. Therefore, the linear constraints $\mathbf{A}$ and $\mathbf{a}$ must be converted to function constraints.

13.3 Data Sources

The first dataset used in this chapter is the `EuStockMarkets` dataset, obtained from the `datasets` package that is part of the standard R installation. `EuStockMarkets` consists of the daily closing prices of major European stock indices from 1991 to 1998, including the German DAX (Ibis), Swiss SMI, French CAC, and UK FTSE. The dataset is readily available and convenient to use, as there is no need to download the data from an external source. The first lines of the dataset are shown in the following code snippet:

```
> head(EuStockMarkets)
         DAX    SMI    CAC   FTSE
[1,] 1628.75 1678.1 1772.8 2443.6
[2,] 1613.63 1688.5 1750.5 2460.2
[3,] 1606.51 1678.6 1718.0 2448.2
[4,] 1621.04 1684.1 1708.1 2470.4
[5,] 1618.16 1686.6 1723.1 2484.7
[6,] 1610.61 1671.6 1714.3 2466.8
```

TABLE 13.1
Symbols of the NASDAQ indices and U.S. Treasury yields composing the dataset.

IXBK	NASDAQ Bank
NBI	NASDAQ Biotechnology
IXK	NASDAQ Computer
IXF	NASDAQ Financial 100
IXID	NASDAQ Industrial
IXIS	NASDAQ Insurance
IXUT	NASDAQ Telecommunications
IXTR	NASDAQ Transportation
FVX	U.S. Treasury yield 5 years
TYX	U.S. Treasury yield 30 years

Although the `EuStockMarkets` dataset is sufficient for the presentation of portfolio optimization problems in this chapter, we also would like to show how a larger dataset can be downloaded from a website, converted to an R object, and saved as a binary file for later use, as such examples are closer to real applications of portfolio optimization in R. Thus, the second dataset used in this chapter represents NASDAQ indices and U.S. treasury yields listed in Table 13.3 and in the following code snippet:

```
> id <- c("IXBK", "NBI", "IXK", "IXF", "IXID", "IXIS", "IXUT",
+ "IXTR", "FVX", "TYX")
```

The NASDAQ/Treasury historical dataset can be downloaded from the Yahoo! Finance website. At the time of writing, the data can be downloaded in a comma-separated value (csv) format, either manually for each financial index or using a small R function to perform the operations, from http://ichart.finance.yahoo.com/table.csv?s=XYZ, where “XYZ” represents a specific query (note that the URL may change at any time). The R function `read.csv()` reads the csv file and creates an R object; `read.csv` is a wrapper function within the general `read.table()` function, with the arguments set for csv files (see also Chapter 1):

```
> args(read.csv)
function (file, header = TRUE, sep = ",", quote = "\"", dec = ".",
    fill = TRUE, comment.char = "", ...)
NULL
```

In the next code snippet, we are using the three dots argument to pass further arguments to the underlying `read.table()` function. The `colClasses` argument specifies which object class R should transform in each column of the dataset. The first column is converted to a `Date` class, whereas the other six columns are converted to numerical vectors, according to

```
> downloadSymbol <- function(symbol) {
+     address <- "http://ichart.finance.yahoo.com/table.csv"
+     url <- paste(address, symbol, sep = "?s=^")
+     read.csv(url, colClasses = c("Date", rep("numeric", 6)))
+ }
```

As an example of the data input procedure, we show the first part of the IXBK index dataset. The dataset is organized in columns; the date is in the first column, followed by open, high, low, and closing prices, followed by the volume and the adjusted closing price,

```
> head(downloadSymbol("IXBK"))
        Date    Open    High     Low   Close Volume Adj.Close
1 2013-05-31 2162.60 2167.93 2144.64 2146.29      0   2146.29
2 2013-05-30 2154.76 2174.71 2154.60 2172.63      0   2172.63
3 2013-05-29 2154.69 2163.77 2150.68 2154.01      0   2154.01
4 2013-05-28 2167.72 2183.75 2158.48 2168.55      0   2168.55
5 2013-05-24 2127.13 2145.25 2122.67 2144.84      0   2144.84
6 2013-05-23 2119.05 2134.75 2118.47 2134.68      0   2134.68
```

To avoid downloading the data at every session, we aggregate the adjusted closing price of the time series into a `data.frame` and save it as a binary R object. The following code snippet downloads datasets for each symbol using the `lapply()` function, which returns a list with a `data.frame` for each symbol. We then use the `Reduce()` function to successively merge the elements of the list. Using the `merge()` function, which is similar to the database "join" operation, the data are merged according to their date stamps:

```
> lprices <- lapply(id, function(symbol) {
+     df <- downloadSymbol(symbol)[, c("Date", "Adj.Close")]
+     names(df) <- c("Date", symbol)
+     df
+ })
> prices <- Reduce(function(x, y)
+                      merge(x, y, all = TRUE, by = "Date"),
+                  lprices)
```

In the downloaded dataset, missing values are represented by NAs in the merge operation. In the examples in this chapter, data were selected for the period 2003–2006, which is easily accomplished as the first column of the dataset contains the Date argument. Note how we reset the row names of the `data.frame` object in the next code snippet.

```
> pos <- (prices$Date >= as.Date("2003-01-01") &
+         prices$Date < as.Date("2007-01-01"))
> prices <- prices[pos, ]
> isNA <- rowSums(sapply(prices[-1], function(x)  is.na(x)))
> isNA <- as.logical(isNA)
> prices <- prices[!isNA, ]
> rownames(prices) <- NULL
```

The obtained `data.frame` is saved as an R object in an "rds" binary file format, according to

```
> saveRDS(prices, file = "data.rds")
```

Binary files created using R have the advantage that their format is independent of the operating system. Thus, datasets can be saved and loaded in different operating systems that support R. Moreover, by default, the binary file is compressed for optimized storage, which can become important when dealing with large datasets. R binary files can be loaded using the `readRDS()` function as follows:

```
> prices <- readRDS("data.rds")
```

Note, however, that the working directory of the R session must either be the same as the directory in which the dataset is saved, or the appropriate path to the saved file must be recalled.

13.4 Portfolio Returns and Cumulative Performance

Using the downloaded dataset, we converted price data into values that can be modeled by a statistical distribution. The most common transformation yields arithmetic returns, defined at time t by

$$r_t = \frac{P_t - P_{t-1}}{P_{t-1}} = \frac{P_t}{P_{t-1}} - 1,$$

where P_t is the price of the financial instrument at time t. Based on the returns at time t, the aggregation of daily returns over period T is

$$r_T = \frac{P_T}{P_0} - 1 = \frac{P_T}{P_{T-1}}\frac{P_{T-1}}{P_{T-2}} \cdots \frac{P_1}{P_0} - 1 = \prod_{t=1}^{T} \frac{P_t}{P_{t-1}}.$$

Given a portfolio wealth at time t (W_t), which corresponds to the sum of the values of its components, $W_t = \sum_i P_{i,t}$, the portfolio return at time t (R_t) becomes

$$\begin{aligned}
R_t &= \frac{1}{W_{t-1}}(W_t - W_{t-1}) \\
&= \frac{1}{W_{t-1}}(P_{1,t} + P_{2,t} + \cdots + P_{N,t-1} - P_{1,t-1} - P_{2,t-1} - \cdots - P_{N,t-1}) \\
&= \frac{1}{W_{t-1}}[(P_{1,t} - P_{1,t-1}) + (P_{2,t} - P_{2,t-1}) + \cdots + (P_{N,t} - P_{N,t-1})) \\
&= \frac{1}{W_{t-1}}(P_{1,t-1}r_1 + P_{2,t-1}r_2 + \cdots + P_{2,t-1}r_2)
\end{aligned}$$

The portfolio return is therefore the sum of its component returns weighted by their allocation:

$$R_t = \sum_i^N \frac{P_{t-1}}{W_{t-1}} r_i = \sum_i^N w_i r_i.$$

These data allow calculation of the daily arithmetic returns for the dataset presented in the previous section, according to

```
> x <- sapply(prices[-1],
+ function(x) x[-1] / x[-length(x)] - 1)
```

Note that the `sapply()` function is used to apply an operation to each column of the object `prices` that is of class `data.frame`.

If, for some reason, the indices cannot be downloaded, the `EuStockMarkets` dataset can be used as an alternative data source:

```
> # x <- apply(EuStockMarkets, 2,
> # function(x) x[-1] / x[-length(x)] - 1)
```

Given the asset returns $\mathbf{r}$ and the portfolio weights $\mathbf{w}$, the cumulative performance at time t can be calculated as

$$W_t = W_0 \prod_i^t (1 + \mathbf{x}_t^\top \mathbf{w}),$$

where W_0 is the initial portfolio wealth. The cumulative portfolio performance can then be implemented as follows, where the default initial portfolio wealth is set at \$1000:

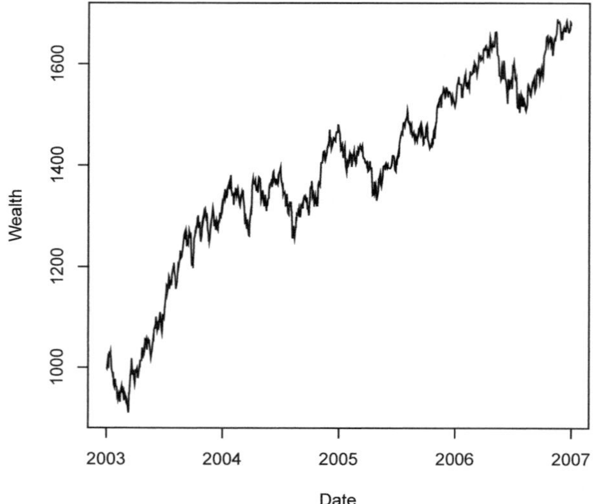

FIGURE 13.1
Cumulated performance of the equally weighted portfolio with an initial wealth of \$1000.

```
> pftPerf <- function(x, w, W0 = 1000) {
+     W0 * cumprod(c(1, 1 + x %*% w))
+ }
```

The method is illustrated by the following code snippet, which plots (Figure 13.1) the cumulative performance for the equally weighted portfolio.

```
> nc <- ncol(x)
> w <- rep(1/nc, nc)
> plot(prices$Date, pftPerf(x, w), type = "l",
+      main = "Portfolio Cumulated Performance",
+      xlab = "Date", ylab = "Wealth")
```

13.5 Portfolio Optimization in R

13.5.1 Introduction

In this chapter, we adopt the general formulation for portfolio optimization, which consists of minimization of a risk measure given a target reward and operational constraints. We first present the mean-variance portfolio for which the risk measure is represented by the covariance matrix of the portfolio. In the second example, the risk is measured as a conditional Value-at-Risk. The third example considers the minimization of drawdowns. In addition to risk measures, an important aspect of portfolio optimization is the formulation of constraints that reflect either operational or decisional constraints. Before presenting the portfolio models, we first review different types of constraints that are implemented in the form of an R function, and which correspond to the canonical optimization problems presented in Section 13.2.

Target Reward

The first constraint is set by the goal to achieve the target reward measure. The target reward constraint $\bar{r}$ is given by the weights $\mathbf{w}$ and average returns μ of each of the components, according to $\mu^\top \mathbf{w} = \bar{r}$. The constraint is given in terms of the matrix and vector coefficients $\mathbf{A}_{\text{eq}}$ and $\mathbf{a}_{\text{eq}}$, respectively, according to

```
> targetReturn <- function(x, target) {
+     list(Aeq = rbind(colMeans(x)), aeq = target)
+ }
```

Full Investment

The full investment constraint states that all capital must be invested in the portfolio. The full investment constraint corresponds to the condition in which the sum of all weights $\mathbf{w}$ is equal to 100%, where the weights correspond to portions of the capital allocated to a given component. We obtain the R function for the full investment constraint for the dataset x:

```
> fullInvest <- function(x) {
+     list(Aeq = matrix(1, nrow = 1, ncol = ncol(x)), aeq = 1)
+ }
```

Long Only

Another type of constraint is related to long only positions, which specify that we can only buy shares and therefore have only position-related weights in contrast to the case of short positions, in which the selling positions we do not own would be reflected as negative weights. Thus, the long only position becomes

```
> longOnly <- function(x) {
+     list(A = diag(1, ncol(x)), a = rep(0, ncol(x)))
+ }
```

Group Constraints

Group constraints, which are also common in portfolio optimization, can be derived from operational restrictions, in which one is obliged to have a minimum portion of shares in a class of instruments. We consider the following three constraints as examples:

- At most, 10% of the portfolio wealth is in the financial or bank sectors;
- At most, 30% of the portfolio wealth is in a single instrument;
- At least 10% of the portfolio wealth is in the U.S. Treasury bonds.

These group constraints can be implemented according to

```
> GroupBudget <- function() {
+
+     # max 10\% in financial and bank sector
+     A1 <- matrix(0, ncol = length(id), nrow = 1)
+     colnames(A1) <- id
```

```
+     A1[1, c("IXBK", "IXF")] <- -1
+     a1 <- -0.1
+
+     # max 30\% in a single instrument
+     A2 <- diag(-1, length(id))
+     a2 <- rep(-0.3, length(id))
+
+     # At least 10\% in trusery
+     A3 <- matrix(0, ncol = length(id), nrow = 1)
+     colnames(A3) <- id
+     A3[1, c("FVX", "TYX")] <- 1
+     a3 <- 0.1
+
+     list(A = rbind(A1, A2, A3), a = c(a1, a2, a3))
+ }
```

13.5.2 Mean–Variance Portfolio

The first case study demonstrates the solution of the mean-variance (MV) portfolio with long only constraints. Markowitz introduced the MV portfolio in 1953, paving the way for modern portfolio optimization. The optimization goal is to determine the best trade-off between return and risk, subject to a set of constraints. The MV model assumes the following: (1) the portfolio consists of both risk assets and risk-free assets; (2) the prices of the instruments are exogenous and given; (3) the investors, who are risk takers, do not influence the price of investments; (4) the returns follow stochastic processes that are elliptically distributed in probability space, meaning that a covariance matrix exists; (5) there are no transaction, tax, or other costs; (6) the markets for all assets are liquid; (7) the assets are infinitely divisible; and (8) full investment is required.

The risk measure developed by Markowitz is an asset-weighted covariance matrix, $\mathbf{w}^\top \Sigma \mathbf{w}$, where Σ is the covariance matrix and $\mathbf{w}$ are the portfolio weights. The optimization solution is obtained by setting a target portfolio return $\bar{r}$ and the long only and full investment conditions, such that

$$\begin{aligned} \underset{\mathbf{w}}{\text{minimize}} \quad & \mathbf{w}^\top \boldsymbol{\Sigma}\, \mathbf{w} && \text{Covariance Risk,} \\ \text{subject to} \quad & \mathbf{w}^\top \hat{\mu} = \bar{x}, && \text{Target Return} \\ & \mathbf{w}^\top \mathbf{1} = 1, && \text{Full Investment} \\ & \mathbf{w} \geq 0, && \text{Long Only Positions,} \end{aligned} \tag{13.4}$$

where $\hat{\mu}$ is the mean return vector of the assets. This problem cannot be solved analytically and therefore the solution requires optimization tools. The MV portfolio is represented as a quadratic programming problem (QP), and because a wrapper function has been defined for the QP solver, implementation of the MV portfolio is straightforward, according to

```
> MV_QP <- function(x, target, Sigma = cov(x), ...,
+                   cstr = c(fullInvest(x),
+                            targetReturn(x, target),
+                            longOnly(x), ...),
+                   trace = FALSE) {
+
+     # quadratic coefficients
```

```
+     size <- ncol(x)
+     c <- rep(0, size)
+     Q <- Sigma
+
+     # optimization
+     sol <- QP_solver(c, Q, cstr, trace)
+
+     # extract weights
+     weights <- sol$solution
+     names(weights) <- colnames(x)
+     weights
+ }
```

where `x` are the asset returns, `target` is the target portfolio return, and `Sigma` is the covariance estimate that is, by default, the classical estimator. The remaining arguments are used to pass the constraints of the optimization problem.

The function can be a tested in a variety of ways. For example, we can optimize the weights that minimize the risk measure using the equally weighted portfolio return (`mean(x)`) as the target return:

```
> w <- MV_QP(x, mean(x))
```

We can also verify that the full investment condition is fulfilled, using

```
> sum(w)
[1] 1
```

and display the optimized weights in a barplot (Figure 13.2, on the left) using the `barplot()` function. Testing of the other constrains is straightforward, using

```
> barplot(w, ylim = c(0, 1), las = 2,
+         main = "Test MV Implementation")
```

Another possible test of our implementation involves testing to see that when we set the target return as the smallest possible mean asset return, the entire allocation is assigned to this asset. In our case, the asset that has the smallest mean return is TYX. As shown in the next code snippet and in Figure 13.2, on the right, the optimized portfolio is fully invested in the asset that yields the smallest return:

```
> w <- MV_QP(x, min(colMeans(x)))
> barplot(w, ylim = c(0, 1), las = 2,
+         main = "Smallest Portfolio Return")
```

Note that the constraints are passed using the `cstr` argument or the three dots arguments. Thus, additional constraints can now be easily added to the MV portfolio. For example, the following code snippet illustrates how the budget constraints defined in the previous section can be added to the optimization problem.

```
> w <- MV_QP(x, mean(x), Sigma = covMcd(x)$cov, GroupBudget())
> barplot(w, ylim = c(0, 1), las = 2,
+         main = "MV with Budget Constraints")
```

The solution of the MV portfolio with group and budget constraints is displayed in Figure 13.4, on the left.

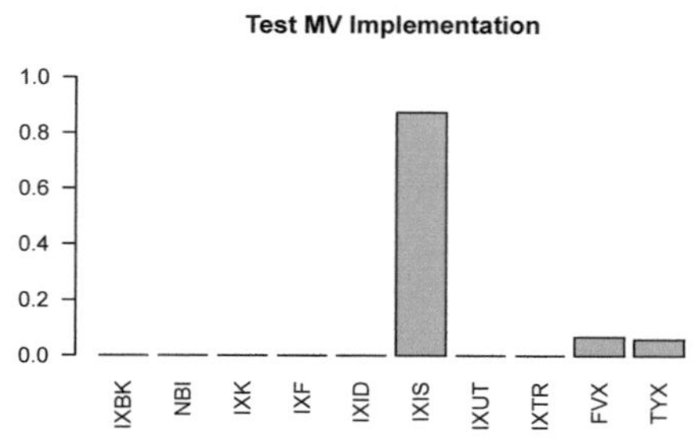

FIGURE 13.2
Solution of the MV portfolio with the equally weighted portfolio return as target return, on the left, and with the smallest mean return as target return, on the right.

13.5.3 Robust Mean–Variance Portfolio

A frequently cited drawback of the MV portfolio model is the use of the covariance matrix to estimate risk. The problem resides in the fact that the sample covariance estimator is sensitive to outliers. However, outliers frequently appear in financial data. Because we pass the covariance estimate as an argument to the `MV_QP()` function, we can easily modify the MV portfolio by using a robust covariance estimator; in this chapter, we use the `covMcd()` function from the `robustbase` (Rousseeuw et al. (2012)) package, which implements the fast minimum covariance determinant method given in Rousseeuw & Driessen (1999). Taking the equally weighted portfolio return as the target return, the robust covariance estimator can be used as in the following code snippet:

```
> w1 <- MV_QP(x, mean(x), Sigma = cov(x))
> barplot(w1, ylim = c(0, 1), las = 2,
+         main = "MV with classical cov")

> w2 <- MV_QP(x, mean(x), Sigma = covMcd(x)$cov)
> barplot(w2, ylim = c(0, 1), las = 2,
+         main = "MV with robust cov")
```

Figure 13.3 shows the weights obtained with both the classical and robust covariance estimators.

13.5.4 Minimum Variance Portfolio

Use of the mean return has been cited as another drawback of the MV portfolio model. It has been shown that the error in the mean estimator can suppress any benefits from optimization, in which case the optimized weights can produce an inappropriate portfolio on account of the error introduced by potential outliers that influence the estimation of the mean. In this regard, one might wish to consider only the minimum variance portfolio in the absence of a target return, which can be easily implemented by removing the target return condition from the constraints of the portfolio, using

```
> w <- MV_QP(x, cstr = c(fullInvest(x), longOnly(x)))
> barplot(w, ylim = c(0, 1), las = 2,
+         main = "Minimum Variance Portfolio")
```

The resultant plot is displayed in Figure 13.4, on the right.

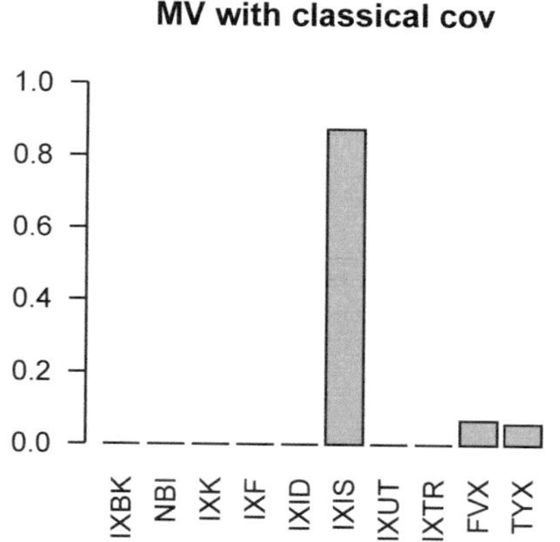

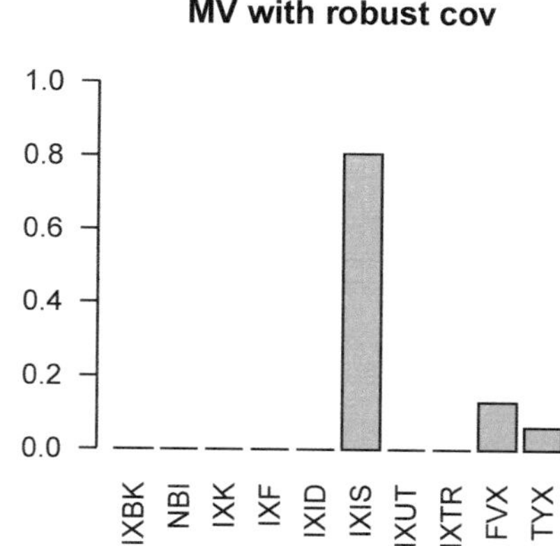

FIGURE 13.3
Weights of the MV portfolio with the classical and robust covariance estimator.

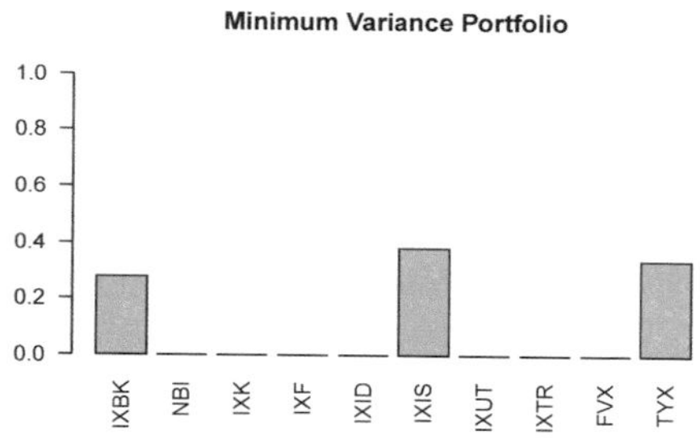

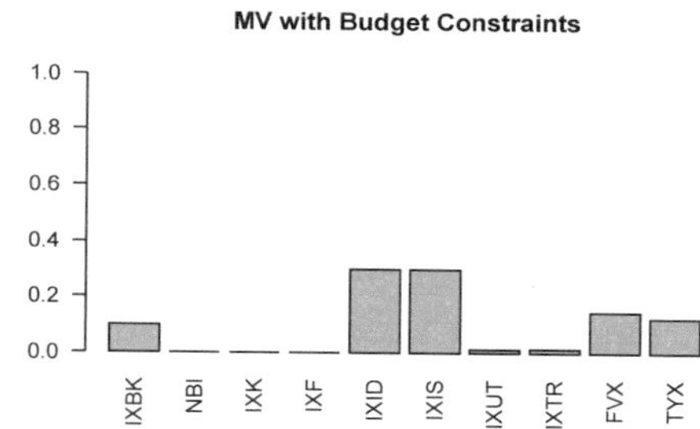

FIGURE 13.4
Solution of the MV portfolio with budget constraints, on the left, and of the minimum variance portfolio, on the right.

13.5.5 Conditional Value-at-Risk Portfolio

Alternative measures of risk have been introduced in addition to the covariance matrix. One popular risk measure is the so-called Value-at-Risk (VaR) measure. The VaR, which is widely used to measure the risk of loss of a portfolio, defines the loss threshold that might be exceeded for a given probability level. For example, a 5% VaR on a \$1000 portfolio indicates a 0.05 probability of a loss of \$1000 or more. In statistical terms, the VaR corresponds to a quantile of the portfolio distribution. For example, if P denotes the probability function of X,

$$\mathrm{VaR}_\alpha(L) = \inf\{l \in \mathbb{R} : P(l) \geq \alpha\}$$

gives the value-at-risk of the portfolio. Figure 13.5 illustrates the 5% value-at-risk of a probability density function.

The drawback of the VaR is that it does not give any information about the maximum loss that can be expected when the VaR has been exceeded, which is especially critical for financial returns that might exhibit a heavy-tailed distribution. The conditional value-at-risk

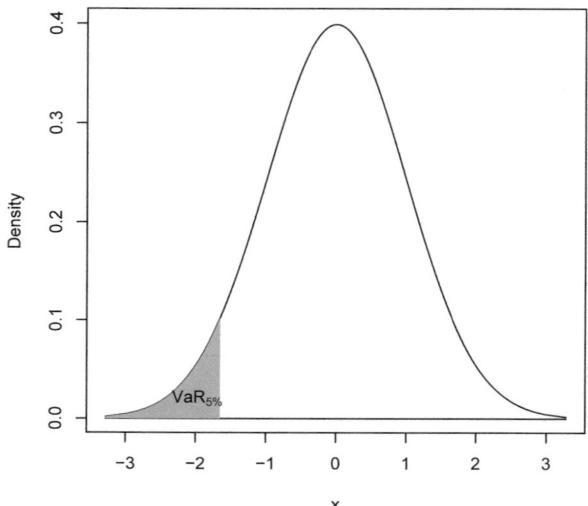

FIGURE 13.5
The 5% value-at-risk of a probability density function.

(CVaR), which was introduced as a modification of the VaR, consists of taking the weighted average between the VaR and the losses exceeding the VaR. The CVaR is the conditionally expected value of the loss under the condition that the VaR has been exceeded. The CVaR can be defined as

$$\text{CVaR}_\alpha = \frac{1}{1-\beta}\int_{f(w,x)}^{VaR_\alpha(\beta)} f(w,x)\,p(w,x)dx,$$

where VaR_α is the value-at-risk, $f(w,x)$ is a loss function defined for the portfolio allocation w and the value of the portfolio components, and p is the probability distribution of the portfolio with weights w. In contrast to the VaR, the CVaR is a coherent risk measure, as explained by Artzner et al. (1999). A coherent risk measure is one that satisfies the properties of translation invariance, subadditivity, monotonicity, and positive homogeneity. These properties can be understood in terms of risk measure ρ on some assets $\mathbf{X}$. Positive homogeneity corresponds to the fact that when the dataset is weighted by a given factor, the risk measure is also weighted according to $\rho(\lambda X) = \lambda\rho(X)$. The subadditivity condition specifies that for two assets X_1 and X_2, the risk associated with both assets combined is less than or equal to the sum of the individual risks, according to $\rho(X_1+X_2) \leq \rho(X_1)+\rho(X_2)$. The translation invariance specifies that adding a risk-free asset to the initial portfolio position decreases the risk measure, according to $\rho(X+\alpha r) = \rho(X) - \alpha$. The monotocity condition specifies that for all X and Y, where $X \leq Y$, the risk of X is less than or equal to the risk of Y, according to $\rho(X) \leq \rho(Y)$.

Initially, the portfolio measure based on the CVaR yields a nonlinear programming problem. However, Uryasev & Rockafellar (2001) showed how to transform the solution into a linear programming problem. The beauty of their approach was to move the VaR_α from the boundaries of the integral into the equation, thus adding it as a new parameter to the optimization problem. The transformed problem then becomes

$$F(X, VaR) = VaR + \frac{1}{1-L}\int_{f(w,x)\geq \text{VaR}} (f(w,x) - VaR)\, p(w,x)dx,$$

which is equivalent to the CVaR when F is minimized ($\min F = CVaR$). The next step is to note that, given the integral boundary conditions, the element $(f(w,x) - \alpha)$ must be positive. The problem can therefore be transformed to the more general problem in which only positive parts are considered:

$$VaR + \frac{1}{1-L}\int_x (f(w,x) - VaR)^+ p(w,x)dy. \tag{13.5}$$

Scenario-based portfolios represent a large class of portfolios for which the CVaR is an illustrative example. In practice, the scenarios can be obtained using actual portfolio returns or returns obtained by simulation, such as by Monte Carlo simulation. The integral can be approximated using the portfolio returns x_t as

$$\approx \frac{1}{J}\sum_{j=1}^{J}(f(w,x) - VaR)^+.$$

Linearization of Equation (13.5) can now be performed, where $(f(w,x_i) - VaR)^+$ is the nonlinear part of the equation. The linearization consists of adding the new variables $z_i = (f(w,x_i) - Var)$ to the objective function and adding new constraints to ensure that the optimized z_i^{opt} are equal to the intended $(f(w,x_i) - VaR)^+$, thus yielding the programming problem

$$\begin{array}{lll}
\underset{\mathbf{w}}{\text{minimize}} & \alpha + \dfrac{1}{(1-\beta)J}\displaystyle\sum_{i=1}^{J} z_j & \\
\text{subject to} & z_j \geq f(w, x_j) - \alpha, & \text{(Z1)} \\
 & z_j \geq 0, & \text{(Z2)} \\
 & \mathbf{w}^\top \hat{\mu} = \bar{x}, & \text{(R)} \\
 & \mathbf{w}^\top \mathbf{1} = 1, & \text{(F)} \\
 & \mathbf{w} \geq 0. & \text{(L)}
\end{array} \tag{13.6}$$

Addition of the target return, long only, and full investment conditions then gives the standard CVaR portfolio model. Linearization of the CVaR portfolio consists of the introduction of $J+1$ additional variables (VaR_α and z_j) and $2J$ linear constraints to the optimization problem. Using the linear programming formulation defined in Equation (13.1), we obtain the objective coefficients

$$c = \begin{bmatrix} w_1 \\ w_2 \\ \vdots \\ w_n \\ \\ \alpha \\ \\ z_1 \\ z_2 \\ \vdots \\ z_j \end{bmatrix} \begin{matrix} \left.\begin{matrix} \\ \\ \\ \\ \end{matrix}\right\} N \\ \\ \}1\,. \\ \\ \left.\begin{matrix} \\ \\ \\ \\ \end{matrix}\right\} J \end{matrix}$$

The number of linear constraints in the CVaR portfolio model are therefore becoming sensitively larger than those in the other models described thus far. Because the number of unknown vectors has been increased and the CVaR problem has been linearized, it is now necessary to express the full investment, long only, and other constraints. The equality linear constraints (F and R in Equation (13.6)) for the full investment and target return become, in terms of the new unknown vectors,

$$A_{\text{eq}} = \left[\begin{array}{cccc c cccc} \multicolumn{4}{c}{\overbrace{}^{N}} & \overbrace{}^{1} & \multicolumn{4}{c}{\overbrace{}^{J}} \\ 1 & 1 & \cdots & 1 & 0 & 0 & 0 & \cdots & 0 \\ w_1 & w_2 & \cdots & w_n & 0 & 0 & 0 & \cdots & 0 \end{array} \right], \quad a_{\text{eq}} = \left[\begin{array}{c} 1 \\ \bar{r} \end{array} \right] \begin{array}{c} \text{(F)} \\ \text{(R)} \end{array}.$$

The inequality linear constraints (Z1 and Z2 in Equation (13.6)), with the long only (L) constraints, become

$$A = \left[\begin{array}{cccc|c|cccc} \multicolumn{4}{c}{\overbrace{\phantom{x_{11} \; x_{12} \; \cdots \; x_{1n}}}^{N}} & \overbrace{}^{1} & \multicolumn{4}{c}{\overbrace{}^{J}} \\ x_{11} & x_{12} & \cdots & x_{1n} & 1 & 1 & & & \\ x_{21} & x_{22} & \cdots & x_{2n} & 1 & & 1 & & \\ \vdots & \vdots & \ddots & \vdots & \vdots & & & \ddots & \\ x_{j1} & x_{j2} & \cdots & x_{jn} & 1 & & & & 1 \\ \hline & & & & & 1 & & & \\ & & & & & & 1 & & \\ & & & & & & & \ddots & \\ & & & & & & & & 1 \\ \hline 1 & & & & & & & & \\ & 1 & & & & & & & \\ & & \ddots & & & & & & \\ & & & 1 & & & & & \end{array} \right], \quad a = \left[\begin{array}{c} \\ \\ \\ \\ \hline \\ \\ \\ \\ \hline \\ \\ \\ \\ \end{array} \right] \begin{array}{l} \left.\vphantom{\begin{array}{c} \\ \\ \\ \\ \end{array}}\right\} J \quad \text{(Z1)} \\ \left.\vphantom{\begin{array}{c} \\ \\ \\ \\ \end{array}}\right\} J \quad \text{(Z1)}, \\ \left.\vphantom{\begin{array}{c} \\ \\ \\ \\ \end{array}}\right\} N \quad \text{(L)} \end{array}$$

where missing entries are equal to zero.

When the integral in Equation (13.5) is approximated using a simulation approach, the number of optimized variables and the size of the matrix for the linear constraints can become very large. However, the constraint matrix is mainly filled with zeros, and can be represented by sparse matrices. A common approach for storage of sparse matrices is the triplet representation, in which each non-zero value is stored with its row and column indices.

Given the matrix representation, the typical linear algebraic routine can be implemented to take advantage of the new representation. The recommended R packages for sparse matrices are `Matrix` (Bates & Maechler (2012)) and the alternative package `slam` (Hornik et al. (2013)), both of which are supported by the linear programming package `Rglpk` used in this chapter.

The following code snippet implements the linearized CVaR portfolio problem (note the use of sparse matrices for constructing the constraint matrix):

```
> CVaR_LP <- function(x, target, alpha = 0.95, ...,
+                     cstr = c(fullInvest(x),
+                              targetReturn(x, target),
+                              longOnly(x)),
+                     trace = FALSE) {
+
```

```
+     # number of scenarios
+     J <- nrow(x)
+
+     # number of assets
+     size = ncol(x)
+
+     # objective coefficients
+     c_weights <- rep(0, size)
+     c_VaR <- 1
+     c_Scenarios <- rep(1 / ((1 - alpha) * J), J)
+     c <- c(c_weights, c_VaR, c_Scenarios)
+
+     # extract values from constraint to extend them
+     # with CVaR constraints
+     Aeq <- Reduce(rbind, cstr[names(cstr) %in% "Aeq"])
+     aeq <- Reduce(c, cstr[names(cstr) %in% "aeq"])
+     A <- Reduce(rbind, cstr[names(cstr) %in% "A"])
+     a <- Reduce(c, cstr[names(cstr) %in% "a"])
+
+     # build first two blocks of the constraint matrix
+     M1 <- cbind(Aeq, simple_triplet_zero_matrix(nrow(Aeq), J + 1))
+     M2 <- cbind(A, simple_triplet_zero_matrix(nrow(A), J + 1))
+
+     # identity matrix and vector of zeros
+     I <- simple_triplet_diag_matrix(1, J)
+
+     # block CVaR constraint (y x + alpha + z_j >= 0)
+     M3 <- cbind(x, rep(1, J), I)
+
+     # block CVaR constraint (z_j >= 0)
+     M4 <- cbind(simple_triplet_zero_matrix(J, size + 1), I)
+
+     # vector of zeros used for the rhs of M3 and M4
+     zeros <- rep(0, J)
+
+     # combine constraints
+     cstr <- list(Aeq = M1,
+                  aeq = aeq,
+                  A = rbind(M2, M3, M4),
+                  a = c(a, zeros, zeros))
+
+     # optimization
+     sol <- LP_solver(c, cstr, trace = trace)
+
+     # extract weights
+     weights <- sol$solution[1:size]
+     names(weights) <- colnames(x)
+
+     # extract VaR and CVaR
+     VaR <- sol$solution[size + 1]
+     CVaR <- c(c %*% sol$solution)
```

```
+     attr(weights, "risk") <- c(VaR = VaR, CVaR = CVaR)
+
+     weights
+ }
```

In the next code snippet, we optimize the weights that minimize the CVaR measure with $\alpha = 0.05$ using the equally weighted portfolio return (`mean(x)`) as the target return:

```
> round(CVaR_LP(x, mean(x)), 3)
 IXBK   NBI   IXK   IXF  IXID  IXIS  IXUT  IXTR   FVX   TYX
0.050 0.000 0.000 0.000 0.026 0.822 0.000 0.000 0.051 0.052
attr(,"risk")
       VaR       CVaR
0.01081098 0.01363397
```

The `CVaR_LP` returns the optimized weights together with the estimated VaR and CVaR.

13.5.6 Minimum Drawdown Portfolio

Only linear constraints have been introduced thus far. However, in some instances, portfolio allocation models may require nonlinear constraints. For example, an investor might be interested in minimizing the maximum drawdown of his portfolio. The drawdown rate of financial instruments at time t corresponds to the rate between the value of the instrument at time t and the latest historical peak:

$$D(t) = \left(P_t - \max_{i\in(1,t)} P_i\right) / \max_{i\in(1,t)} P_i.$$

Historical peaks can be calculated in R using the cumulative maximum function `cummax()`; thus, the cumulative peaks for the IXBK index can obtained using the following code snippet and result is displayed in Figure 13.6, on the left:

```
> plot(prices$Date, cummax(prices$IXBK), type = "l",
+      xlab = "Date", ylab = "Cumulative Maximum",
+      main = "Historical Peaks")
```

Because the returns have already been determined in the previous portfolio optimization problem, the following code snippet implements the drawdown rate from these returns:

```
> drawdown <- function(x) {
+      value <- cumprod(c(1, 1 + x))
+      cummaxValue <- cummax(value)
+      (value - cummaxValue) / cummaxValue
+ }
```

Figure 13.6, on the right, illustrates the drawdown rate for the IXBK index and was created with the next code snippet:

```
> plot(prices$Date, drawdown(x[, "IXBK"]), type = "l",
+      xlab = "Date", ylab = "Drawdowns of IXBK",
+      main = "Max Drawdowns")
```

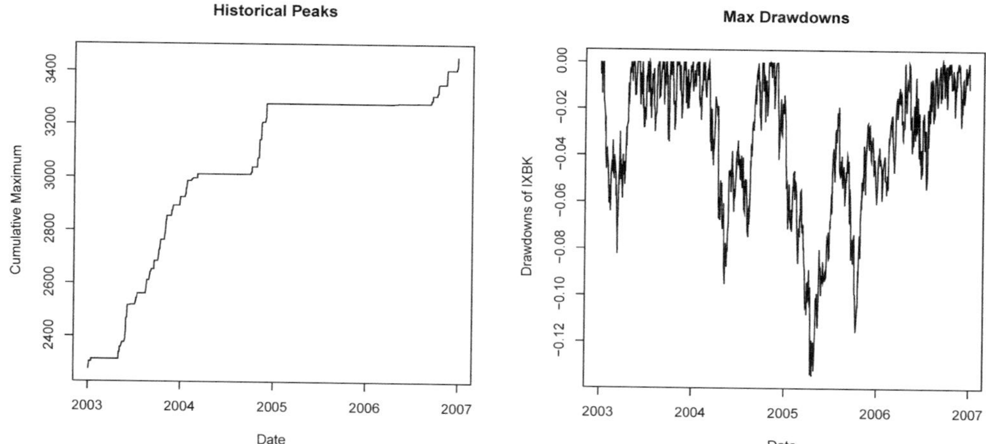

FIGURE 13.6
Historial peaks of the IXBK index, on the left, and drawdown rates for the IXBK index, on the right.

13.6 Display Results

13.6.1 Efficient Frontier

The usual approach for comparing the feasibility of different portfolios with a given set of assets is the so-called efficient frontier approach, which consists of comparing reward measures and risk measures for different feasible sets of asset weightings.

The following example shows, step-by-step, the method to construct the efficient frontier for an MV portfolio. A straightforward approach is to set a target risk measure and maximize the reward measure. However, in Section 13.5.2, we have implemented the MV portfolio as a quadratic programming problem in which the risk measure is minimized for a given level of reward. Given this formulation, we can minimize the variance of the portfolio for the range of feasible returns of the portfolio. When considering long only positions, the feasible portfolio returns range from the smallest to the largest returns of the portfolio constituents. We set the range of feasible returns of the portfolio as

```
> mu <- apply(x, 2, mean)
> reward <- seq(from = min(mu), to = max(mu), length.out = 300)
> sigma <- apply(x, 2, sd)
```

The respective MV portfolios can now be calculated over different return steps. In this example we selected 300 steps so as to achieve a nicely smoothed efficient frontier. Using the function we implemented in Section 13.5.2, we obtain

```
> Sigma <- cov(x)
> riskCov <- sapply(reward, function(targetReturn) {
+     w <- MV_QP(x, targetReturn, Sigma)
+     sd(c(x %*% w))
+ })
```

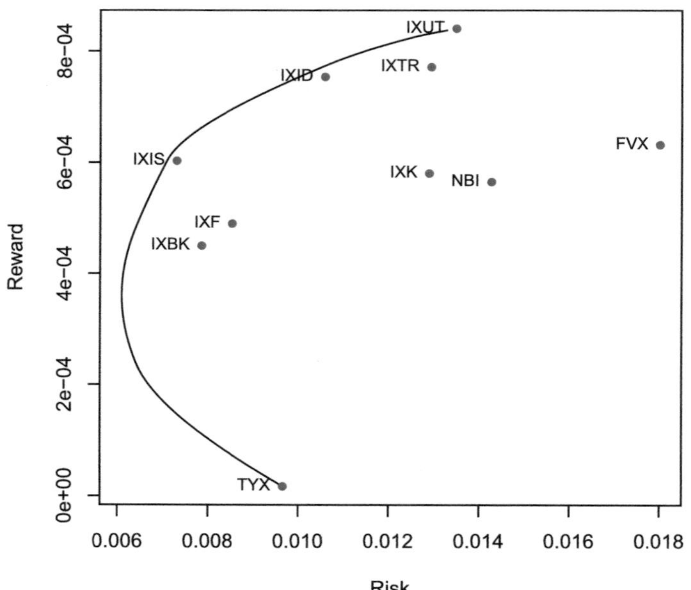

FIGURE 13.7
Efficient frontier of the MV portfolio.

Two points are noteworthy in the previous R snippet. First, the covariance matrix was precalculated and passed to the quadratic optimization problem; this is important from a computational point of view, as otherwise the covariance matrix would be recalculated 300 times, which would increase the processing time. Second, we returned the standard deviation, as this is the measure we have selected for the efficient frontier; this is important if one wishes to compare the efficient frontiers of portfolios optimized with different types of covariance matrices.

The efficient frontier can now be plotted together with the locations of the individual asset returns, using

```
> xlim <- range(c(sigma, riskCov), na.rm = TRUE)
> ylim <- range(mu)
> plot(riskCov, reward, type = "l", xlim = xlim, ylim = ylim,
+      xlab = "Risk", ylab = "Reward",
+      main = "Efficient Frontier")
> points(sigma, mu, col = "steelblue", pch = 19, cex = 0.8)
> text(sigma, mu, labels = colnames(x), pos = 2, cex = 0.8)
```

13.6.2 Weighted Return Plots

In addition to the efficient frontier, the weighted return plot is another common approach for comparing the weights of feasible portfolio sets. The principle of the weighted return plot is to compare the weight diversifications of the different target reward or risk measures.

Weight plots can constructed in R using the `barplot()` function. The matrix weights calculated in the previous section can be passed as the first argument, and each set of weights is then represented by stacked sub-bars, which together make up a single histogram bar. The following code snippet implements a simple weighted return plot:

```
> weightbarplot <- function(weights,
+                           title = "Weighted Return Plot") {
+
+     # color palette
+     size <- nrow(weights)
+     col <- gray(seq(size) / (size + 1))
+
+     # Bar plot
+      len <- ncol(weights)
+     h <- barplot(weights, space = 0, border = col,  col = col,
+                  xlim = c(0, len * 1.2), main = title)
+     idx <- seq(1, len, length.out = 5)
+
+     # Reward label
+     mtext("Reward Target", side = 1, line = 2, at = mean(h))
+     axis(1, at = h[idx], labels = signif(reward[idx], 2),
+          cex = 0.6)
+
+     # Weight label
+     mtext("Weight", side = 2, line = 2)
+
+     # legend
+     legend("topright", legend = rownames(weights), bty = "n",
+            fill = col)
+ }
```

The most important part of the code is the call to the `barplot()` function and the passing of the weights matrix as the first argument. The remainder of the code only adds cosmetic improvements to the default `barplot()` settings. Note that the `gray` function generates a palette of gray levels, based on the number of assets in the portfolio basket. The use of a consistent and pleasing palette (and one that represents a corporation's color palette) is critical for the generation of effective and professional reports.

Given that the R function has been implemented to construct the weighted return plot, we can compare, for example, the weighted diversification of the MV portfolio implemented in Section 13.5.2 using the classical and robust covariance estimators. The following two snippets estimate the set of weights along the range of feasible target returns. The weighted return of the MV portfolio with the classical covariance estimator is calculated as

```
> Sigma <- cov(x)
> weightsCov <- sapply(reward, function(targetReturn) {
+     MV_QP(x, targetReturn, Sigma)
+ })
> weightbarplot(weightsCov,
+               title = "Weighted Return Plot (Sample Cov)")
```

and the one with the robust covariance estimator is

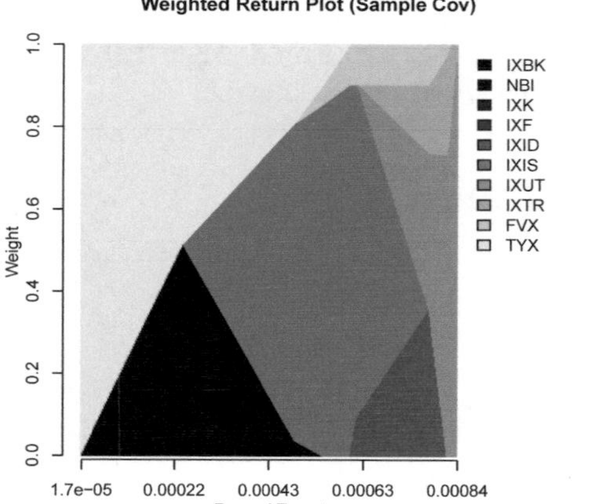

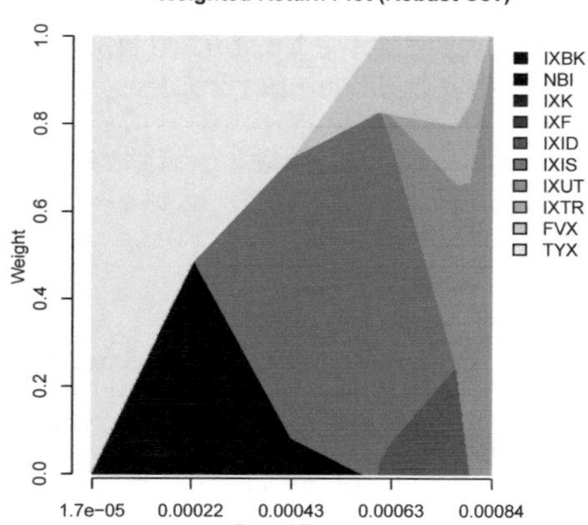

FIGURE 13.8
Weighted return plots of the MV portfolio using the sample and robust covariance estimator.

```
> SigmaRob <- covMcd(x)$cov
> weightsRob <- sapply(reward, function(targetReturn) {
+     MV_QP(x, targetReturn, SigmaRob)
+ })
> weightbarplot(weightsRob,
+               title = "Weighted Return Plot (Robust Cov)")
```

The resultant graphics are displayed in Figure 13.8.

13.7 Conclusion

This chapter described the basic elements of portfolio optimization for problems consisting of minimizing a risk measure under a given set of constraints. The examples were constructed in such a way that they can be extended to more advanced and complex situations. However, the reader should bear in mind that a large number of portfolio models exist that have not been described in this chapter. For example, the extension of the Black–Litterman approach by Meucci, known as the entropy polling approach, is an interesting method for combining information extracted from historical data with forecasts from analysts. Moreover, other types of measures are available, such as the divergence measure from information theory, as introduced in Chalabi (2012).

Finally, it is important to point out that the choice of optimization routine can have an important impact on the optimized weights. We encourage practitioners to compare and choose optimization algorithms that are most appropriate for their portfolio problems. In this regard, an interesting approach to the comparison of several optimization routines that interface with R is by using the programming language AMPL, which offers a common interface for a large set of free and commercial solvers.

Good sources of information for portfolio optimization in R are the works of Würtz et al. (2009) from the Rmetrics Association, and the work of Pfaff (2012), in which readers are guided in a step-by-step approach to portfolio solvers in R; the works are accompanied by the packages `fPortfolio` and `FRAPO`, respectively.

In addition, `BLCOP`, see Gochez (2011), is an implementation of the Black–Litterman model and Meucci's copula opinion pooling framework; `PerformanceAnalytics` (Carl & Peterson (2013)) provides numerous econometric tools for performance and risk analysis; the package `backtest` Enos, Kane, with contributions from Kyle Campbell, Gerlanc, Schwartz, Suo, Colin, & Zhao (2012) provide facilities for exploring portfolio-based conjectures about financial instruments (stocks, bonds, swaps, options, etc.); `crp.CSFP` (Jakob et al. (2013)) implements the program CreditRisk+ (see Boston (1997)); `parma`, see Ghalanos (2013), implements portfolio allocation and risk management applications; `portfolioSim`, see Enos, Kane & with contributions from Kyle Campbell (2012b), is a framework for simulating equity strategies; `portfolio` (Enos, Kane, with contributions from Daniel Gerlanc & Campbell (2012)) provides R classes for analyzing and implementing equity portfolios; `rportfolios` (Novomestky (2012)) offers functions to generate random portfolios; `stockPortfolio`, see Diez & Christou (2012), can be used to build stock models and analyze stock portfolios; and `tawny`, see Rowe (2013), applies random matrix theory and the shrinkage estimator to portfolio problems.

Part IV

Non-Life Insurance

14

General Insurance Pricing

Jean-Philippe Boucher
Université du Québec à Montréal
Montréal, Québec, Canada

Arthur Charpentier
Université du Québec à Montréal
Montréal, Québec, Canada

CONTENTS

In this chapter, we will discuss the use of Generalized Linear Models in (a priori) motor ratemaking. The goal is to propose a premium that an insurance company should charge a client, for a yearly contract, based on a series of characteristics (of the driver, such as the age or the region, or of the car, such as the power, the make, or the type of gas). Those models are described in Kass et al. (2008), Frees (2009), de Jong & Zeller (2008), and Ohlsson & Johansson (2010).

14.1 Introduction and Motivation

14.1.1 Collective Model in General Insurance

A standard premium principle (as least from a theoretical point of view) is the expected value principle: the premium associated to some (annual) risk S is $\pi(S) = (1+\alpha)\mathbb{E}(S)$, where $\alpha > 0$ denotes some loading, and where S is the annual random loss. Let (N_t) be the count process that denotes the number of claims occurred during period $[0,t]$. Let (Y_i) denote the amount of the ith claim. Then the total loss over period $[0,t]$ is

$$S_t = \sum_{i=1}^{N_t} Y_i \text{ with } S_t = 0 \text{ if } N_t = 0.$$

In the case where $\alpha = 0$, this premium is called the *pure premium*. Charging this premium to the insured is justified by the law of large numbers, if we assume losses among the insured to be independent and identically distributed. But from ruin theory, it is known that ruin will be inevitable whenever there is no loading. But the first step is to compute $\mathbb{E}(S)$, where $S = S_1$ is the annual total charge.

If N_1 (which will be denoted N in this chapter) and $Y_1, \cdots, Y_n \cdots$ are independent; and if losses Y_i are i.i.d., then

$$\pi = \mathbb{E}(S) = \mathbb{E}(N) \cdot \mathbb{E}(Y).$$

The (annual) pure premium is the the product of two terms:

- (Annual) claims frequency $\mathbb{E}(N)$,
- Average cost of individual claims $\mathbb{E}(Y)$.

14.1.2 Pure Premium in a Heterogenous Context

In a more realistic world, we should probably take heterogeneity into account. Consider the case where heterogeneity can be observed, through some binary random variable Z (say *low* and *large* risk for, respectively, 50% of the insured). Assume that N has a binomial distribution, with probability either 10% or 20% depending on Z and the loss is deterministic, $Y = 100$. Then, two choices can be made:

- The insurance company charges the same premium for all the insured, $\pi = \mathbb{E}(S) = 15$.
- The insurance company charges a premium taking into account heterogeneity, $\pi(z) = \mathbb{E}(S|Z=z)$ that will be either 10 or 20, depending on z.

From an economic perspective, this might be dangerous if there are two companies with different rating policies. Insured with lower risk will buy insurance from the second company (that charges 10 for an expected risk of 10) while insured with higher risk will buy insurance from the first one (that charges 15 for an expected risk of 20). Thus, the insurance company that does not sell differentiated premiums, taking into account individual risk characteristics, will not survive in a competitive market. Thus, if Z is the (observed) heterogeneity variable, an insurance company should charge

$$\pi(z) = \mathbb{E}(S|Z = z) = \mathbb{E}(N|Z = z) \cdot \mathbb{E}(Y|Z = z).$$

But assuming that heterogeneity is fully observable is a strong assumption. Instead, the insurance company might have available information, summarized in a vector $\boldsymbol{X}$ of information, related to the policyholder, to the car (in the context of car insurance), etc. And some variates can be used to get a good proxy of the unobservable latent factor Z. In that case, with non-partially observable heterogeneity, an insurance company should charge

$$\pi(\boldsymbol{x}) = \mathbb{E}(S|\boldsymbol{X} = \boldsymbol{x}) = \mathbb{E}(N|\boldsymbol{X} = \boldsymbol{x}) \cdot \mathbb{E}(Y|\boldsymbol{X} = \boldsymbol{x}).$$

Thus, the goal in this chapter will be to propose predictive models to estimate $\mathbb{E}(N|\boldsymbol{X} = \boldsymbol{x})$, the annualized claims frequency for some insured with characteristics $\boldsymbol{x}$, and $\mathbb{E}(Y|\boldsymbol{X} = \boldsymbol{x})$, the average cost of (individual) accidents, claimed by the insured with characteristics $\boldsymbol{x}$.

14.1.3 Dataset

The first dataset, `CONTRACTS`, contains contract and client information from a French insurance company, related to some motor insurance portfolio

- Att. 1 (numeric) `ID`, contract number (used to link with the claims dataset)
- Att. 2 (numeric) `NB`, number of claims during the exposure period
- Att. 3 (numeric) `EXPOSURE`, exposure, in years
- Att. 4 (factor) `POWER`, power of the car (ordered categorical)
- Att. 5 (numeric) `AGECAR`, age of car in years
- Att. 6 (numeric) `AGEDRIVER`, age of driver in years (in France, people can drive a car at 18)
- Att. 7 (factor) `BRAND`, brand of the car, `A`: Renaut Nissan, and Citroën; `B`: Volkswagen, Audi, Skoda, and Seat; `C` Opel, General Motors, and Ford; `D` Fiat; `E` Mercedes Chrysler, and BMW; `F` Japanese (except Nissan) and Korean; `G` other.
- Att. 8 (factor) `GAZ`, with `diesel` or `regular`,
- Att. 9 (factor) `REGION`, with different regions, in France (based on a standard French classification)
- Att. 10 (numeric) `DENSITY`, density of inhabitants (number of inhabitants per square kilometer) in the city the driver of the car lives in

Remark 14.1 *In the French insurance market, there is a compulsory no-claim bonus system (see Lemaire (1984) for a description of the system). Insurance pricing in France is then a mix between* a priori *ratemaking, discussed in this chapter, and* a posteriori *ratemaking, which will be discussed in the next chapter. But one should keep in mind that because of this no-claim bonus system, or malus in the case where the insured claims a loss, there might be (financial) incentives not to declare some claims, if the associated loss is smaller than the malus the insured will have in the future.*

As in Chapter 4, it might be more convenient to work with categorized variables,

Att. 5 (factor) `agecar`, `[0,1)`, `[1,4)`, `[4,15)`, and `[15,Inf)`

Att. 6 (factor) `agedriver`, `(17,22]`, `(22,26]`, `(26,42]`, `(42,74]`, and `(74,Inf]`

Att. 10 (factor) `density`, `[0-40)`, `[40,200)`, `[200,500)`, `[500,4500)`, and `[4500,Inf)`

```
> CONTRACTS.f <- CONTRACTS
> CONTRACTS.f$AGEDRIVER <- cut(CONTRACTS$AGEDRIVER,c(17,22,26,42,74,Inf))
> CONTRACTS.f$AGECAR <- cut(CONTRACTS$AGECAR,c(0,1,4,15,Inf),
+ include.lowest = TRUE)
> CONTRACTS.f$DENSITY <- cut(CONTRACTS$DENSITY,c(0,40,200,500,4500,Inf),
+ include.lowest = TRUE)
```

The second dataset, `CLAIMS`, contains claims information, from the same company

Att. 1 (numeric) `ID`, contract number (used to link with the contract dataset)

Att. 2 (numeric) `INDEMNITY`, cost of the claim, seen as at a recent date.

14.1.4 Structure of the Chapter and References

In Sections 14.2 to 14.4, we will discuss modeling of claims numbers, and see how to estimate $\mathbb{E}(N|\boldsymbol{X}=\boldsymbol{x})$ using General Linear Models. In Section 14.2, the Poisson regression will be introduced and discussed, and then extensions will be considered in Section 14.4 (with the Negative Binomial regression, as well as Zero Inflated models). In Section 14.5, the goal will be to model individual losses, using standard models, for example, gamma, log-normal, as well as mixtures in Section 14.6, especially to take into account very large claims. Finally, Tweedie regressions will be discussed in Section 14.7

The theory used in this section can be found in McCullagh & Nelder (1989) for an introduction to Generalized Linear Models, and in Hastie & Tibshirani (1990) for an introduction to Generalized Additive Models. More specific models for counts are described in Cameron & Trivedi (1998) and Hilbe (2011). Computational aspects with R can be found in Faraway (2006). Finally, actuarial applications of these models can be found in Denuit et al. (2007), de Jong & Zeller (2008), and Frees (2009). As claimed in Meyers & Cummings (2009), *goodness of fit* is not the same as *goodness of lift*: the goal here is not to predict observed losses, but to derive accurate estimates of the expected value of losses (which is unobserved) that we will call *price* of the insurance contract (see Goovaerts et al. (1984) for a discussion on premium principles). While goodness-of-fit measures are useful in the estimation of statistical models (see for instance Chapter 4 of this book), it will be less interesting in actuarial pricing.

14.2 Claims Frequency and Log-Poisson Regression

14.2.1 Annualized Claims Frequency

Assume that claims occurrence, for an insured, is driven by a homogeneous Poisson process, with intensity λ, so that the process of claims of occurrence has independent increments, that the number of claims observed during time interval $[t, t+h]$ has a Poisson distribution

$\mathcal{P}(\lambda \cdot h)$. This model will allow us to link yearly frequency and observed frequency on a given period of exposure. For policy holder i

- the annualized number of claims N_i over the period $[0,1]$ is (usually) an unobserved variable,
- the actual number of claims in the database Y_i occurred during period $[0, E_i]$, where E_i is the exposure.

In some sense, we are dealing here with censored data, as we were not able to observe the contract over a full year. Without any explanatory variable, the average annualized frequency and its empirical variance are, respectively,

$$m_N = \frac{\sum_{i=1}^n Y_i}{\sum_{i=1}^n E_i} \text{ and } S_N^2 = \frac{\sum_{i=1}^n [Y_i - m_N \cdot E_i]^2}{\sum_{i=1}^n E_i},$$

```
> vY <- CONTRACTS.f$NB
> vE <- CONTRACTS.f$EXPOSURE
> m <- sum(vY)/sum(vE)
> v <- sum((vY-m*vE)^2)/sum(vE)
> cat("average =",m," variance =",v,"phi =",v/m,"\n")
average = 0.0697  variance = 0.0739 phi = 1.0597
```

where phi is such that $S_N^2 = \varphi \cdot m_X$. For a Poisson distribution, φ should be equal to 1.

Those quantities can also be be computed when taking into account categorial covariates. In that case,

$$m_{N,\boldsymbol{x}} = \frac{\sum_{i,\boldsymbol{X}_i=\boldsymbol{x}} Y_i}{\sum_{i,\boldsymbol{X}_i=\boldsymbol{x}} E_i} \text{ and } S_{N,\boldsymbol{x}}^2 = \frac{\sum_{i,\boldsymbol{X}_i=\boldsymbol{x}} [Y_i - m_N \cdot E_i]^2}{\sum_{i,\boldsymbol{X}_i=\boldsymbol{x}} E_i}.$$

If we consider the case where $\boldsymbol{X}$ is the region of the driver,

```
> vX <- as.factor(CONTRACTS.f$REGION)
> for(i in 1:length(levels(vX))){
+ vEi <- vE[vX==levels(vX)[i]]
+ vYi <- vY[vX==levels(vX)[i]]
+ mi <- sum(vYi)/sum(vEi)
+ vi <- sum((vYi-mi*vEi)^2)/sum(vEi)
+ cat("average =",mi," variance =",vi," phi =",vi/mi,"\n")
+ }
average = 0.0857  variance = 0.0952 phi = 1.1107
average = 0.0692  variance = 0.0738 phi = 1.0672
average = 0.0630  variance = 0.0648 phi = 1.0285
average = 0.0678  variance = 0.0715 phi = 1.0535
average = 0.0821  variance = 0.0920 phi = 1.1207
average = 0.0718  variance = 0.0759 phi = 1.0564
average = 0.0674  variance = 0.0701 phi = 1.0407
average = 0.0716  variance = 0.0755 phi = 1.0548
average = 0.0736  variance = 0.0806 phi = 1.0948
average = 0.0822  variance = 0.0895 phi = 1.0888
```

where again phi is such that $S_{N,\boldsymbol{x}}^2 = \varphi \cdot m_{X,\boldsymbol{x}}$.

14.2.2 Poisson Regression

Let N_i denote the annual claims frequency for insured i, and assume that $N_i \sim \mathcal{P}(\lambda)$ (so far, all insured have the same λ). If insured i was observed during a period of time E_i (called exposure), the number of claims is $Y_i \sim \mathcal{P}(\lambda \cdot E_i)$. To estimate λ, maximum likelihood techniques can be used,

$$\mathcal{L}(\lambda, \boldsymbol{Y}, \boldsymbol{E}) = \prod_{i=1}^{n} \frac{e^{-\lambda E_i}[\lambda E_i]^{Y_i}}{Y_i!},$$

so that the log-likelihood is

$$\log \mathcal{L}(\lambda, \boldsymbol{Y}, \boldsymbol{E}) = -\lambda \sum_{i=1}^{n} E_i + \sum_{i=1}^{n} Y_i \log[\lambda E_i] - \log\left(\prod_{i=1}^{n} Y_i!\right)$$

The first-order condition is

$$\frac{\partial}{\partial \lambda} \log \mathcal{L}(\lambda, \boldsymbol{Y}, \boldsymbol{E})\bigg|_{\lambda=\widehat{\lambda}} = -\sum_{i=1}^{n} E_i + \frac{1}{\widehat{\lambda}} \sum_{i=1}^{n} Y_i = 0,$$

which yields

$$\widehat{\lambda} = \frac{\sum_{i=1}^{n} Y_i}{\sum_{i=1}^{n} E_i} = \sum_{i=1}^{n} \omega_i \frac{Y_i}{E_i} \text{ where } \omega_i = \frac{E_i}{\sum_{i=1}^{n} E_i}.$$

```
> Y <- CONSTRACTS$NB
> E <- CONSTRACTS$EXPOSURE
> (lambda <- sum(Y)/sum(E))
[1] 0.07279295
> weighted.mean(Y/E,E)
[1] 0.07279295
> dpois(0:3,lambda)*100
[1] 92.979  6.768  0.246  0.006
```

As discussed in the introduction, it might be legitimate to assume that λ depends on the insured, and that those λ_i's are functions of some covariates.

Thus, assume that $N_i \sim \mathcal{P}(\lambda_i)$ where N was the annualized claim frequency (which is the variable of interest in ratemaking because we should price a contract for a 1-year period). Unfortunately, N_i is unobservable; only the number of claims Y_i during period E_i has been observed. Thus, assume the $N_i \sim \mathcal{P}(\lambda_i)$ can equivalently be written $Y_i \sim \mathcal{P}(E_i \cdot \lambda_i)$. With a logarithm link function, then $\lambda_i = e^{\boldsymbol{X}'_i \boldsymbol{\beta}}$, and

$$Y_i \sim \mathcal{P}(e^{\boldsymbol{X}'_i \boldsymbol{\beta} + \log E_i}).$$

The exposure here is a particular variable in the regression. The logarithm of the exposure is indeed an explanatory variable, but no coefficient should be estimated (as it has to be equal to 1). This is the idea of the offset variable. Thus, to model the annualized claim frequency variable N, we run a regression on the observed number of claims Y, and the logarithm of the exposure appears as the offset variable. Assume that $\lambda_i = \exp[\boldsymbol{X}'_i \boldsymbol{\beta}]$ (at least to ensure positivity of the parameter). Then, the log-likelihood is written as

$$\log \mathcal{L}(\boldsymbol{\beta}; \boldsymbol{Y}, \boldsymbol{E}) = \sum_{i=1}^{n} [Y_i \log(\lambda_i E_i) - [\lambda_i E_i] - \log(Y_i!)],$$

or, because $\lambda_i = \exp[\boldsymbol{X}_i'\beta]$,

$$\log \mathcal{L}(\boldsymbol{\beta}; \boldsymbol{Y}, \boldsymbol{E}) = \sum_{i=1}^{n} Y_i \cdot [\boldsymbol{X}_i'\boldsymbol{\beta} + \log(E_i)] - \exp[\boldsymbol{X}_i'\boldsymbol{\beta} + \log(E_i)] - \log(Y_i!).$$

The gradient here is

$$\nabla \log \mathcal{L}(\boldsymbol{\beta}; \boldsymbol{Y}, \boldsymbol{E}) = \frac{\partial \log \mathcal{L}(\boldsymbol{\beta}; \boldsymbol{Y})}{\partial \boldsymbol{\beta}} = \sum_{i=1}^{n} (Y_i - \exp[\boldsymbol{X}_i'\boldsymbol{\beta} + \log(E_i)])\boldsymbol{X}_i'$$

while the Hessian matrix is

$$H(\boldsymbol{\beta}) = \frac{\partial^2 \log \mathcal{L}(\boldsymbol{\beta}; \boldsymbol{Y})}{\partial \boldsymbol{\beta} \partial \boldsymbol{\beta}'} = -\sum_{i=1}^{n} (Y_i - \exp[\boldsymbol{X}_i'\boldsymbol{\beta} + \log(E_i)])\boldsymbol{X}_i\boldsymbol{X}_i'.$$

Based on those quantities, it is possible to solve, numerically, the first-order condition, using Newton–Raphson's algorithm (also called Fisher Scoring). Those computations can be performed using the `glm` function, whose generic code is

```
> glm(Y~X1+X2+X3+offset(E),family=poisson(link="log"))
```

We specify the distribution using `family=poisson`, while parameter `link="log"` means that the logarithm of the expected value will be equal to the score, as $\log[\lambda_i] = \boldsymbol{X}_i'\beta$. This is the *link* function in GLM terminology. More generally, it is possible to consider another transformation g such that $g(\lambda_i) = \boldsymbol{X}_i'\beta$. In the context of a Poisson model, $g = \log$ is the canonical link function (see McCullagh & Nelder (1989) for a discussion of GLMs).

For instance, if we consider a regression of `NB` on `GAS`, `AGEDRIVER`, and `DENSITY`, we obtain

```
> reg=glm(NB~GAS+AGEDRIVER+DENSITY+offset(log(EXPOSURE)),
+ family=poisson,data=CONTRACTS.f)
> summary(reg)

Coefficients:
                       Estimate Std. Error z value Pr(>|z|)
(Intercept)            -1.86471    0.04047 -46.079  < 2e-16 ***
GASRegular             -0.20598    0.01603 -12.846  < 2e-16 ***
AGEDRIVER(22,26]       -0.61606    0.04608 -13.370  < 2e-16 ***
AGEDRIVER(26,42]       -1.07967    0.03640 -29.657  < 2e-16 ***
AGEDRIVER(42,74]       -1.07765    0.03549 -30.362  < 2e-16 ***
AGEDRIVER(74,Inf]      -1.10706    0.05188 -21.338  < 2e-16 ***
DENSITY(40,200]         0.18473    0.02675   6.905 5.02e-12 ***
DENSITY(200,500]        0.31822    0.02966  10.730  < 2e-16 ***
DENSITY(500,4.5e+03]    0.52694    0.02593  20.320  < 2e-16 ***
DENSITY(4.5e+03,Inf]    0.63717    0.03482  18.300  < 2e-16 ***
---
Signif. codes:  0 *** 0.001 ** 0.01 * 0.05 . 0.1   1

(Dispersion parameter for poisson family taken to be 1)

    Null deviance: 105613  on 413168  degrees of freedom
Residual deviance: 103986  on 413159  degrees of freedom
AIC: 135263

Number of Fisher Scoring iterations: 6
```

This stucture of the output is very close to the one obtained with the linear regression function `lm`, in R. See Chambers & Hastie (1991) and Faraway (2006) for more details.

In the following sections, we will discuss the interpretation of this regression, on categorical variables (one or two) and on continuous variables (one or two).

14.2.3 Ratemaking with One Categorical Variable

Consider here one regressor: the type of gas (either Diesel or Regular),

```
> vY <- CONTRACTS.f$NB
> vE <- CONTRACTS.f$EXPOSURE
> X1 <- CONTRACTS.f$GAS
> name 1 <- levels (xi)
```

The number of claims per gas type is

```
> tapply(vY, X1, sum)
 Diesel Regular
   8446    7735
```

and the annualized claim frequency is

```
> tapply(vY, X1, sum)/tapply(vE, X1, sum)
    Diesel    Regular
0.07467412 0.06515364
```

The Poisson regression without the `(Intercept)` variable is here

```
> df <- data.frame(vY,vE,X1)
> regpoislog <- glm(vY~0+X1+offset(log(vE)),data=df,
+ family=poisson(link="log"))
> summary(regpoislog)

Call:
glm(formula = vY ~ 0 + X1 + offset(log(vE)), family = poisson(link = "log"),
    data = df)

Deviance Residuals:
    Min       1Q   Median       3Q      Max
-0.5092  -0.3610  -0.2653  -0.1488   6.5858

Coefficients:
          Estimate Std. Error z value Pr(>|z|)
X1Diesel  -2.59462    0.01088  -238.5   <2e-16 ***
X1Regular -2.73101    0.01137  -240.2   <2e-16 ***
---
Signif. codes:  0 *** 0.001 ** 0.01 * 0.05 . 0.1   1

(Dispersion parameter for poisson family taken to be 1)

    Null deviance: 450747  on 413169  degrees of freedom
Residual deviance: 105537  on 413167  degrees of freedom
AIC: 136799

Number of Fisher Scoring iterations: 6
```

The exponential of the coefficients are the observed annualized frequencies, per gas type

```
> exp(coefficients(regpoislog))
  X1Diesel  X1Regular
0.07467412 0.06515364
```

which can be obtained using function `predict()` (with the option `type="response"`)

```
> newdf <- data.frame(X1=names1,vE=rep(1,length(names1)))
> predict(regpoislog,newdata=newdf,type="response")
         1          2
0.07467412 0.06515364
```

With the `(Intercept)`, the regression is

```
> regpoislog <- glm(vY~X1+offset(log(vE)),data=df,
+ family=poisson(link="log"))
> summary(regpoislog)

Call:
glm(formula = vY ~ X1 + offset(log(vE)), family = poisson(link = "log"),
    data = df)

Deviance Residuals:
    Min       1Q   Median       3Q      Max
-0.5092  -0.3610  -0.2653  -0.1488   6.5858

Coefficients:
            Estimate Std. Error  z value Pr(>|z|)
(Intercept) -2.59462    0.01088 -238.454   <2e-16 ***
X1Regular   -0.13639    0.01574   -8.666   <2e-16 ***
---
Signif. codes:  0 *** 0.001 ** 0.01 * 0.05 . 0.1   1

(Dispersion parameter for poisson family taken to be 1)

    Null deviance: 105613  on 413168  degrees of freedom
Residual deviance: 105537  on 413167  degrees of freedom
AIC: 136799

Number of Fisher Scoring iterations: 6
```

where the reference is a car with Diesel type of gas,

```
> exp(coefficients(regpoislog))
(Intercept)   X1Regular
 0.07467412  0.87250624
```

Here, claims frequency for Diesel cars is 0.0746, and for Regular gas cars, it would be 87.25% of the value obtained for the reference one,

```
> prod(exp(coefficients(regpoislog)))
[1] 0.06515364
```

Predictions are similar; only the interpretation is different, here because the `(Intercept)` will be associated with some reference.

14.2.4 Contingency Tables and Minimal Bias Techniques

In Chapter 4, the logistic regression on two categorical covariates was discussed. Similarly when we work with counting variables, it is natural to work with contingency matrices, with two regressors, for instance the GAS (as in the previous section) and the DENSITY (here as a factor):

```
> X2 <- CONTRACTS.f$DENSITY
> names1 <- levels(X1)
> names2 <- levels(X2)
> (P=table(X1,X2))

          X2
X1         [0,40] (40,200] (200,500] (500,4.5e+03] (4.5e+03,Inf]
  Diesel    36626    64966     31436         59797         13120
  Regular   25706    53588     31459         72461         24010
```

Define also the exposure matrix $\boldsymbol{E} = [E_{i,j}]$ (E in R) and the claims count matrix $\boldsymbol{Y}$ (Y in R):

```
> E <- Y <- P
> for(k in 1:length(names1)){
+ E[k,] <- tapply(vE[X1==names1[k]],X2[X1==names1[k]],sum)
+ Y[k,] <- tapply(vY[X1==names1[k]],X2[X1==names1[k]],sum)}
> E
          X2
X1            [0,40]  (40,200] (200,500] (500,4.5e+03] (4.5e+03,Inf]
  Diesel   23049.805 38716.498 17588.139     28573.604      5176.733
  Regular  16943.598 33682.835 19577.038     38011.191     10504.727
> Y
          X2
X1         [0,40] (40,200] (200,500] (500,4.5e+03] (4.5e+03,Inf]
  Diesel     1266     2575      1347          2760           498
  Regular     777     1858      1235          2941           924
```

The annualized (empirical) claims frequency is then

```
> (N <- Y/E)
          X2
X1            [0,40]  (40,200] (200,500] (500,4.5e+03] (4.5e+03,Inf]
  Diesel   0.0549245 0.0665091 0.0765857     0.0965926     0.0961996
  Regular  0.0458580 0.0551616 0.0630841     0.0773719     0.0879604
```

Let $\boldsymbol{N}$ denote the annualized frequency (matrix N in the example above). Consider a multiplicative model for $\boldsymbol{N}$, in the sense that $N_{i,j} = L_i \cdot C_j$ for some vectors $\boldsymbol{L} = [L_i]$ and $\boldsymbol{C} = [C_j]$. Criteria to construct such vectors are usually on of the following three: (weighted) least squares, minimization of some distance (e.g., chi-square), or minimal bias (as introduced by Bailey (1963)). For the least squares techniques, consider $(\boldsymbol{L}, \boldsymbol{C})$ that solve

$$\min\left\{\sum_{i,j} E_{i,j}(N_{i,j} - L_i \cdot C_j)^2\right\},$$

while for the chi-square method, we have to solve

$$\min\left\{\sum_{i,j} E_{i,j}\cdot\frac{(N_{i,j}-L_i\cdot C_j)^2}{L_i\cdot C_j}\right\}.$$

Those two techniques cannot be solved analytically, and iterative algorithms should be used (see Exercise 14.2). Bailey (1963) assumed that predicted sums per row, and per column, should be equal to empirical ones. More precisely, if we sum per column, for a given j,

$$\sum_i Y_{i,j}=\sum_i E_{i,j}\cdot N_{i,j}=\sum_i E_{i,j}\cdot[L_i\cdot C_j],$$

and if we sum per row, for a given i,

$$\sum_j Y_{i,j}=\sum_j E_{i,j}\cdot N_{i,j}=\sum_j E_{i,j}\cdot[L_i\cdot C_j].$$

Again, these equations cannot be solved explicitly. Nevertheless, one can derive the following relationships for L_i:

$$L_i=\frac{\sum_j Y_{i,j}}{\sum_j E_{i,j}C_j},$$

and for C_j,

$$C_j=\frac{\sum_i Y_{i,j}}{\sum_i E_{i,j}L_i}.$$

An iterative algorithm can be used to solve those equations (even if we should keep in mind that there might be identifiability issues because $\boldsymbol{L}$ and $\boldsymbol{C}$ are—under some assumptions—unique up to a multiplicative constant) starting from some initial values for $\boldsymbol{C}$ (say).

```
> L <- matrix(NA,100,length(names1))
> C <- matrix(NA,100,length(names2))
> C[1,] <- rep(sum(vY)/sum(vE),length(names2));colnames(C) <- names2
> for(j in 2:100){
+  for(k in 1:length(names1)) L[j,k] <- sum(Y[k,])/sum(E[k,]*C[j-1,])
+  for(k in 1:length(names2)) C[j,k] <- sum(Y[,k])/sum(E[,k]*L[j,])
+ }
```

After 100 loops, we obtain the following values for $\boldsymbol{L}$ and $\boldsymbol{C}$:

```
> L[100,]
[1] 1.098979 0.907170
> C[100,]
      [0,40]     (40,200]     (200,500] (500,4.5e+03] (4.5e+03,Inf]
  0.05019412  0.06063908   0.06961690    0.08653034    0.09343771
```

That can be used to predict annualized claim frequency:

```
> PredN=N
>  for(k in 1:length(names1)) PredN[k,]<-L[100,k]*C[100,]
> PredN
         X2
X1              [0,40]    (40,200]   (200,500] (500,4.5e+03] (4.5e+03,Inf]
  Diesel  0.05516229 0.06664107 0.07650751    0.09509503    0.10268609
  Regular 0.04553460 0.05500995 0.06315436    0.07849773    0.08476389
```

Given the observed exposure, the prediction of the number of claims for `Diesel` cars (for instance) with this model would be

```
> sum(PredN[1,]*E[1,])
[1] 8446
```

which is exactly the same as the observed total, for the first row,

```
> sum(Y[1,])
[1] 8446
```

This is the minimal bias method, used in Bailey (1963) in motor insurance pricing. The interesting point is that this method coincides with the Poisson regression,

```
> df <- data.frame(vY,vE,X1,X2)
> regpoislog <- glm(vY~X1+X2,offset=log(vE),data=df,
+ family=poisson(link="log"))
> newdf <- data.frame(
+ X1=factor(rep(names1,length(names2))),
+ vE=rep(1,length(names1)*length(names2)),
+ X2=factor(rep(names2,each=length(names1))))
> matrix(predict(regpoislog,newdata=newdf,
+ type="response"),length(names1),length(names2))
           [,1]       [,2]       [,3]       [,4]       [,5]
[1,] 0.05516229 0.06664107 0.07650751 0.09509503 0.10268609
[2,] 0.04553460 0.05500995 0.06315436 0.07849773 0.08476389
```

The interpretation will be that if we use a Poisson regression on categorical variables, the total prediction per modality will equal to the annualized empirical sum of claims.

14.2.5 Ratemaking with Continuous Variables

In Chapter 4, it was mentioned that working with continuous covariates might be interesting, because cutoff (to make variables categorical) levels might be too arbitrary. Thus, categorical variables constructed from continuous ones (age of the driver, age of the car, spatial location, etc.) will create artificial discontinuities, which might be dangerous in the context of highly competitive markets.

```
> reg.cut <- glm(NB~AGEDRIVER+offset(log(EXPOSURE)),
+ family=poisson,data=CONTRACTS.f)
> summary(reg.cut)

Deviance Residuals:
    Min       1Q   Median       3Q      Max
-0.6218  -0.3615  -0.2632  -0.1491   6.5690

Coefficients:
                   Estimate Std. Error z value Pr(>|z|)
(Intercept)        -1.66337    0.03365  -49.43   <2e-16 ***
AGEDRIVER(22,26]   -0.56935    0.04602  -12.37   <2e-16 ***
AGEDRIVER(26,42]   -1.04009    0.03628  -28.67   <2e-16 ***
AGEDRIVER(42,74]   -1.06454    0.03542  -30.05   <2e-16 ***
AGEDRIVER(74,Inf]  -1.17659    0.05177  -22.73   <2e-16 ***
```

```
---
Signif. codes:  0 *** 0.001 ** 0.01 * 0.05 . 0.1   1

(Dispersion parameter for poisson family taken to be 1)

    Null deviance: 105613  on 413168  degrees of freedom
Residual deviance: 104734  on 413164  degrees of freedom
AIC: 136001

Number of Fisher Scoring iterations: 6
```

The standard regrenion on the age of the driver would yield

```
> reg.poisson <- glm(NB~AGEDRIVER+offset(log(EXPOSURE)),
+ family=poisson,data=CONTRACTS)
> summary(reg.poisson)

Deviance Residuals:
    Min       1Q   Median       3Q      Max
-0.5523  -0.3510  -0.2678  -0.1504   6.4415

Coefficients:
              Estimate Std. Error z value Pr(>|z|)
(Intercept) -2.1513378  0.0262347  -82.00   <2e-16 ***
AGEDRIVER   -0.0111060  0.0005579  -19.91   <2e-16 ***
---
Signif. codes:  0 *** 0.001 ** 0.01 * 0.05 . 0.1   1

(Dispersion parameter for poisson family taken to be 1)

    Null deviance: 105613  on 413168  degrees of freedom
Residual deviance: 105206  on 413167  degrees of freedom
AIC: 136467

Number of Fisher Scoring iterations: 6
```

But as in Chapter 4, assuming a linear relationship might be too restrictive.

```
> newdb <- data.frame(AGEDRIVER=18:99,EXPOSURE=1)
> pred.poisson <- predict(reg.poisson,newdata=newdb,type="response",se=TRUE)
> plot(18:99,pred.poisson$fit,type="l",xlab="Age of the driver",
+ ylab="Annualized Frequency", ylim=c(0,.3),col="white")
> segments(18:99,pred.poisson$fit-2*pred.poisson$se.fit,
+ 18:99,pred.poisson$fit+2*pred.poisson$se.fit,col="grey",lwd=7)
> lines(18:99,pred.poisson$fit)
> abline(h=sum(CONTRACTS$NB)/sum(CONTRACTS$EXPOSURE),lty=2)
```

In order to get a nonparametric estimator, it is possible to consider the age (which is here an integer) as a factor variable:

```
> reg.np <- glm(NB~as.factor(AGEDRIVER)+offset(log(EXPOSURE)),
+ family=poisson,data=CONTRACTS)
```

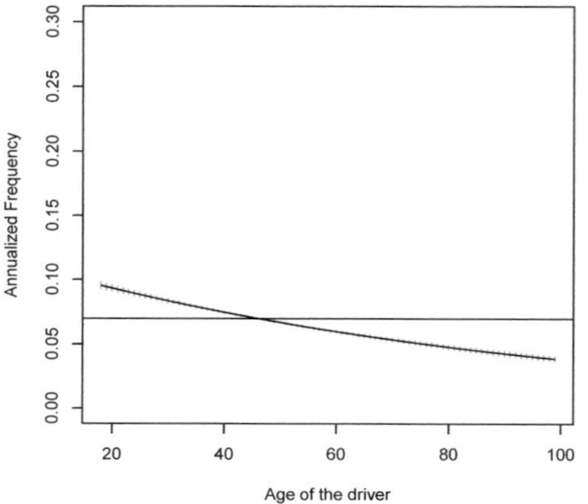

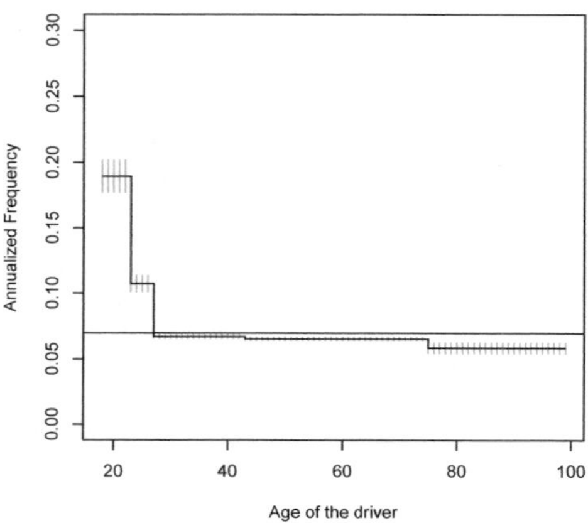

FIGURE 14.1
Poisson regression of the annualized frequency on the age of the driver, with a *linear* model, and when the age of the driver is a categorical variable.

or to use spline regressions, so that we consider a linear model including $s(X)$ instead of X, where $s()$ will be estimated using penalized regression splines (see Hastie & Tibshirani (1990), Bowman & Azzalini (1997), and Wood (2006) for more details):

```
> library(mgcv)
> reg.splines <- gam(NB~s(AGEDRIVER)+offset(log(EXPOSURE)),
+ family=poisson,data=CONTRACTS)
```

Predictions using the four models can be visualized in Figure 14.1 for the (standard) log-Poisson regression, and with arbitrary cuts for the age (to get a categorical variable), and in Figure 14.2 where the age is considered a factor, and using Generalized Additive Models.

14.2.6 A Poisson Regression to Model Yearly Claim Frequency

In order to have significant factors, the age of the car is here only in two classes: less than 15 years old, and more than 15 years old:

```
> CONTRACTS.f$AGECAR <- cut(CONTRACTS$AGECAR,c(0,15,Inf),
+ include.lowest = TRUE)
```

The levels are here

```
> levels(CONTRACTS.f$AGECAR)
 [1] "[0,15]" "(15,Inf]"
```

For the brand, we only distinguish cars of brand `F`,

```
> CONTRACTS.f$brandF <- factor(CONTRACTS.f$BRAND=="F",labels=c("other","F"))
```

and for the power of the car, three classes are considered here,

```
> CONTRACTS.f$powerF <- factor(1*(CONTRACTS.f$POWER%in%letters[4:6])+
+ 2*(CONTRACTS.f$POWER%in%letters[7:8]),
+ labels=c("other","DEF","GH"))
```

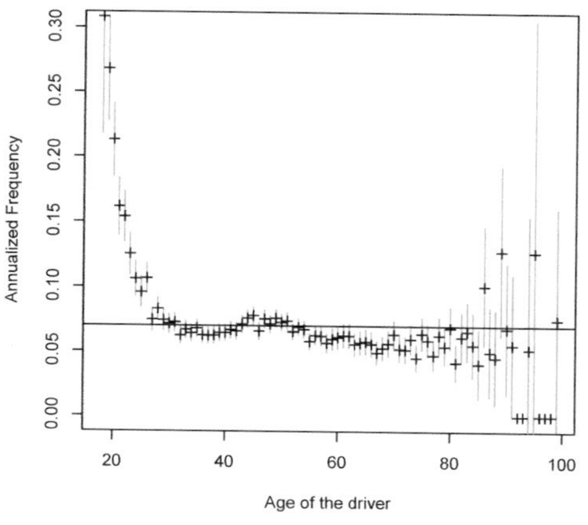

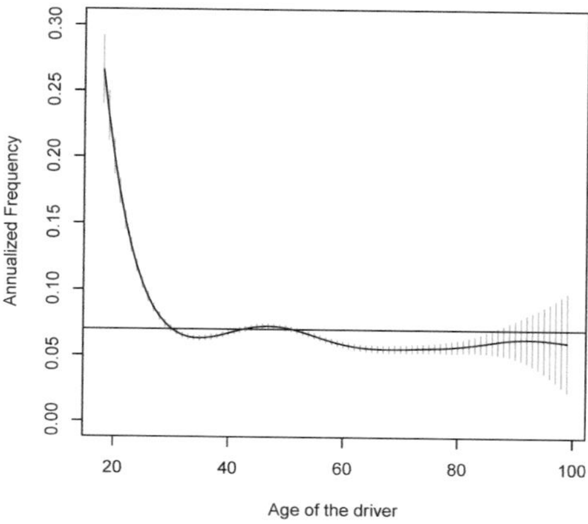

FIGURE 14.2
Poisson regression of the annualized frequency on the age of the driver, per age as a factor (because it is a discrete variable) (left) and using a spline smoother (right).

Our model here is

```
> freg <- formula(NB ~ AGEDRIVER+AGECAR+DENSITY+brandF+powerF+GAS+offset(log(EXPOSURE)))
> regp <- glm(freg,data=CONTRACTS.f,family=poisson(link="log"))
> summary(regp)

Coefficients:
                        Estimate Std. Error z value Pr(>|z|)
(Intercept)             -1.67848    0.04652 -36.082  < 2e-16 ***
AGEDRIVER(22,26]        -0.61506    0.04610 -13.341  < 2e-16 ***
AGEDRIVER(26,42]        -1.08085    0.03652 -29.599  < 2e-16 ***
AGEDRIVER(42,74]        -1.07777    0.03566 -30.223  < 2e-16 ***
AGEDRIVER(74,Inf]       -1.10041    0.05190 -21.203  < 2e-16 ***
AGECAR(15,Inf]          -0.25011    0.03085  -8.108 5.14e-16 ***
DENSITY(40,200]          0.18291    0.02676   6.834 8.26e-12 ***
DENSITY(200,500]         0.31544    0.02968  10.627  < 2e-16 ***
DENSITY(500,4.5e+03]     0.53238    0.02606  20.428  < 2e-16 ***
DENSITY(4.5e+03,Inf]     0.66731    0.03581  18.633  < 2e-16 ***
brandFF                 -0.18756    0.02477  -7.574 3.63e-14 ***
powerFDEF               -0.17892    0.02428  -7.370 1.71e-13 ***
powerFGH                -0.15361    0.02641  -5.816 6.03e-09 ***
GASRegular              -0.19507    0.01620 -12.041  < 2e-16 ***
---
Signif. codes:  0 *** 0.001 ** 0.01 * 0.05 . 0.1   1

(Dispersion parameter for poisson family taken to be 1)

    Null deviance: 105613  on 413168  degrees of freedom
Residual deviance: 103831  on 413155  degrees of freedom
AIC: 135117

Number of Fisher Scoring iterations: 6
```

14.3 From Poisson to Quasi-Poisson

The Poisson assumption is stronger than necessary when we use the first-order condition

$$\mathbb{E}[(Y_i - \exp(\boldsymbol{X}_i'\boldsymbol{\beta})) \cdot \boldsymbol{X}_i] = \boldsymbol{0},$$

which can be rewritten as

$$\sum_{i=1}^{n} \frac{(Y_i - \mu_i)}{g'(\mu_i) \cdot \text{var}[Y_i]} \cdot \boldsymbol{X}_i = \boldsymbol{0}.$$

In this case, various forms of $g(\mu_i)$ (the link function) generate convergent estimators $\boldsymbol{\beta}$. Other forms of $\text{var}[Y_i]$ can also be chosen. In such a case, we can show that $\widehat{\boldsymbol{\beta}}_{PMLE}$ estimators have the following distributions:

$$\widehat{\boldsymbol{\beta}}_{PMLE} \rightarrow \mathcal{N}(\boldsymbol{\beta}, \text{var}_{PMLE}[\widehat{\boldsymbol{\beta}}]],$$

with

$$\text{var}_{PMLE}[\widehat{\boldsymbol{\beta}}] = \Big[\sum_{i=1}^{n} \mu_i \boldsymbol{x}_i \boldsymbol{x}_i'\Big]^{-1} \Big[\sum_{i=1}^{n} \omega_i \boldsymbol{X}_i \boldsymbol{X}_i'\Big] \Big[\sum_{i=1}^{n} \mu_i \boldsymbol{X}_i \boldsymbol{X}_i'\Big]^{-1}$$

and $\omega_i = \text{var}[Y_i | \boldsymbol{X}_i]$.

Obviously, if we suppose $\omega_i = \mu_i$, we obtain the general first-order condition for a Poisson distribution:

$$\text{var}_{PMLE}[\widehat{\boldsymbol{\beta}}] = \Big[\sum_{i=1}^{n} \mu_i \boldsymbol{X}_i \boldsymbol{X}_i'\Big]^{-1} = \text{var}_{MLE}[\widehat{\boldsymbol{\beta}}].$$

However, other forms of ω_i can be explored.

14.3.1 NB1 Variance Form: Negative Binomial Type I

If we suppose an NB1 variance form, such as $\omega_i = \varphi \mu_i$, we have

$$\text{var}_{NB1}[\widehat{\beta}] = \varphi \Big[\sum_{i=1}^{n} \mu_i \boldsymbol{x}_i \boldsymbol{x}_i'\Big]^{-1} = \varphi \cdot \text{var}_{MLE}[\widehat{\beta}].$$

Consequently, a simple way to generalize the variance form is to use the MLE of a Poisson distribution and multiply the variance of $\widehat{\beta}$ by an estimator of ϕ. A natural estimator of ϕ is

$$\widehat{\varphi} = \frac{1}{n-k} \sum_{i=1}^{n} \frac{(y_i - \widehat{\mu}_i)^2}{\widehat{\mu}_i},$$

but nonconstant exposition parameter E_i causes biais in the estimation. Instead, we should use

$$\widehat{\varphi} = \frac{\sum_{i=1}^{n} (y_i - \widehat{\mu}_i)^2}{\sum_{i=1}^{n} \widehat{\mu}_i}.$$

14.3.2 NB2 Variance Form: Negative Binomial Type II

If we suppose an NB2 variance form (the NB2 model is presented in detail in Section 14.4.1) such as $\omega_i = \mu_i + \alpha\mu_i^2$, we have

$$\text{var}_{NB2}[\widehat{\boldsymbol{\beta}}] = \Big[\sum_{i=1}^{n} \mu_i \boldsymbol{X}_i \boldsymbol{X}_i'\Big]^{-1} \Big[\sum_{i=1}^{n} (\mu_i + \alpha\mu_i^2) \boldsymbol{X}_i \boldsymbol{X}_i'\Big] \Big[\sum_{i=1}^{n} \mu_i \boldsymbol{X}_i \boldsymbol{X}_i'\Big]^{-1}.$$

As with the NB1 estimator of overdispersion, the classic estimator of α is

$$\widehat{\alpha} = \frac{1}{n-k} \sum_{i=1}^{n} \frac{(Y_i - \widehat{\mu}_i)^2 - \widehat{\mu}_i}{\widehat{\mu}_i^2},$$

but is unappropriate for non-constant E_i. The following estimator is preferred:

$$\widehat{\alpha} = \frac{\sum_{i=1}^{n} (Y_i - \widehat{\mu}_i)^2 - \widehat{\mu}_i}{\sum_{i=1}^{n} \widehat{\mu}_i^2}.$$

The NB2 regression is obtained using `glm.nb` of library `MASS`.:

```
> library(MASS)
> regnb2 <- glm.nb(freg,data=CONTRACTS.f)
> summary(regnb2)

Coefficients:
                       Estimate Std. Error z value Pr(>|z|)
(Intercept)            -1.65354    0.04858 -34.037  < 2e-16 ***
AGEDRIVER(22,26]       -0.63051    0.04838 -13.033  < 2e-16 ***
AGEDRIVER(26,42]       -1.10163    0.03850 -28.612  < 2e-16 ***
AGEDRIVER(42,74]       -1.09808    0.03765 -29.169  < 2e-16 ***
AGEDRIVER(74,Inf]      -1.12431    0.05410 -20.783  < 2e-16 ***
AGECAR(15,Inf]         -0.24896    0.03159  -7.882 3.23e-15 ***
DENSITY(40,200]         0.18367    0.02742   6.697 2.12e-11 ***
DENSITY(200,500]        0.31690    0.03047  10.399  < 2e-16 ***
DENSITY(500,4.5e+03]    0.53506    0.02676  19.995  < 2e-16 ***
DENSITY(4.5e+03,Inf]    0.66957    0.03690  18.148  < 2e-16 ***
brandFF                -0.19253    0.02540  -7.581 3.42e-14 ***
powerFDEF              -0.17756    0.02507  -7.084 1.40e-12 ***
powerFGH               -0.15312    0.02726  -5.617 1.94e-08 ***
GASRegular             -0.19553    0.01670 -11.707  < 2e-16 ***
---
Signif. codes:  0 *** 0.001 ** 0.01 * 0.05 . 0.1   1

(Dispersion parameter for Negative Binomial(0.8527) family taken to be 1)

    Null deviance: 91316  on 413168  degrees of freedom
Residual deviance: 89623  on 413155  degrees of freedom
AIC: 134743

Number of Fisher Scoring iterations: 1
```

```
           Theta:  0.8527
       Std. Err.:  0.0573

 2 x log-likelihood:  -134712.7800
```

14.3.3 Unstructured Variance Form

More generally, we can estimate the variance $\widehat{\beta}$ without supposing a specific form of ω_i. In this case, the unstructured variance form of $\widehat{\beta}$ is

$$\text{var}_{RS}[\widehat{\boldsymbol{\beta}}] = \left[\sum_{i=1}^{n} \mu_i \boldsymbol{X}_i \boldsymbol{X}_i'\right]^{-1} \left[\sum_{i=1}^{n} (Y_i - \mu_i)^2 \boldsymbol{X}_i \boldsymbol{X}_i'\right] \left[\sum_{i=1}^{n} \mu_i \boldsymbol{X}_i \boldsymbol{X}_i'\right]^{-1},$$

evaluated at $\widehat{\mu}_i$.

14.3.4 Nonparametric Variance Form

A nonparametric way to estimate ω_i is to use the approach of Delgado & Kniesner (1997). They do not suppose a specific form for the conditional variance σ_i^2 of the i-th contract, and estimate it using a consistent estimator of $\boldsymbol{\beta}$, $\widehat{\boldsymbol{\beta}}$, such as the one obtained by a Poisson regression:

$$\widehat{\omega}_i = \sum_{j=1, j\neq i}^{n} [n_i - \widehat{\mu}_j]^2 w_{i,j}, \tag{14.1}$$

where $w_{i,j}$ can be seen as weight applied to the j-th observation to compute the variance of the i-th insured with, obviously, $\sum_{j=1,j\neq i}^{n} w_{i,j} = 1$. Using $w_{i,j} = 1/(n-1)$ leads to an estimated variance almost equal to the empirical variance. Delgado & Kniesner (1997) evaluated the $w_{i,j}$ using nonparametric k nearest neighbor probabilistic weights. His method estimates variance of insured i using its k nearest neighbors, evaluated by its proximity in its normalized covariates. The k nearest neighbors were selected using a neighbor specification k that is proportional to the number of observations, such as $n^{1/2}$ or $n^{3/5}$.

Instead of using the nearest neighbors estimation that gives the same weight on all k observations (which can lead to situations where the $\widehat{\sigma}_i^2$ would be estimated using only insureds having the same profiles, while other $\widehat{\sigma}_i^2$ would be estimated using other insured's profiles), we can rather use kernel estimation which seems to offer a better smooth.

Remark 14.2 *Obviously, we can obtain a robust estimator of the variance of $\widehat{\beta}$ by bootstrapping. By bootstrapping, we can even obtain the distribution of $\widehat{\boldsymbol{\beta}}$, instead of using the asymptotic normality of the $\widehat{\boldsymbol{\beta}}$.*

14.4 More Advanced Models for Counts

A popular way to generalize count distributions is to use a compound sum of the following form:

$$Y = \sum_{j=1}^{M} Z_j. \tag{14.2}$$

With specific choices of distribution M and Z, we obtain the following distributions:

- If $M \sim \mathcal{P}(\mu)$, and $Z \sim Logarithmic(\eta)$, we have a negative binomial distribution.
- If $M \sim \mathcal{B}(\phi)$, and $Z \sim \mathcal{P}(\lambda)$ (or negative binomial), we have a *Zero-inflated Poisson (or Zero-inflated negative binomial)* distribution.
- If $M \sim \mathcal{B}(\phi)$, and $Z \sim$ truncated $\mathcal{P}(\lambda)$ (or negative binomial), meaning that $X_i \in \{1, 2, \cdots\}$, we have a *Hurdle Poisson (or Hurdle-negative binomial)* distribution.

Note that instead of using a truncated at zero distribution, we can also use a shifted at zero distribution.

14.4.1 Negative Binomial Regression

In Chapters 2 and 3, it was mentioned that if $Y|\Theta \sim \mathcal{P}(\lambda \cdot \Theta)$, where Θ has a Gamma distribution, with identical parameters α (so that $\mathbb{E}(\Theta) = 1$), then Y has a negative binomial distribution:

$$\mathbb{P}(Y = y) = \frac{\Gamma(y + \alpha^{-1})}{\Gamma(y+1)\Gamma(\alpha^{-1})} \left(\frac{1}{1+\lambda/\alpha}\right)^{\alpha^{-1}} \left(1 - \frac{1}{1+\lambda/\alpha}\right)^{y}, \forall y \in \mathbb{N}.$$

Let $r = \alpha^{-1}$ and $p = (1 + \alpha\lambda)^{-1}$;

$$f(y) = \mathbb{P}(Y = y) = \binom{y}{y + r - 1} p^r [1 - p]^y, \forall y \in \mathbb{N},$$

which can be written

$$f(y) = \exp\left[y \log(1 - p) + r \log p + \log \binom{y}{y + r - 1}\right], \forall y \in \mathbb{N},$$

which is a distribution of the exponential family (see McCullagh & Nelder (1989)) when $\theta = \log[1 - p]$, $b(\theta) = -r \log(p)$ and $a(\varphi) = 1$, with a known r. The mean of Y is here

$$\mathbb{E}(Y) = b'(\theta) = \frac{\partial b}{\partial p}\frac{\partial p}{\partial \theta} = \frac{r(1-p)}{p} = \lambda,$$

while its variance is

$$\text{var}(Y) = b''(\theta) = \frac{\partial^2 b}{\partial p^2}\left(\frac{\partial p}{\partial \theta}\right)^2 + \frac{\partial b}{\partial p}\frac{\partial^2 p}{\partial \theta^2} = \frac{r(1-p)}{p^2},$$

which can also be written as

$$\text{Var}(Y) = \frac{1}{p}\mathbb{E}(Y) = [1 + \alpha \cdot \lambda] \cdot \lambda.$$

This is so-called Type 2 Negative Binomial regression (NB2, as in Hilbe (2011), discussed in Section 14.3.2):

$$\mathbb{E}(Y) = \lambda = \mu \text{ and } \text{var}(Y) = \lambda + \alpha\lambda^2.$$

The canonical link, such that $g(\lambda) = \theta$, is $g(\mu) = \log(\alpha\mu) - \log(1 + \alpha\mu)$. The generic function in R for an NB2 regression is

```
> library(MASS)
> glm.nb(Y~X1+X2+X3+offset(log(E)))
```

(neither the family nor the link function is specified here). In that case, α is unknown and will be obtained using `summary()`.

In the case where α is known, it is possible to use `family=negative.binomiale` in the standard `glm` function. For instance, a geometric regression will be obtained using

```
> glm(Y~X1+X2+X3+offset(log(E)),family=negative.binomiale(1))
```

In that case, $\text{Var}(Y) = \lambda + \lambda^2$.

Using the context of NB2 regression, it is possible to test if the Poisson regression is a suitable model, the alternative being an NB2 model, with α significant. We must be careful with tests when the null hypothesis is on the boundary of the parameter space. Indeed, in such situations, the MLE are no longer asymptotically normal under H_0. The asymptotic properties of the score test, and also called the Lagrange multiplier test, as shown by Moran (1971) and Chant (1974), are not altered when testing on the boundary of the parameter space.

Using the score test, we assume here that

$$\text{var}(Y|\boldsymbol{X} = \boldsymbol{x}) = \mathbb{E}(Y|\boldsymbol{X} = \boldsymbol{x}) + \alpha \cdot \mathbb{E}(Y|\boldsymbol{X} = \boldsymbol{x})^2,$$

and we would like to test

$$H_0 : \alpha = 0 \text{ against } H_1 : \alpha > 0.$$

A standard statistic is

$$T = \frac{\sum_{i=1}^{n}[(Y_i - \widehat{\mu}_i)^2 - Y_i]}{\sqrt{2\sum_{i=1}^{n}\widehat{\mu}_i^2}}$$

which has a centered Gaussian distribution, with unit variance, under H_0. This test has been implemented in R in the `AER` library,

```
> library(AER)
> dispersiontest(regp)

        Overdispersion test

data:  regp
z = 5.7935, p-value = 3.447e-09
alternative hypothesis: true dispersion is greater than 1
sample estimates:
dispersion
  1.091931
```

This kind of binomial distribution (studied in Santos Silva & Windmeijer (2001)) has also been proposed with another form of parametrization. Using Equation (14.2), the parameter η_i of the logarithmic distribution has been used as

$$\exp(\boldsymbol{X}_i'\gamma) = \frac{\eta_i}{1 - \eta_i},$$

while $M \sim \mathcal{P}(\exp(\boldsymbol{X}_i'\beta))$.

Consequently, N_i is binomial negative with parameters $\lambda_i / \log(1 + \exp(x_i'\gamma))$ and $\exp(x_i'\gamma)$, with a probability function defined as

$$\mathbb{P}(N_i = k|\boldsymbol{X}_i = \boldsymbol{x}_i) = \frac{\Gamma\left(k + \dfrac{\lambda_i}{\log(1 + \exp(\boldsymbol{x}_i'\gamma))}\right)\exp(-\lambda_i)}{\Gamma(k+1)\Gamma\left(\dfrac{\lambda_i}{\log(1 + \exp(x_i'\gamma))}\right)(1 + \exp(-x_i'\gamma))^k}.$$

The first two moments of those models are

$$\begin{aligned}\mathbb{E}[N_i|\boldsymbol{X}=\boldsymbol{x}] &= \frac{\exp(x_i'\beta+\boldsymbol{x}_i'\gamma)}{\log(1+\exp(\boldsymbol{x}_i'\gamma))}\\ \text{var}[N_i|\boldsymbol{X}=\boldsymbol{x}] &= (1+\exp(\boldsymbol{x}_i'\gamma)]\mathbb{E}[N_i].\end{aligned}$$

14.4.2 Zero-Inflated Models

A model is said to be zero-inflated if it can be written as a mixture, with probability π_i a Dirac distribution in 0 and a standard counting regression model (Poisson or Negative Binomial):

$$\mathbb{P}(N_i=k)=\begin{cases}\pi_i+[1-\pi_i]\cdot p_i(0) \text{ if } k=0\\ [1-\pi_i]\cdot p_i(k) \text{ if } k=1,2,\cdots.\end{cases}$$

Thus, for some counting model $p_i(\cdot)$, for instance the Poisson distribution with parameter λ_i, then

$$\mathbb{E}(N_i)=[1-\pi_i]\lambda_i \text{ and } \text{var}(N_i)=\pi_i\lambda_i+\pi_i\lambda_i^2[1-\pi_i]>\mathbb{E}(N_i).$$

Remark 14.3 *There is a class of zero-adapted regression where*

$$\mathbb{P}(N_i=k)=\begin{cases}\pi_i \text{ if } k=0\\ [1-\pi_i]\cdot\dfrac{p_i(k)}{1-p_i(0)} \text{ if } k=1,2,\cdots,\end{cases}$$

which can be used with function `ZENBI` *of library* `gamlss`.

If we suppose that $p_i\geq 0$, testing $p_i=0$ is also a test with the null hypothesis on the boundary of the parameter space. Consequently, using a score test is recommended. Van den Broek (1995) use this test and show that the test statistic of a zero-inflated Poisson against a Poisson distribution can be expressed as

$$LM=\frac{\left(\sum_{i=1}^n(\mathbf{1}(N_i=0))-e^{-\widehat{\lambda}_i})/e^{-\widehat{\lambda}_i}\right)}{\sqrt{\left(\sum_{i=1}^n(1-e^{-\widehat{\lambda}_i})/e^{-\widehat{\lambda}_i}\right)-\sum_{i=1}^n N_i}}. \tag{14.3}$$

Under the null hypothesis, this statistic will have an asymptotic $\mathcal{N}(0,1)$ distribution. Construction of the LM test for heterogeneous models against their zero-inflated modification can be done the same way.

Zero-inflated and zero-adapted models can be estimated either using functions `ZIP` (zero-inflated Poisson) or `ZAP` (zero-adapted Poisson), or functions `ZINI` (zero-inflated Negative Binomial) or `ZIBI` (zero-adapted Negative Binomial) from library `gamlss`, or functions of library `pscl` (described in Zeileis et al. (2008)). For instance, for a zero-inflated model, where probability π_i is function of covariates $\boldsymbol{X}_2$ (with a logistic transformation), while λ_i is function of covariates $\boldsymbol{X}_1$, the generic code will be

```
> reg <- zeroinfl(NB ~ X1 | X2 , data=CONTRACTS.f , dist = "poisson" ,
+  link="logit"))
```

If we consider a zero-inflated Poisson model, with constant probability π, we obtain

```
> fregzi <- formula(NB ~ AGEDRIVER+AGECAR+DENSITY+brandF+powerF+
+ GAS+offset(log(EXPOSURE))|1)
> regzip=zeroinfl(fregzi,data=CONTRACTS.f,dist = "poisson",link="logit")
> summary(regzip)

Pearson residuals:
     Min       1Q   Median       3Q      Max
 -0.4907  -0.2315  -0.1827  -0.1038 101.3696

Count model coefficients (poisson with log link):
                      Estimate Std. Error z value Pr(>|z|)
(Intercept)           -0.92717    0.05953 -15.574  < 2e-16 ***
AGEDRIVER(22,26]      -0.62772    0.04872 -12.885  < 2e-16 ***
AGEDRIVER(26,42]      -1.09943    0.03885 -28.302  < 2e-16 ***
AGEDRIVER(42,74]      -1.09585    0.03799 -28.844  < 2e-16 ***
AGEDRIVER(74,Inf]     -1.11988    0.05431 -20.619  < 2e-16 ***
AGECAR(15,Inf]        -0.24825    0.03167  -7.839 4.54e-15 ***
DENSITY(40,200]        0.18379    0.02746   6.693 2.18e-11 ***
DENSITY(200,500]       0.31679    0.03051  10.382  < 2e-16 ***
DENSITY(500,4.5e+03]   0.53457    0.02679  19.951  < 2e-16 ***
DENSITY(4.5e+03,Inf]   0.67016    0.03692  18.150  < 2e-16 ***
brandFF               -0.19000    0.02539  -7.483 7.25e-14 ***
powerFDEF             -0.17657    0.02507  -7.042 1.90e-12 ***
powerFGH              -0.15223    0.02727  -5.583 2.37e-08 ***
GASRegular            -0.19499    0.01672 -11.660  < 2e-16 ***

Zero-inflation model coefficients (binomial with logit link):
            Estimate Std. Error z value Pr(>|z|)
(Intercept)  0.07414    0.06441   1.151     0.25
---
Signif. codes:  0 '***' 0.001 '**' 0.01 '*' 0.05 '.' 0.1 ' ' 1

Number of iterations in BFGS optimization: 25
Log-likelihood: -6.736e+04 on 15 Df
```

If we consider a probability π_i function of the age of the driver, we obtain

```
> fregzi <- formula(NB ~ AGEDRIVER+AGECAR+DENSITY+brandF+powerF+
+ GAS+offset(log(EXPOSURE))|AGEDRIVER)
> regzip=zeroinfl(fregzi,data=CONTRACTS.f,dist = "poisson",link="logit")
> summary(regzip)

Pearson residuals:
     Min       1Q   Median       3Q      Max
 -0.4946  -0.2306  -0.1827  -0.1038 101.3908

Count model coefficients (poisson with log link):
                      Estimate Std. Error z value Pr(>|z|)
(Intercept)           -0.96211    0.11575  -8.312  < 2e-16 ***
AGEDRIVER(22,26]      -0.39498    0.15828  -2.496   0.0126 *
AGEDRIVER(26,42]      -1.09883    0.12822  -8.570  < 2e-16 ***
AGEDRIVER(42,74]      -1.10280    0.12236  -9.013  < 2e-16 ***
```

```
AGEDRIVER(74,Inf]     -0.74634    0.18353  -4.067 4.77e-05 ***
AGECAR(15,Inf]        -0.24745    0.03167  -7.813 5.57e-15 ***
DENSITY(40,200]        0.18414    0.02746   6.706 1.99e-11 ***
DENSITY(200,500]       0.31706    0.03051  10.391  < 2e-16 ***
DENSITY(500,4.5e+03]   0.53491    0.02679  19.964  < 2e-16 ***
DENSITY(4.5e+03,Inf]   0.67015    0.03691  18.154  < 2e-16 ***
brandFF               -0.18993    0.02538  -7.483 7.27e-14 ***
powerFDEF             -0.17644    0.02506  -7.041 1.91e-12 ***
powerFGH              -0.15233    0.02725  -5.591 2.26e-08 ***
GASRegular            -0.19486    0.01672 -11.655  < 2e-16 ***

Zero-inflation model coefficients (binomial with logit link):
                     Estimate Std. Error z value Pr(>|z|)
(Intercept)         0.0083752  0.2036001   0.041   0.9672
AGEDRIVER(22,26]    0.4143760  0.2695767   1.537   0.1243
AGEDRIVER(26,42]   -0.0003184  0.2382535  -0.001   0.9989
AGEDRIVER(42,74]   -0.0152864  0.2267851  -0.067   0.9463
AGEDRIVER(74,Inf]   0.6404166  0.2952266   2.169   0.0301 *
---
Signif. codes:  0 '***' 0.001 '**' 0.01 '*' 0.05 '.' 0.1 ' ' 1

Number of iterations in BFGS optimization: 61
Log-likelihood: -6.736e+04 on 19 Df
```

14.4.3 Hurdle Models

The Hurdle distribution is based on a dichotomic process, where insureds that report are considered completely different from those who report at least once. For some specific probability mass functions $p_i^{(1)}(.)$ and $p_i^{(2)}(.)$, the probability function of the hurdle is expressed as

$$\mathbb{P}(N_i = k) = \left\{ \begin{array}{l} p_i^{(1)}(0) \text{ if } k = 0 \\ \frac{1-p_i^{(1)}(0)}{1-p_i^{(2)}(0)} p_i^{(2)}(k) \text{ if } k = 1, 2, \cdots \end{array} \right\}.$$

The model collapses to $p_i(.)$ when $p_i^{(1)}(.) = p_i^{(2)}(.) = p_i(.)$, which can be used as a basis for a specification test. The mean and variance corresponding to the model described above are given by

$$\begin{aligned} \mathbb{E}[N_i] &= \frac{1 - p_i^{(1)}(0)}{1 - p_i^{(2)}(.)(0)} \mu_2 \\ \text{var}[N_i] &= \mathbb{P}(N_i > 0)\text{var}[N_i|N_i > 0] + \mathbb{P}(N_i = 0)\mathbb{E}[N_i|N_i > 0], \end{aligned}$$

where μ_2 is the expected value associated with the probability mass function $p_i^{(2)}(.)$. An advantage of the model is its property to have a separable log-likelihood:

$$\begin{aligned} \log \mathcal{L} &= \sum_{i=1}^{n} \mathbf{1}(N_i = 0) \log(p_i^{(1)}(0)) + \mathbf{1}(N_i > 0) \log(1 - p_i^{(1)}(0)) \\ &\quad + \sum_{i=1}^{n} \mathbf{1}(N_i > 0) \log(p_i^{(2)}(N_i)/(1 - p_i^{(2)}(0)). \end{aligned}$$

Then, as done with the number of claims and the cost of the claims, the maximization of the log-likelihood can be done separately for each part (zero case and positive values).

These models can be implemented using the `hurdle` function, from the `pscl` library. If we consider a (truncated) negative binomial distribution for $p^{(2)}$ and a binomial distribution for $p^{(1)}$, on covariates $\boldsymbol{X}_2$ and $\boldsymbol{X}_1$ respectively, the generic code will be

```
> hurdle(NB ~ X1 | X2 , data=CONTRACTS.f, dist = "poisson",
+ zero.dist = "binomial")
```

Here, we obtain

```
> freghrd <- formula(NB ~ AGEDRIVER+AGECAR+DENSITY+brandF+powerF+
+ GAS+offset(log(EXPOSURE))|AGEDRIVER)
> reghrd=hurdle(freghrd,data=CONTRACTS.f,dist = "negbin",zero.
+ dist = "binomial", link = "logit")
> summary(reghrd)

Pearson residuals:
    Min      1Q  Median      3Q     Max
-0.2787 -0.1959 -0.1924 -0.1825 21.7379

Count model coefficients (truncated negbin with log link):
                     Estimate Std. Error z value Pr(>|z|)
(Intercept)          -2.92542    0.94540  -3.094 0.001972 **
AGEDRIVER(22,26]     -0.37124    0.18182  -2.042 0.041171 *
AGEDRIVER(26,42]     -1.10622    0.14957  -7.396 1.40e-13 ***
AGEDRIVER(42,74]     -1.12550    0.14375  -7.829 4.90e-15 ***
AGEDRIVER(74,Inf]    -0.72592    0.20704  -3.506 0.000455 ***
AGECAR(15,Inf]       -0.02225    0.16247  -0.137 0.891071
DENSITY(40,200]       0.45729    0.15381   2.973 0.002949 **
DENSITY(200,500]      0.41196    0.16760   2.458 0.013973 *
DENSITY(500,4.5e+03]  0.69487    0.14832   4.685 2.80e-06 ***
DENSITY(4.5e+03,Inf]  1.14389    0.17446   6.557 5.50e-11 ***
brandFF               0.80303    0.10326   7.777 7.43e-15 ***
powerFDEF             0.19143    0.11868   1.613 0.106739
powerFGH              0.06720    0.12951   0.519 0.603820
GASRegular           -0.10676    0.07777  -1.373 0.169835
Log(theta)           -0.87532    1.33388  -0.656 0.511683
Zero hurdle model coefficients (binomial with logit link):
                 Estimate Std. Error z value Pr(>|z|)
(Intercept)      -2.55504    0.03645  -70.09   <2e-16 ***
AGEDRIVER(22,26] -0.50876    0.04940  -10.30   <2e-16 ***
AGEDRIVER(26,42] -0.82424    0.03908  -21.09   <2e-16 ***
AGEDRIVER(42,74] -0.68664    0.03823  -17.96   <2e-16 ***
AGEDRIVER(74,Inf] -0.59923   0.05535  -10.83   <2e-16 ***
---
Signif. codes:  0 '***' 0.001 '**' 0.01 '*' 0.05 '.' 0.1 ' ' 1

Theta: count = 0.4167
Number of iterations in BFGS optimization: 57
Log-likelihood: -6.869e+04 on 20 Df
```

14.5 Individual Claims, Gamma, Log-Normal, and Other Regressions

In the introduction, we have seen that the pure premium associated to an insured with characteristics $\boldsymbol{x}$ should be $\pi(\boldsymbol{x}) = \mathbb{E}(N|\boldsymbol{X} = \boldsymbol{x}) \cdot \mathbb{E}(Y|\boldsymbol{X} = \boldsymbol{x})$. The first part has been detailed in the previous sections, and now it might be time to propose some models for individual claims losses $\mathbb{E}(Y|\boldsymbol{X} = \boldsymbol{x})$. This section is usually described briefly in actuarial literature, because the tools used are the same as before (Generalized Linear Models); so for statistical reasons, there is no reason to spend too much time here. But also because, in practice, covariates are much less informative to predict amounts than to predict frequency. The dataset contains a claims `ID` (which is a contract number) and an indemnity amount:

```
> tail(CLAIMS)
          ID INDEMNITY
16176 303133       769
16177 302759        61
16178 299443      1831
16179 303389      4183
16180 304313       566
16181 206241      2156
```

It is possible to merge this dataset with the one that contains information about the contracts:

```
> claims   <- merge(CLAIMS,CONTRACTS)
> claims.f <- merge(CLAIMS,CONTRACTS.f)
```

Note that this dataset contains only claims with a strictly positive indemnity (losses claims but filed away are not in this dataset). Thus, what we need to model losses is a distribution on $\mathbb{R}_+$.

14.5.1 Gamma Regression

Y has a Gamma distribution if its density can be written

$$f(y) = \frac{1}{y\Gamma(\varphi^{-1})}\left(\frac{y}{\mu\varphi}\right)^{\varphi^{-1}} \exp\left(-\frac{y}{\mu\varphi}\right), \quad \forall y \in \mathbb{R}_+.$$

This distribution is in the exponential family, as

$$f(y) = \exp\left[\frac{y/\mu - (-\log\mu)}{-\varphi} + \frac{1-\varphi}{\varphi}\log y - \frac{\log\varphi}{\varphi} - \log\Gamma\left(\varphi^{-1}\right)\right], \quad \forall y \in \mathbb{R}_+.$$

The canonical link here is $\theta = \mu^{-1}$, and $b(\theta) = -\log(\mu)$, so that the variance function will be $V(\mu) = \mu^2$. Observe, further, that $\text{var}(Y) = \varphi\mathbb{E}(Y)^2$, and thus the coefficient of variation is constant; here,

$$\frac{\sqrt{\text{var}(Y)}}{\mathbb{E}(Y)} = \varphi.$$

Remark 14.4 *A particular distribution in this family is the exponential distribution, with density*

$$f(y) = \lambda\exp(-\lambda y), \quad \forall y \in \mathbb{R}_+$$

and mean λ^{-1}.

14.5.2 The Log-Normal Model

Y has a log-normal distribution if its density can be written

$$f(y) = \frac{1}{y\sqrt{2\pi\sigma^2}}\, e^{-\frac{(ln \to Pog\, y-\mu)^2}{2\sigma^2}}, \quad \forall y \in \mathbb{R}_+,$$

which is not in the exponential family (so Generalized Linear Models cannot be used to model Y). Nevertheless, recall that Y has a log-normal distribution if $\log Y$ has a Gaussian distribution (which is in the exponential family).

Remark 14.5 *If Y has a log-normal distribution with parameters μ and σ^2, then $Y = \exp[Y^\star]$, where $Y^\star \sim \mathcal{N}(\mu, \sigma^2)$. Thus, μ is not the average of Y, and neither is $\exp[\mu]$. In fact,*

$$\mathbb{E}(Y) = \mathbb{E}(\exp[Y^\star]) \neq \exp\left[\mathbb{E}(Y^\star)\right] = \exp(\mu),$$

from Jensen inequality. One can easily prove that

$$\mathbb{E}(Y) = e^{\mu+\sigma^2/2} = e^{\sigma^2/2}\cdot \exp\left[\mathbb{E}(Y^\star)\right] \text{ and } var(Y) = (e^{\sigma^2}-1)e^{2\mu+\sigma^2}.$$

14.5.3 Gamma versus Log-Normal Models

Consider a Gamma regression for Y_i. Using Taylor's expansion (of order 2),

$$\log(Y_i) \sim \log\mu_i + \frac{1}{\mu_i}[Y_i-\mu_i] - \frac{1}{2\mu_i^2}[Y_i-\mu_i]^2$$

if φ is small. Then, if we take the expected value, it comes that

$$\mathbb{E}(\log Y_i) \sim \log\mu_i - \frac{1}{2}\varphi.$$

Further, one can also prove that $\text{var}(\log Y_i) \sim \varphi$. If we consider a logarithm link function, so that Y_i has variance $\varphi\mu_i^2$ where $\mu_i = \exp[\boldsymbol{X}_i'\boldsymbol{\beta}]$, then

$$\mathbb{E}(\log Y_i) \sim \boldsymbol{X}_i'\boldsymbol{\beta} - \frac{1}{2}\varphi \text{ and } \text{var}(\log Y_i) \sim \varphi.$$

Consider now a log-normal regression, $\log Y_i = \boldsymbol{X}_i'\boldsymbol{\alpha} + \varepsilon_i$, where $\varepsilon_i \sim \mathcal{N}(0,\sigma^2)$ i.i.d. Then

$$\mathbb{E}(\log Y_i) \sim \boldsymbol{X}_i'\boldsymbol{\alpha} \text{ and } \text{var}(\log Y_i) = \sigma^2.$$

Except for the intercept, coefficients $\boldsymbol{\alpha}$ and $\boldsymbol{\beta}$ should be rather close if the coefficient of variation is small

```
> reg.logn <- lm(log(INDEMNITY)~AGECAR+GAS,
+ data=claims[claims$INDEMNITY<15000,])
> reg.gamma <- glm(INDEMNITY~AGECAR+GAS,family=Gamma(link="log"),
+ data=claims[claims$INDEMNITY<15000,])
> summary(reg.gamma)

Coefficients:
              Estimate Std. Error t value Pr(>|t|)
(Intercept)    7.26172    0.02568 282.772  < 2e-16 ***
AGECAR(1,4]    0.02127    0.03095   0.687  0.49194
```

```
AGECAR(4,15]    -0.07575     0.02713  -2.792  0.00525 **
AGECAR(15,Inf]  -0.08397     0.04048  -2.074  0.03807 *
GASRegular      -0.02307     0.01719  -1.342  0.17950
---
Signif. codes:  0 *** 0.001 ** 0.01 * 0.05 . 0.1   1

(Dispersion parameter for Gamma family taken to be 1.166754)

    Null deviance: 13410  on 16002  degrees of freedom
Residual deviance: 13377  on 15998  degrees of freedom
AIC: 261934

Number of Fisher Scoring iterations: 5

> summary(reg.logn)

Coefficients:
                 Estimate Std. Error t value Pr(>|t|)
(Intercept)     6.8475524  0.0247901 276.221  < 2e-16 ***
AGECAR(1,4]     0.0004851  0.0298728   0.016  0.98704
AGECAR(4,15]   -0.0803790  0.0261929  -3.069  0.00215 **
AGECAR(15,Inf] -0.0306088  0.0390797  -0.783  0.43350
GASRegular     -0.0240781  0.0165916  -1.451  0.14674
---
Signif. codes:  0 *** 0.001 ** 0.01 * 0.05 . 0.1   1

Residual standard error: 1.043 on 15998 degrees of freedom
Multiple R-squared: 0.001459,       Adjusted R-squared: 0.00121
F-statistic: 5.845 on 4 and 15998 DF,  p-value: 0.000107
```

Remark 14.6 *One can prove that for a Gamma regression, the inverse of Fisher information for $\widehat{\beta}$ (which will be the asymptotic variance) is $\varphi(\boldsymbol{X}'\boldsymbol{X})^{-1}$, which is exactly the variance of $\widehat{\alpha}$ in the log-normal model.*

14.5.4 Inverse Gaussian Model

Y has an inverse Gaussian distribution if its density can be written

$$f(y) = \left[\frac{\lambda}{2\pi y^3}\right]^{1/2} \exp\left(\frac{-\lambda(y-\mu)^2}{2\mu^2 y}\right), \quad \forall y \in \mathbb{R}_+$$

with expected value μ.

14.6 Large Claims and Ratemaking

If claims are not too large, then the log-normal and the gamma regressions should be quite close, as mentioned in the section above. But consider the following two regressions, on the age of the driver (as a continuous variate):

```
> reg.logn <- lm(log(INDEMNITY)~AGEDRIVER,data=claims)
> reg.gamma <- glm(INDEMNITY~AGEDRIVER,family=Gamma(link="log"),data=claims)
> summary(reg.gamma)

Coefficients:
             Estimate Std. Error t value Pr(>|t|)
(Intercept)  8.095074   0.203485  39.782   <2e-16 ***
AGEDRIVER   -0.009926   0.004304  -2.307   0.0211 *
---
Signif. codes:  0 *** 0.001 ** 0.01 * 0.05 . 0.1   1

(Dispersion parameter for Gamma family taken to be 66.85011)

> summary(reg.logn)

Coefficients:
             Estimate Std. Error t value Pr(>|t|)
(Intercept) 6.7361807  0.0277458 242.782  < 2e-16 ***
AGEDRIVER   0.0020374  0.0005868   3.472 0.000518 ***
---
Signif. codes:  0 *** 0.001 ** 0.01 * 0.05 . 0.1   1

Residual standard error: 1.115 on 16179 degrees of freedom
```

Here, coefficients are significant, but with opposite signs. With a Gamma regression, the younger the driver, the less expensive the claims, while it is the reverse with a log-normal regression. The interpretation is related to large claims: outliers will affect the Gamma regression more than the log-normal one (because it is a regression on the logarithm of losses). On the other hand, on average, predictions obtained with a log-normal model may not be consistent with observed losses: here, the average cost of the claim was

```
> mean(claims$INDEMNITY)
[1] 2129.972
```

and the average predictions were, with the two models,

```
> mean(predict(reg.gamma,type="response"))
[1] 2130.361
> sigma <- summary(reg.logn)$sigma
> mean(exp(predict(reg.logn))*exp(sigma^2/2))
[1] 1718.981
```

So, in order to have a more robust pricing method, we should find a way to deal with large claims (see also Teugels (1982) for a discussion of the concept of *large* claims, or Beirlant & Teugels (1992)). A natural technique is to consider that differentiating premiums should be valid for standard claims, while extremely large ones might be spread between all the insureds, without differentiating (pooling extremely large losses among the insureds).

```
> M <- claims[order(-claims$INDEMNITY),c("INDEMNITY","NB","POWER",
+ "AGECAR","AGEDRIVER","GAS","DENSITY")]
> M$SUM <- cumsum(M$INDEMNITY)/sum(M$INDEMNITY)*100
> head(M)
      INDEMNITY NB POWER AGECAR AGEDRIVER     GAS DENSITY    SUM
```

```
5033    2036833  1    i    13     19 Regular    93  5.909
1854    1402330  2    f    13     20 Regular   203  9.978
4948     306559  1    g     1     21  Diesel   108 10.868
2637     301302  1    f     3     46  Diesel    10 11.742
11566    281403  1    f     4     61  Diesel  1064 12.558
4512     254944  1    d    12     27 Regular   319 13.298
```

The largest claim cost more than 2 million euros, almost 6% of the total loss. The three largest (almost 11% of the total loss) were caused by young drivers (19, 20, and 21 years old, respectively).

14.6.1 Model with Two Kinds of Claims

A standard result in probability theory is that $\mathbb{E}(Y) = \mathbb{E}(\mathbb{E}(Y|Z))$ for any Z. Thus, given a multinomial random variable Z, taking values z_i's,

$$\mathbb{E}(Y) = \sum_i \mathbb{E}(Y|Z = z_i) \cdot \mathbb{P}(Z = z_i)$$

on any probabilistic space. Thus, we can consider any conditional expectation

$$\mathbb{E}(Y|\boldsymbol{X}) = \sum_i \mathbb{E}(Y|Z = z_i, \boldsymbol{X}) \cdot \mathbb{P}(Z = z_i|\boldsymbol{X}).$$

Consider the case where Z is some information about the size of the claim, namely Z belongs either to $\{Y > s\}$ or $\{Y \leq s\}$, for some high amount s. Therefore,

$$\mathbb{E}(Y) = \mathbb{E}(Y|Y > s) \cdot \mathbb{P}(Y > s) + \mathbb{E}(Y|Y \leq s) \cdot \mathbb{P}(Y \leq s).$$

As before, in the case where the probability is computed under conditional probability, given X, the relationship above becomes

$$\mathbb{E}(Y|\boldsymbol{X}) = \underbrace{\mathbb{E}(Y|\boldsymbol{X}, Y \leq s)}_{A} \cdot \underbrace{\mathbb{P}(Y \leq s|\boldsymbol{X})}_{C} + \underbrace{\mathbb{E}(Y|Y > s, \boldsymbol{X})}_{B} \cdot \underbrace{\mathbb{P}(Y > s|\boldsymbol{X})}_{C},$$

where

- A is the average cost of *normal* claims (excluding claims that exceed s)
- B is the average cost of *large* claims (those that exceed s)
- C is the probability of having a large or a normal claim

For part C, a logistic regression can be run. For parts A and B, regressions are on subsets of the dataset. Consider a threshold `s <- 10000`, reached by less than 2% of the claims:

```
> claims$STANDARD <- (claims$INDEMNITY<s)
> mean(claims$STANDARD)
[1] 0.982943
```

Consider a logistic regression to model the probability that a claim will be a standard one:

```
>  library(splines)
>  age <- seq(18,100)
>  regC <- glm(STANDARD~bs(AGEDRIVER),data=claims,family=binomial)
>  ypC <- predict(regC,newdata=data.frame(AGEDRIVER=age),type="response",
     se=TRUE)
```

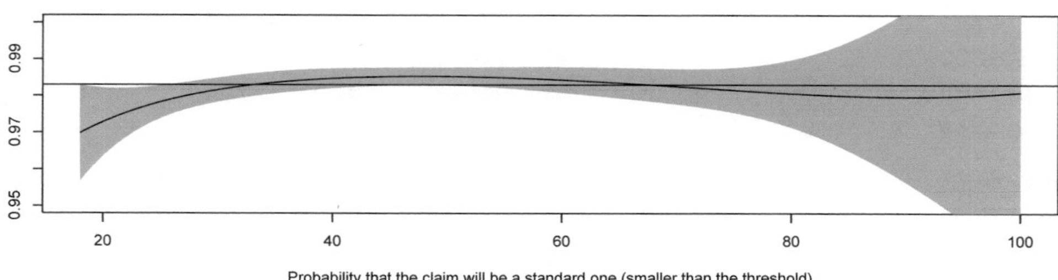

FIGURE 14.3
Probability of having a standard claim, given that a claim occurred, as a function of the age of the driver. Logistic regression with a spline smoother.

The probability can be visualized in Figure 14.3:

```
> plot(age,ypC$fit,ylim=c(.95,1),type="l",)
> polygon(c(age,rev(age)),c(ypC$fit+2*ypC$se.fit,rev(ypC$fit-2*ypC$se.fit)),
+ col="grey",border=NA)
> abline(h=mean(claims$STANDARD),lty=2)
```

For standard and large claims, consider two gamma regressions on the two subsets,

```
>  indexstandard <- which(claims$INDEMNITY<s)
>  mean(claims$INDEMNITY[indexstandard])
[1] 1280.085
>  mean(claims$INDEMNITY[-indexstandard])
[1] 51106.23

> regA <- glm(INDEMNITY~bs(AGEDRIVER),data=claims[indexstandard,],
+ family=Gamma(link="log"))
> ypA <- predict(regA,newdata=data.frame(AGEDRIVER=age),type="response")

> regB <- glm(INDEMNITY~bs(AGEDRIVER),data=claims[-indexstandard,],
+ family=Gamma(link="log"))
> ypB <- predict(regB,newdata=data.frame(AGEDRIVER=age),type="response")
```

In order to compare, let us fit one model on the overall dataset:

```
> reg <- glm(INDEMNITY~bs(AGEDRIVER),data=claims,family=Gamma(link="log"))
> yp <- predict(reg,newdata=data.frame(AGEDRIVER=age),type="response")
```

```
and let us compare the two approaches, on Figure \ref{Fig:GLM:age-small-large}
```

```
> ypC <- predict(regC,newdata=data.frame(AGEDRIVER=age),type="response")
> plot(age,yp,type="l",lwd=2,ylab="Average cost",xlab="Age of the driver")
> lines(age,ypC*ypA+(1-ypC)*ypB,type="h",col="grey",lwd=6)
> lines(age,ypC*ypA,type="h",col="black",lwd=6)
> abline(h= mean(claims$INDEMNITY),lty=2)
```

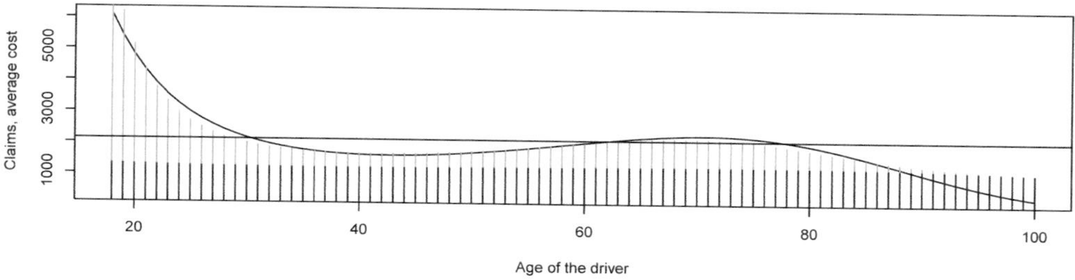

FIGURE 14.4
Average cost of a claim, as a function of the age of the driver, s =10,000.

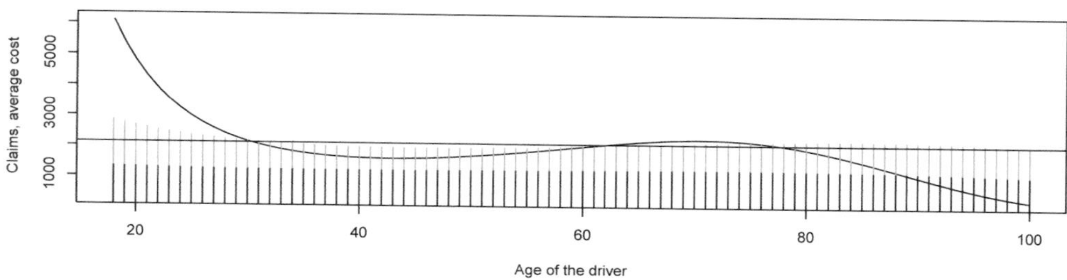

FIGURE 14.5
Average cost of a claim, as a function of the age of the driver, s =10,000, when extremely large claims are pooled among the insured.

The dotted horizontal line in Figure 14.3 is the average cost of a claim. The dark line, in the back, is the prediction on the whole dataset `reg`. The dark part is the part of the average claim related to standard claims (smaller than s) and the lighter area is the part of the average claim due to possible large claims (exceeding s).

For standard and large claims, consider two gamma regressions, on the two subsets. As mentioned earlier, it is possible to substitute constant parts in the pricing model, for example,

$$\mathbb{E}(Y|\boldsymbol{X}) \sim \underbrace{\mathbb{E}(Y|\boldsymbol{X}, Y \leq s)}_{A} \cdot \underbrace{\mathbb{P}(Y \leq s|\boldsymbol{X})}_{C} / + \underbrace{\mathbb{E}(Y|Y > s)}_{B_0} \cdot \underbrace{\mathbb{P}(Y > s|\boldsymbol{X})}_{C}$$

where B_0 is obtained by considering the average cost of large claims, without any explanatory variable.

It is possible to focus not on large claims, but on extremely large claims, `s <- 100000`, as on Figure 14.6, where three smoothed regressions are considered.

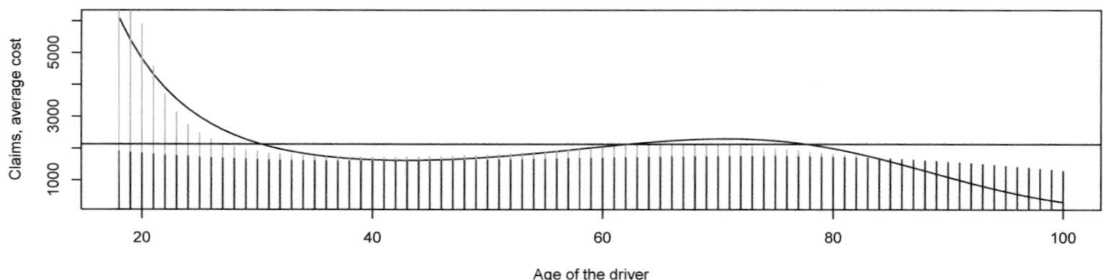

FIGURE 14.6
Average cost of a claim, as a function of the age of the driver, s =100,000.

14.6.2 More General Model

A more general model can be considered here:

$$
\begin{aligned}
\mathbb{E}(Y|\boldsymbol{X}) \quad = \quad & \underbrace{\mathbb{E}(Y|\boldsymbol{X}, Y \leq s_1)}_{A} \cdot \underbrace{\mathbb{P}(Y \leq s_1|\boldsymbol{X})}_{D,\pi_1(\boldsymbol{X})} \\
& + \underbrace{\mathbb{E}(Y|Y \in (s_1, s_2], \boldsymbol{X})}_{B} \cdot \underbrace{\mathbb{P}(Y \in (s_1, s_2]|\boldsymbol{X})}_{D,\pi_2(\boldsymbol{X})} \\
& + \underbrace{\mathbb{E}(Y|Y > s_2, \boldsymbol{X})}_{C} \cdot \underbrace{\mathbb{P}(Y > s_2|\boldsymbol{X})}_{D,\pi_3(\boldsymbol{X})},
\end{aligned}
$$

where mixing probabilities $(\pi_1(\boldsymbol{X}), \pi_2(\boldsymbol{X}), \pi_3(\boldsymbol{X}))$ are associated to a multinomial random variable.

A multinomial regression model can be written, as an extension of the logistic regression one. Assume here that

$$(\pi_1, \pi_2, \pi_3) \propto (\exp(\boldsymbol{X}'\boldsymbol{\beta}_1), \exp(\boldsymbol{X}'\boldsymbol{\beta}_2), 1).$$

The estimation of coefficients $\boldsymbol{\beta}_1$ and $\boldsymbol{\beta}_2$ (including standard errors) can be obtained using regression `multinom` of `library(nnet)`:

```
> library(nnet)
> threshold <- c(0,1150,10000,Inf)
> regD <- multinom(cut(claims$INDEMNITY,breaks=threshold)~bs(AGEDRIVER),data=claims)
# weights:  15 (8 variable)
initial  value 17776.645443
iter  10 value 12391.379124
final  value 12389.058985
converged
> summary(regD)
Call:
multinom(formula = cut(claims$INDEMNITY, breaks = threshold) ~
    bs(AGEDRIVER), data = claims)

Coefficients:
                  (Intercept) bs(AGEDRIVER)1 bs(AGEDRIVER)2 bs(AGEDRIVER)3
(1.15e+03,1e+04]     0.126849     -0.5866635      0.4435400      0.2998493
```

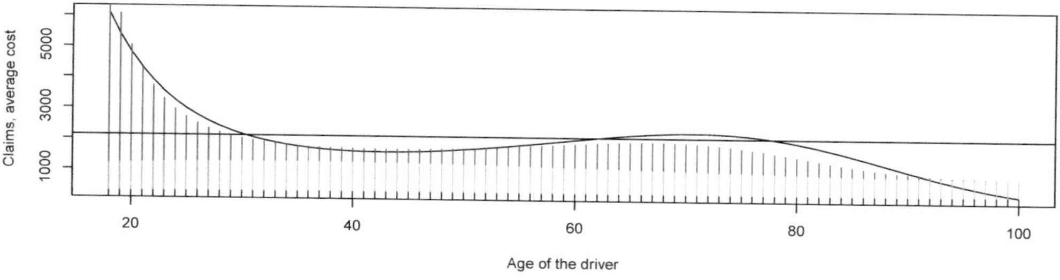

FIGURE 14.7
Probability of having a standard claim, given that a claim occurred, as a function of the age of the driver. Logistic regression with a spline smoother.

```
(1e+04,Inf]         -2.716344     -1.8494277       0.1237384      -0.2909207

Std. Errors:
                  (Intercept) bs(AGEDRIVER)1 bs(AGEDRIVER)2 bs(AGEDRIVER)3
(1.15e+03,1e+04]   0.06893703      0.2353106      0.2562023      0.3562258
(1e+04,Inf]        0.22592318      0.8219250      0.9575998      1.2674017

Residual Deviance: 24778.12
AIC: 24794.12
```

For instance, for drivers with age 20 or 50, given that a claim has occurred, the probability that it could be in one of the three tranches will be

```
> predict(regD,newdata=data.frame(AGEDRIVER=c(20,50)),type="probs")
  (0,1.15e+03] (1.15e+03,1e+04] (1e+04,Inf]
1    0.4655077        0.5074693  0.02702298
2    0.4885641        0.4967026  0.01473323
```

If we plot the parts due to small claims (less than s_1), medium claims (from s_1 to s_2), and large claims (more than s_2), we obtain the graph in Figure 14.7

14.7 Modeling Compound Sum with Tweedie Regression

As a conclusion, we can mention a large class of Generalized Linear Models, called Tweedie models in Jørgensen (1987), studied in Jørgensen (1997). A Tweedie distribution is in the exponential family, and it satisfies

$$\text{var}(Y) = \varphi[\mathbb{E}(Y)]^p.$$

If $p = 0$ is the distribution with constant variance (normal distribution), if $p = 1$, then the variance function is linear (Poisson distribution); and if $p = 2$, it has a quadratic variance function (Gamma distribution).

An interesting case is when p lies in the interval $(1, 2)$. In that case, Y is a compound Poisson-Gamma distribution. A natural idea is then to use a Tweedie model on the aggregated sum, per insured.

```
> A <- tapply(CLAIMS$INDEMNITY,CLAIMS$ID,sum)
> ADF <- data.frame(ID=names(A),INDEMNITY=as.vector(A))
> CT <- merge(CONTRACTS,ADF,all.x = TRUE)
> CT$INDEMNITY[is.na(CT$INDEMNITY)] <- 0
> tail(CT)
        NB EXPOS POWER AGECAR AGEDRIVER BRAND     GAS REGION DENSITY IND
 413164  0 0.005     d      0        61     F Regular     25     205   0
 413165  0 0.002     j      0        29     F  Diesel     11    2471   0
 413166  0 0.005     d      0        29     F Regular     11    5360   0
 413167  0 0.005     k      0        49     F  Diesel     11    5360   0
 413168  0 0.002     d      0        41     F Regular     11    9850   0
 413169  0 0.002     g      6        29     F  Diesel     72      65   0
> CT.f=merge(CONTRACTS.f,ADF,all.x = TRUE)
> CT.f$INDEMNITY[is.na(CT.f$INDEMNITY)] <- 0
> tail(CT.f)
        NB EXPOS POWER AGECAR AGEDRIVER BRAND     GAS REGION          DENSITY IND
 413164  0 0.005     d  [0,1]    (42,74]     F Regular     25       (200,500]   0
 413165  0 0.002     j  [0,1]    (26,42]     F  Diesel     11 (500,4.5e+03]   0
 413166  0 0.005     d  [0,1]    (26,42]     F Regular     11 (4.5e+03,Inf]   0
 413167  0 0.005     k  [0,1]    (42,74]     F  Diesel     11 (4.5e+03,Inf]   0
 413168  0 0.002     d  [0,1]    (26,42]     F Regular     11 (4.5e+03,Inf]   0
 413169  0 0.002     g (4,15]    (26,42]     F  Diesel     72        (40,200]   0
```

Using library tweedie we obtain

```
> out <- tweedie.profile(INDEMNITY~POWER+AGECAR+
+ AGEDRIVER+BRAND+GAS+DENSITY,data=CT.f,
+ p.vec=seq(1.05,1.95,by=.05) )
```

We can then plot the profile likelihood function associated to the Tweedie regression:

```
> out$p.max
[1] 1.564286
> plot(out,type="b")
> abline(v=out$p.max,lty=2)
```

In order to ensure convergence of the algorithm, we can use coefficients obtained from a Poisson regression:

```
> reg1 <- glm(INDEMNITY~POWER+AGECAR+AGEDRIVER+BRAND+GAS+DENSITY,
+ data=CT.f, family = tweedie(var.power = 1, link.power = 0))
```

The following function returns $\widehat{\boldsymbol{\beta}}^{(p)}$ as a function of p:

```
> coef <- function(p){
+ glm(INDEMNITY~POWER+AGECAR+AGEDRIVER+BRAND+GAS+DENSITY,
+ data=CT.f, family = tweedie(var.power = p, link.power = 0),
+ start=reg1$coefficients)$coefficients}
```

It is also possible to visualize the evolution of the coefficients estimates as a function of p (without the (Intercept)), with a logarithm link function:

```
> vp <- seq(1,2,by=.1)
> Cp <- Vectorize(coef)(vp)
> matplot(vp,t(Cp[-1,]),type="l")
> text(2,Cp[-1,length(vp)],rownames(Cp[-1,]),cex=.5,pos=4)
```

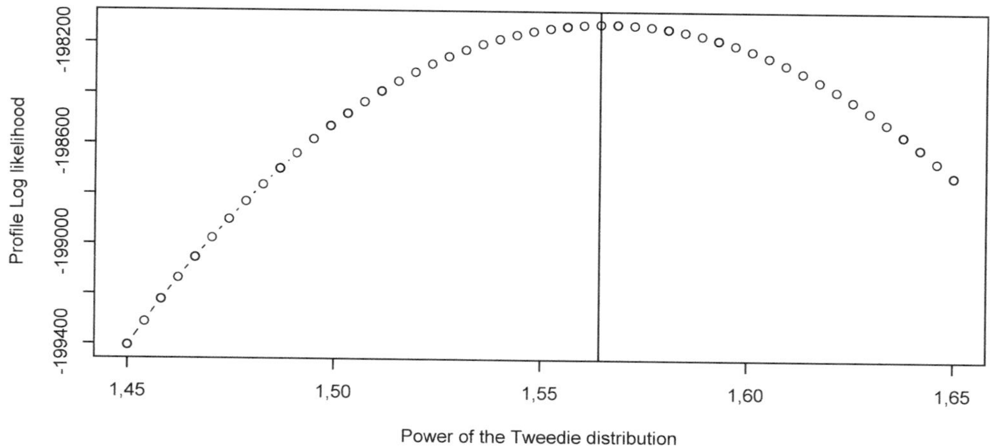

FIGURE 14.8
Profile likelihood of a Tweedie regression, as a function of p.

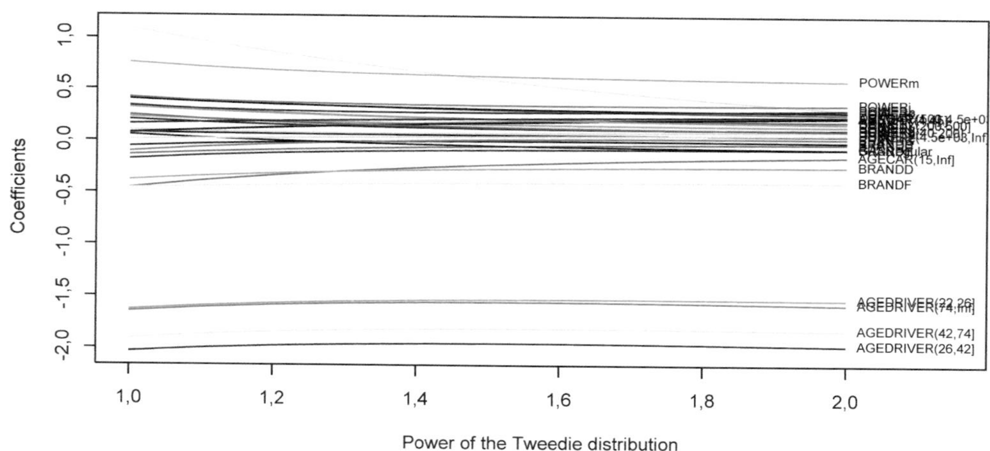

FIGURE 14.9
Evolution of $\widehat{\boldsymbol{\beta}}^{(p)}$'s from a Tweedie regression, as a function of p

One can also use library `cplm` for fitting Tweedie compound Poisson linear models.
Smyth & Jørgensen (2002) suggested a Tweedie model using a joint likelihood function (for claim cost and frequency).

14.8 Exercises

14.1. In the minimum bias method, based on the least squares approach we have to minimize

$$D = \sum_{i,j} E_{i,j}(N_{ij} - L_i \cdot C_j)^2.$$

Write the first-order condition, and solve this equation, numerically, on the same variables as in Section 14.2.4.

14.2. In the minimum bias method, based on the chi-square approach, we have to minimize

$$Q = \sum_{i,j} \frac{E_{i,j}(N_{i,j} - L_i \cdot C_j)^2}{L_i \cdot C_j}.$$

Write the first-order condition, and solve this equation, numerically, on the same variables as in Section 14.2.4.

15

Longitudinal Data and Experience Rating

Katrien Antonio
Universiteit van Amsterdam
Amsterdam, Netherlands

Peng Shi
University of Wisconsin Madison
Madison, Wisconsin, USA

Frank van Berkum
Universiteit van Amsterdam
Amsterdam, Netherlands

CONTENTS

15.1 Motivation

15.1.1 A Priori Rating for Cross-Sectional Data

In Chapter 14 on "General Insurance Pricing," Boucher & Charpentier discuss regression techniques suitable for pricing with cross-sectional data. A cross-sectional dataset observes each subject in the sample (for instance, a policy(holder)) only once. Each subject is therefore described by a single response, say Y_i for subject i, and vector with covariate information, $\boldsymbol{x}_i$. Assuming independence between subjects, (Generalized) Linear Models [(G)LMs] are directly available for statistical modeling and explain the response as a function of the risk factors, within an appropriate distributional framework. For pricing in general insurance, the actuary builds such models for a (cross-sectional) dataset with claim counts on the one hand and claim severities on the other hand as response variables, obtained by following policyholders during a single time period. The result is a tariff based on risk classification through regression modeling. When the explanatory variables used as rating factors express *a priori* correctly measurable information about the policyholder (or, for instance, the vehicle or the insured building), the system is called an *a priori* classification scheme. The examples in Chapter 14 illustrate this idea and use, for instance, age of the driver and age of the car to explain the number of claims registered by a policyholder.

15.1.2 Experience Rating for Panel Data

However, despite the presence of an *a priori* rating system, some important risk factors remain unmeasurable or unobservable. For example, in automobile insurance, the insurer is unable to detect the driver's aggressiveness behind the wheel or the quickness of his reflexes to avoid a possible accident (see Denuit et al. (2007) for further motivation). This motivates the presence of inhomogeneous tariff cells within an *a priori* rating system. An *a posteriori* or *experience rating* system is necessary to allow for the reevaluation of the premium (established *a priori*) based on the history of claims as reported by the insured. One can argue that an important predictor for the future number of claims reported by an insured will be the number of claims reported in the past. Predictive modeling for experience rating will confront analysts with data structures going beyond the cross-sectional design dealt with in (G)LMs. Longitudinal (or panel) data arise when the claim history of policyholders (or, in general, a group of "subjects") is registered repeatedly over time. Thus, with longitudinal data, the variables will have double subscripts, indicating the subject and observation period, respectively. Specifically, let Y_{it} denote the response for the i-th subject in the t-th time period, and let $\mathbf{x}_{it}$ denote the associated vector of explanatory variables. Assuming that there are n subjects and following the i-th subject over $t = 1, \cdots, T_i$ time periods, we observe

$$\begin{array}{ll} \text{1st subject} & \{(y_{11}, \mathbf{x}_{11}), (y_{12}, \mathbf{x}_{12}), \cdots, (y_{11}, \mathbf{x}_{1T_1})\} \\ \text{2nd subject} & \{(y_{21}, \mathbf{x}_{21}), (y_{22}, \mathbf{x}_{22}), \cdots, (y_{21}, \mathbf{x}_{2T_2})\} \\ \vdots & \vdots \\ n\text{-th subject} & \{(y_{n1}, \mathbf{x}_{n1}), (y_{n2}, \mathbf{x}_{n2}), \cdots, (y_{n1}, \mathbf{x}_{nT_n})\} \end{array}$$

Longitudinal data have several potential advantages. First, longitudinal data are a hybrid of cross-sectional and time series data. On the one hand, they allow for the examination of the effects of covariates on the response, as in usual regression. On the other hand, similar to time series analysis, they also permit the identification of *dynamic* relations over time. Because they share subject-specific characteristics, observations on the same subject over time are correlated and require an adjusted toolkit for statistical modeling. In this chapter we study regression models incorporating these dynamics, among others, by extending *a priori* rating with so-called random effects. These random effects structure correlation between observations registered on the same subject, and also take heterogeneity among subjects, due to unobserved characteristics, into account.

15.1.3 From Panel to Multilevel Data

The panel data setting has two layers (or levels) of data: the *time* level on the one hand and the *subject* level on the other hand. However, insurers may have several other layers of data at their disposal. For example, Antonio et al. (2010) discuss experience rating for a dataset on fleet covers, registered for multiple insurance companies. Fleet policies are umbrella-type policies issued to customers whose insurance covers more than a single vehicle. The hierarchical or multilevel structure of the data is as follows: vehicles (v) observed over time (t), nested within fleets (f), with policies issued by insurance companies (c). Multilevel models allow for incorporating the hierarchical structure of the data by specifying random effects at the various levels in the data. Once again, these random effects represent unobservable characteristics at each level. Moreover, random effects allow *a posteriori* updating of an *a priori* tariff, by taking into account the past performance of—in the case of intercompany fleet contracts—the vehicle, fleet, and company.

15.1.4 Structure of the Chapter

Section 15.2 of this chapter considers linear models for longitudinal data. We discuss three approaches to capture unobserved heterogeneity in the longitudinal data context. Section 15.2.2 introduces the basic fixed effects model and describes the model specification and diagnostics. Section 15.2.3 extends these models to incorporate serial correlation in error terms. Section 15.2.4 presents models with random effects and generalizes the framework to linear mixed models. Section 15.2.5 covers the prediction for the linear mixed effects model and points out its connection to the actuarial credibility theory. The computational aspects are illustrated using a dataset introduced in Section 15.2.1. In Section 15.3 we leave the framework of linear models and switch to a distributional framework that is probably more appealing to actuaries, namely the generalized linear models and their random effects extensions. *Actuarial credibility* systems are examples of *a posteriori* rating systems accounting for the history of claims as it emerges for an individual risk. Commercial versions of these experience rating schemes are more widely known in practice as *Bonus–Malus scales*. A case study (using R) with such rating schemes is the topic of Section 15.3.2. The theory on longitudinal data models is based on Diggle et al. (2002), Frees (2004), Hsiao (2003), Wooldridge (2010), and the references therein. Section 15.3 of this chapter is based on Antonio & Valdez (2012), Antonio & Zhang (2014), and Denuit et al. (2007) but focus now on implementation with R. We refer to these papers and the references therein for more technical background. This chapter only covers examples with panel data. We refer to Antonio et al. (2010) for examples with multilevel data structures.

15.2 Linear Models for Longitudinal Data

15.2.1 Data

For linear longitudinal data models, we demonstrate the theory and computational aspects using a dataset of automobile bodily injury liability claims that was described and employed in Frees & Wang (2005). The dataset contains claims of 6 years from 1993 to 1998 for a random sample of twenty-nine towns in the state of Massachusetts. All variables in monetary values are rescaled using the consumer price index to mitigate the effect of time trends. We are interested in the behavior of average claims per unit of exposure, that is, the pure premium, for each town and each year. Two explanatory variables are available for the

TABLE 15.1
Description of variables in the auto claim dataset.

Variable	Description
TOWNCODE	The index of Massachusetts towns
YEAR	The calendar year of the observation
AC	Average claims per unit of exposure
PCI	Per-capita income of the town
PPSM	Population per square mile of the town

regression analysis, the per-capita income (PCI) and the population per square mile (PPSM) of each town. The variables and their descriptions are summarized in Table 15.1.

```
> #  File name is AutoClaimData.txt
> AutoClaim = read.table(choose.files(), sep ="", quote = "",header=TRUE)

> names(AutoClaim)
[1] "TOWNCODE" "YEAR"     "AC"         "PCI"        "PPSM"

> AutoClaim[1:12,]  # Check longitudinal structure
   TOWNCODE YEAR       AC      PCI      PPSM
1        10 1993 160.8522 18134.04 1475.5515
2        10 1994 158.3382 18495.88 1461.8110
3        10 1995 156.8098 18778.29 1488.9911
4        10 1996 168.9899 18740.46 1502.9322
5        10 1997 171.8229 18809.62 1534.4251
6        10 1998 153.7644 19034.59 1557.6937
7        11 1993 149.3873 15597.56  855.4350
8        11 1994 137.5546 15908.79  877.2725
9        11 1995 169.9164 16151.69  872.8024
10       11 1996 169.0598 16119.15  898.7802
11       11 1997 161.3425 16178.64  929.2647
12       11 1998 138.0516 16372.14  940.9162
```

We use the data in the first 5 years, namely 1993–1997, to develop the model and keep the observations in the final year for validation purposes. To explore relations among variables, the techniques used for usual regressions such as histogram and correlations statistics are ready to apply for longitudinal data. In addition, we introduce several more specialized techniques. The first is the multiple time series plot as exhibited in Figure 15.1, where the average claims in multiple years for each town are joined using straight lines. The plot shows the development of claims over time and helps visualize town-specific effects.

```
# Use year 1993-1997 as trainning data and reserve year 1998 for validation
AutoClaimIn <- subset(AutoClaim, YEAR < 1998)

> # Multiple time series plot
> plot(AC ~ YEAR, data = AutoClaimIn, ylab="Average Claim", xlab="Year")
>   for (i in AutoClaimIn$TOWNCODE) {
+   lines(AC ~ YEAR, data = subset(AutoClaimIn, TOWNCODE == i)) }
```

One can also use scatterplots to help detect the relation between the response and explanatory variables. Figure 15.2 displays the scatterplot for variables `PCI` and `PPSM`, suggesting

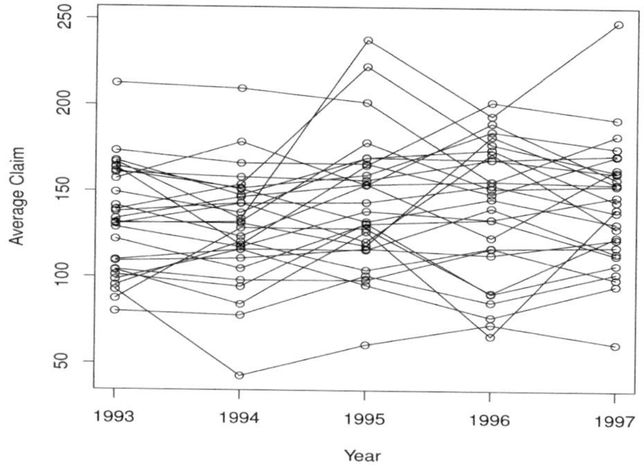

FIGURE 15.1
Multiple time series plot of average claims.

the negative relation between AC and PCI and the positive relation between AC and PPSM. Note that we use both PCI and PPSM in log scale, and logarithmic values will be used in the following analysis. In addition, we also serially connect the observations to identify potential patterns in each covariate. In this case, we observe that PCI varies over time and PPSM is relatively statable.

```
> # Scatter plot to explore relations
> AutoClaimIn$lnPCI <- log(AutoClaimIn$PCI)
> AutoClaimIn$lnPPSM <- log(AutoClaimIn$PPSM)

> plot(AC ~ lnPCI, data = AutoClaimIn, ylab="Average Claim", xlab="PCI")
>  for (i in AutoClaimIn$TOWNCODE) {
+  lines(AC ~ lnPCI, data = subset(AutoClaimIn, TOWNCODE == i)) }

> plot(AC ~ lnPPSM, data = AutoClaimIn, ylab="Average Claim", xlab="PPSM")
>  for (i in AutoClaimIn$TOWNCODE) {
+  lines(AC ~ lnPPSM, data = subset(AutoClaimIn, TOWNCODE == i)) }
```

As a preliminary analysis, we consider a pooled cross-sectional regression model Pool.fit assuming all observations are independent, that is,

$$y_{it} = \alpha + \mathbf{x}_{it}^{'}\boldsymbol{\beta} + \epsilon_{it}. \tag{15.1}$$

Here, α is the homogeneous intercept for all towns and $\boldsymbol{\beta}$ is the vector of regression coefficients. Variables PCI, PPSM, and YEAR are included as covariates. As expected, we observe a significant negative effect of PCI and a positive effect of PPSM. We also observe an increasing trend in claims after purging off the inflation. Functions such as lm and anova are used to fit and analyze the ordinary least squares regression:

```
> AutoClaimIn$YEAR <- AutoClaimIn$YEAR-1992
> Pool.fit <- lm(AC ~ lnPCI+lnPPSM+YEAR, data=AutoClaimIn)
```

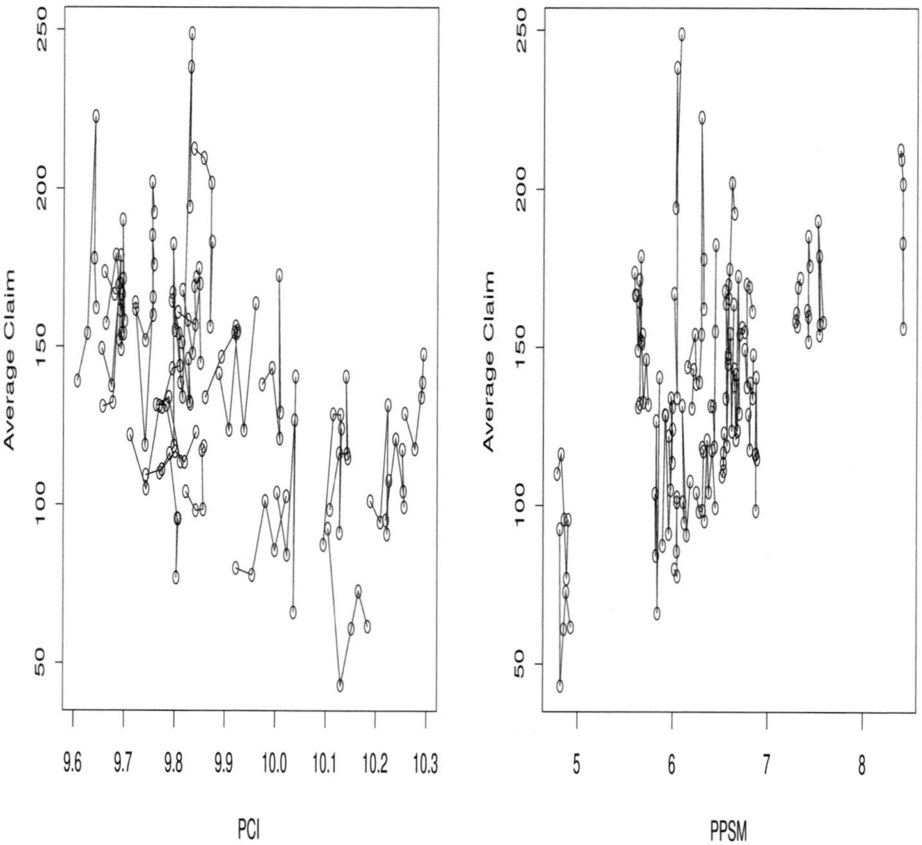

FIGURE 15.2
Scatterplot between average claims and explanatory variables.

```
> summary(Pool.fit)

Call:
lm(formula = AC ~ lnPCI + lnPPSM + YEAR, data = AutoClaimIn)

Residuals:
    Min      1Q  Median      3Q     Max
-49.944 -16.154  -1.759  14.300 104.468

Coefficients:
             Estimate Std. Error t value Pr(>|t|)
(Intercept)  899.569    120.150   7.487 6.98e-12 ***
lnPCI        -92.604     11.855  -7.812 1.17e-12 ***
```

```
lnPPSM        22.305     2.933   7.606 3.64e-12 ***
YEAR           3.923     1.519   2.583  0.01082 *
---
Signif. codes:  0 '***' 0.001 '**' 0.01 '*' 0.05 '.' 0.1 ' ' 1

Residual standard error: 25.75 on 141 degrees of freedom
Multiple R-squared:  0.4908,    Adjusted R-squared:   0.48
F-statistic:  45.3 on 3 and 141 DF,  p-value: < 2.2e-16

> anova(Pool.fit)
Analysis of Variance Table

Response: AC
           Df Sum Sq Mean Sq F value    Pr(>F)
lnPCI       1  46355   46355 69.9028 5.402e-14 ***
lnPPSM      1  39344   39344 59.3302 2.141e-12 ***
YEAR        1   4423    4423  6.6704   0.01082 *
Residuals 141  93502     663
---
Signif. codes:  0 '***' 0.001 '**' 0.01 '*' 0.05 '.' 0.1 ' ' 1
```

15.2.2 Fixed Effects Models

Repeated observations allow one to study the heterogeneity, be it either of subject or of time. We begin with the basic fixed effects model by introducing subject-specific intercepts in the model

$$y_{it} = \alpha_i + \mathbf{x}_{it}^{'}\boldsymbol{\beta} + \epsilon_{it}. \tag{15.2}$$

Hereby, α_i is a town-specific intercept ($i = 1, \cdots, n$); $\mathbf{x}_{it} = (x_{it,1}, \cdots, x_{it,K})^{'}$ is the vector of covariates; and $\boldsymbol{\beta} = (\beta_1, \cdots, \beta_K)^{'}$ is the vector of regression coefficients to be estimated. There are alternative methods to treat the heterogeneous intercepts. In this section, we assume $\{\alpha_i\}$ are fixed parameters to be estimated along with $\boldsymbol{\beta}$. Here, $\boldsymbol{\beta}$ is known as the population parameter capturing the common effects of explanatory variables. $\{\alpha_i\}$, called *nuisance* parameters, vary by subject (here, town) and account for the subject heterogeneity. In the following, we will be using notations $T = \max\{T_1, \cdots, T_n\}$ and $N = \sum_i^n T_i$.

The basic fixed effects model assumes that there is no within-subject serial correlation, that is, ϵ_{it} are i.i.d. random variables with mean zero and variance σ^2. Thus, by the Gauss-Markov theorem, the OLS estimates are the best linear unbiased estimates with

$$\hat{\boldsymbol{\beta}} = \left(\sum_{i=1}^{n}\sum_{t=1}^{T_i}(\mathbf{x}_{it} - \bar{\mathbf{x}}_i)(\mathbf{x}_{it} - \bar{\mathbf{x}}_i)^{'}\right)^{-1}\left(\sum_{i=1}^{n}\sum_{t=1}^{T_i}(\mathbf{x}_{it} - \bar{\mathbf{x}}_i)(y_{it} - \bar{y}_i)^{'}\right) \tag{15.3}$$

and

$$\hat{\alpha}_i = \bar{y}_i - \bar{\mathbf{x}}_i^{'}\hat{\boldsymbol{\beta}}. \tag{15.4}$$

Here, $\bar{y}_i$ and $\bar{\mathbf{x}}_i$ are averages of $\{y_{it}\}$ and $\{\mathbf{x}_{it}\}$ over time, respectively. The above is also known as within estimator because it uses the time variation within each cross section. In addition, the variance of $\hat{\boldsymbol{\beta}}$ is shown to be

$$\text{Var}\,\hat{\boldsymbol{\beta}} = s^2\left(\sum_{i=1}^{n}\sum_{t=1}^{T_i}(\mathbf{x}_{it} - \bar{\mathbf{x}}_i)(\mathbf{x}_{it} - \bar{\mathbf{x}}_i)^{'}\right)^{-1}, \tag{15.5}$$

where s^2 is the unbiased estimate of σ^2 using residuals. In deriving the large sample property, one assumes $n \to \infty$ and T remaining fixed. Under regular conditions, one can show that $\hat{\boldsymbol{\beta}}$ is consistent and asymptotically normally distributed. However, $\{\alpha_i\}$ are not consistent and are not even approximately normal if the responses are not normally distributed.

We fit this basic fixed effects model `FE.fit` using `lm` by treating `TOWNCODE` as a categorical variable. The t- and F-statistics are constructed in the same way as in classical regression models. Note that the above model could be easily modified to account for time-specific heterogeneity by replacing α_i with λ_t. Similarly, using categorical variables for the time dimension, least squares estimation is readily applied.

```
> # Basic fixed-effects model
> FE.fit <- lm(AC ~ factor(TOWNCODE)+lnPCI+lnPPSM+YEAR-1, data=AutoClaimIn)
> summary(FE.fit)

Call:
lm(formula = AC ~ factor(TOWNCODE) + lnPCI + lnPPSM + YEAR -
    1, data = AutoClaimIn)

Residuals:
    Min      1Q  Median      3Q     Max
-55.645  -8.900   0.177   8.995  50.141

Coefficients:
                   Estimate Std. Error t value Pr(>|t|)
factor(TOWNCODE)10 1660.321   1846.793   0.899    0.371
factor(TOWNCODE)11 1558.851   1794.617   0.869    0.387
factor(TOWNCODE)12 1554.375   1884.831   0.825    0.411
factor(TOWNCODE)13 1360.128   1731.874   0.785    0.434
factor(TOWNCODE)14 1443.895   1780.094   0.811    0.419
factor(TOWNCODE)15 1681.983   1841.401   0.913    0.363
(et cetera)
lnPCI               -22.631    159.268  -0.142    0.887
lnPPSM             -176.831    107.240  -1.649    0.102
YEAR                  5.947      2.738   2.172    0.032 *
---
Signif. codes:  0 '***' 0.001 '**' 0.01 '*' 0.05 '.' 0.1 ' ' 1

Residual standard error: 18.88 on 113 degrees of freedom
Multiple R-squared:  0.9863,    Adjusted R-squared:  0.9824
F-statistic: 254.4 on 32 and 113 DF,  p-value: < 2.2e-16

> anova(FE.fit)
Analysis of Variance Table

Response: AC
                  Df  Sum Sq Mean Sq  F value  Pr(>F)
factor(TOWNCODE)  29 2897069   99899 280.3677 < 2e-16 ***
lnPCI              1    2231    2231   6.2621 0.01377 *
lnPPSM             1      34      34   0.0967 0.75638
YEAR               1    1681    1681   4.7168 0.03196 *
Residuals        113   40263     356
```

```
---
Signif. codes:  0 '***' 0.001 '**' 0.01 '*' 0.05 '.' 0.1 ' ' 1
```

We further discuss three specific tests for model specification and diagnostics. The first is the pooling test, where one wishes to test whether the subject-specific effect is significant. The null hypothesis is

$$H_0 : \alpha_1 = \alpha_2 = \cdots = \alpha_n = \alpha.$$

This can be done using the partial $F-$ (Chow) test (see Chow (1960)) by calculating

$$\text{F-ratio} = \frac{(ErrorSS)_{Pooled} - ErrorSS}{(n-1)s^2}.$$

Here, $ErrorSS$ and s^2 are from the heterogeneous model (i.e., `FE.fit`) and $(ErrorSS)_{Pooled}$ are from the homogeneous model (i.e., `pool.fit`). It can be shown that $F{-}ratio$ follows an F-distribution with degrees of freedom $df_1 = n-1$ and $df_2 = N-(n+K)$. In this example, the F-statistic is equal to $(93,502 - 40,263)/(29-1)/18.88^2 = 5.33$, so we reject the null hypothesis.

```
> anova(Pool.fit,FE.fit)
Analysis of Variance Table

Model 1: AC ~ lnPCI + lnPPSM + YEAR
Model 2: AC ~ factor(TOWNCODE) + lnPCI + lnPPSM + YEAR - 1
  Res.Df   RSS Df Sum of Sq      F    Pr(>F)
1    141 93502
2    113 40263 28     53238 5.3362 7.214e-11 ***
---
Signif. codes:  0 '***' 0.001 '**' 0.01 '*' 0.05 '.' 0.1 ' ' 1
```

15.2.3 Models with Serial Correlation

An alternative approach to capture heterogeneity is to use serial correlation. The intuition is that if there are some unobserved time constant variables affecting the response, they will introduce correlation among repeated observations. To motivate this approach, we examine the serial correlation of residuals from `Pool.fit`. The results show strong temporal correlation among AC after removing the effects of explanatory variables. This suggests that the i.i.d. assumption used in the homogeneous model is not appropriate.

```
> # Correlation among residuals
> AutoClaimIn$rPool <- resid(Pool.fit)
> rvec <- cbind(subset(AutoClaimIn,YEAR==1)$rPool,subset(AutoClaimIn,YEAR==2)$rPool,
+ subset(AutoClaimIn,YEAR==3)$rPool,subset(AutoClaimIn,YEAR==4)$rPool,
+ subset(AutoClaimIn,YEAR==5)$rPool)
> cor(rvec)
          [,1]      [,2]      [,3]      [,4]      [,5]
[1,] 1.0000000 0.5862895 0.5187797 0.4207831 0.5424555
[2,] 0.5862895 1.0000000 0.3911814 0.2164202 0.2555096
[3,] 0.5187797 0.3911814 1.0000000 0.3955654 0.7890728
[4,] 0.4207831 0.2164202 0.3955654 1.0000000 0.4778912
[5,] 0.5424555 0.2555096 0.7890728 0.4778912 1.0000000
```

To relax the i.i.d. assumption, we first consider a homogeneous model with serial correlation. For subject i, the matrix presentation of the model is

$$\mathbf{y}_i = \mathbf{X}_i\boldsymbol{\beta} + \boldsymbol{\epsilon}_i, \tag{15.6}$$

where

$$\mathbf{y}_i = \begin{pmatrix} y_{i1} \\ y_{i2} \\ \vdots \\ y_{iT_i} \end{pmatrix}, \; \mathbf{X}_i = \begin{pmatrix} x_{i1,1} & x_{i1,2} & \cdots & x_{i1,K} \\ x_{i2,1} & x_{i2,2} & \cdots & x_{i2,K} \\ \vdots & \vdots & \cdots & \vdots \\ x_{iT_i,1} & x_{iT_i,2} & \cdots & x_{iT_i,K} \end{pmatrix} = \begin{pmatrix} \mathbf{x}_{i1}^{'} \\ \mathbf{x}_{i2}^{'} \\ \vdots \\ \mathbf{x}_{iT_i}^{'} \end{pmatrix}, \; \boldsymbol{\epsilon}_i = \begin{pmatrix} \epsilon_{i1} \\ \epsilon_{i2} \\ \vdots \\ \epsilon_{iT_i} \end{pmatrix}. \tag{15.7}$$

Now we assume that $\boldsymbol{\epsilon}_i$ are correlated with $\text{Var}(\boldsymbol{\epsilon}_i) = \mathbf{R}_i$. Let $\mathbf{R} = \mathbf{R}(\boldsymbol{\tau})$ denote the $T \times T$ temporal covariance matrix for a vector of T observations. Unknown parameters in this covariance matrix are denoted with $\boldsymbol{\tau}$. Note there are at most $T(T+1)/2$ unknown elements in $\mathbf{R}$. Commonly used special cases of $\mathbf{R}$ are (using $T=5$):

$$\text{Independent } \mathbf{R} = \begin{pmatrix} \sigma^2 & 0 & 0 & 0 & 0 \\ 0 & \sigma^2 & 0 & 0 & 0 \\ 0 & 0 & \sigma^2 & 0 & 0 \\ 0 & 0 & 0 & \sigma^2 & 0 \\ 0 & 0 & 0 & 0 & \sigma^2 \end{pmatrix} \quad \text{Compound Symmetry } \mathbf{R} = \sigma^2 \begin{pmatrix} 1 & \rho & \rho & \rho & \rho \\ \rho & 1 & \rho & \rho & \rho \\ \rho & \rho & 1 & \rho & \rho \\ \rho & \rho & \rho & 1 & \rho \\ \rho & \rho & \rho & \rho & 1 \end{pmatrix} \quad \text{AR(1) } \mathbf{R} = \sigma^2 \begin{pmatrix} 1 & \rho & \rho^2 & \rho^3 & \rho^4 \\ \rho & 1 & \rho & \rho^2 & \rho^3 \\ \rho^2 & \rho & 1 & \rho & \rho^2 \\ \rho^3 & \rho^2 & \rho & 1 & \rho \\ \rho^4 & \rho^3 & \rho^2 & \rho & 1 \end{pmatrix}$$

$$\text{Toeplitz } \mathbf{R} = \begin{pmatrix} \sigma^2 & \sigma_1 & \sigma_2 & \sigma_3 & \sigma_4 \\ \sigma_1 & \sigma^2 & \sigma_1 & \sigma_2 & \sigma_3 \\ \sigma_2 & \sigma_1 & \sigma^2 & \sigma_1 & \sigma_2 \\ \sigma_3 & \sigma_2 & \sigma_1 & \sigma^2 & \sigma_1 \\ \sigma_4 & \sigma_3 & \sigma_2 & \sigma_1 & \sigma^2 \end{pmatrix} \quad \text{Banded Toeplitz } \mathbf{R} = \begin{pmatrix} \sigma^2 & \sigma_1 & \sigma_2 & 0 & 0 \\ \sigma_1 & \sigma^2 & \sigma_1 & \sigma_2 & 0 \\ \sigma_2 & \sigma_1 & \sigma^2 & \sigma_1 & \sigma_2 \\ 0 & \sigma_2 & \sigma_1 & \sigma^2 & \sigma_1 \\ 0 & 0 & \sigma_2 & \sigma_1 & \sigma^2 \end{pmatrix} \quad \text{Unstructured } \mathbf{R} = \begin{pmatrix} \sigma^2 & \sigma_{12} & \sigma_{13} & \sigma_{14} & \sigma_{15} \\ \sigma_{12} & \sigma^2 & \sigma_{23} & \sigma_{24} & \sigma_{25} \\ \sigma_{13} & \sigma_{23} & \sigma^2 & \sigma_{34} & \sigma_{35} \\ \sigma_{14} & \sigma_{24} & \sigma_{34} & \sigma^2 & \sigma_{45} \\ \sigma_{15} & \sigma_{25} & \sigma_{35} & \sigma_{45} & \sigma^2 \end{pmatrix}.$$

For the i-th observation, the covariance matrix $\text{Var}(\boldsymbol{\epsilon}_i) = \mathbf{R}_i(\boldsymbol{\tau})$ is a $T_i \times T_i$ matrix. Here, $\mathbf{R}_i(\boldsymbol{\tau})$ is positive definite and depends on i only through its dimension; thus it can be determined by removing certain rows and columns of the matrix $\mathbf{R}(\boldsymbol{\tau})$. This set of notations allows us to easily handle missing data and incomplete observations.

The model can be estimated using either moment-based or likelihood-based methods. With known $\mathbf{R}_i$, the generalized least squares (GLS) estimates are obtained by minimizing

$$\sum_{i=1}^{n} (\mathbf{y}_i - \mathbf{X}_i\boldsymbol{\beta})^{'}\mathbf{R}_i^{-1}(\mathbf{y}_i - \mathbf{X}_i\boldsymbol{\beta}),$$

and we have

$$\hat{\boldsymbol{\beta}} = \left(\sum_{i=1}^{n} \mathbf{X}_i^{'}\mathbf{R}_i^{-1}\mathbf{X}_i\right)^{-1} \sum_{i=1}^{n} \mathbf{X}_i^{'}\mathbf{R}_i^{-1}\mathbf{y}_i.$$

We can estimate such a model using the `R` package `nlme`. Two types of likelihood-based methods are provided to estimate regression parameter $\boldsymbol{\beta}$ and variance components $\boldsymbol{\tau}$, the full maximum likelihood (ML) estimation and the restricted maximum likelihood (REML) estimation. Based on the assumption of multivariate normality of the response $\mathbf{y}_i$, the full log-likelihood function ($l = \log(\mathcal{L})$) for the model is

$$l_{ML}(\boldsymbol{\beta}, \boldsymbol{\tau}) \equiv -\frac{1}{2}\left(\sum_{i=1}^{n} \log \det \mathbf{R}_i(\boldsymbol{\tau}) + \sum_{i=1}^{n} (\mathbf{y}_i - \mathbf{X}_i\boldsymbol{\beta})' \mathbf{R}_i(\boldsymbol{\tau})^{-1} (\mathbf{y}_i - \mathbf{X}_i\boldsymbol{\beta})\right). \quad (15.8)$$

The MLE follows by maximizing the above likelihood function over $\boldsymbol{\beta}$ and $\boldsymbol{\tau}$ simultaneously. It is also easy to show that for fixed covariance parameter $\boldsymbol{\tau}$, the MLE of $\boldsymbol{\beta}$ are the same as the generalized least squares estimators. It is known that the MLE of $\boldsymbol{\tau}$ is biased downward. To mitigate the bias, the restricted maximum likelihood maximizes the following log-likelihood function:

$$l_{REML}(\boldsymbol{\beta}, \boldsymbol{\tau}) \equiv l_{ML}(\boldsymbol{\beta}, \boldsymbol{\tau}) - \frac{1}{2} \log \det \left(\sum_{i=1}^{n} \mathbf{X}_i' \mathbf{R}_i^{-1} \mathbf{X}_i\right). \quad (15.9)$$

The REML estimation will be discussed in more detail in the section on random-effects models.

In our application, we fit the linear model with three types of serial correlation: the compound symmetry, the AR(1), and the unstructured. See Table 15.2 for the results. We denote the resulting models by `SCex.fit`, `SCar.fit`, and `SCun.fit`, respectively. The models are fit using the function `gls()` in the `nlme` package. The argument `correlation` is used to specify matrix $\mathbf{R}(\boldsymbol{\tau})$, and the argument `method` is used to specify the estimation method. The default estimation approach is the REML. The estimation results are displayed in Table 15.2. The estimates of regression coefficients are similar and are consistent with the pooled cross-sectional regression model. The estimates of variance components suggest significant within-subject temporal correlation. Note that when unstructured covariance is specified, the model is not identifiable in its most general form due to the nonuniqueness of $\mathbf{R}(\boldsymbol{\tau})$. Thus, additional constraints are necessary for identification purposes. The `gls()` function estimates the model under the parameterization $\mathbf{R} = \sigma^2 \boldsymbol{\Sigma}$, where σ^2 is a scale parameter and $\boldsymbol{\Sigma}$ is the correlation matrix.

For inference, the estimation error of population parameter $\boldsymbol{\beta}$ is based on

$$\widehat{\text{Var}\hat{\boldsymbol{\beta}}} = \left(\sum_{i=1}^{n} \mathbf{X}_i' R(\hat{\boldsymbol{\tau}})_i^{-1} \mathbf{X}_i\right)^{-1}.$$

The estimation error of $\hat{\boldsymbol{\tau}}$ can follow in different ways. The approach implemented in the `gls()` is to use the inverse of the observed Fisher information. The confidence interval for the scale parameter σ and correlation parameter ρs are obtained based on the approximate normal distribution of the ML or REML estimators of a transformation of parameters. Specifically, the 95% confidence interval of σ is

$$\left[\exp\left(\widehat{\sigma^*} - 1.645 s_{\widehat{\sigma^*}}\right), \exp\left(\widehat{\sigma^*} + 1.645 s_{\widehat{\sigma^*}}\right)\right],$$

where $\widehat{\sigma^*} \equiv \ln \hat{\sigma}$ and $s_{\widehat{\sigma^*}}$ is the associated standard error derived from the Fisher information. Similarly, the 95% confidence interval of ρ is

$$\left[\frac{\exp(\widehat{\rho^*} - 1.645 s_{\widehat{\rho^*}}) - 1}{\exp(\widehat{\rho^*} - 1.645 s_{\widehat{\rho^*}}) + 1}, \frac{\exp(\widehat{\rho^*} + 1.645 s_{\widehat{\rho^*}}) - 1}{\exp(\widehat{\rho^*} + 1.645 s_{\widehat{\rho^*}}) + 1}\right],$$

where $\widehat{\rho^*} \equiv \ln \frac{1-\hat{\rho}}{1+\hat{\rho}}$ and $s_{\widehat{\rho^*}}$ is the corresponding standard error. In the package `nlme`, function `intervals` can be used to call for the 95% confidence interval of $\boldsymbol{\tau}$, and function `getVarCov` can be used to call for the estimates of $\mathbf{R}(\hat{\boldsymbol{\tau}})$.

```
> library(nlme)
> # Compound symmetry
> SCex.fit <- gls(AC ~ lnPCI+lnPPSM+YEAR, data=AutoClaimIn,
+ correlation=corCompSymm(form=~1|TOWNCODE))
> summary(SCex.fit)
> intervals(SCex.fit,which = "var-cov")
> getVarCov(SCex.fit)
Marginal variance covariance matrix
       [,1]   [,2]   [,3]   [,4]   [,5]
[1,] 688.50 326.07 326.07 326.07 326.07
[2,] 326.07 688.50 326.07 326.07 326.07
[3,] 326.07 326.07 688.50 326.07 326.07
[4,] 326.07 326.07 326.07 688.50 326.07
[5,] 326.07 326.07 326.07 326.07 688.50
  Standard Deviations: 26.239 26.239 26.239 26.239 26.239

> # AR(1)
> SCar.fit <- gls(AC ~ lnPCI+lnPPSM+YEAR, data=AutoClaimIn,
+ correlation=corAR1(form=~1|TOWNCODE))
> summary(SCar.fit)
> intervals(SCar.fit,which = "var-cov")
> getVarCov(SCar.fit)
Marginal variance covariance matrix
        [,1]    [,2]   [,3]    [,4]    [,5]
[1,] 673.210 292.350 126.96  55.132  23.942
[2,] 292.350 673.210 292.35 126.960  55.132
[3,] 126.960 292.350 673.21 292.350 126.960
[4,]  55.132 126.960 292.35 673.210 292.350
[5,]  23.942  55.132 126.96 292.350 673.210
  Standard Deviations: 25.946 25.946 25.946 25.946 25.946

> # Unstructured
> SCun.fit <- gls(AC ~ lnPCI+lnPPSM+YEAR, data=AutoClaimIn,
+ correlation=corSymm(form=~1|TOWNCODE))
> summary(SCun.fit)
> intervals(SCun.fit,which = "var-cov")
> getVarCov(SCun.fit)
Marginal variance covariance matrix
       [,1]   [,2]   [,3]   [,4]   [,5]
[1,] 696.15 485.50 324.79 315.06 374.16
[2,] 485.50 696.15 227.51 179.88 190.11
[3,] 324.79 227.51 696.15 284.12 522.68
[4,] 315.06 179.88 284.12 696.15 351.96
[5,] 374.16 190.11 522.68 351.96 696.15
  Standard Deviations: 26.385 26.385 26.385 26.385 26.385
```

The usual t- or F-test statistics follow as for the i.i.d. case. Caution is needed for the tests based on the likelihood function. For example, the likelihood ratio test relies on the value of log-likelihood function rather than the restricted likelihood. One can use `method="ML"` in the `gls()` function to implement maximum likelihood estimation. We perform the test

TABLE 15.2
Estimation for models with serial correlation.

	SCex.fit		SCar.fit		SCun.fit	
Parameter	*Est.*	*S.E.*	*Est.*	*S.E.*	*Est.*	*S.E.*
(Intercept)	887.89	206.81	891.45	168.25	878.68	200.85
lnPCI	−91.20	20.41	−91.33	16.61	-90.81	19.81
lnPPSM	21.96	5.08	21.76	4.11	23.70	4.95
YEAR	3.91	1.14	3.55	1.66	1.82	1.03
	Est.	*95%CI*	*Est.*	*95%CI*	*Est.*	*95%CI*
CS	0.47	(0.29,0.64)				
AR(1)			0.43	(0.26,0.58)		
UN						
corr(1,2)					0.70	(0.46,0.84)
corr(1,3)					0.47	(0.15,0.70)
corr(1,4)					0.45	(0.06,0.72)
corr(1,5)					0.54	(0.19,0.76)
corr(2,3)					0.33	(−0.00,0.59)
corr(2,4)					0.26	(−0.16,0.60)
corr(2,5)					0.27	(−0.13,0.60)
corr(3,4)					0.41	(0.11,0.64)
corr(3,5)					0.75	(0.57,0.86)
corr(4,5)					0.51	(0.20,0.72)
Scale	26.24	(22.22,30.98)	25.95	(22.62,29.76)	26.38	(22.27 ,31.26)
log−REML	−645.96		−654.25		−635.93	
log-ML	−655.61		−663.67		−645.38	
AIC	1323.21		1339.34		1320.75	
BIC	1341.07		1357.21		1365.40	

using **anova** for the models with serial correlation and the pooled cross–sectional regression. The results support the evidence of positive serial correlation.

```
> # Likelihood ratio test
> SCex.fit.ml <- gls(AC ~ lnPCI+lnPPSM+YEAR, data=AutoClaimIn,
+ correlation=corCompSymm(form=~1|TOWNCODE), method="ML")
> SCar.fit.ml <- gls(AC ~ lnPCI+lnPPSM+YEAR, data=AutoClaimIn,
+ correlation=corAR1(form=~1|TOWNCODE), method="ML")
> SCun.fit.ml <- gls(AC ~ lnPCI+lnPPSM+YEAR, data=AutoClaimIn,
+ correlation=corSymm(form=~1|TOWNCODE), method="ML")

> anova(SCex.fit.ml, Pool.fit)
            Model df      AIC      BIC    logLik   Test  L.Ratio p-value
SCex.fit.ml     1  6 1323.212 1341.073 -655.6062
Pool.fit        2  5 1359.497 1374.381 -674.7487 1 vs 2 38.28505  <.0001
> anova(SCar.fit.ml, Pool.fit)
            Model df      AIC      BIC    logLik   Test  L.Ratio p-value
SCar.fit.ml     1  6 1339.344 1357.205 -663.6721
Pool.fit        2  5 1359.497 1374.381 -674.7487 1 vs 2 22.15326  <.0001
> anova(SCun.fit.ml, Pool.fit)
```

```
             Model df      AIC      BIC    logLik   Test  L.Ratio p-value
SCun.fit.ml      1 15 1320.753 1365.404 -645.3763
Pool.fit         2  5 1359.497 1374.381 -674.7487 1 vs 2 58.74476  <.0001
```

Finally, we extend the above model to allow for heterogeneity. We consider a more general model where not only subject specific intercepts, but also subject-specific slopes are incorporated in the linear model as

$$\mathbf{y}_i = \mathbf{Z}_i\boldsymbol{\alpha}_i + \mathbf{X}_i\boldsymbol{\beta} + \boldsymbol{\epsilon}_i, \tag{15.10}$$

with explanatory matrix

$$\mathbf{Z}_i = \begin{pmatrix} z_{i1,1} & z_{i1,2} & \cdots & z_{i1,q} \\ z_{i2,1} & z_{i2,1} & \cdots & z_{i2,1} \\ \vdots & \vdots & \cdots & \vdots \\ z_{iT_i,1} & z_{iT_i,2} & \cdots & z_{iT_i,q} \end{pmatrix} = \begin{pmatrix} \mathbf{z}_{i1}^{'} \\ \mathbf{z}_{i2}^{'} \\ \vdots \\ \mathbf{z}_{iT_i}^{'} \end{pmatrix}$$

and subject-specific parameters $\boldsymbol{\alpha}_i = (\alpha_{i1}, \cdots, \alpha_{iq})^{'}$. The temporal correlation is allowed through the assumption Var $\boldsymbol{\epsilon}_i = \mathbf{R}_i(\boldsymbol{\tau})$. This is known as the fixed-effects linear longitudinal data model. The GLS of parameters can be shown as

$$\hat{\boldsymbol{\beta}} = \left(\sum_{i=1}^{n} \mathbf{X}_i^{'}\mathbf{R}_i^{-1/2}\boldsymbol{\Omega}_i\mathbf{R}_i^{-1/2}\mathbf{X}_i^{'}\right)^{-1} \sum_{i=1}^{n} \mathbf{X}_i^{'}\mathbf{R}_i^{-1/2}\boldsymbol{\Omega}_i\mathbf{R}_i^{-1/2}\mathbf{y}_i^{'}$$

and

$$\widehat{\alpha_i} = (\mathbf{Z}_i^{'}\mathbf{R}_i^{-1}\mathbf{Z}_i)^{-1}\mathbf{Z}_i^{'}\mathbf{R}_i^{-1}(\mathbf{y}_i - \mathbf{X}_i\hat{\boldsymbol{\beta}}),$$

with

$$\boldsymbol{\Omega}_i = \mathbf{I}_i - \mathbf{R}_i^{-1/2}\mathbf{Z}_i(\mathbf{Z}_i^{'}\mathbf{R}_i^{-1}\mathbf{Z}_i)^{-1}\mathbf{Z}_i\mathbf{R}_i^{-1/2}$$

The above model can also be easily implemented using `gls()` by modifying the R code. For example, in the special case of $\mathbf{z}_{it} = 1$, the model reduces to the subject-specific intercept model with serial correlation. One could simply add `factor(TOWNCODE)` in the `SCar.fit`.

15.2.4 Models with Random Effects

Consider the linear longitudinal data model

$$y_{it} = \mathbf{z}_{it}^{'}\boldsymbol{\alpha}_i + \mathbf{x}_{it}^{'}\boldsymbol{\beta} + \epsilon_{it}. \tag{15.11}$$

Instead of treating $\boldsymbol{\alpha}_i$ as fixed parameters, another approach to study heterogeneity is to view $\boldsymbol{\alpha}_i$ as random variables. This model, containing fixed effects parameter $\boldsymbol{\beta}$ and random effects $\boldsymbol{\alpha}_i$, is known as the Linear Mixed-Effects Model (LMM). In its most general form, we assume that $\mathrm{E}(\boldsymbol{\alpha}_i) = \mathbf{0}$ and $\mathrm{Var}(\boldsymbol{\alpha}_i) = \mathbf{D}$, a $q \times q$ positive definite matrix. Furthermore, the subject effects and error term are assumed to be uncorrelated, that is, $\mathrm{Cov}(\boldsymbol{\alpha}_i, \boldsymbol{\epsilon}_i^{'}) = 0$. Under these assumptions, the variance of each subject can be expressed as

$$\mathrm{Var}(\mathbf{y}_i) = \mathbf{Z}_i\mathbf{D}\mathbf{Z}_i^{'} + \mathbf{R}_i = \mathbf{V}_i(\boldsymbol{\tau}),$$

where vector $\boldsymbol{\tau}$ determines the covariance matrix.

For inference purposes, the GLS estimator of population parameter $\boldsymbol{\beta}$ is

$$\hat{\boldsymbol{\beta}}_{GLS} = \left(\sum_{i=1}^{n} \mathbf{X}_i^{'}\mathbf{V}_i^{-1}\mathbf{X}_i\right)^{-1} \sum_{i=1}^{n} \mathbf{X}_i^{'}\mathbf{V}_i^{-1}\mathbf{y}_i$$

and its variance is

$$\text{Var}\ \hat{\boldsymbol{\beta}}_{GLS} = \left(\sum_{i=1}^{n} \mathbf{X}_i^{'} \mathbf{V}_i^{-1} \mathbf{X}_i \right)^{-1}.$$

Similar to the fixed-effects model, it is easy to show that the MLE under multivariate normality is the same as the GLS estimators of $\boldsymbol{\beta}$. For feasible estimates, we discuss likelihood-based methods for the estimation of variance components. Using $\hat{\boldsymbol{\beta}}_{GLS}$, the concentrated log-likelihood function is shown as

$$\begin{aligned} & l_{ML}(\hat{\boldsymbol{\beta}}_{GLS}(\boldsymbol{\tau}), \boldsymbol{\tau}) \\ & \equiv -\frac{1}{2} \left(\sum_{i=1}^{n} \log \det \mathbf{V}_i(\boldsymbol{\tau}) + \sum_{i=1}^{n} (\mathbf{y}_i - \mathbf{X}_i \hat{\boldsymbol{\beta}}_{GLS}(\boldsymbol{\tau}))^{'} \mathbf{V}_i(\boldsymbol{\tau})^{-1} (\mathbf{y}_i - \mathbf{X}_i \hat{\boldsymbol{\beta}}_{GLS}(\boldsymbol{\tau})) \right). \end{aligned}$$

Viewing $\hat{\boldsymbol{\beta}}_{GLS}$ as a function of $\boldsymbol{\tau}$, one can maximize the log-likelihood with respect to $\boldsymbol{\tau}$. This can be done using either Newton–Raphson or the Fisher scoring method. As in the OLS regression, the MLEs of variance component are biased downward. To mitigate the bias, one could employ restricted maximum likelihood by modifying the concentrated log-likelihood function:

$$l_{REML}(\hat{\boldsymbol{\beta}}_{GLS}(\boldsymbol{\tau}), \boldsymbol{\tau}) \equiv l_{ML}(\hat{\boldsymbol{\beta}}_{GLS}(\boldsymbol{\tau}), \boldsymbol{\tau}) - \frac{1}{2} \log \det \left(\sum_{i=1}^{n} \mathbf{X}_i^{'} \mathbf{V}_i(\boldsymbol{\tau})^{-1} \mathbf{X}_i \right). \tag{15.12}$$

Now we examine the so-called *error components* model (or, random intercept model), a special case that is important in actuarial science where $\mathbf{z}_{it} = 1$ and Var $\boldsymbol{\epsilon}_i = \sigma^2 \mathbf{I}_i$. See Sections 15.3 and 15.3.2 for more examples of this specification. The model becomes

$$y_{it} = \alpha_i + \mathbf{x}_{it}^{'} \boldsymbol{\beta} + \epsilon_{it}.$$

The model has the same presentation as the basic fixed-effects model and assumes no serial correlation within each subject. The difference is that the subject-specific intercept α_i is assumed to be random with zero mean and variance σ_α^2. The error components model corresponds to the random sampling scheme where subjects consist of a random subset from a population. One can show that the variance of subject i is

$$\text{Var}\ \mathbf{y}_i = \sigma_\alpha^2 \mathbf{J}_i + \sigma^2 \mathbf{I}_i = \mathbf{V}_i = \sigma^2 \begin{pmatrix} 1 & \rho & \cdots & \rho \\ \rho & 1 & \cdots & \rho \\ \vdots & \vdots & \cdots & \vdots \\ \rho & \rho & \cdots & 1 \end{pmatrix},$$

where $\mathbf{J}_i$ is a $T_i \times T_i$ matrix with all elements equal to one, $\mathbf{J}_i$ is a T_i-dimensional identity matrix, and $\rho = \sigma_\alpha^2/(\sigma^2 + \sigma_\alpha^2)$. Thus, the error components model is equivalent to the model with exchangeable serial correlation.

We implement the error components model `EC.fit` using function `lme()` in the `nlme` package. The argument `random` is used to specify the random effects in the mixed-effects model. Comparing with Table 15.2, we notice that estimates of $\boldsymbol{\beta}$ are the same as the model with the exchangeable serial correlation. The default uses the REML to estimate model parameters. The confidence intervals of variance components are calculated in a similar way as for models with serial correlation (see Section 15.2.3) and can be called by function `intervals()`.

```
> library(nlme)
```

```
> # Error-components model
> EC.fit <- lme(AC ~ lnPCI+lnPPSM+YEAR, data=AutoClaimIn, random=~1|TOWNCODE)
> summary(EC.fit)
Linear mixed-effects model fit by REML
 Data: AutoClaimIn
       AIC      BIC    logLik
  1303.913 1321.606 -645.9566

Random effects:
 Formula: ~1 | TOWNCODE
        (Intercept) Residual
StdDev:    18.05746 19.03756

Fixed effects: AC ~ lnPCI + lnPPSM + YEAR
               Value Std.Error  DF   t-value p-value
(Intercept) 887.8878 206.81071 113  4.293239   0e+00
lnPCI       -91.1979  20.41210 113 -4.467833   0e+00
lnPPSM       21.9614   5.07913 113  4.323844   0e+00
YEAR          3.9119   1.14457 113  3.417801   9e-04
 Correlation:
       (Intr) lnPCI  lnPPSM
lnPCI  -0.988
lnPPSM -0.249  0.096
YEAR    0.197 -0.205 -0.082

Standardized Within-Group Residuals:
        Min          Q1         Med          Q3         Max
-2.53017784 -0.61089180  0.01099886  0.50082006  2.91907172

Number of Observations: 145
Number of Groups: 29
> intervals(EC.fit, which="var-cov")
Approximate 95% confidence intervals

 Random Effects:
  Level: TOWNCODE
                   lower     est.    upper
sd((Intercept)) 12.93758 18.05746 25.20347

 Within-group standard error:
   lower     est.    upper
16.72928 19.03756 21.66434
```

A relevant question to ask is whether the subject-specific effects are significant or the intercepts take a common value. Because α_i is random, we wish to test the null hypothesis $H_0: \sigma_\alpha^2 = 0$. We consider the following procedure:

- Run the pooled cross-sectional model $y_{it} = \mathbf{x}_{it}^{'}\boldsymbol{\beta} + \epsilon_{it}$ and then calculate residuals e_{it}.
- For each subject, compute an estimator of σ_α^2,

$$s_i = \frac{1}{T_i(T_i-1)}\left(T_i^2\bar{e}_i^2 - \sum_{t=1}^{T_i} e_{it}^2\right)$$

- Compute test statistic and compare it with a quantile of an $\chi^2(1)$:

$$TS = \frac{1}{2n}\left(\frac{\sum_{i=1}^{n} s_i\sqrt{T_i(T_i-1)}}{N^{-1}\sum_{i=1}^{n}\sum_{t=1}^{T_i} e_{it}}\right)^2$$

In our example, the test statistic is equal to 56.82 and thus we reject the null hypothesis of constant intercept.

```
> # Pooling test
> tcode = unique(AutoClaimIn$TOWNCODE)
> n = length(tcode)
> N = nrow(AutoClaimIn)
> T <- rep(NA,n)
> s <- rep(NA,n)
> for (i in 1:n){
+ T[i] <- nrow(subset(AutoClaimIn,TOWNCODE==tcode[i]))
+ s[i] <- (sum(subset(AutoClaimIn,TOWNCODE==tcode[i])$rPool)^2
+          - sum(subset(AutoClaimIn,TOWNCODE==tcode[i])$rPool^2))/T[i]/(T[i]-1)
+ }
> TS <- (sum(s*sqrt(T*(T-1)))*N/sum(AutoClaimIn$rPool^2))^2/2/n
> TS
[1] 56.85278
```

To implement the mixed-effects model, one could use `correlation` in the `lme()` function to specify serial correlation. For example, in the model `RE.fit`, we use `update()` to include $AR(1)$ temporal correlation in the error components model. Here we see that with subject-specific intercept, the serial correlation (-0.014) is not significant. The function `getVarCov()` can be used to output the variance-covariance matrix. The argument `type="conditional"` provides the estimate of $\mathbf{R}_i$ and the argument `type="marginal"` provides the estimate of $\mathbf{V}_i$. We further perform a likelihood ratio test to test for the serial correlation using `anova`. Consistently, the large p-value does not show support for serial correlation in the error components model. Note: we use `method="ML"` to get the true log-likelihood value for this test.

```
> # Error component with AR1
> RE.fit <- update(EC.fit, correlation=corAR1(form=~1|TOWNCODE))
> summary(RE.fit)
Linear mixed-effects model fit by REML
 Data: AutoClaimIn
      AIC      BIC    logLik
  1305.897 1326.538 -645.9484

Random effects:
 Formula: ~1 | TOWNCODE
        (Intercept) Residual
StdDev:    18.10974  18.9826

Correlation Structure: AR(1)
 Formula: ~1 | TOWNCODE
 Parameter estimate(s):
        Phi
-0.01444735
Fixed effects: AC ~ lnPCI + lnPPSM + YEAR
```

```
                Value Std.Error  DF   t-value p-value
(Intercept) 887.8789 206.74423 113  4.294577   0e+00
lnPCI       -91.2038  20.40536 113 -4.469601   0e+00
lnPPSM       21.9669   5.07795 113  4.325938   0e+00
YEAR          3.9237   1.13499 113  3.457055   8e-04
 Correlation:
       (Intr) lnPCI  lnPPSM
lnPCI  -0.988
lnPPSM -0.249  0.096
YEAR    0.198 -0.207 -0.082

Standardized Within-Group Residuals:
        Min          Q1         Med          Q3         Max
-2.55033919 -0.60887177  0.02008323  0.49759528  2.91281638

Number of Observations: 145
Number of Groups: 29
> intervals(RE.fit, which="var-cov")
Approximate 95% confidence intervals

 Random Effects:
  Level: TOWNCODE
                   lower     est.    upper
sd((Intercept)) 12.96079 18.10974 25.30422

 Correlation structure:
         lower         est.    upper
Phi -0.2431935 -0.01444735 0.215821
attr(,"label")
[1] "Correlation structure:"

 Within-group standard error:
   lower     est.    upper
16.55969 18.98260 21.76003

> # Get variance components
> getVarCov(RE.fit)
Random effects variance covariance matrix
            (Intercept)
(Intercept)      327.96
  Standard Deviations: 18.11
> getVarCov(RE.fit, type="conditional")
TOWNCODE 10
Conditional variance covariance matrix
            1           2          3           4           5
1  3.6034e+02  -5.2059000   0.075212  -0.0010866  1.5699e-05
2 -5.2059e+00 360.3400000  -5.205900   0.0752120 -1.0866e-03
3  7.5212e-02  -5.2059000 360.340000  -5.2059000  7.5212e-02
4 -1.0866e-03   0.0752120  -5.205900 360.3400000 -5.2059e+00
5  1.5699e-05  -0.0010866   0.075212  -5.2059000  3.6034e+02
  Standard Deviations: 18.983 18.983 18.983 18.983 18.983
```

```
> getVarCov(RE.fit, type="marginal")
TOWNCODE 10
Marginal variance covariance matrix
       1      2      3      4      5
1 688.30 322.76 328.04 327.96 327.96
2 322.76 688.30 322.76 328.04 327.96
3 328.04 322.76 688.30 322.76 328.04
4 327.96 328.04 322.76 688.30 322.76
5 327.96 327.96 328.04 322.76 688.30
  Standard Deviations: 26.236 26.236 26.236 26.236 26.236
```

```
> # Likelihood ratio test
> EC.fit.ml <- lme(AC ~ lnPCI+lnPPSM+YEAR, data=AutoClaimIn,
+ random=~1|TOWNCODE, method="ML")
> RE.fit.ml <- update(EC.fit, correlation=corAR1(form=~1|TOWNCODE), method="ML")
> anova(EC.fit.ml, RE.fit.ml)
          Model df      AIC      BIC    logLik   Test    L.Ratio p-value
EC.fit.ml     1  6 1323.212 1341.073 -655.6062
RE.fit.ml     2  7 1325.171 1346.009 -655.5857 1 vs 2 0.04087198  0.8398
```

We conclude this section with the Hausman test. We have discussed the linear fixed-effects panel data model and the linear mixed-effects model. Both allow for subject specific heterogeneity but with different assumptions. An interesting question is how to choose from the two classes, that is, whether to treat $\boldsymbol{\alpha}_i$ as fixed or random. A possible solution is to refer to the Hausman test (see Hausman (1978)) with test statistic given by

$$TS = \left(\hat{\boldsymbol{\beta}}_{FE} - \hat{\boldsymbol{\beta}}_{GLS}\right)' \left(\text{Var}\ \hat{\boldsymbol{\beta}}_{FE} - \text{Var}\ \hat{\boldsymbol{\beta}}_{GLS}\right)^{-1} \left(\hat{\boldsymbol{\beta}}_{FE} - \hat{\boldsymbol{\beta}}_{GLS}\right),$$

where $\boldsymbol{\beta}_{FE}$ and $\boldsymbol{\beta}_{GLS}$ denote the fixed-effects estimator and the random-effects estimator, respectively. We compare the test statistic with a quantile of a $\chi^2(q)$. A large value supports the fixed-effects estimator. As an example, we compare the basic fixed-effects model with the error components model. The test statistic's observed value is 3.97, supporting the error components formulation.

```
> # Hausman test
> Var.FE <- vcov(FE.fit)[-(1:n),-(1:n)]
> Var.EC <- vcov(EC.fit)[-1,-1]
> beta.FE <- coef(FE.fit)[-(1:n)]
> beta.EC <- fixef(EC.fit)[-1]
> ChiSq <- t(beta.FE-beta.EC)%*%solve(Var.FE-Var.EC)%*%(beta.FE-beta.EC)
> ChiSq
         [,1]
[1,] 3.970489
```

15.2.5 Prediction

This section reviews prediction for longitudinal data mixed-effects models (as discussed in Section 15.2.4). In previous sections, we discussed the estimation and inference of fixed parameters $\boldsymbol{\beta}$ in the model. It is also of interest to summarize the subject-specific effects described by random variable $\boldsymbol{\alpha}_i$. For example, in credibility theory, one is interested in the prediction of expected claims for a policyholder given his risk class. In doing so, we develop the best linear unbiased predictor (BLUP) of a random variable. Predictors are said to be linear if they are formed from a linear combination of the response and the BLUPs are constructed by minimizing the mean square error.

In a linear mixed-effects model where we have $\mathrm{E}(\mathbf{y}_i) = \mathbf{X}_i\boldsymbol{\beta}$ and $\mathrm{Var}(\mathbf{y}_i) = \mathbf{Z}_i\mathbf{D}\mathbf{Z}_i^{'}+\mathbf{R}_i = \mathbf{V}_i$, we wish to predict a random variable η with $\mathrm{E}(\eta) = \mathbf{c}^{'}\boldsymbol{\beta}$ and $\mathrm{Var}(\eta) = \sigma_\eta^2$. Let $\hat{\boldsymbol{\beta}}_{GLS}$ to be the generalized least squares estimator of $\boldsymbol{\beta}$, then the BLUP of η is

$$\eta_{BLUP} = \mathbf{c}^{'}\hat{\boldsymbol{\beta}}_{GLS} + \sum_{i=1}^{n}\left\{\mathrm{Cov}(\eta, \mathbf{y}_i)^{'}\mathbf{V}_i^{-1}\left(\mathbf{y}_i - \mathbf{X}_i\hat{\boldsymbol{\beta}}_{GLS}\right)\right\}$$

and the mean squared error is

$$\begin{aligned}\mathrm{Var}(\eta_{BLUP} - \eta) = &\left(\mathbf{c}^{'} - \sum_{i=1}^{n}\mathrm{Cov}(\eta, \mathbf{y}_i)^{'}\mathbf{V}_i^{-1}\mathbf{X}_i\right)\left(\sum_{i=1}^{n}\mathbf{X}_i^{'}\mathbf{V}_i^{-1}\mathbf{X}_i\right) \\ &\times\left(\mathbf{c}^{'} - \sum_{i=1}^{n}\mathrm{Cov}(\eta, \mathbf{y}_i)^{'}\mathbf{V}_i^{-1}\mathbf{X}_i\right)^{'} - \sum_{i=1}^{n}\mathrm{Cov}(\eta, \mathbf{y}_i)^{'}\mathbf{V}_i^{-1}\mathrm{Cov}(\eta, \mathbf{y}_i) + \sigma_\eta^2.\end{aligned}$$

For example, consider a special case $\eta = \mathbf{w}_1^{'}\boldsymbol{\alpha}_i + \mathbf{w}_2^{'}\boldsymbol{\beta}$, a linear combination of population parameters and subject-specific effects. Using the above relation, we can show that

$$\hat{\eta}_{BLUP} = \mathbf{w}_1^{'}\mathbf{D}\mathbf{Z}_i^{'}\mathbf{V}_i^{-1}(\mathbf{y}_i - \mathbf{X}_i\hat{\boldsymbol{\beta}}_{GLS}) + \mathbf{w}_2^{'}\hat{\boldsymbol{\beta}}_{GLS}.$$

Taking $\mathbf{w}_2 = 0$, we further have the BLUP of $\boldsymbol{\alpha}_i$:

$$\hat{\boldsymbol{\alpha}}_{i,BLUP} = \mathbf{D}\mathbf{Z}_i^{'}\mathbf{V}_i^{-1}(\mathbf{y}_i - \mathbf{X}_i\hat{\boldsymbol{\beta}}_{GLS}).$$

Another special case that is useful for diagnostics is the residual $\eta = \epsilon_{it}$. In this case, we have $\mathbf{c} = \mathbf{0}$ and its BLUP is straightforwardly shown as

$$\hat{e}_{it,BLUP} = y_{it} - \left(\mathbf{z}_{it}^{'}\hat{\boldsymbol{\alpha}}_{i,BLUP} + \mathbf{x}_{it}^{'}\hat{\boldsymbol{\beta}}_{GLS}\right).$$

Some special cases of BLUPs are available in package `nlme`. For the example of the error-components model `EC.fit`, function `ranef()` could be used to get the BLUP of random intercept $\hat{\boldsymbol{\alpha}}_{i,BLUP}$, and function `residuals()` could be used get the BLUP of residuals $\hat{e}_{it,BLUP}$ and its standardized version.

```
> # BLUP
> alpha.BLUP <- ranef(EC.fit)
> beta.GLS <- fixef(EC.fit)
> resid.BLUP <- residuals(EC.fit, type="response")
> rstandard.BLUP <- residuals(EC.fit, type="normalized")
> alpha.BLUP
   (Intercept)
10  -0.2049993
11  -6.9197373
12  17.7349235
13  20.9538588
14  -0.1942180
15  -5.6464625
et cetera
```

To conclude this section, we compare the performance of alternative models using the data of automobile insurance. Our interest is to predict the expected claims of each policyholder in the next year. So the quantity of interest is $\eta = \mathbb{E}(y_{i,T_i+1}|\alpha_i)$. The corresponding BLUP is $\eta_{BLUP} = \mathbf{z}_{it}^{'}\hat{\boldsymbol{\alpha}}_{i,BLUP} + \mathbf{x}_{it}\hat{\boldsymbol{\beta}}_{GLS}$. Recall that we developed various longitudinal data models using data of years 1993–1997, and use the data of year 1998 to validate the prediction. Table 15.3 presents the performance of various longitudinal data models based on both in-sample and out-of-sample data. For in-sample data, we report the information-based model selection criteria AIC and BIC. For out-of-sample, we report the sum of squared prediction error (SSPE) and the sum of absolute prediction error (SAPE). The results show that models that account for subject-specific effects perform better, regardless of the way that heterogeneity is accommodated.

```
> # Use data of year 1998 for validation
> AutoClaimOut <- subset(AutoClaim, YEAR == 1998)

> # Define new variables
> AutoClaimOut$lnPCI <- log(AutoClaimOut$PCI)
> AutoClaimOut$lnPPSM <- log(AutoClaimOut$PPSM)
> AutoClaimOut$YEAR <- AutoClaimOut$YEAR-1992

> # Compare models Pool.fit, SCar.fit, FE.fit, EC.fit, RE.fit and FEar.fit
> # Fixed-effects model with AR(1)
> FEar.fit <- gls(AC ~ factor(TOWNCODE)+lnPCI+lnPPSM+YEAR-1,
+ data=AutoClaimIn, correlation=corAR1(form=~1|TOWNCODE))
> FEar.fit.ml <- gls(AC ~ factor(TOWNCODE)+lnPCI+lnPPSM+YEAR-1,
+ data=AutoClaimIn, correlation=corAR1(form=~1|TOWNCODE), method="ML")

# Prediction
> Xmat <- cbind(rep(1,nrow(AutoClaimOut)),AutoClaimOut$lnPCI,
+ AutoClaimOut$lnPPSM,AutoClaimOut$YEAR)
beta.Pool <- coef(Pool.fit)
pred.Pool <- Xmat%*%beta.Pool
MSPE.Pool <- sum((pred.Pool - AutoClaimOut$AC)^2)
MAPE.Pool <- sum(abs(pred.Pool - AutoClaimOut$AC))

beta.SCar <- coef(SCar.fit)
pred.SCar <- Xmat%*%beta.SCar
MSPE.SCar <- sum((pred.SCar - AutoClaimOut$AC)^2)
MAPE.SCar <- sum(abs(pred.SCar - AutoClaimOut$AC))

beta.FE <- coef(FE.fit)[-(1:29)]
pred.FE <- coef(FE.fit)[1:29] + Xmat[,-1]%*%beta.FE
MSPE.FE <- sum((pred.FE - AutoClaimOut$AC)^2)
MAPE.FE <- sum(abs(pred.FE - AutoClaimOut$AC))

beta.FEar <- coef(FEar.fit)[-(1:29)]
pred.FEar <- coef(FEar.fit)[1:29] + Xmat[,-1]%*%beta.FEar
MSPE.FEar <- sum((pred.FEar - AutoClaimOut$AC)^2)
MAPE.FEar <- sum(abs(pred.FEar - AutoClaimOut$AC))

alpha.EC <- ranef(EC.fit)
```

TABLE 15.3
Comparison of alternative models.

	In-Sample		Out-of-Sample	
	AIC	BIC	SSPE	SAPE
Pooled cross-sectional model	1359.50	1374.38	22201.78	681.25
Pooled cross-sectional with AR(1)	1339.34	1357.21	21242.64	658.98
Fixed-effects model	1293.33	1391.56	21506.07	660.59
Fixed-effects with AR(1)	1286.03	1387.24	21573.79	662.04
Error-components model	1323.21	1341.07	19515.86	619.44
Error-components with AR(1)	1325.17	1346.01	19572.94	620.64

```
beta.EC <- fixef(EC.fit)
pred.EC <- alpha.EC+Xmat%*%beta.EC
MSPE.EC <- sum((pred.EC - AutoClaimOut$AC)^2)
MAPE.EC <- sum(abs(pred.EC - AutoClaimOut$AC))

alpha.RE <- ranef(RE.fit)
beta.RE <- fixef(RE.fit)
pred.RE <- alpha.RE+Xmat%*%beta.RE
MSPE.RE <- sum((pred.RE - AutoClaimOut$AC)^2)
MAPE.RE <- sum(abs(pred.RE - AutoClaimOut$AC))
```

15.3 Generalized Linear Models for Longitudinal Data

As in the previous section, we have a dataset at our disposal consisting of n subjects, where for each subject i, $(1 \leq i \leq n)$ T_i observations are available. Relevant examples in experience rating are (among others) a dataset with n policyholders followed over time, and for which claim counts and severities are registered during each time period under consideration. As explained in Section 15.1 and demonstrated in Section 15.2 for linear models, we extend the GLMs discussed in Chapter 14 by including subject- (or, policyholder-) specific random effects. The random effects structure correlation between observations registered on the same subject, and also take heterogeneity among subjects, due to unobserved characteristics, into account. Therefore, our approach is in line with the random effects approach discussed in Section 15.2.4. Other methods exist for the analysis of longitudinal data in the framework of generalized linear models (the so-called marginal and conditional models; see Verbeke & Molenberghs (2000) and Antonio & Zhang (2014) for a discussion), but those will not be covered here.

15.3.1 Specifying Generalized Linear Models with Random Effects

Given the vector $\boldsymbol{\alpha}_i$ with the random effects for subject i, the repeated measurements $Y_{i1}, \ldots, Y_{iT_i}$ are assumed to be independent with a density from the exponential family

$$f(y_{it}|\boldsymbol{\alpha}_i, \boldsymbol{\beta}, \phi) = \exp\left(\frac{y_{it}\theta_{it} - \psi(\theta_{it})}{\phi} + c(y_{it}, \phi)\right), \; t = 1, \ldots, T_i. \tag{15.13}$$

Some explicit examples follow in the illustrations discussed below. Similar to expressions obtained in Chapter 14, the following (conditional) relations hold:

$$\mu_{it} = \mathbb{E}[Y_{it}|\boldsymbol{\alpha}_i] = \psi^{'}(\theta_{it}) \text{ and } \text{Var}[Y_{it}|\boldsymbol{\alpha}_i] = \phi\psi^{''}(\theta_{it}) = \phi V(\mu_{it}), \tag{15.14}$$

where $g(\mu_{it}) = \boldsymbol{z}_{it}^{'}\boldsymbol{\alpha}_i + \boldsymbol{x}_{it}^{'}\boldsymbol{\beta}$. As before, $g(\cdot)$ is called the link and $V(\cdot)$ is the variance function. $\boldsymbol{\beta}$ $(p \times 1)$ denotes the fixed-effects parameter vector (governing *a priori* rating) and $\boldsymbol{\alpha}_i$ $(q \times 1)$ the random-effects vector. $\boldsymbol{x}_{it}$ $(p \times 1)$ and $\boldsymbol{z}_{it}$ $(q \times 1)$ contain subject i's covariate information for the fixed and random effects, respectively. The specification of the GLMM is completed by assuming that the random effects, $\boldsymbol{\alpha}_i$ $(i = 1, \ldots, n)$, are mutually independent and identically distributed with a density function $f(\boldsymbol{\alpha}_i|\boldsymbol{\nu})$. Herewith, $\boldsymbol{\nu}$ denotes the unknown parameters in the density. In general statistics, the random effects often have a (multivariate) normal distribution with zero mean and covariance matrix determined by $\boldsymbol{\nu}$. Observations on the same subject are dependent because they share the same random effects $\boldsymbol{\alpha}_i$.

The likelihood function for the unknown parameters $\boldsymbol{\beta}$, $\boldsymbol{\nu}$, and ϕ then becomes

$$\mathcal{L}(\boldsymbol{\beta}, \boldsymbol{\nu}, \phi; \boldsymbol{y}) = \prod_{i=1}^{n} f(\boldsymbol{y}_i|\boldsymbol{\alpha}, \boldsymbol{\beta}, \phi) = \prod_{i=1}^{n} \int \prod_{t=1}^{T_i} f(y_{it}|\boldsymbol{\alpha}_i, \boldsymbol{\beta}, \phi) f(\boldsymbol{\alpha}_i|\boldsymbol{\nu}) d\boldsymbol{\alpha}_i, \tag{15.15}$$

where $\boldsymbol{y} = (\boldsymbol{y}_1^{'}, \ldots, \boldsymbol{y}_n^{'})^{'}$ and the integral is with respect to the q-dimensional vector $\boldsymbol{\alpha}_i$. For instance, with normally distributed data and random effects (our setting in Section 15.2), the integral can be worked out analytically and explicit expressions follow for the maximum likelihood estimator of $\boldsymbol{\beta}$ and the Best Linear Unbiased Predictor ('BLUP') for $\boldsymbol{\alpha}_i$. For more general GLMMs, however, approximations to the likelihood or numerical integration techniques are required to maximize Equation (15.15) with respect to the unknown parameters. Such techniques are discussed (and demonstrated) in Antonio & Zhang (2014) (and references therein).

To illustrate the concepts described above, we now consider a Poisson GLMM with normally distributed random intercept, that is, a Poisson error components model. This GLMM allows explicit calculation of the marginal mean and covariance matrix. In this way, one can clearly see how the inclusion of the random effect leads to overdispersion and within-subject covariance.

Example 15.1 (A Poisson GLMM) *Let N_{it} denote the claim frequency registered in year t for policyholder i. Assume that, conditional on α_i, N_{it} follows a Poisson distribution with mean $\mathbb{E}[N_{it}|\alpha_i] = \exp(\boldsymbol{x}_{it}^{'}\boldsymbol{\beta} + \alpha_i)$ and that $\alpha_i \sim N(0, \sigma_b^2)$. Straightforward calculations lead to*

$$\begin{aligned} Var(N_{it}) &= Var(\mathbb{E}(N_{it}|\alpha_i)) + \mathbb{E}(Var(N_{it}|\alpha_i)) \\ &= \mathbb{E}(N_{it})(\exp(\boldsymbol{x}_{it}^{'}\boldsymbol{\beta})[\exp(3\sigma_b^2/2) - \exp(\sigma_b^2/2)] + 1), \end{aligned} \tag{15.16}$$

and

$$\begin{aligned} Cov(N_{it_1}, N_{it_2}) &= Cov(E(N_{it_1}|\alpha_i), \mathbb{E}(N_{it_2}|\alpha_i)) + \mathbb{E}(Cov(N_{it_1}, N_{it_2}|\alpha_i)) \\ &= \exp(\boldsymbol{x}_{it_1}^{'}\boldsymbol{\beta})\exp(\boldsymbol{x}_{it_2}^{'}\boldsymbol{\beta})(\exp(2\sigma_b^2) - \exp(\sigma_b^2)). \end{aligned} \tag{15.17}$$

Hereby, we used the expressions for the mean and variance of a log-normal distribution. In the expression for the covariance, we used the fact that, given the random effect α_i, N_{it_1} and N_{it_2} are independent. We see that the expression inside the parentheses in Equation (15.16) is always bigger than 1. Thus, although $N_{it}|\alpha_i$ follows a regular Poisson distribution, the marginal distribution of N_{it} is overdispersed. According to Equation (15.17), due to the random intercept, observations on the same subject are no longer independent.

Example 15.2 (A Poisson GLMM – continued) *Let N_{it} again denote the claim frequency for policyholder i in year t. Assume that, conditional on α_i, N_{it} follows a Poisson distribution with mean $E[N_{it}|\alpha_i] = \exp(\boldsymbol{x}_{it}^{'}\boldsymbol{\beta} + \alpha_i)$ and that $\alpha_i \sim N(-\frac{\sigma_b^2}{2}, \sigma_b^2)$. This re-parameterization is commonly used in ratemaking. Indeed, we now get*

$$\mathbb{E}[N_{it}] = \mathbb{E}[\mathbb{E}[N_{it}|\alpha_i]] = \exp\left(\boldsymbol{x}_{it}^{'}\boldsymbol{\beta} - \frac{\sigma_b^2}{2} + \frac{\sigma_b^2}{2}\right) = \exp(\boldsymbol{x}_{it}^{'}\boldsymbol{\beta}), \tag{15.18}$$

and

$$\mathbb{E}[N_{it}|\alpha_i] = \exp(\boldsymbol{x}_{in}^{'}\boldsymbol{\beta} + \alpha_i). \tag{15.19}$$

This specification shows that the a priori premium, given by $\exp(\boldsymbol{x}_{it}^{'}\boldsymbol{\beta})$, is correct on the average. The a posteriori correction to this premium is determined by $\exp(\alpha_i)$.
Besides the log-normal distribution from the above examples, other mixing distributions can be used. In the Poisson-Gamma framework, for instance, the conjugacy of these distributions allows for explicit calculation of the predictive premium. Example 15.3 (A Poisson-Gamma rating model).

$$N_{it} \sim Poi(b_i\lambda_{it}), \text{where } \lambda_{it} = \exp(\boldsymbol{x}_{it}^{'}\boldsymbol{\beta}) \text{ and } b_i \sim \Gamma(a, a).$$

It follows that $\mathbb{E}[b_i] = 1$ and the resulting joint, unconditional distribution then becomes

$$\begin{aligned} & Pr(N_{i1} = n_{i1}, \ldots, N_{iT_i} = n_{iT_i}) \\ = & \left(\prod_{t=1}^{T_i} \frac{\lambda_{it}^{n_{it}}}{n_{it}!}\right) \frac{\Gamma(\sum_{t=1}^{T_i} n_{it} + \alpha)}{\Gamma(\alpha)} \left(\frac{\alpha}{\sum_{t=1}^{T_i} n_{it} + \alpha}\right)^{\alpha} \times \left(\sum_{t=1}^{T_i} n_{it} + \alpha\right)^{-\sum_{t=1}^{T_i} n_{it}} \end{aligned} \tag{15.20}$$

with $\mathbb{E}[N_{it}] = \mathbb{E}[\mathbb{E}[N_{it}|b_i]] = \lambda_{it}$ and $Var[N_{it}] = \mathbb{E}[Var[N_{it}|b_i]] + Var[\mathbb{E}[N_{it}|b_i]] = \lambda_{it} + \frac{1}{\alpha}\lambda_{it}^2$.

For the specification in Equation (15.20), the posterior distribution of the random intercept b_i has again a Gamma distribution with

$$f(b_i|N_{i1} = n_{i1}, \ldots, N_{iT_i} = n_{iT_i}) \propto \Gamma(\sum_{t=1}^{T_i} n_{it} + a, \sum_{t=1}^{T_i} \lambda_{it} + a). \tag{15.21}$$

The (conditional) mean and variance of this posterior distribution are given, respectively, by

$$\mathbb{E}[b_i|N_{it} = n_{it},\ t = 1, \ldots, T_i] = \frac{a + \sum_{t=1}^{T_i} n_{it}}{a + \sum_{t=1}^{T_i} \lambda_{it}} \text{ and} \tag{15.22}$$

$$Var[b_i|N_{it} = n_{it},\ t = 1, \ldots, T_i] = \frac{a + \sum_{t=1}^{T_i} n_{it}}{\left(a + \sum_{t=1}^{T_i} \lambda_{it}\right)^2}. \tag{15.23}$$

This leads to the following a posteriori premium

$$\begin{aligned} \mathbb{E}[N_{i,T_i+1}|N_{it} = n_{it},\ t = 1, \ldots, T_i] &= \lambda_{i,T_i+1}\mathbb{E}[b_i|N_{it} = n_{it},\ t = 1, \ldots, T_i] \\ &= \lambda_{i,T_i+1}\left\{\frac{\alpha + \sum_{t=1}^{T_i} n_{it}}{\alpha + \sum_{t=1}^{T_i} \lambda_{it}}\right\}. \end{aligned} \tag{15.24}$$

The above credibility premium is optimal when a quadratic loss function is used. Indeed, as is known in mathematical statistics, the conditional expectation minimizes a mean squared error criterion.

Experience rating based on multilevel (panel or higher order) models poses a challenge to the insurer when it comes to communicating the predictive results of these models to the policyholders. Customers may find it difficult to understand. It is not readily transparent to an ordinary policyholder how the surcharges (*maluses*) for reported claims and the discounts (*bonuses*) for claim-free periods are evaluated. In order to establish an experience rating system where insureds can easily understand the effect of reported claims or periods without claims, Bonus–Malus scales have been developed. We develop a case study (using R) of such scales in Section 15.3.2.

15.3.2 Case Study: Experience Rating with Bonus–Malus Scales in R

We now demonstrate how the statistical models from Section 15.3 allow us to develop a specific type of experience rating system, namely a Bonus–Malus ([BM]) scale. This type of experience rating is very common in motor (or vehicle) insurance. See Lemaire (1984) and Denuit et al. (2007) for detailed discussions. In a BM scale, an *a priori* tariff is adjusted based on the claim history of a policyholder. A "good" history will create a *bonus*, and therefore premium reduction. A 'bad' performance causes a *malus*, and penalizes the policyholder by a premium increase. We closely follow Denuit et al. (2007) in this section, and extend the discussion in Antonio & Valdez (2012) with an implementation in R of a simple BM scale.

Experience rating with a BM scale is appealing from a commercial and communication point of view. An insurer can easily explain to a customer how his claims reported in year t will change the premium applicable to year $t+1$ for automobile insurance. To discuss the probabilistic, statistical, as well as computational aspects of Bonus–Malus scales, a **credibility model** similar to the one in Example 15.3 is assumed. Let N_{it} denote the number of claims registered for policyholder i in year t. Our credibility model is structured as follows:

- Policy(holder) i of the portfolio ($i = 1, \ldots, n$) is represented by a sequence $(\Theta_i, \boldsymbol{N}_i)$ where $\boldsymbol{N}_i = (N_{i1}, N_{i2}, \ldots)$ and Θ_i represents unexplained heterogeneity and has mean 1;
- Given $\Theta_i = \theta$, the random variables N_{it} ($t = 1, 2, \ldots$) are independent and $\mathrm{P}(\lambda_{it}\theta)$ distributed; and
- The sequences $(\Theta_i, \boldsymbol{N}_i)$ ($i = 1, \ldots, n$) are assumed to be independent.

15.3.2.1 Bonus–Malus Scales

A BM scale consists of a certain number of levels, say $s+1$, that are numbered from $0, \ldots, s$, with 0 being the best scale. Let ℓ_0 be the entrance level of a new driver. According to the number of claims reported during the insured period, drivers will move up and down the scale. A claim-free year results in a *bonus* point, which implies that the driver goes one level down. Claims are penalized by *malus* points, meaning that for each claim filed, the driver goes up a certain number of levels, denoted with pen (for *penalty*). We introduce a set of random variables that allows us to describe the technicalities of a BM scale. L_k represents the level occupied by the driver in the time interval $(k, k+1)$. Thus, L_k takes a value in $\{0, \ldots, s\}$, and $\{L_1, L_2, \ldots\}$ is the driver's trajectory over time. With N_k the number of claims reported by the insured in the period $(k-1, k)$, the future level of an insured L_k is obtained from the present level L_{k-1} and the number of claims reported during the present year N_k. We recognize the so-called Markov property: the future depends on the present but not on the past. The *relativity* r_ℓ associated with each level ℓ in the scale determines the premium discount/penalty awarded to the driver. A policyholder who has at present

TABLE 15.4
Transitions in the (−1/top scale) BM system.

Starting Level	Level 0 Claim	Occupied if ≥ 1 is Reported
0	0	5
1	0	5
2	1	5
3	2	5
4	3	5
5	4	5

a priori premium λ_{it} (determined using the techniques from Chapter 14) and is in scale ℓ, has to pay $r_\ell \times \lambda_{it}$. With $r_\ell < 1$, the driver receives a discount based on a favorable record of past claims. When $r_\ell > 1$, the driver is penalized for his past performance. The relativities, together with the transition rules in the scale, are the commercial alternative for the credibility-type corrections to an *a priori* tariff, as discussed above. We want to demonstrate in this section the calculation of these relativities for a given portfolio and BM scale.

Example 15.3 (-1/Top Scale) *We consider a very simple example of a BM scale to illustrate the concepts : the (−1/Top Scale). See Denuit et al. (2007) for more realistic examples. This scale has six levels, numbered 0,1,. . . ,5. Starting class is level 5. Each claim-free year is rewarded by one bonus class. When an accident is reported, the policyholder is transferred to scale 5. Table 15.4 represents these transitions.*

15.3.2.2 Transition Rules, Transition Probabilities and Stationary Distribution

To enable the calculation of the relativity corresponding with each level ℓ, some probabilistic concepts associated with BM scales must be introduced. The **transition rules** corresponding with a certain BM scale are indicator variables $t_{ij}(k)$ such that

$$t_{ij}(k) \quad = \quad \begin{cases} 1 \text{ if the policy transfers from } i \text{ to } j \text{ when } k \text{ claims are reported,} \\ 0 \text{ otherwise.} \end{cases} \tag{15.25}$$

We define the transition matrix $\boldsymbol{T}(k)$, with k the number of claims reported by the driver,

$$\boldsymbol{T}(k) \quad = \quad \begin{pmatrix} t_{00}(k) & t_{01}(k) & \dots & t_{0s}(k) \\ t_{10}(k) & t_{11}(k) & \dots & t_{1s}(k) \\ \vdots & \vdots & \ddots & \vdots \\ t_{s0}(k) & t_{s1}(k) & \dots & t_{ss}(k) \end{pmatrix}, \tag{15.26}$$

where

$$t_{ij}(k) \quad = \quad \begin{cases} 1 \text{ if the policy transfers from } i \text{ to } j \text{ when } k \text{ claims are reported,} \\ 0 \text{ otherwise.} \end{cases} \tag{15.27}$$

Thus, this matrix is a $0-1$ matrix and each row has exactly one 1.

Assuming $N_1,\ N_2,\dots$ are independent and $\mathcal{P}(\theta)$ distributed, the trajectory this driver follows through the scale will be represented as $\{L_1(\theta), L_2(\theta), \ldots\}$. The transition probability of this driver to go from level ℓ_1 to ℓ_2 in a single step is

$$\begin{aligned} p_{\ell_1\ell_2}(\theta) &= \mathrm{P}[L_{k+1}(\theta) = \ell_2 | L_k(\theta) = \ell_1] \\ &= \sum_{n=0}^{+\infty} \mathrm{P}[L_{k+1}(\theta) = \ell_2 | N_{k+1} = n, L_k(\theta) = \ell_1] \mathrm{P}[N_{k+1} = n | L_k(\theta) = \ell_1] \\ &= \sum_{n=0}^{+\infty} \frac{\theta^n}{n!} \exp(-\theta) t_{\ell_1\ell_2}(n), \end{aligned} \tag{15.28}$$

where we used the independence of N_{k+1} and L_k. In matrix form, the **one-step transition matrix $\boldsymbol{P}(\theta)$** is given by

$$\boldsymbol{P}(\theta) = \begin{pmatrix} p_{00}(\theta) & p_{01}(\theta) & \ldots & p_{0s}(\theta) \\ p_{10}(\theta) & p_{11}(\theta) & \ldots & p_{1s}(\theta) \\ \vdots & \vdots & \ddots & \vdots \\ p_{s0}(\theta) & p_{s1}(\theta) & \ldots & p_{ss}(\theta) \end{pmatrix}. \tag{15.29}$$

The probability of being transferred from level i to level j in n steps is expressed by the n-step transition probability $p_{ij}^{(n)}$:

$$\begin{aligned} p_{ij}^{(n)}(\theta) &= \mathrm{P}[L_{k+n}(\theta) = j | L_k(\theta) = i] \\ &= \sum_{i_1=0}^{s} \sum_{i_2=0}^{s} \cdots \sum_{i_{n-1}=0}^{s} p_{ii_1}(\theta) p_{i_1i_2}(\theta) \ldots p_{i_{n-1}j}(\theta), \end{aligned} \tag{15.30}$$

which composes the **$\boldsymbol{n}$-step transition matrix $\boldsymbol{P}^{(n)}(\theta)$**

$$\boldsymbol{P}^{(n)}(\theta) = \begin{pmatrix} p_{00}^{(n)}(\theta) & p_{01}^{(n)}(\theta) & \ldots & p_{0s}^{(n)}(\theta) \\ p_{10}^{(n)}(\theta) & p_{11}^{(n)}(\theta) & \ldots & p_{1s}^{(n)}(\theta) \\ \vdots & \vdots & \ddots & \vdots \\ p_{s0}^{(n)}(\theta) & p_{s1}^{(n)}(\theta) & \ldots & p_{ss}^{(n)}(\theta) \end{pmatrix}. \tag{15.31}$$

The following relation holds between the 1 and n-step transition matrices: $\boldsymbol{P}^{(n)}(\theta) = \boldsymbol{P}^n(\theta)$.

Ultimately, the BM system will stabilize and the proportion of policyholders occupying each level of the scale will remain unchanged. These proportions are captured in the **stationary distribution** $\boldsymbol{\pi}(\theta) = (\boldsymbol{\pi}_0(\theta), \ldots, \boldsymbol{\pi}_s(\theta))^{'}$, which are defined as

$$\boldsymbol{\pi}_{\ell_2}(\theta) = \lim_{n \to +\infty} p_{\ell_1\ell_2}^{(n)}(\theta). \tag{15.32}$$

Correspondingly, $\boldsymbol{P}^{(n)}(\theta)$ converges to $\boldsymbol{\Pi}(\theta)$ defined as

$$\lim_{n \to +\infty} \boldsymbol{P}^{(n)}(\theta) = \boldsymbol{\Pi}(\theta) = \begin{pmatrix} \boldsymbol{\pi}^{'}(\theta) \\ \boldsymbol{\pi}^{'}(\theta) \\ \vdots \\ \boldsymbol{\pi}^{'}(\theta) \end{pmatrix}. \tag{15.33}$$

For the BM scale introduced in Illustration 15.3 the transition and one-step probability matrices are given as follows:

$$\boldsymbol{T}(0) = \begin{pmatrix} 1 & 0 & 0 & 0 & 0 & 0 \\ 1 & 0 & 0 & 0 & 0 & 0 \\ 0 & 1 & 0 & 0 & 0 & 0 \\ 0 & 0 & 1 & 0 & 0 & 0 \\ 0 & 0 & 0 & 1 & 0 & 0 \\ 0 & 0 & 0 & 0 & 1 & 0 \end{pmatrix} \text{ and } \boldsymbol{T}(1) = \begin{pmatrix} 0 & 0 & 0 & 0 & 0 & 1 \\ 0 & 0 & 0 & 0 & 0 & 1 \\ 0 & 0 & 0 & 0 & 0 & 1 \\ 0 & 0 & 0 & 0 & 0 & 1 \\ 0 & 0 & 0 & 0 & 0 & 1 \\ 0 & 0 & 0 & 0 & 0 & 1 \end{pmatrix} \quad (15.34)$$

$$\boldsymbol{P}(\theta) = \begin{pmatrix} \exp(-\theta) & 0 & 0 & 0 & 0 & 1-\exp(-\theta) \\ \exp(-\theta) & 0 & 0 & 0 & 0 & 1-\exp(-\theta) \\ 0 & \exp(-\theta) & 0 & 0 & 0 & 1-\exp(-\theta) \\ 0 & 0 & \exp(-\theta) & 0 & 0 & 1-\exp(-\theta) \\ 0 & 0 & 0 & \exp(-\theta) & 0 & 1-\exp(-\theta) \\ 0 & 0 & 0 & 0 & \exp(-\theta) & 1-\exp(-\theta) \end{pmatrix}. \quad (15.35)$$

In R, we specify this one-step transition matrix $\boldsymbol{P}$ as follows:

```
Pmatrix =
function(th) {
P = matrix(nrow=6,ncol=6,data=0)
P[1,1]=P[2,1]=P[3,2]=P[4,3]=P[5,4]=P[6,5]= exp(-th)
P[,6] = 1-exp(-th)
return(P)}
```

Using a result from Rolski et al. (1999) (also see Denuit et al. (2007)), the stationary distribution $\boldsymbol{\pi}(\theta)$ can be obtained as $\boldsymbol{\pi}'(\theta) = \boldsymbol{e}'(\boldsymbol{I}-\boldsymbol{P}(\theta)+\boldsymbol{E})^{-1}$, with $\boldsymbol{E}$ the $(s+1)\times(s+1)$ matrix with all entries 1. For the (−1/Top Scale), this results in

$$\boldsymbol{\pi}'(\theta) = (1,1,1,1,1,1)\times \begin{pmatrix} 2-\exp(-\theta) & 1 & 1 & 1 & 1 & \exp(-\theta) \\ 1-\exp(-\theta) & 2 & 1 & 1 & 1 & \exp(-\theta) \\ 1 & 1-\exp(-\theta) & 2 & 1 & 1 & \exp(-\theta) \\ 1 & 1 & 1-\exp(-\theta) & 2 & 1 & \exp(-\theta) \\ 1 & 1 & 1 & 1-\exp(-\theta) & 2 & \exp(-\theta) \\ 1 & 1 & 1 & 1 & 1-\exp(-\theta) & 1+\exp(-\theta) \end{pmatrix}^{-1}. \quad (15.36)$$

We specify the stationary distribution of the (−1/Top Scale) in R:

```
lim.distr =
function(matrix) {
et  = matrix(nrow=1, ncol=dim(matrix)[2], data=1)
E = matrix(nrow=dim(matrix)[1], ncol=dim(matrix)[2], data=1)
mat  = diag(dim(matrix)[1]) - matrix + E
inverse.mat = solve(mat)
p  = et %*% inverse.mat
return(p)}
```

For instance, with $\theta = 0.1$ (as in the example of Denuit et al. (2007), page 180, Example 4.9), the stationary distribution becomes

$$\boldsymbol{\pi}'(0.1) = \begin{pmatrix} 0.6065307 & 0.06378939 & 0.07049817 & 0.07791253 & 0.08610666 & 0.09516258 \end{pmatrix}. \quad (15.37)$$

In R, we use the following instructions:

```
> P = Pmatrix(0.1)
> P
          [,1]      [,2]      [,3]      [,4]      [,5]       [,6]
[1,] 0.9048374 0.0000000 0.0000000 0.0000000 0.0000000 0.09516258
[2,] 0.9048374 0.0000000 0.0000000 0.0000000 0.0000000 0.09516258
[3,] 0.0000000 0.9048374 0.0000000 0.0000000 0.0000000 0.09516258
[4,] 0.0000000 0.0000000 0.9048374 0.0000000 0.0000000 0.09516258
[5,] 0.0000000 0.0000000 0.0000000 0.9048374 0.0000000 0.09516258
[6,] 0.0000000 0.0000000 0.0000000 0.0000000 0.9048374 0.09516258
> pi = lim.distr(P)
> pi
          [,1]       [,2]       [,3]       [,4]       [,5]       [,6]
[1,] 0.6065307 0.06378939 0.07049817 0.07791253 0.08610666 0.09516258
```

15.3.2.3 Relativities

The calculation of the relativities in a BM scale reveals some similarities with explicit credibility-type calculations. Following Norberg (1976) with the number of levels and transition rules being fixed, the optimal relativity r_ℓ, corresponding with level ℓ, is determined by maximizing the asymptotic predictive accuracy. This implies that one tries to minimize

$$\mathbb{E}[(\Theta - r_L)^2], \tag{15.38}$$

the difference between the relativity r_L and the "true" relative premium Θ, under the assumptions of our credibility model. Simplifying the notation in this model, the *a priori* premium of a random policyholder is denoted with Λ and the residual effect of unknown risk characteristics with Θ. The policyholder then has (unknown) annual expected claim frequency $\Lambda\Theta$, where Λ and Θ are assumed to be independent. The weights of different risk classes follow from the *a priori* system with $\mathrm{P}[\Lambda = \lambda_k] = w_k$.

Calculation of the r_ℓ's goes as follows:

$$\begin{aligned}
\min \mathbb{E}[(\Theta - r_L)^2] &= \sum_{\ell=0}^{s} \mathbb{E}[(\Theta - r_\ell)^2 | L = \ell]\mathrm{P}[L = \ell] &(15.39)\\
&= \sum_{\ell=0}^{s} \int_0^{+\infty} (\theta - r_\ell)^2 \mathrm{P}[L = \ell | \Theta = \theta] dF_\Theta(\theta)\\
&= \sum_k w_k \int_0^{+\infty} \sum_{\ell=0}^{s} (\theta - r_\ell)^2 \pi_\ell(\lambda_k \theta) dF_\Theta(\theta), &(15.40)
\end{aligned}$$

where $\mathrm{P}[\Lambda = \lambda_k] = w_k$. In the last step of the derivation, conditioning is on Λ. It is straightforward to obtain the optimal relativities by solving

$$\frac{\partial \mathbb{E}[(\Theta - r_L)^2]}{\partial r_j} = 0 \text{ with } \quad j = 0, \ldots, s. \tag{15.41}$$

Alternatively, from mathematical statistics it is well known that for a quadratic loss function (see Equation (15.39)) the optimal $r_\ell = \mathbb{E}[\Theta|L=\ell]$. This is calculated as follows:

$$\begin{aligned} r_\ell &= \mathbb{E}[\Theta|L=\ell] \\ &= \mathbb{E}[\mathbb{E}[\Theta|L=\ell,\Lambda]|L=\ell] \\ &= \sum_k \mathbb{E}[\Theta|L=\ell,\Lambda=\lambda_k]\mathrm{P}[\Lambda=\lambda_k|L=\ell] \\ &= \sum_k \int_0^{+\infty} \theta \frac{\mathrm{P}[L=\ell|\Theta=\theta,\Lambda=\lambda_k]w_k}{\mathrm{P}[L=\ell,\Lambda=\lambda_k]} dF_\Theta(\theta) \frac{\mathrm{P}[\Lambda=\lambda_k, L=\ell]}{\mathrm{P}[L=\ell]}, \end{aligned} \tag{15.42}$$

where the relation $f_{\Theta|L=\ell,\Lambda=\lambda_k}(\theta|\ell,\lambda_k) = \frac{\mathrm{P}[L=\ell|\Theta=\theta,\Lambda=\lambda_k]\times w_k \times f_\Theta(\theta)}{\mathrm{P}[\Lambda=\lambda_k,L=\ell]}$ is used. The optimal relativities are given by

$$r_\ell = \frac{\sum_k w_k \int_0^{+\infty} \theta \pi_\ell(\lambda_k\theta)dF_\Theta(\theta)}{\sum_k w_k \int_0^{+\infty} \pi_\ell(\lambda_k\theta)dF_\Theta(\theta)}. \tag{15.43}$$

When no *a priori* rating system is used, all the λ_k's are equal (estimated by $\hat{\lambda}$) and the relativities reduce to

$$r_\ell = \frac{\int_0^{+\infty} \theta \pi_\ell(\hat{\lambda}\theta)dF_\Theta(\theta)}{\int_0^{+\infty} \pi_\ell(\hat{\lambda}\theta)dF_\Theta(\theta)}. \tag{15.44}$$

Calculation of these relativities in R goes as follows. We replicate Example 4.11 from Denuit et al. (2007) where no *a priori* rating is used. This example uses a $\Gamma(a,a)$ distribution for the policyholder-specific random effect Θ_i (as in Illustration 46), with $\hat{a} = 0.888$ and $\hat{\lambda} = 0.1474$. Those estimates are obtained by calibrating a Negative Binomial distribution on the data from Portfolio A in Denuit et al. (2007) (see Section 1.6, pages 44–45 , in the book). Data in Portfolio A are the claim counts registered on 14,505 policies during calendar year 1997.

```
### Without a priori ratemaking
a.hat          = 0.8888
lambda.hat         = 0.1474

int1 =
function(theta, s, a, lambda) {
        a         = a.hat
        lambda = lambda.hat
        f.dist = gamma(a)^(-1) * a^a * theta^(a-1) * exp(-a*theta)
        p          = lim.distr(Pmatrix((lambda*theta)))
        return(theta*p[1,s+1]*f.dist)}
P1 = matrix(nrow=1, ncol=6, data=0)
for (i in 0:5) P1[1,i+1] = integrate(Vectorize(int1),lower=0,upper=Inf,s=i)$value

int2 =
function(theta, s, a, lambda) {
        a         = a.hat
        lambda = lambda.hat
        f.dist = gamma(a)^(-1) * a^a * theta^(a-1) * exp(-a*theta)
        p          = lim.distr(Pmatrix((lambda*theta)))
        return(p[1,s+1]*f.dist)}

P2 = matrix(nrow=1, ncol=6, data=0)
```

```
for (i in 0:5) P2[1,i+1] = integrate(Vectorize(int2),lower=0,upper=Inf,s=i)$value
R = P1 / P2
> R   # relativities without a priori rating
          [,1]     [,2]     [,3]     [,4]     [,5]     [,6]
[1,] 0.5466848 1.21958 1.348203 1.507254 1.709032 1.973534
```

To demonstrate the calculation of relativities when accounting for *a priori* rating, we use the Portfolio A data from Denuit et al. (2007) again with the $\hat{\lambda}_k$'s and w_k's printed in Table 2.7 (page 91) of the book. $\hat{\lambda}_k$ is the *a priori* annually expected claim frequency for risk class k, as determined by a set of *a priori* observed risk factors. The selection of risk factors and estimated annual claim frequencies are obtained by fitting a Negative Binomial regression model to the Portfolio A data. Negative Binomial regression for a single year of data on observed claim counts, say k_i with $i = 1 \ldots, N$ is based on the following likelihood

$$\mathcal{L}(\boldsymbol{\beta}, \alpha) = \prod_{i=1}^{n} \frac{\lambda_i^{k_i}}{k_i!} \left(\frac{a}{a+\lambda_i} \right)^a (a+\lambda_i)^{-k_i} \frac{\Gamma(a+k_i)}{\Gamma(a)}, \tag{15.45}$$

where $\lambda_i = d_i \exp(\boldsymbol{x}_i^t \boldsymbol{\beta})$ (with d_i the exposure registered for policyholder i). Negative Binomial regression is available in R from the `glm.nb()` function.

```
lambda  = c(0.1176,0.1408,0.1897,0.2272,0.1457,0.1746,0.2351,0.2816,
            0.1761,0.2109,0.2840,0.3402,0.2182,0.2614,0.3520,0.0928,
            0.1112,0.1498,0.1794,0.1151,0.1378,0.1856,0.2223)
weights = c(0.1049,0.1396,0.0398,0.0705,0.0076,0.0122,0.0013,0.0014,
            0.0293,0.0299,0.0152,0.0242,0.0007,0.0009,0.0002,0.1338,
            0.1973,0.0294,0.0661,0.0372,0.0517,0.0025,0.0044)
a = 1.065
n=length(weights)

int3 =
function(theta, lambda, a, l) {
        p = lim.verd(Pmatrix(lambda*theta))
        f.dist = gamma(a)^(-1) * a^a * theta^(a-1) * exp(-a*theta)
        return(theta*p[1,l+1]*f.dist)}

int4 =
function(theta, lambda, a, l) {
        p = lim.verd(Pmatrix(lambda*theta))
        f.dist = gamma(a)^(-1) * a^a * theta^(a-1) * exp(-a*theta)
        return(p[1,l+1]*f.dist)}

teller1 = teller2 = noemer = array(dim=6, data=0)
result1 = result2 = array(dim=6, data=0)

for (i in 0:5) {
        b = c = array(dim=n,data=0)
        for (j in 1:n) {
                b[j] = integrate(Vectorize(int3),lower=0, upper=Inf,lambda=lambda[j],a=a,l=i)$value
                c[j] = integrate(Vectorize(int4),lower=0, upper=Inf,lambda=lambda[j],a=a,l=i)$value}
        teller1[i+1] = b %*% weights
        noemer[i+1]  = c %*% weights
        R     = teller1/noemer
        }

> R   # relativities with a priori rating
[1] 0.6118907 1.2088841 1.3124752 1.4388207 1.5985014 1.8123074
```

Summarizing, we obtain the relativities displayed in Table 15.5 (with and without *a priori* rating) for the (−1/Top Scale) and the Portfolio A data from Denuit et al. (2007). The *a posteriori* corrections are less severe when *a priori* rating is taken into account.

TABLE 15.5
Numerical characteristics for the (−1/top scale) and portfolio A data from Denuit et al. (2007), without and with *a priori* rating taken into account.

Level ℓ	$r_\ell = E[\Theta \mid L = \ell]$ without *a priori*	$r_\ell = E[\Theta \mid L = \ell]$ with *a priori*
5	197.3%	181.2%
4	170.9%	159.9%
3	150.7%	143.9%
2	134.8 %	131.3%
1	122.0%	120.9%
0	54.7%	61.2%

16

Claims Reserving and IBNR

Markus Gesmann
ChainLadder project
London, United Kingdom

CONTENTS

16.1 Introduction

16.1.1 Motivation

The insurance industry, unlike other industries, does not sell products as such, but rather promises. An insurance policy is a promise by the insurer to the policyholder to pay for future claims for an upfront received premium.

As a result, insurers do not know the upfront cost for their service, but rely on historical data analysis and judgment to predict a sustainable price for their offering. In General Insurance (or Non-Life Insurance, e.g. motor, property and casualty insurance); most policies run for a period of 12 months. However, the claims payment process can take years or even decades. Therefore, often not even the delivery date of their product is known to insurers.

In particular, losses arising from casualty insurance can take a long time to settle and even when the claims are acknowledged, it may take time to establish the extent of the claims settlement cost. Claims can take years to materialize. A complex and costly example involves the claims from asbestos liabilities, particularly those in connection with mesothelioma and lung damage arising from prolonged exposure to asbestos. A research report by a working party of the Institute and Faculty of Actuaries estimated that the undiscounted cost of U.K. mesothelioma-related claims to the U.K. Insurance Market for the period 2009 to 2050 could be around £10bn; see Gravelsons et al. (2009). The cost for asbestos-related claims in the United States for the worldwide insurance industry was estimated to be around \$120bn in 2002, see Michaels (2002).

Thus, it should come as no surprise that the biggest item on the liabilities side of an insurer's balance sheet is often the provision or reserves for future claims payments. Those reserves can be broken down into case reserves (or outstanding claims), which are losses already reported to the insurance company and losses that are incurred but not reported (IBNR) yet.

Historically, reserving was based on deterministic calculations with pen and paper, combined with expert judgment. Since the 1980s, with the arrival of personal computer, spreadsheet software has become very popular for reserving. Spreadsheets not only reduced the calculation time, but allowed actuaries to test different scenarios and the sensitivity of their forecasts.

As the computer became more powerful, ideas of more sophisticated models started to evolve. Changes in regulatory requirements, for example, Solvency II[1] in Europe, have fostered further research and promoted the use of stochastic and statistical techniques. In particular, for many countries, extreme percentiles of reserve deterioration over a fixed time period have to be estimated for the purpose of capital setting.

Over the years, several methods and models have been developed to estimate both the level and variability of reserves for insurance claims; see Schmidt (2012) or P.D. England & R.J. Verrall (2002) for an overview.

In practice, the Mack chain-ladder and bootstrap chain-ladder models are used by many actuaries, along with stress testing / scenario analysis and expert judgment to estimate ranges of reasonable outcomes; see the surveys of U.K. actuaries in 2002, Lyons et al. (2002), and across the Lloyd's market in 2012, Orr (2012).

16.1.2 Outline and Scope

In this chapter we can only give an introduction to some reserving models and the focus will be on the practical implementation in R. For a more comprehensive overview, see Wüthrich & Merz (2008). The remainder of this chapter is structured as follows. Section 16.2 gives an overview of the data structure used for a typical reserving exercise and introduces the example dataset used throughout this chapter. We discuss the classical deterministic chain-ladder reserving method in Section 16.3 and introduce the concept of a tail factor. In Section 16.4 we show first that the chain-ladder algorithm can be considered a weighted linear regression through the origin, and move on from there to introduce stochastic reserving models. We start with the Mack model, which provides a stochastic framework for the chain-ladder methods and allows the estimation of the mean squared error of the payment predictions. Following this, we discuss the Poisson model, a generalised linear model that replicates the chain-ladder forecasts. To estimate the full distribution of the reserve, we consider a bootstrap approach. We finish the section with a log-incremental reserving model that is particularly suited to identify changing trends in the data. Finally, Section 16.5 will briefly

[1]See http://ec.europa.eu/internal_market/insurance/solvency/index_en.htm.

discuss the differences between the ultimo and one-year reserve risk measurements in the context of Solvency II.

16.2 Development Triangles

Historical insurance data are often presented in the form of a triangle structure, showing the development of claims over time for each exposure (origin) period. An origin period could be the year the policy was written or earned, or the loss occurrence period. Of course, the origin period does not have to be yearly, for example, quarterly or monthly origin periods are also often used. The development period of an origin period is also called age or lag. Data on the diagonals present payments in the same calendar period. Note: Data of individual policies are usually aggregated to homogeneous lines of business, division levels or perils.

As an example, we present a claims payment triangle from a U.K. Motor Non-Comprehensive account as published by Christofides (1997). For convenience we set the origin period from 2007 to 2013.

The following dataframe presents the claims data in a typical form as it would be stored in a database. The first column holds the origin year, the second column the development year and the third column has the incremental payments / transactions.

```
> n <- 7
> Claims <- data.frame(originf = factor(rep(2007:2013, n:1)),
+                      dev=sequence(n:1),
+                      inc.paid=
+                      c(3511, 3215, 2266, 1712, 1059, 587,
+                        340, 4001, 3702, 2278, 1180,  956,
+                        629, 4355, 3932, 1946, 1522, 1238,
+                        4295, 3455, 2023, 1320, 4150, 3747,
+                        2320, 5102, 4548, 6283))
```

To present the data in a triangle format, we can use the `matrix` function:

```
> (inc.triangle  <- with(Claims, {
+     M <- matrix(nrow=n, ncol=n,
+                 dimnames=list(origin=levels(originf), dev=1:n))
+     M[cbind(originf, dev)] <- inc.paid
+     M
+   }))

      dev
origin    1    2    3    4    5   6   7
  2007 3511 3215 2266 1712 1059 587 340
  2008 4001 3702 2278 1180  956 629  NA
  2009 4355 3932 1946 1522 1238  NA  NA
  2010 4295 3455 2023 1320   NA  NA  NA
  2011 4150 3747 2320   NA   NA  NA  NA
  2012 5102 4548   NA   NA   NA  NA  NA
  2013 6283   NA   NA   NA   NA  NA  NA
```

It is the objective of a reserving exercise to forecast the future claims development in the bottom right corner of the triangle and potential further developments beyond development age 7. Eventually all claims for a given origin period will be settled, but it is not always obvious to judge how many years or even decades it will take. We speak of long and short tail business depending on the time it takes to pay all claims.

Often it is helpful to consider the cumulative development of claims as well, which is presented below:

```
> (cum.triangle <- t(apply(inc.triangle, 1, cumsum)))

origin    1    2     3     4     5     6     7
  2007 3511 6726  8992 10704 11763 12350 12690
  2008 4001 7703  9981 11161 12117 12746    NA
  2009 4355 8287 10233 11755 12993    NA    NA
  2010 4295 7750  9773 11093    NA    NA    NA
  2011 4150 7897 10217    NA    NA    NA    NA
  2012 5102 9650    NA    NA    NA    NA    NA
  2013 6283   NA    NA    NA    NA    NA    NA
```

The latest diagonal of the triangle presents the latest cumulative paid position of all origin years:

```
> (latest.paid <- cum.triangle[row(cum.triangle) == n - col(cum.triangle) + 1])

[1]  6283  9650 10217 11093 12993 12746 12690
```

We add the cumulative paid data as a column to the data frame as well:

```
> Claims$cum.paid <- cum.triangle[with(Claims, cbind(originf, dev))]
```

To start the reserving analysis, we plot the data:

```
> op <- par(fig=c(0,0.5,0,1), cex=0.8, oma=c(0,0,0,0))
> with(Claims, {
+     interaction.plot(x.factor=dev, trace.factor=originf, response=inc.paid,
+                      fun=sum, type="b", bty='n', legend=FALSE); axis(1, at=1:n)
+     par(fig=c(0.45,1,0,1), new=TRUE, cex=0.8, oma=c(0,0,0,0))
+     interaction.plot(x.factor=dev, trace.factor=originf, response=cum.paid,
+                      fun=sum, type="b", bty='n'); axis(1,at=1:n)
+   })
> mtext("Incremental and cumulative claims development",
+         side=3, outer=TRUE, line=-3, cex = 1.1, font=2)
> par(op)


> library(lattice)
> xyplot(cum.paid ~ dev | originf, data=Claims, t="b", layout=c(4,2),
+  as.table=TRUE, main="Cumulative claims development")
```

Figures 16.1 and 16.2 present the incremental and cumulative claims development by origin year. The triangle appears to be fairly well behaved. The past two years, 2012 and 2013, appear to be slightly higher than years 2008 to 2011, and the values in 2007 are lower in comparison to the later years, for example, the book changed over the years. The last payment of 1,238 for the 2009 origin year stands out a bit as well.

Other claims information can provide valuable insight into the reserving process too, such as claims numbers, transition timings between different claims settlement stages and

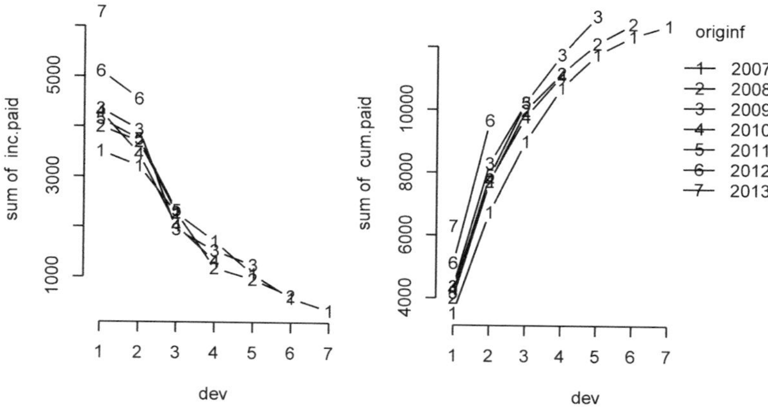

FIGURE 16.1
Plot of incremental and cumulative claims payments by origin year using base graphics, using `interaction.plot` of the `stats` package in R.

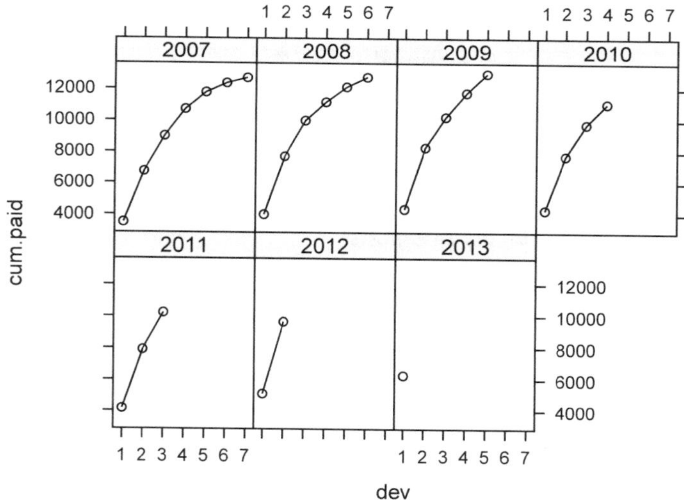

FIGURE 16.2
Cumulative claims developments by origin year using the `lattice` package, with one panel per origin year.

earning patterns. See, for example, Miranda et al. (2012), Orr (2007), Murray & Lauder (2011), respectively. A deep understanding of the whole business process from pricing, to underwriting, claims handling and data management will guide the actuary to interpret the claims data at hand. The Claims Reserving Working Party Paper, Lyons et al. (2002), outlines the different aspects in more detail.

Notation

Using terminology in Wütherich & Merz (2008), cumulative payments are noted as $C_{i,j}$, for origin period i (or period of occurrence) seen after j periods of development and incremental payments $X_{i,j}$. The outstanding liabilities, or reserves, for accident year i at time j is given by

$$R_{i,j} = \sum_{k>j} X_{i,j} = \left(\lim_{k\to\infty} C_{i,k}\right) - C_{i,j} \tag{16.1}$$

Because this quantity involves unobserved data (i.e. amounts that will be paid in the future), $R_{i,j}$ will be the estimated claims reserves.

Remark 16.1 *For convenience, we will assume that we work on square matrices, with n rows and n columns. For a more general setting, see Wütherich & Merz (2008).*

Remark 16.2 *Throughout this chapter we note the first development period as 1. Other authors use 0 for the first development period. This is more a matter of taste than having any practical implications.*

Remark 16.3 *Many of the methods and models presented here can be applied to paid and reported (often also called 'incurred') data. We either have to estimate the reserve or incurred but not reported (IBNR) claims. For the purpose of this chapter, we assume*

$$\begin{aligned} \textit{Ultimate loss cost} &= \textit{paid} + \textit{reserve} && (16.2)\\ &= \textit{paid} + \textit{case reserve} + \textit{IBNR} && (16.3)\\ &= \textit{incurred} + \textit{IBNR} && (16.4) \end{aligned}$$

Remark 16.4 *Some methods and models require data > 0, which for paid claims should be given (insurers rarely receive money back from the insured after a claim was paid[2]), but case reserves can show negative adjustments over time; therefore incremental incurred triangles do show negatives occasionally. To ensure that data have only positive values, it can be temporarily shifted, or, in a given context, be ignored.*

16.3 Deterministic Reserving Methods

The most established and probably oldest method or algorithm for estimating reserves is the so-called chain-ladder method or loss development factor (LDF) method.

The classical chain-ladder method is a deterministic algorithm to forecast claims based on historical data. It assumes that the proportional developments of claims from one development period to the next is the same for all origin periods.

16.3.1 Chain-Ladder Algorithm

Most commonly as a first step, the age-to-age link ratios f_k are calculated as the volume weighted average development ratios of a cumulative loss development triangle from one age period to the next C_{ik} for $i, k = 1, \ldots, n$.

[2]Examples are late salvage or subrogation payments.

$$f_k = \frac{\sum_{i=1}^{n-k} C_{i,k+1}}{\sum_{i=1}^{n-k} C_{i,k}} \tag{16.5}$$

```
> f <- sapply((n-1):1, function(i) {
      sum( cum.triangle[1:i, n-i+1] ) / sum( cum.triangle[1:i, n-i] )
  })
```

Initially we expect no further development after year 7. Hence, we set the last link ratio (often called the tail factor) to 1:

```
> tail <- 1
> (f <- c(f, tail))

[1] 1.889 1.282 1.147 1.097 1.051 1.028 1.000
```

These factors f_k are then applied to the latest cumulative payment in each row ($C_{i,n-i+1}$) to produce stepwise forecasts for future payment years $k \in \{n-i+1, \dots, n\}$:

$$\hat{C}_{i,k+1} = f_k \hat{C}_{i,k}, \tag{16.6}$$

starting with $\hat{C}_{i,n+1-i} = C_{i,n+1-i}$. The *squaring* of the claims triangle is calculated below:

```
> full.triangle <- cum.triangle
> for(k in 1:(n-1)){
     full.triangle[(n-k+1):n, k+1] <- full.triangle[(n-k+1):n,k]*f[k]
  }
> full.triangle

origin    1     2     3     4     5     6     7
  2007 3511  6726  8992 10704 11763 12350 12690
  2008 4001  7703  9981 11161 12117 12746 13097
  2009 4355  8287 10233 11755 12993 13655 14031
  2010 4295  7750  9773 11093 12166 12786 13138
  2011 4150  7897 10217 11720 12854 13509 13880
  2012 5102  9650 12375 14195 15569 16362 16812
  2013 6283 11870 15222 17461 19151 20126 20680
```

The last column contains the forecast ultimate loss cost:

```
> (ultimate.paid <- full.triangle[,n])

 2007  2008  2009  2010  2011  2012  2013
12690 13097 14031 13138 13880 16812 20680
```

The cumulative products of the age-to-age development ratios provide the loss development factors for the latest cumulative paid claims for each row to ultimate:

```
> (ldf <- rev(cumprod(rev(f))))

[1] 3.291 1.742 1.359 1.184 1.080 1.028 1.000
```

The inverse of the loss development factor estimates the proportion of claims developed to date for each origin year, often also called the gross up factors or growth curve:

```
> (dev.pattern <- 1/ldf)

[1] 0.3038 0.5740 0.7361 0.8444 0.9261 0.9732 1.0000
```

The total estimated outstanding loss reserve with this method is

```
> (reserve <- sum (latest.paid * (ldf - 1)))

[1] 28656
```

or via

```
> sum(ultimate.paid - latest.paid)

[1] 28656
```

Remark 16.5 *The basic chain-ladder algorithm has the implicit assumption that each origin period has its own unique level and that development factors are independent of the origin periods; or equivalently, there is a constant payment pattern. Therefore, if a_i is the ultimate (cumulative) claim for origin period i and b_j is the percentage of ultimate claims in development period j, with $\sum b_j = 1$, then the incremental payment $\hat{X}_{ij}$ can be described as $\hat{X}_{ij} = a_i b_j$; see Christofides (1997).*

```
> a <- ultimate.paid
> (b <- c(dev.pattern[1], diff(dev.pattern)))

[1] 0.30382 0.27017 0.16208 0.10828 0.08170 0.04716 0.02679

> (X.hat <- a %*% t(b))

     [,1] [,2] [,3] [,4] [,5]  [,6]  [,7]
[1,] 3855 3428 2057 1374 1037 598.4 340.0
[2,] 3979 3538 2123 1418 1070 617.6 350.9
[3,] 4263 3791 2274 1519 1146 661.6 375.9
[4,] 3992 3549 2129 1423 1073 619.5 352.0
[5,] 4217 3750 2250 1503 1134 654.5 371.9
[6,] 5108 4542 2725 1820 1374 792.8 450.4
[7,] 6283 5587 3352 2239 1690 975.2 554.1
```

Remark 16.6 *As the chain-ladder method is a deterministic algorithm and does not regard the observations as realizations of random variables but absolute values, the forecast of the most recent origin periods can be quite unstable. To address this issue, Bornhuetter & Ferguson (1972) suggested a credibility approach, which combines the chain-ladder forecast with prior information on expected loss costs, for example, from pricing data. Under this approach the chain-ladder development to ultimate pattern is used as weighting factors between the pure chain-ladder and expected loss cost estimates.*

Suppose the expected loss cost for the 2013 origin year is 20,000; then the BF method would estimate the ultimate loss cost as

```
> (BF2013 <- ultimate.paid[n] * dev.pattern[1] + 20000 * (1 - dev.pattern[1]))

 2013
20207
```

16.3.2 Tail Factors

In the previous section we implicitly assumed that there are no claims payment after 7 years, or in other words, that the oldest origin year is fully developed.

However, often it is not suitable to assume that the oldest origin year is fully settled. A typical approach to overcome this shortcoming is to extrapolate the development ratios, for example, assuming a linear model of the log development ratios minus one, which reflects the incremental changes on the previous cumulative payments; see also Figure 16.3.

```
> dat <- data.frame(lf1=log(f[-c(1,n)]-1), dev=2:(n-1))
> (m <- lm(lf1 ~ dev , data=dat))

Call:
lm(formula = lf1 ~ dev, data = dat)

Coefficients:
(Intercept)          dev
     -0.131       -0.572

> plot(lf1 ~ dev, main="log(f - 1) ~ dev", data=dat, bty='n')
> abline(m)

> sigma <- summary(m)$sigma
> extrapolation <- predict(m, data.frame(dev=n:100))
> (tail <- prod(exp(extrapolation + 0.5*sigma^2) + 1))

[1] 1.037
```

We have not carried out any sense checks apart from the plot in Figure 16.3; however, the ratio analysis presented above would suggest that we can expect another 3.7% claims development after year 7 and therefore we should consider increasing our reserve to 29,728.

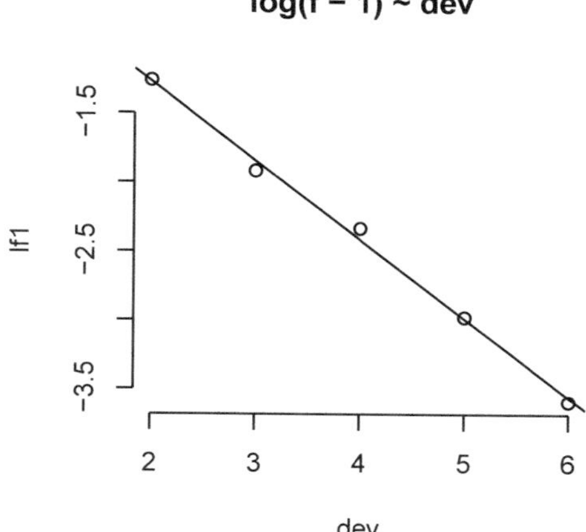

FIGURE 16.3
Plot of the loss development factors −1 on a log-scale against development period.

More generally, the factors used to project the future payments need not always be drawn from the dollar weighted averages of the triangle. Other sources of factors from which the actuary may *select* link ratios include simple averages from the triangle, averages weighted toward more recent observations or adjusted for outliers, and benchmark patterns based on related, more credible loss experience. Also, because the ultimate value of claims is simply the product of the most current diagonal and the cumulative product of the link ratios, the completion of the interior of the triangle is usually not displayed; instead, the eventual value of the claims, or ultimate value is shown.

For example, suppose the actuary decides that the volume weighted factors from the claims triangle are representative of expected future growth, but discards the tail factor derived from the linear fit in favor of a tail based on data from a larger book of similar business. The LDF method might be displayed in R as follows:

```
> library(ChainLadder)
> ata(cum.triangle)

origin   1-2   2-3   3-4   4-5   5-6   6-7
  2007 1.916 1.337 1.190 1.099 1.050 1.028
  2008 1.925 1.296 1.118 1.086 1.052    NA
  2009 1.903 1.235 1.149 1.105    NA    NA
  2010 1.804 1.261 1.135    NA    NA    NA
  2011 1.903 1.294    NA    NA    NA    NA
  2012 1.891    NA    NA    NA    NA    NA
  smpl 1.890 1.284 1.148 1.097 1.051 1.028
  vwtd 1.889 1.282 1.147 1.097 1.051 1.028
```

16.4 Stochastic Reserving Models

As the provision for outstanding claims is often the biggest item on the liabilities side of an insurer's balance sheet, it is important not only to estimate the mean but also the uncertainty of the reserve.

Over the years many statistical techniques have been developed to embed the reserving analysis into a stochastic framework. The key idea is to regard the observed data as one realization of a random variable, rather than absolutes. Statistical techniques also allow for more formal testing, make modelling assumptions more explicit and, in particular, help to monitor actual versus expected claims developments (A versus E). It is the regular A versus E exercise which can help to drive management actions. Hence, as a minimum, not only the mean reserve, or *best estimate liabilities*[3], should be estimated but also the volatility of reserves.

In this section we first show that the deterministic chain-ladder algorithm of the previous section can be considered a weighted linear regression through the origin. Indeed, the following Mack model provides a stochastic framework for the chain-ladder method and allows us to estimate the mean squared error of future payments, using many estimators from the linear regression output. An alternative to the Mack model is the Poisson model, a generalized linear model that replicates the chain-ladder forecasts as well. Yet, the Poisson model is often not directly applicable to insurance data, as the variance of the data is

[3]Note: Actuaries distinguish between best estimate liabilities undiscounted and discounted to present value.

frequently greater than the mean and hence we consider a quasi-Poisson model to estimate uncertainty metrics. Following this we present a bootstrap technique to estimate the full reserve distribution. For many triangles it is reasonable to assume that the incremental payments follow a log-normal distribution and hence we finish this section with a parametric reserving model that is particularly suited to identify and model changing trends in data.

16.4.1 Chain-Ladder in the Context of Linear Regression

Since the early 1990s, several papers have been published to embed the deterministic chain-ladder method into a statistical framework. Barnett & Zehnwirth (2000) and Murphy (1994) were not the only ones to point out that the chain-ladder age-to-age link ratios could be regarded as coefficients of a linear regression through the origin. To illustrate this concept, we follow Barnett & Zehnwirth (2000).

Let $C_{\cdot,k}$ denote the k-th column in the cumulative claims triangle. The chain-ladder algorithm can be seen as

$$C_{\cdot,k+1} = f_k\, C_{\cdot,k} + \varepsilon(k) \text{ with } \varepsilon_k \sim N(0, \sigma_k^2 C_{\cdot,k}^{\delta}) \tag{16.7}$$

The parameter f_k describes the slope or the 'best' line through the origin and data points $[C_{\cdot,k}, C_{\cdot,k+1}]$, with δ as a 'weighting' parameter. Barnett & Zehnwirth (2000) distinguish the cases:

- $\delta = 0$ ordinary regression with intercept 0
- $\delta = 1$ historical chain ladder age-to-age link ratios
- $\delta = 2$ straight averages of the individual link ratios

Indeed, we can demonstrate the different cases by applying different linear models to our data. First, we add columns to the original dataframe `Claims`, to have payments of the current and previous development period next to each other; additionally we add a column with the development period as a factor.

```
> names(Claims)[3:4] <- c("inc.paid.k", "cum.paid.k")
> ids <- with(Claims, cbind(originf, dev))
> Claims <- within(Claims,{
     cum.paid.kp1 <- cbind(cum.triangle[,-1], NA)[ids]
     inc.paid.kp1 <- cbind(inc.triangle[,-1], NA)[ids]
     devf <- factor(dev)
     }
  )
```

In the next step we apply the linear regression function `lm` to each development period, vary the weighting parameter δ from 0 to 2 and extract the slope coefficients:

```
> delta <- 0:2
> ATA <- sapply(delta, function(d)
     coef(lm(cum.paid.kp1 ~ 0 + cum.paid.k : devf,
        weights=1/cum.paid.k^d, data=Claims))
  )
> dimnames(ATA)[[2]] <- paste("Delta = ", delta)
> ATA
```

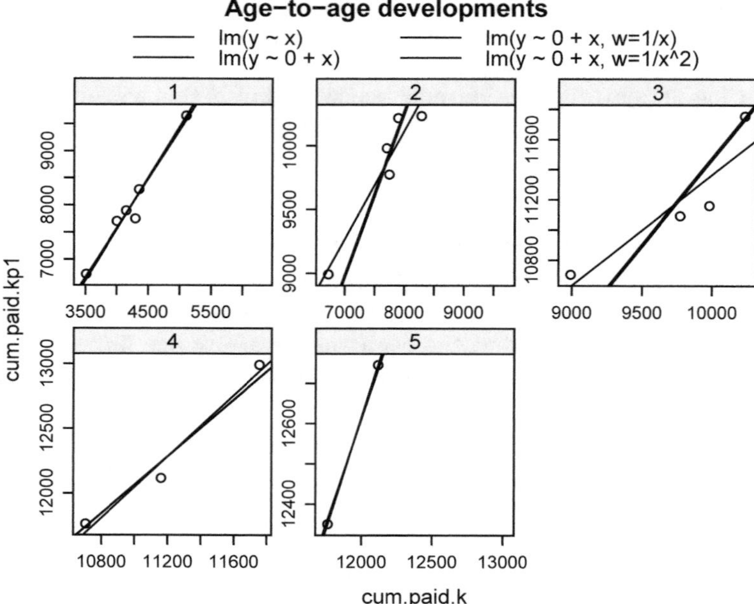

FIGURE 16.4
Plot of the cumulative development positions from one development year to the next for each development year, including regression lines of different linear models.

```
                 Delta =  0 Delta =  1 Delta =  2
cum.paid.k:devf1      1.888      1.889      1.890
cum.paid.k:devf2      1.280      1.282      1.284
cum.paid.k:devf3      1.146      1.147      1.148
cum.paid.k:devf4      1.097      1.097      1.097
cum.paid.k:devf5      1.051      1.051      1.051
cum.paid.k:devf6      1.028      1.028      1.028
```

Indeed, the development ratios for $\delta = 1$ and $\delta = 2$ tally with those of the previous section. Let us plot the data again, with the cumulative paid claims of one period against the previous one, including the regression output for each development period; see Figure 16.4.

```
> xyplot(cum.paid.kp1 ~ cum.paid.k | devf,
         data=subset(Claims, dev < (n-1)),
         main="Age-to-age developments", as.table=TRUE,
         scales=list(relation="free"),
         key=list(columns=2, lines=list(lty=1:4, type="l"),
                  text=list(lab=c("lm(y ~ x)",
                                  "lm(y ~ 0 + x)",
                                  "lm(y ~ 0 + x, w=1/x)",
                                  "lm(y ~ 0 + x, w=1/x^2)"))),
         panel=function(x,y,...){
           panel.xyplot(x,y,...)
           if(length(x)>1){
             panel.abline(lm(y ~ x), lty=1)
```

```
            panel.abline(lm(y ~ 0 + x), lty=2)
            panel.abline(lm(y ~ 0 + x, weights=1/x), lty=3)
            panel.abline(lm(y ~ 0 + x, , weights=1/x^2), lty=4)
        }
    }
)
```

Note that for development periods 2 and 3, we observe a difference in the slope of the linear regression with and without an intercept. Of course we could test the significance of the intercept via the usual tests.

16.4.2 Mack Model

Mack (1993, 1999) suggested a model to estimate the first two moments (mean and standard errors) of the chain-ladder forecast, without assuming a distribution under three conditions.

In order to forecast the amounts $\hat{C}_{ik}$ for $k > n+1-i$, the Mack chain-ladder model assumes

$$\text{CL1: } \mathbb{E}\left(F_{ik} \middle|\, C_{i,1}, C_{i,2}, \ldots, C_{i,k}\right) = f_k \text{ with } F_{ik} = \frac{C_{i,k+1}}{C_{i,k}} \tag{16.8}$$

$$\text{CL2: } Var\left(F_{i,k} \middle|\, C_{i,1}, C_{i,2}, \ldots, C_{ik}\right) = \frac{\sigma_k^2}{w_{ik} C_{ik}^{\alpha}} \tag{16.9}$$

$$\text{CL3: } \left\{C_{i,1}, \ldots, C_{i,n}\right\}, \left\{C_{j,1}, \ldots, C_{j,n}\right\} \text{ are independent for origin period } i \neq j \tag{16.10}$$

with $w_{ik} \in [0;1], \alpha \in \{0,1,2\}$. Note that Mack uses the following notation for the weighting parameter $\alpha = 2 - \delta$, with δ defined as in the previous section. In other words, the Mack model assumes that the link ratios for each development period are consistent across all origin periods (CL1), the volatility decreases as losses are paid (CL2) and all origin periods are independent; for example, there is no structural change or market cycle (CL3). If these assumptions hold, the Mack model gives an unbiased estimator for future claims. Thus,

$$\widehat{f}_k = \frac{\sum_{i=1}^{n-k} w_{ik} C_{i,k}^{\alpha} F_{i,k}}{\sum_{i=1}^{n-k} w_{ik} C_{i,k}^{\alpha}} \tag{16.11}$$

is an unbiased estimator for f_k, given past observations in the triangle, and $\widehat{f}_k$ and $\widehat{f}_j$ are non-correlated for $k \neq j$. Hence, an unbiased estimator for $\mathbb{E}(C_{i,k}|C_{i,1}, \ldots, C_{i,n+1-i})$ is

$$\widehat{C}_{i,k} = \widehat{f}_{n-i} \cdot \widehat{f}_{n-i+1} \cdots \widehat{f}_{k-2} \left(\widehat{f}_{k-1} - 1\right) \cdot C_{i,n+1-i}. \tag{16.12}$$

Recall that $\widehat{f}_k$ is the estimator with minimal variance among all linear estimators obtained from the $F_{i,k}$'s. Finally, if $\alpha = 1$ and $\omega_{i,k} = 1$, then[4].

$$\widehat{\sigma}_k^2 = \frac{1}{n-k-1} \sum_{i=1}^{n-k} \left(F_{i,k} - \widehat{f}_k\right)^2 \cdot C_{i,k} \tag{16.13}$$

is an unbiased estimator of σ_k^2, given past observations in the triangle. Based on these estimators, it is possible to compute the mean squared error of prediction for reserve $\widehat{R}_i$, given past observations $\mathcal{F}$ in the triangle

$$MSE_i = \underbrace{\widehat{Var}(\widehat{R}_i|\mathcal{F})}_{\text{process variance}} + \underbrace{\mathbb{E}\left([R_i - \widehat{R}_i]^2|\mathcal{F}\right)}_{\text{estimation error}}. \tag{16.14}$$

[4]For the general case, see Mack (1999)

The process variance originates from the stochastic movement of the process, whereas the estimation error reflects the uncertainty in the estimation of the parameters.

From Mack (1999) (see also Chapter 3 in Wüthrich & Merz (2008)), the process variance can be estimated using

$$\widehat{Var}(\widehat{R}_i|\mathcal{F}) = \widehat{R}_i \sum_{k=n+1-i}^{n-1} \frac{\widehat{\sigma}_k^2}{\widehat{f}_k^2 \widehat{C}_{i,k}}, \tag{16.15}$$

and the estimation error estimated by

$$\mathbb{E}\left([R_i - \widehat{R}_i]^2|\mathcal{F}\right) = \widehat{R}_i^2 \sum_{k=n+1-i}^{n-1} \frac{\widehat{\sigma}_k^2}{\widehat{f}_k^2}\left(\frac{1}{C_{i,k}} + \frac{1}{\sum_{l=1}^{n-k} C_{l,k}}\right). \tag{16.16}$$

In order to derive the conditional mean squared error of total reserve prediction $\widehat{R}$, define the covariance term, for $i < j$, as

$$MSE_{i,j+1} = \widehat{C}_{i,j}^2 \sum_{k=j+2-i}^{l} \frac{\widehat{\sigma}_k^2}{\widehat{f}_k^2}\left(\frac{1}{C_{i,k}} + \frac{1}{\sum_{l=1}^{j+1-k} C_{l,k}}\right) + MSE_{i,j}, \tag{16.17}$$

with $MSE_{i,n+1-i} = 0$. Then the conditional mean squared error of reserves (all years) is

$$MSE = \sum_{i=1}^{n} MSE_i + 2\sum_{j>i} MSE_{i,j}. \tag{16.18}$$

These formulas are implemented in the `ChainLadder` package, Gesmann, Murphy & Zhang (2013), via the function `MackChainLadder`. As an example, we apply the `MackChainLadder` function[5] to our triangle:

```
> library(ChainLadder)
> (mack <- MackChainLadder(cum.triangle, weights=1, alpha=1,
                           est.sigma="Mack"))

MackChainLadder(Triangle = cum.triangle, weights = 1, alpha = 1,
    est.sigma = "Mack")

     Latest Dev.To.Date Ultimate   IBNR Mack.S.E CV(IBNR)
2007 12,690       1.000   12,690      0     0.00      NaN
2008 12,746       0.973   13,097    351     3.62   0.0103
2009 12,993       0.926   14,031  1,038    22.90   0.0221
2010 11,093       0.844   13,138  2,045   141.98   0.0694
2011 10,217       0.736   13,880  3,663   426.70   0.1165
2012  9,650       0.574   16,812  7,162   692.39   0.0967
2013  6,283       0.304   20,680 14,397   900.58   0.0626

               Totals
Latest:     75,672.00
Dev:             0.73
Ultimate:  104,327.77
```

[5]The interested reader may want to review the help page of the `MackChainLadder` function in more detail and investigate the source code of the function, which utilises `lm` and its output.

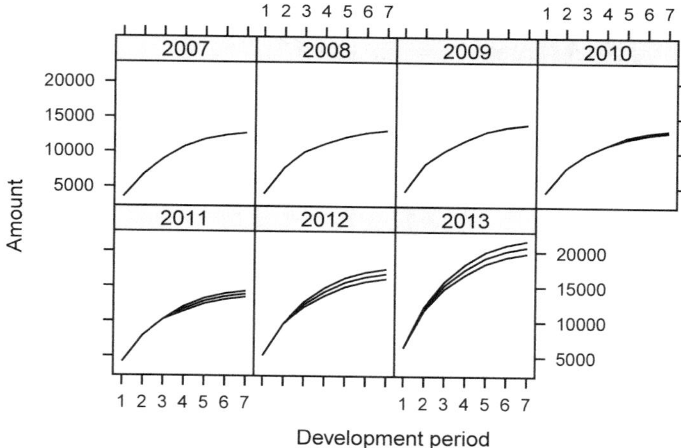

FIGURE 16.5
Plot of the actual and expected cumulative claims development and estimated standard error of the Mack model forecast.

```
IBNR:          28,655.77
Mack S.E.:      1,417.27
CV(IBNR):           0.05
```

The output provides immediate access to various statistics of the Mack model, including the forecast future payments (here labelled as IBNR), its estimated mean squared error, at individual and across all origin period levels, here ±5%. Hence, the predicted future payments and their errors can be used for an A versus E exercise in the following development period; see Figure 16.5, which was produced using the following command:

```
> plot(mack, lattice=TRUE, layout=c(4,2))
```

To check if the assumptions for the Mack model are held, we review the residual plots of the Mack model; see Figure 16.6.

```
> plot(mack)
```

From the residual plots we note that smaller values appear under-fitted and larger values slightly over-fitted. The effect of under-fitting appears to be particularly pronounced for data from earlier calendar years, where fewer data are available.

Remark 16.7 *One way to address the above shortcomings might be to find appropriate weights and to review the choice of the α parameter in the Mack model. The function* `CLFMdelta`*, following Bardis et al. (2012), can help to find consistent α values based on a set of selected age-to-age ratios.*

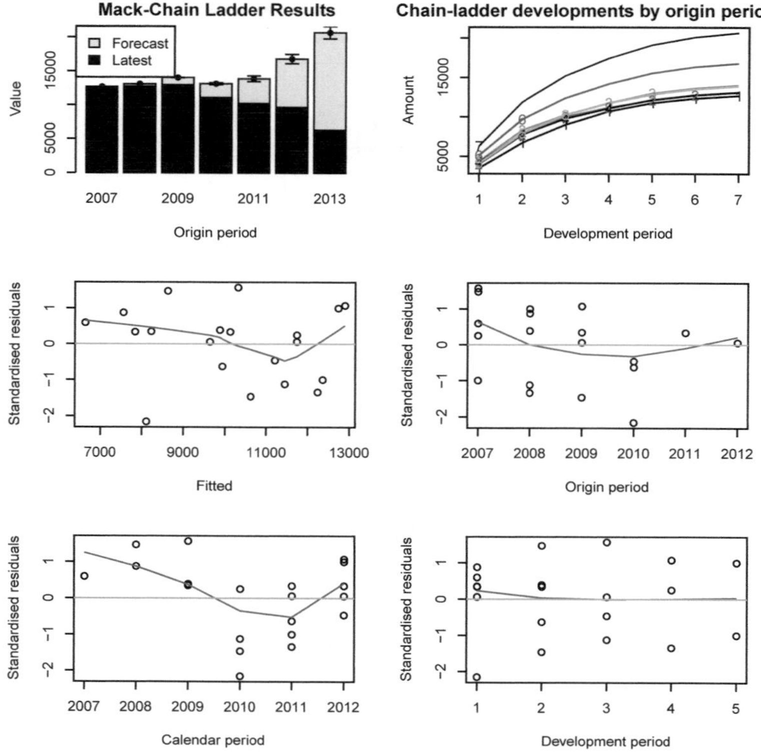

FIGURE 16.6
Plot of `MackChainLadder` output. The top-left panel shows the latest actual position with the forecasts stacked on top and whiskers indicating the estimated standard error. The top-right panel presents the claims developments to ultimate for each origin year. The four residual plots show the standardized residuals against fitted values, origin, calendar and development period. The residual plots should not show any obvious patterns and about 95% of the standardized residuals should be contained in the range of −2 to 2 for the Mack model to be strictly applicable.

Multivariate Chain-Ladder Models

The Mack chain-ladder technique can be generalized to the multivariate setting where multiple reserving triangles are modelled and developed simultaneously. The advantage of the multivariate modelling is that correlations among different triangles can be modelled, which will lead to more accurate uncertainty assessments.

Reserving methods that explicitly model the between-triangle contemporaneous correlations can be found in Pröhl & Schmidt (2005) and Merz & Wüthrich (2008b).

Another benefit of multivariate loss reserving is that structural relationships between triangles can also be reflected, where the development of one triangle depends on past losses from other triangles. For example, there is generally a need for the joint development of the paid and incurred losses; Quarg & Mack (2004).

Most of the chain-ladder-based multivariate reserving models can be summarised as sequential, seemingly unrelated regressions; Zhang (2010). We note another strand of multivariate loss reserving builds a hierarchical structure into the model to allow estimation of one triangle to *borrow strength* from other triangles, reflecting the core insight of actuarial credibility; Zhang et al. (2012).

The ChainLadder package provides implementation of multivariate chain-ladder models via the functions MunichChainLadder, MultiChainLadder and MultiChainLadder2. See the package vignette and help files for more details.

16.4.3 Poisson Regression Model for Incremental Claims

Hachemeister & Stanard (1975), Kremer (1982) and finally Mack (1991) examined the validity of treating incremental paid claims as Poisson distributed random variables, noting that the model gives the same forecast as the volume weighted chain-ladder method. Renshaw & Verrall (1998) presented the technique in the context of generalized linear models.

The idea is to assume that incremental payments X_{ij} are Poisson distributed, given the factors origin period a_i and development period b_j with an intercept c and corner constraints $a_1 = 0$, $b_1 = 0$:

$$\log \mathbb{E}(X_{i,j}) = \eta_{i,j} = c + a_i + b_j \tag{16.19}$$

The Poisson model can be directly implemented via the glm function in R:

```
> preg <- glm(inc.paid.k ~ originf + devf,
              data=Claims, family=poisson(link = "log"))
> summary(preg)

Call:
glm(formula = inc.paid.k ~ originf + devf, family = poisson(link = "log"),
    data = Claims)

Deviance Residuals:
   Min      1Q  Median      3Q     Max
-7.057  -1.776   0.036   1.675   8.776

Coefficients:
            Estimate Std. Error z value Pr(>|z|)
(Intercept)  8.25725    0.01064  776.17  < 2e-16 ***
originf2008  0.03156    0.01263    2.50   0.0124 *
originf2009  0.10042    0.01265    7.94  2.0e-15 ***
originf2010  0.03468    0.01326    2.62   0.0089 **
originf2011  0.08966    0.01367    6.56  5.4e-11 ***
originf2012  0.28129    0.01408   19.98  < 2e-16 ***
originf2013  0.48835    0.01650   29.59  < 2e-16 ***
devf2       -0.11739    0.00914  -12.84  < 2e-16 ***
devf3       -0.62832    0.01170  -53.70  < 2e-16 ***
devf4       -1.03172    0.01498  -68.89  < 2e-16 ***
devf5       -1.31341    0.01910  -68.78  < 2e-16 ***
devf6       -1.86298    0.02991  -62.29  < 2e-16 ***
devf7       -2.42831    0.05527  -43.94  < 2e-16 ***
---
Signif. codes:  0 *** 0.001 ** 0.01 * 0.05 . 0.1   1

(Dispersion parameter for poisson family taken to be 1)
```

```
    Null deviance: 26197.85  on 27  degrees of freedom
Residual deviance:   322.95  on 15  degrees of freedom
AIC: 615.6

Number of Fisher Scoring iterations: 4
```

The intercept term estimates the first log-payment of the first origin period. The other coefficients are then additive to the intercept. Thus, the predictor for the second payment of 2008 would be $\exp(8.25725+0.03156-0.11739) = 3538$. The second column in the output above gives us immediate access to the standard errors. Other test statistics are provided by `summary` as well, such as deviance and AIC (see also Chapter 14 for an introduction to the Poisson regression).

Based on those estimated coefficients, we can predict the incremental claims payments as $\widehat{X}_{i,j}$, but first we have to create a dataframe that holds the future time periods:

```
> allClaims <- data.frame(origin = sort(rep(2007:2013, n)),
                          dev = rep(1:n,n))
> allClaims <- within(allClaims, {
     devf <- factor(dev)
     cal <- origin + dev  - 1
     originf <- factor(origin)
   })
> (pred.inc.tri <- t(matrix(predict(preg,type="response",
                                 newdata=allClaims), n, n)))

     [,1] [,2] [,3] [,4] [,5]  [,6]  [,7]
[1,] 3855 3428 2057 1374 1037 598.4 340.0
[2,] 3979 3538 2123 1418 1070 617.6 350.9
[3,] 4263 3791 2274 1519 1146 661.6 375.9
[4,] 3992 3549 2129 1423 1073 619.5 352.0
[5,] 4217 3750 2250 1503 1134 654.5 371.9
[6,] 5108 4542 2725 1820 1374 792.8 450.4
[7,] 6283 5587 3352 2239 1690 975.2 554.1
```

The total amount of reserves is the sum of incremental predicted payments for calendar years beyond 2013:

```
> sum(predict(preg,type="response", newdata=subset(allClaims, cal > 2013)))

[1] 28656
```

Observe not only that the total amount of reserves is the same as the chain-ladder method, but also the predicted triangle; see Remark 16.5 on page 550.

From the regression coefficients we can also calculate the chain-ladder age-to-age ratios again:

```
> df <- c(0, coef(preg)[(n+1):(2*n-1)])
> sapply(2:7, function(i) sum(exp(df[1:i]))/sum(exp(df[1:(i-1)])))

[1] 1.889 1.282 1.147 1.097 1.051 1.028
```

While it is interesting to assume a Poisson model, since the output is the same as the one obtained using the chain-ladder technique, we have to test if this model is appropriate from a statistical perspective. For instance, assuming equi-dispersion is clearly not valid here:

```
> library(AER)
> dispersiontest(preg)

        Overdispersion test

data:  preg
z = 2.966, p-value = 0.001508
alternative hypothesis: true dispersion is greater than 1
sample estimates:
dispersion
     11.57
```

A quasi-Poisson model, with the variance proportional to the mean, should be more reasonable. We will investigate this further in the next section.

Remark 16.8 *Note that the so-called Poisson regression — namely* `glm(Y ..., family=poisson)` *—can be used on non-integers. Recall that in* R*, output from a generalised linear model is obtained using iterated least squares in a standard linear regression on* `log(Y)`*, if the link function is logarithmic. The only important point in this section is thus to have (strictly) positive incremental payments, not integers; even this can be loosened to the sum of incremental payments for each development period to be positive for a quasi-Poisson model; see P.D. England & R.J. Verrall (2002) and Firth (2003).*

Quantifying Uncertainty in GLMs

We continue our analysis with an over-dispersion Poisson model. As mentioned in Kaas et al. (2008), there are closed forms for the variance of any quantity, in generalized linear models. With notations of Chapter 14, in the case of Poisson regression with a logarithmic link function, we have

$$\mathbb{E}(X_{i,j}|\mathcal{F}) = \mu_{i,j} = \exp[\eta_{i,j}] \text{ and } \widehat{\mu}_{i,j} = \exp[\widehat{\eta}_{i,j}]. \tag{16.20}$$

Using Taylor series expansion, we can approximate $Var(\widehat{x}_{i,j})$:

$$Var(\widehat{x}_{i,j}) \approx \left|\frac{\partial \mu_{i,j}}{\partial \eta_{i,j}}\right|^2 \cdot Var(\widehat{\eta}_{i,j}), \tag{16.21}$$

which, with a logarithmic link function, can be simplified to

$$\frac{\partial \mu_{i,j}}{\partial \eta_{i,j}} = \mu_{i,j}. \tag{16.22}$$

Thus, the mean squared error of the total amount of reserve is here

$$\mathbb{E}\left([R-\widehat{R}]^2\right) \approx \left(\sum_{i+j>n+1} \widehat{\phi}\cdot\widehat{\mu}_{i,j}\right) + \widehat{\boldsymbol{\mu}}'\cdot\widehat{Var}(\widehat{\boldsymbol{\eta}})\cdot\widehat{\boldsymbol{\mu}}. \tag{16.23}$$

With this preparation done we can carry out the regression, assuming a quasi-Poisson distribution:

```
> summary(odpreg <- glm(inc.paid.k ~ originf + devf, data=Claims,
                        family=quasipoisson))
```

```
Call:
glm(formula = inc.paid.k ~ originf + devf, family = quasipoisson,
    data = Claims)

Deviance Residuals:
   Min       1Q   Median       3Q      Max
-7.057   -1.776    0.036    1.675    8.776

Coefficients:
            Estimate Std. Error t value Pr(>|t|)
(Intercept)   8.2573     0.0494  166.99  < 2e-16 ***
originf2008   0.0316     0.0587    0.54  0.59861
originf2009   0.1004     0.0588    1.71  0.10819
originf2010   0.0347     0.0616    0.56  0.58185
originf2011   0.0897     0.0635    1.41  0.17850
originf2012   0.2813     0.0654    4.30  0.00063 ***
originf2013   0.4883     0.0767    6.37  1.3e-05 ***
devf2        -0.1174     0.0425   -2.76  0.01452 *
devf3        -0.6283     0.0544  -11.55  7.2e-09 ***
devf4        -1.0317     0.0696  -14.82  2.3e-10 ***
devf5        -1.3134     0.0888  -14.80  2.4e-10 ***
devf6        -1.8630     0.1390  -13.40  9.4e-10 ***
devf7        -2.4283     0.2569   -9.45  1.0e-07 ***
---
Signif. codes:  0 *** 0.001 ** 0.01 * 0.05 . 0.1   1

(Dispersion parameter for quasipoisson family taken to be 21.6)

    Null deviance: 26197.85  on 27  degrees of freedom
Residual deviance:   322.95  on 15  degrees of freedom
AIC: NA

Number of Fisher Scoring iterations: 4
```

Note that the coefficients are the same as for the Poisson model without over-dispersion. However, the dispersion parameter is 21.6 and the errors changed as well. Now we can compute all the components of the mean squared error:

```
> mu.hat <- predict(odpreg, newdata=allClaims, type="response")*(allClaims$cal>2013)
> phi <- summary(odpreg)$dispersion
> Sigma <- vcov(odpreg)
> model.formula <- as.formula(paste("~", formula(odpreg)[3]))
> # Future design matrix
> X <- model.matrix(model.formula, data=allClaims)
> Cov.eta <- X%*% Sigma %*%t(X)
```

Hence, the mean squared error is

```
> sqrt(phi * sum(mu.hat) + t(mu.hat) %*% Cov.eta %*% mu.hat)

     [,1]
[1,] 1708
```

Observe that this is comparable with Mack's mean squared error of 1417.

Nevertheless, the method we just described might not be valid since that expression was obtained using asymptotic theory on generalised linear models, which might not be valid here since we have less than 50 observations.

```
> op <- par(mfrow=c(2,2), oma = c(0, 0, 3, 0))
> plot(preg)
> par(op)
```

```
Warning messages:
1: not plotting observations with leverage one:
  7, 28
2: not plotting observations with leverage one:
  7, 28
```

Still, the residual plots, see Figure 16.7, look reasonable, but R gives us a warning in respect of two data points that have a leverage greater than one. Lines 7 and 28 refer to the development years 2 and 1 for the origin years 2009 and 2013 respectively.

Indeed, the first payment in 2013 is considerably higher than those in previous years and also the payment for the 2009 year after 24 months is higher in relation to the payment in year 3. Again, this should prompt further investigations into the data.

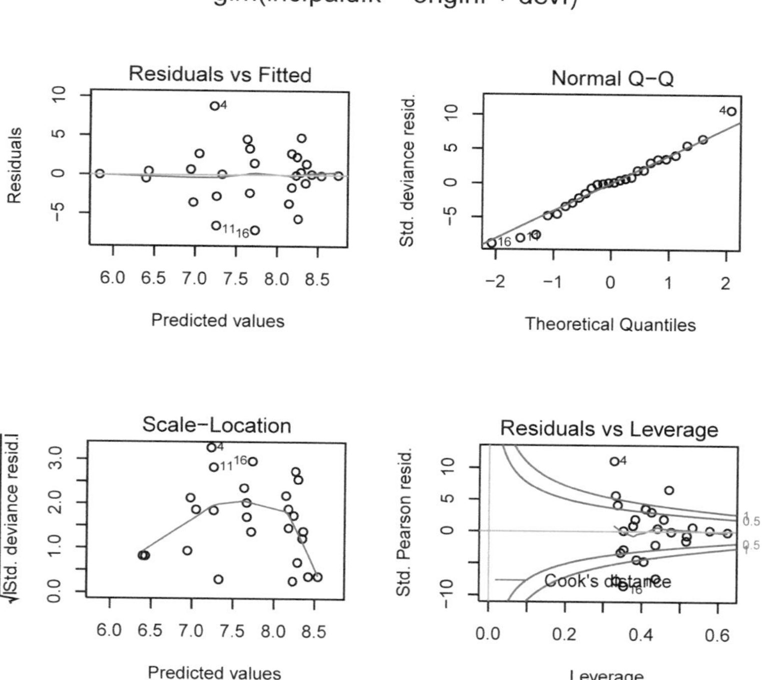

FIGURE 16.7
The output of the Poisson model appears to be well behaved. However, the last plot produces a warning that should be investigated.

Remark 16.9 *A wider range of generalized reserving models is provided in the* `ChainLadder` *package via the function* `glmReserve`*. It takes origin period and development period as mean predictors in estimating the ultimate loss reserves, and provides both analytical and bootstrapping methods to compute the associated prediction errors. The bootstrapping approach also generates the full predictive distribution for loss reserves. To replicate the results of this section we use:*

```
> (odp <- glmReserve(as.triangle(inc.triangle), var.power=1, cum=FALSE))

      Latest Dev.To.Date Ultimate  IBNR    S.E      CV
2008   12746      0.9732    13097   351  125.8 0.35843
2009   12993      0.9260    14031  1038  205.1 0.19757
2010   11093      0.8443    13138  2045  278.9 0.13636
2011   10217      0.7361    13880  3663  386.8 0.10559
2012    9650      0.5740    16812  7162  605.3 0.08451
2013    6283      0.3038    20680 14397 1158.1 0.08044
total  62982      0.6873    91638 28656 1708.2 0.05961
```

Remark 16.10 *The use of generalised linear models in insurance loss reserving has many compelling aspects; for example,*

- *When the over-dispersed Poisson model is used, it reproduces the estimates from the chain-ladder method.*
- *It provides a more coherent modelling framework than the Mack model.*
- *All the relevant established statistical theory can be directly applied to perform hypothesis testing and diagnostic checking.*

However, the user should be cautious of some of the key assumptions that underlie the generalised linear model, in order to determine whether this model is appropriate for the problem considered:

- *The generalised linear model assumes no tail development, and it only projects losses to the latest time point of the observed data. To use a model that enables tail extrapolation, consider the growth curve model* `ClarkLDF` *or* `ClarkCapeCod` *in the* `ChainLadder` *package; see also Clark (2003).*
- *The model assumes that each incremental loss is independent of all the others. This assumption may not be valid, in that cells from the same calendar year are usually correlated due to inflation or business operating factors (e.g. catastrophe losses can effect policies from multiple origin periods).*
- *The model tends to be over-parametrized, which may lead to inferior predictive performance.*

16.4.4 Bootstrap Chain-Ladder

An alternative to asymptotic econometric relationships can be to use the bootstrap methodology. Here we present a two-stage simulation approach, following P.D. England & R.J. Verrall (2002).

In the first stage, a quasi-Poisson model is applied to the claims triangle to forecast future payments. From this we calculate the scaled-Pearson residuals, assuming that they are approximately independent and identical distributed. These residuals are re-sampled with replacement many times to generate bootstrapped (pseudo) triangles and to forecast future claims payments to estimate the parameter error. Recall that the predictions of the quasi-Poisson model are the same as those from the chain-ladder method, hence we use the latter faster algorithm.

In the second stage, we simulate the process error with the bootstrap value as the mean and an assumed process distribution, here a quasi-Poisson. The set of reserves obtained in this way forms the predictive distribution, from which summary statistics such as mean, prediction error or quantiles can be derived.

In a Poisson regression, the Pearson's residuals are

$$\varepsilon_{i,j} = \frac{Y_{i,j} - \widehat{Y}_{i,j}}{Var(Y_{i,j})}. \tag{16.24}$$

In order to have a proper estimator of the variance (to have residuals with unit variance), we have to adjust the residuals for the number of regression parameters k (i.e. $2n-1$) and observations n:

$$\tilde{\varepsilon}_{i,j} = \sqrt{\frac{n}{n-k}} \cdot \frac{Y_{i,j} - \widehat{Y}_{i,j}}{\sqrt{\widehat{Y}_{i,j}}}. \tag{16.25}$$

The strategy is to bootstrap among those residuals to get a sample $\tilde{\varepsilon}^b_{i,j}$, and to generate a pseudo triangle

$$Y^b_{i,j} = \widehat{Y}_{i,j} + \sqrt{\widehat{Y}_{i,j}} \cdot \tilde{\varepsilon}^b_{i,j}. \tag{16.26}$$

Then we can use standard techniques to complete the triangle, and extrapolate the lower part. As mentioned in the introduction to Mack's approach, there are two kinds of uncertainty: uncertainty in the estimation of the model, and uncertainty in the process of future payments.

If we use the predictions from a quasi-Poisson model in this new triangle, we will predict the expected value of future payments. In order to quantify uncertainty, it is necessary to generate scenarios of payments.

This two-stage bootstrapping/simulation approach is implemented in the `BootChainLadder` function as part of the `ChainLadder` package.

As input parameters we provide the cumulative triangle, the number of bootstraps and the process distribution to be assumed:

```
> set.seed(1)
> (B <- BootChainLadder(cum.triangle, R=1000, process.distr="od.pois"))

BootChainLadder(Triangle = cum.triangle, R = 1000, process.distr = "od.pois")

     Latest Mean Ultimate Mean IBNR SD IBNR IBNR 75% IBNR 95%
2007 12,690        12,690         0       0        0        0
2008 12,746        13,099       353     127      430      569
2009 12,993        14,025     1,032     202    1,165    1,368
2010 11,093        13,133     2,040     270    2,216    2,525
2011 10,217        13,866     3,649     392    3,900    4,336
2012  9,650        16,815     7,165     630    7,559    8,268
2013  6,283        20,677    14,394   1,185   15,203   16,298
```

```
                  Totals
Latest:           75,672
Mean Ultimate:   104,305
Mean IBNR:        28,633
SD IBNR:           1,721
Total IBNR 75%:   29,787
Total IBNR 95%:   31,312
```

The first two moments are, not surprisingly, similar to the Poisson model. In contrast to the Mack model, which only provides the first two moments for the underlying distributions, here we have also access to the various percentiles of the estimated distribution of future payments.

The default plot of the model output, see Figure 16.8, presents the distribution of the simulated future payments (here labelled *IBNR*) and an initial sense check of the model by comparing the latest actual payments against simulated data.

```
> plot(B)
```

For capital setting purposes, it is desirable to look at the extreme percentiles of the simulated data. The `BootChainLadder` function also gives us access to the simulated triangles, which allows us to extract percentiles using the `quantile` function:

```
> quantile(B, c(0.75,0.95,0.99, 0.995))

$ByOrigin
     IBNR 75% IBNR 95% IBNR 99% IBNR 99.5%
2007        0      0.0        0        0.0
2008      430    569.1      716      746.2
2009     1165   1368.2     1514     1580.2
2010     2216   2525.0     2757     2809.4
2011     3900   4336.1     4641     4803.1
2012     7559   8268.0     8858     9014.0
2013    15203  16298.3    17215    17482.1

$Totals
            Totals
IBNR 75%:    29787
IBNR 95%:    31312
IBNR 99%:    32859
IBNR 99.5%:  33278
```

For many lines of business in non-life insurance, it is not unreasonable that losses follow a log-normal distribution. We can test this idea for our data by fitting a log-normal distribution to the predicted future payments. The `fitdistrplus` package by Delignette-Muller et al. (2013) makes it a one liner in R:

```
> library(fitdistrplus)
> (fit <- fitdist(B$IBNR.Totals[B$IBNR.Totals>0], "lnorm"))

Fitting of the distribution ' lnorm ' by maximum likelihood
Parameters:
        estimate Std. Error
```

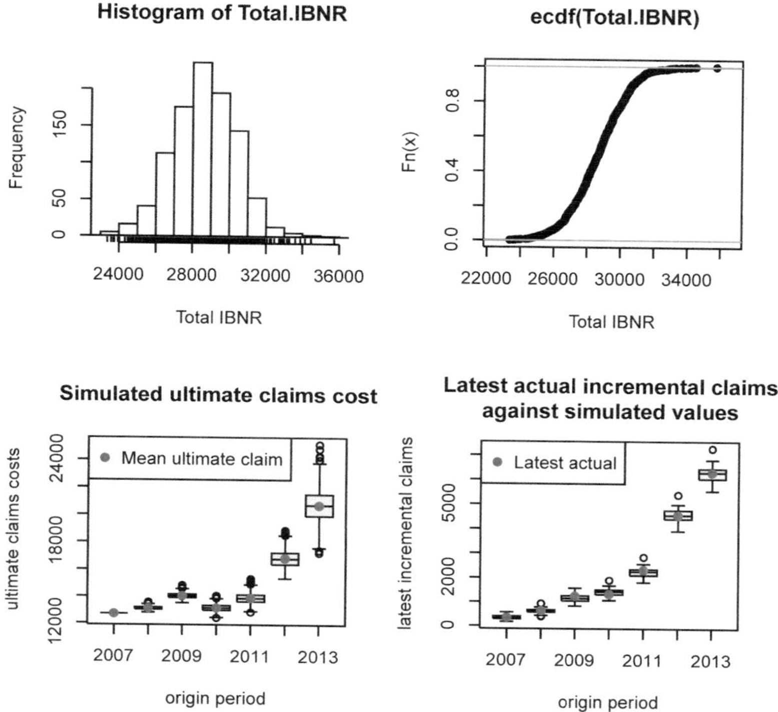

FIGURE 16.8
The top-left chart shows a histogram of the simulated future payments (here labelled *IBNR*) across all origin years. The line chart in the top right presents the empirical cumulative distribution of those simulated future payments. The bottom row shows a breakdown by origin year. The first box-whisker plot on the left displays the simulations by year, with the mean highlighted as a dot. The box-whisker plot to the right can be used to test the model, as it shows the latest actual claim for each origin year against the simulated distribution.

```
meanlog 10.26051   0.001908
sdlog    0.06033   0.001347
```

```
> plot(fit)
```

The fit looks very reasonable indeed, see Figure 16.9, and the 99.5 percentile of the fitted log-normal is close to the sample percentile above:

```
> qlnorm(0.995, fit$estimate['meanlog'], fit$estimate['sdlog'])
```

```
[1] 33387
```

Payment distribution over the next year

As we have access to all simulated triangles, we can also estimate percentiles for payments in the following year. For the 99.5 percentile payment over the next 12 months, we get

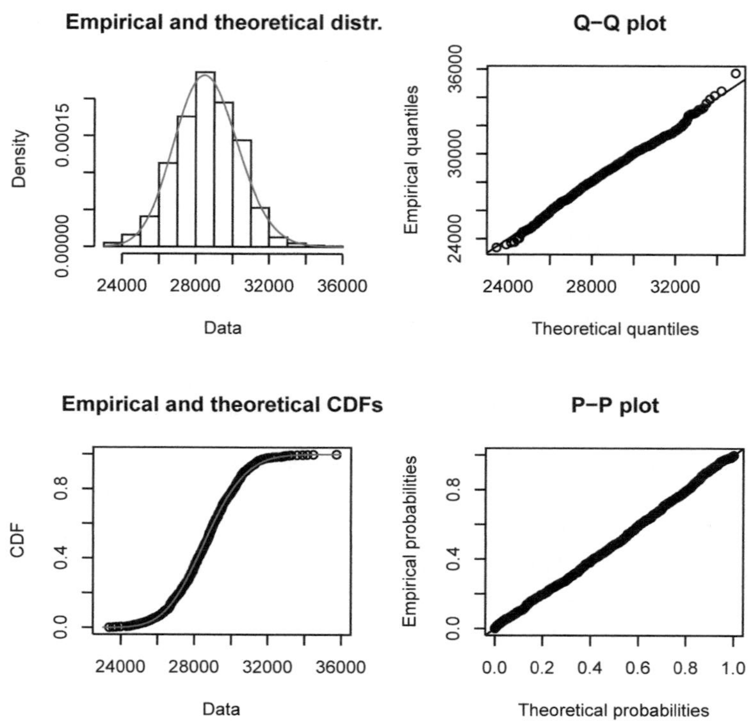

FIGURE 16.9
Plots of simulated data from `BootChainLadder` and a fitted log-normal distribution.

```
> ny <- (col(inc.triangle) == (nrow(inc.triangle) - row(inc.triangle) + 2))
> paid.ny <- apply(B$IBNR.Triangles, 3,
                   function(x){
                     next.year.paid <- x[col(x) == (nrow(x) - row(x) + 2)]
                     sum(next.year.paid)
                   })
> paid.ny.995 <- B$IBNR.Triangles[,,order(paid.ny)[round(B$R*0.995)]]
> inc.triangle.ny <- inc.triangle
> (inc.triangle.ny[ny] <- paid.ny.995[ny])

[1] 7076 3308 1352  970  868  338
```

This would reflect a 49% reserve utilisation over the next year (sum of payments next year devided by total reserve).

16.4.5 Reserving Based on Log-Incremental Payments

We noted in the previous section that the claims appear to follow a log-normal distribution. Zehnwirth (1994) was not the first to consider modelling the log of the incremental claims

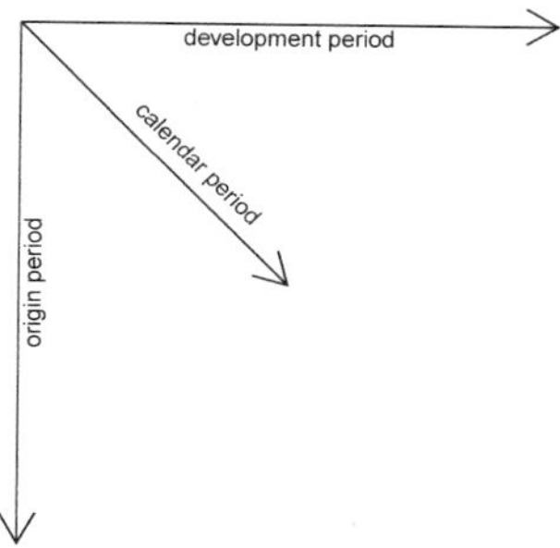

FIGURE 16.10
Structure of a typical claims triangle and the three time directions: origin, development, and calendar periods.

payments, but his papers and software ICRFS[6] have popularised this approach. Here we present the key concepts of what Zehnwirth (1994) calls the probabilistic trend family (PTF).

Zehnwirth's model assumes the following structure for the incremental claims $X_{i,j}$:

$$\ln(X_{i,j}) = Y_{i,j} = \alpha_i + \sum_{k=1}^{j} \gamma_k + \sum_{t=1}^{i+j} \iota_t + \varepsilon_{i,j}, . \tag{16.27}$$

The errors are assumed to be normal with $\varepsilon_{i,j} \sim \mathcal{N}(0, \sigma^2)$. The parameters $\alpha_i, \gamma_j, \iota_t$ model trends in three time directions, namely origin year, development year and calendar (or payment) year, respectively; see Figure 16.10.

Christofides (1997) examines a very similar model, but uses the following notation:

$$\ln(X_{i,j}) = Y_{i,j} = a_i + d_j + \varepsilon_{i,j}, \tag{16.28}$$

with a, d representing the parameters in origin and development period direction (a parameter p_{i+j-1} for the payment year direction could be added). Although the models in Equations (16.27) and (16.28) are essentially the same, the design matrices differ and therefore the coefficients and their interpretation.

Remark 16.11 *Note that the above model is not a GLM, that is,* $\log(y+\varepsilon) = X\beta$. *Instead, it models* $\log(y) = X\beta + \varepsilon$, *although both models assume* $\varepsilon \sim \mathcal{N}(0, \sigma^2)$. *Hence, we will use least squares regression to fit the coefficients via* `lm` *again.*

Before we apply the log-linear model to the data, and we will follow Christofides (1997), we shall plot it again on a log-scale.

```
> Claims <- within(Claims, {
      log.inc <- log(inc.paid.k)
      cal <- as.numeric(levels(originf))[originf] + dev - 1
  })
```

The interaction plot, Figure 16.11, suggests a linear relationship after the second development year on a log-scale. The lines of the different origin years are fairly closely grouped, but the last 2 years, labelled 6 and 7, do stand out. We shall test if this is significant. We

[6]Interactive Claims Reserving and Forecasting System.

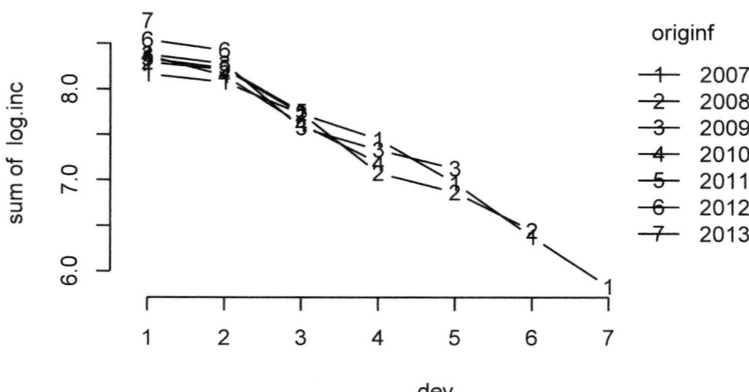

FIGURE 16.11
The interaction plot shows the developments of the origin years on a log-scale. From the second development year, the decay appears to be linear.

start with a model using all levels of the origin factor and two dummy parameters for the development year, with $d_1 = d1$ and $d_j = (j-1) \cdot d27$ for $j > 1$. Hence, we add two dummy variables to our data:

```
> Claims <- within(Claims, {
      d1 <- ifelse(dev < 2, 1, 0)
      d27 <- ifelse(dev < 2, 0, dev - 1)
  })
```

The dummy variable $d1$ is 1 for the first development period and 0 otherwise, while $d27$ is 0 for the first development period and counts up from 1 then onwards. Hence, we will estimate one parameter for the first payment and a constant trend (decay) for the following periods:

```
> summary(fit1 <- lm(log.inc ~  originf + d1 + d27, data=Claims))

Call:
lm(formula = log.inc ~ originf + d1 + d27, data = Claims)

Residuals:
    Min      1Q  Median      3Q     Max 
-0.2214 -0.0397  0.0112  0.0329  0.1962 

Coefficients:
             Estimate Std. Error t value Pr(>|t|)    
(Intercept)  8.572835   0.075690  113.26  < 2e-16 ***
originf2008  0.000956   0.063935    0.01  0.98822    
originf2009  0.092037   0.068675    1.34  0.19600    
originf2010 -0.018715   0.075261   -0.25  0.80629    
```

```
originf2011  0.063828   0.084302    0.76  0.45825
originf2012  0.272668   0.098245    2.78  0.01205 *
originf2013  0.468983   0.131593    3.56  0.00207 **
d1          -0.296215   0.069903   -4.24  0.00045 ***
d27         -0.434960   0.018488  -23.53  1.6e-15 ***
---
Signif. codes:  0 *** 0.001 ** 0.01 * 0.05 . 0.1   1

Residual standard error: 0.114 on 19 degrees of freedom
Multiple R-squared:  0.983,      Adjusted R-squared:  0.976
F-statistic:  139 on 8 and 19 DF,  p-value: 3.29e-15
```

The model output confirms what we had noticed from the interaction plot already; apart from the origin years 2012 and 2013 there is no significant difference between the years; the p-values are all greater than 5% and the coefficients are less than twice their standard errors. Therefore we reduce the model and replace the origin variable with two dummy columns for those years:

```
> Claims <- within(Claims, {
     a6 <- ifelse(originf == 2012, 1, 0)
     a7 <- ifelse(originf == 2013, 1, 0)
  })
> summary(fit2 <- lm(log.inc ~  a6 + a7 + d1 + d27, data=Claims))

Call:
lm(formula = log.inc ~ a6 + a7 + d1 + d27, data = Claims)

Residuals:
     Min       1Q   Median       3Q      Max
-0.21567 -0.04910  0.00654  0.05137  0.27199

Coefficients:
            Estimate Std. Error t value Pr(>|t|)
(Intercept)   8.6079     0.0515  167.14  < 2e-16 ***
a6            0.2435     0.0852    2.86  0.00887 **
a7            0.4411     0.1217    3.62  0.00142 **
d1           -0.3035     0.0678   -4.48  0.00017 ***
d27          -0.4397     0.0167  -26.39  < 2e-16 ***
---
Signif. codes:  0 *** 0.001 ** 0.01 * 0.05 . 0.1   1

Residual standard error: 0.112 on 23 degrees of freedom
Multiple R-squared:  0.98,      Adjusted R-squared:  0.977
F-statistic:  288 on 4 and 23 DF,  p-value: <2e-16
```

The reduction in parameters from 9 to 5 seems sensible, all coefficient are significant and the model error reduced from 0.114 to 0.112 as well. Further we can read off the coefficient for $d27$ that claims payments are predicted to reduce by 44% each year after year 1. Next, we plot the model:

```
> op <- par(mfrow=c(2,2), oma = c(0, 0, 3, 0))
> plot(fit2)
> par(op)
```

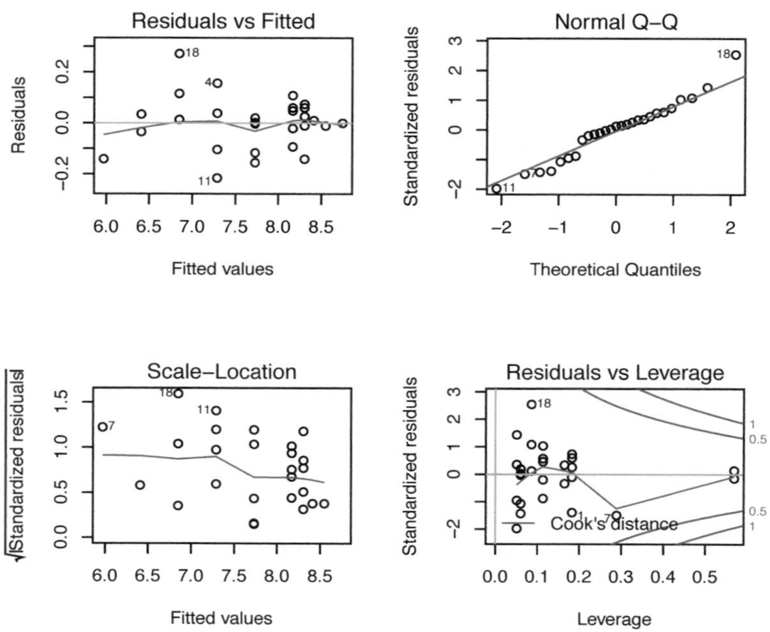

FIGURE 16.12
Residual plots of the log-incremental model `fit2`. The last payment of 2009 (row 18) is highlighted again as a potential outlier, and so are rows 4, 7 and 11.

Reviewing the residual plots in Figure 16.12 highlights again the latest payment for the 2009 origin year (the 18th row of the Claims data) as a potential outlier.

The error distribution appears to follow a normal distribution, top right QQ-plot in Figure 16.12, confirmed by the Shapiro–Wilk normality test:

```
> shapiro.test(fit2$residuals)

        Shapiro-Wilk normality test

data:  fit2$residuals
W = 0.9654, p-value = 0.4638
```

To investigate the residuals further, we shall plot them against the fitted values and the three trend directions. The following function will create those four plots for our model.

```
> resPlot <- function(model, data){
     xvals <- list(
       fitted = model[['fitted.values']],
       origin = as.numeric(levels(data$originf))[data$originf],
       cal=data$cal, dev=data$dev
     )
```

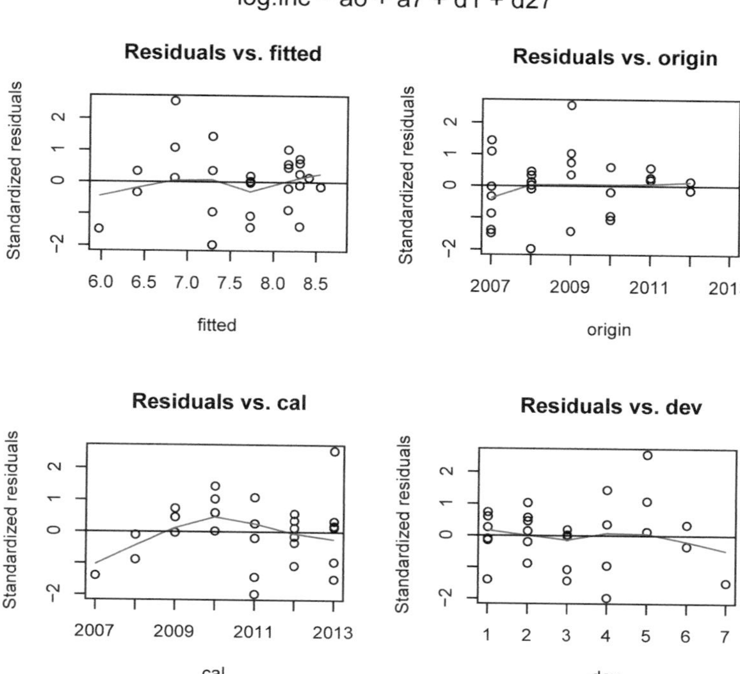

FIGURE 16.13
Residual plots of the log-incremental model `fit2` against fitted values and the three trend directions.

```
    op <- par(mfrow=c(2,2), oma = c(0, 0, 3, 0))
    for(i in 1:4){
    plot.default(rstandard(model) ~ xvals[[i]] ,
                 main=paste("Residuals vs", names(xvals)[i] ),
                 xlab=names(xvals)[i], ylab="Standardized residuals")
    panel.smooth(y=rstandard(model), x=xvals[[i]])
    abline(h=0, lty=2)
    }
    mtext(as.character(model$call)[2], outer = TRUE, cex = 1.2)
    par(op)
  }

> resPlot(fit2, Claims)
```

Again, the residual plots all look fairly well behaved; however, we notice from the bottom-left plot in Figure 16.13 that claims for the payment years 2007, 2008 are slightly over-fitted and 2009, 2010 are under-fitted. Hence, we introduce an additional parameter for that period and update our model:

```
> Claims <- within(Claims, {
    p34 <- ifelse(cal < 2011 & cal > 2008, cal-2008, 0)
```

```
  })
> summary(fit3 <- update(fit2, ~ . + p34, data=Claims))

Call:
lm(formula = log.inc ~ a6 + a7 + d1 + d27 + p34, data = Claims)

Residuals:
    Min      1Q  Median      3Q     Max
-0.1941 -0.0595  0.0164  0.0511  0.2840

Coefficients:
            Estimate Std. Error t value Pr(>|t|)
(Intercept)   8.5576     0.0540  158.51  < 2e-16 ***
a6            0.2822     0.0819    3.45  0.00230 **
a7            0.4777     0.1152    4.15  0.00042 ***
d1           -0.2897     0.0638   -4.54  0.00016 ***
d27          -0.4301     0.0163  -26.45  < 2e-16 ***
p34           0.0603     0.0292    2.07  0.05074 .
---
Signif. codes:  0 *** 0.001 ** 0.01 * 0.05 . 0.1   1

Residual standard error: 0.105 on 22 degrees of freedom
Multiple R-squared:  0.984,       Adjusted R-squared:  0.98
F-statistic:  264 on 5 and 22 DF,  p-value: <2e-16

> resPlot(fit3, Claims)
```

The residual plot against calendar years, Figure 16.14, has improved and the parameter $p34$ could be regarded significant. The coefficient $p34$ describes a 6% increase of claims payments in those 2 years. An investigation should clarify if this effect is the result of a temporary increase in claims inflation, a change in the claims settling process, other causes or just random noise. Observe that the new model has a slightly lower residual standard error of 0.105 compared to 0.112.

Within the linear regression framework we can forecast the claims payments and estimate the standard errors. We follow the paper by Christofides (1997) again. Recall that for a log-normal distribution, the mean is $E(X) = \exp(\mu + 1/2\sigma^2)$ and the variance is $Var(X) = \exp(2\mu + \sigma^2)(\exp(\sigma^2) - 1)$, where μ and σ are the mean and standard deviation of the logarithm, respectively

```
> log.incr.predict <- function(model, newdata){
    Pred <- predict(model, newdata=newdata, se.fit=TRUE)
    Y <- Pred$fit
    VarY <- Pred$se.fit^2 + Pred$residual.scale^2
    P <- exp(Y + VarY/2)
    VarP <-  P^2*(exp(VarY)-1)
    seP <- sqrt(VarP)
    model.formula <- as.formula(paste("~", formula(model)[3]))
    mframe <- model.frame(model.formula, data=newdata)
    X <- model.matrix(model.formula, data=newdata)
    varcovar <- X %*% vcov(model) %*% t(X)
    CoVar <-  sweep(sweep((exp(varcovar)-1), 1, P, "*"), 2, P, "*")
    CoVar[col(CoVar)==row(CoVar)] <- 0
```

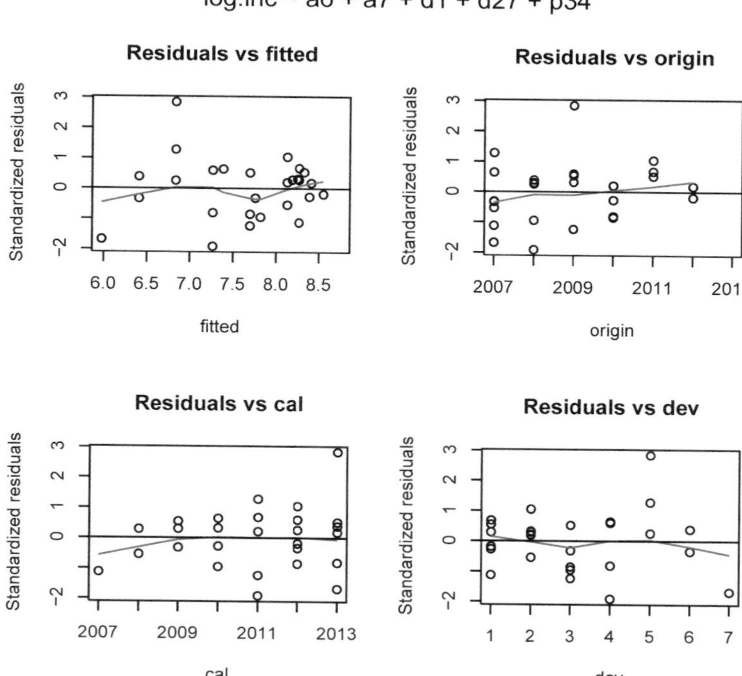

FIGURE 16.14
Residual plot of the log-incremental model `fit3`.

```
    Total.SE <- sqrt(sum(CoVar) + sum(VarP))
    Total.Reserve <- sum(P)
    Incr=data.frame(newdata, Y, VarY, P, seP, CV=seP/P)
    out <- list(Forecast=Incr,
                Totals=data.frame(Total.Reserve,
                                  Total.SE=Total.SE,
                                  CV=Total.SE/Total.Reserve))
    return(out)
  }
```

With the above function it is straightforward to carry out the prediction for future claims payment and standard errors. As a bonus we can estimate payments beyond the available data.

To forecast the future claims, we prepare a data frame with the predictors for those years, here with 6 years beyond age 7:

```
> tail.years <-6
> fdat <- data.frame(
     origin=rep(2007:2013, n+tail.years),
     dev=rep(1:(n+tail.years), each=n)
     )
> fdat <- within(fdat, {
```

```
    cal <- origin + dev - 1
    a7 <- ifelse(origin == 2013, 1, 0)
    a6 <- ifelse(origin == 2012, 1, 0)
    originf <- factor(origin)
    p34 <- ifelse(cal < 2011 & cal > 2008, cal-2008, 0)
    d1 <- ifelse(dev < 2, 1, 0)
    d27 <- ifelse(dev < 2, 0, dev - 1)
  })
```

So, here are the results for the two models:

```
> reserve2 <- log.incr.predict(fit2, subset(fdat, cal>2013))
> reserve2$Totals

  Total.Reserve Total.SE      CV
1         33847     2545 0.07519

> reserve3 <- log.incr.predict(fit3, subset(fdat, cal>2013))
> reserve3$Totals

  Total.Reserve Total.SE      CV
1         34251     2424 0.07078
```

The two models produce very similar results and it should not be much of a surprise as they are quite similar indeed. The third model has proportionally a slightly smaller standard error and may hence be the preferred choice.

The future payments can be displayed with the xtabs function:

```
> round(xtabs(P ~ origin + dev, reserve3$Forecast))

      dev
origin    2    3    4    5    6    7    8    9   10   11   12   13
  2007    0    0    0    0    0    0  259  168  110   71   47   30
  2008    0    0    0    0    0  397  259  168  110   71   47   30
  2009    0    0    0    0  610  397  259  168  110   71   47   30
  2010    0    0    0  937  610  397  259  168  110   71   47   30
  2011    0    0 1441  937  610  397  259  168  110   71   47   30
  2012    0 2946 1916 1247  812  529  344  224  146   95   62   40
  2013 5529 3595 2338 1521  990  645  420  273  178  116   76   49
```

The model structure is clearly visible in the above future claims triangle; as the origin years 2007 to 2011 share the same parameter, the predicted future payments for those years have the same identical mean expectations.

For comparison, here is the output of the Mack chain-ladder model, assuming a tail factor of 1.05 and standard error of 0.02:

```
> round(summary(MackChainLadder(cum.triangle, est.sigma="Mack",
                              tail=1.05, tail.se=0.02))$Totals,2)

               Totals
Latest:      75672.00
Dev:             0.69
Ultimate:   109544.16
```

```
IBNR:         33872.16
Mack S.E.:     2563.40
CV(IBNR):         0.08
```

The chain-ladder method provides a forecast similar to the log-incremental regression model, but at the price of many more parameters and hence potential instability.

A model with few parameters is potentially more robust and can be analysed by back testing the model with fewer data points.

The log-incremental regression model provides an intuitive and elegant stochastic claims reserving model and can help to investigate trends in the calendar/payment year direction, such as claims inflation, which is challenging to define and measure; Gesmann, Rayees & Clapham (2013). Additionally, the tail extrapolation is part of the model design and not an artificial add-on.

See Christofides (1997) and Zehnwirth (1994) for a more detailed discussion of the log-incremental model.

16.5 Quantifying Reserve Risk

The most frequent reason insurance companies failed in the past was insufficient reserves; Massey et al. (2002). Hence, as mentioned earlier, monitoring claims development is one of main purposes of the models presented in the previous section. Models, unlike deterministic point estimators, allow the actuary to judge the materiality of payment deviation from modelled expected claims development.

Yet, we have to acknowledge that a model cannot be proven right, or, to put it more bluntly: *You make money, until you don't.*

Therefore it is important to understand the underlying model assumptions and to quantify the uncertainty of the predicted mean ultimate loss costs. Typical risk measures are mean squared error, value at risk (VaR) and tail value at risk (TVaR). These risk metrics can be defined over different time horizons and play a key role in assessing capital requirements for reserve risk.

16.5.1 Ultimo Reserve Risk

The uncertainty around the ultimate loss cost, also called *ultimo* reserve risk, estimates the risk that the reserve is not sufficient to cover claims payments for the full run-off of today's liabilities. It is an important metric for capital setting and when pricing the transfer of run-off books of business and has been in use for many years.

Estimators for the ultimo risk were given for all the stochastic reserving models of the previous section. The log-incremental and bootstrap models provide direct access to various percentiles, while the Mack model only provides estimation for the mean and mean squared error and hence requires a distribution assumption, for example, log-normal to estimate extreme percentiles.

16.5.2 One-Year Reserve Risk

The reserve risk over a 1-year time horizon measures the change required in the estimate of ultimate loss cost conditioned on the claims development over the following year. This metric is used to assess the reserve risk under the proposed Solvency II regime.

The Solvency II framework is consistent with a mark to market basis, and therefore focuses on the 1-year view of the balance sheet, which requires the discounting of future cash flows based on an assumed interest rate(s). Reserve risk is specifically defined as the difference between the best estimate reserve at $t = 0$ and after 12 months ($t = 1$), following claims deterioration or a claims shock with 0.5% probability.

Merz & Wüthrich (2008a) analysed the Mack model and derived analytical formulas for the *claims development result (CDR)*. Ohlsson & Lauzeningks (2009) also described a simulation approach to quantify the 1-year risk.

Estimating the 1-year reserve risk can be computationally demanding, and many aspects such as the inclusion of expert judgement and tail factors are fields of active research.

Remark 16.12 *Similar ideas to the 1-year reserve risk are applied to back test reserving models. By removing the latest calendar year data from a claims triangle and comparing its prediction with the forecast based on the complete data, we can test the robustness of the model and its parameters.*

Illustrative 1-year reserve risk example

To clarify the concept, we present an illustrative example of how to estimate the 1-year reserve risk. We follow the ideas of Felisky et al. (2010) and simplify those even further.

From our bootstrap model we extracted the future claims triangle that has shown the highest payment in the following calendar year at the 99.5 percentile, see page 567. We add the shock payment of the next year to the original triangle and re-forecast the extended triangle to ultimate, using the chain-ladder algorithm (note that the age-to-age link ratios will change, but no tail factor is assumed), and compare the newly predicted reserve with the original forecast.

```
> (cum.triangle.ny <- t(apply(inc.triangle.ny,1,cumsum)))

origin    1     2     3     4     5     6     7
  2007 3511  6726  8992 10704 11763 12350 12690
  2008 4001  7703  9981 11161 12117 12746 13084
  2009 4355  8287 10233 11755 12993 13861    NA
  2010 4295  7750  9773 11093 12063    NA    NA
  2011 4150  7897 10217 11569    NA    NA    NA
  2012 5102  9650 12958    NA    NA    NA    NA
  2013 6283 13359    NA    NA    NA    NA    NA

> f.ny <- sapply((n-1):1, function(i){
     sum(cum.triangle.ny[1:(i+1), n-i+1])/sum(cum.triangle.ny[1:(i+1), n-i])
  })
> (f.ny <- c(f.ny[-(n-1)],1))

[1] 1.936 1.295 1.144 1.094 1.057 1.000

> full.triangle.ny <- cum.triangle.ny
> for(k in 2:(n-1)){
     full.triangle.ny[(n-k+2):n, k+1] <- full.triangle.ny[(n-k+2):n,k]*f[k]
  }
> (sum(re.reserve.995 <- full.triangle.ny[,n] - rev(latest.paid)))

[1] 31951
```

We observe that after a 1 in 200 payment shock year, the reserve would increase from 28,656 to 31,951 (+11%). Therefore the one-year reserve risk is 3,295, assuming that the actuary would not change further assumptions.

To put this metric into context, suppose the reserve is twice the volume of net earned premiums in the following year; then this would suggest a prior year deterioration of 22% on the combined ratio.

This simplified re-reserving approach has of course its limitation; it is only a point estimator. The payment shock movement is purely dependent on the historical data volatility and cannot capture events such as changes in legislation or other exogenous influences.

Note that the reserve after the 1-year shock development is less than the predictions that include a tail factor. This demonstrates that estimating the 1-year reserve risk is far more complex than illustrated with the toy example above and that the actuary would have to consider carefully how she would re-reserve the same book of business in the following year, that is, she may want to re-consider trends in the calendar year direction.

In reality, if we had a massive spike in paid losses, unless this related to a single claim, which seems unlikely at the 1 in 200 level, chances are we have issues in many areas. Therefore we will probably cross check claims with material case reserves and review them as well. The influence of case reserves is ignored in this example.

Remark 16.13 *Long tail classes of business may have a lower reserve risk over a 1-year horizon than short tail lines of business. Their claims development can be much slower and hence changes can take longer to materialize.*

On the other hand, it is also more difficult to detect reserve deterioration for long tail lines. Exemplified in the past, by the soft U.S.-casualty cycle of the late 1990s that showed prior year deteriorations reported over several years.

16.6 Discussion

This chapter gave a brief overview of how some of the more popular claims reserving methods and models can be applied in R. For more details about the models and their mathematical derivation, see the original papers, Wüthrich & Merz (2008) or Kaas et al. (2008).

Claims reserving is a complex and evolving subject. Changes in regulation, e.g. Solvency II in Europe, are expected to accelerate the adoption of stochastic reserving frameworks. Although the Mack and Bootstrap models, which are simple to implement in spreadsheet software, are popular with actuaries today, we demonstrated that many other stochastic models are straightforward to apply in a statistical environment such as R.

There are many other aspects that need to be considered in a full reserving analysis and which were not covered here; to name but a few:

- Claims inflation and other exogenous influences such as legislation, social environment, climate change
- Changes in business processes, earning patterns, accounting practice and data quality in general
- Changes to exposure and underlying policy terms and conditions

- Aggregation of reserves across multiple lines of business
- Treatment of large and catastrophic losses
- Treatment of reinsurance

Reserving is always mixture of art and science, and requires a combination of sound data analysis with business knowledge and judgement. In this sense, a Bayesian approach would lead naturally to the inclusion of expert judgement and the full reserve distribution.

Using hierarchical or multilevel models as presented by Guszcza (2008) appears a natural next step for claims reserving. The Clark LDF model, Clark (2003), mentioned by Guszcza (2008), has already been implemented by Daniel Murphy in R as part of the `ChainLadder` package. Additionally, the double-chain-ladder approach of Miranda et al. (2012) looks like a promising extension of the chain-ladder family of models.

Also worth mentioning is the `lossDev` package by Laws & Schmid (2012) that implements robust loss development using Markov Chain Monte Carlo (MCMC) using `rjags`, Plummer (2013).

So, which reserving model should I use?

It depends. Unfortunately, there is no easy answer.

It depends on the data, the context, the type of business, the tail characteristic, the time available, your statistical knowledge and, of course, the aim of the analysis.

Remember that the purpose of the data is to prove the model wrong. Hence, it is often easier to start by reviewing which models not to consider.

16.7 Exercises

Exercise 16.1 Christofides (1997) provides data on exposure changes and inflation for the example triangle used in this chapter:

```
> exposure <- data.frame(origin=factor(2007:2013),
    volume.index=c(1.43, 1.45, 1.52, 1.35, 1.29, 1.47, 1.91))
> inflation <- data.frame(cal=2007:2013,
    earning.index=c(1.55, 1.41, 1.3, 1.23, 1.13, 1.05, 1))
```

Adjust the historical data for those exposure changes and inflationary effects, using

$$\text{Normalised claims} = \frac{\text{Claims} \times \text{Inflation}}{\text{Exposure}}. \tag{16.29}$$

Carry out a reserving analysis. Which changes do you observe? How do you scale the output back to the original scale?

Exercise 16.2 Consider different scenarios of future claims inflation. By how much does the reserve change if you set claims inflation at 7.5%?

Exercise 16.3 Add two new arguments to the function `log.incr.predict` to take into account exposure and inflation assumptions.

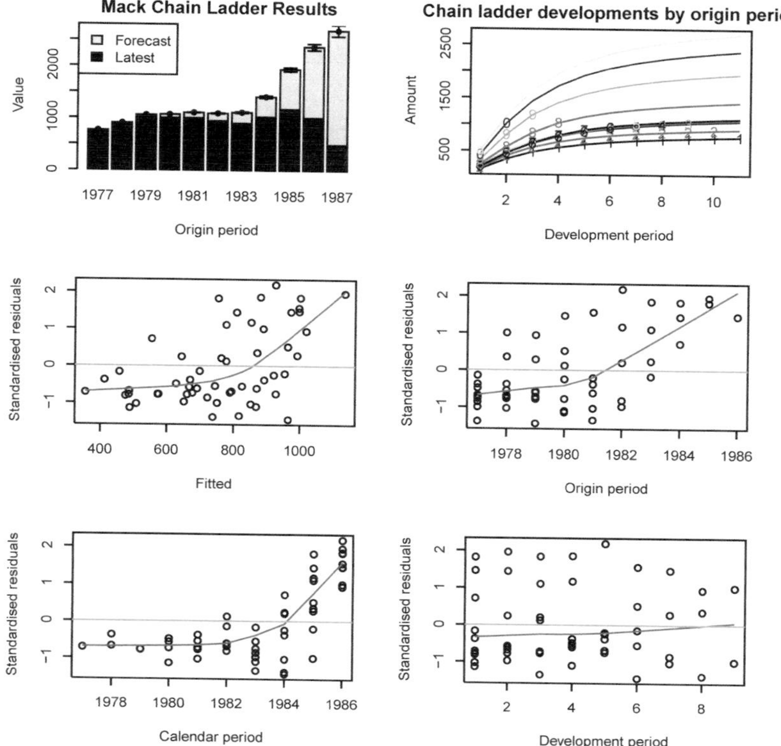

FIGURE 16.15
Mack chain-ladder output for the `ABC` triangle.

Exercise 16.4 Test the stability of your reserve. Follow the approach by Barnett & Zehnwirth (2000): remove the latest calendar year information from your data and re-forecast the reserve. Re-fit your model and discuss if the changes in the parameters are significant.

Exercise 16.5 The `ABC` triangle of the `ChainLadder` package shows significant calendar year trends, see Figure 16.15.

```
> library(ChainLadder)
> data(ABC)
> M <- MackChainLadder(ABC/1000, est.sigma="Mack")

> plot(M)
```

Investigate the data with the log-incremental model. Can a calendar year trend parameter be found?

Exercise 16.6 The log-incremental model estimates the first two moments of the reserve, assuming the incremental claims data follow a log-normal distribution. Estimate the 99.5 percentile movement of the reserve and compare against the output of the bootstrap chain-ladder model.

Exercise 16.7 The `glm` function has the argument `offset`. Discuss how it can be used to take into account exposure information. See also the `glmReserve` function in the `ChainLadder` function.

Exercise 16.8 Glenn Meyers and Peng Shi provide loss reserving data pulled from NAIC Schedule P via the CAS website: http://www.casact.org/research/index.cfm?fa=loss_reserves_data. Read the data into R and analyse it.

Investigate how multivariate chain-ladder models can be used to estimate the reserve for several of these triangles simultaneously.

Bibliography

Aalen, O. (1978), Non-parametric inference for a family of counting processes, *Annals of Statistics* **6**, 710–726.

Abramowitz, M. & Stegun, I. A. (1970), *Eds., Handbook of Mathematical Functions with Formulas, Graphs, and Mathematical Tables*, Applied Mathematics Series 55, National Bureau of Standards, Washington, DC. ninth printing.

Andersen, E. (1970), Sufficiency and exponential families for discrete sample spaces, *Journal of the American Statistical Association* **65**, 1248–1255.

Andersen, S. (1957), On the collective theory of risk in case of contagion between claims, *Bulletin of the Institute of Mathematics and its Applications* **12**, 2775–279.

Anderson, R. (2007), *The Credit Scoring Toolkit: Theory and Practice for Retail Credit Risk Management and Decision Automation*, Oxford University Press, New York.

Ang, A. & Piazzesi, M. (2003), A no-arbitrage vector autoregression of term structure dynamics with macroecconomic and latent variables, *Journal of Monetary Economics* **50**(4), 745–787.

Antonio, K., Frees, E. & Valdez, E. (2010), A multilevel analysis of intercompany claim counts, *ASTIN Bulletin: The Journal of the International Actuarial Association* **40**(1), 151–177.

Antonio, K. & Valdez, E. (2012), Statistical aspects of *a priori* and *a posteriori* risk classification in insurance, *Advances in Statistical Analysis* **96**(2), 187–224.

Antonio, K. & Zhang, W. (2014), *Predictive Modelling in Actuarial Science*, Cambridge University Press, chapter Mixed models for predictive modelling in actuarial science.

Archer, K. J. (2010), `rpartOrdinal`: An R package for deriving a classification tree for predicting an ordinal response, *Journal of Statistical Software* **34**(7), 1–17.

Arnold, B. C. (1983), *Pareto Distributions*, International Co-operative Publishing House, Fairland, MD.

Artzner, P., Delbaen, F., Eber, J.-M. & Heath, D. (1999), Coherent measures of risk, *Mathematical Finance* **9**(3), 203–228.

Asmussen, S. & Albrecher, H. (2010), *Ruin Probabilities*, 2nd edn., World Scientific, Hackensack, NJ.

Asmussen, S. & Rolski, T. (1991), Computational methods in risk theory: A matrix-algorithmic approach, *Insurance: Mathematics and Economics* **10**, 259–274.

Baier, T. & Neuwirth, E. (2003), High-level interface between R and Excel, *in* K. Hornik, F. Leisch & A. Zeileis, Eds., *Proceedings of the 3rd International Workshop on Distributed Statistical Computing (DSC 2003)*, Vienna, Austria.

Bailey, R. (1963), Insurance rates with minimum bias, *Proceedings of the Society of Actuaries* **50**, 4–11.

Baillie, R. T. & Bollerslev, T. (1989), Common stochastic trends in a system of exchange rates, *Journal of Finance* **44**(1), 167–181.

Banerjee, S., Carlin, B. P. & Gelfand, A. E. (2004), *Hierachical Modelling and Analysis for Spatial Data*, Chapman and Hall, Boca Raton, Florida.

Bank for International Settlements (2005), Zero-coupon yield curves: Technical documentation, BIS papers 25, Monetary and Economic Department.

Barbi, E. (1999), Eterogeneità della Popolazione e Sopravvivenza Umana: Prospettive Metodologiche ed Applicazioni alle Generazioni Italiane 1870–1895, PhD thesis, Florence.

Bardis, M., Majidi, A. & Murphy, D. (2012), A family of chain-ladder factor models for selected link ratios, *Variance* **2**, 143–160.

Barnett, G. & Zehnwirth, B. (2000), Best estimates for reserves, *Proceedings of the CAS* LXXXVII(167).

Bates, D. (2003), Converting packages to S4, *R News* **3**(1), 6–8.

Bates, D. (2005), Fitting linear mixed models in R, *R News* **5**(1), 27–30.

Bates, D. & Maechler, M. (2012), *Matrix: Sparse and Dense Matrix Classes and Methods*. R package version 1.0-6.

Becker, R. A. (1994), A brief history of S, Technical report, AT&T Bell Laboratories, Murray Hil, NJ.

Becker, R. A. & Chambers, J. M. (1984), *S. An Interactive Environment for Data Analysis and Graphics*, Wadsworth and Brooks/Cole, Monterey, CA. [the 'brown book'].

Beirlant, J., Goegebeur, Y., Teugels, J. & Segers, J. (2004), *Statistics of Extremes*, Wiley Series in Probability and Statistics, John Wiley & Sons Ltd., Chichester. Theory and applications, With contributions from Daniel De Waal and Chris Ferro.

Beirlant, J. & Teugels, J. (1992), Modelling large claims in non-life insurance, *Insurance: Mathematics and Economics* **11**, 17–29.

Ben-Israel, A. & Greville, T. N. E. (2003), *Generalized Inverses: Theory and Applications*, Springer, New York.

Berger, J. (1985), *Bayesian Inference in Statistical Analysis*, Springer-Verlag, Berlin.

Berger, J. O., Bernardo, J. M. & Sun, D. (2009), The formal definition of reference priors, *Annals of Statistics* **37**, 905–938.

Bernegger, S. (1997), The Swiss Re exposure curves and the MBBEFD distribution class, *ASTIN Bulletin* **27**(1), 99–111.

Besag, J. (1974), Spatial interaction and the statistical analysis of lattice data systems (with discussion), *Journal of the Royal Statistical Society, Series B* **36**, 192–225.

Bickel, P. & Doksum, K. (2001), *Mathematical Statistics: Basic Ideas and Selected Topics*, Vol. 1, Prentice Hall, Englewood Cliffs, NJ.

Biller, B. & Nelson, B. (2003), Modeling and generating multivariate time-series input processes using a vector autoregressive technique, *ACM Transactions on Modeling and Computer Simulation* **13**(3), 211–237.

Bivand, R. S., Pebesma, E. & Gomez-Rubio, V. (2013), *Applied Spatial Data Analysis with R, Second edition*, Springer, New York.

Black, F. & Litterman, R. (1992), Global portfolio optimization, *Financial Analysts Journal* **48**(5), 28–43.

Bladt, M. (2005), A review on phase-type distributions and their use in risk theory, *ASTIN Bulletin* **35**(1), 145–161.

Blanpain, N. & Chardon, O. (2010), Projections de populations 2007-2060 pour la France métropolitaine: Méthode et principaux résultats, Série des Documents de Travail de la direction des statistiques Démographiques et Sociales F1008, Institut National de la Statistique et des Études Économiques.

Blue, J., Fox, P., Fullerton, W., Gay, D., Grosse, E., Hall, A., Kaufman, L., Petersen, W. & Schryer, N. (1978), *PORT Mathematical Subroutine Library*, Lucent Technologies Murray Hill, NJ.

Bollerslev, T. (1986), Generalized autoregressive conditional heteroskedasticity, *Journal of Econometrics* **31**(3), 307–327.

Bollerslev, T. (1987), A conditionally heteroskedastic time series model for speculative prices and rates of return, *The Review of Economics and Statistics* pp. 542–547.

Bollerslev, T., Chou, R. Y. & Kroner, K. F. (1992), ARCH modeling in finance: A review of the theory and empirical evidence, *Journal of Econometrics* **52**(1), 5–59.

Bollerslev, T. & Ghysels, E. (1996), Periodic autoregressive conditional heteroscedasticity, *Journal of Business & Economic Statistics* **14**(2), 139–151.

Booth, H. (2006), Demographic forecasting: 1980 to 2005 in review, *International Journal of Forecasting* **22**(3), 547–581.

Booth, H., Hyndman, R. J., Tickle, L. & de Jong, P. (2006), Lee-Carter mortality forecasting: A multi-country comparison of variants and extensions, *Demographic Research* **15**(9), 289–310.

Booth, H., Maindonald, J. & Smith, L. (2002), Applying Lee-Carter under conditions of variable mortality decline, *Population Studies* **56**(3), 325–336.

Booth, H. & Tickle, L. (2008), Mortality modelling and forecasting: A review of methods, *Annals of Actuarial Science* **3**(1-2), 3–43.

Booth, H., Tickle, L. & Smith, L. (2005), Evaluation of the variants of the Lee-Carter method of forecasting mortality: A multi-country comparison, *New Zealand Population Review* **31**(1), 13–37. Special Issue on Stochastic Population Projections, edited by A. Dharmalingam & I. Pool.

Bornhuetter, R. L. & Ferguson, R. E. (1972), The actuary and IBNR, *Proceedings of the Casualty Actuarial Society* pp. 181–195.

Boston, C. S. F. (1997), Creditrisk+: A credit risk management framework, Technical report, Credit Suisse First Boston, New York.

Bowers, N. L., Jones, D. A., Gerber, H. U., Nesbitt, C. J. & Hickman, J. C. (1997), *Actuarial Mathematics, 2nd edition*, Society of Actuaries, Schaumburg, IL.

Bowman, A. & Azzalini, A. (1997), *Applied Smoothing Techniques for Data Analysis*, Oxford University Press, London.

Breiman, L. (1996), Bagging predictors, *Machine Learning* **24**(2), 123–140.

Breiman, L., Friedman, J., Olshen, R. & Stone, C. (1984), *Classification and Regression Trees*, Wadsworth, New York.

Brillinger, D. R. (1986), The Natural variability of vital rates and associated statistics, *Biometrics* **42**, 693–734.

Brouhns, N., Denuit, M. & Vermunt, J. K. (2002), A Poisson log-bilinear regression approach to the construction of projected lifetables, *Insurance: Mathematics and Economics* **31**(3), 373–393.

Broyden, C. G. (1970), The convergence of a class of double-rank minimization algorithms, *Journal of the Institute of Mathematics and Its Applications* **6**, 76–90.

Bühlmann, H. (1969), Experience rating and credibility, *ASTIN Bulletin* **5**, 157–165.

Bühlmann, H. & Gisler, A. (2005), *A Course in Credibility Theory and Its Applications*, Springer, Berlin.

Butt, Z. & Haberman, S. (2009), ilc: A collection of R functions for fitting a class of Lee-Carter mortality models using iterative fitting algorithms, Actuarial Research Paper 190, Cass Business School, London.

Cairns, A. (2009), A quantitative comparison of stochastic mortality models using data from England and Wales and the United States, *North American Actuarial Journal* **13**(1), 1–35.

Cairns, A. J., Blake, D. & Dowd, K. (2006*a*), A two-factor model for stochastic mortality with parameter uncertainty: Theory and calibration, *Journal of Risk and Insurance* **73**(4), 687–718.

Cairns, A. J. G., Blake, D. & Dowd, K. (2006*b*), Pricing death: Frameworks for the valuation and securitization of mortality risk, *ASTIN Bulletin* **36**, 79–120.

Camarda, C. G. (2012), MortalitySmooth: An R package for smoothing Poisson counts with P-splines, *Journal of Statistical Software* **50**(1), 1–24.

Cameron, A.C. & Trivedi, P. (1998), *Regression Analysis of Counts Data*, Cambridge University Press, New York.

Cardot, H., Ferraty, F. & Sarda, P. (2003), Spline estimators for the functional linear model, *Statistica Sinica* **13**(3), 571–591.

Carl, P. & Peterson, B. G. (2013), `PerformanceAnalytics`*: Econometric tools for performance and risk analysis.* R package version 1.1.0. *http://CRAN.R-project.org/package=PerformanceAnalytics*

Casella, G. & Berger, R. (2002), *Statistical Inference*, Duxbury Thomson Learning, N. Scituate, MA.

Chalabi, Y. (2012), New Directions in Statistical Distributions, Parametric Modeling and Portfolio Selection, PhD thesis, Eidgenössische Technische Hochschule ETH Zürich.

Chambers, J. M. & Hastie, T. (1991), *Statistical Models in S*, Chapman & Hall, London.

Chambers, J. M. & Lang, D. T. (2001), Object-oriented programming in R, *R News* **1**(3), 17–19.

Chant, D. (1974), On asymptotic tests of composite hypotheses in nonstandard conditions, *Biometrika* **61**, 291–298.

Cherubini, U., Luciano, E. & Vecchiato, W. (2004), *Copula Methods in Finance*, Wiley, New York.

Chiang, C. L. (1984), *The Life Table and Its Applications*, Robert E. Krieger Publishing, Malabar, FL.

Ching, W. & Ng, M. (2006), *Markov Chains: Models, Algorithms and Applications: Models, Algorithms and Applications*, International series in Operations Research & Management Science, Springer Verlag, Berlin.

Chow, G. (1960), Tests of equality between sets of coefficients in two linear regressions, *Econometrica* pp. 591–605.

Christofides, S. (1997), Regression models based on log-incremental payments, *Claims Reserving Manual* **2**, D5.1–D5.53.

Clark, D. R. (2003), *LDF Curve-Fitting and Stochastic Reserving: A Maximum Likelihood Approach*, Casualty Actuarial Society. CAS Fall Forum.

Clarke, A. (1973), *Profiles of the Future: An Inquiry into the Limits of the Possible*, Millennium Edition.

Cleveland, W. S. (1993), *Visualizing Data*, Hobart Press, Summit, NJ.

Coale, A., Demeny, P. & Vaughan, B. (1983), *Regional Model Life Tables and Stable Populations*, Studies in Population, Academic Press, New York.

Cohen, Y. & Cohen, J. (2008), *Statistics and Data with R: An Applied Approach through Examples*, Wiley, New York.

Coles, S. G. (2001), *An Introduction to Statistical Modelling of Extreme Values*, Springer, London.

Consiglio, A. & Guirreri, S. (2011), Simulating term structure of interest rates with arbitrary marginals, *International Journal of Risk Assessment and Management* **15**(4).

Cormen, T. H., Leiserson, C. E. & Rivest, R. L. (1989), *Introduction to Algorithms*, MIT Press, Cambridge, MA.

Coughlan, G. D., Epstein, D., Ong, A., Sinha, A., Hevia-Portocarrero, J., Gingrich, E., Khalaf-Allah, M. & Joseph, P. (2007), *LifeMetrics: A Toolkit for Measuring and Managing Longevity and Mortality Risks*. Version 1.1: 13 March 2007. JP Morgan Pension Advisory Group.

Courant, R. & Hilbert, D. (2009), *R through Excel*, Springer Verlag, Berlin.

Cox, J., Ingersoll, J. & Ross, S. (1985), A theory of the term structure of interest rates, *Econometrica* **53**, 385–407.

Cox, J. R. (1972), Regression models and life-tables (with discussion), *Journal of Royal Stat.* **Ser. B**, 187–220.

Cramér, H. (1946), *Mathematical Methods of Statistics*, Princeton University Press, Princeton, NJ.

Crawley, M. (2012), *The R Book*, Wiley Interscience, New York.

Csardi, G. & Nepusz, T. (2006), The igraph software package for complex network research, *InterJournal* p. 1695.

Cullen, A. & Frey, H. (1999), *Probabilistic Techniques in Exposure Assessment*, Springer, New York.

Culp, M., Johnson, K. & Michailidis, G. (2006), ada: An R package for stochastic boosting, *Journal of Statistical Software* **17**(2), 1–27.

Currie, I. D. (2006), Smoothing and forecasting mortality rates with P-splines, Heriot Watt University, London.

Currie, I. D., Durbán, M. & Eilers, P. H. C. (2004), Smoothing and forecasting mortality rates, *Statistical Modelling* **4**(4), 279–298.

D'Agostino, R. & Stephens, M. (1986), *Goodness-of-Fit Techniques, first edn.*, Dekker, New York.

Dalgaard, P. (2009), *Introductory Statistics with R*, Springer Verlag, Berlin.

Daniel, J.W. (2004), Multi-state transition models with actuarial applications, http://www.casact.org/library/studynotes/daniel.pdf.

De Jong, P. & Tickle, L. (2006), Extending LeeCarter Mortality Forecasting, *Mathematical Population Studies* **13**(1), 1–18.

De Jong, P. & Zeller, G. (2008), *Generalized Linear Models for Insurance Data*, Cambridge University Press, New York.

Debón, A., Montes, F. & Sala, R. (2006), A comparison of nonparametric methods in the graduation of mortality: Application to data from the Valencia region (Spain), *International Statistical Review* **74**(2), 215–233.

Delgado, M. & Kniesner, T. (1997), Count data models with variance of unknown form: An application to a hedonic model of worker absenteeism, *The Review of Economics and Statistics* **79**(1), 41–49.

Delignette-Muller, M. & Dutang, C. (2013), `fitdistrplus`: An R Package for Fitting Distributions. Working paper.

Delignette-Muller, M. L., Pouillot, R., Denis, J.-B. & Dutang, C. (2013), *fitdistrplus: Help to Fit of a Parametric Distribution to Non-censored or Censored Data.* R package version 1.0-1.

Delwarde, A., Denuit, M., Olié, L. & Kachakhidze, D., (2004), Modèles linéaires et additifs géneralisés, maximum de vraisemblance local et méthodes relationelles en assurance sur la vie, *Bulletin Français d'Actuariat* **6**(12), 77–102.

Delwarde, A. & Denuit, M. (2005), Construction de tables de mortalite periodiques et prospectives. *Assurance Audit Actuariat. Economica.*

Denuit, M. & Goderniaux, A. C. (2005), Closing and projecting life tables using log-linear models, *Bulletin of the Swiss Association of Actuaries* (1), 29–48.

Denuit, M., Marechal, X., Pitrebois, S. & Walhin, J.-F. (2007), *Actuarial Modelling of Claim Counts: Risk Classification, Credibility and Bonus-Malus Systems*, Wiley, New York.

Deshmukh, S. (2012), *Multiple Decrement Models in Insurance: An Introduction Using R*, SpringerLink : Bücher, Springer Verlag.

Dey, D. K., Chen, M. H. & Chang, H. (1997), Bayesian approach for nonlinear random effects models, *Biometrics* **53**, 1239–1252.

Dickson, D. (2010), *Insurance Risk and Ruin*, Cambridge University Press, London.

Dickson, D., Hardy, M. & Waters, H. (2009), *Actuarial Mathematics for Life Contingent Risks*, International Series on Actuarial Science, Cambridge University Press, London.

Diebold, X. & Li, C. (2006), Forecasting the term structure of government bond yields, *Journal of Econometrics* **130**(2), 337–364.

Diez, D. & Christou, N. (2012), *stockPortfolio: Build stock models and analyze stock portfolios.* R package version 1.2.

Diggle, P., Heagerty, P., Liang, K. & Zeger, S. (2002), *Analysis of Longitudinal Data*, 2nd edn, Oxford University Press, London.

Duffie, D. & Kan, R. (1996), A yield-factor model of interest rates, *Mathematical Finance* **6**, 379–406.

Dutang, C., Goulet, V. & Pigeon, M. (2008), actuar: An R package for actuarial science, *Journal of Statistical Software* **25**(7), 1–37.

Dwyer, P. S. (1951), *Linear Computations*, Wiley, New York.

Eddelbuettel, D. & François, R. (2011), Rcpp: Seamless R and C++ integration, *Journal of Statistical Software* **40**(8), 1–18.

Embrechts, P., Klüppelberg, C. & Mikosch, T. (1997), *Modelling Extremal Events*, Springer Verlag, Berlin.

Embrechts, P., Lindskog, F. & McNeil, A. (2001), Modelling Dependence with Copulas and applications to Risk Management, Technical report, ETH Zurich.

England P.D. & Verrall R.J. (2002), Stochastic claims reserving in general insurance, *British Actuarial Journal* **8**, 443–544.

Engle, R. F. (1982), Autoregressive conditional heteroscedasticity with estimates of the variance of United Kingdom inflation, *Econometrica: Journal of the Econometric Society* pp. 987–1007.

Enos, J., Kane, D., with contributions from Gerlanc, D. & Campbell, K. (2012a), *portfolio: Analysing equity portfolios.* R package version 0.4-5.

Enos, J., Kane, D. with contributions from Campbell, K. (2012b), *portfolioSim: Framework for simulating equity portfolio strategies.* R package version 0.2-6.

Enos, J., Kane, D., with contributions from Campbell, K. Gerlanc, D., Schwartz, A., Suo, D., Colin, A., & Zhao, L. (2012c), **backtest**: *Exploring portfolio-based conjectures about financial instruments.* R package version 0.3-1.

Estrella, A. & Mishkin, F. S. (1996), The Term Structure of Interest Rates and its Role in Monetary Policy for the European Central Bank, Research Paper 9526, Federal Reserve Bank of New York.

Estrella, A. & Trubin, M. R. (2006), The yield curve as a leading indicator: Some practical issues, *Current Issues in Economics and Finance* **12**(5).

Fang, K.-T., Kotz, S. & Ng, K.-W. (1990), *Symmetric Multivariate and Related Distributions*, Chapman & Hall, London.

Faraway, J. (2006), *Extending the Linear Model with R: Generalized Linear, Mixed Effects and Nonparametric Regression Models*, CRC Press, Boca Raton, FL.

Feinerer, I. (2008), An introduction to text mining in R, *R News* **8**(2), 19–22.

Felipe, A., Guillén, M. & Pérez-Marín, A. (2002), Recent mortality trends in the Spanish population, *British Actuarial Journal* **8**(4), 757–786.

Felisky, K., Akoh-Arrey, A. & Cabrera, E. (2010), *Solvency II and Technical Provisions Dealing with the Risk Margin*, Institute of Actuaries. GIRO conference and exhibition.

Finan, M. (n.d.), A reading of the theory of life contingency models: A preparation for exam mlc, http://faculty.atu.edu/mfinan/actuarieshall/MLCbook2.pdf. Accessed: 01/11/2012.

Firth, D. (2003), *Overdispersed Poisson - negative observation.* Discussion on R-help. *https://stat.ethz.ch/pipermail/r-help/2003-January/028743.html*

Fisher, R. (1940), The precision of discriminant functions, *Annals of Eugenics* **10**, 422–429.

Fishman, G. & Moore, L. (1982), A statistical evaluation of multiplicative congruential random number generators with modulus 2311, *Journal of the American Statistical Association* **77**, 129–136.

Fletcher, R. (1970), A new approach to variable metric algorithms, *Computer Journal* **13**, 317–322.

Forfar, D., McCutcheon, J. & Wilkie, A. (1988), On graduation by mathematical formula, *Journal of the Institute of Actuaries* **115**(1), 1–459.

Fox, J. (2009), Aspects of the Social Organization and Trajectory of the R Project, *The R Journal* **1**, 5–13.

Fox, J. & Weisberg, S. (2011), *An R Companion to Applied Regression*, 2nd edn., Sage, Thousand Oaks CA.

Frees, E. (2004), *Longitudinal and Panel Data: Analysis and Applications in the Social Sciences*, Cambridge University Press, London.

Frees, E. (2009), *Regression Modeling with Actuarial and Financial Applications*, Cambridge University Press, New York.

Frees, E. W. & Valdez, E. (1998), Understanding relationships using copulas, *North American Actuarial Journal* **2**(1), 1–25.

Frees, E. W. & Wang, P. (2006), Copula credibility for aggregate loss models, *Insurance: Mathematics and Economics* **38**, 360–373.

Frees, E. & Wang, P. (2005), Credibility using copulas, *North American Actuarial Journal* **9**(2), 31–48.

Frost, P. A. & Savarino, J. E. (1986), An empirical Bayes approach to efficient portfolio selection, *The Journal of Financial and Quantitative Analysis* **21**, 293–305.

Fu, W. J. (1998), Penalized regressions: The bridge versus the lasso, *Journal of Computational and Graphical Statistics* **7**(3), 397416.

Furnival, G. M. & Wilson, Robert W., J. (1974), Regression by leaps and bounds, *Technometrics* **16**, 499–511.

Galimberti, G., Soffritti, G. & Maso, M. D. (2012), Classification trees for ordinal responses in R: The `rpartScore` package, *Journal of Statistical Software* **47**(10), 1–25.

Gandrud, C. (2013), *Reproducible Research with R and RStudio*, Chapman & Hall / CRC Press, Boca Raton, FL.

Geisser, S. & Eddy, W. F. (1979), A predictive approach to model selection (Corr: V75 p765), *Journal of the American Statistical Association* **74**, 153–160.

Gelman, A. & Hill, J. (2007), *Data Analysis Using Regression and Multilevel/Hierarchical Models*, Cambridge University Press, New York.

Gendron, M. & Crepeau, H. (1989), On the computation of the aggregate claim distribution when individual claims are inverse Gaussian, *Insurance: Mathematics and Economics* **8**, 251–258.

Genolini, C. (2008), *A (Not So) Short Introduction to S4*.
URL: christophe.genolini.free.fr/Tutorial/notSoShort.php

Genton, M., ed. (2004), *Skew-Elliptical Distributions and Their Applications*, Chapman & Hall/CRC, Boca Raton, FL.

Gesmann, M., Murphy, D. & Zhang, Y. (2013), *ChainLadder: Mack-, Bootstrap and Munich-Chain-Ladder Methods for Insurance Claims Reserving.* R package version 0.1.7.

Gesmann, M., Rayees, R. & Clapham, E. (2013), Claims inflation – A known unknown, *The Actuary*, May 2013 pp. 30–31.

Ghalanos, A. (2013), *parma: Portfolio Allocation and Risk Management Applications.* R package version 1.03.

Ghalanos, A. & Theussl, S. (2012), *Rsolnp: General Non-linear Optimization Using Augmented Lagrange Multiplier Method.* R package version 1.14.

Giles, T. L. (1993), Life insurance application of recursive formulas, *Journal of Actuarial Practice* **1**(2), 141–151.

Gochez, F. (2011), *BLCOP: Black-Litterman and Copula-Opinion Pooling Frameworks.* R package version 0.2.6.

Goldberg, D. (1991), What every computer scientist should know about floating-point arithmetic, *ACM Comput. Surveys* **23**(1), 5–48.

Goldfarb, D. (1970), A family of variable metric updates derived by variational means, *Mathematics of Computation* **24**, 23–26.

Goldfarb, D. & Idnani, A. (1982), Dual and primal-dual methods for solving strictly convex quadratic programs, in *Numerical Analysis*, Springer Verlag, Berlin, pp. 226–239.

Goldfarb, D. & Idnani, A. (1983), A numerically stable dual method for solving strictly convex quadratic programs, *Mathematical Programming* **27**(1), 1–33.

Goovaerts, M., De Vylder, F. & Haezendonck, J. (1984), *Insurance Premiums*, North Holland Publishing, Amsterdam.

Graham, R. L., Knuth, D. E. & Patashnik, O. (1989), *Concrete Mathematics*, Addison-Wesley, Reading, MA.

Gravelsons, B., Ball, M., Beard, D., Brooks, R., Couchman, N., Kefford, C., Michaels, D., Nolan, P., Overton, G., Robertson-Dunn, S., Ruffini, E., Sandhouse, G., Schilling, J., Sykes, D., Taylor, P., Whiting, A., Wilde, M. & Wilson, J. (2009), B12: UK asbestos working party update 2009', http://www.actuaries.org.uk/research-and-resources/documents/b12-uk-asbestos-working-party-update-2009-5mb. Presented at the *General Insurance Convention*.

Greenwood, A. J. & Durand, D. (1960), Aids for fitting the gamma distribution by maximum likelihood, *Technometrics* **2**, 55–65.

Grothendieck, G. & Petzoldt, T. (2004), R Help Desk: Date and time classes in R, *R News* **4**(1), 29–32.

Guirreri, S. (2010), Simulating Term Structure of Interest Rates with Arbitrary Marginals, PhD thesis, University of Palermo.

Guirreri, S. (2012), *YieldCurve: Modelling and Estimation of the Yield Curve.* R package version 4.0.
http://www.guirreri.host22.com

Guszcza, J. (2008), Hierarchical Growth Curve Models for Loss Reserving, *in CAS Forum*, pp. 146–173.

Haberman, S. & Pitacco, E. (1999), *Actuarial Models for Disability Insurance*, Chapman & Hall/CRC, Boca Raton, FL.

Haberman, S. & Renshaw, A. E. (2011), A comparative study of parametric mortality projection models, *Insurance: Mathematics and Economics* **48**(1), 35–55.

Hachemeister, C. A. & Stanard, J. N. (1975), IBNR Claims Count Estimation with Static Lag Functions, *in 12th ASTIN Colloquium*, Portimao, Portugal.

Halton, J. H. (1960), On the efficiency of certain quasi-random sequences of points in evaluating multi-dimensional integrals, *Numerische Mathematik* **2**, 84–90.

Hand, D. J. (2005), Good practice in retail credit score-card assessment, *Journal of the Operational Research Society* **56**, 1109–1117.

Harrell, F. (2006), Problems caused by categorizing continuous variables, *available online at http:// biostat.mc.vanderbilt.edu/ twiki/ bin/ view/Main/.*

Hastie, T. J. & Tibshirani, R. J. (1990), *Generalized Additive Models*, Chapman & Hall, London.

Hastie, T., Tibshirani, R. & Friedman, J. (2009), *Elements of Statistical Learning: Data Mining, Inference and Prediction*, Springer-Verlag, Berlin.

Hausman, J. (1978), Specification tests in econometrics, *Econometrica* pp. 1251–1271.

He, X. & Ng, P. T. (1999), COBS: Qualitatively constrained smoothing via linear programming', *Computational Statistics* **14**(3), 315–337.

Heiberger, R. M. & Neuwirth, E. (1953), *Methodes of Mathematical Physics, volume 1*, Wiley, New York.

Hemmerle, W. J. (1967), *Statistical Computations on a Digital Computer*, Blaisdell, Waltham, MA.

Henley, W. & Hand, D. (1996), A k-nearest-neighbour classifier for assessing consumer credit risk, *The Statistician* **45**(1), 77–95.

Herzog, T. N. (1996), *Introduction to Credibility Theory*, second edn., ACTEX, Winsted, CT.

Hilbe, J. (2009), *Logistic Regression Models*, Chapman & Hall/CRC Press, Boca Raton FL.

Hilbe, J. (2011), *Negative Binomial Regression*, Wiley, New York.

Hill, B. M. (1975), A simple general approach to inference about the tail of a distribution, *Annals of Statistics* **3**(5), 1163–1174.

Hoerl, A. E. & Kennard, R. W. (1970), Ridge regression: Biased estimation for non-orthogonal problems, *Technometrics* **42(1)**, 80–86.

Hoffmann, T. J. (2011), Passing in command line arguments and parallel cluster/multicore batching in R with `batch`, *Journal of Statistical Software, Code Snippets* **39**(1), 1–11.

Højsgaard, S., Edwards, D. & Lauritzen, S. (2012), *Graphical Models with R*, Springer Verlag, Berlin.

Holmes, S. (2006), Review of Fionn Murtagh's book: Correspondence analysis and data coding with Java and R, *R News* **6**(4), 41–43.

Hornik, K., Meyer, D. & Buchta, C. (2013), *slam: Sparse Lightweight Arrays and Matrices*. R package version 0.1-28.

Hornik, K. & Theussl, S. (2012), *Rglpk: R/GNU Linear Programming Kit Interface*. R package version 0.3-10.

Hosmer, D. & Lemeshow, S. (2000), *Applied Logistic Regression*, Wiley, New York.

Hothorn, T., Bretz, F. & Genz, A. (2001), On multivariate t and Gauß probabilities in R, *R News* **1**(2), 27–29.

Hsiao, C. (2003), *Analysis of Panel Data*, 2nd edn., Cambridge University Press, New York.

Hsieh, D. A. (1989), Testing for nonlinear dependence in daily foreign exchange rates, *Journal of Business* **62**, 339–368.

Hubert, M., Rousseeuw, P. J. & Verboven, S. (2002), A fast method of robust principal components with applications to chemometrics, *Chemometrics and Intelligent Laboratory Systems* **60**, 101–111.

Human Mortality Database (2013), University of California, Berkeley (USA), and Max Planck Institute for Demographic Research (Germany). Downloaded on 22 February 2013.
www.mortality.org

Hyndman, R. & Fan, Y. (1996), Sample quantiles in statistical packages, *American Statistician* **50**, 361–365.

Hyndman, R. J. (2012), *Demography: Forecasting mortality, fertility, migration and population data.* With contributions from Heather Booth, Leonie Tickle and John Maindonald.

Hyndman, R. J. & Booth, H. (2008), Stochastic population forecasts using functional data models for mortality, fertility and migration, *International Journal of Forecasting* **24**(3), 323–342.

Hyndman, R. J., Booth, H. & Yasmeen, F. (2013), Coherent mortality forecasting: The product-ratio method with functional time series models, *Demography* **50**(1), 261–283.

Hyndman, R. J. & Khandakar, Y. (2008), Automatic time series forecasting: The forecast package for R, *Journal of Statistical Software* **27(3)**, 1-22.

Hyndman, R. J., Koehler, A. B., Ord, J. K. & Snyder, R. D. (2008), *Forecasting with Exponential Smoothing: The State Space Approach*, Springer-Verlag, Berlin.

Hyndman, R. J. & Shang, H. L. (2009), Forecasting functional time series (with discussion), *Journal of the Korean Statistical Society* **38**(3), 199–221.

Hyndman, R. J. & Ullah, S. (2007), Robust forecasting of mortality and fertility rates: A functional data approach, *Computational Statistics and Data Analysis* **51**(10), 4942–4956.

Ihaka, R. & Gentleman, R. (1996), R: A language for data analysis and graphics, *Journal of Computational and Graphical Statistics* **5**(3), 299–314.

Jagger, T. & Elsner, J. (2008), Modelling tropical cyclone intensity with quantile regression, *International Journal of Climatology* **29**, 1351–1361.

Jakob, K., Fischer, D. M. & Kolb, S. (2013), *crp.CSFP: CreditRisk+ portfolio model.* R package version 1.2.1.

Jewell, W. (2004), Bayesian statistics, *in* J. L. Teugels & B. Sundt, Eds., 'Encyclopedia of Actuarial Science', Vol. 1, Wiley.

Joe, H. (1997), Multivariate models and dependence concepts, *in Monographs on Statistics and Applied Probability*, Vol. 73, Chapman & Hall, Boca Raton, FL.

Johnson, N. (1949), System of frequency curves generated by methods of translation, *Biometrika* **36**, 149–176.

Johnson, N., Kemp, A. & Kotz, S. (2005), *Univariate Discrete Distributions*, 3rd ed., Wiley-Interscience, New York.

Jones, O., Maillardet, R. & Robinson, A. (2009), *Introduction to Scientific Programming and Simulation Using R*, first ed., Chapman and Hall/CRC Press, Boca Raton, FL.

Jørgensen, B. (1987), Exponential dispersion models, *Journal of the Royal Statistical Society, Series B* **49**, 127–162.

Jørgensen, B. (1997), *The Theory of Dispersion Models*, Chapman & Hall, London.

Kaas, R., Goovaerts, M., Dhaene, J. & Denuit, M. (2008), *Modern Actuarial Risk Theory – Using R*, second ed., Springer-Verlag, Heidelberg.

Kabacoff, R. (2011), *R in Action*, Manning Publications, Greenwich, CT.

Kahle, D. & Wickham, H. (2013), *ggmap: A Package for Spatial Visualization with Google Maps and OpenStreetMap*. R package version 2.3.

Kendrick, D., Mercado, P. & Amman, H. (2006), Computational economics: Help for the underestimated undergraduate, *Computational Economics* **77**, 261–271.

Kernighan, B. W. (1988), *C/C++Programming Language*, Prentice Hall, Englewood Cliff NJ.

Kilibarda, M. (2013), *plotGoogleMaps: Plot SP or SPT(STDIF,STFDF) Data as HTML Map Mashup over Google Maps*. R package version 2.0.

Kleiber, C. & Zeileis, A. (2008), *Applied Econometrics with R*, Springer-Verlag, New York.

Kleinman, K. & Horton, N. J. (2010), *SAS and R : Data Management, Statistical Analysis, and Graphics*, Chapman & Hall/CRC, Press Boca Raton, FL.

Klugman, S. (1992), *Bayesian Statistics in Actuarial Science: With Emphasis on Credibility*, Kluwer, Boston.

Klugman, S. A., Panjer, H. H. & Willmot, G. E. (2009), *Loss Models: From Data to Decisions*, Wiley Series in Probability and Statistics, New York.

Klugman, S. & Parsa, R. (1999), Fitting bivariate loss distributions with copulas, *Insurance: Mathematics and Economics* **24**, 139–148.

Knorr-Held, L. & Best, N. (2001), A shared component model for detecting joint and selective clustering of two diseases, *Journal of the Royal Statistical Society, Series B* **164**, 73–85.

Knuth, D. E. (1973), *Fundamental Algorithms*, Vol. 1 of *The Art of Computer Programming*, second ed., Addison-Wesley, Reading, MA, section 1.2, pp. 10–119. A full INBOOK entry.

Koissi, M.-C., Shapiro, A. F. & Högnäs, G. (2006), Evaluating and extending the Lee-Carter model for mortality forecasting: Bootstrap confidence interval, *Insurance: Mathematics and Economics* **38**(1), 1–20.

Krause, A. (2009), *The Basics of S-PLUS*, Springer Verlag, Berlin.

Kremer, E. (1982), IBNR claims and the two-way model of ANOVA, *Scandinavian Actuarial Journal* pp. 47–55.

Kuhn, M. (2008), Building predictive models in R using the caret package, *Journal of Statistical Software* **28**(5), 1–26.

Lang, D. T. (2001), In search of C/C++ & FORTRAN routines, *R News* **1**(3), 20–23.

Lawrence, M. & Verzani, J. (2012), *Programming Graphical User Interfaces in R*, Chapman & Hall / CRC, Press, Boca Raton, FL.

Laws, C. W. & Schmid, F. A. (2012), *lossDev: Robust Loss Development Using MCMC*. R package version 3.0.0-4.

Lazar, D. & Denuit, M. M. (2009), A multivariate time series approach to projected life tables, *Applied Stochastic Models in Business and Industry* **25**, 806–823.

Lee, R. D. & Carter, L. R. (1992), Modeling and forecasting U.S. mortality, *Journal of the American Statistical Association* **87**(419), 659–671.

Lee, R. D. & Miller, T. (2001), Evaluating the performance of the Lee-Carter method for forecasting mortality, *Demography* **38**(4), 537–49.

Leisch, R. (2009), *Creating R Packages: A Tutorial.*
URL: cran.c project.org/doc/contrib/Leisch-CreatingPackages.pdf.

Lemaire, J. (1984), *Bonus-Malus Systems in Automobile Insurance*, North-Holland Publishing, Amsterdam.

Lemieux, C. (2009), *Monte Carlo and Quasi-Monte Carlo Sampling*, Springer-Verlag, New York.

Liaw, A. & Wiener, M. (2002), Classification and regression by randomForest, *R News* **2**(3), 18–22.

Liddell, F. D. K. (1984), Simple exact analysis of the standardized mortality ratio, *Journal of Epidemiology and Community Health* **38**, 85–88.

Ligges, U. & Fox, J. (2008), R Help Desk: How can I avoid this loop or make it faster?, *R News* **8**(1), 46–50.

Lindley, D. V. (1983), Theory and practice of Bayesian statistics, *The Statistician* **32**, 1–11.

Litterman, R. & Scheikman, J. (1991), Common factors affecting bond returns, *Journal of Fixed Income* **1**, 54–61.

Liu, Y.-H., Makov, U. E. & Smith, A. F. M. (1996), Bayesian methods in actuarial science, *The Statistician* **45**, 503–515.

Loader, C. R. (1999), *Local Regression and Likelihood*, Statistics and Computing Series, Springer Verlag, New York.

Loecher, M. (2013), *RgoogleMaps: Overlays on Google Map Tiles in R.* R package version 1.2.0.3.

Lumley, T. (2004), Programmers' niche: A simple class, in S3 and S4, *R News* **4**(1), 33–36.

Lunn, D., Spiegelhalter, D., Thomas, A. & Best., N. (2009), The BUGS project: Evolution, critique and future directions (with discussion), *Statistics in Medicine* **28**, 3049–3082.

Lyons, G., Forster, W., Kedney, P., Warren, R. & Wilkinson, H. (2002), *Claims Reserving Working Party paper*, Institute of Actuaries, London.

MacAdie, C. J., Landsea, C. W., Neumann, C. J., David, J. E., Blake, E. & Hammer, G. R. (2009), Tropical Cyclones of the North Atlantic Ocean, 1851–2006, Technical memo, National Climatic Data Center in cooperation with the National Hurricane Center.

Mack, T. (1991), A simple parametric model for rating automobile insurance or estimating IBNR claims reserves, *ASTIN Bulletin* **21**, 93–109.

Mack, T. (1993), Distribution-free calculation of the standard error of chain ladder reserve estimates, *ASTIN Bulletin* **23**, 213–225.

Mack, T. (1999), The standard error of chain ladder reserve estimates: Recursive calculation and inclusion of a tail factor, *ASTIN Bulletin* **29**(2), 361–266.

Maindonald, J. & Braun, W. J. (2007), *Data Analysis and Graphics Using R: An Example-Based Approach*, Cambridge University Press, Berlin.

Markowitz, H. (1952), Portfolio selection, *The Journal of Finance* **7**(1), 77–91.

Marshall, A. W. & Olkin, I. (1988), Families of multivariate distributions, *Journal of the American Statistical Association* **83**(403), 834–841.

Martinussen, T. & Scheike, T. (2006), *Dynamic Regression Models for Survival Data*, Springer, New York.

Massey, R., Widdows, J., Bhattacharya, K., Shaw, R., Hart, D., Law, D. & Hawes, W. (2002), *Insurance company failure.* General Insurance Convention Working Party. *http://www.actuaries.org.uk/research-and-resources/documents/insurance-company-failure*

Matloff, N. (2011), *The Art of R Programming* , No Starch Press, San Francisco.

May, E. (2004), *Credit Scoring for Risk Managers: The Handbook for Lenders*, South-Western Publishing, Independence, KY.

McCullagh, P. & Nelder, J. A. (1989), *Generalized Linear Models (Second edition)*, Chapman & Hall, London.

McGrayne, S. (2012), *The Theory That Would Not Die*, Yale University Press, New Haven, CT.

McNeil, A. (1997), Estimating the tails of loss severity distributions using extreme value theory, *ASTIN Bulletin* **27**(1), 117–137.

McNeil, A. J. & Frey, R. (2000), Estimation of tail-related risk measures for heteroscedastic financial time series: An extreme value approach, *Journal of Empirical Finance* **7**, 271–300.

Merz, M. & Wüthrich, M. V. (2008*a*), Modelling the claims development result for solvency purposes, *CAS E-Forum* pp. 542–568.

Merz, M. & Wüthrich, M. V. (2008*b*), Prediction error of the multivariate chain ladder reserving method, *North American Actuarial Journal* **12**, 175–197.

Metropolis, N. & Ulam, S. (1949), The Monte Carlo method, *Journal of the American Statistical Association* **44**(247), 335–341.

Meucci, A. (2009), *Risk and Asset Allocation*, Springer Verlag, Berlin.

Meyers, G. & Cummings, D. (2009), "Goodness of fit" vs. "goodness of lift", *Actuarial Review* **36**, 16–17.

Michaels, D. (2002), APH: How the Love Carnal and silicone implants nearly destroyed Lloyd's (slides)', http://www.actuaries.org.uk/research-and-resources/documents/aph-how-love-carnal-and-silicone-implants-nearly-destroyed-lloyds-s. Presented at the Younger Members' Convention.

Miranda, M. D. M., Nielsen, J. P. & Verrall, R. (2012), *Double Chain Ladder*, ASTIN Bulletin, **42**(1), P. 59-76.

Moler, C. & Van Loan, C. (1978), Nineteen dubious ways to compute the matrix exponential, *SIAM Review* **20**(4), 801–836.

Moran, P. (1971), Maximum-likelihood estimation in non-standard conditions, *Mathematical Proceedings of the Cambridge Philosophical Society* **70**(3), 441–450.

Murdoch, D. (2002), Reading foreign files, *R News* **2**(1), 2–3.

Murell, P. (2012), *R Graphics*, Chapman & Hall / CRC Press, Boca Raton, FL.

Murphy, D. (1994), Unbiased loss development factors, *PCAS* **81**, 154 – 222.

Murray, R. & Lauder, A. (2011), *Modelling the Claims Process — An Alternative to Development Factor Modelling*, Institute of Actuaries. GIRO Conference and Exhibition.

Murrell, P. & Ripley, B. (2006), Non-standard fonts in PostScript and PDF graphics, *R News* **6**(2), 41–47.

Nelsen, R. B. (2006), *An Introduction to Copulas*, Springer, New York.

Nelson, C. & Siegel, A. (1987), Parsimonious modelling of yield curves, *Journal of Business* **60**.

Niederreiter, H. (1992), *Random Number Generation and Quasi-Monte-Carlo Methods*, Society for Industrial and Applied Mathematics, Philadelphia.

Nisbet, R., Elder, J. & Miner, G. (2011), *Handbook of Statistical Analysis and Data Mining Applications*, Academic Press, New York.

Norberg, R. (1976), A credibility theory for automobile bonus systems, *Scandinavian Actuarial Journal* pp. 92–107.

Norberg, R. (1979), The credibility approach to ratemaking, *Scandinavian Actuarial Journal* **1979**, 181–221.

Norris, J. (1997), *Markov Chains*, Cambridge University Press, New York.

Novomestky, F. (2012), *rportfolios: Random Portfolio Generation.* R package version 1.0.

O'Cinneide, C. (1990), Characterization of phase-type distributions, *Stochastic Models* **6**(1), 1–57.

Ohlsson, E. & Johansson, B. (2010), *Non-Life Insurance Pricing with Generalized Linear Models*, Springer Verlag, Berlin.

Ohlsson, E. & Lauzeningks, J. (2009), The one-year non-life insurance risk, *Insurance: Mathematics and Economics* **45**(2), 203–208.

Olver, F. W. J., Lozier, D. W., Boisvert, R. F. & Clark, C. W., Eds. (2010), *NIST Handbook of Mathematical Functions*, Cambridge University Press, New York.

Orr, J. (2007), *A Simple Multi-State Reserving Model*, Colloqiua Orlando ed., ASTIN Colloquium in Orlando.

Orr, J. (2012), *GIROC Reserving Research Workstream*, Institute of Actuaries, London.

Pagan, A. (1996), The econometrics of financial markets, *Journal of Empirical Finance* **3**(1), 15–102.

Palm, F. C. (1996), 7 GARCH models of volatility, *Handbook of Statistics* **14**, 209–240.

Panjer, H. H. (1981), Recursive evaluation of a family of compound distributions, *ASTIN Bulletin* **12**, 22–26.

Pasupathy, R. (2010), Generating nonhomogeneous Poisson processes, in *Wiley Encyclopedia of Operations Research and Management Science*, Wiley & Sons, New York.

Pearson, K. (1895), Contributions to the mathematical theory of evolution, II: Skew variation in homogeneous material, *Philosophical Transactions of the Royal Society of London.* **186**, 343–414.

Pfaff, B. (2012), *Financial Risk Modelling and Portfolio Optimization with R*, John Wiley & Sons, Ltd., New York.

Pickle, L. W., Mungiole, M., Jones, G. K. & White, A. A. (1999), Exploring spatial patterns of mortality: The new atlas of United States mortality, *Statistics in Medicine* **18**(23), 3211–3220.

Pielke, J. R. A., Gratz, J., Landsea, C. W., Collins, D., Saunders, M. A. & Musulin, R. (2008), Normalized hurricane damages in the United States: 1900–2005, *Natural Hazards Review* **9**, 29–42.

Pinheiro, J. C. & Bates, D. M. (2000), *Mixed-Effects Models in S and S-Plus*, 1 ed., Springer-Verlag, Berlin.

Planchet, F. & Thérond, P. (2011), *Modélisation Statistique des Phénomènes de Durée—Applications Actuarielles*, Assurance Audit Actuariat, Economica Paris.

Plat, R. (2009), On stochastic mortality modeling, *Insurance: Mathematics and Economics* **45**(3), 393–404.

Plummer, M. (2003), JAGS: a program for analysis of Bayesian graphical models using Gibbs sampling, *in Proceedings of the 3rd International Workshop on Distributed Statistical Computing*, Vienna, Austria.

Plummer, M. (2011), *JAGS Version 3.1.0 User Manual.*

Plummer, M. (2013), *rjags: Bayesian Graphical Models Using MCMC.* R package version 3-10.

Press, W., Teukolsky, S., Vetterling, W. & Flannery, B. (2007), *Numerical Recipes: The Art of Scientific Computing*, Cambridge University Press, London.

Pröhl, C. & Schmidt, K. D. (2005), Multivariate chain-ladder, *Dresdner Schriften zur Versicherungsmathematik* .

Quarg, G. & Mack, T. (2004), Munich chain ladder, Munich Re Group. *Casualty Actuarial Society*, Vol. 2, Issue 2, 266–299.

Quenouille, M. H. (1949), Approximate tests of correlation in time series, *Journal of the Royal Statistical Society Series B* **11**, 68–84.

Raftery, A. E., Chunn, J. L., Gerland, P. & Sevčíková, H. (2013), Bayesian probabilistic projections of life expectancy for all countries, *Demography* **50**, 777–801.

Ramsay, J. O. & Silverman, B. W. (2005), *Functional Data Analysis*, 2nd ed., Springer-Verlag, New York.

Rebonato, R., Sukhdeep, M., Mark, J., Lars-Dierk, B. & Ken, N. (2005), Evolving yield curves in the real-world measures: A semi-parametric approach, *The Journal of Risk* **7**(3), 29–61.

Renshaw, A. E. & Haberman, S. (2003), Lee-Carter mortality forecasting with age-specific enhancement, *Insurance: Mathematics and Economics* **33**(2), 255–272.

Renshaw, A. E. & Haberman, S. (2006), A cohort-based extension to the Lee–Carter model for mortality reduction factors, *Insurance: Mathematics and Economics* **38**(3), 556–570.

Renshaw, A. E. & Verrall, R. J. (1998), A stochastic model underlying the chain-ladder technique, *British Actuarial Journal* **4**, 903–923.

Ripley, B. D. (2001), Installing R under Windows, *R News* **1**(2), 11–14.

Ripley, B. D. (2005), Packages and their management in R 2.1.0, *R News* **5**(1), 8–11.

Ripley, B. D. & Hornik, K. (2001), Date-time classes, *R News* **1**(2), 8–11.

Robert, C. P. & Casella, G. (2010), *Introducing Monte Carlo Methods with R*, Springer, New York.

Rolski, T., Schmidli, H., Schmidt, V. & Teugels, J. (1999), *Stochastic Processes for Insurance and Finance*, Wiley, Chichester.

Ross, S. (2009), *First Course in Probability, A (8th Edition)*, 8 ed., Prentice Hall, Englewood Cliffs, NJ.

Rousseeuw, P., Croux, C., Todorov, V., Ruckstuhl, A., Salibian-Barrera, M., Verbeke, T., Koller, M. & Maechler, M. (2012), *robustbase: Basic Robust Statistics.* R package version 0.9-7.

Rousseeuw, P. J. & Driessen, K. V. (1999), A fast algorithm for the minimum covariance determinant estimator, *Technometrics* (41), 212–223.

Rowe, B. L. Y. (2013), *tawny: Provides Various Portfolio Optimization Strategies Including Random Matrix Theory and Shrinkage Estimators.* R package version 2.1.0.

Royston, R., Altman, D. & Sauerbrei, W. (2006), Dichotomizing continuous predictors in multiple regression: a bad idea, *Statistics in Medicine* **25**(1), 49–56.

Rubin, D. (1976), Inference and missing data, *Biometrika* **63**, 581–592.

Ruckdeschel, P., Kohl, M., Stabla, T., & Camphausen, F. (2006), S4 classes for distributions, *R News* **6**(2), 2–6.

Ruckman, C. & Francis, J. (n.d.), *Financial Mathematics: A Practical Guide for Actuaries and Other Business Professionals*, BPP Professional Education, Phoenix, AZ.

Rue, H., Martino, S. & Chopin, N. (2009), Approximate Bayesian inference for latent Gaussian models by using integrated nested Laplace approximations, *Journal of the Royal Statistical Society: Series B (Statistical Methodology)* **71**(2), 319–392.

Santos Silva, J. & Windmeijer, F. (2001), Two-part multiple spell models for health care demand, *Journal of Econometrics* **104**(1), 67–89,.

Sarkar, D. (2002), Lattice, *R News* **2**(2), 19–23.

Sarkar, D. (2008), *Lattice: Multivariate Data Visualization with R*, Springer Verlag, Berlin.

Satchell, S. & Scowcroft, A. (2000), A demystification of the Black-Litterman model: Managing quantitative and traditional portfolio construction, *Journal of Asset Management* **1**, 138–150.

Scarsini, M. (1984), On measures of concordance, *Stochastica* **8**, 201–218.

Schmidberger, M., Morgan, M., Eddelbuettel, D., Yu, H., Tierney, L. & Mansmann, U. (2009), State of the art in parallel computing with R, *Journal of Statistical Software* **31**(1), 1–27.

Schmidt, K. D. (2012), A bibliography on loss reserving, http://www.math.tu-dresden.de/sto/schmidt/dsvm/reserve.pdf.

Schmock, U. (1999), Estimating the value of the WINCAT coupons of the Winterthur insurance convertible bond, *ASTIN Bulletin* **29**(1), 101–163.

Shang, H. L., Booth, H. & Hyndman, R. J. (2011), Point and interval forecasts of mortality rates and life expectancy: A comparison of ten principal component methods, *Demographic Research* **25**(5), 173–214.

Shanno, D. F. (1970), Conditioning of quasi-newton methods for function minimization, *Mathematics of Computation* **24**, 647–656.

Shao, J. & Tu, D. (1995), *The Jackknife and Bootstrap*, Springer, New York.

Silverman, B. (1986), *Density Estimation for Statistics and Data Analysis*, Chapman & Hall, London.

Sklar, A. (1959), Fonctions de répartition à n dimensions et leurs marges, *Publications de l'ISUP de Paris 8* **8**, 229–231.

Smyth, G. & Jørgensen, B. (2002), Fitting Tweedie's Compound Poisson Model to Insurance Claims Data: Dispersion Modelling, *ASTIN Bulletin* **32**, 143–157.

Sobol, I. (1967), On the distribution of points in a cube and the approximate evaluation of integrals, *USSR Computational Mathematics and Mathematical Physics* **7**, 86–112.

Spector, P. (2008), *Data Manipulation with R*, Springer Verlag, Berlin.

Spedicato, G. A. (2013*a*), *Lifecontingencies: An R package to perform life contingencies actuarial mathematics*. R package version 0.9.8.

Spedicato, G. A. (2013*b*), *`markovchain`: An R package to easily handle discrete Markov chains*. R package version 0.0.2.

Spiegelhalter, D. J., Best, N. G., Carlin, B. P. & Linde, A. (2002), Bayesian measures of model complexity and fit, *Journal of the Royal Statistical Society, Series B* **64**, 583–639.

Stan Development Team (2012), Stan: A C++ Library for Probability and Sampling, version 1.0.

Steehouwer, H. (2005), Macroeconomic Scenarios and Reality: A Frequency Domain Approach for Analyzing Historical Time Series and Generating Scenarios for the Future, PhD thesis, Free University of Amsterdam.

Stroustrup, B. (2013), *The C/C++ Programming Language*, Addison-Wesley.

Sturtz, S., Ligges, U. & Gelman, A. (2005), R2winbugs: A package for running winbugs from R, *Journal of Statistical Software* **12**(3), 1–16.

Svensson, L. (1994), Estimating and Interpreting Forward Interest Rates: Sweden 1992–1994, International Monetary Fund, Washington, D.C.

Teetor, P. (2011), *R Cookbook*, O'Reilly Media, Sebastopol, CA.

Teugels, J. (1982), Large claims in insurance mathematics, *ASTIN Bulletin* **13**, 81–88.

Theussl, S. (2013), *Optimization and Mathematical Programming.* CRAN Task View.

Thomas, A., O'Hara, B., Ligges, U. & Sturtz, S. (2006), Making bugs open, *R News* **6**(1), 12–17.

Thomas, L. (2000), A survey of credit and behavioural scoring: Forecasting financial risk of lending to consumers, *International Journal of Forecasting* **16**, 149–172.

Tibshirani, R. (1996), Regression shrinkage and selection via the lasso, *Journal of the Royal Statistical Society B* **58**(1), 267–288.

Tomas, J. (2011), A local likelihood approach to univariate graduation of mortality, *Bulletin Français d'Actuariat* **11**(22), 105–153.

Tomas, J. (2013), Quantifying Biometric Life Insurance Risks with Non-Parametric Methods, PhD thesis, Amsterdam School of Economics Research Institute.

Tomas, J. & Planchet, F. (2014), Constructing entity specific prospective mortality table: Adjustment to a reference, *European Actvarial Journal* to appear.

Tomas, J. & Planchet, F. (2013), Multidimensional smoothing by adaptive local kernel-weighted log-likelihood with application to long-term care insurance, *Insurance: Mathematics & Economics* **52**(3), 573–589.

Tse, Y. (2009), *Nonlife Actuarial Models: Theory, Methods and Evaluation*, International Series on Actuarial Science, Cambridge University Press, London.

Tufféry, S. (2011), *Data Mining and Statistics for Decision Making*, Wiley.

Tukey, J. W. (1958), Bias and confidence in not quite large samples (Abstract), *Annual of Mathematical; Statistics* **29**, 614.

Tuljapurkar, S., Li, N. & Boe, C. (2000), A universal pattern of mortality decline in the G7 countries, *Nature* **405**, 789–792.

Uryasev, S. & Rockafellar, R. T. (2001), Conditional value-at-risk: Optimization approach, in *Stochastic Optimization: Algorithms and Applications*, Springer Verlag, Berlin, pp. 411–435.

Van den Broek, J. (1995), A score test for zero inflation in a Poisson distribution, *Biometrics* **51**, 738–743.

Vance, A. (2009), Data analysts captivated by R's power, *New York Times* (January 7th), B6. *http://www.newyorktimes.com/2009/01/07/technology/business-computing/07program.html*

Vasicek, O. (1977), An equilibrium characterization of the term structure, *Journal of Financial Economics* **5**, 177–188.

Vaupel, J., Manton, K. & Stallard, E. (1979), The impact of heterogeneity in individual frailty on the dynamics of mortality, *Demography* **16**, 439–454.

Venables, W. N. & Ripley, B. D. (2002), *Modern Applied Statistics with S*, 4th ed., Springer, New York.

Venzon, D. J. & Moolgavkar, S. H. (1988), A method for computing profile-likelihood-based confidence intervals, *Applied Statistics* **37**(1), 87–94.

Verbeke, G. & Molenberghs, G. (2000), *Linear Mixed Models for Longitudinal Data*, Springer Series in Statistics, Springer, Berlin.

von Neumann, J. (1951), Various techniques used in connection with random digits, *National Bureau of Standards, Applied Mathematics Series* **12**, 36–38.

Wainer, H. (2006), Finding what is not there through the unfortunate binning of results: The Mendel effect, *Chance* **19**, 127–141.

Weingessel, A. (2011), *quadprog: Functions to Solve Quadratic Programming Problems.* R package version 1.5-4; S original by Berwin A. Turlach.

Wei, X. (2012), PROC_R: A SAS macro that enables native R programming in the base SAS environment, *Journal of Statistical Software, Code Snippets* **46**(2), 1–13.

Whittle, P. (1954), On stationary processes in the plane, *Biometrika* **41**, 434–439.

Wickham, H. (2009), *ggplot2: Elegant Graphics for Data Analysis*, Springer Verlag, Berlin.

Wickham, H. (2011), The split-apply-combine strategy for data analysis, *Journal of Statistical Software* **40**(1), 1–29.

Wilkinson, G. N. & Rogers, C. E. (1977), Symbolic description of factorial models for analysis of variance, *Journal of the Royal Statistical Society. Series C (Applied Statistics)* **22**, 392–399.

Wilkinson, J. H. (1963), *Rounding Errors in Algebraic Processes*, Prentice-Hall, Englewood Cliffs, NJ.

Wilkinson, L. (1999), *The Grammar of Graphics*, Springer Verlag, Berlin.

Willets, R. C. (2004), *The cohort effect: insights and explanations*, Cambridge Univversity Press, New York.

Williamson, R. (1989), Probabilistic Arithmetic. PhD thesis, University of Queensland.

Wood, S. N. (1994), Monotonic smoothing splines fitted by cross validation, *SIAM Journal on Scientific Computing* **15**(5), 1126–1133.

Wood, S. N. (2003), Thin plate regression splines, *Journal of the Royal Statistical Society, Series B* **65**(1), 95–114.

Wood, S.N. (2006), *Generalized Additive Models: An Introduction with R*, Chapman and Hall/CRC Press, Boca Raton, FL, London.

Wooldridge, J. (2010), *Econometric Analysis of Cross Section and Panel Data*, 2nd ed., The MIT Press, Cambridge, MA.

Würtz, D., Chalabi, Y., Chen, W. & Ellis, A. (2009), *Portfolio Optimization with R/Rmetrics*, Finance Online GmbH, Zurich.

Wütherich, M. V. & Merz, M. (2008), *Stochastic Claims Reserving Methods in Insurance*, Wiley Finance, New York.

Yan, J. & Prates, M. (2012), *rbugs: Fusing R and OpenBugs.* R package version 0.5-6.

Yang, R. & Berger, J. O. (1998), A catalog of noninformative priors. working paper, Duke University, Durham, NC.

Ye, Y. (1987), Interior Algorithms for Linear, Quadratic, and Linearly Constrained Non-Linear Programming, PhD thesis, Department of EES, Stanford University.

Yee, T. W. (2008), The VGAM package, *R News* **8**(2), 28–39.

Yu, H. (2002), Rmpi: Parallel statistical computing in R, *R News* **2**(2), 10–14.

Zehnwirth, B. (1994), Probabilistic development factor models with applications to loss reserve variability, prediction intervals and risk based capital, *Casualty Actuarial Society Forum* **2**, 447–605.

Zeileis, A. (2005), 'CRAN task views', *R News* **5**(1), 39–40.

Zeileis, A., Kleiber, C. & Jackman, S. (2008), Regression models for count data in R, *Journal of Statistical Software* **27**, 1–25.

Zhang, J. & Gentleman, R. (2004), Tools for interactively exploring R packages, *R News* **4**(1), 20–25.

Zhang, Y. (2010), A general multivariate chain ladder model, *Insurance: Mathematics and Economics* **46**, 588–599.

Zhang, Y., Dukic, V. & Guszcza, J. (2012), A Bayesian nonlinear model for forecasting insurance loss payments, *Journal of the Royal Statistical Society, Series A* **175**, 637–656.

Zhu, C., Byrd, R. H., Lu, P. & Nocedal, J. (1997), Algorithm 778: L-bfgs-b: Fortran subroutines for large-scale bound-constrained optimization, *ACM Transactions on Mathematical Software (TOMS)* **23**(4), 550–560.

Zuur, A. F., Ieno, E. N. & Meesters, E. (2009), *A Beginner's Guide to R*, Springer Verlag, Berlin.

Index

R Command Index

For Product Safety Concerns and Information please contact our
EU representative GPSR@taylorandfrancis.com Taylor & Francis
Verlag GmbH, Kaufingerstraße 24, 80331 München, Germany